FORTSCHRITTE DER CHEMIE ORGANISCHER NATURSTOFFE

PROGRESS IN THE CHEMISTRY OF ORGANIC NATURAL PRODUCTS

BEGRÜNDET VON · FOUNDED BY

L. ZECHMEISTER

HERAUSGEGEBEN VON · EDITED BY

W. HERZ   H. GRISEBACH   G. W. KIRBY
TALLAHASSEE, FLA.   FREIBURG i. BR.   GLASGOW

VOL. 30

VERFASSER · AUTHORS

M. J. CORMIER · H. FLASCH · B. FRANCK · K. HORI · L. JAENICKE
W. KELLER-SCHIERLEIN · H. D. LOCKSLEY · D. G. MÜLLER
JUDITH POLONSKY · R. TSCHESCHE · J. E. WAMPLER · G. WULFF

1973

WIEN · SPRINGER-VERLAG · NEW YORK

Mit 28 Abbildungen   With 28 Figures

ISBN-13:978-3-7091-7103-5   e-ISBN-13:978-3-7091-7102-8
DOI:10.1007/978-3-7091-7102-8

Inhaltsverzeichnis

Contents

Quassinoid Bitter Principles. By JUDITH POLONSKY, Institut de Chimie des Substances Naturelles, Centre Nationale de la Recherche Scientifique, Gif-sur-Yvette, France . 101

Inhaltsverzeichnis — Contents   V

Die Ergochrome (Physiologie, Isolierung, Struktur und Biosynthese). Von B. Franck und H. Flasch, Organisch-Chemisches Institut der Universität Münster, BRD .. 151

Chemie und Biologie der Saponine. Von R. Tschesche und G. Wulff, Organisch-Chemisches Institut, Universität Bonn, BRD ... 461

Bioluminescence: Chemical Aspects

By M. J. CORMIER, J. E. WAMPLER, and K. HORI

Athens, Georgia

With 9 Figures

Contents

Note on Abbreviations

The abbreviations used in this review are as follows: LH and LH_2 commonly refer to the luciferin molecule under discussion, L refers to dehydroluciferin and $L=O$ to oxyluciferin. E commonly refers to the luciferase molecule under discussion. $FMNH_2$ and FMN are the reduced and oxidized forms of flavin mononucleotide respectively. DPNH and DPN are the reduced and oxidized forms of nicotinamide adenine dinucleotide. ATP and PP_i represent adenosine triphosphate and inorganic pyrophosphate respectively, ANS is anilinonaphthalenesulfonate and RCHO is used for an unspecified long chain aldehyde.

I. Introduction

Bioluminescence may be defined simply as the emission of visible radiation by an enzyme-catalyzed reaction. An analogous phenomenon, chemiluminescence, may be defined as the emission of visible radiation by a nonenzymic (purely chemical) reaction. Generally bioluminescence represents an efficient reaction with quantum yields (number of photons emitted divided by the number of molecules of substrate oxidized) ranging from about 0.1 to nearly 1.0 (*34, 113, 120, 141*). On the other hand chemiluminescent reactions cover a broad spectrum of quantum yields.

In general chemiluminescent systems are often limited either by an emitter molecule with an intrinsically low quantum yield or by quenching effects both of which are selected against in most bioluminescence reactions. Thus the enzymes involved in catalyzing bioluminescent reactions must play an important role in increasing the efficiency of such reactions, probably both by decreasing the rates of excited state quenching and by increasing the efficiency of excited state formation. Bioluminescence takes place in an aqueous environment at near neutral pH, conditions which do not allow chemiluminescence to occur. The environmental requirements for chemiluminescence generally include strong base and an aprotic solvent. Presumably the enzyme provides the appropriate basic and aprotic environment to allow not only for an efficient reaction but for the appropriate transition levels as well. Certainly this concept must apply to enzymic catalysis in general.

The terms luciferin and luciferase have become part of the terminology in the field of bioluminescence primarily due to the classic work of Dubois (*28*). In the classic sense luciferase refers to an enzyme that catalyzes the oxidation of the substrate, luciferin, with light emission. During this oxidative reaction a large amount of energy is released (50—72 kcal per mole quanta). This energy is utilized to create an electronically excited state either directly or by energy transfer to some fluorescent species formed, or present, during the reaction. This fluorescent species, or emitter, may be a product of luciferin, luciferin itself, or some

Table 1. *A List of Known Bioluminescent Reactions Separated According to Type Reaction*

Type Reaction	Examples	Emission Maxima (nm)	
		in vivo	*in vitro*
A. *Pyridine-nucleotide linked:*			
1. $DPNH + H^+ + FMN \xrightarrow{\text{Dehydrogenase}} FMNH_2 + DPN$	Bacteria	470—505	490
$FMNH_2 + RCHO + O_2 \xrightarrow{\text{Luciferase}} Light$			
2. $DPNH + H^+ + L \xrightarrow{\text{Dehydrogenase}} LH + DPN$			
$LH + O_2 \xrightarrow{\text{Luciferase}} Light$	Fungi	528	—
B. *Adenine-nucleotide linked:*			
1. $LH + ATP + O_2 \xrightarrow[\text{Mg}^{2+}]{\text{Luciferase}} Light$	Firefly	552—582	562 (ph 7.6)
2. $LH\text{-}Sulfate + DPA \xrightarrow{\text{Luciferin Sulfokinase}} LH + PAPS$	*Renilla* (coelenterate)	509	488
	Cavernularia (coelenterate)	509	488
$LH + O_2 \xrightarrow{\text{Luciferase}} Light$	*Yarella* (fish)	—	—
C. *Enzyme-substrate systems:*	*Cypridina* (crustacean)	—	460
	Porichthys (fish)	—	460
	Apogon (fish)	—	460
	Parapriacanthus (fish)	—	460
	Pholas (clam)	—	480
$LH + O_2 \xrightarrow{\text{Luciferase}} Light$	*Gonyaulax* (protozoan) sol. sys.	470	470
	Odontosyllis (annelid)	—	507
	Latia (limpet)	—	535

Table 1 (continued)

Type Reaction	Examples	Emission Maxima (nm)	
		in vivo	*in vitro*
	Euphausia (crustacean)	—	476
	Diaphus (myctophid fish)	—	—
	Hoplophorus (decapod)	—	—
D. Peroxidation systems:			
$LH + H_2O_2 \xrightarrow{\text{Luciferase}} Light$ }	*Balanoglossus* (acorn worm)	—	—
	Diplocardia (earthworm)	507	507
E. Activation of "precharged" systems:			
1. Precharged Particle $\xrightarrow{H^+;O_2} Light$ }	*Gonyaulax* (protozoan)	470	470
	Aequorea (coelenterate)	509	460
	Halistaura (coelenterate)	—	460
	Obelia (coelenterate)	509	475
	Clytia (coelenterate)	509	—
2. Precharged Protein $\xrightarrow{Ca^{2+}} Light$ }	*Phialidium* (coelenterate)	509	—
	Ptilosarcus (coelenterate)	509	—
	Pelagia (coelenterate)	—	—
	Campanularia (coelenterate)	—	—
	Mnemiopsis (coelenterate)	485	485
F. Unclassified systems:	*Chaetopterus* (annelid)	—	460
	Meganyctiphanes (crustacean)	—	476

protein bound chromophore in close association with luciferase, depending upon the system being investigated.

It is important to note that luciferin is a generic term which does not apply to any specific chemical structure. Although luciferin structures may vary widely, as will be shown subsequently in this chapter, we also find in a number of instances that luciferins isolated from widely diverse animals may be identical. Similar observations have been made for the luciferases.

The basic problem in bioluminescence may be stated simply: how is energy, derived from chemical oxidation, efficiently coupled to electronic excitation? One of the very interesting properties of luciferase proteins that catalyze bioluminescent reactions is their ability to manipulate large amounts of energy and to catalyze the creation of an electronically excited state. Sufficient work has been done on the chemistry of certain of these systems to allow us to make a start in the formulation of mechanism and to ponder the answer to the question posed above.

As will be shown subsequently there is a great deal of similarity in the mechanisms of bioluminescent reactions. This might be expected since the requirements for mechanisms which allow for the manipulation of a large amount of energy and release of this energy in a single reaction step may be extremely selective as to both the substrate requirements and the details of the oxidation reaction.

Work done by numerous investigators over the past decade has made this chapter possible. Many of the interesting observations made on bioluminescence will not be touched upon since it is beyond the scope of this chapter to do so. We have concentrated rather on those systems about which some chemical information is available. Additional aspects of bioluminescence have been covered in recent reviews (*18, 19, 42, 45, 55, 68, 96*).

CORMIER and TOTTER (*18, 19*) have developed a useful classification of bioluminescent reactions in which a particular system falls under one of several "type reactions". Table 1 gives an expanded version of their original classification based on recent information.

II. Renilla (Sea Pansy) Bioluminescence

1. General Comments

Renilla reniformis (commonly referred to as the sea pansy) is a marine animal (coelenterate) whose bioluminescence is controlled by a nerve network. As the result of a mechanical or electrical stimulus, concentric waves of greenish luminescence can be seen to move over the animal surface. Light from the living animal is green and exhibits a very narrow, structured emission (*141*) having a maximum at 19,640 cm^{-1} (509 nm). At low enzyme concentration the *in vitro* reaction produces a blue structureless emission (*141*) with a maximum at 20,500 cm^{-1} (488 nm).

A good deal of information is now available regarding the requirements for this light emission process and some information is now at hand which allows us to speculate on the control mechanisms involved. As described below, studies on *Renilla* will provide an understanding of the bioluminescent reaction in a number of marine organisms since several of these have been shown to have essentially identical biochemical requirements for light emission.

2. Chemical Requirements for Light Emission

The tentative structure of the major portion of *Renilla* luciferin (**1**) has recently been reported (*23, 66*). Some uncertainties in the structure still remain regarding groups R_2 and R_3. Note the indole-pyrazine ring structure and the similarity of this structure to that of *Cypridina* luciferin (**22**). Unlike the situation which obtains in *Cypridina* luciferin (*18*) groups R_2 and R_3 are not amino acid side chains (*78*). Free luciferin is easily autooxidized (*65*), but the living animal stabilizes it and stores luciferin as a sulfonated derivative (*15*) which we term luciferyl sulfate (**2**). The position of the sulfate group has been recently established by infra-red data and chemical synthesis (*66*). For example luciferin exhibits an absorption at 1630 cm^{-1} which is assigned to an amide linkage present in the fused pyrrole ring. This absorption is absent in luciferyl sulfate but new absorption bands appear at 1215 cm^{-1} and 1265 cm^{-1} which are assigned to an acid sulfate linkage equivalent to such linkages found in indolyl-3-sulfate and ascorbic acid sulfate. We have also been able to synthesize luciferyl sulfate from luciferin and sulfamic acid. This methods result in the synthesis of sulfate esters of alcoholic and phenolic hydroxyls. The synthesized product is identical with the natural compound in chromatographic behavior and absorption characteristics. In addition requirements for light production with synthetic luciferyl sulfate are identical with those reported for the natural compound and the kinetics are the same (*11*). Thus the structure of luciferyl sulfate is **2**.

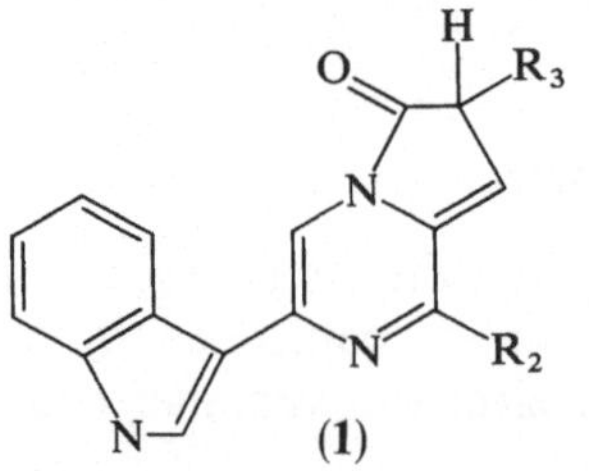

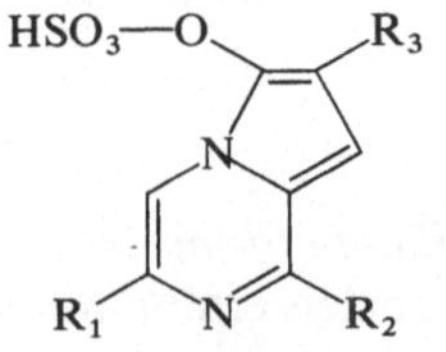

Since luciferyl sulfate does not react with luciferase to produce light the animal must possess a means for effecting its conversion to luciferin. We have isolated an enzyme from *Renilla* that does catalyze the conversion of luciferyl sulfate to luciferin which we have named luciferin sulfokinase (*15*). As shown by CORMIER *et al.* (*15*) luciferin sulfokinase catalyzes the conversion of luciferyl sulfate to luciferin in the presence of 3′,5′-diphosphoadenosine (DPA) (**3**). The activity of partially purified luciferin

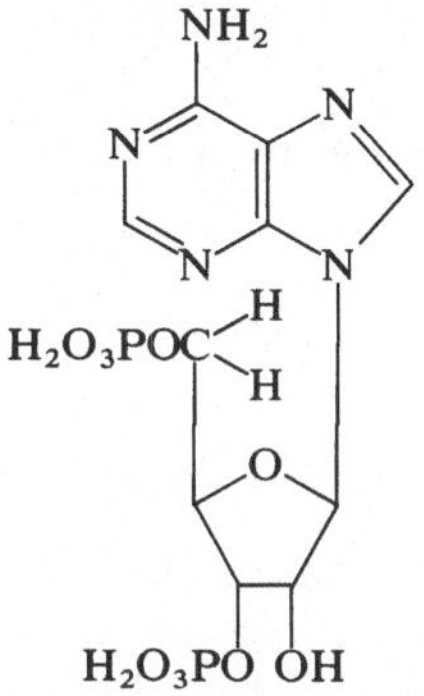

$$NH_2$$

$$H_2O_3POC$$

$$H_2O_3PO \quad OH$$

(**3**)

sulfokinase is stimulated by the addition of calcium ion but that of the purified enzyme is not (*11, 86*). As shown in Scheme 1 the other reaction product is 3′-phosphoadenylyl sulfate (PAPS). S^{35} experiments have shown that the reaction is reversible (*15*).

$$\text{Luciferyl sulfate} + \text{DPA} \underset{\text{Sulfokinase}}{\overset{\text{Luciferin}}{\rightleftharpoons}} \text{Luciferin} + \text{PAPS}$$

Scheme 1

Most of the luciferin in *Renilla* exists in the form of luciferyl sulfate except for small amounts which we find bound to protein. Furthermore the conversion of luciferyl sulfate to luciferin by luciferin sulfokinase is a very slow reaction (*15*). In addition it is well known that repeated stimulation of *Renilla* results in a bioluminescence fatigue phenomena (*109*) which requires considerable time (1—2 hours) for complete recovery. These observations suggest that the low rate of synthesis of luciferin by luciferin sulfokinase plays an important role in this recovery process and is thus one of the controlling features of bioluminescence in *Renilla*.

Once luciferin is formed the only requirement for bioluminescence *in vitro* are luciferin, oxygen and luciferase. As shown in Scheme 2 luciferase catalyzes the oxidation of luciferin to produce light, CO_2 and oxyluciferin which represents the product excited state in the reaction (*23, 141*).

$$\text{Luciferin} + O_2 \xrightarrow{\quad \text{Luciferase} \quad} \text{Oxyluciferin*} + CO_2$$
$$\downarrow$$
$$\text{Oxyluciferin} + h\nu$$

Scheme 2

Luciferase is an energy conversion enzyme of low molecular weight, approximately 24,000. The evidence suggest that it is a dimer of identical subunits with a molecular weight of 12,000 (*73*). Its Stokes radius is 27 Å, its diffusion coefficient, $D_{20,w}$, is 8×10^{-7} cm^2 sec^{-1} and its extinction coefficient, $\frac{0,1\%}{280}$, at pH 7.5 is 1.04. Its amino acid composition has also been determined (*73*).

3. Mechanism of the Light Reaction

During the bioluminescent oxidation of luciferin by luciferase (Scheme 2) one mole of CO_2 is produced per mole of luciferin present (*23*). The only other products are light and oxyluciferin (**8**, Scheme 3) whose fluorescence emission spectrum essentially matches that of the bioluminescence emission (Fig. 1). This same compound is released into the surrounding sea water by *Renilla* when they are stimulated to luminesce (*141*). Thus this fluorescence product must represent the *in vitro* emitter.

When light emission proceeded in an $^{18}O_2$ atmosphere there was negligible incorporation of ^{18}O into the CO_2. However when the reaction was carried out in the presence of $H_2{}^{18}O$, both oxygens in the released CO_2 were labelled (*23*). The production of CO_2 and the production of the *in vitro* product (**8**) as well as labelling of the CO_2 by $H_2{}^{18}O$ requires the presence of both luciferase and luciferin (**1**). A mechanism that predicts the ^{18}O labelling data during *Renilla* bioluminescence is shown in Scheme 3. One of the two labelled oxygens found in the CO_2 most probably arises from a non-enzymic exchange of water oxygen with the ketone group of luciferin (**1**) where R_1 is the indolyl moiety. We postulate the formation of a carbanion (**5**) followed by oxygen attack to give a linear hydroperoxide (**6**). Hydroxyl ion then adds at the carbonyl carbon to form **7** which then rearranges to form the electronically excited state product **8** and CO_2. Compound **8** then returns to the ground state with the release of a photon. For convenience this mechanism will be referred to as the "hydroxyl ion mechanism".

When luciferase catalyzes the oxidation of luciferin according to Scheme 3, blue light is observed with a maximum at 20,500 cm^{-1} (488 nm) as shown in Fig. 1. The quantum yield of this reaction with respect to luciferin is low, i.e. about 0.04 (*141*). The color of light produced from

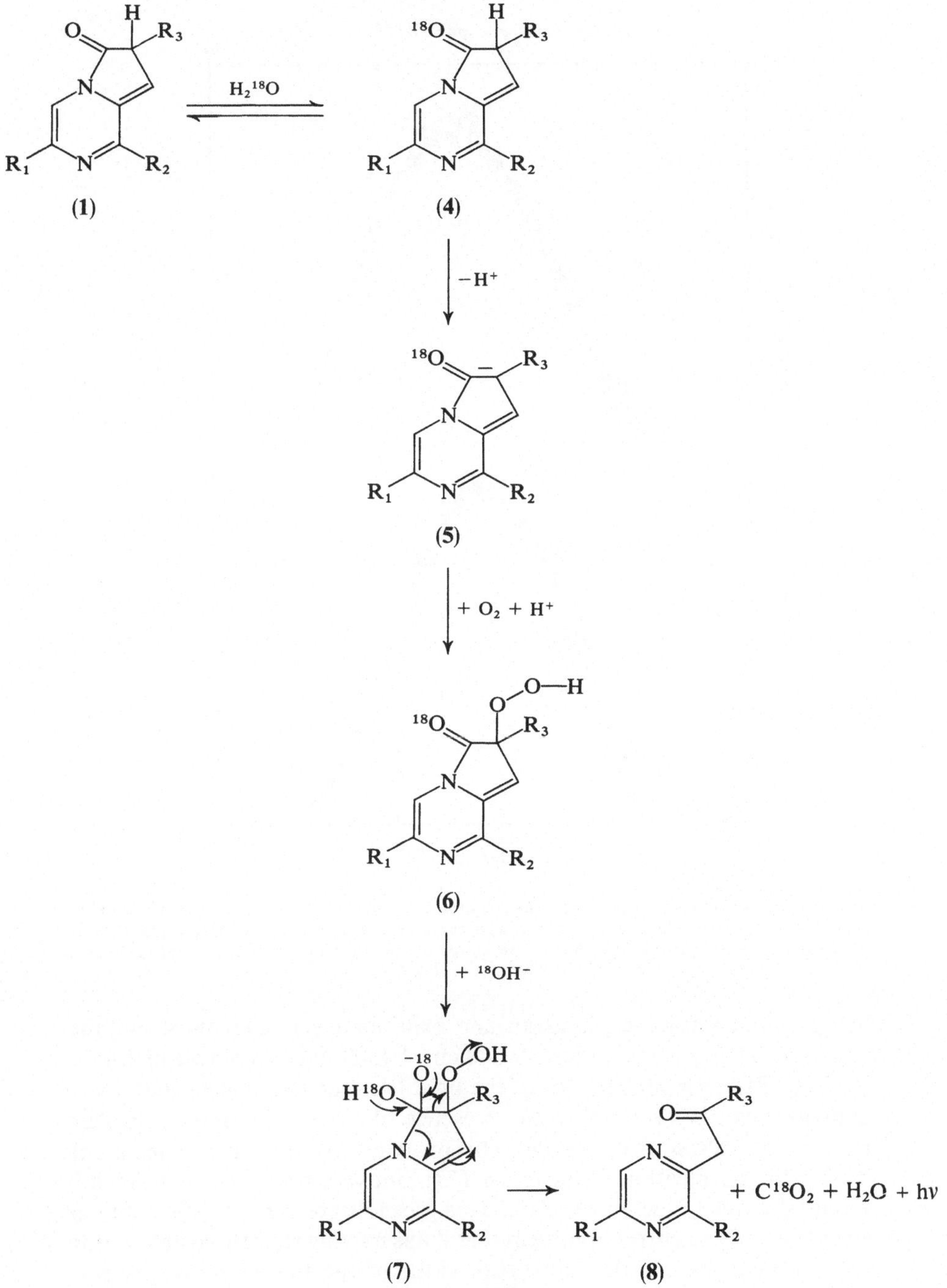

$$(1) \xrightarrow{H_2{}^{18}O} (4) \xrightarrow{-H^+} (5) \xrightarrow{+ O_2 + H^+} (6) \xrightarrow{+ {}^{18}OH^-} (7) \longrightarrow (8) + C^{18}O_2 + H_2O + h\nu$$

Scheme 3

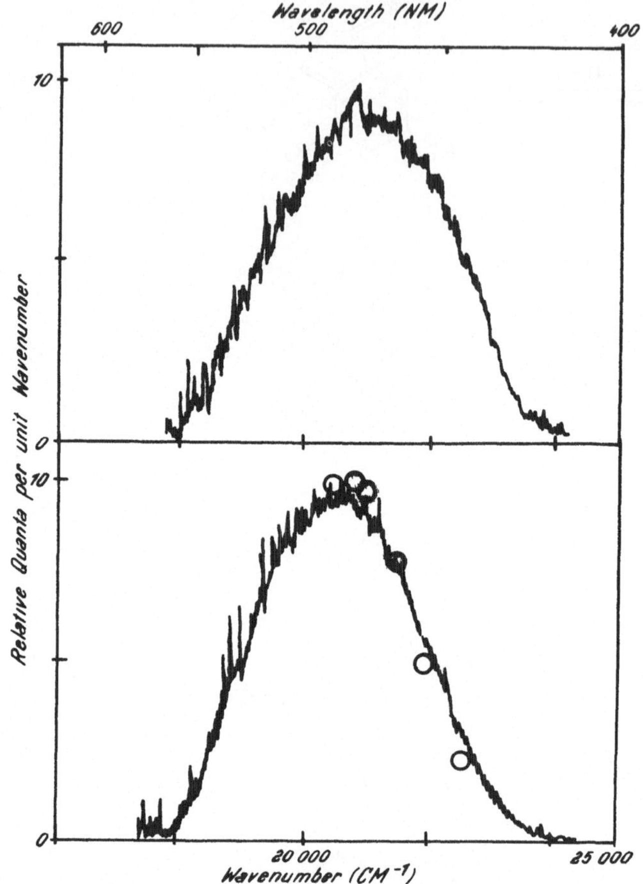

Fig. 1. (Upper) Fluorescence spectrum of *Renilla* oxyluciferin isolated from *in vitro* reaction mixtures. (Lower) *In vitro* bioluminescence of *Renilla reniformis* system using 1 mm slits. Open circles show the comparison of this spectrum with the blue region of the *in vivo* bioluminescence. Data taken from WAMPLER *et al.* (*141*)

living animals however, is green and exhibits a very narrow structured emission having a maximum 19,640 cm^{-1} (509 nm) as shown in Fig. 3.

The difference in the color of the light between the *in vitro* and *in vivo* emissions can be ascribed to the presence of a green fluorescent protein that can be isolated from *Renilla*. This protein was originally carried along during the purification of luciferase (*73*), but we have recently isolated it essentially free of luciferase (*142*). This green fluorescent protein contains a very interesting chromophore in that it has narrow structured absorption and emission bands, the half-band width of the fluorescence spectrum being 800 cm^{-1} and exhibits only a small Stokes shift (360 cm^{-1}) as shown in

Fig. 2. The fluorescence excitation spectrum shown in Fig. 2 is quite similar to the absorption band. There is clearly a mirror image relationship between the absorption and emission spectra, and the shape and position of the emission spectrum is excitation wavelength independent over the whole excitation spectrum. When the polarization of the emission was determined as a function of excitation (Fig. 2) the values were high over the entire excitation spectrum. The quantum yield of this chromophore determined relative to quinine was 0.30. Taken together these data suggest that the absorption and emission transitions investigated here arise from the same electronic transition and that the structure seen is of a vibrational nature and not due to electronic or species differences.

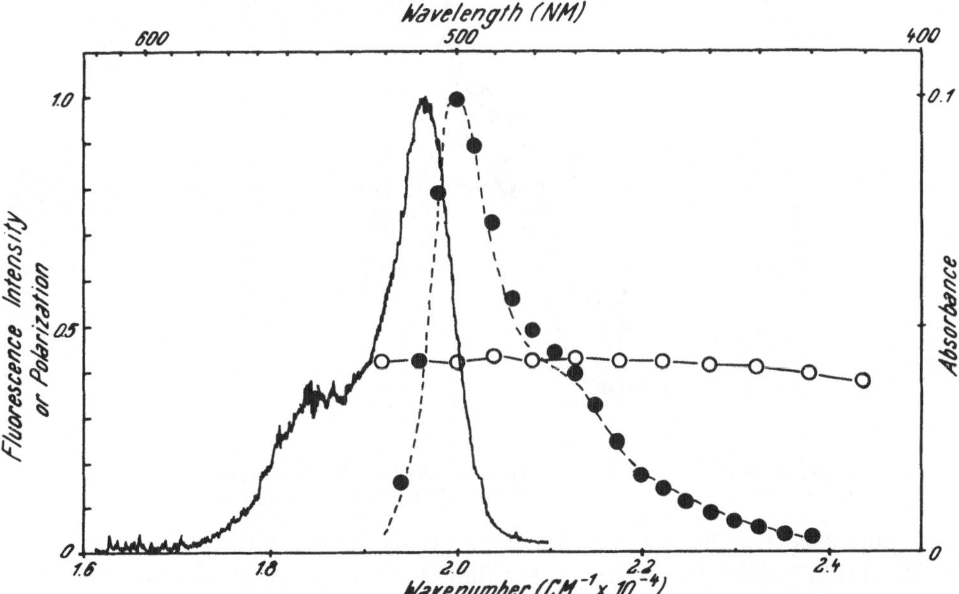

Fig. 2. Spectral characteristics of the protein-bound green chromophore isolated from *Renilla*. The absorption spectrum (dashed line) reflects a mirror image of the structured fluorescence spectrum (solid line). The excitation spectrum is indicated by closed circles. The polarization of fluorescence excitation spectrum is given by the open circles. Data taken from WAMPLER *et al.* (*141*)

The green fluorescent chromophore is apparently covalently bound to the protein (*141*). The importance of this green fluorescent protein is indicated by comparison of its fluorescence with the *in vivo* bioluminescence of *Renilla* (Fig. 3). In both cases the spectra are identical and it is concluded that this protein bound chromophore is the *in vivo* emitter. Although the fluorescence and bioluminescence emissions shown in Fig. 3 are identical over most regions of the spectrum, there is one important exception where

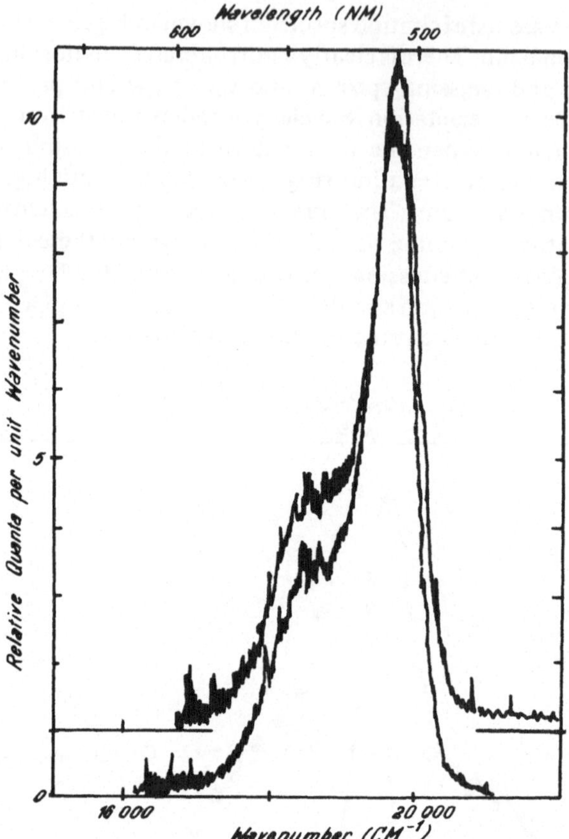

Fig. 3. Comparison between the *in vivo* bioluminescence of *Renilla mülleri* (upper) and the fluorescence of the protein-bound green chromophone (lower). Data taken from Wampler *et al.* (*141*)

a difference can be noted in the blue region of the spectrum, i.e. at 20,500 cm^{-1} (488 nm). Whereas the fluorescence cuts off in this region, the bioluminescence emission does not. When the difference spectrum in this blue region is normalized to the *in vitro* bioluminescence emission a striking similarity in the two spectra is observed (Fig. 1). Thus during normal *in vivo* bioluminescence in *Renilla* there are two emitters. One is derived from the electronically excited state of **8** and the other from the electronically excited state of the green fluorescent protein.

In the absence of the green fluorescent protein, the luciferase-catalyzed oxidation of luciferin produces blue light as shown in Fig. 4. However, when sufficient concentrations of both luciferase and green fluorescent protein (about 2 and 3 mg/ml respectively) are present the characteristic green emission is observed (Fig. 4). When this occurs the quantum yield also increases and approaches the fluorescent quantum

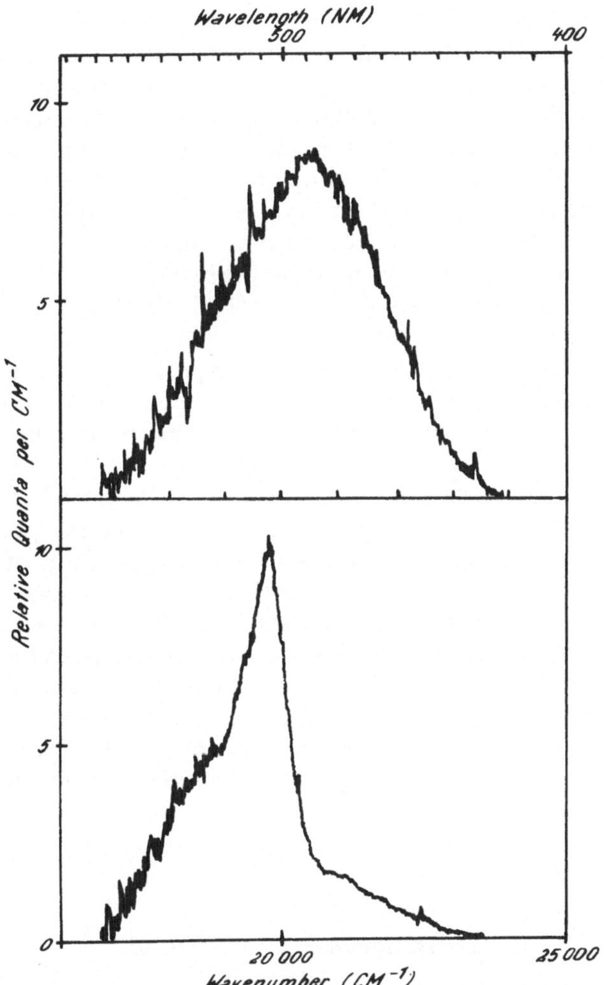

Fig. 4. Effect of addition of isolated green-protein on the *in vitro* bioluminescence of *Renilla reniformis*. (Upper) *In vitro* bioluminescence spectrum obtained upon addition of an oxygen free luciferin solution to 1 ml of luciferase solution. (2 mg/ml in phosphate buffer, pH 7.2). (Lower) *In vitro* bioluminescence spectrum under the same conditions including 3 mg/ml of isolated green fluorescent protein

yield of the green chromophore (*141*). When such a mixture of luciferase and green fluorescent protein is diluted to a protein concentration of about 1 mg/ml or less, blue light is again observed upon the addition of luciferin. That is, the production of green light is highly concentration dependent both with respect to luciferase and to the green fluorescent protein. These and other observations suggest that there is protein-protein interaction between luciferase and the green fluorescent protein. Such an association between these two proteins could result in intramolecular energy transfer between the electronically excited state product **8** derived from luciferin oxidation and the green fluorescent chromophore. Since the overlap of the product emission with the green fluorescent protein absorption is very good, and since this absorption appears to be a strongly allowed transition, the conditions for intramolecular energy transfer are excellent. Förster's theory (*30, 36, 74*) requires that such energy transfer could not come about from homogeneous solution since the concentrations of donnor and acceptor are several orders of magnitude too low to account for the observed transfer efficiency (*141*). This objection is overcome by assuming concentration-dependent association of luciferase and the green fluorescent protein. Associated proteins of this molecular size (ca. 30,000 daltons) could easily provide interaction distances within the 28 Å critical distance calculated for this energy transfer model (*141*).

We also have evidence that all components of the *Renilla* bioluminescent system are packaged within a membrane-enclosed subcellular particle which is triggered to luminesce after lysis in the presence of calcium ion (*1*). Presumably it is a nerve-mediated control of calcium flux across the membrane of these organelles which results in control of the luminous flash *in vivo*. A similar particle has been isolated from the bioluminescent coelenterate, *Obelia* (*104*). We have examined the color of the light emitted by these *Renilla* particles by means of a computerized spectrofluorimeter (*140*). Once again the characteristic green *in vivo* type emission is observed. This is not surprising since one would expect localized high concentrations of luciferase and green fluorescent protein within the isolated particles.

4. Related Bioluminescent Systems

Approximately ten years ago SHIMOMURA *et al.* (*125, 126, 127*) isolated a most interesting type of bioluminescent system from the bioluminescent jelly fish, *Aequorea* and *Halistaura*. When the luminous tissues were extracted with EDTA-containing buffers a protein component was isolated which exhibited a brillant blue (emission maximum = 460 nm) luminescence upon the addition of calcium ion. Dissolved oxygen was not required for luminescence. Such proteins were termed "photoproteins". JOHNSON *et al.* (*71*) also observed that light from the living animal appeared green and that a green fluorescent protein could also be isolated in addition to the "photoprotein". They also suggested that this green fluorescent protein might be involved during bioluminescence *in vivo*. The similarity of the fluorescence of this green protein to the *in vivo* emission of *Renilla* was noted recently by WAMPLER *et al.* (*141*) and by MORIN and HASTINGS (*104*).

More recently "photoproteins" with properties similar to those described above have been extracted from a variety of coelenterates by MORIN and HASTINGS (*103, 104*). These include *Obelia, Campanularia, Clytia, Lovenella, Phialidium, Aequorea, Diphyes, Renilla, Ptilosarcus, Stylatula, Pelagia* and *Mnemiopsis.* In all cases examined the bioluminescence emission of the photoproteins were blue, ranging from 460 to 485 nm. However, in all but two cases (*Mnemiopsis* and *Pelagia*) the *in vivo* bioluminescence was observed to be green and to exhibit the characteristic narrow emission described for *Renilla* (*104, 141*), and to exhibit the same 509 nm emission maximum (see Fig. 3). In the case of *Obelia* a green fluorescent material (presumably the green fluorescent protein) is localized within the photocytes (*105*). A similar green fluorescent material has been observed to occur in the photocytes of other coelenterates whose *in vivo* bioluminescence is in the green (*104*). *Mneiopsis* and *Pelagia* apparently do not contain the green fluorescent protein (*104*) thus accounting for their blue emission during bioluminescence *in vivo.*

Thus it appears that the green fluorescent protein recently studied in some detail by WAMPLER *et al.* (*141*) is widely distributed among the coelenterates and represents the energy transfer acceptor during bioluminescence of those animals.

In view of the observations made above it seems probable that the biochemical requirements for light emission among the coelenterates may be identical with or very similar to those already described for *Renilla* by CORMIER and associates. As supporting evidence for the proposal that the biochemical requirements for light emission among the coelenterates may be the same it should be pointed out that we have examined yet another coelenterate, *Cavernularia obesa,* and indeed have found requirements identical with those described for *Renilla* (*66*). For example, a luciferyl sulfate has been isolated from *Cavernularia obesa* which appears to be identical with the luciferyl sulfate isolated from the coelenterate *Renilla* in chromatographic behavior and absorption characteristics. Furthermore, luciferyl sulfate isolated from *Cavernularia* can replace *Renilla* luciferyl sulfate as the substrate for light production in the presence of *Renilla* luciferase, 3',5'-diphosphoadenosine, *Renilla* luciferin sulfokinase, calcium ion, and oxygen. Also, luciferase and luciferin sulfokinase obtained from *Cavernularia* will produce light in the presence of *Renilla* luciferyl sulfate, 3',5'-diphosphoadenosine, calcium ion, and oxygen. Thus the bioluminescence systems of these two coelenterates appear to be very similar in their biochemical requirements. *Cavernularia* has also been found to contain the characteristic green fluorescent protein recently described as being responsible for the *in vivo* light emission of *Renilla.*

What then we might ask is a photoprotein? It was previously proposed by McElroy and Seliger (*95*) and by Cormier and Totter (*18, 19*) that the photoprotein luminescent reaction be viewed as a calcium-triggered release of an EDTA-inhibited, oxygen containing, luciferin-luciferase complex. It has also been suggested that photoproteins represent a stabilized peroxide intermediate of some sort (*56*). In this regard a protein complex which contains a stabilized luciferin hydroperoxide (such as **6**, Scheme 3) that could be triggered to luminesce upon the addition of calcium ion would seem to be an attractive candidate for a photoprotein. This proposal suggests a very straightforward experiment, i.e. an attempt to isolate the carbon dioxide evolved during the calcium triggered photoprotein reaction in the presence of ^{18}O labelled oxygen and alternatively in the presence of ^{18}O labelled water. Knowledge of the labelling patterns in the carbon dioxide would contribute greatly to elucidation of the nature of photoproteins.

The fact that a variety of coelenterates appear to possess requirements for bioluminescence essentially identical with those already described for *Renilla* is interesting. However it is perhaps even more interesting that such requirements have also been observed in the bioluminescence of a fish, *Yarella* (K. Hori and M. J. Cormier, unpublished results). This fish has distinct intradermal photophores (*53*). From the photophores we have been able to extract a luciferin and luciferase which will cross react with *Renilla* luciferase and luciferin to produce light. This is another example of an interphylum bioluminescent cross reaction suggesting that a *Renilla* type bioluminescent system, which occured early in evolution, has been maintained throughout evolution such that it is now found in fish, the most advanced evolutionary example of bioluminescence.

It now seems obvious that the characterizations of Table 1 may be arbitrary. For instance many of the precharged systems (Type E) may in fact have chemical mechanisms analogous to those already described for *Renilla* (Type B).

III. Firefly Bioluminescence

1. General Comments

Among the terrestrial bioluminescence systems the firefly system is the best defined. It represents one of those few systems, terrestrial or marine, where the developments of many scientific disciplines have been pooled to produce a unified picture. Studies have been made of behavior (*9*), anatomy (*8*), physiology (*7*), and chemistry (*94, 96*) as they relate to the problem of bioluminescence in the firefly.

In general the light produced by any particular firefly species originates in one or more luminous organs called lanterns. The color of light varies between species and in some cases within a species from individual to individual. In at least one instance the color varies between the lanterns of a single individual (*4*). The variations in color range from green to red. In several species examined the function of bioluminescence appears to be in mating. The luminous signal is often manifest as a characteristic flash pattern (*9*).

The firefly bioluminescence reaction is again a simple redox reaction involving molecular oxygen and a complex organic substrate, luciferin (**9**). The mechanism of this reaction is fairly well defined and the enzyme which catalyzes it has been examined extensively. It should be noted that the following discussion is somewhat restricted since most of the studies have involved a single firefly species, *Photinus pyralis*. Therefore generalizations unless supported by comparative experiments — which will be mentioned where available — cannot be justified. For general reviews of this subject the reader is referred to references *94*, *96*, and *145*.

2. Chemical Requirements for Light Emission

Like *Renilla*, the firefly bioluminescence system requires an activation reaction proceeding the bioluminescence reaction. However, unlike the *Renilla* system (Section II), this reaction is catalyzed by luciferase itself. Since it is a reversible reaction, its consideration cannot be separated from the light yielding reaction. The overall chemical requirements can then be expressed by the following scheme:

$$\text{Luciferin } (\mathbf{9}) + \text{ATPMg}^{2+} + \text{O}_2 \rightarrow \text{CO}_2 + \text{Oxyluciferin } (\mathbf{10}) + \text{AMP} + \text{PP}_i\text{Mg}^{2+} + h\nu$$

Scheme 4

The quantum yield of bioluminescence is approximately one both with respect to oxygen (*94*) and luciferin (*113*). Again, as in the *Renilla* case, an adenine nucleotide is involved in the activation reaction. This reaction is shown by the next equation.

$$\text{Luciferin } (\mathbf{9}) + \text{ATPMg}^{2+} \leftrightarrows \text{Luciferyl-adenylate } (\mathbf{11}) + \text{PP}_i\text{Mg}^{2+}$$

Scheme 5

Rhodes and McElroy (*111*) have demonstrated the reversibility of this reaction. The actual bioluminescence reaction is as follows:

$$\text{Luciferyl-adenylate } (\mathbf{11}) + O_2 \rightarrow \text{AMP} + \text{Oxyluciferin}(\mathbf{10}) + CO_2 + h\nu$$

Scheme 6

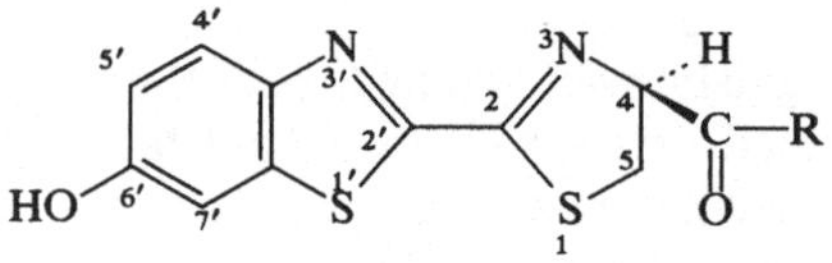

(9)

d(−) Luciferin

R = OH

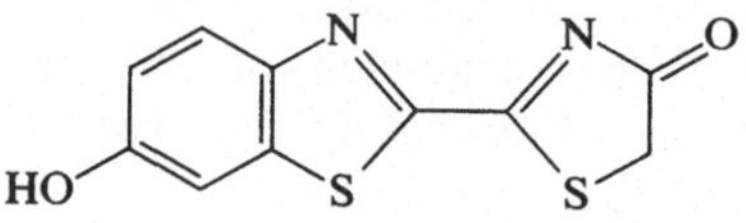

(10)

Oxyluciferin

(11)

Luciferyl adenylate

R = AMP

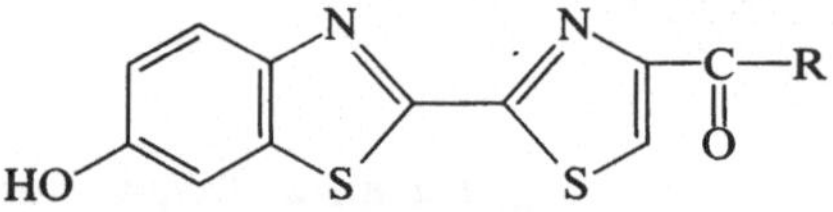

(12)

Dehydroluciferyl adenylate

R = AMP

(13)

Dehydroluciferin

R = OH

The catalyst for the firefly bioluminescence reaction, firefly luciferease, has been studied primarily in the form of an enzyme isolated from one species, *Photinus pyralis* (*46*). Hence, generalizations about catalytic mechanism, binding, molecular weight, and control must be made with caution, a warning which has frequently not been stressed sufficiently in the recent literature. Such a warning is particularly *à propos* if it is remembered that there exists a wide range from species to species in the color of firefly bioluminescence. This variation has been demonstrated to be directly related to differences in the enzyme, not the substrate (*115*).

Although the enzyme from *Photinus* has been studied for many years, a definite chemical and physical description can not yet be given. The enzyme is routinely isolated and purified by recrystallization by the procedure of GREEN and MCELROY (*46*). The molecular weight of the minimum active species of the enzyme is 50,000. However, it tends to aggregate and is active as an aggregate right up to the limit of precipitation (*24*). The aggregation phenomenon is partially explained by the large number of hydrophobic amino acids which make up the enzyme (*25*).

The isolated enzyme is heterogeneous by several criteria (*27*). Representative data (*27*) indicate the following interesting and puzzling properties:

1. One "active site" sulfhydryl is found per 50,000 daltons, but only one site per 100,000 daltons for the luciferyl-adenylate analog dehydroluciferyl-adenylate (**12**) (L–AMP).

2. Only one $ATPMg^{2+}$ site per 100,000 daltons, but one ATP site per 50,000.

3. Two different C-terminal amino acids.

4. Two active fractions by isoelectric focusing.

In addition, the inhibitor dehydroluciferin (**13**) is found to bind to one site per 50,000 daltons (*26*). The contradictions here are obvious and DENBURG and MCELROY (*27*) have not been able to hypothesize a mechanism consistent with all of these findings. In contrast to the direct experiments, DELUCA's (*20*) data on the reversal of dye inhibition indicates that dehydroluciferyl-adenylate binds to two sites per 100,000 daltons, not one. The model of LEE and MCELROY (*84*) suggests that most of the above data can be explained by postulating that all of the binding sites rest on one of two non-identical subunit or monomer species. The role of the other inactive monomer species remains to be explained.

Much of the collected data indicate that the active site, in fact luciferase in general, is hydrophobic. The dye studies of DELUCA (*20*) give results consistent with a nonpolar binding site, as do the lifetime measurements on luciferase-bound dehydroluciferin (*21*). Many non-polar molecules act as inhibitors of firefly luciferin (**9**) (*96*).

The indications are strong that one sulfhydryl group per 50,000 daltons is involved in catalysis (*26*). Recent results of DENBURG and MCELROY (*27*) show that while this sulfhydryl group is protected from sulfhydryl reagents by L-AMP (**12**), competitive inhibitors, such as 1,5 ANS, do not offer this same protection (*20*). Hence, the sulfhydryl group may not directly be involved in the binding site as had been previously postulated (*24*). TRAVIS and MCELROY (*135*) using radioactive $[1-^{14}C]N$-ethylmaleimide were able to isolate and sequence the peptide containing the sulfhydryl group which is protected from sulfhydryl reagents by dehydroluciferin binding. Their results indicate that the minimum molecular weight of luciferase is 50,000 daltons, but again they implicated a dimer as the active species. The number of peptides formed by tryptic digestion supported the presence of identical subunit species.

Firefly luciferin has been shown by both chemical and physical measurements as well as by synthesis to have structure (9) (*116, 144*). This luciferin is the common substrate for the following genera: *Photuris*, *Pyrophorus*, *Diphotus*, *Lecontae*, and *Luciola* (*80, 115*). Native luciferin is the d(–) form and not the l(+) form (*144*). Synthetic l-luciferin is not active in the light reaction but can be adenylated (*116*). The relative three dimensional orientation of d-luciferin in a crystalline structure has been investigated by X-ray analysis by Blank *et al.* (*5*).

The products of the activation reaction as indicated in Scheme 5 are luciferyl-adenylate (**11**) and inorganic pyrophosphate (*111*). The reversal of this reaction has been demonstrated by the exchange of radioactive labels.

The bioluminescence reaction itself (Scheme 6) produces carbon dioxide (*110*), AMP (*111*), and oxyluciferin (*133*). Until recently the structure of oxyluciferin has been elusive. Hopkins *et al.* (*64*) postulated structure **10** on the basis of chemiluminescence experiments using luciferin analogs; subsequently Suzuki and coworkers (*133, 134*) synthesized this substance and verified its identity with oxyluciferin by isolating the same material from firefly lanterns and from chemiluminescence reaction mixtures (*133*).

3. Mechanism of the Light Reaction

In summary, firefly bioluminescence involves two steps, an activation reaction and a redox reaction. The activation reaction results in the reversible adenylation of firefly luciferin using $ATPMg^{2+}$ and produces pyrophosphate in addition to luciferin adenylate (**11**). The Michaelis constant for the firefly reaction measured by Rhodes and McElroy (*111*) was 2.4×10^{-6} M and gave a lower limit for the equilibrium constant of the activation step as 4×10^5. The same workers have measured the activation equilibrium constant for dehydroluciferin (**12**) (Scheme 7).

$$L\ (13) + ATPMg\ ^+ \leftrightarrows L - AMP\ (12) + PP_i\,Mg\ ^{2+}$$

Scheme 7

Since $L - AMP$ formed by this reaction cannot react with oxygen to produce light, a true equilibrium can be established. The measured equilibrium constant varied between 2 and 3×10^5 M. They were also able to measure the rate of dissociation of enzyme bound $L - AMP$

$(k_1 = 1.3 \times 10^{-3} \text{ sec}^{-1})$ and its dissociation constant $(K = \dfrac{k_1}{k_2} = 5.1 \times 10^{-10})$ (Scheme 8).

$$E - L - AMP \underset{k_2}{\overset{k_1}{\rightleftharpoons}} E + L - AMP \quad (12)$$

Scheme 8

These values allow calculation of the value of k_2 as $2.5 \times 10^6 \text{ M}^{-1} \text{ sec}^{-1}$. The upper limit for the $LH_2 - AMP$ (**11**) dissociation constant from the Michaelis constant was $2 \times 10^{-7} \text{ M}$.

The light producing oxidation reaction as outlined above involves molecular oxygen and produces oxyluciferin, carbon dioxide, and AMP. The rate constant for this step from the kinetic measurements of RHODES and MCELROY (*111*) varied between 2×10^{-1} and $2 \times 10^{-2} \text{ sec}^{-1}$.

The simplicity of this two step reaction is disturbed in actual *in vitro* systems by a rapid inhibition of the overall reaction. MCELROY and SELIGER (*94*) have explained this effect by assuming that dehydroluciferyl adenylate acts as an inhibitor after having been produced by enzymic adenylation of dehydroluciferin which was in turn formed by a side reaction. The overall reaction sequence of Fig. 5 would then represent the reactions and competing reactions of firefly bioluminescence. In addition to these reactions oxyluciferin undergoes rapid non-enzymic chemical breakdown resulting in at least three products (*110*). These same products are also produced in the chemiluminescence reactions (*110*).

$$E + LH_2 + ATPMg^{2+} \rightleftharpoons E - LH_2 - AMP + PPMg^{2+}$$

Scheme 9

$$E - LH_2 - AMP \rightleftharpoons E + LH_2 - AMP$$

Scheme 10

$$E - LH_2 - AMP + O_2 \Rightarrow E + L = O + CO_2 + AMP + Light$$

Scheme 11

$$LH_2 + O_2 \rightarrow L + H_2O_2$$

Scheme 12

$$E + L + ATPMg^{2+} \rightleftharpoons E - L - AMP + PPMg^{2+}$$

Scheme 13

$$E - L - AMP \rightleftharpoons E + L - AMP$$

Scheme 14

Fig. 5. Reactions of firefly bioluminescence

This sequence of reactions (Fig. 5) explains the stimulation produced by added pyrophosphate on the inhibited part of the reaction time course and the added inhibition seen when pyrophosphatase is incorporated into the reaction (*93*). In the first case reversal of Scheme 13 would lead to free enzyme, and in the second cleavage of pyrophosphate would pull the reaction of Scheme 13 to completion.

The typical *in vitro* reaction using, for example, *Photinus* luciferase has a pH dependence characterized by green light with a high quantum yield in basic solutions, and by red light with a low quantum yield in acid solutions. This pH profile is illustrated by Fig. 6. Both types of emission are seen in chemiluminescence experiments (*114*).

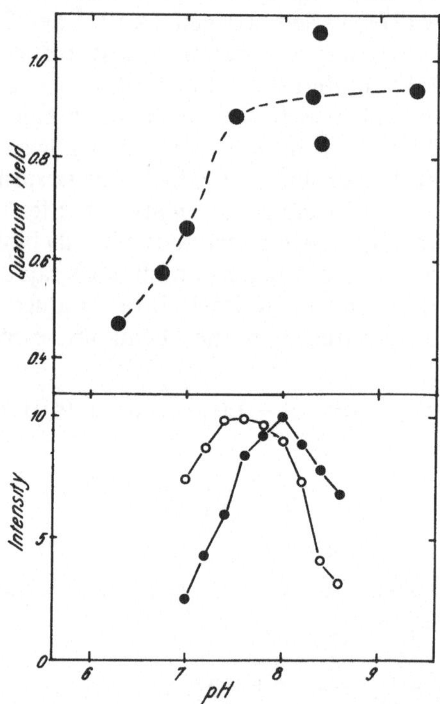

Fig. 6. Relationship between color and quantum yield for firefly bioluminescence. (Upper) pH dependence of the quantum yield of firefly *in vitro* bioluminescence taken from SELIGER and MCELROY (*113*). (Lower) pH dependence of the color of light obtained from the *in vitro* firefly reactions. Red light intensity (open circles). Green light intensity (closed circles). Data are from SELIGER and MCELROY (*115*)

Structure **10** was postulated for the reaction product, oxyluciferin by HOPKINS *et al.* (*64*) prior to its synthesis. Responsibility for the pH dependent color change was assigned to the phenolic hydroxyl at position 6′, an assignment which rested on experiments with luciferin analogs. Of the many analogs synthesized (*146, 147, 148*) only two, 4′-hydroxyluciferin (**14**) and

6'-aminoluciferin (**15**), were acive in *in vitro* bioluminescence reactions. Both of these substrate analogs give red light and lack the pH dependence seen with native luciferin. Hence, the reactive species of luciferyl adenylate in the green light reaction was postulated as the 6'-phenolate ion. Experiments on the chemiluminescence of various analogs (*64*) indicated that the phenolate of oxyluciferin was indeed the red luminescence emitter, but that green luminescence required deprotonation of both the phenol and enol hydroxyls in the enol tautomer (**16**) of oxyluciferin. These conclusions were supported by the chemiluminescence experiments of WHITE *et al.* (*145*). Finally SUZUKI *et al.* (*133, 134*) showed that synthetic oxyluciferin in strongly basic DMSO has a green fluorescence ($\lambda_{max}=557$) consistent with the observed chemiluminescence of luciferin.

(14)

(15)

(16)

On the basis of the above evidence the emitters in chemiluminescence and in *Photinus* bioluminescence appear to be the mono- and dianions of oxyluciferin (**10**). Since the same luciferin molecule is involved in all the firefly systems (*115, 80*) which have been investigated so far differences in the color of light emission in different species are probably due to variations in the environment which the emitter occupies on the individual enzymes. This conclusion is supported by the cross reaction experiments of SELIGER and McELROY (*115*). MORTON *et al.* (*106*) have shown that the spectroscopy of luciferin analogs can be made to depend on environmental conditions making it likely that similar factors account for the variations in color *in vivo*.

The mechanism for the firefly light reaction (*64*) shown in Scheme 15 is based upon the dioxetane mechanism developed by McCAPRA (*87*) as applied to the chemiluminescence of model systems (*90, 143*) and was proposed for firefly luminescence by HOPKINS *et al.* (*64*). In this mechanism AMP is lost after a concerted attack of molecular oxygen on the 4-position of luciferyl adenylate (11) which results in formation of the four-membered peroxide ring of 18. The oxyluciferin excited state is then produced by cleavage of 18 to form CO_2. The highest energy emission observed for the firefly species examined by BIGGLEY et al. (4) corresponded to a wavelength maximum at 546 nm with a half band width of about 70 nm. This places the minimum energy requirement for the energy-yielding step of the reaction at about 56 kcal/mole quanta.

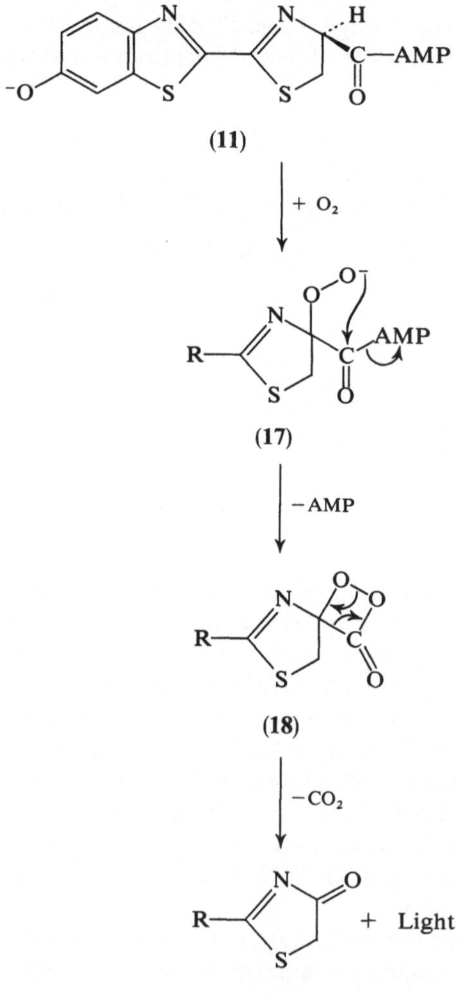

(11)

+ O_2

(17)

− AMP

(18)

− CO_2

(10) + Light

Scheme 15

One of the predictions of the above mechanism is that the CO_2 liberated by this reaction would include one atom of oxygen acquired from molecular oxygen. However, DELUCA and DEMPSEY (*22*) report experiments using $^{18}O_2$ which show that this prediction is not fulfilled.

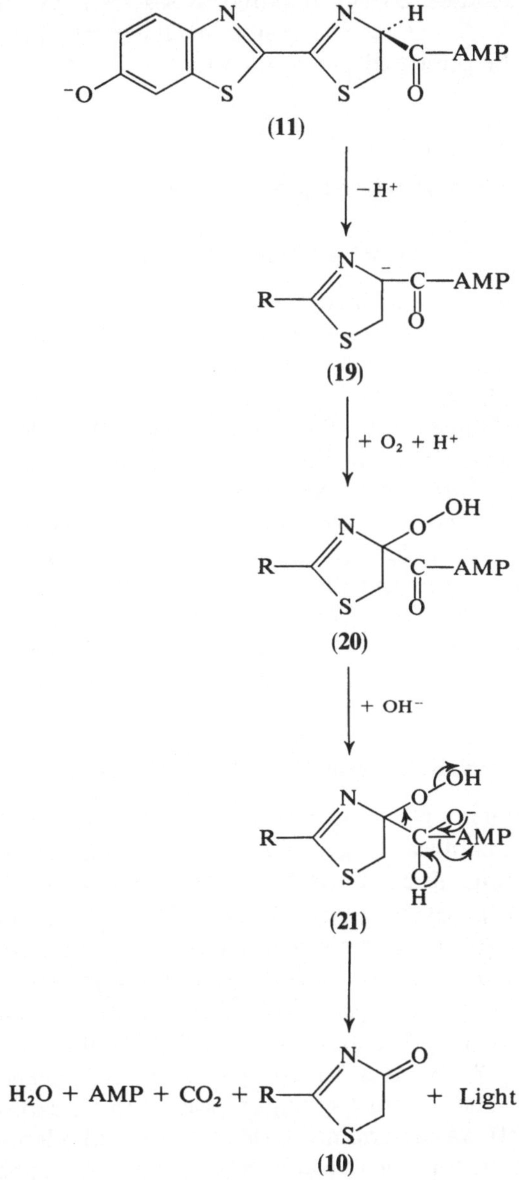

Scheme 16

Instead they demonstrate that the added oxygen originates from water. In the light of these results they propose the mechanism illustrated in Scheme 16. As in *Renilla* this requires, in addition to oxygen, the attack of a hydroxyl ion to account for the oxygen in CO_2 originating from water. The excitation energy required arises from the concerted bond changes of **21**. A further discussion of the general aspects of this mechanism will be presented in section VIII.

IV. Cypridina Bioluminescence

1. General Comments

Cypridina hilgendorfii is a small (3 – 3.9 mm in length) luminous ostracod that is abundant in Japanese waters. The luciferin and luciferase of this animal are stored in separate gland cells. One type of gland cell is filled with small (2 – 3 micron diameter) colorless granules while the second is filled with larger (10 micron diameter) yellow granules (*52*). Upon stimulation the animal simultaneously secretes both types of granules into the sea water where they mix and dissolve, thus forming a localized concentration of luciferase and luciferin which results in a brilliant bluish luminescence. The emission is broad and unstructured with an emission maximum at about 460 nm (*68*).

The live animals when appropriately dried retain their ability to luminescence when moistened (*52*). Thus the dried material has provided a ready source of luciferase and luciferin for biochemical investigations.

2. Chemical Requirements for Light Emission

The only requirements for light emission in *Cypridina* are luciferin (**22**), luciferase, and dissolved oxygen. The isolation and crystallization of pure luciferin from dried *Cypridina* was reported in 1957 by Shimomura *et al.* (*117*). Several years later a method was reported for obtaining luciferin from live animals which resulted in increased yields such that about 20 milligrams of luciferin was obtained per kilogram wet weight of *Cypridina* (*49*). This provided sufficient material for chemical studies which culminated in 1966 in the structural elucidation of luciferin (**22**) by Kishi *et al.* (*77*, *78*) as a 3-indolopyrazine with isoleucine and arginine side chains. The absolute configuration of the β-carbon atom of the isoleucine moiety was determined by enzymatic methods and found to be l (*45*). A methanolic solution of luciferin has the following absorption bands with molar absorptivities in parentheses: 218 nm (27,500), 270 nm

(17,000), 310 nm (11,500), and 435 nm (9,000) (*78*). Luciferin shows strong fluorescence (*45, 123*) in aqueous solution ($\lambda_{max} = 535$ nm) as well as in methanol ($\lambda_{max} = 514$ nm).

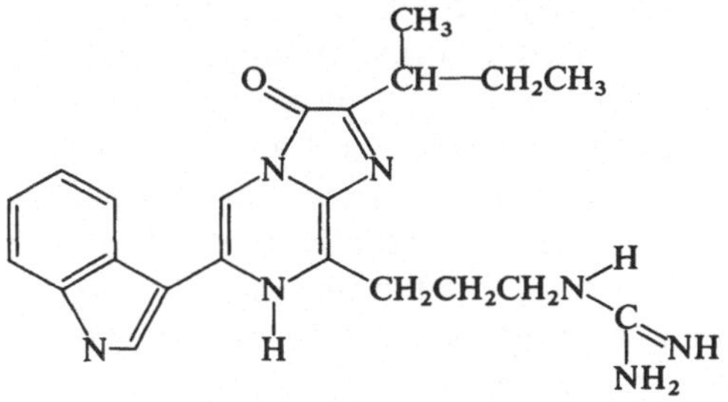

(22)

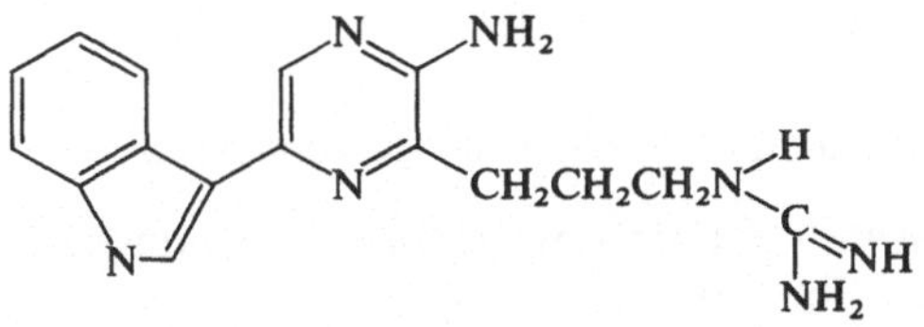

(23)

Structure **22** was confirmed by total synthesis of luciferin (*79*). This was accomplished by the condensation of synthetic etioluciferin (**23**) with α-keto-β-methylvaleric acid followed by reduction and ring closure. The yield was less than 1% but the synthetic material appeared to be identical to the naturally occurring substance. A much improved synthetic method was reported recently by INOUE *et al.* (*67*). This method involves the condensation of the dihydrobromide of synthetic etioluciferin (**23**) with α-keto-β-methylvaleraldehyde to give crude luciferin in 70% yield. Purification and crystallization of the crude product resulted in a final yield of 30%.

The effect of chemical modifications of the luciferin structure on bioluminescence and chemiluminescence has been summarized in two recent reviews (*42, 45*). For chemiluminescence the indole ring can be replaced by phenyl and the amino acid side chains by hydrogen or methyl without markedly changing the rate. This is not so for bioluminescence. For example replacement of the indole moiety by phenyl or removal of the guanidino group reduces the bioluminescence rate by a factor of about 10^2 to 10^3. On the other hand the isoleucine side chain may be replaced by other alkyl groups without producing such a large effect on the bioluminescence rate. These results suggest that both the indole moiety and the arginine side chain are important for proper binding to luciferase.

Luciferase was isolated from *Cypridina* in 1961 and many physical constants of the purified enzyme have been determined (*123, 139, 124*). A molecular weight in the range of 52,000—57,000 was found by Shimomura *et al.* (*123, 124*) for *Cypridina* luciferase whereas Tsuju and Sowinski have reported a value of approximately 80,000 (*139*). The sedimentation constant, $S^{\circ}_{20,\,w}$, was determining to be about 4.6. The diffusion constant, $D_{20,\,w}$, is 7.9×10^{-7} cm^2 sec^{-1} and the isoelectric point of the enzyme is 4.35. The Michaelis constant for luciferin is 5.2×10^{-7} M. It has recently been shown that *Cypridina* luciferase is a metalloprotein containing one to two gram atoms of tightly bound calcium per molecular weight of 70,000 (*85*).

According to Shimomura *et al.* (*124*) the luciferase content per animal is about one microgram which is equivalent to 1.2×10^{13} molecules. From the amount of extractable luciferin its concentration per animal was estimated to be about one hundred times greater than that of luciferase.

3. Mechanism of the Luminescent Reaction

The bioluminescent oxidation of luciferin (**22**) catalyzed by luciferase can be illustrated by Scheme 17.

$$\text{Luciferin (\textbf{22})} + O_2 \xrightarrow{\text{Luciferase}} \text{Oxyluciferin (\textbf{24})} + CO_2 + \text{Light}$$

Scheme 17

During this reaction 1 mole of luciferin reacts with 1 mole of oxygen to yield 1 mole of CO_2 and presumably 1 mole of oxyluciferin (*78, 130*). A recent determination of the bioluminescence quantum yield gave a value of 0.3 (*120*). The molecular activity (micromoles of luciferin transformed per minute per micromole of luciferase) was found to be 1,600. Thus luciferase "turns over" as had been predicted from kinetic data (*10*).

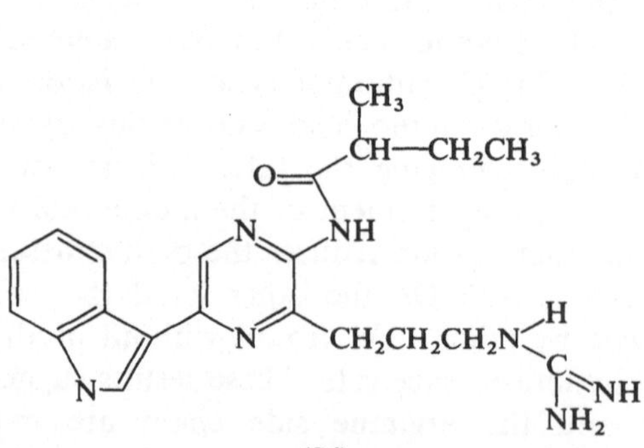

(24)

The correct structure of oxyluciferin (**24**) was recently deduced (*44, 130*). Acid hydrolysis of oxyluciferin leads to the formation of etioluciferin (**23**) and α-methylbutyric acid. It is interesting that *Cypridina* luciferase also possesses hydrolase activity for oxyluciferin, the reaction again leading to the production of etioluciferin and α-methylbutyric acid. The molecular activity for luciferase in this reaction is 2 (*123*).

The nature of the product excited state during *Cypridina* bioluminescence was predicted from studies on the chemiluminescence of luciferin and appropriate analogs (*43, 44, 88*). For example McCapra and Chang (*88*) showed that **25** (Scheme 18) is chemiluminescent in DMSO in the presence of oxygen and a base, and that the product excited state is **26**. Goto *et al.* (*43, 44*) showed that **27** (Scheme 19), as well as luciferin (**22**) produced an intense chemiluminescence in diethylene glycol dimethyl ether (diglyme). The product excited state produced from the chemiluminescence of **27** was found to be **28**. In both cases emission occurred from the anions. The similarities between **26** and **28** and the structure of oxyluciferin (**24**) should be noted. Thus it is no surprise that studies on the chemiluminescence of luciferin (22) showed that oxyluciferin (**24**) was the product excited state during the chemiluminescent reaction (*44*). In this case however it appears that photon emission from the electronically excited state of oxyluciferin occurs from

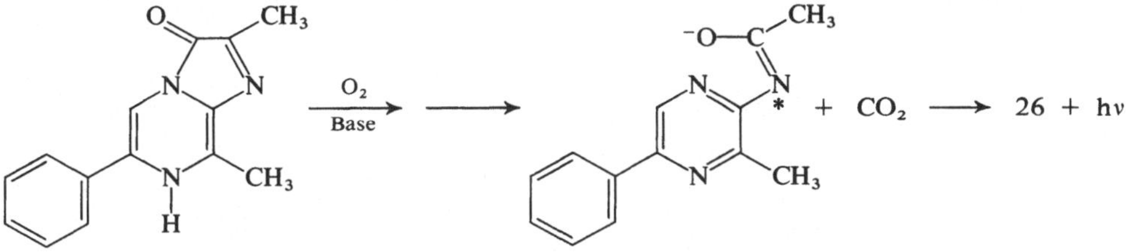

Scheme 18

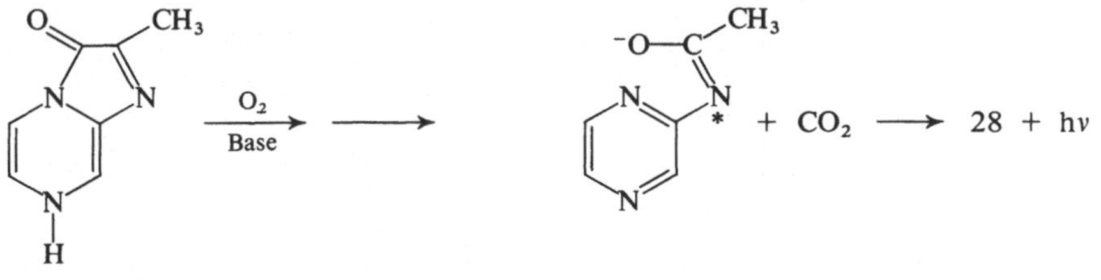

Scheme 19

the neutral species rather than from the anion. It also appears that the electronically excited state of the anion is formed first and that this is followed by protonation prior to photon release.

Oxyluciferin (**24**) is only very weakly fluorescent in aqueous solution (*123*). However, it is strongly fluorescent in aprotic solvents; this results in pronounced blue shifts which depend upon the solvent (*123, 44*). Thus it was predicted (*44*) that a luciferase-oxyluciferin complex represented the electronically excited state product during *Cypridina* bioluminescence. Luciferase was thought to provide the proper aprotic environment leading to the high bioluminescence quantum yield observed. This is quite analogous to the situation which prevails during bacterial bioluminescence where again luciferase provides the proper environment for high quantum yield photon emission from the electronically excited state of the flavin cation (*34*). Shimomura *et al.* (*123*) have provided recent evidence that the light emitter during *Cypridina* bioluminescence is indeed a luciferase-oxyluciferin complex. Oxyluciferin (**24**) combines with luciferase in a 1:1 ratio. Upon binding to the enzyme the fluorescence intensity of oxyluciferin increases about 100-fold and the fluorescence emission matches that of the luciferase catalyzed bioluminescent oxidation of luciferin (**22**) as shown in Fig. 7.

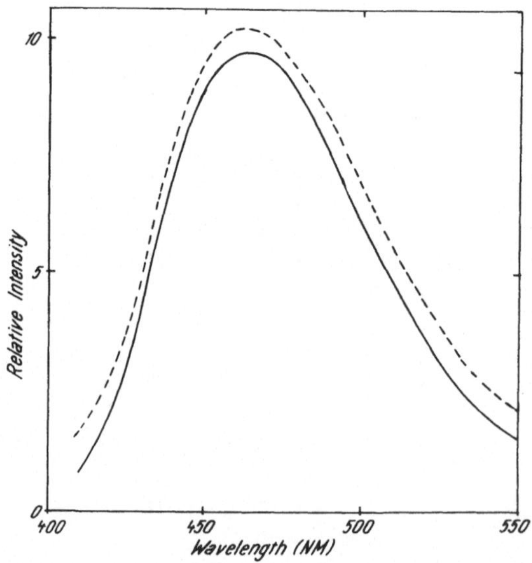

Fig. 7. Comparison between *in vitro* bioluminescence of *Cypridina* and fluorescence of the oxyluciferin-luciferase complex. Data are from Shimomura *et al.* (*123*)

When the bioluminescent reaction is allowed to proceed in the presence of $^{18}O_2$ approximately one oxygen atom of the CO_2 produced is labelled (*121*). Much less labelling occurs when the reaction is carried out in the presence of $H_2^{18}O$. These labelling patterns are in direct contrast to those observed for *Renilla* and firefly bioluminescence as outlined in previous sections. Thus the mechanism proposed for *Cypridina* bioluminescence is somewhat different and presumably operates as shown in Scheme 20. The first part of the mechanism is analogous to that postulated for the *Renilla* (Scheme 3) and firefly (Scheme 16) cases, *i.e.*, the linear hydroperoxide (**29**) probably arises by oxygen attack on a carbanion. The major difference from Schemes 3 and 16 lies in postulating the formation of the dioxetane intermediate **30** rather than assuming attack by hydroxyl ion on an intermediate corresponding to **29**. The dioxetane intermediate is presumed to decompose in a concerted fashion leading to the formation of oxyluciferin (**24**) in an electronically excited state. Its return to the ground state results in photon emission (see Section VIII for a more detailed discussion of mechanisms in bioluminescence and chemiluminescence).

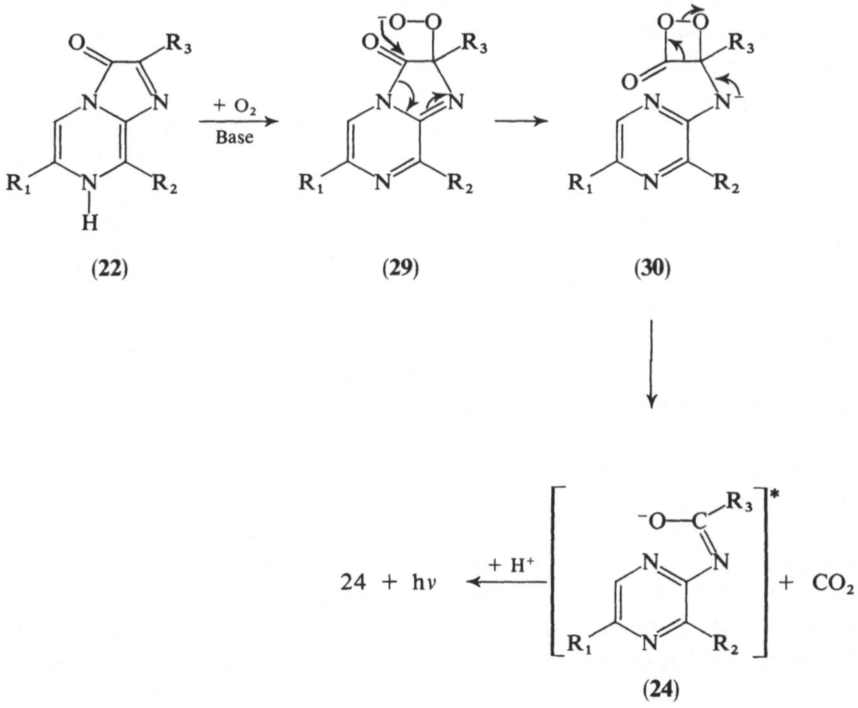

Scheme 20

4. Related Bioluminescent Systems

It has been shown (69, 72) that the biochemical requirements for light emission in certain types of shallow water bioluminescent fish (*Apogon ellioti* and *Parapriacanthus beryciformes*) are closely related to those found in the crustacean *Cypridina*. In all these cases the luciferases and luciferins have been shown to be interchangeable. There is good evidence that crystalline luciferin obtained from *Parapriacanthus* (72) has chemical and physical properties identical with those described for *Cypridina* luciferin. Thus the structures of the luciferins are assumed to be identical. The emission spectra of the *in vitro* bioluminescent reaction of *Cypridina* and the fish systems are identical as are those involving an interchange of luciferase and luciferins (128). On the other hand there do seem to be some subtile differences in the luciferases. These differences include chromatographic, kinetic, and immunological behavior (136).

When the "interphylum luciferin-luciferase reaction" described above was discovered it was also realized that luciferin and luciferase in these fish were stored in two luminous glands, one of which was connected to the gut *via* a duct. Since these fish are known to ingest live *Cypridina* the question arose as to the origin of luciferin and luciferase in these glands. This is a difficult question to answer.

Of possible importance to a solution of this problem is the observation that the luciferin and luciferase from another bioluminescent fish, *Porichthys porosissmus*, will cross-react with *Cypridina* luciferin and luciferase to produce light (12). Here again the evidence suggests close similarities in the luciferin structures and identical bioluminescence emissions were observed. In addition it has recently been observed that *Porichthys* luciferin and luciferase will cross-react to produce light with luciferin and luciferase obtained from a number of other species including the fish *Apogon* and *Parapriacanthus* and the crustacean *Euphansia* (138). Furthermore the luciferin and luciferase from a bioluminescent myctophid fish has been shown to cross-react with the luciferin and luciferase from *Cypridina* (137). Now the luciferins and luciferases of *Porichthys* and the mytophid are located within intradermal photophores. Since these photophores have no connection to the gut or other body organs, it seemed rather unlikely that the source of luciferase was their food supply. However, Cormier et al. (12) have shown that *Porichthys* luciferin can be found in all tissues of its body while Tsuji et al. (138) have shown that *Porichthys* ingest luminous euphansiids (138). These observations suggest the possibility that luciferin could be derived from the food supply of *Porichthys* after all.

Certainly such observations on similarities in the luminescent systems among animals of different phyla will aid in gaining an understanding of the evolutionary relationships in bioluminescence.

V. Bacterial Bioluminescence

1. General Comments

Several species of both symbiotic and free living bacteria exhibit bioluminescence (*50*). Symbiotic forms have been isolated (*47, 48, 50*) from several species of fish. In the host species there are often highly evolved culture organs, lens, and reflector systems whereby the host can control the luminescence. Biochemical research has been carried out using *Photobacterium javanescence, Photobacterium fischeri,* and *Photobacterium phospherium* (*19, 50, 54*), but *P. fischeri* has been studied to the largest extent. The luciferase responsible for catalyzing the bioluminescence reaction has been isolated and purified from two strains of *P. fischeri* and from *P. phosphoreum*. This enzyme carries out a complicated reaction involving oxidation of reduced flavin mononucleotide (**31**) by molecular oxygen and, according to current arguments, is also involved in the concomitant chemical alteration of long chain aldehyde. All three substrates are required for full bioluminescence activity, but the classical role of a luciferin more nearly fits the flavin molecule.

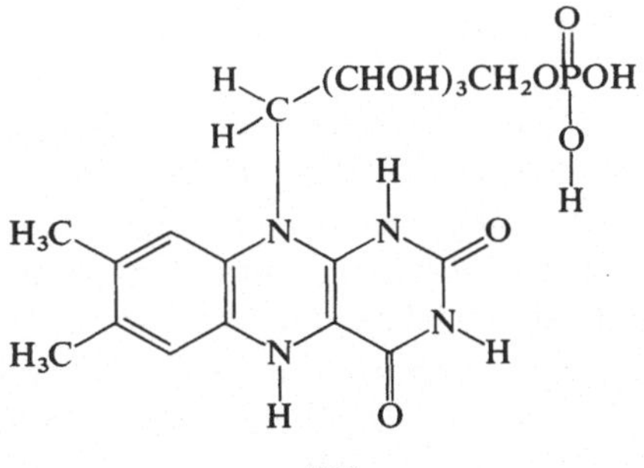

$$\text{(31)}$$

Again the primary problem in analyzing this system is the energetics and species involved in the mechanism. The nature of the excited state species responsible for emission has been particularly elusive, as has been the role which long chain aldehydes play in the chemistry of the reaction. We will attempt here to analyze only the more recent data in order to postulate possible mechanisms and to suggest some experimental approaches to the solution of this problem. Recent reviews (*19, 54*) can give the reader an excellent background for understanding the development of research on bacterial bioluminescence.

2. Chemical Requirements for the Light Reaction

The purified bacterial bioluminescence system exhibits four requirements for light emission: a luciferase with a molecular weight of approximately 80,000 (*60*), reduced flavin mononucleotide (*132*), molecular oxygen (*131*), and long chain aldehyde (*131*). The products of the reaction are oxidized flavin mononucleotide (*17*) and an as yet unidentified aldehyde derivative or derivatives. The overall reaction can then be represented by Scheme 21.

$$FMNH_2 + RCHO + O_2 \rightarrow FMN + Products + Light$$

Scheme 21

The enzymes from both *P. fischeri* and *P. phosphoreum* strains have been purified (*60, 62, 81, 107*). The *P. fischeri* enzymes have been obtained in crystalline form. To date, the molecular weights are all near 80,000; the luciferase from the two strains of *P. fischeri* which have been studied in detail is composed of two non-identical subunits designated α and β (*38, 100*). While the subunits can be separated and the native enzyme regenerated from them with the MAV luciferase (*99*), attempts to form hybrid enzyme from the two *fischeri* strains have been unsuccessful (*62*).

Meighen *et al.* (*98, 99*) have approached the question of subunit function in *fischeri* luciferase (from the MAV strain) by means of chemical alteration studies. They succinylated luciferase to a level of between 30 and 50% of the available acceptor groups (35 lysine residues per luciferase molecule). The inactive succinylated luciferase formed (designated $\alpha_s\beta_s$) was employed in a series of hybridization experiments. These results indicate that while the hybrid $\alpha\beta_s$ was at least half active, the $\alpha_s\beta_s$ hybrid was not active at all. Further studies (*99*) revealed that the catalytic role rests primarily with the α subunit and that the β subunit does not participate directly in binding of substrates or catalysis.

These enzymes are fairly specific in their substrate requirements. Various flavin analogues will substitute for FMN with reduced activity levels (*101*), and aldehydes of various chain lenghts can be used

(*131*); however the rate of reaction is very dependent upon which aldehyde is used.

$FMNH_2$ is by far the best substrate of the flavin analogs tried. $FMNH_2$ (**31**) is oxidized during the bioluminescence reaction, but does not appear to be changed in any other way since, with an appropriate reducing system, the reaction has been shown to recycle FMN. For example, individual determinations of the quantum yield with respect to FMN have resulted in values ranging from 4 to 100 (*17, 83*).

The requirement for long chain aldehyde has been most controversial. The source of the conflict has been the endogenous bioluminescence activity of luciferase preparations, which led some workers (*56, 58, 59*) to conclude that aldehyde was not an absolute requirement for light emission. Support for this view was obtained from experiments where the bioluminescence reaction was allowed to take place in ice at low temperatures. Under these conditions the stimulating effect of aldehyde was reduced and the endogenous activity increased (*58*). These results were explained by ELEY and CORMIER (*32*) as being due to the effect of low levels of aldehyde impurities and the puddling of substrates in ice crystals. They demonstrated that the bioluminescence of ice systems is stimulated remarkably by small concentrations of aldehydes. For instance using the *phosphoreum* enzyme, the amount of aldehyde which gave a 5-fold stimulation at room temperature gave a 35-fold stimulation at low temperatures. They conclude that the endogenous bioluminescence activity is due to a low level aldehyde contamination. Thus far, maximal activity has been obtained with dodecanal for the *phosphoreum* enzyme (*31*) and tetradecanal for the *fischeri* enzyme (*61*).

3. Mechanism of the Light Reaction

As pointed out above, until recently the direct participation of long chain aldehyde in the chemistry of bacterial bioluminescence was in question. It was proposed that aldehyde acted only to change the conformation of the enzyme and thereby the quantum yield of the reaction (*56, 58, 59*). However early inhibition experiments (*129*) showed that 2-decenal was a competitive inhibitor of the light stimulating effect of decanal by affecting the rate of reaction but not the quantum yield. Taken together with the observation that 9-undecenal was not an inhibitor of undecanal (in fact it is a better substrate) these data suggested that the α-carbon atom of the aldehyde molecule was implicated in the reaction chemistry. More recent studies (*32*) on frozen bioluminescence and on the stimulatory effects of added aldehyde with purified preparations of enzyme have given further support to an absolute requirement for aldehyde. Similar results and conclusions can be seen in the work of Erlanger *et al.* (*35*) who using resin bound enzyme systems could recycle the *fischeri* enzyme reducing the endogenous level. Under these conditions they observed a 135-fold stimulation upon addition of aldehyde compared with the 45-fold stimulation reported by Hastings *et al.* (*61*). Finally the recent data of Meighen and Hastings (*97*) which indicate that only one $FMNH_2$ molecule is involved in the reaction point out the necessity of aldehyde involvement in order to explain the energetics of the reaction.

The nature of the emitting species has been elusive. The most obvious canditate has been the flavin molecule itself. The typical bioluminescence spectrum (Fig. 8) is, however, considerably different

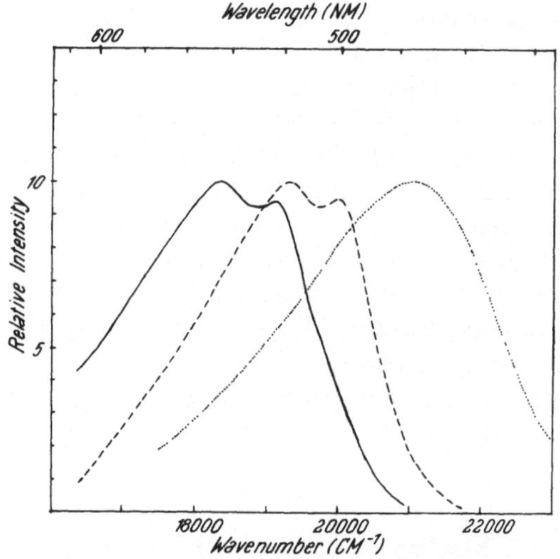

Fig. 8. Comparison between flavin fluorescence in fluid aqueous solution (solid line) and the rigid glass (dashed line) and bioluminescence of *Photobacterium phosphoreum* (dotted line). Data taken from Eley *et al.* (*34*)

from the fluorescence spectrum of FMN (*61*) and environmental effects (polarity, etc.) have not been successful in accounting for the observed emission spectrum. For this reason some workers have at one time or another postulated that various enzyme chromophores are responsible for the emission (*14, 54*). However, more recent data supports the hypothesis that flavin or a flavin derivative is the emitting species (*34, 101*). These experiments have shown that the emission spectrum of bioluminescence can be varied by using different flavin analoges as substrates, thus indicating that the flavin ring system is related directly to the emitting species. However, the high quantum yields with respect to flavin in these systems (*17, 83*) show turnover of flavin to such a degree that the emitter species must be a transient derivative; in fact, it is probably an enzyme bound species formed during the reaction. ERLANGER *et al.* (*35*), again using the resin bound enzyme, have demonstrated that light emission occurs from an enzyme bound intermediate. This conclusion is also supported by the observation that while the flavin analogs alter the bioluminescence spectrum, there is no correlation between the bioluminescence in the presence of a specific analog and its fluorescence spectrum.

The nature of the emitter is more clearly indicated by experiments of ELEY *et al.* (*34*) which show that under appropriate conditions of restricted molecular movement, the flavin cation (**32**) protonated at

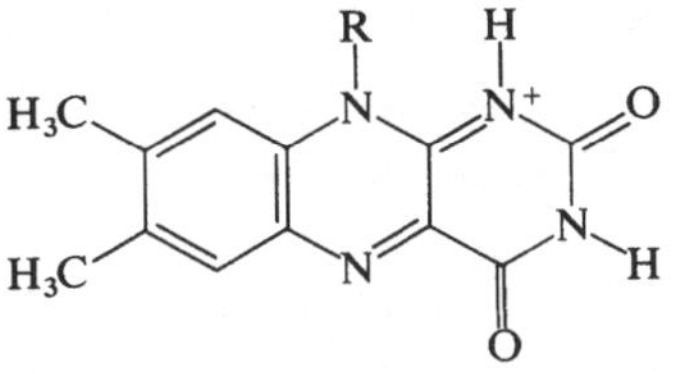

(32)

position N-1 can give a fluorescence which is nearly identical with the bioluminescence. Moreover this fluorescence has a quantum yield of 20% which is more than adequate to account for the observed bioluminescence quantum yield which is less than 10% (*34*). In addition a rough correspondence was seen between the fluorescence maxima of the analog cations and the corresponding bacterial bioluminescence (*in vitro*) in the presence of the analogs. This is illustrated by the data of Table 2 Fig. 9 illustrates the comparison between cation fluorescence and bacterial bioluminescence.

Table 2. *Comparison of Intensity Maxima of Flavin Cation Fluorescence and Bioluminescence Emission*

Flavin	Absorption Maximum (25 C) of the Cation	Flavin Cation Fluorescence, 77 K		Bioluminescence*, 9° C
	Ethanon-HCl	Ethanol-HCl (1:1)	18 N H_2SO_4	
	λ_A (nm)	λ_F (nm)	λ_F (nm)	λ_B (nm)
FMN	393	480	493	492
FAD	393	480	493	484
Iso-FMN	383	530	530	472
2-Thio-FMN	245	510—530	510—530	534

* Bioluminescence data are uncorrected values (MITCHELL and HASTINGS (*101*)); λ_F represents corrected values as described in ELEY *et al.* (*34*).

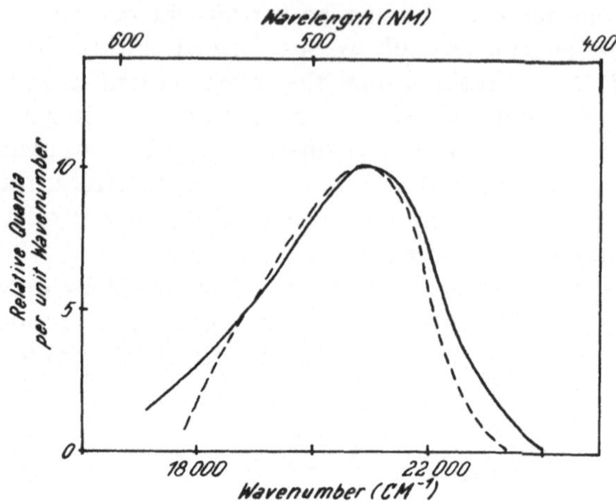

Fig. 9. Comparison between fluorescence of flavin cation in rigid glass (dashed line) and bioluminescence of *Photobacterium phosphoreum* (solid line). Data taken from Eley *et al.* (*34*)

Hastings and coworkers (*40, 57*) have demonstrated the existence of a light-inducible bioluminescence in bacterial luciferase preparations which requires both aldehyde and molecular oxygen. The action spectrum of this activity corresponds to the absorption spectrum of these enzyme preparations and this LIP (light-inducible protein) fraction has been isolated from luciferase (*102*). The spectral studies of Eley *et al.* (*33*) also indicate the presence of a chromophoric group in their enzyme preparations which is tightly bound, but which exhibits fluorescence properties similar to the flavin cation (**32**). Binding studies of FMN suggest that this species is chemically distinct from the cation, since the cation appears to be poorly bound.

The reaction sequence shown in Scheme 22 is a hypothetical scheme based primarily on kinetic data and is proposed here merely to aid in summarizing the possibilities for the reactions which might be involved and for a mechanism of bacterial bioluminescence.

For some time the autooxidation of flavin (reaction ①) has clouded the results of kinetic studies. Gibson and Hastings (*39*) studied this reaction and found that it was self-catalyzing. The rate constant for autooxidation as measured by them was $k_1 = 24$ sec^{-1}. The equilibrium of reaction ② is of course disturbed by the autooxidation reaction. However, Gibson *et al.* (41) have estimated the rate constants involved ($k_3 = .4$ sec^{-1}; $k_2 = 1 \times 10^7$ M^{-1} sec^{-1}; $k_{eq} = 2 \times 10^{-6}$). Recently Meighen and Hastings (*97*) have reported that the stoichiometry of this reaction requires one FMNH$_2$ per enzyme molecule. This finding contradicts previous experiments which indicated a two to one stoichiometry (*92*).

②
FMNH₂ + E ⇌ E−FMNH₂ + O₂
K₂
K₃
O₂ ① K₁
K₄ ③
FMN + H₂O   E−FMNH₂−O₂ + RCHO   E−FMNX
K₇ K₆ ⑤
K₅ ④
⑥ K₈ K₉   ⑬ K₁₆ O₂   K₁₅ ⑫
E−FMN + H₂O₂   E−FMNXH
(E−FMNH₂−O₂−RCHO)
K₁₀ ⑦
K₁₄ hν ⑪
Products + E−FMNH⁺* →(K₁₃) E−FMNX
⑩
K₁₁ ⑧
E−FMNH⁺
K₁₂ ⑨
E + FMN + H⁺

Scheme 22

The order of addition of the additional reactants, oxygen and aldehyde, was determined by Hastings and Gibson (56). They have also indicated that the addition of aldehyde is a reversible step (reaction ⑥).

The slow oxidation of the enzyme-flavin-oxygen intermediate (reaction ④) has been measured (56). The rate constant for this step is $k_5 = 0.17\ sec^{-1}$.

The rate limiting step of bacterial bioluminescence is reaction ⑦ which yields the product excited state. The details of the mechanism of this reaction are unknown. McCapra and Wrigglesworth have suggested that a Schiff's base intermediate derived from an aldehyde might be involved in this reaction (91), because of the chemiluminescence produced by oxidation of model Schiff's bases. However, until the products of the reaction have been identified, the mechanism will remain in doubt. The rate constant does vary with aldehyde chain length and has been found to be maximal with dodecanal for the *phosphoreum* enzyme (31) and with tetradecanal as $k_{10} = 1.4\ sec^{-1}$ for the *fischeri* enzyme (59). The cation-like excited state formed by this reaction decays with emission. This is in all probability a singlet transition with a rate constant of around $10^8\ sec^{-1}$. The dissociation of the ground state cation species from the enzyme appears to be rapid and highly favored since essentially no binding reaction has been observed between FMN and the enzyme (41). The lack of affinity of the enzyme for oxidized flavin is not too surprising in light of recent X-ray studies of Kierkegaard *et al.* (76) which show that oxidized flavins are planar structures while reduced flavins have a folded structure. Hence, the binding site may recognize only the folded reduced flavin.

In Scheme 22 the reactions corresponding to rate constants $k_{13} - k_{16}$ are representative of the light-induced bioluminescence phenomenon reported by Gibson *et al.* (40). Here we have postulated that a side reaction involving utilization of the excited state (reaction ⑩) results in formation of an altered oxidized flavin-enzyme adduct, LIP (E-FMNX). Since flavin has been shown to turn over in the bacterial reaction (17, 83), it seems unlikely that LIP is involved in the light reaction pathway.

Several interesting studies have been carried out to investigate the nature of this adduct (40, 57, 102, 82). The reaction steps shown (⑩, ⑪, ⑫, ⑬) summarize the results. Absorption of a photon (reaction ⑪) results in photoreduction of this adduct to form an enzyme-bound, reduced flavin-enzyme intermediate which can then react with oxygen to recycle in the light reaction (reaction ⑬). This reduced intermediate spontaneously decomposes (Step ⑫) at a rate measured by Hastings and Gibson (57) as $k_{15} = 0.9\ sec^{-1}$. The physical constants descriptive of light-inducible bioluminescence are analogous to those describing the $FMNH_2$ initiated biolumnescence. The luminescence spectra are similar as are the overall reaction rate constants with various aldehydes. The similarities extend to frozen state bioluminescence (*40*) and the temperature dependence of this phenomenon (*82*).

VI. Latia Bioluminescence

1. General Comments

Latia neritoides is an unusual bioluminescent animal found in New Zealand. Whereas most bioluminescent species are of either terrestrial or marine origin, *Latia* spends its entire life cycle in fresh water. *Latia* is a fresh water mollusc, a limpet that upon stimulation secretes a brilliantly luminous mucus having an emission maximum at 535 nm (*68, 119*).

2. Chemical Requirements for Light Emission

The production of light by cell-free extracts of *Latia* was demonstrated many years ago (*6*). More recently an oxygen requirement for bioluminescence was reported and isolation and properties of *Latia* luciferin were described (*122*). The luciferin has a molecular weight of 236 and exhibits a single absorption band at 207 nm. Its structure (**33**), reported by SHIMOMURA and JOHNSON (*118*) is most unusual. Note that it contains a β-ionone ring like that found in vitamin A, and a side chain with an enol formate linkage. The structure of this luciferin has been confirmed by chemical synthesis (*37*), natural and synthetic material

(**33**)

(**34**)

being identical in chromatographic behavior, mass spectrum, and light emission in the presence of luciferase. Recently an alternate synthetic method has been reported (*108*) which results in much higher yields (50%). The authors (*108*) were also able to determine the geometrical configuration of the double bond located in the side chain of *Latia* luciferin which, as shown in **34** is *trans*.

Purified luciferase has a molecular weight of 173,000 and exhibits a typical protein absorption at about 280 nm (*119*).

In addition to luciferin, luciferase and O_2 an unusual requirement for bioluminescence in *Latia* was reported (*119*). This is the requirement for a so-called "purple protein" of molecular weight 39,000. Solutions of this protein exhibit absorption bands in the visible at 568 and 620 nm in addition to the usual 280 nm absorption.

3. Mechanism of the Light Reaction

In attempting to arrive at a reasonable mechanism for the light reaction in *Latia* several observations should be kept in mind. Firstly, one of the products of the light reaction has been identified (*118, 119*) as β-7,8-dihydroionone (**35**) which in analogy to other bioluminescent systems we have named oxyluciferin. Secondly, the kinetics of the reaction show an initial rapid phase followed by a progressively slower phase. This and other evidence suggest that luciferase is strongly product-inhibited (*119*). Thirdly, total light production is proportional to the luciferin concentration suggesting that luciferin acts as a typical substrate (*119*). Fourthly, the purple protein has been observed to have a quantum yield as high as 8 showing that it recycles in the reaction (*119*). Fifthly, an initial lag in the reaction rate of the early phase of the reaction is eliminated when luciferin (**33**) is preincubated with purple protein, but not when it is preincubated with luciferase (*119*). Hence it was suggested that luciferin (**33**) first combines with the purple protein prior to its involvement with luciferase in the light reaction.

On the basis of the evidence outlined above we view the reaction sequence as shown in Schemes 23 and 24. That is, luciferin (**33**) combines

$$\text{Purple protein (P)} + \text{Luciferin (LH)} \longrightarrow \text{P} - \text{LH}$$

Scheme 23

$$\text{P} - \text{LH} + O_2 \xrightarrow{\text{Luciferase}} \text{Oxyluciferin (35)} + \text{P} + h\nu$$

Scheme 24

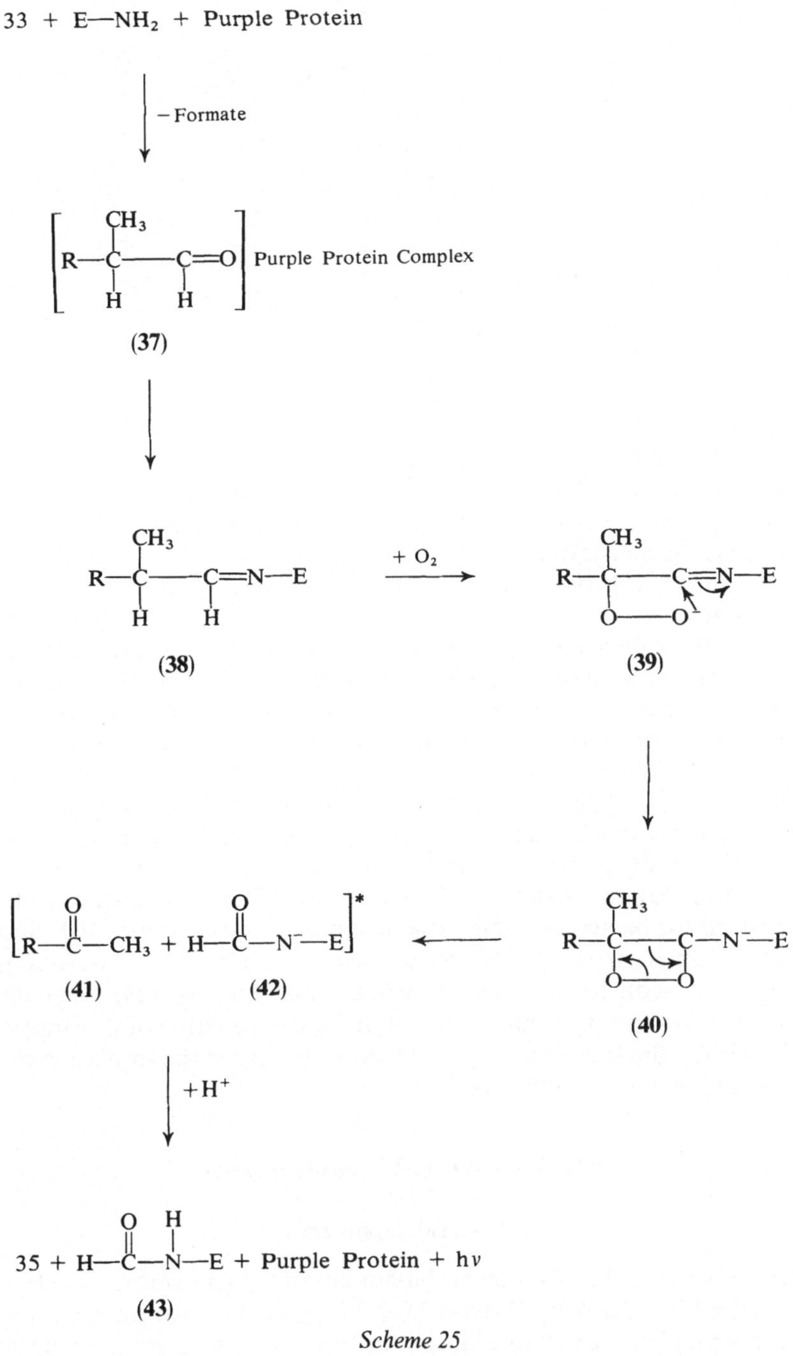

33 + E—NH$_2$ + Purple Protein

$\downarrow$ $-$ Formate

$\left[\text{R—}\overset{\overset{\displaystyle CH_3}{|}}{\underset{\underset{\displaystyle H}{|}}{C}}\text{—}\overset{}{\underset{\underset{\displaystyle H}{|}}{C}}\text{=O} \right]$ Purple Protein Complex

(37)

(38) $\xrightarrow{+ \text{O}_2}$ (39)

(40)

$\left[\text{R—}\overset{\overset{\displaystyle O}{\parallel}}{C}\text{—CH}_3 + \text{H—}\overset{\overset{\displaystyle O}{\parallel}}{C}\text{—N}^-\text{—E} \right]^*$

(41)   (42)

$\downarrow$ $+\text{H}^+$

35 + H—$\overset{\overset{\displaystyle O}{\parallel}}{C}$—$\overset{\overset{\displaystyle H}{|}}{N}$—E + Purple Protein + h$\nu$

(43)

Scheme 25

(35)

(36)

with the purple protein to form the complex $P - LH$. This complex becomes the true substrate for luciferase and in the presence of O_2 reacts to yield the products shown. In Schemes 23 and 24 the purple protein would recycle as has been demonstrated.

It is interesting that neither luciferin (33) nor oxyluciferin (35) are fluorescent.

McCapra and Wrigglesworth (91) have postulated a mechanism for the *Latia* reaction which is based on the observation that a chemiluminescent Schiff's base can be formed from aldehydes and 2-aminopyridine. A modified mechanism is illustrated in Scheme 25 in which it is assumed that luciferin is first hydrolyzed to luciferin aldehyde (36) by either luciferase $(E - NH_2)$ or the purple protein. This is followed by the formation of a Schiff's base with a free amino group of luciferase to give an intermediate such as 38. The decomposition of 40 by a concerted mechanism would yield oxyluciferin (35) and formylated luciferase (43). This is analogous to the mechanism for bacterial bioluminescence proposed in section VIII. Such a mechanism is attractive because it accounts for several observations. Firstly, it accounts for the missing carbons in oxyluciferin (35). Secondly, it accounts for "product inhibition" of luciferase by virtue of its formylation. Thirdly, the lack of product fluorescence suggest the formation of a transient fluorescent intermediate.

VII. Earthworm Bioluminescence

1. General Comments

Some general observations on bioluminescent earthworms have been summarized in a book by Harvey (51). Many of these worms are rather large, ranging between 10 to 28 inches in length (2, 70). Light is produced by the coelomic fluid of these worms which is ejected from the mouth, anus and/or dorsal pores of the worms following mechanical, electrical, or chemical stimulation. When first ejected the coelomic fluid has a milky appearance due to the presence of large coelomic cells. However, within a few seconds after ejection the coelomic cells lyse producing a luminous slime. In the instance of the bioluminescent earthworm, *Diplocardia longa*, it has been shown that all the bioluminescence can be accounted for by the coelomic cells which were isolated by sucrose gradient techniques (2). These coelomic cells contain numerous granules which dissolve upon lysis of the cells and it is thought that the bioluminescent machinery is located within these granules.

2. Chemical Requirements for Light Emission

JOHNSON *et al.* (*70*) have reported the existence of an interesting bioluminescent earthworm from New Zealand whose coelomic fluid produces a bright orange-yellow luminescence. Working with the crude coelomic fluid they reported that oxygen was required for this bioluminescence. Using whole worms and observing the secreted coelomic fluid HARVEY reported an oxygen requirement for bioluminescence from another species of earthworm (*51*). Under similar conditions an oxygen requirement for bioluminescence has also been observed using *Diplocardia longa* (*2, 3*). If, for example, whole worms are placed in an argon atmosphere and subsequently forced to eject their coelomic fluid no luminescence is observed. If oxygen is now admitted a brillant bluish-green luminescence is observed.

In the case of *Diplocardia longa* we now know oxygen is required for bioluminescence only in the crude coelomic fluid (*2, 3*). We have isolated and purified both the luciferin and the luciferase from this animal. When purified preparations of luciferin and luciferase are employed the only additional requirement for bioluminescence is hydrogen peroxide and oxygen is actually an inhibitor of the light reaction (*2*). Thus the overall *in vitro* reaction can be viewed as shown in Scheme 26.

$$\text{Luciferin} + H_2O_2 \xrightarrow{\quad \text{Luciferase} \quad} \text{Product} + \text{Light}$$

Scheme 26

The colors of the light from crude coelomic fluid and from the H_2O_2-dependent *in vitro* systems are the same. There is an emission maximum at 19,700 cm^{-1} (507 nm) and a half-band width of 3,070 cm^{-1} (78 nm) in each case.

Unfortinately no structural information pertaining to the luciferin is as yet available. However luciferase has been brought to a state of apparent homogeneity and many of its properties have been studied (*2, 3*). Luciferase represents about 5% of the total soluble protein of coelomic fluid. Its molecular weight is 320,000, its sedimentation constant ($S_{20, w}$) is 7.3 and its Stokes radius is 100 angstroms. At low salt concentration lucierase aggregates and this process is reversible. Luciferase consists of three non-identical subunits of molecular weight 71,000, 58,000, and 14,500. The native protein presumably consists of two pairs of the larger subunits and four of the low molecular weight one. Its frictional ratio (f/f_0) is 2.25 which suggest that the protein is highly asymmetric. An analysis for metal content has revealed the presence of approximately 1 mole of cupric copper per molecular weight of 320,000 (*2, 3*).

Although H_2O_2 is required for light emission (*2, 3*) it is of interest that luciferase does not contain a heme prosthetic group. Solutions of luciferase exhibit only a single absorption band at 278 nm. Thus the protein should be considered a non-heme peroxidase. The role of copper is not known at the present time.

The participation of H_2O_2 rather than O_2 in the *in vitro* reaction is of interest since most bioluminescent reactions studied are known to require O_2. Another H_2O_2-linked bioluminescence system however has been reported (*13, 29*). In addition a model bioluminescence system has been developed using horseradish peroxidase, H_2O_2, and luminol in which it was shown that radical intermediates are involved in the pathway to light production (*16*). In the case of *Diplocardia*, H_2O_2 is not only an absolute requirement for light production but also irreversibly inactives luciferase. The data show that one molecule of H_2O_2 per luciferase site is required for this inactivation. During the light reaction therefore luciferase does not exhibit enzymatic catalysis and "turnover". Metal binding agents such as cyanide and *o*-phenanthroline will inhibit light production but will also protect luciferase from inactivation by H_2O_2. However we cannot be sure that H_2O_2 inactivation of luciferase involves H_2O_2 attack on a metal in the protein.

Because of the destructive effect of H_2O_2 on luciferase the requirements for O_2 in the crude coelomic fluid must play an important role in the bioluminescence of earthworms. We believe that the crude coelomic fluid contains an oxidase which acts as an H_2O_2 generating system. That is, H_2O_2 must not be present in the coelomic cells until they lyse in the presence of O_2. Thus luciferase would be protected from inactivation by H_2O_2 until there were a need for the worm to bioluminescence. It seems possible that the observations concerning oxygen requirements for bioluminescence in other earthworms (*51, 70*) may involve the same kind of phenomenon.

References, pp. 53—60

VIII. Comments on General Mechanisms Involved in Bioluminescence

Beginning in the mid-sixties numerous papers appeared on possible mechanisms leading to light emission which involved a variety of chemiluminescent compounds. In many of these cases an unusual compound, consisting in part of a four-membered oxygen-containing ring (dioxetane **44**) was proposed as a key intermediate. It is beyond the scope of this

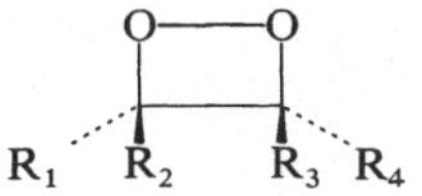

$$R_1 = R_4 = H$$
$$R_2 = R_3 = OC_2H_5$$

(44) (53)

chapter to discuss all the experiments. However it is pertinent to an understanding of the mechanisms of bioluminescence to mention those papers dealing with the chemiluminescence of luciferin analogs. For example the mechanism of chemiluminescence of a number of analogs of *Cypridina* luciferin was studied (*43, 44, 88*). In such cases the product excited state was identified and CO_2 was predicted (but not demonstrated) to be the other product. As discussed earlier (p. 29ff.) a mechanism was proposed which involves a dioxetane as the key intermediate (Scheme 27). At about the same time similar work on analogs of firefly luciferin (*64, 89*) led (p. 23ff.) to equivalent results and predictions (Scheme 28).

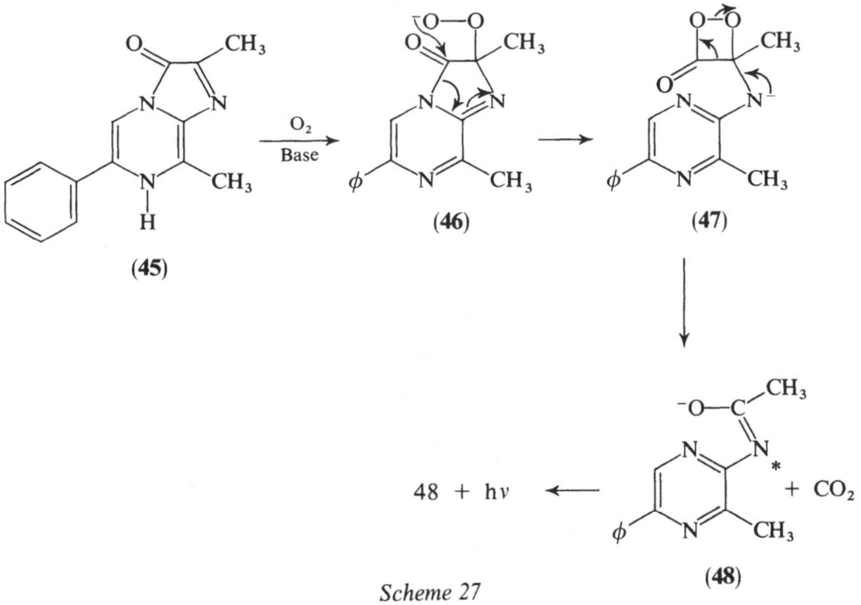

Scheme 27

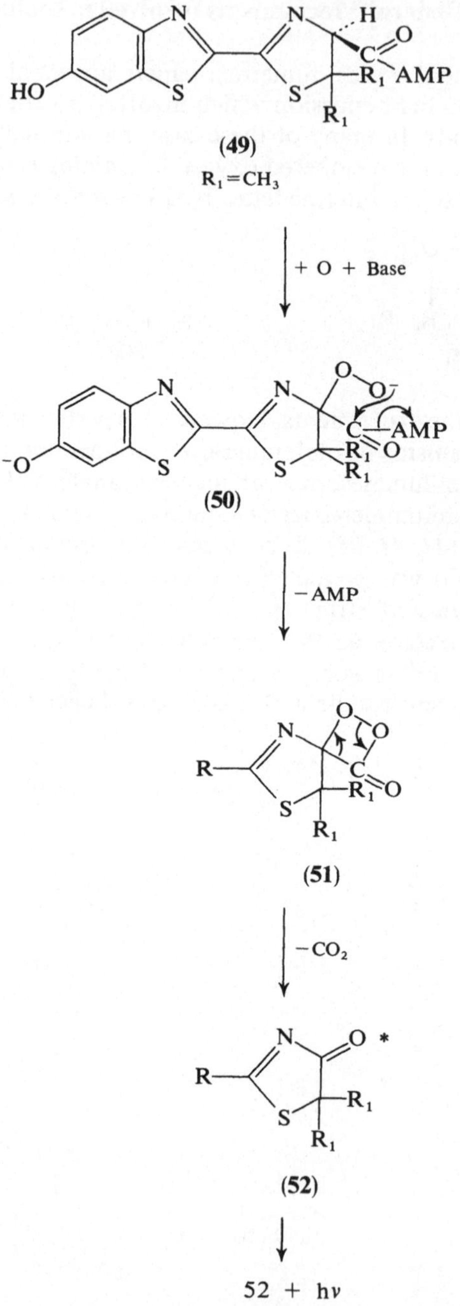

M. J. Cormier, J. E. Wampler, and K. Hori:

(49)

$R_1 = CH_3$

$+ O + Base$

(50)

$- AMP$

(51)

$- CO_2$

(52)

$52 + h\nu$

Scheme 28

These studies were not only important to an understanding of luminescence mechanisms but provided for the first time a theoretical basis for
the formation of electronically excited states in these reactions. It was
pointed out (*87*) that if the decomposition of the dioxetane is concerted then
conservation of orbital symmetry during product formation necessarily
leaves one of the carbonyl products in an electronically excited state.

Such studies stimulated a series of experiments on the mechanisms
of bioluminescent reactions. Using the mechanisms worked out for
chemiluminescence one could for the first time attempt to predict the
products of bioluminescent reactions. Indeed such mechanisms do predict
the correct excited state products for *Cypridina* (*78, 123, 130*), firefly
(*64, 133*) and *Renilla* bioluminescence (*23, 141*). Furthermore CO_2 has
for the first time been demonstrated as a product of the *in vitro* bioluminescence of *Cypridina*, firefly and *Renilla* (*22, 23, 110, 130*).

More importantly, certain of the experiments on bioluminescence were
designed so as to test the validity of the dioxetane mechanism. Since both
the dioxetane (Section IV) and hydroxyl ion mechanisms (Section II)
predict the same reaction products it is difficult to distinguish between them
except by the use of ^{18}O labelling. As seen in Schemes 27 and 28 if the
reactions were to proceed in the presence of $^{18}O_2$ one of the oxygen atoms
of the liberated CO_2 should be labeled, but ^{18}O from $H_2^{18}O$ should not
be incorporated. The same arguments should hold true for the *in vitro*
bioluminescence reactions of *Renilla*, firefly and *Cypridina*. In contrast
with this postulate, however, it was found in experiments with *Renilla* and
firefly (Sections II, III), that the label in the evolved CO_2 came from the
oxygen of H_2O and not O_2, while on the other hand the incorporation of
label into CO_2 in *Cypridina* (Section IV) was in accordance with
Scheme 27.

It is difficult to draw generalizations from these data. Obviously these
experiments are important and should be continued and expanded. For
example it will be most interesting to carry out such experiments on
chemiluminescence using the *Cypridina* and firefly luciferin analogs
previously described (*64, 43, 44, 88, 89*). It should also be mentioned
that when experiments of this type are carried out in so-called "dry
DMSO" it is difficult to reduce the water concentration below 10^{-2} to
10^{-3} molar. Thus the "hydroxyl ion mechanism" (see Section II) could
easily be operative during such chemiluminescent reactions.

Whether one uses the "hydroxyl ion mechanism" or the "dioxetane
mechanism", the same products are predicted from the bioluminescent
and chemiluminescent reactions mentioned above. Aside from this, two
questions remain in analysis of these reaction mechanisms. First, is the
energy yield sufficient to account for the quantum of energy emitted, and
second, is electronic excited state formation an efficient process? The effi-

ciency requirement is a very important one since for example the firefly reaction has an overall quantum yield of nearly unity. For the dioxetane mechanism several recent studies indicate that the answer to both of these questions is positive. Theoretical calculations (75) predict that dioxetane decomposition results in an electronic excited state, probably of singlet multiplicity. In addition calculations of the energy yield expected for a concerted dioxetane decomposition (149) result in a value of about 100 kcal/mole. This is certainly enough energy for even the bluest of the chemiluminescent reactions. The efficiency of excited state production in simple model systems such as the decomposition of *cis*-diethoxy-1,2-dioxetane (53) has been calculated as nearly unity (149).

As yet the "hydroxyl ion" model has not been considered from a theoretical viewpoint. The energy yield for this mechanism depends upon what bond changes actually occur during the concerted, excited state generating step and the degree of participation of other chemical species. Depending upon these details of the concerted reaction and its catalysis the energy generated may or may not be sufficient for the excited state. Most model systems would not distinguish between the two mechanisms. However, the experiments of Richardson and Hodge (112) using chloro-*t*-butyl hydroperoxide (54) and (hydroxyl-*t*-butyl-*t*-butyl peroxide) (55)

(54) (55)

indicate that while the former compound can undergo basic decomposition to form electronic excited states and give light in the presence of a fluorescent dye (Scheme 29), the *t*-butyl peroxide (55) which is prevented from forming a dioxetane intermediate gives no light under the same conditions. While this is certainly not proof that the hydroxyl mechanism is inefficient in excited state production, it does point out the need for both chemical and theoretical investigations into the possibilities of this mechanism. The fact remains that the ^{18}O labeling experiments (22, 23) must be explained.

These two mechanisms are useful in predicting the chemical pathways leading to light emission during bioluminescence and chemiluminescence reactions. With this in mind we have applied these mechanisms to the case of bacterial bioluminescence and arrived at a mechanism (Scheme 30) which represents a modification of one proposed earlier (91).

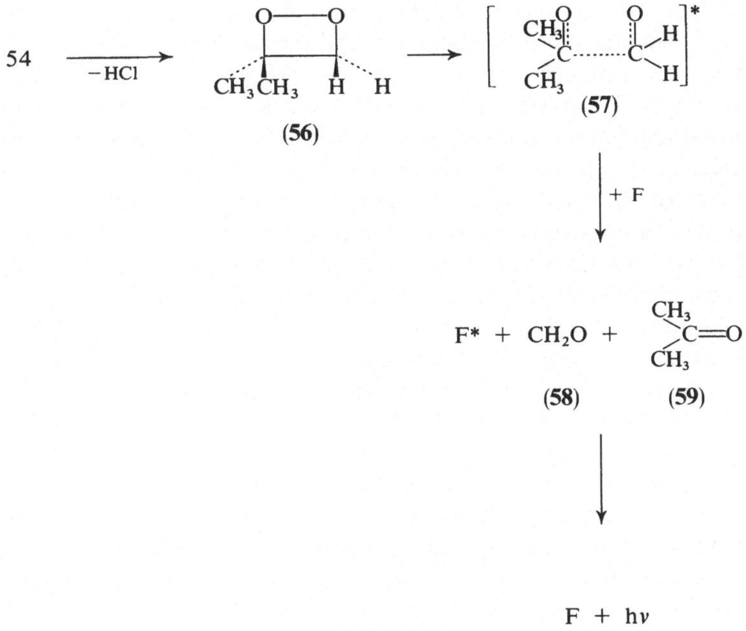

Scheme 29

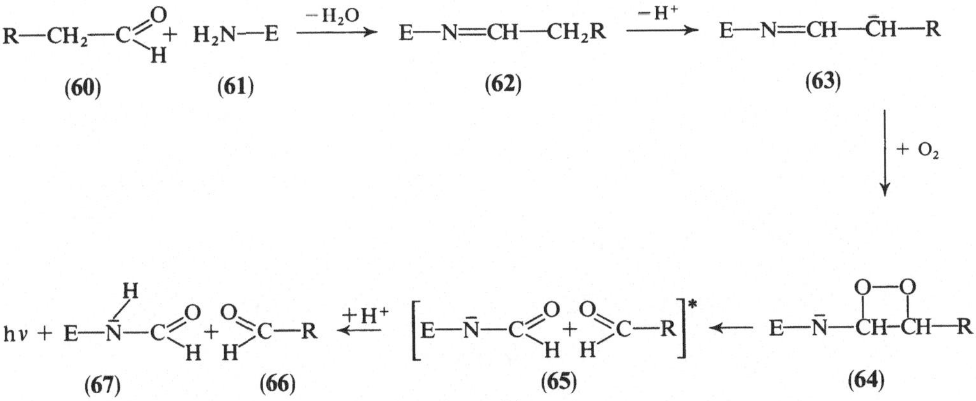

Scheme 30

As pointed out in Section V there has been much scientific debate as to whether aldehyde is actually utilized during bacterial bioluminescence (*14, 17, 32, 55, 58, 59, 92*), but the evidence indicates that it must be (*14, 17, 32, 92, 97*). It has also been mentioned that McCapra and Wrigglesworth (*91*) have described a highly chemiluminescent Schiff's base formed from isobutyraldehyde and 2-aminopyridine. As shown in Scheme 30 if one assumes that long chain aldehydes react with an amino group of the luciferase-$FMNH_2 - O_2$ complex ($E - NH_2$) to produce the Schiff's base **62**, then O_2 (presumably bound) could react to yield intermediates **63** and **64** analogous to those described in Schemes 27 and 28. For convenience we are using the "dioxetane mechanism" but the same final products are predicted if one utilizes the "hydoxyl ion mechanism".

Since nothing is known about the nature of the luciferase-$FMNH_2 - O_2$ complex we will focus our attention only upon what may happen during oxidation of the aldehyde. We are assuming that the oxidant is O_2, presumably bound in accord with the data of HASTINGS and GIBSON (*56*). The predicted products are 1. the lower homolog of the aldehyde used initially and 2. formylated luciferase, rather than CO_2. The energy released during the decomposition of **64** could be transferred to a luciferase-bound flavin cation (*34*) with subsequent photon emission.

Finally we would like to present a general observation on the luciferases, the proteins that catalyze bioluminescent reactions. Many bioluminescent reactions appear to represent a variation of the type of reaction catalyzed by monooxygenases. In the case of monooxygenase (*63*) one atom each of O_2 is incorporated into a substrate (S) and into the H_2O produced in the presence of another hydrogen donor (AH_2) as shown in Scheme 31.

$$S + O_2 + AH_2 \longrightarrow SO + H_2O + A$$

Scheme 31

Certain luciferases (*Renilla* and firefly; see Sections II and III) catalyze a slight variation of this reaction requiring the participation of hydroxyl ion in which one atom each of O_2 is apparently incorporated into the hydrogen donor itself (luciferin in this case) and into the H_2O produced. This is illustrated in Scheme 32.

$$LH + O_2 + OH^- \longrightarrow L'O + H_2O + CO_2$$

Scheme 32

where L' refers to LH (luciferin) minus CO. On the other hand *Cypridina* luciferase apparently catalyzes a reaction during which one atom each of oxygen is incorporated into the hydrogen donor (luciferin) and into the CO_2 which is produced rather than into H_2O. Water is not a product according to the proposed mechanism (*121*).

Since two other bioluminescent forms (*Cavernularia* and *Yarella*), and possibly certain coelenterates as well (see Section II), are known to have luciferase and luciferins that must be closely related in structure, if not identical, to those found in *Renilla*, it is logical to expect that the bioluminescent mechanisms in these other forms might also be of the modified monooxygenase type found in *Renilla* and firefly. Furthermore there are numerous bioluminescent organisms that have been shown to "cross react" with the *Cypridina* system. This implies that those organisms contain *Cypridina*-like luciferases and luciferins (see Section IV). Again in these organisms the mechanism might also be expected to follow the modified monooxygenase type outlined above.

Thus one can begin to envision a classification of bioluminescent reactions based upon mechanism. This could be based upon the catalytic nature of the particular luciferase involved. For example we might have a class of oxygenase-type bioluminescent reactions in which the individual members might be referred to as firefly luciferin monooxygenase, *Renilla* luciferin monooxygenase, *Cypridina* luciferin dioxygenase, etc. In addition we could have a class of peroxidase-type bioluminescent reactions such as *Diplocardia* luciferin peroxidase, *Balanoglossid* luciferin peroxidase, etc.

It would appear that among the higher marine forms there exist at least two basic bioluminescent types which contain structurally related luciferins and where emission is produced by similar mechanisms. These are the *Renilla* type and the *Cypridina* type. Not too many years ago it was felt that most bioluminescent forms produce their light by widely differing chemical mechanisms. Although the reactions in some of the systems studied so far differ chemically, the basic chemical mechanism may be the same for most of those bioluminescent reactions requiring molecular oxygen.

References

1. ANDERSON, J., J. E. WAMPLER, and M. J. CORMIER: Unpublished results.
2. BELLISARIO, R. L.: Studies on the Bioluminescent Earthworm, *Diplocardia longa*. Thesis, Univ. of Georgia, Athens, U.S.A. (1971).
3. BELLISARIO, R., and M. J. CORMIER: Peroxide-Linked Bioluminescence Catalyzed by a Cooper-containing, Non-heme Luciferase Isolated from a bioluminescent Earthworm. Biochem. Biophys. Res. Comm. **43**, 800 (1971).
4. BIGGLEY, W. H., J. E. LLOYD, and H. H. SELIGER: The Spectral Distribution of Firefly Light II. J. Gen. Physiol. **50**, 1681 (1967).
5. BLANK, G. E., J. PLETCHER, and M. SAX: The Molecular Structure of Firefly D-(-)-luciferin: A Single Crystal X-ray Analysis. Biochem. Biophys. Res. Comm. **42**, 583 (1971).
6. BOWDEN, B. J.: Some Observations on a Luminescent Freshwater-Limpet from New Zealand. Biol. Bull. **99**, 373 (1950).

7. Buck, J. B.: The Anatomy and Physiology of the Light Organ in Fireflies. Ann. New York Acad. Sci. **49**, 397 (1948).

8. — Unit Activity in Firefly Lanterns. In Bioluminescence in Progress (F. H. Johnson and Y. Haneda, eds.), p. 459. Princeton, N. J.: Princeton University Press. 1966.

9. Buck, J., and E. Buck: Mechanism of Rhythmic Synchronous Flashing of Fireflies. Science **159**, 1319 (1968).

10. Chase, A. M.: Activity and Inhibition of *Cypridina* Luciferase. In Bioluminescence in Progress (F. H. Johnson and Y. Haneda, eds.), p. 119. Princeton, N. J.: Princeton University Press. 1966.

11. Cormier, M. J.: Studies on the Bioluminescence of *Renilla reniformis*. II. Requirement for 3′,5′-Diphosphoadenosine in the Luminescent Reaction. J. Biol. Chem. **237**, 2032 (1962).

12. Cormier, M. J., J. M. Crane, Jr., and Y. Nakano: Evidence for the Identity of the Luminescent Systems of *Porichthys porosissimus* (fish) and *Cypridina hilgendorfii* (Crustacean). Biochem. Biophys. Res. Comm. **29**, 747 (1967).

13. Cormier, M. J., and L. S. Dure: Studies on the Bioluminescence of *Balanoglossus biminiensis* Extracts. I. Requirements for Hydrogen Peroxide and Characteristics of the System. J. Biol. Chem. **238**, 785 (1963).

14. Cormier, M. J., M. Eley, S. Abe, and Y. Nakano: On the Requirement and Mode of Action of Long Chain Aldehydes During Bacterial Bioluminescence. Photochem. Photobio. **9**, 351 (1969).

15. Cormier, M. J., K. Hori, and Y. D. Karkhanis: Studies on the Bioluminescence of *Renilla reniformis*. VII. Conversion of Luciferin into Luciferyl Sulfate by Luciferin Sulfokinase. Biochemistry **9**, 1184 (1970).

16. Cormier, M. J., and P. M. Prichard: An Investigation of the Mechanism of the Luminiscent Peroxidation of Luminol by Stopped Flow Techniques. J. Biol. Chem. **243**, 4706 (1968).

17. Cormier, M. J., and J. R. Totter: Quantum Efficiency Determinations on Components of the Bacterial Luminiscence System. Biochim. Biophys. Acta **25**, 229 (1957).

18. — — Bioluminescence. Ann. Rev. Biochem. **33**, 431 (1964).

19. — — Bioluminescence. Enzymic Aspects. Photophysiology **4**, 315 (1968).

20. Deluca, M.: Hydrophobic Nature of the Active Site of Firefly Luciferase. Biochemistry **8**, 160 (1969).

21. Deluca, M., L. Brand, T. A. Cabula, H. H. Seliger, and A. F. Makula: Nanosecond Time-Resolved Proton Transfer Studies with Dehydroluciferin and its Complex with Luciferase. J. Biol. Chem. **246**, 6702 (1971).

22. Deluca, M., and M. Dempsey: Mechanism of Oxidation in Firefly Luminescence. Biochem. Biophys. Res. Comm. **40**, 117 (1970).

23. Deluca, M., M. E. Dempsey, K. Hori, J. E. Wampler, and M. J. Cormier: Mechanism of Oxidative Carbon Dioxide Production During *Renilla reniformis* Bioluminescence. Proc. Nat. Acad. Sci. (USA) **68**, 1658 (1971).

24. Deluca, M., and W. D. McElroy: The Hydrolase Properties of Firefly Luciferase. Biochem. Biophys. Res. Comm. **18**, 836 (1965).

25. Deluca, M., G. W. Wirtz, and W. D. McElroy: Role of Sulfhydryl Groups in Firefly Luciferase. Biochemistry **3**, 935 (1964).

26. Denberg, J. L., R. T. Lee, and W. D. McElroy: Substrate Binding Properties of Firefly Luciferase. I. Luciferin Binding Site. Arch. Biochem. Biophys. **134**, 381 (1969).

27. Denberg, J. L., and W. D. McElroy: Catalytic Subunit of Firefly Luciferase. Biochemistry **9**, 4619 (1970).

28. Dubois, R.: Note sur la Physiologie des Pyrophores. Compt. Rend. Soc. Biol. **37**, 559 (1885).

29. Dure, L. S., and M. J. Cormier: Studies on the Bioluminescence of *Balanoglossus*

biminiensis Extracts. II. The Peroxidase Nature of Luciferase. J. Biol. Chem **238**, 790 (1963).

30. DUYSENS, L. N. M.: Transfer of Excitation Energy in Photosynthesis, Keminka and Son, Utrecht (1952).

31. ELEY, M.: Studies on the Bioluminescence of *Photobacterium phosphoreum*. M. S. Thesis, Univ. of Georgia, Athens, USA (1968).

32. ELEY, M., and M. J. CORMIER: On the Function of Aldehyde in Bacterial Bioluminescence: Evidence for an Aldehyde Requirement During Luminescence from the Frozen State. Biochem. Biophys. Res. Comm. **32**, 454 (1968).

33. ELEY, M. H., M. J. CORMIER, and J. LEE: Studies on the Mechanism of Bacterial Bioluminescence: Enzyme Bound Intermediates and Reaction Products. Am. Chem. Soc. Paper Abstracts (SE-SW), 20 (1970).

34. ELEY, M., J. LEE, J.-M. LHOSTE, C. Y. LEE, M. J. CORMIER, and P. HEMMERICH: Bacterial Bioluminescence, Comparisons of Bioluminescence Emission Spectra, the Fluorescence of Luciferase Reaction Mixtures and the Fluorescence of Flavin Cations. Biochemistry **9**, 2902 (1970).

35. ERLANGER, B. F., M. ISAMBERT, and A. M. MICHELSON: Insoluble Bacterial Luciferase: A New Approach to Some Problems in Bioluminescence. Biochem. Biophys. Res. Comm. **40**, 70 (1970).

36. FORSTER, T.: Experimentelle und theoretische Untersuchung des zwischenmolekularen Übergangs von Elektronenanregungsenergie. Zeit. Naturforsch. **4a**, 321 (1949).

37. FRACHEBOUD, M. G., O. SHIMOMURA, R. K. HILL, and F. H. JOHNSON: Synthesis of *Latia* Luciferin. Tetrahedron Letters **45**, 3951 (1969).

38. FRIEDLAND, J., and J. W. HASTINGS: Nonidentical Subunits of Bacterial Luciferase: Their Isolation and Recombination to Form Active Enzyme. Proc. Nat. Acad. Sci. (USA) **58**, 2336 (1967).

39. GIBSON, Q. H., and J. W. HASTINGS: The Oxidation of Reduced Flavin Mononucleotide by Monecular Oxygen. Biochem. J. **83**, 368 (1962).

40. GIBSON, Q. H., J. W. HASTINGS, and C. GREENWOOD: On the Molecular Mechanism of Bioluminescence. II. Light-Induced Luminescence. Proc. Nat. Acad. Sci. (USA) **53**, 187 (1965).

41. GIBSON, Q. H., J. W. HASTINGS, G. WEBER, W. DUANE, and J. MASSA: The Interaction of Flavin Mononucleotide with an Enzyme System from *Photobacterium fischeri*. In Flavins and Flavoproteins (E. C. Slater, ed.), p. 341. Amsterdam: Elsevier Publishing Co. 1966.

42. GOTO, T.: Chemistry of Bioluminescence. Pure and Applied Chemistry **17**, 421 (1968).

43. GOTO, T., S. INOUE, and S. SUGIURA: *Cypridina* Bioluminescence. IV. Synthesis and Chemiluminescence of 3,7-Dihydroimidazo[1,2-a]pyrazin-3-zone and its 2-Methyl Derivative. Tetrahedron Letters 3873 (1968).

44. GOTO, T., S. INOUE, S. SUGIURA, K. NISHIKUWA, M. ISOBE, and Y. ABE: *Cypridina* Bioluminescence. V. Structure of Emitting Species in the Luminescence of *Cypridina* Luciferin and its Related Compounds. Tetrahedron Letters 4035 (1968).

45. GOTO, T., and Y. KISHI: Luciferins, Bioluminescent Substances. Angew. Chem. **7**, 407 (1968).

46. GREEN, A. A., and W. D. MCELROY: Crystalline Firefly Luciferase. Biochim. Biophys. Acta **20**, 170 (1956).

47. HANEDA, Y.: The Luminiscence of Some Deep-Sea Fishes of the Families *Gadidae* and *Macrouridae*. Pacific Science **4**, 372 (1951).

48. — Observations on Luminescence in the Deep-Sea Fish, *Paratrachichthys prosthemius*. Science Report of the Yokosuka City Museum 15 (1957).

49. HANEDA, Y., F. H. JOHNSON, Y. MASUDA, Y. SAIGA, O. SHIMOMURA, H. C. SIE, N. SUGIYAMA, and I. TAKATSUKI: Crystalline Luciferin from Live *Cypridina*. J. Cell. Comp. Physiol. **57**, 55 (1961).

50. Harvey, E. N.: Bioluminescence, pp. 1—95. New York: Academic Press. 1952.
51. — Bioluminescence, p. 233. New York: Academic Press. 1952.
52. — Bioluminescence, p. 302. New York: Academic Press. 1952.
53. — Bioluminescence, p. 512. New York: Academic Press. 1952.
54. Hastings, J. W.: The Chemistry of Bioluminescence. In Current Topics in Bioenergetics 1, 113 (1966).
55. — Bioluminescence. Annu. Rev. Biochem. 37, 597 (1968).
56. Hastings, J. W., and Q. H. Gibson: Intermediates in the Bioluminescent Oxidation of Reduced Flavin Mononucleotide. J. Biol. Chem. 238, 2537 (1963).
57. — — The Role of Oxygen in the Photoexcited Luminescence of Bacterial Luciferase. J. Biol. Chem. 242, 720 (1967).
58. Hastings, J. W., Q. H. Gibson, and C. Greenwood: The Molecular Mechanism of Bioluminescence. I. The Role of Long-Chain Aldehyde. Proc. Nat. Acad. Sci. (USA) 52, 1529 (1964).
59. Hastings, J. W., Q. H. Gibson, J. Friedland, and J. Spudich: Molecular Mechanisms in Bacterial Bioluminescence. On Energy Storage Intermediates and the Role of Aldehyde in the Reaction. In Bioluminescence in Progress (F. H. Johnson and Y. Haneda, eds.), p. 151. Princeton, N. J.: Princeton Univ. Press. 1966.
60. Hastings, J. W., W. H. Riley, and J. Massa: The Purification, Properties, and Chemiliuminescent Quantum Yield of Bacterial Luciferase. J. Biol. Chem. 240, 1473 (1965).
61. Hastings, J. W., J. A. Spudlich, and G. Malnic: The Influence of Aldehyde Chain Length Upon the Relative Quantum Yield of the Bioluminescent Reaction of *Achromobacter fischeri*. J. Biol. Chem. 238, 3100 (1963).
62. Hastings, J. W., K. Weber, J. Friedland, A. Eberhard, G. W. Mitchell, and A. Gunsalus: Structurally Distinct Bacterial Luciferases. Biochemistry 8, 4681 (1969).
63. Hayaishi, O.: Oxygenases (Oxygen-Transferring Enzymes). In Biological Oxidations (T. Singer, ed.), p. 585. New York: Interscience. 1968.
64. Hopkins, T. A., H. H. Seliger, E. H. White, and M. W. Cass: The Chemiluminescence of Firefly Luciferin. A Model for the Bioluminescence Reaction and Identification of the Product Excited State. J. Amer. Chem. Soc. 89, 7148 (1967).
65. Hori, K., and M. J. Cormier: Studies on the Bioluminescence of *Renilla reniformis*. VI. Some Chemical Properties and the Tentative Partial Structure of Luciferin. Biochim. Biophys. Acta 130, 420 (1966).
66. Hori, K., Y. Nakano, and M. J. Cormier: Studies on the Bioluminescence of *Renilla reniformis*. XI. Location of the Sulfate Group in Luciferyl Sulfate. Biochim. Biophys. Acta, in press.
67. Inoue, S., S. Sugiura, H. Kakoi, and T. Gota: *Cypridina* Bioluminescence. VI. A New Route for the Synthesis of *Cypridina* Luciferin and its Analogs. Tetrahedron Letters 1609 (1969).
68. Johnson, F. H.: Bioluminescence. Comprehensive Biochemistry 27, 79 (1967).
69. Johnson, F. H., Y. Haneda, and E. H. C. Sie: An Interphylum Luciferin-Luciferase Reaction. Science 132, 422 (1960).
70. Johnson, F. H., O. Shimomura, and Y. Haneda: A Note on the Large Luminescent Earthworm, *Octochaetus multiporus*, of New Zealand. In Bioluminescence in Progress (F. H. Johnson and Y. Haneda, eds.), p. 385. Princeton, N. J.: Princeton Univ. Press. 1966.
71. Johnson, F. H., O. Shimomura, Y. Saiga, L. C. Gershman, G. T. Renolds, and J. R. Waters: Quantum Efficiency of *Cypridina* Luminescence, with a Note on that of *Aequorea*. J. Cell. Comp. Physiol. 60, 85 (1962).
72. Johnson, F. H., N. Sugiyama, O. Shimomura, Y. Saiga, and Y. Haneda: Crystalline Luciferin from a Luminescent Fish, *Parapriacanthus beryciformes*. Proc. Nat. Acad. Sci. (USA) 47, 486 (1961).

73. KARKHANIS, Y. D., and M. J. CORMIER: Isolation and Properties of *Renilla reniformis* Luciferase, a Low- Molecular Weight Energy Conversion Enzyme. Biochemistry **10**, 317 (1971).
74. KARREMAN, G., and R. H. STEEL: On the Possibility of Long Distance Energy Transfer by Resonance in Biology. Biochim. Biophys. Acta **25**, 280 (1957).
75. KEARNS, D. R.: Selection Rules for Singlet-Oxygen Reactions. Concerted Addition Reactions. J. Amer. Chem. Soc. **91**, 6554 (1969).
76. KIERKEGAARD, P., R. NORRESTAM, P.-E. WERNER, I. CSÖREGH, M. VON GLEHN, R. KARLSSON, M. LEIJONMARCK, O. RÖNNQUIST, B. STENSLAND, O. TILLBERG, and L. TORBJÖRNSSON: X-ray Structure Investigation of Flavin Derivatives. In Flavins and Flavoproteins (H. Kamin, ed.), p. 1. Baltimore, Md.: University Park Press. 1971.
77. KISHI, Y., T. GOTO, S. EGUCHI, Y. HIRATA, E. WATANABE, and T. AOYAMA: *Cypridina* Bioluminescence. II. Structural Studies of *Cypridina* Luciferin by Means of a High Resolution Mass Spectrometer and an Amino Acid Analyzer. Tetrahedron Letters 3437 (1966).
78. KISHI, Y., T. GOTO, Y. HIRATA, O. SHIMOMURA, and F. H. JOHNSON: *Cypridina* Bioluminescence. I. Structure of *Cypridina* Luciferin. Tetrahedron Letters 3427 (1966).
79. KISHI, Y., T. GOTO, S. INOUE, S. SUGIURA, and H. KISHIMOTO: *Cypridina* Bioluminescence. III. Total Synthesis of *Cypridina* Luciferin. Tetrahedron Letters 3445 (1966).
80. KISHI, Y., S. MATSUURA, S. INOUE, O. SHIMOMURA, and T. GOTO: Luciferin and Luciopterin Isolated from the Japanese Firefly *Luciola cruciata*. Tetrahedron Letters 2847 (1968).
81. KUWABARA, S., M. J. CORMIER, L. S. DURE, P. KREISS, and P. PFUDERER: Crystalline Bacterial Luciferase from *Photobacterium fischeri*. Proc. Nat. Acad. Sci. (USA) **53**, 822 (1965).
82. LAVELLE, F., J.-P. HENRY, and M. MICHELSON: E'tudes de Bioluminescence. Comp. Rend. Acad. Sci. Paris **270**, 2126 (1970).
83. LEE, J., and H. H. SELIGER: Absolute Spectral Sensitivity of Phototubes and the Application to the Measurement of the Absolute Quantum Yields of Chemiluminescence and Bioluminescence. Photochem. Photobio. **4**, 1046 (1965).
84. LEE, R., and W. D. MCELROY: Effects of s'-Adenylic Acid on Firefly Luciferase. Arch. Biochem. Biophys. **145**, 78 (1971).
85. LYNCH, R. V., F. I. TSUJI, and D. H. DONALD: Personal communication from Dr. Tsuji.
86. MATTHEWS, J. C.: Purification and Properties of *Renilla* Luciferin Sulfokinase. Thesis, Univ. of Georgia, Athens (USA), 1971.
87. MCCAPRA, F.: An Application of the Theory of Electrocyclic Reactions to Bioluminescence. Chem. Commun. **1968**, 155 (1968).
88. MCCAPRA, F., and Y. C. CHANG: The Chemiluminescence of *Cypridina* Luciferin Analogue. Chem. Commun. **1967**, 1011 (1967).
89. MCCAPRA, F., Y. C. CHANG, and V. P. FRANCOIS: The Chemiluminescence of a Firefly Luciferin Analogue. Chem. Commun. **1968**, 22 (1968).
90. MCCAPRA, F., D. G. RICHARDSON, and Y. C. CHANG: Chemiluminescence Involving Perioxide Decompositions. Photochem. Photobio. **4**, 1111 (1965).
91. MCCAPRA, F., and R. WRIGGLESWORTH: A Chemiluminescent Schiff's Base: A Possible Mechanism for *Latia* and Bacterial Bioluminescence. Chem. Commun. **1969**, 91 (1969).
92. MCELROY, W. D., and A. A. GREEN: Enzymic Properties of Bacterial Luciferase. Arch. Biochem. Biophys. **56**, 240 (1955).
93. MCELROY, W. D., J. HASTINGS, J. COULOMBRE, and V. SONNENFELD: The Mechanism of Action of Pyrophosphate in Firefly Luminescence. Arch. Biochem. Biophys. **46**, 399 (1953).
94. MCELROY, W. D., and H. H. SELIGER: Mechanism of Action of Firefly Luciferase. Fed. Proc. **21**, 1006 (1962).

95. McElroy, W. D., and H. H. Seliger: The Chemistry of Light Emission. Adv. Enzymology **25**, 119 (1963).
96. McElroy, W. D., H. H. Seliger, and E. H. White: Mechanism of Bioluminescence, Chemiluminescence and Enzyme Function in the Oxidation of Firefly Luciferin. Photochem. Photobio. **10**, 153 (1969).
97. Meighen, E. A., and J. W. Hastings: Binding Site Determination from Kinetic Data. Reduced Flavin Mononucleotide Binding to Bacterial Luciferase. J. Biol. Chem. **246**, 7666 (1971).
98. Meighen, E. A., M. Z. Nicoli, and J. W. Hastings: Hybridization of Bacterial Luciferase with a Variant Produced by Chemical Modification. Biochemistry **10**, 4062 (1971).
99. — — — Functional Differences of the Nonidentical Subunits of Bacterial Luciferase. Properties of Hybrids of Native and Chemically Modified Bacterial Luciferase. Biochemistry **10**, 4069 (1971).
100. Meighen, E. A., L. B. Smillie, and J. W. Hastings: Subunit Homologies in Bacterial Luciferase. Biochemistry **9**, 4949 (1970).
101. Mitchell, G., and J. W. Hastings: The Effect of Flavin Isomers and Analogues Upon the Color of Bacterial Bioluminescence. J. Biol. Chem. **244**, 2572 (1969).
102. Light-Induced Bioluminescence. Isolation and Characterization of a Specific Protein Involved in the Absorption and Delayed Emission of Light. Biochemistry **9**, 2699 (1970).
103. Morin, J. G., and J. W. Hastings: Biochemistry of the Bioluminescence of Colonial Hydroids and other Coelenterates. J. Cell. Physiol. **77**, 305 (1971).
104. — — Energy Transfer in a Bioluminescence System. J. Cell. Physiol. **77**, 313 (1971).
105. Morin, J. G., and G. T. Reynolds: The Cellular Origin of Bioluminescence in the Colonial Hydroid, *Obelia*. J. Exp. Biology, in press.
106. Morton, R. A., T. A. Hopkins, and H. H. Seliger: The Spectroscopic Properties of Firefly Luciferin and Related Compounds. An Approach to Product Emission. Biochemistry **8**, 1598 (1969).
107. Nakamura, T., and K. Mutsuda: Studies on Luciferase from *Photobacterium phosphoreum*. I. Purification and Physiochemical Properties. J. Biochem. **70**, 35 (1971).
108. Nakatsubo, F., Y. Kishi, and T. Goto: Synthesis and Stereochemistry of *Latia* Luciferin. Tetrahedron Letters **5**, 381 (1970).
109. Nicol, J. A. C.: In the Luminescence of Biological Systems (F. H. Johnson, ed.), American Association for the Advancement of Science, Washington, D. C., p. 303 (1955).
110. Plant, P. J., E. H. White, and W. D. McElroy: The Decarboxylation of Luciferin in Firefly Bioluminescence. Biochem. Biophys. Res. Comm. **31**, 98 (1968).
111. Rhodes, W. C., and W. D. McElroy: The Synthesis and Function of Luciferyl-Adenylate and Oxyluciferyl-adenylate. J. Biol. Chem. **233**, 1528 (1958).
112. Richardson, W. H., and V. F. Hodge: The Neighboring Perioxide Anion and the 1,2-Dioxetane Intermediate. A Kinetic and Product Study of the Basic Decomposition of Chloro-*tert*-butyl-Hydroperoxide. J. Amer. Chem. Soc. **93**, 3996 (1971).
113. Seliger, H. H., and W. D. McElroy: Spectral Emission and Quantum Yield of Firefly Bioluminescence. Arch. Biochem. Biophys. **88**, 136 (1960).
114. — — Chemiluminescence of Firefly Luciferin Without Enzyme. Science **138**, 683 (1962).
115. — — The Colors of Firefly Bioluminescence: Enzyme Configuration and Species Specificity. Proc. Nat. Acad. Sci. (USA) **52**, 75 (1964).
116. Seliger, H. H., W. D. McElroy, E. H. White, and G. F. Field: Stereospecificity and Firefly Bioluminescence, A Comparison of Natural and Synthetic Luciferins. Proc. Nat. Acad. Sci. (USA) **47**, 1129 (1961).
117. Shimomura, O., T. Goto, and Y. Hirata: Crystalline *Cypridina* Luciferin. Bull. Chem. Soc. Japan **30**, 928 (1957).

118. SHIMOMURA, O., and F. H. JOHNSON: The Structure of *Latia* Luciferin. Biochemistry **7**, 1734 (1968).

119. — — Purification and Properties of the Luciferase and of a Protein Cofactor in the Bioluminescence System of *Latia neritoides.* Biochemistry **7**, 2574 (1968).

120. — — Mechanisms in the Quantum Yield of *Cypridina* Bioluminescence. Photochem. Photobio. **12**, 291 (1970).

121. — — Mechanism of the Luminescent Oxidation of *Cypridina* Luciferin. Biochem. Biophys. Res. Comm. **44**, 340 (1971).

122. SHIMOMURA, O., F. H. JOHNSON, and Y. HANEDA: Isolation of the Luciferin of the New Zealand Fresh-water Limpet, *Latia neritoides* Gray. In Bioluminescence in Progress (F. H. Johnson and Y. Haneda, eds.), p. 391. Princeton, N. J.: Princeton Univ. Press. 1966.

123. SHIMOMURA, O., F. H. JOHNSON, and T. MASUGI: *Cypridina* Bioluminescence: Light Emitting Oxyluciferin-Luciferase Complex. Science **164**, 1299 (1969).

124. SHIMOMURA, O., F. H. JOHNSON, and Y. SAIGA: Purification and Properties of *Cypridina* Luciferase. J. Cell. Comp. Physiol. **58**, 113 (1961).

125. — — — Extraction, Purification and Properties of Aequorian a Bioluminescent Protein from the Luminous Hydromedusan, *Aequorea.* J. Cell. Comp. Physiol. **59**, 223 (1962).

126. — — — Further Data on the Bioluminescent Protein, Aequorian. J. Cell. Comp. Physiol. **62** 1 (1963).

127. — — — Extraction and Properties of Halistaulin, a Bioluminescent Protein from the Hydromedusan, *Halistaura.* J. Cell. Comp. Physiol. **62**, 9 (1963).

128. SIE, H.-C. E., W. D. McELROY, F. H. JOHNSON, and Y. HANEDA: Spectroscopy of the *Apogon* Luminescent System and of its Cross-reaction with the *Cypridina* System. Arch. Biochem. Biophys. **93**, 286 (1961).

129. SPUDICH, J., and J. W. HASTINGS: Inhibition of the Bioluminescent Oxidation of Reduced Flavin Mononucleotide by 2-Decenal. J. Biol. Chem. **238**, 3106 (1963).

130. STONE, H.: The Enzyme Catalyzed Oxidation of *Cypridina* Luciferin. Biochem. Biophys. Res. Comm. **31**, 386 (1968).

131. STREHLER, B. L., and M. J. CORMIER: Isolation, Identification, and Function of Long Chain Aldehydes Affecting the Bacterial Luciferin-Luciferase Reaction. J. Biol. Chem. **211**, 213 (1954).

132. STREHLER, B. L., E. N. HARVEY, J. J. CHANGE, and M. J. CORMIER: The Luminescent Oxidation of Reduced Riboflavin or Reduced Riboflavin Phosphate in the Bacterial Luciferin-Luciferase Reaction. Proc. Nat. Acad. Sci. (USA) **40**, 10 (1954).

133. SUZUKI, N., and T. GOTO: Firefly Bioluminescence. II. Identification of 2(6'-hydroxy-benzothiazol-2'-yl)-4 hydroxythiazole as a Product in the Bioluminescence of Firefly Lanterns and as a Product in the Chemiluminescence of Firefly Luciferin in DMSO. Tetrahedron Letters 2021 (1971).

134. SUZUKI, N., M. SATO, K. NISHIKANA, and T. GOTO: Synthesis and Spectral Properties of 2-(6'-hydroxylbenzothiazol-2'-yl)-4 hydroxythiazole, a Possible Emitting Species in the Firefly Bioluminescence. Tetrahedron Letters 4683 (1969).

135. TRAVIS, J., and W. D. McELROY: Isolation and Sequence of an Essential Sulfhydryl Peptide at the Active Site of Firefly Luciferase. Biochemistry **5**, 2170 (1966).

136. TSUJI, F. I., and Y. HANEDA: Chemistry of the Luciferase of *Cypridina hilgendorfii* and *Apogon ellioti.* In Bioluminescence in Progress (F. H. Johnson and Y. Haneda, eds.), p. 137. Princeton, N. J.: Princeton Univ. Press. 1966.

137. — — Studies on the Luminiscence Reaction of a Myctophid Fish, *Diaphus etucens* Brauer. Science Report of Yokosuka City Museum, 104 (1971).

138. TSUJI, F. I., Y. HANEDA, R. V. LYNCH, III, and N. SUGIYAMA: Luminescence Cross-reactions of *Porichthys* Luciferin and Theories on the Origin of Luciferin in Some Shallow-Water Fishes. Comp. Biochem. Physiol. **40 A**, 163 (1971).

139. Tsuji, F. I., and R. Sowinski: Purification and Molecular Weight of *Cypridina* Luciferase. J. Cell. Comp. Physiol. **58,** 125 (1961).

140. Wampler, J. E., and R. J. Desa: An On-line Spectrofluorimeter System for Rapid Collection of Absolute Luminescence Spectra. Applied Spectroscopy **25,** 623 (1971).

141. Wampler, J. E., K. Hori, J. W. Lee, and M. J. Cormier: Structured Bioluminescence. Two Emitters During Both the *In Vitro* and the *In Vivo* Bioluminescence of the Sea Pansy, *Renilla.* Biochemistry **10,** 2903 (1971).

142. Wampler, J. E., Y. D. Karkhanis, K. Hori, and M. J. Cormier: Protein-Protein Interactions Between *Renilla* Luciferase and a Green Fluorescent Protein, the Energy Transfer Acceptor During *Renilla* Bioluminescence. Federat. Proc., in press.

143. White, E. H., and M. J. C. Harding: The Chemiluminescence of Lophine and its Derivates. J. Amer. Chem. Soc. **86,** 5686 (1964).

144. White, E. H., F. McCapra, G. F. Field, and W. D. McElroy: The Structure and Synthesis of Firefly Luciferin. J. Amer. Chem. Soc. **83,** 2402 (1961).

145. White, E. H., E. Rapabert, T. H. Hopkins, and H. H. Seliger: Chemi- and Bioluminescence of Firefly Luciferin. J. Amer. Chem. Soc. **91,** 2178 (1969).

146. White, E. H., and H. Worther: Analogs of Firefly Luciferin. III. J. Organ. Chem. (USA) **31,** 1484 (1966).

147. White, E. H., H. Worther, G. F. Field, and W. D. McElroy: Analogs of Firefly Luciferin. J. Organ. Chem. (USA) **30,** 2344 (1965).

148. White, E. H., H. Worther, H. H. Seliger, and W. D. McElroy: Amino Analogs of Firefly Luciferin and Biological Activity Thereof. J. Amer. Chem. Soc. **88,** 2015 (1966).

149. Wilson, T., and A. P. Schaap: The Chemiluminescence from *cis*-Diethoxy-1,2--dioxetane. An Unexpected Effect of Oxygen. J. Amer. Chem. Soc. **93,** 4126 (1971).

(Received February 28, 1972)

Gametenlockstoffe
bei niederen Pflanzen und Tieren

Von L. JAENICKE und D. G. MÜLLER, Köln

Mit 5 Abbildungen

Inhaltsübersicht

I. Einleitung

Die Befruchtung spielt eine bedeutende Rolle im Entwicklungsgang der Organismen. In ihrem Verlauf werden die Gametenkerne, und damit das genetische Material zweier Individuen, vereinigt. Durch die Meiose (Reduktionsteilung) wird der Ausgangszustand wiederhergestellt, jedoch in einer solchen Weise, daß die elterlichen Erbfaktoren in den verschiedensten Kombinationen neu verteilt werden können. Die Beobachtung von Befruchtungsvorgängen zeigt, daß der eigentlichen Verschmelzung der Geschlechtszellen Reaktionen vorausgehen, an denen offensichtlich chemische Wechselwirkungen beteiligt sind. Besonders deutlich lassen sich solche stofflichen Wechselwirkungen bei niederen Pflanzen beobachten, bei denen beispielsweise Geschlechtsorgane nur gebildet werden, wenn der entsprechende Partner oder Filtrate seines Kulturmediums vorhanden sind. In anderen Fällen kann man beobachten, daß Geschlechtsorgane durch Luft oder Wasser gerichtet aufeinander zuwachsen oder, daß freibewegliche männliche Gameten sich gezielt zu stationären weiblichen Gameten hinbewegen. Solche biologischen Reaktionen sind außerordentlich empfindlich, und die Konzentration der beteiligten Stoffe ist entsprechend gering. Dies macht die Ansammlung von genügenden Mengen für die Strukturaufklärung zu einem äußerst mühsamen Unternehmen. Obwohl auch in der älteren Literatur zahlreiche Hinweise auf chemische Wechselwirkungen zwischen Sexualpartnern vorhanden sind (*42, 43*), konnten bisher erst wenige derartige Geschlechtsstoffe identifiziert werden. Erst die hochleistungsfähigen modernen Analysenmethoden, wie Massen- und Kernresonanz-Spektroskopie, haben hierbei die Forschung wesentlich vorangetrieben. Man kennt heute die Molekularstruktur von vier Sexualhormonen, die nunmehr auch synthetisch zugänglich sind.

Es handelt sich um zwei Induktionsstoffe, die bei niederen Pilzen von einem Geschlechtspartner gebildet werden und beim anderen Geschlecht die Bildung von Sexualorganen bewirken:

Das Steroid Antheridiol bei *Achlya* und die Terpenoid-Carbonsäure Trisporsäure C bei *Mucor* (*6*).

In zwei weiteren Fällen wurden Substanzen identifiziert, die von weiblichen Gameten abgegeben werden und die die zugehörigen männlichen Gameten chemotaktisch anlocken (29a).

Es handelt sich hierbei um das Sequiterpen Sirenin bei dem Pilz *Allomyces* und um den Lockstoff der weiblichen Gameten der Meeresbraunalge *Ectocarpus,* den Cycloheptadien-Kohlenwasserstoff Ectocarpen.

In dieser Übersicht wird in erster Linie über diese beiden nunmehr bekannten Lockstoffe und daran anschließend über einige weitere, weniger gut charakterisierte Substanzen mit Lockstoffwirkung berichtet.

Zunächst soll kurz auf einige terminologische Schwierigkeiten eingegangen werden. Im deutschen Sprachgebrauch hat es sich eingebürgert, ganz allgemein Substanzen, die an der chemischen Wechselwirkung zwischen Sexualpartnern (von der Ausbildung von Sexualorganen bis hin zur Verschmelzung der verschiedengeschlechtlichen Kerne) beteiligt sind, als Gamone zu bezeichnen (*71*). Dabei wird jedoch die ursprüngliche Definition von Gamon als „Gametenhormon" (*26*) erweitert.

Im amerikanischen Sprachgebrauch hat der Begriff „Gamon" keinen Eingang gefunden. Hier wird die Bezeichnung „sexual hormone" vorgezogen, wobei auf eine sehr weit gefaßte Definition von „Hormon" Bezug genommen wird: „A substance produced in one portion of the organism and transported by any means, including diffusion to other portions of the same individual or to other individuals of the same species, where it induces specific responses", (*70*). Die Verwendung dieses erweiterten Hormonbegriffs hat den Vorteil, daß dadurch in übersichtlicher Weise alle diffusiblen Wechselwirkungen zusammengefaßt und deutlich gegen Vorgänge an stationären Strukturen, wie z. B. spezifische Kontakt- und Membranreaktionen, abgegrenzt werden.

KÖHLER (*32*) unterscheidet nun zur Klärung der Funktion solcher hormoneller Faktoren „Fernwirkungen" einerseits und „Agglutinations- und Zellkontaktwirkung" andererseits. Als generischen Terminus für chemotaktisch und chemotropisch wirksame Substanzen schlägt er die klangvolle und einleuchtende Bezeichnung „Sirenin" vor. Der Name Sirenin allerdings wurde zuerst von MACHLIS für die von den weiblichen Gameten des Pilzes *Allomyces* produzierte Substanz eingeführt (*37*). Dies geschah in Anlehnung an die mythischen Sirenen, die durch ihren Gesang die vorbeisegelnden Seefahrer anlockten. Da MACHLIS es ablehnt, die Bezeichnung „Sirenin" in erweitertem Sinne für chemotaktisch und chemotropisch wirksame Sexualstoffe zu verwenden (39a), wollen wir zunächst keine neuen terminologischen Begriffe einführen, um die bereits bestehende Verwirrung nicht noch zu vergrößern*.

Ferner ist eine Abgrenzung der hier behandelten Gametenlockstoffe gegen tierische Pheromone notwendig. Die von Gametenlockstoffen ausgelösten Erscheinungen sind in der Regel Reaktionen zwischen Geschlechtszellen, also Vorgänge auf zellulärer Ebene, während der Name „Pheromon" bisher nur für Substanzen verwendet wurde, die Kommunikation zwischen höher organisierten Tieren bewirken. Allerdings schließt die

---

* In einer Korrespondenz schlägt Professor L. MACHLIS (Berkeley) als generische Begriffe für derartige Sexualstoffe Ableitungen vom griechischen „Eros" vor. Für chemotaktisch wirksame Hormone wäre damit die Bezeichnung „Erototaktine", für chemotropisch wirksame „Erototropine" und für organinduzierende Substanzen „Erotogene" zu bilden.

Definition der Pheromone, wie sie von Karlson und Lüscher gegeben wurde (*31*), Wechselwirkungen auf zellulärer Ebene nicht ausdrücklich aus.

II. Verbreitung der Gameten-Lockstoffe

Gameten-Chemotaxis ist im Prinzip überall da möglich, wo mindestens ein selbständig beweglicher Gamet am Befruchtungsprozeß beteiligt ist. Dies trifft im Pflanzenreich zu für die meisten Algen, niederen Pilze, Moose, Farne, bis hin zu einigen Gymnospermen, wie z. B. *Gingko biloba*. Bei sämtlichen Angiospermen sowie den Nadelhölzern werden jedoch keine freien Spermatozoiden mehr gebildet. Die Eizelle kann in diesem Fall durch gerichtete Wachstumsreaktionen (Chemotropismus) der Pollenschläuche aufgesucht werden. Im Tierreich scheint chemotaktische Anlockung von Spermien nur in vereinzelten Fällen vorzukommen. Hier werden die Tiere als Geschlechtspartner zusammengeführt, und bei Balz- und Kopulationsvorgängen geraten dann Eier und Spermien in enge räumliche Beziehung, so daß Reaktionen zwischen Spermium und stationären Strukturen der Eioberfläche für die Befruchtung auszureichen

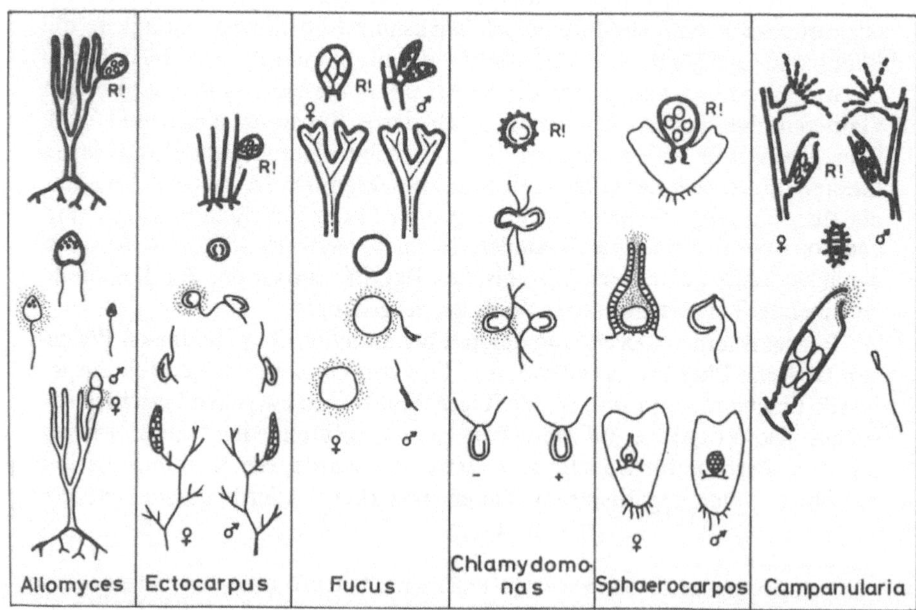

Abb. 1. Schematische Darstellung des Entwicklungsganges der besprochenen Organismen. Dicke Konturen: diploide Phase; dünne Konturen: haploide Phase. *R!* = Reduktionsteilung. Geschlechtszellen stark vergrößert, Produktion der Lockstoffmoleküle durch Punktierung angedeutet

Literaturverzeichnis: SS. 96—100

scheinen. Der einzige gesicherte Fall von tierischer Spermien-Chemotaxis liegt bei marinen Hydroidpolypen (s. S. 92) vor, bei denen die Spermien zur Öffnung des Eibehälters gelockt werden. Auch bei den mit einer derben Hülle versehenen Eiern einiger Fische, bei denen der Zugang für die Spermien nur an einer vorgeformten Stelle möglich ist, spielt vielleicht chemotaktische Anlockung der Spermien zu dieser Stelle hin eine Rolle (5).

III. Spezifität chemotaktisch wirksamer Substanzen

Da Gameten-Lockstoffe bei der Einleitung der Befruchtung von Bedeutung sind, ist die Frage nach dem Zusammenhang von Bastardierungsfähigkeit nahestehender Arten und der Spezifität der chemotaktischen Reaktionen von besonderem Interesse. Die Anlockung der männlichen Gameten durch die nativen Substanzen ist in den meisten bisher daraufhin untersuchten Arten interspezifisch; eine Ausnahme ist möglicherweise *Allomyces* (39).

In einigen Fällen reagieren die Gameten auf ein breites Spektrum von „Pseudolockstoffen". Ob allerdings die nativen Lockstoffe diesen nicht doch überlegen sind, läßt sich zur Zeit mangels quantitativer Vergleichsverfahren nicht entscheiden. Da Artbastarde in der Natur nur selten auftreten, die chemotaktischen Reaktionen aber, nach unserem gegenwärtigen Wissensstand, nicht streng spezifisch sind, wird die Gametenchemotaxis lediglich als einer unter mehreren Faktoren am Erfolg oder Nichterfolg von Bastardierungen beteiligt sein.

IV. Topo- und phobotaktische Reaktion

Die Anhäufung beweglicher Zellen an einer Reizstoff-Quelle kann auf zwei Reaktionsweisen beruhen:

1. Der Organismus ist in der Lage, die Quelle gerichtet direkt anzusteuern (topotaktische Reaktion), oder
2. der Organismus bewegt sich nicht gerichtet, sondern der Eintritt in Bereiche niedrigerer Konzentration wirkt als Reiz. Dieser löst eine „Schreckreaktion" aus, durch die die Bewegung in eine neue Richtung gelenkt wird. Auf diese Weise wird durch Versuch und Irrtum ebenfalls die Ansammlung der Gameten am Ort der optimalen Reizstoffkonzentration erreicht (phobotaktische Reaktion) (87).

Es läßt sich jedoch auch vorstellen, daß die topische Reaktion durch eine Folge sinnvoll aneinandergereihter phobischer Einzelschritte zustandekommt, wie dies etwa bei der Pseudophototopotaxis (27) der Fall ist. Topotaktische Reaktionen sind durch mikrokinematographische Methoden einwandfrei für die Spermatozoiden von Farnen (*75*) und *Campanularia* (*47*) erwiesen. In den anderen hier behandelten Fällen von Gameten-Chemotaxis ist die Reaktionsweise nicht genau untersucht.

V. Sirenin

a) Vorkommen, Funktion und Bestimmung

Zu der Gattung *Allomyces* (Phycomycetes) gehören Pilze, die in feuchten Böden, Wassergräben, an Teichufern und ähnlichen Orten vorkommen. *Allomyces* hat einen Generationswechsel, bei dem sich eine Geschlechtsgeneration (Gametophyt) mit einer ungeschlechtlichen Generation (Sporophyt) abwechselt. Beide Generationen sind gleichgestaltet; sie bestehen aus einem mehrere Zentimeter großen weißen, verzweigten Mycel. Der Gametophyt trägt männliche und weibliche Fortpflanzungsorgane (Gametangien) paarweise nebeneinander. Wird das Mycel überflutet, so treten die Gameten aus und bewegen sich mit einer Geißel frei im Wasser. Die farblosen weiblichen Gameten werden von den kleineren orangeroten männlichen Gameten befruchtet. Die bewegliche Zygote setzt sich auf einer Unterlage fest und entwickelt sich zum Sporophyten. In den braunen Dauersporangien dieses diploiden Sporophyten findet die Meiose statt. Die Meiosporen entwickeln sich wieder zu Gametophyten und schließen somit den Entwicklungskreislauf (s. Abb. 1).

Schon im Gametangium produzieren die weiblichen Gameten eine Substanz, die in das umgebende Wasser abgegeben wird und männliche Gameten anlockt. Machlis und Vegis (*37*) gaben ihr den Namen „Sirenin" (s. S. 63). Sie ist neutral, hitzebeständig und schwer flüchtig. Sirenin läßt sich auf folgende Weise biologisch nachweisen (*38*): Der Boden eines zylindrischen Gefäßes wird durch eine semipermeable Membran abgeschlossen und in eine Suspension mit männlichen Gameten getaucht. Auf die Membran werden die auf ihre Sireninwirkung zu prüfenden Lösungen gebracht. Die männlichen Gameten setzen sich auf der Unterseite der Membran fest, und zwar in umso größerer Zahl, je höher die angebotene Sireninkonzentration ist. Durch mikroskopische Auszählung konnte gefunden werden, daß noch Konzentrationen von 10^{-10} m Sirenin einen deutlichen Effekt zeigen (*10*).

b) Isolierung

Eine wichtige Voraussetzung für Massenkulturen zur Sireningewinnung war die experimentelle Herstellung eingeschlechtlich männlicher oder weiblicher Stämme. Dies wurde durch Artkreuzungen (*A. arbuscula* x *A. javanicus*) und gezielte Auslese der Nachkommenschaft erreicht. Auf diese Weise wurden Stämme erhalten, die zu über 95% entweder nur männliche oder nur weibliche Gametangien bildeten (*36*). *Allomyces* wächst auf Nähragar bei 25° C im Dunkeln, läßt sich aber auch in belüfteten, aseptisch gehaltenen Submerskulturen (0,2% Glukose, 1,2% autolysierter Hefeextrakt, Natriumacetat, pH 7,5) heranziehen. Nach mehrtägiger Vorkultur erhielt man Suspensionen, die durch massenhafte Entlassung weiblicher Gameten milchig trübe waren. Die Filtrate solcher Ansätze enthielten Sirenin in Konzentrationen bis zu 10^{-6} m. Insgesamt wurden aus derartigen Ansätzen durch Extraktion mit Methylenchlorid und Chromatographie an Aluminiumoxid 2,5 g Sirenin gewonnen (*40*). Die Substanz ist ungesättigt und hat alkoholische Hydroxyl-Funktionen. Zur Lokalisierung und Identifizierung in den Eluaten macht man von diesen Eigenschaften Gebrauch, indem man entweder Proben nach direkter Chromatographie mit Joddampf behandelt (Rf 0,6—0,7) oder die Eluate mit 4(4-Nitrophenylazo)benzoylchlorid (*28*) umsetzt. Dabei entstehen als Hauptprodukte Sirenin-Mono- (Rf 0,2) und -Diester (Rf 0,5), die zur Analyse auf Kieselgel mit Chloroform dünnschichtchromatographisch getrennt werden. Das vorgereinigte Sirenin wird nun präparativ insgesamt in den bis-p-Nitrophenylazobenzoylester umgewandelt und dieser rechromatographiert. Der reine Diester läßt sich aus Benzol-Cyclohexan umkristallisieren (F 186—187° C, $\varepsilon_{330\,mm} = 32\,000$) und durch Alkali spalten. Man erhält so das Sirenin (**1**) als farblose, viskose Flüssigkeit, $[\alpha]_D^{22°}$—45° (c = 1,0, Chloroform) der Formel $C_{15}H_{24}O_2$, M 236. Aus der Formel geht hervor, daß es ein vierfach ungesättigtes Sesquiterpen ist, dessen beide Sauerstoff-Funktionen Ester bilden können.

c) Strukturaufklärung

Am Di-4-(4-Nitrophenylazo)benzoylester wurde die Struktur durch partielle Ozonolyse bei −60° C und spektroskopische, insbesondere NMR-Verfahren, aufgeklärt (*41, 59*). Der Diester (M 744) gibt dabei

1. den p-Nitrophenylazobenzoylester des Hydroxyacetons (**2**),
2. einen p-Nitrophenylazobenzoylester eines Aldehyds $C_{12}H_{19}O_2$ (**3**),
3. einen Cyclopropylcarboxaldehyd (**4**) mit einem α-Proton, einer tertiären Methylgruppe *cis* zur Aldehydgruppe und einer Allylalkohol-Seitenkette.

Wird das Sirenin mit MnO_2 oxydiert, entsteht ein β-Cyclopropyl-α-β-ungesättigter Dialdehyd (**5**).

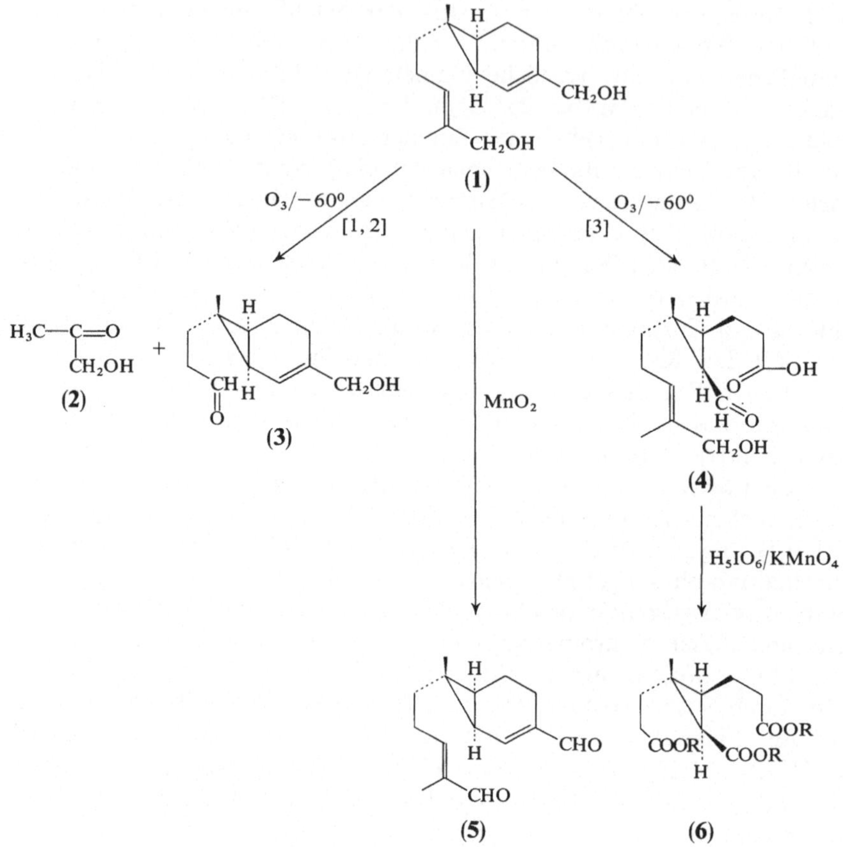

Schema I. Strukturaufklärung von Sirenin

Damit sind die 15C, 2O und 23 der 24H im Molekül festgelegt in der Struktur

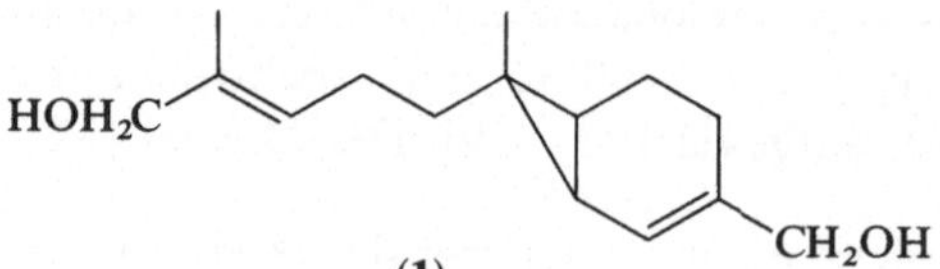

(1)

Die Oxydation mit Perjodat/$KMnO_4$ gibt eine Cyclopropyltricarbonsäure (**6**, R=H), die mit Diazomethan verestert wurde. Dieser Triester (**6**, $R=CH_3$) sowie dessen drei weitere mögliche Isomere wurden synthetisiert. Dadurch wurde bewiesen, daß die Substanz die bicyclische Caren-Struktur hat. Läßt man die Reaktion in deuteriertem Medium ab-

Literaturverzeichnis: SS. 96—100

laufen, entsteht ein Tricarbonsäureester (**6**, R = CH$_3$), in den kein Deuterium eingebaut ist. Das heißt, es ist keine Epimerisierung eingetreten, und die Tricarbonsäure hat die gleiche Stereochemie wie das Ausgangsmaterial. Daß in diesem die acyclische Doppelbindung trans steht, folgt aus den spektralen Daten im Vergleich zum *cis*- und *trans*-2-Methyl-2-penten-1-on (**7**) (*59*).

$$
\begin{array}{ccc}
\text{cis-} & \text{(7)} & \text{trans-}
\end{array}
$$

Tabelle 1. *Eigenschaften des Sirenins*

Formel: C$_{15}$H$_{24}$O$_2$ (M 236)

Farblose viskose Flüssigkeit [α]$_D^{22°}$ − 45° (c 1.0, CHCl$_3$)

Diacetat C$_{19}$H$_{28}$O$_4$ (M 320) b$_{0.3\text{mm}}$ 190° (Bad-Temperatur) n$_D^{29}$ 1.4870

Mono-p-Nitrophenylazo-benzylester C$_{28}$H$_{31}$O$_3$N$_5$ (M 490) F 115—116°

Bis-p-Nitrophenylazo-benzylester C$_{41}$H$_{38}$N$_6$O$_8$ (M 744) F 186—187°

NMR: (60—100 MHz) δ	0,88	1,62	3,97	5,39	5,80
in CDCl$_3$	3H	3H	4H	1H	1H
TMS = 0	s	s	s	t	s
		breit		breit	breit
	$\geq$C−CH$_3$	C=C−CH$_3$	2 × CH$_2$−C(C)=C	C=CH−CH$_2$	CH=C
	Cyclopropyl-CH$_3$	Vinyl-CH$_3$	Carbinyl-methylene	Vinyl-H β zu Carbinylgruppen	Vinyl-H

komplexe Serie von
peaks im 1—2,2-ppm-
Bereich für restliche Protonen

IR (λ CHCl$_3$)   3600, 1660, 1601, 1380, 990 cm^{-1}
(alkohol.
Hydroxyl)

Massenspektrum M$^+$ 236; m/e 187   148 135 133 131 119 109 107 105 95 93 91 79
Basis-
peak

R$_f$ (6% Methanol in Benzol) auf Kieselgel: 0,12

d) Biosynthese

Das Cyclopropan-Derivat (*34*) Sirenin gehört somit zu den Sesquiterpenen, die Isoprenhomologe bekannter Monoterpene, hier des 4-Carens (**8**), sind.

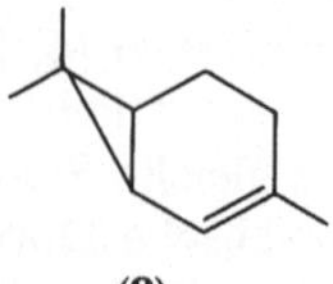

(8)

Daraus läßt sich die Biosynthese aus *cis*-Farnesylpyrophosphat **(9)** über ein Carbonium-Ion **(10)** zum Desoxy-Sirenin (Sesquicaren) **(11)** *(63)* durch eine 1,3-Deprotonierung ableiten *(59)*.

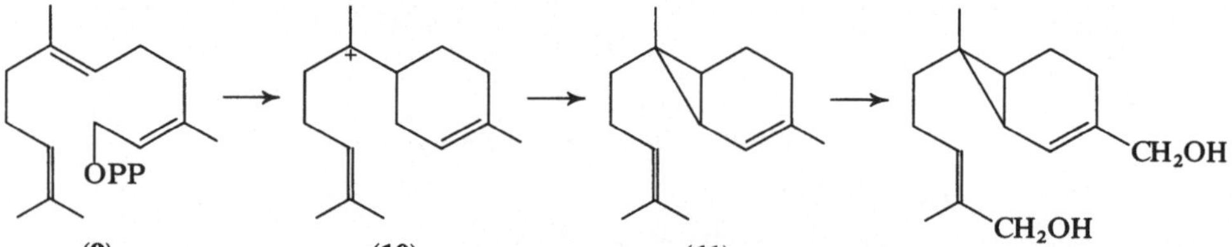

(9)   **(10)**   **(11)**   **(1)**

Schema II. Mutmaßliche Biogenese von Sirenin

Experimente zur Biosynthese liegen noch nicht vor.

e) Total-Synthesen von Sirenin

In kurzem Zeitabstand sind mehrere Total-Synthesen von Sirenin *(9, 18, 19, 25, 52, 53, 69)* ausgearbeitet worden, die aber sämtlich auf ähnlichem Grundprinzip beruhen, wie sie auch für die Synthese von Sesquicaren *(51, 53)* verwendet wurden, nämlich der Bildung des kondensierten Cyclopropansystems durch stereospezifische intramolekulare α-Ketocarben-Addition nach Stork *(23, 67, 81, 82)*: „Δ^5-Diazoketon-Weg"; vgl. a. *(51)*.

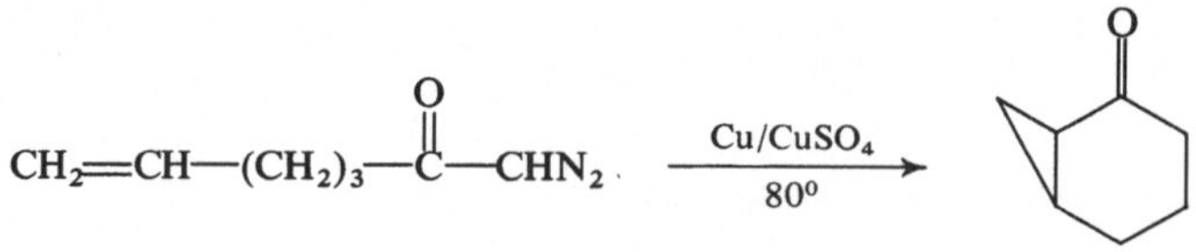

$$CH_2=CH-(CH_2)_3-\overset{O}{\overset{\|}{C}}-CHN_2 \cdot \xrightarrow[80°]{Cu/CuSO_4}$$

Stork-*Reaktion*

Durch dieses synthetische Verfahren sind Bicyclo[4.1.0]Heptan-Ring-systeme generell zugänglich, und es lassen sich Sirenin **(1)** und sein *cis*-Isomeres (Iso-Sirenin) **(12)** sowie Homologe darstellen. Das zentrale Problem ist also die Darstellung geeigneter olefinischer Diazoketone, die sich zum Caren-System cyclisieren lassen und anschließende

Funktionalisierung der Seitenketten sowie Einfügung der restlichen Gruppen nach der Cyclisierung. Die acyclischen Diene werden hierbei zumeist durch Reaktionen nach Art der WITTIG-Synthese (8) gewonnen.

Dies soll am Beispiel der Synthese von dl-Sirenin (1) und dl-Isosirenin (12) von RAPOPORT (9) beschrieben werden: Überschüssiges 6-Methyl-5-hepten-2-on (13) wird mit dem Ylid der Triphenylphosphonovaleriansäure (14) in Dimethylsulfoxid/Tetrahydrofuran umgesetzt. Das Gemisch der isomeren 6,10-Dimethyl-5,9-Undekadiensäuren (15) wird mit Dimethylsulfat verestert. Die Ester können durch präparative Gaschromatographie an Carbowax 20 M in reinen *trans*- und *cis*-Ester getrennt werden.

Ersterer führt bei den weiteren beschriebenen Reaktionen zum dl-Sirenin (1); der andere zum dl-Isosirenin (12).

Der Ester wird verseift, über das Natriumsalz mit Oxalylchlorid in das Säurechlorid umgewandelt und dieses mit Diazomethan in das Diazoketon (16) übergeführt. Beim Kochen in Cyclohexan in Gegenwart von suspendiertem Kupfersulfat verläuft die stereospezifische STORK-Cyclisierung zum bicyclischen 7-Ring-Keton (17) mit der Δ^5-Doppelbindung in ca. 65% Ausbeute. Daneben entsteht etwa 1% des 11-Ring-Ketons durch Cyclisierung mit der Δ^9-Doppelbindung. Durch Kochen unter Rückfluß in Dimethylcarbonat in Gegenwart von überschüssigem Natriumhydrid wird in quantitativer Ausbeute eine $COOCH_3$-Gruppe α-ständig zur Carbonylgruppe einkondensiert, und man erhält einen β-Ketoester (18), der mit Natriumboranat bei $-20°$C den entsprechenden β-Hydroxyester ergibt. Diese wird über den Pivaloylester mit K-t-Butylat in Toluol dehydratisiert (19). Die nun folgende Oxydation der terminalen *trans*-Methylgruppe mit SeO_2, anschließend mit MnO_2, gibt den ω-*trans*-Aldehyd (20), der mit Lithiumalanat/$AlCl_3$ bei $0°$C zu dl-Sirenin (1) bzw. dl-Isosirenin (12) reduziert wird.

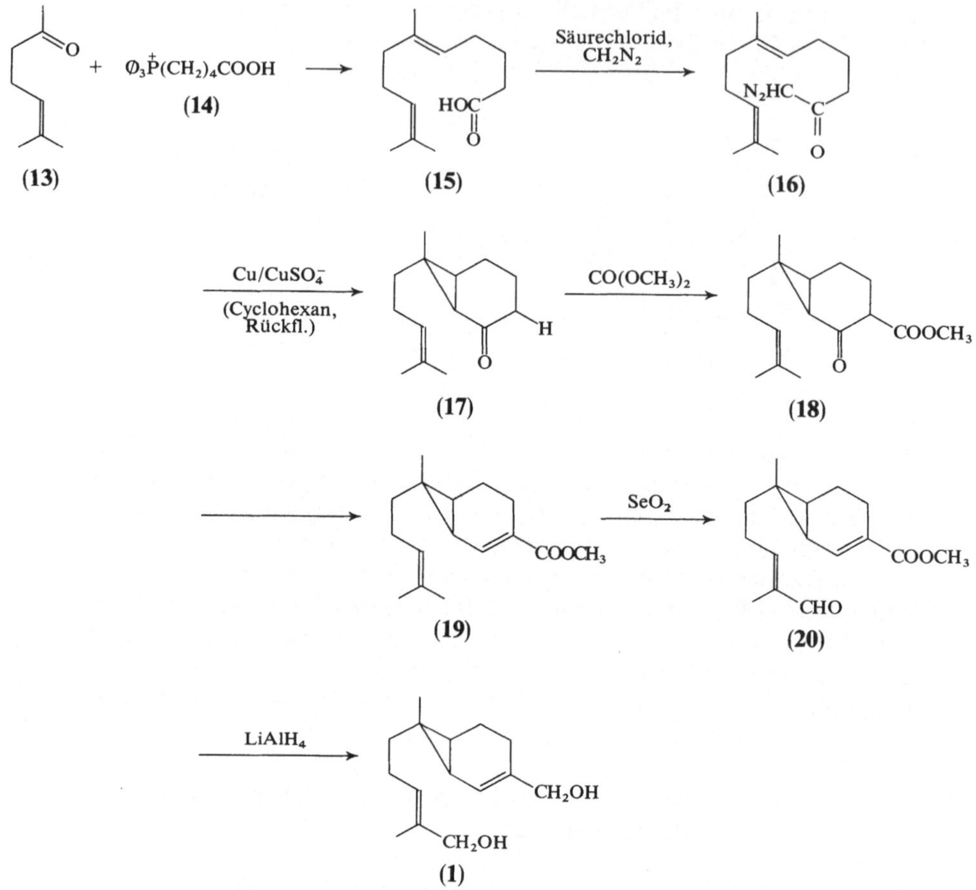

Schema III. Synthese von (dl)-Sirenin nach Rapoport (*9, 68*)

Die Trennung der optischen Antipoden kann bereits auf der Stufe der bicyclischen Ketone (**17**) in Form ihrer Ketale mit optisch aktivem 2,3-Butandiol erfolgen (*69*).

Da l-Sirenin mittels Toluolsulfochlorid in Pyridin und anschließender LiAlH$_4$-Reduktion in mono-Desoxy-Sirenin umgewandelt werden kann (**9**) und dieses wiederum in ähnlicher Reaktionsfolge (SO$_3$/Pyridin; LiAlH$_4$) in 85% Ausbeute in (-)-Sesquicaren (**11**) (*17*), läßt sich die absolute Konfiguration des l-Sirenins als derjenigen des (-)-Sesquicarens identisch erschließen (*69*). Die Stereochemie von (-)-Sesquicaren ist durch CD-Messungen eruiert worden (*63*). In analoger Weise, aber unabhängig davon, wurde bewiesen, daß in l-Sirenin der Cyclopropan-Ring in der angegebenen Weise (**1**) räumlich umgeben ist, also die Methylgruppe *endo* steht (*69*).

Die drei weiteren bisher ausgearbeiteten Synthesen verlaufen in prinzipiell ähnlicher Weise und nehmen ihren Ausgang vom Undekadienylbromid (*51—53*) oder vom Geraniol (*18, 19, 25*), die in geeigneter und eleganter Weise zu einem ungesättigten Diazoketon verlängert und anschließend cyclisiert werden. Die Einführung der funktionellen Gruppen erfolgt durch die hochstereoselektiven Reaktionen der Terpenchemie. Ihr Verlauf ist in den folgenden Formelschemata skizziert.

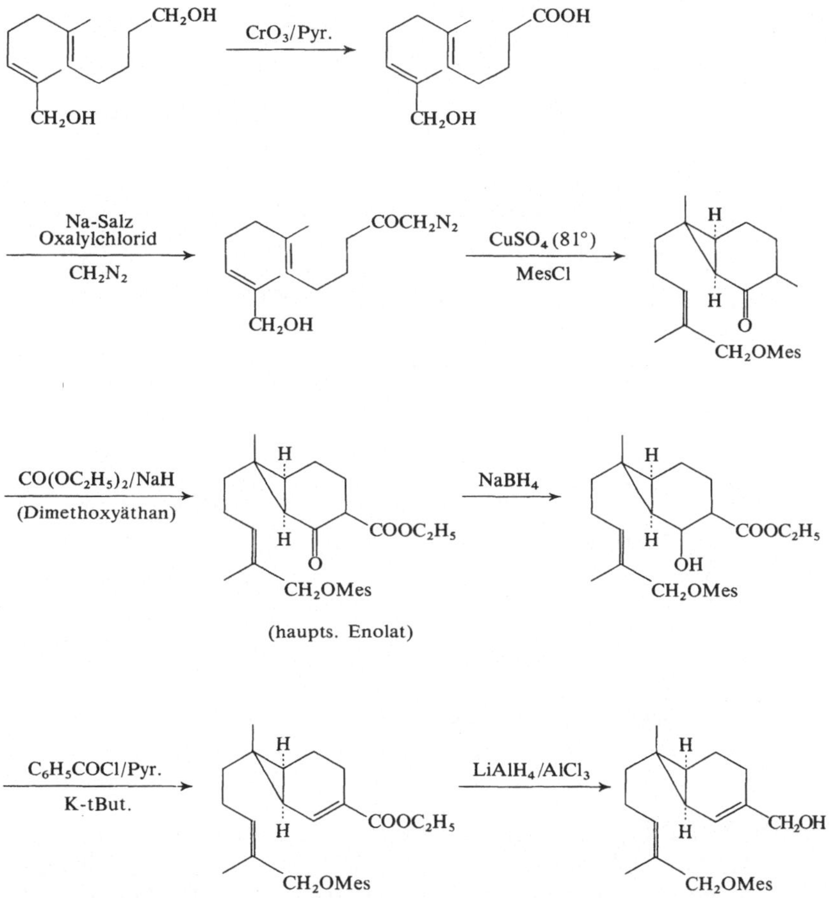

Schema IV. Synthese von dl-Sirenin nach Corey (*18*)

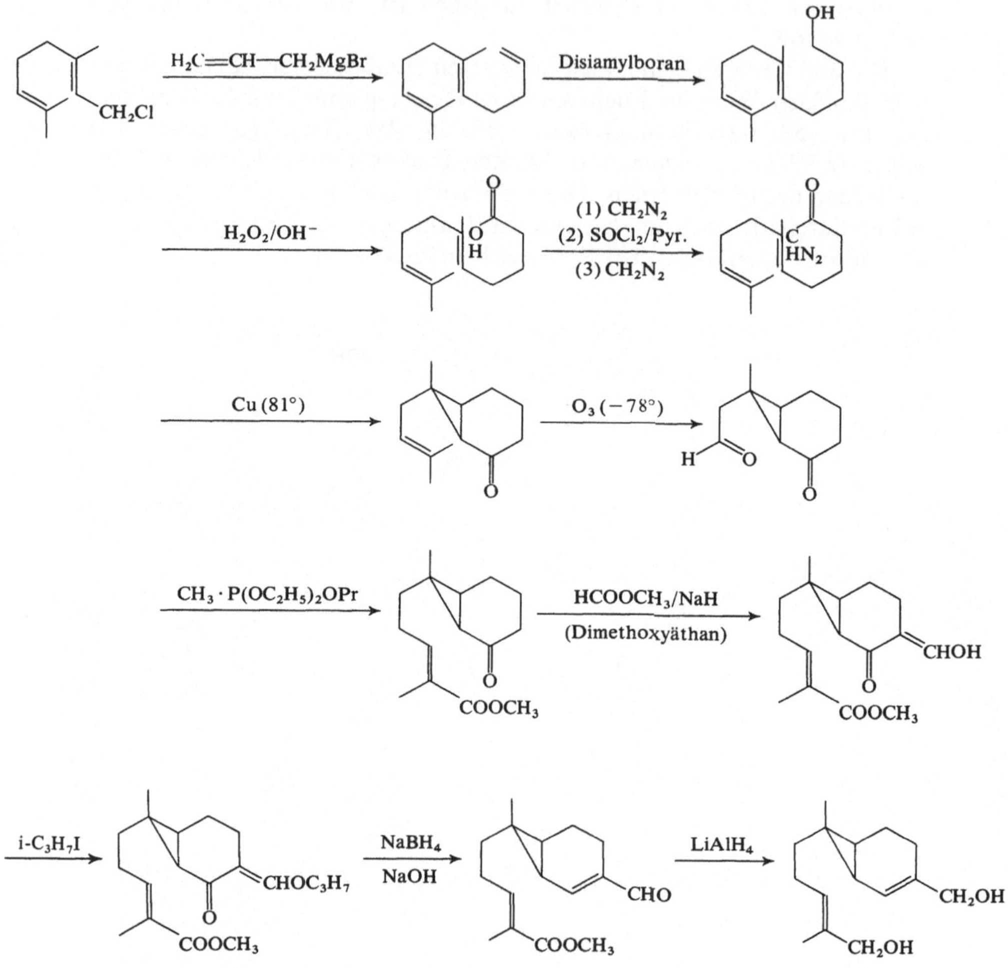

Schema V. Synthese von dl-Sirenin nach Grieco (*25*)

Von den Zwischenprodukten der Synthese ist keines als Lockstoff wirksam: dl-Iso-Sirenin hat nur 8% Aktivität, dl-Sirenin ist dagegen voll aktiv. Damit ist gleichzeitig erwiesen, daß für die biologische Funktion die Allylalkoholgruppierung in der Seitenkette notwendig ist, die Stereochemie der Seitenketten am Cyclopropanring ein weiterer Wirkungsfaktor ist und d-Sirenin nicht hemmt.

Literaturverzeichnis: SS. 96—100

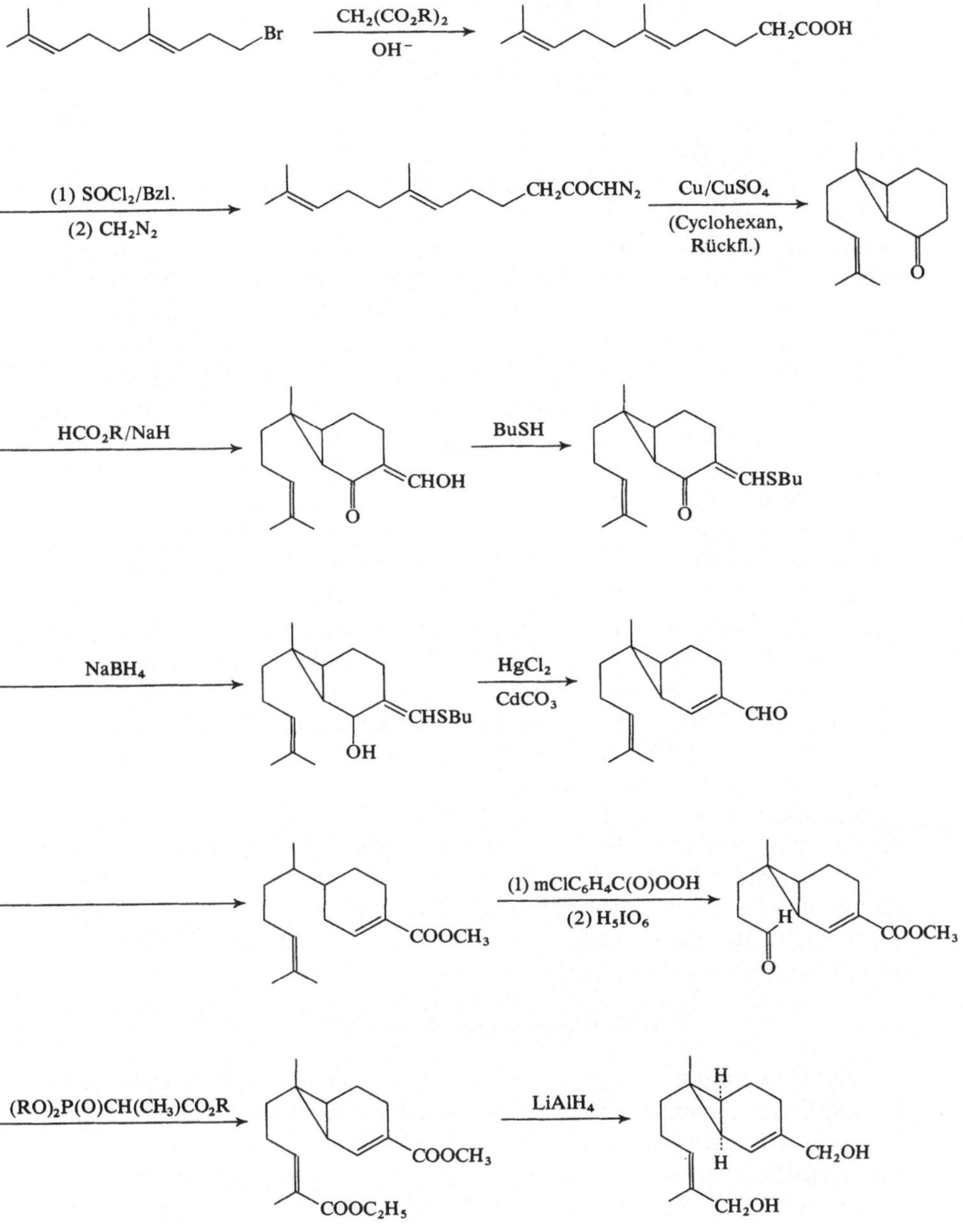

Schema VI. Synthese von dl-Sirenin nach MORI (*52, 53*)

VI. Ectocarpen

a) Vorkommen und Wirkung

Der zweite nunmehr in seiner Struktur aufgeklärte Gameten-Lockstoff ist Ectocarpen. Wir gaben dieser Substanz zunächst den Namen *Ectocarpus*-Sirenin (*57*), da wir es für sinnvoll hielten, derartige Substanzen generell nach ihrer Herkunft und Funktion zu benennen. Da MACHLIS jedoch neuerdings die Verwendung von „Sirenin" als generischem Begriff ablehnt und ihn für die *Allomyces*-Substanz beschränkt wissen will (s. S. 63), wollen wir seinem Vorschlag folgen und der von den weiblichen Gameten von *Ectocarpus siliculosus* produzierten Substanz den neutraleren Namen Ectocarpen zuweisen. *Ectocarpus siliculosus* (Dillw.) Lyngb. ist eine an vielen Meeresküsten verbreitete, aus bis zu 30 cm langen büscheligen Fäden bestehende Braunalge, die sich oft als Bewuchs auf Schiffsrümpfen unliebsam bemerkbar macht. Die Art ist getrenntgeschlechtig; männliche und weibliche Pflanzen sind gleich gebaut. Auch hier liegt ein Generationswechsel vor (s. Abb. 1); die Gametophyten können sich mit einem ähnlich gebauten diploiden Sporophyten, auf dem die Meiose stattfindet, abwechseln. Die ins Seewasser abgegebenen, frei beweglichen Gameten sind zunächst nicht zu unterscheiden und zeigen keinerlei gegenseitige Beeinflussung. Sobald sich jedoch ein weiblicher Gamet auf einer Unterlage festsetzt, wird eine deutliche Reaktion der männlichen Gameten sichtbar. Sie schwimmen auf die weibliche Zelle zu, heften sich mit der Spitze der Vordergeißel an dieser fest und beginnen sie in heftiger Bewegung zu umkreisen. Auf diese Weise können sich Klumpen von bis zu 30 männlichen Gameten um eine weibliche Zelle bilden (Abb. 2a). Sobald ein männlicher Gamet die Befruchtung vollzogen hat, hört die Reizung der restlichen männlichen Gameten auf, sie lösen sich los und schwimmen davon (*54*). Es zeigte sich, daß der von den weiblichen Gameten gebildete Stoff leicht flüchtig ist und als aromatischer, terpenartiger Duft wahrgenommen werden kann (*55*).

b) Gewinnung und Nachweis

Zur Gewinnung des Ectocarpens wurden die Gameten einer weiblichen Einzelpflanze verwendet, die über mehrere Jahre hinweg durch Zerschneiden vegetativ vermehrt wurde (*56*). Die Kultur erfolgte im 14-Std.-Tag bei Kunstlicht in mit Nährsalzen und Erddekokt angereichertem Seewasser bei 20°C. Nach etwa 11—17 Tagen Wachstum haben die Pflanzen weibliche Geschlechtsorgane gebildet und entlassen täglich 1—2 Std. nach Lichtbeginn ihre Gameten. Dabei tritt ein aromatischer, an Wacholder erinnernder Geruch auf. Vegetative weibliche Pflanzenteile, männliche Gameten und asexuelle Fortpflanzungszellen besitzen diesen

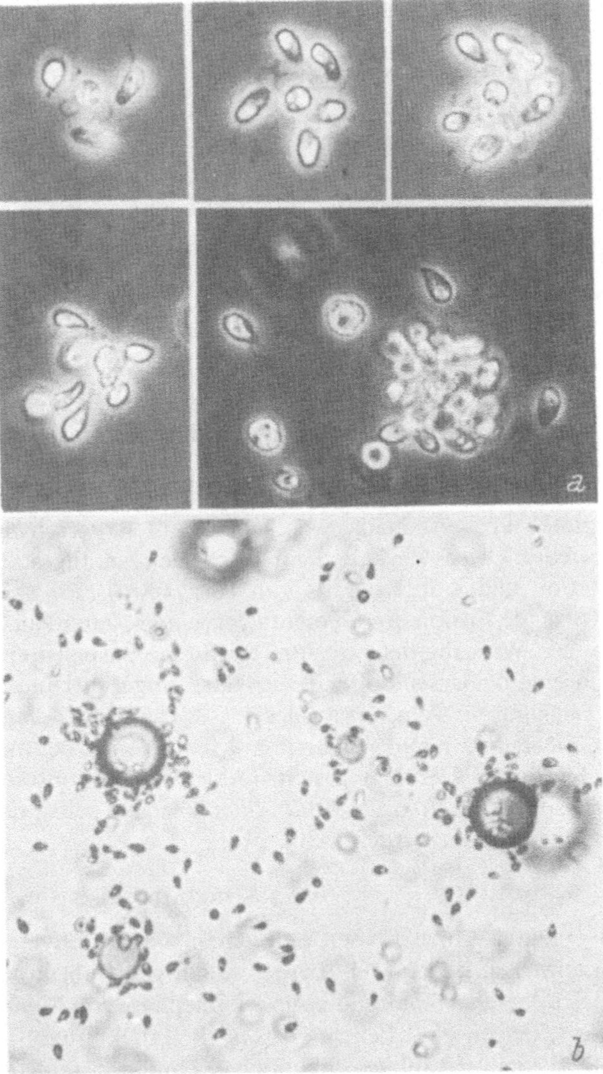

Abb. 2. *Ectocarpus siliculosus.*
a Befruchtung. Ein festgesetzter weiblicher Gamet (Mitte) wird von männlichen Gameten umschwärmt. L. u.: Zellverschmelzung, r. u.: Anlockung besonders zahlreicher männlicher Gameten; in der linken Bildhälfte: 2 Zygoten. Phasenkontrast, Vergr. 600×, Blitzlicht (Original).
b Ectocarpin-haltige Paraffinöltropfen in einer Suspension männlicher Gameten. Hellfeld, Blitzlicht (Original)

Geruch nicht. Die Weibchen-spezifische Substanz, die sehr leicht in den Luftraum übergeht, kann mit einem Luftstrom abgeblasen, in Kältefallen kondensiert und angereichert werden. Sie läßt sich auch in schwerflüchtigen Fettsubstanzen (z. B. Paraffinöl oder Vaseline) auffangen. Der auf diese Weise gewonnene Weibchen-spezifische Stoff ruft bei Suspensionen von männlichen Gameten die charakteristischen chemotaktischen und chemo-kinetischen Reaktionen hervor, die der Befruchtung vorausgehen.

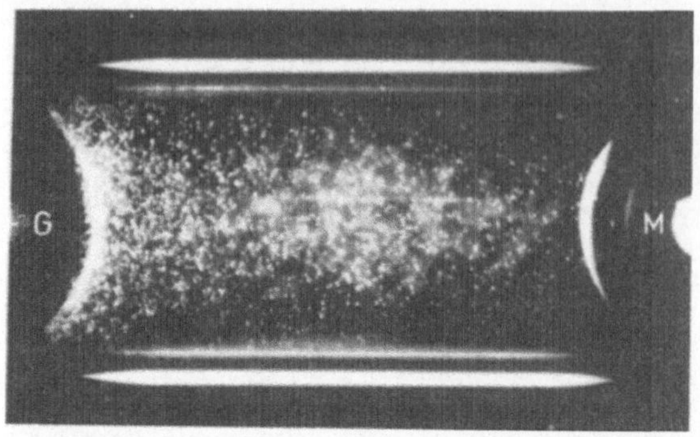

Abb. 3. *Ectocarpus siliculosus.* Männliche Gameten in einer Kapillare. Bei M grenzt an den Meniskus die Dampfphase über reinem Kulturmedium; bei G dagegen die Dampfphase über einer Suspension weiblicher Gameten. Blitzlicht, Dunkelfeld; aus (*56*)

Die chemokinetische Reaktion läßt sich leicht beobachten, wenn einer Suspension männlicher Gameten in einem Hängetropfen der von weiblichen Gameten gebildete Stoff in der Gasphase angeboten wird. An der Seewasser-Luft-Grenze bilden sich Zusammenballungen, in denen bis zu 20 oder mehr männliche Gameten in heftiger Bewegung um einen Mittelpunkt kreisen. Nach wenigen Minuten verschwinden diese Gruppen wieder. Dieses Verhalten entspricht den Anfangsstadien der echten Kopulation und wurde bereits von Hartmann (*26* a) als Indiz für die nachfolgende Fusion der Gameten verwendet. Zum Nachweis der chemotaktischen Wirkung der Substanz wird eine Suspension männlicher Gameten in eine beiderseits offene Kapillare eingeführt und deren eines Ende der Wirkung der Substanz ausgesetzt, das andere jedoch nicht. Die männlichen Gameten sammeln sich an dem Meniskus an, der mit dem Substanz-Luft-Gemisch in Berührung steht (Abb. 3). Die Substanz kann, außer in der Gasphase, auch in Paraffinöl oder Vaseline gelöst den männlichen Gameten angeboten werden (Abb. 2 b).

c) Strukturbeweis

Durch Anzucht der weiblichen Gametophyten in großen Ansätzen konnten größere Mengen der flüchtigen Substanzen gewonnen werden (*57*). Aus dem Rohkondensat der Gasphase von 14 900 Kulturschalen (154 g Trockengewicht) ließen sich nach Trocknung und präparativer Gas-Rechromatographie schließlich 92 mg der reinen Substanz gewinnen. Ectocarpen ist nach Auskunft der NMR- und IR-Spektren sowie des Massenspektrums ein alicyclischer Kohlenwasserstoff $C_{11}H_{16}$ (M 148) ohne funktionelle Gruppen (**21**). Es ist optisch aktiv:

$$[\alpha]_D^{22°} = +72° \; (4)$$

(21)

Literaturverzeichnis: SS. 96—100

Im NMR-Spektrum lassen sich die folgenden Details erkennen: Eine Methylgruppe, ein tertiäres Proton, 6 olefinische Protonen, 1 Divinylmethan-Methylengruppierung und 2 Methylengruppen. Damit sind bereits sämtliche 16 Wasserstoffe definiert. Man kann daraus ableiten, daß die 4 Unsättigungen in der Form von 3 Doppelbindungen und einem Ringsystem vorliegen, wobei das tertiäre Proton an der Verzweigungsstelle steht.

Im charakteristischen Bereich des IR-Spektrums finden sich lediglich Banden für *cis*-Doppelbindungen (730—690 cm^{-1}), keine für *trans*-Doppelbindungen (970 cm^{-1}). Eine UV-Absorption oberhalb 210 nm oder NMR-Banden im $\delta = 0{,}8$- und 1,2-Bereich werden weder vor noch nach der gaschromatographischen Reinigung gemessen. Es hat also während der Reinigung keine Ring-Isomerisierung stattgefunden.

Tabelle 2. *Eigenschaften des Ectocarpens*

Formel: $C_{11}H_{16}$ (M 148)

farblose Flüssigkeit b$_{15mm}$ 80°, n$_D^{25}$ 1,4882, $[\alpha]_D^{22}$ + 72° (c = 0,03, CCl$_4$)

NMR (100 MHz): δ	1,0	3,4	5,6—5,2	2,8	2,1—2,0
in CCl$_4$ TMS = 0	3H	1H	6H	2H	4H
	$\geqslant$C–CH$_3$	$\geqslant$CH	–C=CH	(C=C)$_2$–CH$_2$	$>$CH$_2$
	-Methyl	tertiärer H	Vinyl-H	Divinylmethan	Methylen

keine NMR-Bande im 0,3- und 1,2-δ-Bereich (Cyclopropanring)

IR (λ) 730, 690 cm^{-1}

Massenspektrum M$^+$ 148 m/e 133 119 105 91 79 66 44; $\dfrac{149 \cdot 100}{148} = 10{,}9$

UV keine Absorption oberhalb 210 nm

Bei der katalytischen Hydrierung werden, entsprechend den drei Doppelbindungen, 6 Wasserstoffe aufgenommen (M$^+$ 154). Das Hexahydro-Ectocarpen hat einen fettig-seifigen Geruch. Es ist in allen Eigenschaften mit n-Butyl-cycloheptan (**22**) identisch (*3*).

(**22**)

Demnach ist Ectocarpen n-Butenylcycloheptadien. Die Lage der Doppelbindung an 1, 2′ und 5′ ergibt sich aus den NMR-Daten, so daß der natürliche Lockstoff die Struktur eines all-*cis*-(1-Cycloheptadien-2′,5′-yl)-buten-1 (**21**) besitzt.

Auf Grund der leichten Flüchtigkeit des Ectocarpens wurde auf eine geschlossene Konformation des asymmetrischen Moleküls geschlossen, wie sie das S-Isomere aufweist. Sie konnte auch durch partielle Reduktion und Ozonolyse zu S-Butylbernsteinsäure bewiesen werden (*4, 46*).

d) Synthese

Die Synthese des Ectocarpens (*29*) folgt der Cycloheptadien-Synthese von Vogel (*84, 85*), deren zentraler Schritt die Cope-Umlagerung (*14, 21, 22, 80*) eines geeigneten Divinylcyclopropans ist: *cis-trans*-2-Vinylcyclo-propancarbonsäureester wird durch Carben-Addition aus Diazoessigester und Butadien dargestellt. Reduktion des Esters mit Lithiumalanat gibt das 2-Carbinol; dessen Oxydation mit MnO_2 (*61*) den *cis, trans*-2-Vinyl-cyclopropan-Aldehyd (**23**). Mit dem Wittig-Reagens aus 1-Brompentin-2 und Triphenylphosphin (**24**) erhält man die vier zu erwartenden isomeren Divinylpropan-Kohlenwasserstoffe (**25**). Durch Cope-Umlagerung entsteht Dehydro-Ectocarpen (**26**), dessen Acetylen-Bindung mittels Lindlar-Katalysator stereospezifisch zur *cis*-Doppelbindung reduziert wird. Das Reaktionsprodukt (**21**) wird durch präparative Gaschromatographie abgetrennt und erweist sich als identisch mit dem natürlichen Lockstoff in allen physikalischen und chemischen Eigenschaften mit Ausnahme der optischen Drehung. Die Ausbeute beträgt 31%. Die Substanz zeigt biologische Aktivität gegenüber *Ectocarpus*-Androgameten.

Abwandlungen dieser Synthese, die auch zur Darstellung von Homologen und der stereospezifischen Synthese der optischen Antipoden adaptiert werden kann, bestehen in unterschiedlichen Wegen zur Darstellung des Divinylcyclopropanaldehyds (*24, 45*) oder in der primären Synthese von dl-Cycloheptadiencarbonsäure und deren Antipodenspaltung (*4*).

Auf grundsätzlich gleichem Weg lassen sich auch ^{3}H- und ^{14}C-markierte Ectocarpene, sowie Homologe und Analoge der Alkylcyclohepta-

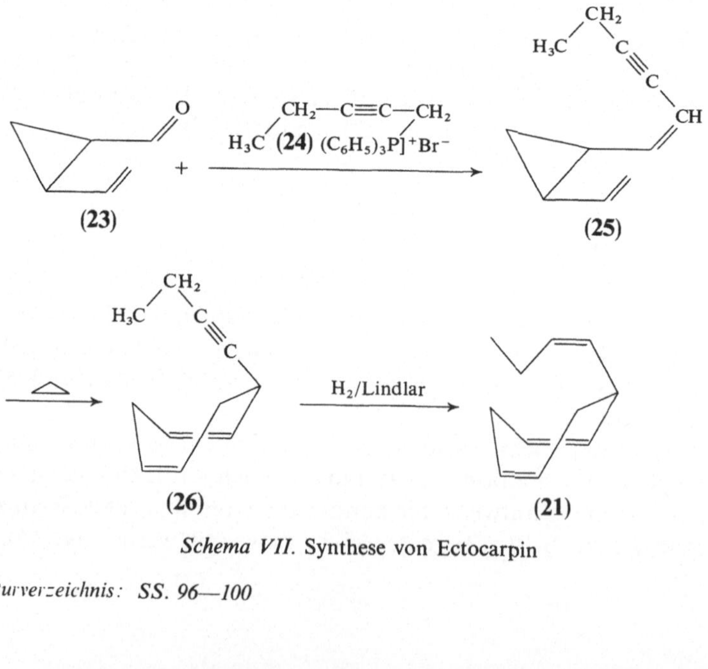

Schema VII. Synthese von Ectocarpin

Literaturverzeichnis: SS. 96—100

diene herstellen. Damit öffnen sich Möglichkeiten, das Stoffwechselschicksal und die Biosynthese der Substanz zu verfolgen sowie ihre Aufnahme durch die Androgameten und den Wirkungsmechanismus zu untersuchen, durch den deren gerichtete Bewegung zustandekommt.

e) Biosynthese

Die Biosynthese des Ectocarpens ist noch unklar. Es könnte sich aus einem konjugierten all-*cis*-Deka-tetraen (**27**) durch Einführung einer Methylengruppe aus aktivem Methyl (S-Adenosyl-Methionin) und anschließende COPE-Umlagerung bilden (*73*) oder durch analoge Reaktion einer polyungesättigten Fettsäure mit nachfolgender Decarboxylierung; aber man könnte sie auch von dem ω-Ende der Linolensäure durch oxydativen Abbau zu einem C_{11}-Bruchstück (**28**) und dessen Cyclisierung ableiten. Von MOORE wurden in *Dictyopteris* mehrere interessante schwefelhaltige Kohlenwasserstoffe isoliert (s. u.), die ebenfalls als Zwischenstufen der Biosynthese diskutiert werden können.

$[CH_3]^+$

(**27**)

(**28**)

VII. Dictyopteren und andere Inhaltsstoffe von Dictyopteris

Mit dem Ectocarpen sehr nahe verwandte Substanzen sind die Dictyopterene, die kürzlich von MOORE (*50*) beschrieben wurden.

Dictyopterene sind Duftstoffe, die von der an den Felsriffen vor der Küste von Hawaii wachsenden Meeresalge *Dictyopteris plagiogramma* abgegeben werden und Bestandteil des Amber-artigen „Geruchs des Pazifiks" (vgl. a. *60*) sind. Ihrer Struktur nach handelt es sich um C_{11}-Cyclopropan-Kohlenwasserstoffe.

MOORE macht keine Angaben über den biologischen Status der extrahierten Pflanzen. Bei *Dictyopteris* sind männliche und weibliche Geschlechtspflanzen sowie die asexuellen Sporophyten gleich gebaut. Außerdem läßt sich bei *Dictyopteris* der europäischen Küsten zeigen, daß sowohl die Sporophyten als auch vegetative Pflanzen ohne Fortpflanzungsorgane den charakteristischen *Dictyopteris*-Geruch produzieren. Deshalb scheint es unwahrscheinlich, daß den Dictyopterenen eine spezifische Rolle bei der sexuellen Fortpflanzung zukommt.

82   L. Jaenicke und D. G. Müller:

Das essentielle Öl von *Dictyopteris* besteht zu etwa 40% aus C_{11}-Kohlenwasserstoffen. Ihre präparative Aufarbeitung erfolgt an 25% $AgNO_3$-Kieselgelsäulen durch Elution mit Äther-Hexan. Die Hauptanteile sind (+)-Dictyopteren A: (+)R,R-*trans*-1-(*trans*-1-hexenyl)-2-vinylcyclopropan (**29**) und (−)-Dictyopteren B: (−)R,R-*trans*-1-(*trans*, *cis*-hexa-1′3′,-dienyl)-2-vinylcyclopropan (**30**); weiter enthält es kleine Mengen (−)-Dictyopteren C′: (−)R-6-Butylcyclohepta-1,4-dien (**32**), dem Antipoden von Dihydro-Ectocarpen (*65*), Dictyopteren D′: (+)S-6-(*cis*-But-1′-enyl)-cyclohepta-1,4-dien (**22**) = Ectocarpen, *trans, cis, cis*-Undeka-1,3,5,8-tetraen (**33**), *trans, trans, cis*-Undeka-1,3,5,8-tetraen (**34**), *trans, cis*-Undeka-1,3,5-trien (**35**) und *trans, trans*-Undeka-1,3,5-trien (*64, 66*) (**36**)

Ein weiterer Anteil sind schwefelhaltige Lipide (*73*). Ihre Abtrennung gelingt an Permutit/Decalso oder Kieselgel mit 1—5% Äther/Hexan und anschließende Gelfiltration an Sephadex LH-20 mit 50% Chloroform/Methanol bzw. durch Dünnschichtchromatographie auf Kieselgel mit Benzol.

Die wichtigsten bisher erhaltenen Thio-Verbindungen sind: Undekan-3-onyl-thiolacetat (**37**) und Di-undekan-3-onyl-disulfid (**38**). In kleinen Mengen wurden weiter isoliert: 3-Hexenyl-4,5-dithiacylcycloheptanon (**39**), sowie *trans*-Undec-4-en-3-onyl-thiolacetat (**40**) und Undec-1-en-3-on (**41**), das vermutlich ein bei der Chromatographie entstehendes Artefakt ist.

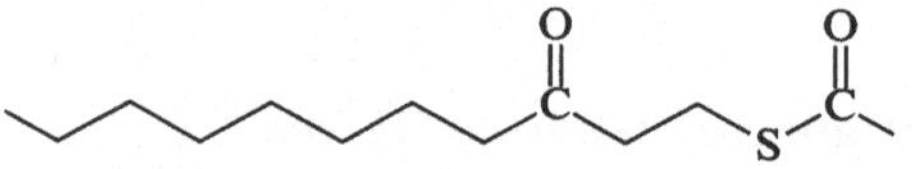

(**37**)

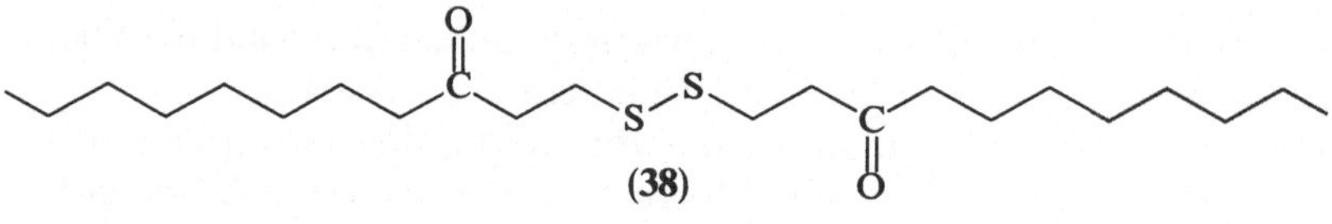

(**38**)

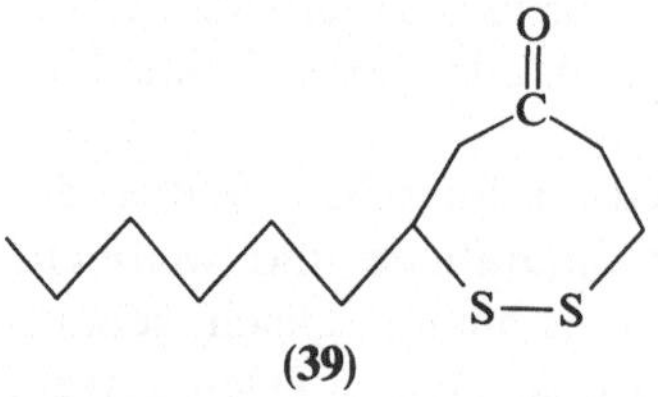

(**39**)

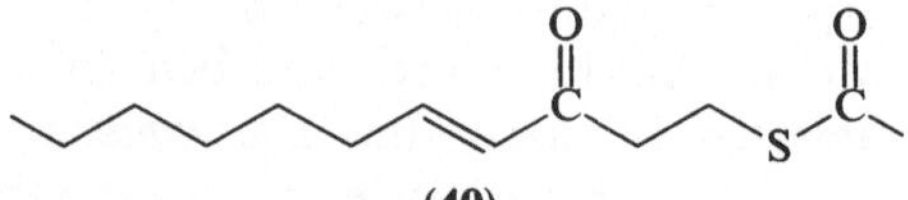

(**40**)

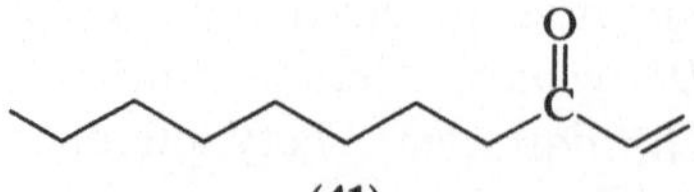

(**41**)

Literaturverzeichnis: SS. 96—100

Da im essentiellen Öl von *Dictyopteris* keine Methyl-verzweigten Verbindungen vorkommen, erscheint es nicht ausgeschlossen, daß diese schwefelhaltigen Lipide die Vorläufer der C_{11}-Kohlenwasserstoffe in *Dictyopteris* sind. Aus ihnen könnten die Undekatriene und Undeka-tetraene entstehen, und der optisch aktive Cyclopropanring der Dictyopterene wiederum könnte durch enzymkatalysierte und licht-in-duzierte (*86*) (angeregter Singulett- oder Triplett-Zustand) Di-π-Methan-Umlagerung (*88, 89*) einer polyungesättigten linearen Fettsäure und an-schließendem Abbau zum C_{11}-Kohlenwasserstoff gebildet werden.

Man kennt derartige Reaktionen bei ungesättigten Kohlenwasser-stoffen mit geminalen Dimethylgruppen, die das intermediäre Bi-Radikal stabilisieren können. So lagert sich Artemisia-trien (**42**) (*trans*-2,5,5-Trime-thyl-1,3,6-heptatrien) leicht in *trans*-1-Isopropenyl-2-vinyl-3,3-dimethyl-cyclopropan um (*76*) (**43**). Allerdings scheint es Hinweise dafür (*35*) zu geben, daß diese Di-π-Methan-Umlagerung von freiem *cis, trans*-1,4,6-Undekatrien zu Dictyopteren A nicht verläuft; jedoch wäre die Triplett-Stabilisierung am Enzym oder durch einen Cofaktor möglich.

(42)   →   (43)

Die Cope-Umlagerung der homologen Divinylcyclopropane verläuft stereospezifisch (*65*). So gibt Dictyopteren A (**29**) nahezu ausschließlich S-Butylcycloheptadien ($[\alpha]_D^{25°} + 2°$ ($CHCl_3$)), den Antipoden des natür-lichen Dictyopteren C′ (**32**). Die absolute Konfiguration dieser Substanz wurde durch Ozonolyse zu (+)R-Butylbernsteinsäure erwiesen (*46*). Dictyopteren C′ (**32**) entsteht vermutlich durch eine entsprechende Umla-gerung aus Dictyopteren C (**31**) bei niedriger Temperatur (*62*), obgleich Dictyopteren C selbst nicht im essentiellen Öl gefunden wurde. Dagegen enthält es in kleiner Menge Dictyopteren D′ (**22**): (+)-6-(*cis*-but-1′-enyl)-cyclo-hepta-1,4-dien, $[\alpha]_D^{25°} + 75°$, das offenbar mit dem Ectocarpen identisch ist. Es sollte in Analogie zum Dictyopteren C′ die S-Konfigura-tion am C_6 haben und der Antipode des Cope-Umlagerungsproduktes aus Dictyopteren B (**30**) sein. Diese Umlagerung erfordert höhere Tem-peraturen (*62*).

Die Synthese von dl-Dictyopteren A (*20, 62*) geht den nächstliegen-den Weg über das Gemisch der stereoisomeren 1-Vinyl-2-(1-hexenyl)-cyclopropane, die durch Reaktion von 2-Vinylcyclopropylcarboxaldehyd (**23**) mit dem Wittig-Reagens n-Pentyl-triphenylphosphoniumbromid in Gegenwart von n-Butyllithium in Äther bei 10° C erhalten wird. Daraus wird durch präparative Gaschromatographie an Polyäthylenglykol-

Tabelle 3. *Übersicht über die Dictyopterene*

Verbindung	Formel	Nr.
Dictyopteren A	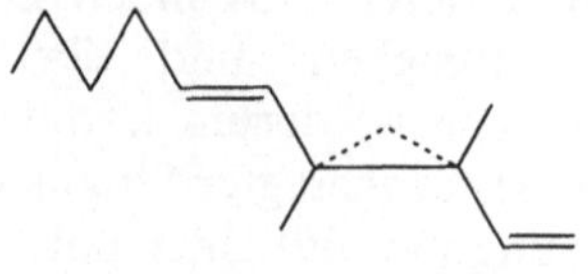	(29)
Dictyopteren B	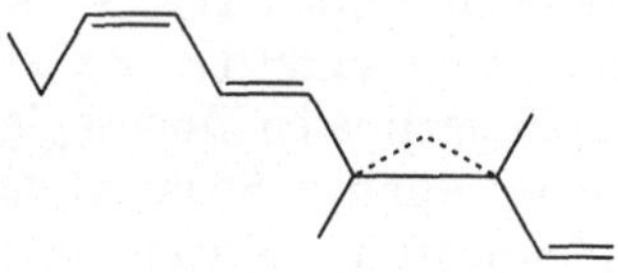	(30)
Dictyopteren C nicht im essentiellen Öl von Dictyopteris	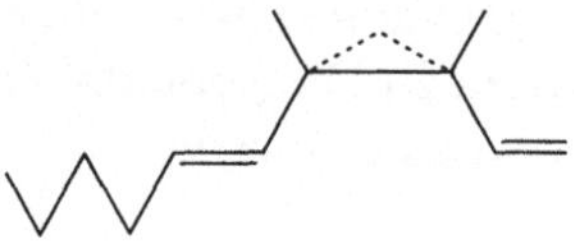	(31)
Dictyopteren C′	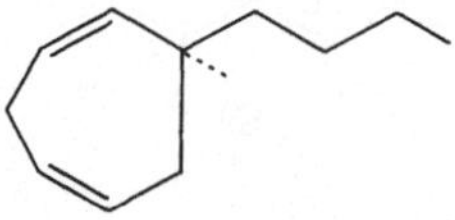	(32)
Dictyopteren D′	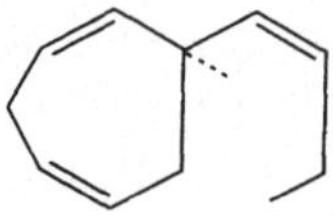	(22)
		(33)
	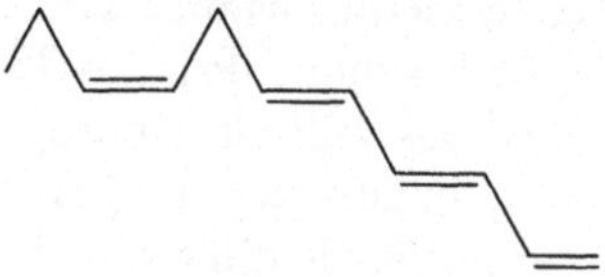	(34)
		(35)
	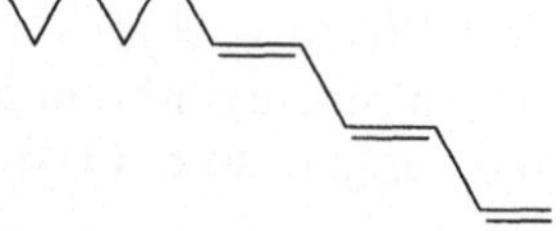	(36)

Literaturverzeichnis: SS. 96—100

Tabelle 3 (Fortsetzung)

Systematischer Namen	Eigenschaften	Durch Cope-Umlagerung bei °C entsteht (65)	
(+) R,R-*trans*-1-(*trans*-1-Hexenyl)-2-vinylcyclopropan (50)	$[\alpha]_D + 77° \pm 5°$ (0,5; A) $+72°$ (1,0; $CHCl_3$) λ_{max} 206 nm ($\varepsilon = 16\,000$)	170°	Antipode von C′ ($[\alpha]_D + 2°$ ($CHCl_3$))
(−) R,R-*trans*-1-(*trans, cis*-Hexa-1′,3′-dienyl)-2-vinyl-cyclopropan (64)	$[\alpha]_D - 43°$ (A) λ_{max} 247 nm	90°	Antipode von D′ ($[\alpha]_D - 24°$ (C_2H_5OH))
cis-1-(*trans*-1-Hexenyl)-2-vinylcyclopropan 162)	instabil 15°	RT	Dictyopteren C′ ($[\alpha]_D - 13°$ ($CHCl_3$))
(−) R-6-Butyl-cyclohepta-1,4-dien (65)	$[\alpha]_D - 13°$ ($CHCl_3$)		
(+) S-6-(*cis*-But-1′-enyl-)cyclohepta-1,4-dien (65)	identisch mit Ectocarpen $[\alpha]_D + 75°$ ($CHCl_3$) $+72°$ (CCl_4)		
trans,cis,cis-Undeca-1,3,5,8-tetraen (64)	λ_{max} 257,266,277 nm		
trans,trans,cis-Undeca-1,3,5,8-tetraen (66)			
trans,cis-Undeca-1,3,5-trien (66)			
trans,trans-Undeca-1,3,5-trien (66)			

succinat oder Apiezon L das gewünschte *trans, trans*-Isomere (**29**) isoliert. Es macht nach Angaben von Das und Weinstein (*20*) 48% des Gemisches aus, das weitere je 2% der *cis, cis*- (**29b**) und der *trans, cis*-Verbindung (**29a**) enthält und 48% des *cis, trans*-Kohlenwasserstoffs (**31**), der sich nach diesen Angaben erst beim Kochen unter Rückfluß in Xylol durch Cope-Umlagerung   in   6-Butyl-*cis,cis*-cyclohepta-1,4-dien(Dihydro-Iso-

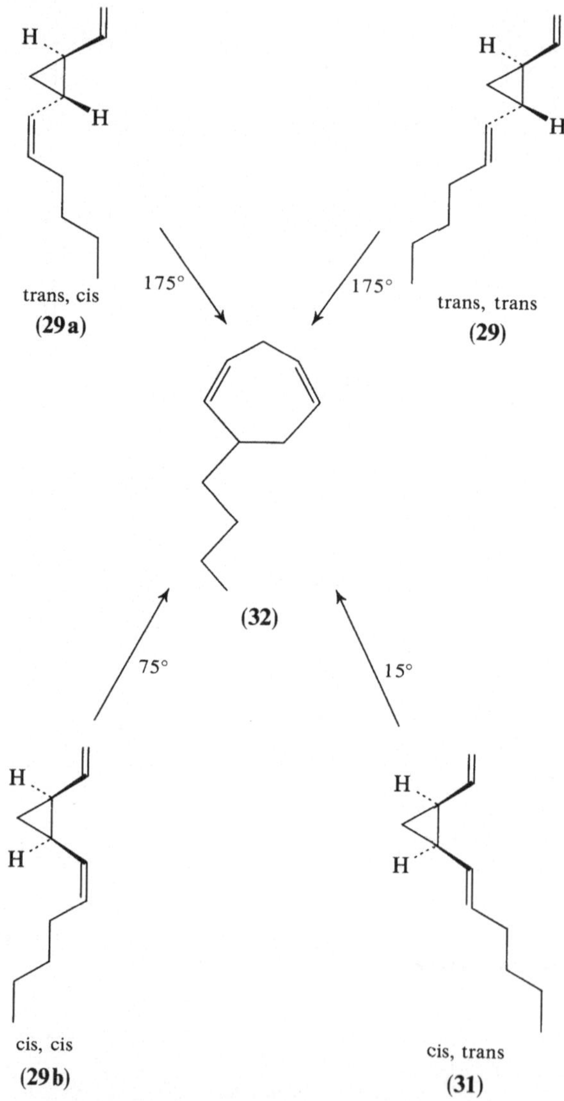

Schema VIII. Umlagerungen von Dictyopterenen

Ectocarpen (**32**) umlagert. Beim Erhitzen soll sich auch der *trans-cis*-Kohlenwasserstoff über eine allylische Diradikal-Zwischenstufe in weiteres *trans*-Vinyl-(*trans*-1-hexenyl)-cyclopropan (Dictyopteren A) umlagern. Eine analoge Umisomerisierung scheint Dictyopteren A in der Hitze zu erleiden, da die Massenspektren von Dictyopteren A und Butylcyclo-heptadien praktisch identisch sind.

Auf im wesentlichen identischen Prinzipien beruht die Synthese von Dic-tyopterenen durch OHLOFF und PICKENHAGEN (*62*). Jedoch wird hierbei vom reinen *trans*-Vinylcyclopropylcarbinol (*33*) ausgegangen. Das Reak-tionsprodukt der WITTIG-Kondensation enthält demnach nur das *trans-, cis-* (**29a**) und das *trans, trans*-Divinylcyclopropan-Derivat (**29**) (dl-Dictyopteren A). Das Mischungsverhältnis wird als 60:40 angegeben. Beide Isomeren valenzisomerisieren bei 175°C zum 6-Butylcyclohepta-dien (**32**). Die Synthese der beiden *cis*-isomeren Kohlenwasserstoffe (**29b** und **31**), die sich vom *cis*-Vinylcyclopropylcarbinol ableiten, erfolgt in gleicher Weise. Das Reaktionsgemisch nach der WITTIG-Umsetzung enthält aber nur ein, allerdings thermolabiles, *cis*-Divinylcyclopropyl-Derivat (**29b**) neben 6-Butylcyclopheptadien (**32**). Es wird angenommen, daß letzteres bereits unterhalb Raumtemperatur aus dem primären *cis, cis*-Divinylcyclopropyl-Kohlenwasserstoff (**31**) entstanden ist, wäh-rend der *cis, trans*-Kohlenwasserstoff (**29b**) erst bei 70° C in das Cyclo-heptadien-Derivat übergeht. Modellbetrachtungen (*74*) zeigen tatsäch-lich, daß die sterische Behinderung der COPE-Isomerisierung durch einen der geminalen Wasserstoffatome des Cyclopropanrings bei einer *cis*-Doppelbindung in der Seitenkette größer ist (*62*).

Die Massenspektren aller dieser Verbindungen sind weitgehend iden-tisch. Die NMR- und IR-Spektren zeigen die zu erwartenden Charakteri-stika.

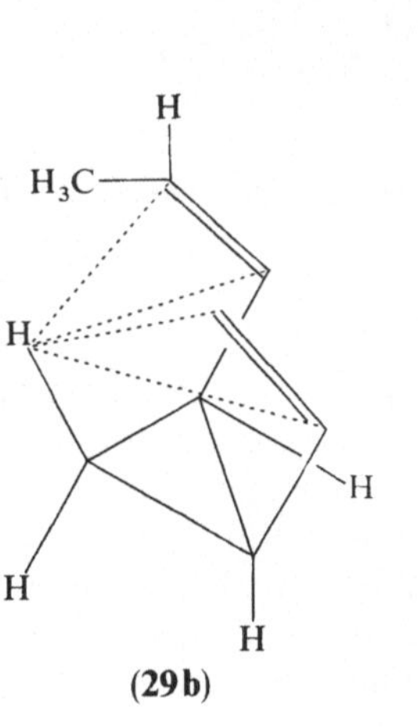

(29b)

(29 und 29a)

R = H   CH₃
R₁ = CH₃   H

VIII. Einige weitere Fälle von Gameten-Chemotaxis

a) Fucus

Die Fucus-Arten gehören zu den derberen Tangen an den Küsten der gemäßigten Klimazone. Die sexuelle Fortpflanzung geschieht bei ihnen durch bis zu 0,1 mm große Eier, die, in großen Massen gebildet, von sehr viel kleineren Spermatozoiden befruchtet werden (vgl. Abb. 1). Da sich Eier und Spermatozoiden leicht und in großen Mengen gewinnen lassen, ist die Befruchtung bei *Fucus* schon sehr lange bekannt. Die mikroskopische Beobachtung zeigt in eindrucksvoller Weise, daß von den Eiern eine Lockwirkung auf die Spermatozoiden ausgehen muß (Abb. 4). Wird Seewasser, in dem sich längere Zeit befruchtungs-

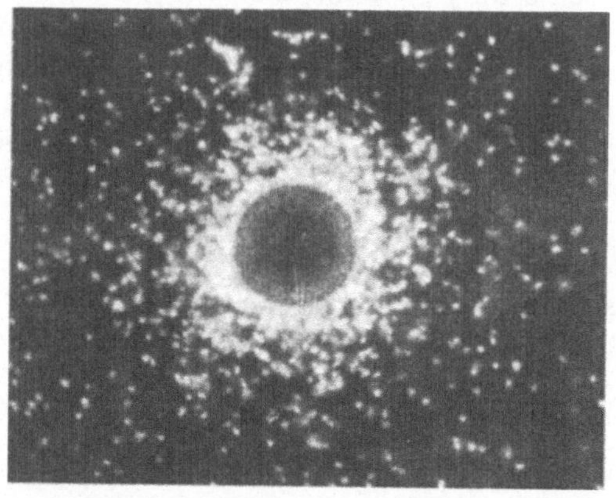

Abb. 4. Ei von *Fucus serratus* in einer Spermatozoidensuspension. Durchmesser der Eizelle 70 μ. Dunkelfeld, Blitzlicht (Original)

fähige Eier befunden haben, in eine Kapillare eingeführt und den Spermatozoiden angeboten, schwimmen sie auf die Mündung der Kapillare zu und sammeln sich dort an. Es läßt sich zeigen, daß die Wirkung der Eier auch durch eine Cellophanmembran hindurch erfolgt. Cook, Elvidge und Heilbron (*16*) fanden, daß die Lockwirkung durch Durchblasen eines inerten Gases (z. B. N_2) aus Eisuspensionen entfernt werden kann. Die chemotaktische Aktivität ist dann im Kondensat (Kältefalle bei $-78°$ C) wieder zu finden. Dieselben Autoren machten auch einen ersten Versuch, die Struktur der wirksamen Substanz aufzuklären (*15*). Dazu wurden Kondensate eines durch Eisupensionen von F. *vesiculosus* und F. *serratus* geführten Stickstoffstromes („Eiwasser-Kondensate") in ein Massenspektrometer eingeführt und untersucht. Die Hauptmengen des Kondensats bestanden aus Wasser und CO_2; Massenfragmente von 74, 73 und 59 schienen möglicherweise spezifisch für die Eiwasser-Kondensate zu sein. Die technischen Schwierigkeiten dieser Analysen waren jedoch groß, so daß die Autoren bemerkten: „there was no guarantee, that the samples did not deteriorate on storage prior to analysis, or that they contained any of the natural chemotactic material".

Im Verlauf einer Prüfung von flüchtigen Substanzen fanden die Autoren eine große Zahl auf *Fucus*-Spermatozoiden chemotaktisch wirkender Chemikalien, u. a. die n-Alkane C_5 bis C_9 sowie mehrere Äther, Alkohole, Ester und Ketone. Die Versuche wurden allerdings nicht weitergeführt. Neuere Beobachtungen (*58*) bestätigen das Vorhandensein eines flüchtigen chemotaktisch wirkenden Lockstoffs in Ei-Suspensionen von *Fucus serratus*. Durch Ausfrieren aus dem über Fucus-Eier geleiteten Luftstrom in einer an das Verfahren bei Ectocarpus angelehnten Methode konnte der Lockstoff als gaschromatographisch einheitliche Fraktion gewonnen werden. Er ist ein dreifach ungesättigter linearer C_8-Kohlenwasserstoff (M = 108) mit dem UV-Spektrum des Hexatriens (λ_{max} 255, 265, 275 nm; ε_{max} ca. 40 000). Bei der Hydrierung entsteht n-Octan. Massenspektrum und gaschromatographisches Verhalten lassen in der Substanz Octatrien-1,3,5 (=1-Äthyl-hexatrien-1,3,5,) vermuten (*29a*). Die Stereochemie der Doppelbindungen ist zur Zeit noch unbekannt.

b) Chlamydomonas

Einen weiteren Fall von Gameten-Chemotaxis fand Tsubo bei *Chlamydomonas moewusii* var. *rotunda* (*83*). Es handelt sich dabei um einzellige begeißelte Grünalgen, die sich bei entsprechender Vorbehandlung in Gameten umwandeln können. Die Geschlechtsverteilung ist diözisch; man unterscheidet einen (+)- und einen (−)-Stamm (vgl. Abb. 1). Gewinnt man ein zellfreies Filtrat des (−)-Stammes und bietet es in einer Kapillare den (+)-Gameten an, so kann man eine deutliche Lockwirkung feststellen. Umgekehrt läßt sich kein solcher Effekt eines (+)-Filtrates auf (−)-Gameten ausmachen. Die Wirksamkeit von aktiven Filtraten verschwand beim Durchperlen mit N_2 sowie beim offenen Erwärmen. Präparate blieben jedoch beim halbstündigen Erhitzen auf $130°C$ aktiv, wenn sie in einer Ampulle luftdicht eingeschmolzen waren. Die von den

($-$)-Gameten produzierte chemotaktisch aktive Substanz ist demnach hitzestabil und leicht flüchtig. Als chemotaktisch wirksam erwiesen sich auch folgende Gase: Äthylen, Acetylen, Kohlenmonoxyd und Leuchtgas. Die chemische Struktur des *Chlamydomonas*-Lockstoffs ist bislang unbekannt.

c) *Sphaerocarpos*

Bei den Archegoniaten (Moosen und Farnen) liegt ein Generationswechsel zwischen einem haploiden Gametophyten und einem diploiden Sporophyten vor. Das weibliche Geschlechtsorgan (Archegonium) zeichnet sich durch einen charakteristischen Bau aus: Die Eizelle ist von einer flaschenartigen Hülle aus vegetativen Zellen umgeben und erst im reifen Zustand durch einen Kanal, den „Flaschenhals", zugänglich. Die männlichen Gameten sind in Regen- oder Tautropfen frei beweglich, sammeln sich an der Archegoniummündung an und dringen durch den Kanal zur Eizelle vor (vgl. Abb. 1). Die Beobachtung zeigt, daß die Spermatozoiden von der Archegoniummündung bzw. der Eizelle angelockt werden und gerichtet darauf zu schwimmen. Die chemotaktische Reaktion von Moos- und Farnspermatozoiden wurde sehr eingehend untersucht. Durch empirische Suche wurde eine große Anzahl von Substanzen gefunden, die bei Spermatozoiden verschiedener Arten positive Chemotaxis hervorrufen (*87*). Als Beispiele seien erwähnt: Casein und andere Eiweißstoffe, sowie verschiedene Kalium-, Rubidium- und Caesiumsalze bei dem Lebermoos *Marchantia polymorpha;* Rohrzucker bei *Funaria* und anderen Laubmoosen; Äpfelsäure und andere organische Säuren bei verschiedenen Farnen. Es ist jedoch klar, daß in keinem dieser Fälle das natürliche Sexual-Chemotaktikum bekannt ist.

Der erste und bislang einzige Ansatz für die Identifizierung eines nativen Lockstoffes bei Archegoniaten sind die Arbeiten von Schieder (*77, 78*). Reife Archegonien des Lebermooses *Sphaerocarpos donnellii* bilden eine Substanz, auf die die Spermatozoiden positiv chemotaktisch reagieren. Sie sammeln sich bei intakt abpräparierten Archegonien an der Mündung des Kanals an (Abb. 5a), bei Archegonien, deren basale Hüllzellen bei der Präparation zerstört wurden, jedoch auch in verstärktem Maß in der Nähe der Eizelle (Abb. 5b).

Die wirksame Substanz kann aus den ganzen fertilen weiblichen Pflanzen extrahiert werden. Die Beurteilung der Reinigungsschritte kann mit dem Kapillartest erfolgen (Abb. 5c), oder aber mit Hilfe des „Trockenrandtests". Dabei wird die zu prüfende Lösung auf einem Objektträger eingedunstet und der Rückstand sodann mit einer Spermatozoiden-Suspension überschichtet. Besaß die zu prüfende Lösung chemotaktische Aktivität, so konzentriert sich diese während des Eindunstens

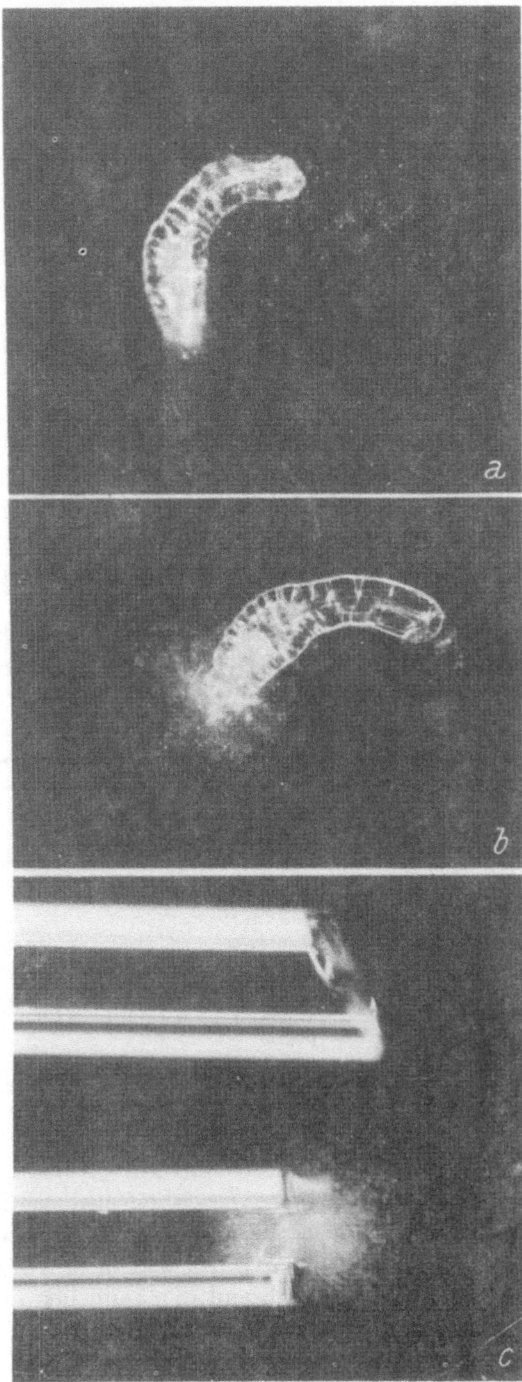

Abb. 5. *Sphaerocarpos donnellii.*
a, b freipräpariertes Archegonium in einer Spermatozoidensuspension. a Archegonium intakt, b basale Hüllzellen des Archegoniums zerstört. Öffnung des Archegoniums jeweils rechts oben. c Kapillar-Test. Obere Kapillare enthält dest. Wasser, untere Kapillare enthält Lösung des extrahierten Lockstoffes. Alle Aufnahmen Dunkelfeld, Blitzlicht. a, b Original O. SCHIEDER; c aus (77)

in der äußersten Randzone des Tropfens. Im Test führt dies zu einer deutlichen Anhäufung der Spermatozoiden in diesem Bereich. Es zeigte sich, daß die wirksame Substanz wasserlöslich und bei trockenem Erwärmen bis 120°C beständig ist. Im sauren Bereich ist sie ebenfalls beständig, im alkalischen Milieu verschwindet die Aktivität. Sephadex-Gel-Filtration deutet auf ein Molekulargewicht zwischen 2 000 und 4 000 hin. Die Substanz ist durch das peptidspaltende Enzym Pronase inaktivierbar. Nach saurer Hydrolyse konnten einige Aminosäuren identifiziert werden. Auffällig ist ein hoher Gehalt an Serin (18%). Diese Ergebnisse weisen darauf hin, daß das Sexual-Chemotaktikum von *Sphaerocarpos donnellii* ein Eiweißstoff bzw. daß für seine Wirksamkeit eine Peptidstruktur unerläßlich ist.

In Tabelle 4 ist die Reizschwelle des natürlichen *Sphaerocarpos*-Chemotaktikums mit den Minimalkonzentrationen bekannter Stoffe für andere Archegoniatenspermatozoiden verglichen. Die *Sphaerocarpos*-Substanz ist etwa 100fach wirksamer als die wirksamsten empirisch gefundenen Chemotaktika. Dies, sowie das zum Teil sehr breite Spektrum wirksamer Substanzen und deren allgemeine Verbreitung in pflanzlichen Geweben spricht dafür, daß die mit dem Griff in den Chemikalienschrank entdeckten Archegoniaten-Chemotaktika nicht mit den Substanzen identisch sind, die im natürlichen Geschehen an der chemischen Wechselwirkung zwischen den Gameten beteiligt sind.

Tabelle 4. *Grenzkonzentrationen chemotaktisch wirkender Stoffe bei Archegoniaten*

Organismus	wirksamstes Chemotaktikum	MG	niedrigste nachweisbare Konzentration	Methode
Verschiedene Farne	Na-Malat	134	6×10^{-5} m	Kapillartest (87)
Lycopodium	Citronensäure	192	5×10^{-6} m	Kapillartest (87)
Sphaerocarpos	Spezifisches Polypeptid	2000 bis 4000	4 bis 8×10^{-8} m	Trockenrandtest (77)

d) Campanularia

Campanularia calceolifera ist ein etwa zentimetergroßer mariner stockbildender Hydroid-Polyp. Die Geschlechtsorgane sind getrennt und auf verschiedene Tierstöcke verteilt. Weibliche Kolonien bilden in krugförmigen Behältern Eier aus, während an männlichen Tierstöcken in ähnlich gebauten Behältern Spermien produziert werden. Die Spermien werden ins Wasser entlassen. Bei genauerer Beobachtung kann man feststellen, daß

Literaturverzeichnis: SS. 96—100

Spermien gerichtet auf die Öffnung des Eibehälters zuschwimmen und sich dort ansammeln (*47*) (vgl. Abb. 1).

Es läßt sich nachweisen, daß die Lockwirkung von dem die Öffnung des Eibehälters umgebenden Gewebe ausgeht. Die Eier selbst scheinen an der Produktion der Locksubstanzen nicht direkt beteiligt zu sein. Die Substanz ist in Wasser und Äthanol löslich und kann aus Präparationen weiblicher Organe isoliert werden (*48*). Dreistündiges Kochen bei 100°C erzeugt keinen Aktivitätsverlust. Die Substanz diffundiert durch Agar und Dialysiermembranen; sie ist unlöslich in Äther und Chloroform-Methanol-Gemischen. Chromatographie an BioGel P2 ergab, daß die Lockwirkung an eine einheitliche Fraktion gebunden ist. Chromatographischer Vergleich mit Substanzen bekannten Molekulargewichts erlaubte eine Schätzung des Molekulargewichtes von etwa 600. Beim Chemotaktikum einer verwandten Art *(Tubularia crocea)* lag das MG bei etwa 240. Weitere chemische Einzelheiten sind bisher nicht bekannt geworden.

IX. Betrachtungen zum Wirkungsmechanismus der Gametenlockstoffe

Mit großer Wahrscheinlichkeit ist die molekulare Basis der beschriebenen intraspezifischen Lockwirkungen zwischen Gynogameten und Androgameten der niederen Organismen durch chemische Signale eine Lockstoff-Rezeptor-Wechselwirkung, an die sich molekulare Umwandlungen, die zur Reizbildung und Reizverarbeitung führen, anschließen. Es handelt sich also, neben dem chemischen, um ein physiologisches und kybernetisches Kommunikationsproblem zwischen Sender und Empfänger. Wir können aber weder erklären, wodurch ein bestimmtes chemisches Molekül zu einem Signalstoff wird, noch wissen wir Sicheres über die elementaren Vorgänge der Rezeptorfunktionen. Das Problem der Anlockung gliedert sich in eine ganze Reihe von Einzelfragen:

1. Woraus entstehen die Signalmoleküle?
2. Was regelt ihre Produktion?
3. Wie werden sie ausgeschieden?
4. Wodurch werden sie aufgenommen?
5. Weshalb bewirken sie bestimmte und gerichtete Reaktionen des Empfängers?
6. Wie werden sie inaktiviert oder abgebaut?

Die Biosynthese, ihre Lokalisation und ihre Kontrolle wurden bereits in den vorhergehenden Abschnitten gestreift. In allen Fällen liegen diese Dinge zwar noch in völligem Dunkel, es lassen sich aber plausible Vorläufer-Beziehungen und Regelungs-Hypothesen entwickeln.

Für die Frage der physiologischen Wirkung der Gametenstoffe wird man vorerst den Erkenntnissen und Axiomen beim Riech- und Schmeckvorgang folgen, wie sie bei dem Bakterium *Escherichia coli* (2) und besonders an Insekten (79) erarbeitet wurden.

Bei Bakterien und Protozoen werden die chemotaktisch wirkenden Substanzen (Carbonsäure-, Zucker- und Aminosäure-Derivate) angeblich weder aufgenommen noch verarbeitet, vielmehr durch wenig spezifische Chemorezeptoren der Außenstruktur erkannt, von denen dann die taktischen Reaktionen gesteuert werden. Adler (2) konnte bei *E. coli* mindestens 5 derartige Rezeptorqualitäten unterscheiden, die allerdings leicht durch Fremdstoffe blockiert werden. Dagegen ist bei den Gameten eine sehr enge Spezifität, hohe Wechselzahl und große Affinität der Rezeptoren zu fordern. So verschwindet das Sirenin von *Allomyces* aus wäßriger Lösung nach Zugabe einer Suspension von Androgameten (11), entweder durch Adsorption oder durch Abbau, vollständig.

Funktionsbezogene elektronenmikroskopische Untersuchungen der weiblichen und männlichen Gameten und ihrer Flagellen liegen noch kaum vor. Die morphologische Grundlage des Empfängersystems stellt vermutlich die Geißel dar, die als Kanal von etwa 0,4 μ Durchmesser und 25 μ Länge (44) die Lockstoffmoleküle festhalten und fortleiten kann, bis sie die empfindliche Rezeptormembran treffen. Dort scheinen sie eine Ja-Nein-Reaktion zu bewirken, die zur Auslösung eines Reizes, zu seiner Verarbeitung und damit zur Einstellung einer bestimmten Reaktionsrichtung führt.

Als Rezeptoren sind Proteine oder Lipoproteine am wahrscheinlichsten, da diese sowohl die große Vielzahl eines hochspezifischen Bindungszentrums (Substratspezifität) als auch die notwendige Stabilität und Elastizität der Tertiärstruktur (Reaktionsspezifität) aufweisen.

Die zur Zeit am besten experimentell belegten Vorstellungen (30) über die Chemorezeption von Molekular-Signalen durch die Zellmembranen sind solche, die sich primär an einfache bimolekulare Protein/Ligand-Wechselwirkungen anlehnen. Wie bei diesen wird der Botenstoff des Senders durch Rezeptormoleküle des Empfängers, ähnlich einem Enzym-Substrat-Komplex, nicht-kovalent gebunden. Die nachfolgenden Vorgänge werden auch insofern komplizierten Enzymmechanismen gleichen, als die Chemorezeption im Sekundärprozeß vermutlich eine verarbeitende physikalische Komponente hat, durch die die Zelle verschiedenen Gebrauch von den auftreffenden Signalen machen kann. Die molekular einfache physikalisch-chemische Wechselwirkung ist daher mechanistisch bereits durchaus komplex mit höher organisierten zellulären Strukturen verbunden.

Zu der Bindung des Lockstoffs durch räumlich ungerichtete Dispersionskräfte, d. h., bildlich, durch Lösung ihres Kohlenstoffgerüstes in der

Literaturverzeichnis: SS. 96—100

hydrophoben Phase des makromolekularen Akzeptors, treten gerichtete Komponenten durch dipolare Wechselwirkungen, wie Wasserstoffbrücken, oder Polarisationsbindungen mittels funktioneller Gruppen (Hydroxylgruppen, Doppelbindungen usw.). Die Dipolkräfte bewirken gerichtete Aggregation der Reaktionspartner und dadurch Konformationsänderung im Akzeptorsystem (*12, 13*). Dies kann wiederum reaktive sekundäre Messenger (vgl. *72*) bilden oder das milieukontrollierte labile Gleichgewicht der Zellmembran so umkippen, daß sich deren Ionen- oder elektrische Permeabilität ändert. Schließlich erfolgt durch diese allosterischen Effektoren (*49*) die Stimulierung des Bewegungszentrums, die Kontrolle der Schlaggeschwindigkeit der Geißeln und damit die gerichtete Reaktion der Zelle.

Für einen derartigen, der Enzymfunktion und Regulation durch Konformationsänderung, Cooperativität und Allosterie ähnlichen Erkennungs- und Verarbeitungsvorgang bei der Lockstoffwirkung lassen sich eine Reihe von Fakten anführen, z. B. die Geschwindigkeit, mit der die Reizaufnahme-Verarbeitung und -Beantwortung erfolgt, die geringe Anzahl von „Treffern", die für den Erfolg erforderlich zu sein scheint, sowie die Spezifität für bestimmte Struktur- und sterische Erfordernisse, wie Mol-Volumen, Fähigkeit zur Wasserstoffbrückenbildung und Polarisierbarkeit. Struktur- und Aktivitäts-Beziehungen sind bei den Sexuallockstoffen allerdings nur wenig untersucht, da noch kein brauchbarer quantativer Routinenachweis ausgearbeitet wurde, man sich dabei wohl auch vor dic gleichen Schwierigkeiten wie bei höheren Tieren gestellt sehen wird, nämlich, eine exakte Aussage über die Anzahl der an der richtigen Stelle auftreffenden Moleküle machen zu können (*11*). Auch wurden Homologe dieser Lockstoffe noch nicht eingesetzt, obgleich ihre Synthese naheliegend ist.

Das Schicksal der Lockstoffe ist ebenfalls völlig unbekannt. Es muß gefordert werden, daß sie abgebaut werden, jedoch nicht als Voraussetzung oder im Verlauf der von ihnen ausgelösten Reaktionen, sondern um sekundäre Fehlinformationen zu verhindern, durch die etwa Androgameten nutzlos zusammengeführt werden würden. Der Abbau braucht nicht rasch oder tiefgreifend zu sein; *cis-trans*-Umlagerung von Doppelbindungen oder Oxydoreduktionen würden ausreichen. Man weiß, daß geringe Strukturumwandlungen Sirenin als Lockstoff unwirksam machen (s. S. 74).

Inaktivierung oder Wegtransport vom Akzeptor ist aber auch erforderlich, um zu gewährleisten, daß der Gamet tatsächlich im Stoffgradienten zum Signalort gelangt. Ein solcher Abtransport des benutzten Signalstoffs würde durch Diffusion oder durch aktiven Transport in Compartimente, in denen er nicht mehr mit dem Rezeptor reagieren kann, durch Abbau oder durch sterische Umlagerung möglich sein. Danach müssen die Akzeptorstrukturen unverändert sein, so daß stets ein frischer unbesetzter Trefferort für neu eintretenden Lockstoff zur Verfügung steht.

Offen bleibt die Frage nach der Abstufung der Wirksamkeit der Lockstoffe, die auf Modifikation der Bindungsstärken im primären oder sekundären Trefferort beruhen mag, und das damit zusammenhängende Problem, weshalb die Zelle im Gradienten gerichtet bleibt. Sie muß erkennen können, daß der Gradient zu- oder abnimmt, und auf den negativen Gradienten durch Richtungsänderung reagieren. Dies bedeutet, daß es verschiedenartig strukturierte oder verschiedenartig reagierende (anisotrope) Akzeptormoleküle in den Zellen gibt.

Es ist zu hoffen, daß die einzelligen, auf so hochspezifische Lockstoffe verhältnismäßig einfach reagierenden Gameten niederer Organismen zur Lösung der Probleme der Chemorezeption beitragen können, zumal sie auch genetischen Manipulationen zugänglich sein dürften, durch die sich die molekularen Vorgänge verdeutlichen und die stofflichen Grundlagen verstärken ließen.

Literaturverzeichnis

1. ADAM, G., and M. DELBRÜCK: Reduction of dimensionality in biological processes. In: Structural Chemistry and Molecular Biology, p. 198. (A. Rich and N. Davidson, eds.) San Francisco-London: Freeman. 1968.

2. ADLER, J.: Chemoreceptors in bacteria. Science (Wash.) **166**, 1588 (1969).

3. AKINTOBI, T.: Die Strukturaufklärung des Sirenins der Meeresbraunalge *Ectocarpus siliculosus*. Diplomarbeit, Köln 1970.

4. AKINTOBI, T.: (1971) unveröff.

5. AUSTIN, C. R.: Fertilization. Englewood Cliffs, N. J.: Prentice-Hall. 1965.

6. BARKSDALE, A. W.: Sexual hormones of *Achlya* and other fungi. Science (Wash.) **166**, 831 (1969).

7. BARKSDALE, A. W., M. J. CARLILE, and L. MACHLIS: A comparative study of hormone A and sirenin. Mycologia (N.Y.) **57**, 138 (1965).

8. BERGELSON, L. D., und M. M. SCHEMJAKIN: Synthese natürlich vorkommender Fettsäuren durch sterisch kontrollierte Carbonyl-Olefinierung. Angew. Chemie **76**, 113 (1964); Angew. Chemie Intern. Edit. **3**, 250 (1964).

9. BHALERAO, U. T., J. J. PLATTNER, and H. RAPOPORT: Synthesis of dl-Sirenin and dl-Isosirenin. J. Amer. Chem. Soc. **92**, 3429 (1970).

10. CARLILE, M. J., and L. MACHLIS: The response of male gametes of *Allomyces* to the sexual hormone Sirenin. Amer. J. Bot. **52**, 478 (1965).

11. — — The response of male gametes of Allomyces to the sexual hormone Sirenin. Amer. J. Bot. **52**, 484 (1965).

12. CHANGEUX, J. P., and J. THIÉRY: On the excitability and cooperativity of biological membranes. In: Regulatory function of biological membranes. (J. Järnefelt, ed.) BBA-Library, Vol. **11**, p. 116. Amsterdam: Elsevier Publ. Co. 1968.

13. CHANGEUX, J. P., J. THIÉRY, Y. TUNG, and C. KITTEL: On the cooperativity of biological membranes. Proc. Natl. Acad. Sci. U.S. **57**, 335 (1967).

14. COPE, A. C., and E. M. HARDY: The introduction of substituted vinyl groups V. A rearrangement involving the migration of an allyl group in a three-carbon system. J. Amer. Chem. Soc. **62**, 441 (1940).

15. COOK, A. H., and J. A. ELVIDGE: Fertilization in the Fucaceae: Investigations on the nature of the chemotactic substance produced by eggs of *Fucus serratus* and *F. vesiculosus*. Proc. Roy. Soc. B. **138**, 97 (1951).

16. COOK, A. H., J. A. ELVIDGE, and I. HEILBRON: Fertilization, including chemotactic phenomena in the Fucaceae. Proc. Roy. Soc. B. **135**, 293 (1947/48).

17. COREY, E. J., and K. ACHIWA: A method for deoxygenation of allylic and benzylic alcohols. J. Org. Chem. **34**, 3667 (1969).

18. — — Simple synthetic route to dl-Sirenin. Tetrahedron Lett., 2245 (1970).

19. COREY, E. J., K. ACHIWA, and J. A. KATZENELLENBOGEN: Total synthesis of dl-Sirenin. J. Amer. Chem. Soc. **91**, 4318 (1969).

20. DAS, K. G., and B. WEINSTEIN: The synthesis of ($\pm$)-Dictyopterene A. Tetrahedron Lett. 3459 (1969).

21. DOERING, W. v. E., and W. R. ROTH: The overlap of two allyl-radicals or a four-centered transition state in the Cope rearrangement. Tetrahedron **18**, 67 (1962).

22. — — Thermische Umlagerungsreaktionen. Angew. Chemie **75**, 27 (1963); Angew. Chemie Intern. Edit. **2**, 115 (1963).

23. FAWZI, M. M., and C. D. GUTSCHE: Intramolecular reactions of olefinic diazo ketones. J. Org. Chem. **31**, 1390 (1966).

24. FETZION, M., and M. GOLFIER: Selective oxidation of alcohols by silver carbonate. C. R. Acad. Sci. Paris, Ser. C. **267**, 900 (1968).

25. GRIECO, P. A.: Total synthesis of dl-Sirenin. J. Amer. Chem. Soc. **91**, 5660 1969).

26. HARTMANN, M., und O. SCHARTAU: Untersuchungen über die Befruchtungsstoffe der Seeigel I. Biol. Zbl. **59**, 571 (1939).

26a. HARTMANN, M.: Allgemeine Biologie, 3. Aufl., S. 581. Jena: G. Fischer. 1947.

27. HAUPT, W.: Die Phototaxis der Algen. In: Handbuch der Pflanzenphysiologie (W. Ruhland, ed.), Bd. 17, Teil 1, S. 318. Berlin-Göttingen-Heidelberg: Springer. 1959.

28. HECKER, E.: 4'-Nitroazobenzolcarbonsäure-(4)-chlorid als Reagens auf Alkohole. Chem. Ber. **88**, 1666 (1955).

29. JAENICKE, L., T. AKINTOBI und D. G. MÜLLER: Synthese des Sexuallockstoffs von *Ectocarpus siliculosus*. Angew. Chemie **83**, 537 (1971); Angew. Chemie Intern. Edit. **10**, 492 (1971).

29a. JAENICKE, L.: Sexuallockstoffe im Pflanzenreich. Köln-Opladen: Westdeutscher Verlag. 1972.

30. KAFKA, W. A.: Molekulare Wechselwirkungen bei der Erregung einzelner Riechzellen. Z. vergl. Physiol. **70**, 105 (1970).

31. KARLSON, P., and M. LÜSCHER: 'Pheromones': a new term for a class of biologically active substances. Nature (London) **183**, 55 (1959).

32. KÖHLER, K.: Die chemischen Grundlagen der Befruchtung (Gamone). In: Handbuch der Pflanzenphysiologie. (W. Ruhland, ed.) Bd. 18, S. 282. Berlin-Heidelberg-New York: Springer. 1967.

33. LANDGREBE, J. A., and L. W. BECKER: Synthesis, stereochemistry and solvolysis of 2-bromobicyclopropyl and closely related structures. J. Org. Chem. **33**, 1173 (1968).

34. LAW, J. H.: Natural products with cyclopropane ring. Accts. of Chem. Res. **4**, 199 (1971).

35. LIU, R. S. H., and G. S. HAMMOND: Photosensitized internal addition of dienes to olefins. J. Amer. Chem. Soc. **89**, 4936 (1967).

36. MACHLIS, L.: Evidence for a sexual hormone in *Allomyces*. Physiol. Plantarum **11**, 181 (1958).

37. — A procedure for the purification of Sirenin. Nature (London) **181**, 1790 (1958).

38. — Evidence for the hormonal control of sexual reproduction in *Oedogonium* and *Allomyces*. In: Physiology of Reproduction, p. 79. (F. L. Hisaw, Jr., Ed.) Corvallis, Oregon: Oregon State University Press. 1963.

39. Machlis, L.: The response of wild-type male gametes of *Allomyces* to Sirenin. Plant Physiol. **43**, 1319 (1968).

39a. — The coming of age of sex hormones in plants. Mycologia **64**, 235 (1972).

40. Machlis, L., W. H. Nutting, M. W. Williams, and H. Rapoport: Production, isolation and characterization of Sirenin. Biochemistry **5**, 1247 (1966).

41. Machlis, L., W. H. Nutting, and H. Rapoport: The structure of Sirenin. J. Amer. Chem. Soc. **90**, 1674 (1968).

42. Machlis, L., and E. Rawitscher-Kunkel: Mechanisms of gametic approach in plants. In: Fertilization. (C. B. Metz and A. Monroy, eds.) Vol. **1**, p. 117. New York: Academic Press. 1967.

43. Machlis, L., E. Rawitscher-Kunkel, and J. R. Raper: Sex hormones: Thallophytes. In: Growth. (P. L. Altman and D. S. Dittmer, eds.) Federation of Am. Soc. for Exper. Biol., Washington, D. C. 1962, 532.

44. Manton, I., B. Clarke, and A. D. Greenwood: Further observations with the electron microscope on spermatozoids in the brown algae. J. Exp. Botany **4**, 319 (1953).

45. Marner, F.-J.: (1971) unveröff.

46. Matell, M.: Über Bildungsbedingungen von partiellen racemischen Verbindungen. 2. Mitt. Alkylbernsteinsäuren und S-Alkylmerkaptobernsteinsäuren. Arkiv Kemi **5**, 17 (1953).

47. Miller, R. L.: Chemotaxis during fertilization in the hydroid *Campanularia*. J. Exp. Zool. **162**, 23 (1966).

48. — Gel filtration of the chemotactants of the hydroids *Campanularia*, *Tubularia* and *Gonothyrea*. Amer. Zoologist **6**, 611 (1966).

49. Monod, J., J. P. Changeux, and F. Jacob: Allosteric proteins and cellular control systems. J. Mol. Biol. **6**, 302 (1963).

50. Moore, R. E., J. A. Pettus, Jr., and M. S. Doty: Dictyopterene A, an odoriferous constituent from algae of the genus *Dictyopteris*. Tetrahedron Lett. 4787 (1968).

51. Mori, K., and M. Matsui: Synthesis of racemic sesquicarene. Tetrahedron Lett. 2729 (1969).

52. — —Synthesis of racemic Sirenin, a plant sex hormone. Tetrahedron Lett. 4435 (1969).

53. — — Synthesis of mono- and sesquiterpenoids. II. ($\pm$)-Sesquicarene and a mixture of ($\pm$)-Sirenin and its C-7 epimer. Tetrahedron Lett. 2801 (1970).

54. Müller, D. G.: Generationswechsel, Kernphasenwechsel und Sexualität der Braunalge *Ectocarpus siliculosus* im Kulturversuch. Planta (Berl.) **75**, 39 (1967).

55. — Ein leicht flüchtiges Gyno-Gamon der Braunalge *Ectocarpus siliculosus*. Naturwiss. **54**, 496 (1967).

56. — Versuche zur Charakterisierung eines Sexual-Lockstoffes bei der Braunalge *Ectocarpus siliculosus*. Planta (Berl.) **81**, 160 (1968).

57. Müller, D. G., L. Jaenicke, M. Donike, and T. Akintobi: Sex attractant in a brown alga: Chemical structure. Science (Wash.) **171**, 815 (1971).

58. Müller, D. G.: Chemotaxis in brown algae: detection and isolation of the attractant released by eggs of *Fucus serratus* L. Naturwiss. **59**, 166 (1972).

59. Nutting, W. H., H. Rapoport, and L. Machlis: The structure of Sirenin. J. Amer. Chem. Soc. **90**, 6434 (1968).

60. Ohloff, G.: Chemie der Geruchs- und Geschmacksstoffe. Fortschr. Chem. Forschung **12**, 185 (1969).

61. Ohloff, G., H. Farnow, W. Phillipp und G. Schade: Germacron und seine pyrolytische Umwandlung. Liebigs Ann. Chem. **625**, 206 (1959).

62. Ohloff, G., und W. Pickenhagen: Synthese von ($\pm$)-Dictyopteren A. Helv. Chim. Acta **52**, 880 (1969).

63. OHTA, Y., and Y. HIROSE: Structure of sesquicarene. Tetrahedron Lett. 1251 (1968).
64. PETTUS, J. A., JR., and R. E. MOORE: Isolation and structure determination of an undeca-1,3,5,8-tetraene and dictyopterene B from algae of the genus *Dictyopteris*. Chem. Comm. 1093 (1970).
65. — — The isolation and structure determination of dictyopterenes C′ and D′ from *Dictyopteris*. Stereospecificity in the Cope rearrangement of dictyopterenes A and B. J. Amer. Chem. Soc. **93**, 3487 (1971).
66. — — Isolation and structure determination of Dictyopterene C, two undeca-1,3,5-trienes and undeca-1,3,5,8-tetraenes from *Dictyopteris*. 161st National Meeting of the ACS, Los Angeles, Calif., March 1971, Abstr. 113, Org. Section.
67. PIERS, E., W. DEWAAL, and R. W. BRITTON: The total synthesis of (+)-4-Demethyl-aristolone and related compounds. Chem. Comm. 188 (1968).
68. PLATTNER, J. J., U. T. BHALERAO, and H. RAPOPORT: Synthesis of dl-Sirenin. J. Amer. Chem. Soc. **91**, 4933 (1969).
69. PLATTNER, J. J., and H. RAPOPORT: Synthesis of d- and l-Sirenin and their absolute configurations. J. Amer. Chem. Soc. **93**, 1758 (1971).
70. RAPER, J. R.: Chemical regulation of sexual processes in the Thallophytes. Botan. Review **18**, 447 (1952).
71. RESCHKE, T.: Die Gamone aus *Blakeslea trispora*. Tetrahedron Lett. 3435 (1969).
72. ROBINSON, G. A., R. W. BUTCHER, and E. W. SUTHERLAND: Cyclic AMP. Ann. Rev. Biochem. **37**, 149 (1968).
73. ROLLER, P., K. AU, and R. E. MOORE: Isolation of S-(3-oxoundecyl)-thioacetate, bis-(3-oxoundecyl)-disulphide, (−)-3-hexyl-4,5-dithiacycloheptanone, and S-(*trans*-3-oxo-undec-4-enyl)-thioacetate from *Dictyopteris*. Chem. Comm. 503 (1971).
74. ROTH, W. R., und B. PELTZER: Thermische und photochemische Isomerisierung von Bicyclo[4.2.0]octadien-(2.4). Cyclo-octatrien-(1.3.5) und -(1.3.6). Liebigs Ann. Chem. **685**, 56 (1965).
75. ROTHSCHILD, Lord: The behavior of spermatozoa in the neighborhood of eggs. Intern. Rev. Cytol. **1**, 257 (1952).
76. SASAKI, T., S. EGUCHI, M. OHNO, and T. UMEMURA: Chrysanthemic acid IV. Di-π-methane rearrangement of Artemisia-triene. Tetrahedron Lett. 3895 (1970).
77. SCHIEDER, O.: Untersuchungen über das chemotaktisch wirksame Gamon von *Sphaerocarpos donnellii* Aust. Z. f. Pflanzenphysiol. **59**, 258 (1968).
78. — Über das Chemotaktikum des Lebermooses *Sphaerocarpos donnellii* Aust. Natur-wiss. **58**, 568 (1971).
79. SCHNEIDER, D.: Insect olfacting receiving systems for chemical messages. Science (Wash.) **163**, 1031 (1969).
80. SCHRÖDER, G., J. F. M. OTH und R. MERENYI: Moleküle mit schneller und reversibler Valenzisomerisierung. Angew. Chemie **77**, 774 (1965); Angew. Chemie Intern. Edit. **4**, 752 (1965).
81. STORK, G., and J. FICINI: Intramolecular cyclization of unsaturated diazoketones. J. Amer. Chem. Soc. **83**, 4678 (1961).
82. STORK, G., and M. GREGSON: Acyl participation in concerted cyclization of cyclo-propyl ketones. J. Amer. Chem. Soc. **91**, 2373 (1969).
83. TSUBO, Y.: Chemotaxis and sexual behavior in *Chlamydomonas*. J. Protozool. **8**, 114 (1961).
84. VOGEL, E., R. ERB, G. LENZ und A. A. BOTHNER-BY: Kleine Kohlenstoffringe. 8. Mitt. Valenzisomerisierung von cis- und trans-2-Vinylcyclopropylisocyanat. Liebigs Ann. Chem. **682**, 1 (1965).
85. VOGEL, E., K. H. OTT und K. GAJEK: Kleine Kohlenstoffringe. 6. Mitt. Valenz-isomerisierung von cis-1,2-Divinylcycloalkanen. Liebigs Ann. Chem. **644**, 172 (1961).
86. WHITE, E. H., J. WIESKO, and D. F. ROSWELL: Photochemistry without light. J. Amer. Chem. Soc. **91**, 5194 (1969).

87. Ziegler, H.: Chemotaxis. In: Handbuch der Pflanzenphysiologie, Bd. 17, Teil 2, S. 484. (W. Ruhland, ed.) Berlin-Göttingen-Heidelberg-New York: Springer. 1962.
88. Zimmerman, H. E., and P. S. Mariano: The di-π-methane rearrangement. Interaction of electronically excited vinyl chromophores. Mechanistic and exploratory organic photochemistry. XLI. J. Amer. Chem. Soc. **91,** 1718 (1969).
89. Zimmermann, H. E., and A. C. Pratt: Organic photochemistry. LIV. Concertedness, stereochemistry and energy dissipation in the di-π-methane rearrangement. Source of singlet-triplet reactivity differences. J. Amer. Chem. Soc. **92,** 1409 (1970).

(Eingelaufen am 20. September 1971)

Quassinoid Bitter Principles

By JUDITH POLONSKY, Gif-sur-Yvette

Contents

                    JUDITH POLONSKY:

I. Introduction

The Simaroubaceae form a large botanical family of mostly pantropical distribution; only a few representatives (such as *Ailanthus*) extend into temperate regions. Many species of this family, for example *Quassia amara* L. (Surinam quassia) and *Picraena excelsa* Lindt. (= *Picrasma excelsa* Planch or *Aeschrion excelsa* Kuntze, *Jamaica quassia,* have been known, for over a century, to contain bitter substances which are called collectively "quassin". However, the isolation of individual constituents and even more, the elucidation of their structures was not accomplished until modern physical techniques of investigation were available. Amongst these, thin layer chromatography, nuclear magnetic resonance spectroscopy, circular dichroism and mass spectroscopy should be mentioned. An example of the power of the new physical techniques in this field, and particularly that of NMR spectroscopy, is provided by the chemistry of quassin (**1**) and neoquassin (**2**), the first Simaroubaceae constituents whose structures were established by VALENTA and his co-workers (*74, 75*) in the early 1960's. During the intervening years a number of genera of the family Simarubaceae have been investigated and many new bitter constituents have been isolated and had their structures elucidated (see Tables). These Simaroubaceae constituents, which occur either as esters or in the free state, are all closely related chemically and form the bitter principles of the quassin group, which are called *quassinoids* (*13*)*.

By establishing the structure of quassin, VALENTA and his co-workers resolved one of the few remaining classical problems of plant structural chemistry. It is interesting to note that another problem which engaged the interest and ingenuity of natural product chemists for well over a century was solved in the same year. This was limonin (**3**), the characteristic bitter principle of citrus species which are included in the Rutaceae family. As is now known, a great number of compounds isolated from the same or a botanically related family, the Meliaceae, have a structure similar to limonin and comprise the bitter principles of the limonin group, which are called *limonoids* (*13, 16*). It could be added that those bitter principles which have been isolated from the Meliaceae family, also called *meliacins*, have an intact, not cleaved ring A; this is exemplified by gedunin (**4**) (Chart 1).

Ten years after recognition of the quassin structure one can see the extent of this new area of terpenoid chemistry. A comprehensive review

---

* The Simaroubaceae bitter constituents are also designated by the name "Simarou-. lides" (*16*).

References, pp. 147—150

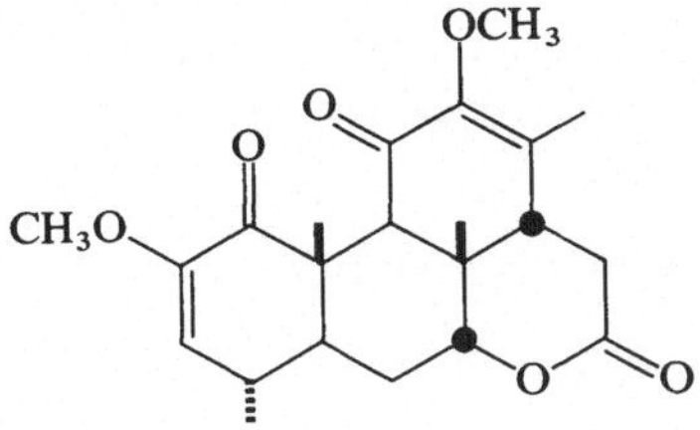

(**1**) Quassin

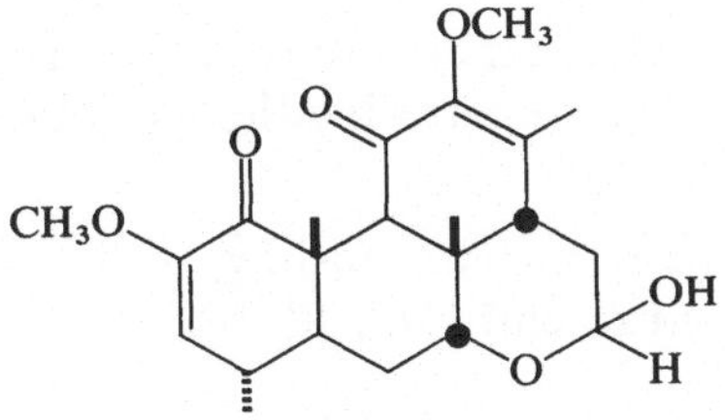

(**2**) Neoquassin

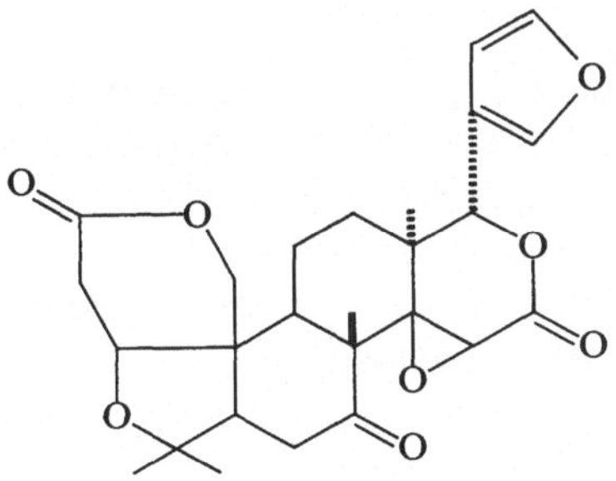

(**3**) Limonin

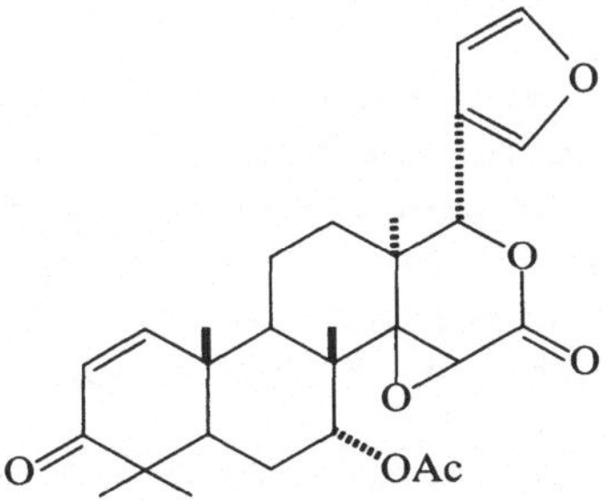

(**4**) Gedunin

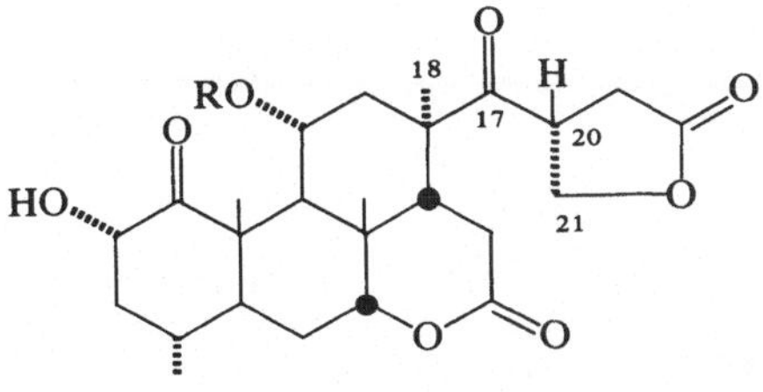

Simarolide (**5**) R=Ac

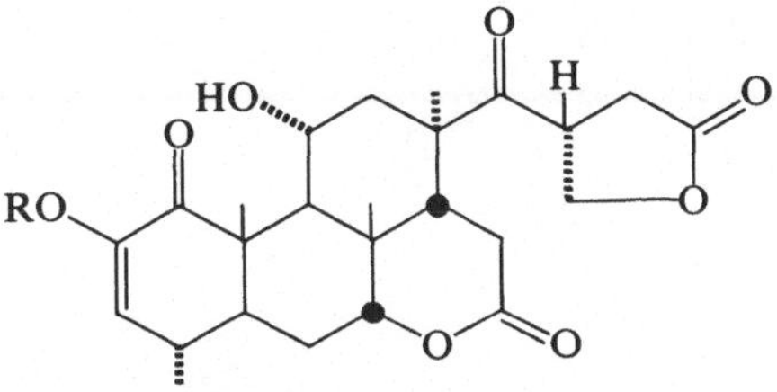

Picrasin A (**6**) R=CH3

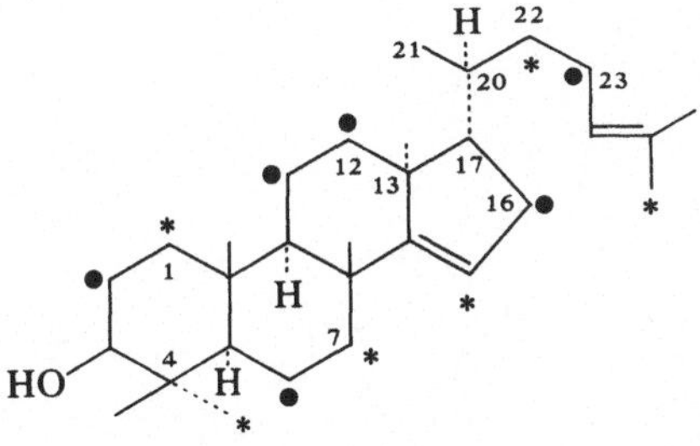

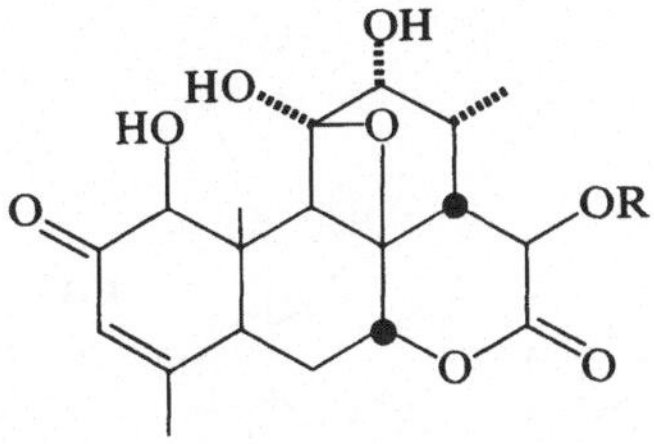

(**7**) Apo-tirucallol

Glaucarubinone (**8a**) R=COC(CH3)OH · C2H5

Glaucarubolone (**8b**) R=H

Chart 1

of this subject therefore seems appropriate at this time when a coherent pattern has emerged although much work still remains yet to be done.

Several reviews dealing with certain aspects of this subject have appeared recently (*13, 50, 58*).

II. Quassinoid General Features

Most of the quassinoids are fundamentally C_{20}-compounds and have, like quassin, the basic skeleton (A) (Chart. 2). Some are C_{19}-compounds with the basic skeleton (B) and only two, simarolide (**5**) (*7, 57*) and picrasin A (**6**) (*33*) are fundamentally C_{25}-compounds with the basic skeleton (C); the two latter compounds have a γ-lactone group which is reminiscent of the furan ring present in the limonoids.

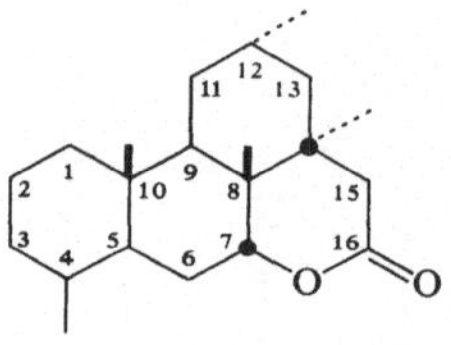

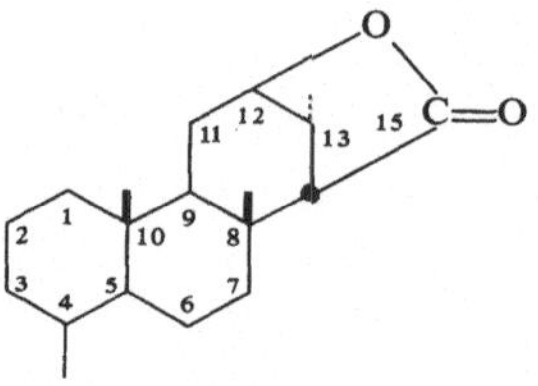

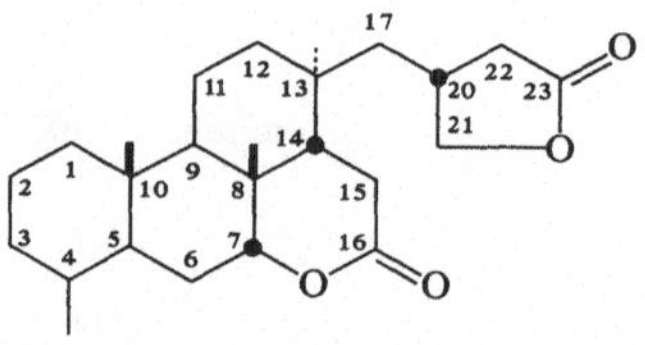

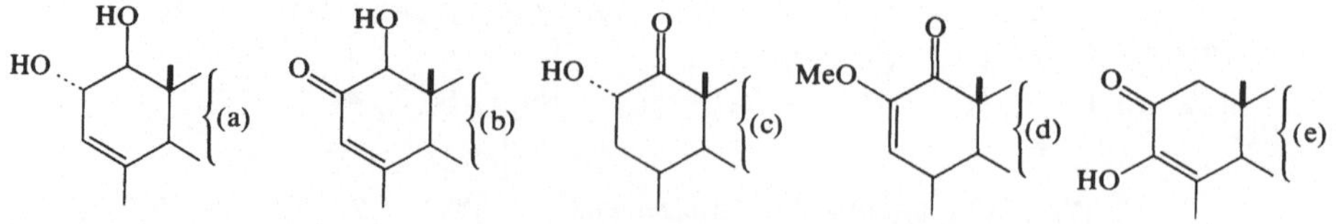

Chart 2

Simultaneously with the chemical studies, there was a development of interest in the biogenesis of the quassinoids. This led to the proposal that these compounds originated, as probably do the limonoids, in a tetracyclic triterpene such as apo-euphol or its 20-α isomer (apo-tirucallol) (7). Direct evidence for this hypothesis came from the incorporation of specifically labelled mevalonate into glaucarubinone and glaucarubolone (8a, 8b). As this problem will be discussed later (p. 132), it is sufficient to note here that the labelling pattern is fully consistent with the triterpene origin of the quassinoids (46, 49). The numbering system of quassinoids is the same as that of their biogenetic precursors, the tetracyclic triterpenes.

All the quassinoids so far known have only one methyl group at C-4. With regard to the proto-triterpene (7), the C_{25}-, C_{20}- and C_{19}-compounds may be considered, respectively, as (seco-16/17; 23/24) pentanor-, (seco-16/17; 13/17) decanor- and (seco-15/16; 13/17) undecanor-triterpenoids. Cleavage of ring A, B or C which is characteristic of the limonoids has not yet been observed in the quassinoids.

The quassinoids are heavily oxygenated lactones (δ-lactones in the C_{20}- and γ-lactones in the C_{19}-compounds) possessing rarely more than one double bond. They have varying numbers of oxygen-containing groups (e. g. hydroxyl or esterified hydroxyl, carbonyl, oxide, methoxyl, or carboxymethyl). In the C_{20}-compounds, these oxygenated functions may be found at most of the carbon atoms, although the methyl groups at C-4 and C-10 and the carbons C-5 and C-9 have never yet been found to have an oxygen function. Despite this widespread pattern of oxygenation, certain positions seem to be preferred. Thus, all the quassinoids, except four, have an oxygenated group at positions 1, 2, 7, 11 and 12. The exceptions are: the two C_{25}-compounds which possess no oxygenated function at position 12 and the bruceins A, B, C and brusatol which have oxygenated functions at positions 2 and 3 but not at 1 and 2.

The variations in the structures of quassinoids are principally as follows:

Ring A may have the structures (a), (b), (c), (d) or (e) depicted in Chart 2.

Ring C may possess at position 8 either a methyl group or a hydroxymethyl group which forms a hemiketal bridge to C-11 or an oxide bridge to C-13.

Ring D may have at C-15 a hydroxyl group which is generally esterified with various small fatty acids (acetic, 2-methyl-butyric, isovaleric, senecioic, 2-hydroxy-2-methylbutyric or 3,4-dimethyl-4-hydroxyvaleric acid).

It is convenient to comment here on certain chemical and physical properties that arise from structural features common to the quassinoids.

1. Chemical Properties

a) All quassinoids containing ring A in the form of partial structure (*a*), readily undergo an acid-catalysed dehydration which eventually leads to two compounds in which the structure of ring A alone is modified. The major product has the A-ring aromatized as in (*f*) and the minor one is an unsaturated ketone ($\lambda_{max} = 230$ mμ) which has partial structure (*g*) (Chart 3).

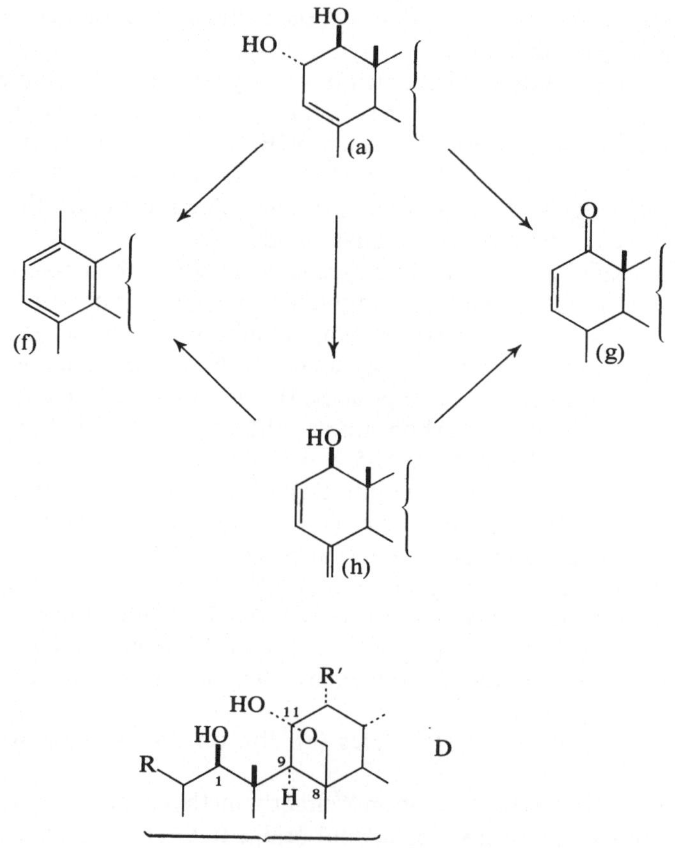

Chart 3

This dehydration has been shown to take place in two stages. Initially loss of one molecule of water occurs to give dienes with structure (*h*). Such dienes have been isolated in crystalline form from several quassinoids with partial structure (*a*) by reaction with very mild acids. On subsequent treatment with mineral acids the dienes may be induced to isomerise to the unsaturated ketones (*g*) or to lose a second molecule of water which results in generation of a benzenoid ring A (*f*). The latter conversion (*h*) → (*f*) is, essentially, a dienol-benzene rearrangement. When ring C is bridged by a hemiketal function [glaucarubol (**55**), chaparrin (**72**)] the B/C *trans*-fusion is maintained during the aromatization as evidenced by lack of deuterium incorporation from DCl. When this feature is absent, partial epimerization at C-9 takes place.

The acid-catalysed dehydration of this type of quassinoids was important in work leading to their structure elucidation. The partially aromatized compounds thus formed were suitable starting materials for structural investigation, the 1-keto-compounds of type g were useful substrates for optical rotatory dispersion and circular dichroism measurements during the stereochemical studies.

b) A number of quassinoids possesses a hemiketal grouping in ring C which involves the carbonyl function at C-11 and the hydroxymethyl at C-8 (partial structure D) (Chart 3), because in the absence of additional carbonyl groups, these compounds do not have the typical U.V.-absorption of saturated ketones, nor do their optical rotatory dispersion or dichroism curves exhibit a Cotton effect. This preferred hemiketal formation can be clearly rationalized in terms of their conformation. Models show that with the A/B and B/C ring junctions *trans* fused, the molecule is well oriented for hemiketal formation, and that this decreases the non-bonded interaction between the α-axial hydroxyl at C-12 and carbon atom C-15 (*37*).

Methylation of these compounds by KUHN's method results in methylation of the hemiketal hydroxyl and all other hydroxyls, except the one which is β and equatorial at C-1 (*65*). Because of steric hindrance caused by the presence of a ketal function at C-11, the hydroxyl at C-1 cannot be acetylated either.

Unlike methylation, acetylation gives principally derivatives which possess a ketone function at C-11 and a primary acetoxy group at C-8. Valuable information about the nature of the function present at C-11 (ketone or ketal) is provided by circular dichroism measurements as well as by the N.M.R. signal of H_9 and the signals of the $-CH_2O$-grouping at C-8.

c) Reaction with diazomethane of this type of compound reveals an unexpected functional group interaction which merits mention (*24*). All the quassinoids with partial structure D afford on treatment in methanol with ethereal diazomethane two monomethylated derivatives in the ratio of about 9 to 1. In the minor product the hydroxyl of the hemiketal has been methylated whereas the hydroxyl at C-1 has been methylated in the major product. It has been shown that this reaction is independent of the nature of the substituent at C-2 or C-12. The formation of these products, particularly that with the methoxyl at C-1, may be explained by assuming strong hydrogen bonding between the hydroxyls at C-1 and C-11; this view is supported by the high value of Δv (OH) observed in the I.R. spectra of some of these compounds.

d) Compounds with partial structure (b) in which the allylic hydroxyl group is oxidised show characteristic U.V. absorption ($\lambda_{max} = 240$ mμ). Quassinoids with partial structures (a) and (b), which very often occur together in the same plant, may be easily interrelated. Thus, oxidation of the compounds of the first type with manganese dioxide readily leads to those of the second type. Reduction of the latter with sodium borohydride gives compounds with structure (a); however, it should be noted that simultaneous reduction of the δ-lactone group to a hemiacetal also occurs. The facile reduction of the lactone group with sodium borohydride, and to a lesser extent by catalytic hydrogenation, has proved to be a characteristic feature of all quassinoids having a six membered lactone ring.

2. Spectral Properties

N.M.R. spectroscopy was without doubt the most important physical tool in the structural elucidation of the quassinoids. The presence of numerous and varied functional groups in these molecules causes the protons to be shielded to different extent. Thus, many of the protons appear in distinctive regions of the spectrum, a circumstance which produce quite characteristic N.M.R. patterns. A vast body of evidence deduced from N.M.R. data was built up which may be of potential aid in the structure determination of quassinoids of as yet unassigned structure. Extensive N.M.R. data have been recorded in most of the papers dealing with constituents of Simaroubaceae; a detailed N.M.R. study of glaucarubine and its congeners has been published separately (23). Therefore, only a few general features of the N.M.R. spectra of the quassinoids are here summarized.

The C-4-methyl resonance in the compounds with A-ring structure (a) occurs consistently at 1.69 ppm; it is shifted downfield to 1.98 ppm in those compounds with A-ring structure (b). The $-CH_2O$-grouping at C-8 generally gives rise to a quartet characteristic of two non-equivalent protons (AB system); the chemical shift of these protons, as well as their coupling constant, is dependant on the nature of the function in which they are involved (hemiketal, ketal, acetoxy or oxide). The β-equatorial H-7 proton appears as a characteristic and easily recognizable triplet ($J \approx 5$ c/s) in the spectra of those quassinoids which have a δ-lactone. In the compounds having a hydroxyl or esterified hydroxyl at position 15, the α-axial proton H-15 ($J_{15, 14} \approx 10$ Hz) is found downfield from the range normally accepted for such protons; this is probably due to the oxygen function located in close proximity at C-12. Moreover, the chemical shift of H-15 depends upon the nature of the group at C-11: the H-15-resonance is shifted downfield when a hemiketal or a hydroxyl at C-11 is replaced by a keto group.

It may also be noted that in the partially aromatized compounds with A-ring structure (*f*), the two aromatic methyl groups appear as two distinctive singlets whereas the two aromatic protons appear as a single (2H) peak.

In contrast to N.M.R. spectroscopy, little use was made of mass spectroscopy during early structural studies of the quassinoids. Although the potentialities of the method increased upon introduction of improved inlet systems which permit the vaporisation of such relatively non-volatile substances as the quassinoids, because of ignorance concerning the cracking pattern, mass spectroscopy was still used mainly for molecular weight determinations. Recently, a systematic study of the mass spectra of several quassinoids has been carried out and fragmentation patterns characteristic of some of their structural features have been observed (*21*). This is illustrated in the discussion of some individual quassinoids later in this chapter.

III. Structure Determination of Quassinoids

1. Quassin (1) and Neoquassin (2)

Serious investigations of the structures of these compounds may be said to have begun in 1950 when ROBERTSON and his co-workers (*42*) were able to achieve a clean separation of quassin and neoquassin from the wood of *Quassia amara* L. and to define most of their chemical properties. Despite the large amount of information accumulated by these authors in the early 1950's (*4, 5, 31*) no structures for quassin and neoquassin could be advanced at that time. In 1960—1962, VALENTA and his co-workers (*74, 75*), using to a great extent the then new technique of nuclear magnetic resonance spectroscopy, were able to correlate the previous work with their own and to establish structures (**1**) and (**2**) for quassin and neoquassin respectively.

The structural evidence obtained by both of the groups is summarized below.

Quassin was found to be a δ-lactone and neoquassin the corresponding hemiacetal; they can be interconverted by reduction and oxidation at C-16. The hemiacetal ring of neoquassin (**2**) was further defined by

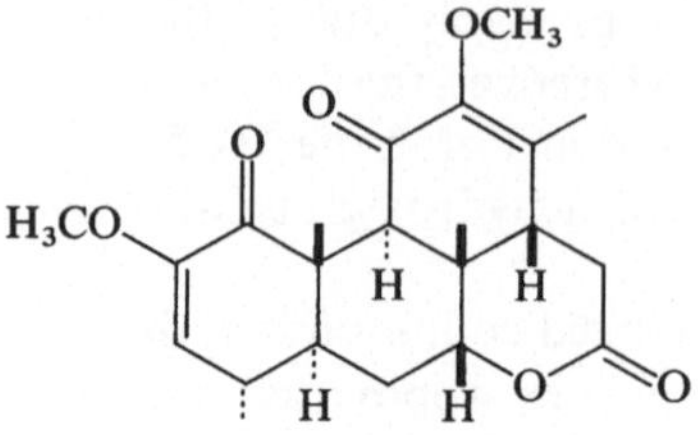

(1)   Quassin

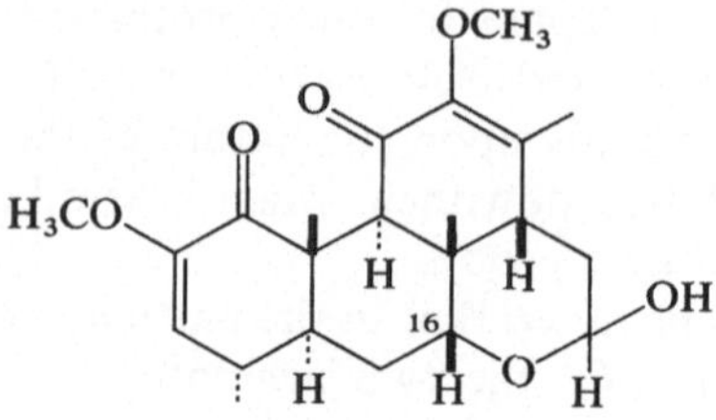

Neoquassin   **(2)**

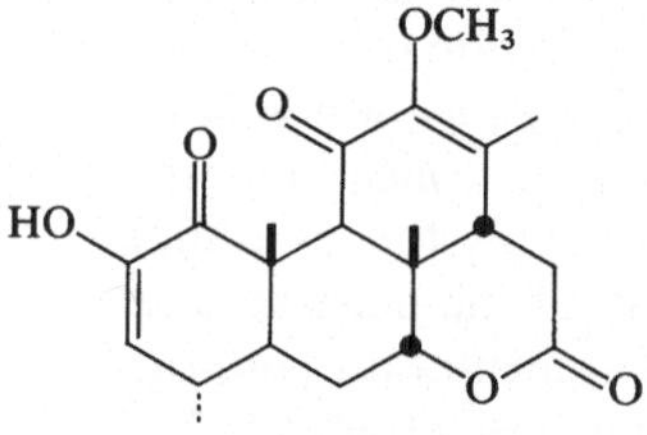

Norquassin   **(13)**

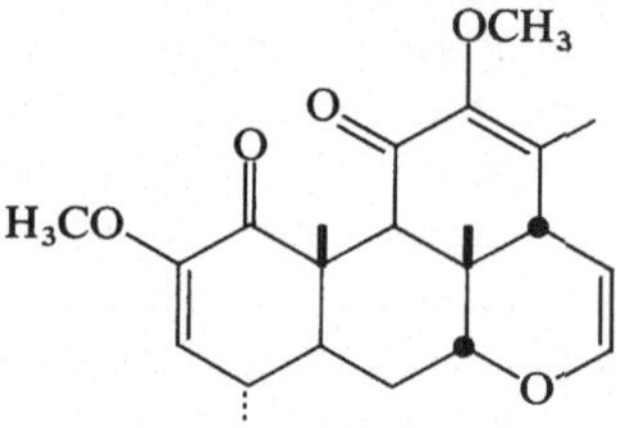

Anhydroneoquassin   **(9)**

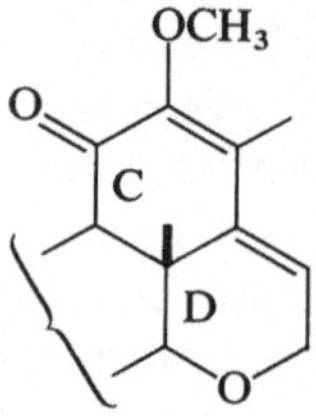

(10)
Dehydroneoquassin
(Isoanhydroneoquassin)

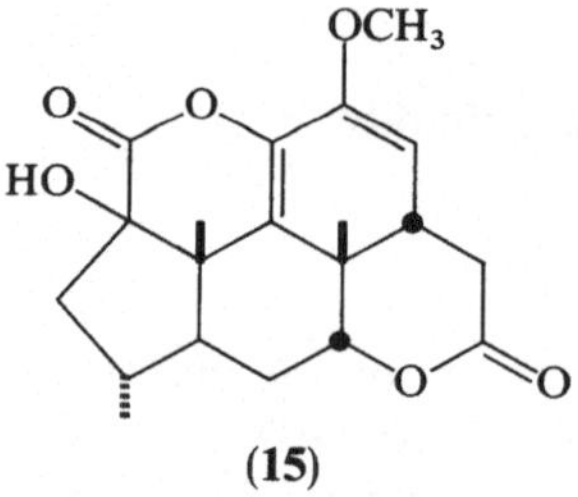

(15)

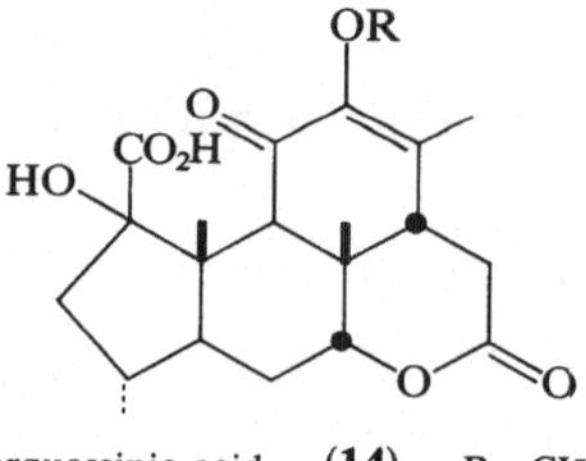

Norquassinic acid   **(14)**   R=CH₃

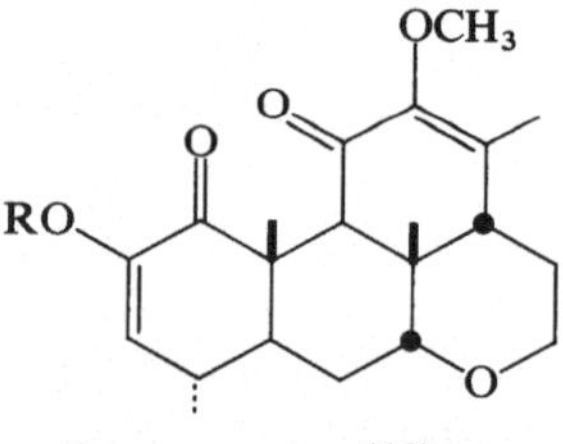

Desoxoquassin   **(11)**   R=CH₃
Desoxonorquassin   **(12)**   R=H

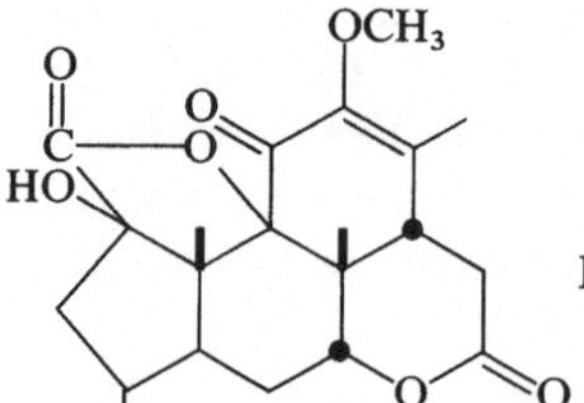

(16)

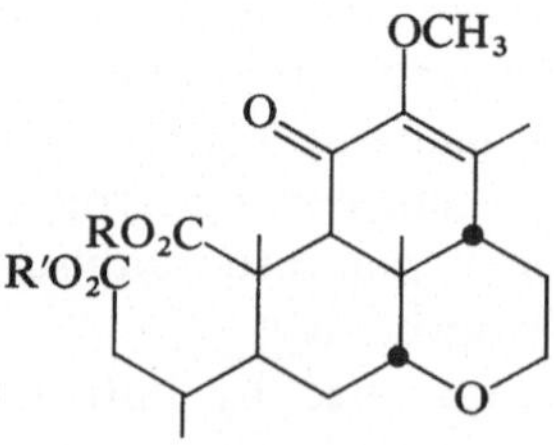

(17)   R=R'=H
(18)   R=H; R'=CH₃

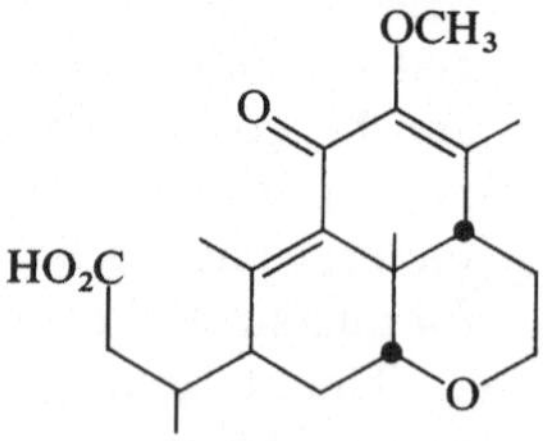

(19)

Chart 4

References, pp. 147—150

reaction with acetic anhydride, which yields the cyclic enol ether, anhydroneoquassin (**9**), and an unusual dehydration product, dehydroneoquassin (**10**). Anhydroneoquassin (**9**) may be reconverted to neoquassin (**2**) on treatment with aqueous acetic acid or reduced to desoxoquassin (**11**) by catalytic hydrogenation.

The infrared and ultraviolet spectra show evidence of two α,β-unsaturated ketone functions; the acidic degradation indicates that both chromophores are diosphenol methyl ethers. Actually, acidic hydrolysis of quassin (**1**) yields norquassin (**13**) which on treatment with diazomethane again gives quassin. Similarly, desoxoquassin (**11**) gives desoxonorquassin (**12**).

The two diosphenol ether systems of ring A und C were related (*74*) in a series of interesting reactions. Thus norquassinic acid (**14**). the benzilic acid rearrangement product of norquassin (**13**), formed the enollactone (**15**) with acetic anhydride-sodium acetate and the γ-lactone (**16**) with lead tetraacetate in acetic acid.

Another important piece of evidence for the structure of quassin came from the study of the desoxodicarboxylic acid (**17**) obtained by alkaline hydrogen peroxide oxidation of desoxonorquassin (**12**). The monoester (**18**), which is stable under the usual decarboxylating conditions, undergoes oxidative decarboxylation on treatment with lead tetraacetate to give the doubly unsaturated ketone (**19**) having a methyl group on the newly-created double bond. This interesting γ-keto acid oxidation defines the substitution at C-9 and C-10 in quassin.

The detailed structure of the lactone ring of quassin was deduced from an oxidative study of anhydroneoquassin (**9**). Only its strongly nucleophilic enol-ether double bond is attacked during mild potassium permanganate oxidation in acetone. This leads to three products: the diol (**20**), hydroxyquassin (**21**) and its non-lactonic isomer, the oxidoenone (**22**). Treatment of hydroxyquassin (**21**) with base eliminates two carbon atoms and leads to the non-lactonic compound (**23**). This reaction which can be visualized as shown in (**21**) establishes the relative position of the lactone ring of quassin and the ketone at C-11.

The fact that ring C in dehydroneoquassin (**10**) does not aromatize provides a strong indication that the second angular methyl group is located at C-8. Confirmation for this assignment comes from the study of the epoxide (**22**), one of the permanganate oxidation products of anhydroneoquassin (**9**).

Treatment of (**22**) with formic acid hydrolyses the normally very stable methoxyl group at C-12, which has become activated by the epoxide function, and leads to the yellow compound (**24**). Treatment of the latter with base yields the phenol (**27**) whose structure rests on detailed NMR, IR and UV spectroscopic evidence. This remarkable transforma-

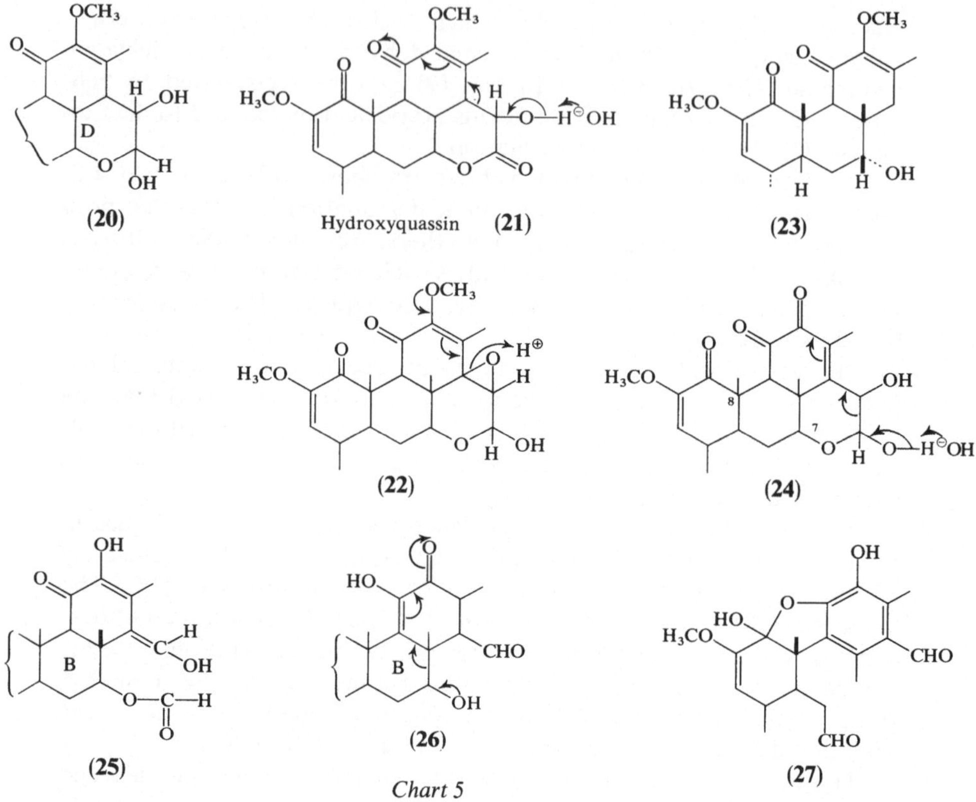

(20)

Hydroxyquassin   (21)

(23)

(22)

(24)

(25)

(26)

(27)

Chart 5

tion which may be rationalized as proceeding through the stages
$(24 \rightarrow 25 \rightarrow 26 \rightarrow 27)$, shows that aromatization is possible only after
cleavage of the $C_7 - C_8$ bond and confirms the location of the angular
methyl group at C-8.

A most informative feature in the chemistry of quassin is its facile
isomerisation and transformation by alkali. Quassin readily dissolves in
aqueous base (lactone opening) and can be recovered unchanged after
saturation with carbon dioxide. However, more vigorous basic treatment
of quassin causes, in addition to the lactone opening, epimerization of the
asymmetric centre at C-9 and ether ring closure between C-7 and C-13
leading to pseudo-quassinolic acid (28)*. The reactions of this acid were
important for deducing the stereochemistry of quassin. On treatment
with acetic anhydride-sodium acetate, pseudoquassinolic acid (28) gives
pseudoquassin (29)*, the product of acylation at C-12. Pyrolysis of
pseudoquassinolic acid (28) furnishes alloquassin (30)*, the formation

---

* Shown as the enantiomer.

References, pp. 147—150

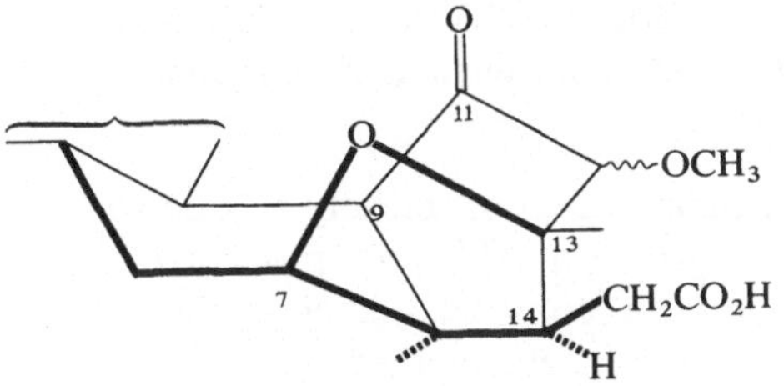

(28)
Pseudoquassinolic acid

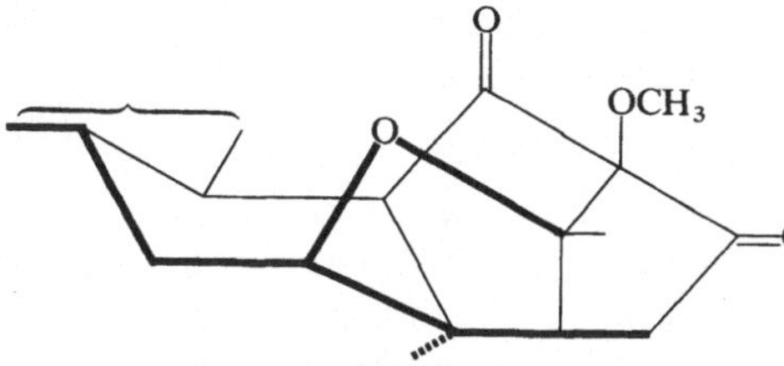

(29)
Pseudoquassin

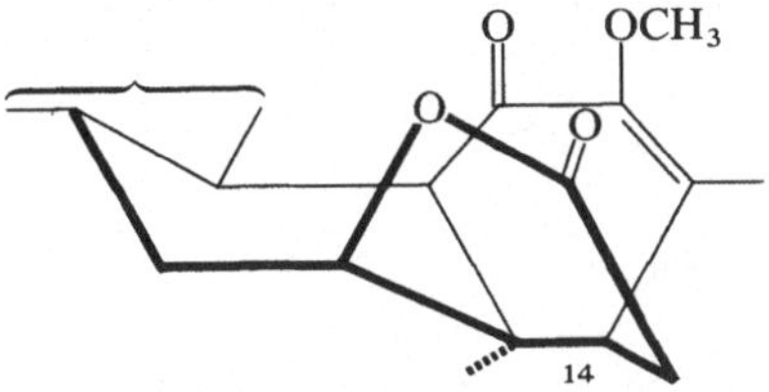

(30)
Alloquassin

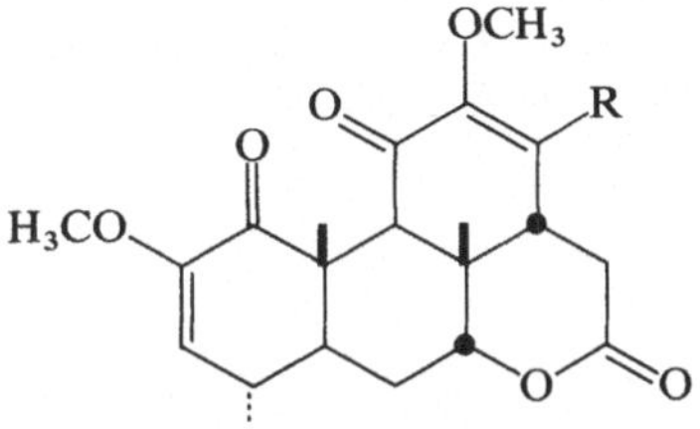

(31) R = CH$_2$OH
(32) R = CHO

Chart 6

of which involves β-elimination of the ether function, epimerization at C-14, and lactone ring closure.

Important information about the stereochemistry of quassin was obtained through the study of another isomerisation. Boiling acetic acid or cold methanolic hydrochloric acid converts quassin to an equilibrium mixture of roughly equal amounts of quassin and a new isomer, isoquassin. NMR studies show clearly that the C$_7$-hydrogen atom of both quassin and isoquassin is equatorial in relation to ring B. The two compounds differ most strikingly in their respective rotatory dispersion or circular dichroism curves; quassin and isoquassin differ by the configuration at C-14.

Quassin and neoquassin also occur in the bark of *Picrasma crenata* Engl. (*67*) and *Ailanthus glandulosa* Desf. Quite recently nigakilactone D and nigakihemiacetal B, isolated from *Picrasma ailanthoides* Planchon (= *P. quassioides* Bennett) have been identified as quassin and neoquassin respectively (*52, 54, 55*).

Commercial quassin from *Quassia amara* L. has been found to be a mixture of quassin (**1**), neoquassin (**2**), isoquassin and 18-hydroxyquassin (**31**). The latter has been oxidised with manganese dioxide to the aldehyde (**32**); the structures of these two compounds rest on NMR evidence (*12*).

2. Nigakilactones A, B, C, E, F, H (33—38), Simalikalactone C (39a), Picrasin D, E and F (39b, 39c and 39d), and Nigakihemiacetals A and C (40, 41)

These twelve simaroubaceous constituents are most closely related to quassin, since they also have a methyl group at C-8 and possess the same A-ring structure. They all have been converted chemically to quassin. The nigakilactones and the nigakihemiacetals have been isolated from *Picrasma ailanthoides* Pl. (*P. quassioides* B.) (*51—54*). Nigakilactone B (=simalikalactone A) has also been found in *Quassia africana* Baill. (*73*) from which simalikalactone C (**39a**) has been isolated as well (*73*).

Quite recently Picrasin D (**39b**), E (**39c**) and F (**39d**) have isolated from *Picrasma quassioides* (*34, 35*).

(**33**)	R_1=OH	R_2=OH	R_3=H	R_4=O	Nigakilactone A
(**34**)	R_1=OH	R_2=OCH$_3$	R_3=H	R_4=O	Nigakilactone B
(**35**)	R_1=OAc	R_2=OCH$_3$	R_3=H	R_4=O	Nigakilactone C
(**36**)	R_1=OAc	R_2=OCH$_3$	R_3=OH	R_4=O	Nigakilactone E
(**37**)	R_1=OH	R_2=OCH$_3$	R_3=OH	R_4=O	Nigakilactone F
(**38**)	R_1=OH	R_2=OCH$_3$	R_3=OH 14βOH	R_4=O	Nigakilactone H
(**39a**)	R_1=OH	R_2=O	R_3=H	R_4=O	Simalikalactone C
(**40**)	R_1=OH	R_2=OCH$_3$	R_3=OH	R_4=H, OH	Nigakihemiacetal A
(**41**)	R_1=OH	R_2=OH	R_3=H	R_4=H, OH	Nigakihemiacetal C

R =R$_1$=H	Picrasin D	(**39b**)
R$_1$=H; R=OH	Picrasin E	(**39c**)
R =R$_1$=OH	Picrasin F	(**39d**)

Chart 7

3. Amarolide (42), 11-Acetylamarolide (43), Chaparrolide (44), Castelanolide (45), Klaineanone (46), 11-Dehydroklaineanone (47), 15-Hydroxyklaineanone (48), Picrasin B (49), and 6-Hydroxypicrasin B (50a), Picrasin C (50b) (*36*)

These quassinoides have also a methyl group at C-8. Amarolide (42) and its acetyl derivative (43) were originally isolated from the bark (*10*) and from the seeds (*20*) of *Ailanthus altissima* (Mill.) Swingle ($=$ *Ailanthus glandulosa* Desf.). Recently the revised structure (42) has been proposed (*72*) for this substance after its discovery in *Castela nicholsoni* Hook. Chaparrolide (44) and castelanolide (45) are minor constituents of this plant (*44*), Klaineanone (46) and its 11-dehydroderivative (47) have been found in the seeds of *Hannoa klaineana* PIERRE and ENGLER (*59, 81*) while picrasin B ($=$ simalikalactone B) (49) has been extracted from *Picrasma quassioides* Benett (*32—36, 55*), from *Quassia africana* Baill. (*73*) and from *Soulamea pancheri* Brongn. (*76*). From the latter 6-hydroxypicrasin B (50a) has also been isolated (*76*); it is one of the rare quassinoids which has a substituent at C-6. 15-Hydroxyklaineanone (48) has been found in *Pierriera orientalis* Courchet (*60*).

The structure assignment of these compounds rests largely upon their spectroscopic properties and the conversion of most of them to quassin (1). Thus, for instance, amarolide (42) has been oxidised by means of Bi_2O_3 to the bis-diosphenol (51) which on methylation with diazomethane yields quassin (1); the last step can be reversed with boron tribromide (*10*). Klaineanone (46) and its 11-dehydroderivative (47) have been converted to quassin (1) in the following way (*81*). Alkaline treatment of (52), which is easily obtained from klaineanone (46) or from (47), leads to the mono-diosphenol (53). Thus, in addition to the hydrolysis of the acetoxy groups, basic treatment of compound (52) causes oxidation at C-12 and a α-ketol rearrangement in ring A. Amarolide (42) or acetyl-amarolide (43), when treated with base, also give the mono-diosphenol (53). Oxidation of the latter with Bi_2O_3 followed by methylation with diazomethane then leads to a mixture of quassin and isoquassin which can be separated by chromatography.

4. Glaucarubin (54), Glaucarubinone (70a), Glaucarubol (55), Glaucarubolone (70b), 15-Isovalerylglaucarubol, α-Methyl-α-acetoxybutyric Ester of Glaucarubolone, Holocanthone, Ailanthinone (70c)

Glaucarubin (54) is the major bitter constituent of the seeds of *Simarouba glauca* from which it was first isolated and studied, in 1954, by HAM *et al.* (*30*). However, detailed structural studies of glaucarubin were carried out several years later and culminated in the elucidation of its complete structure (*64, 65*).

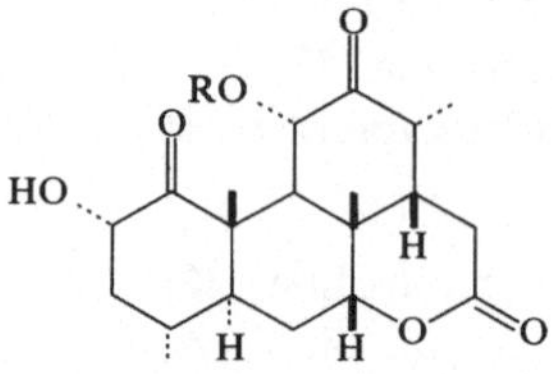

116   JUDITH POLONSKY:

Amarolide   (**42**)   R=H
            (**43**)   R=Ac

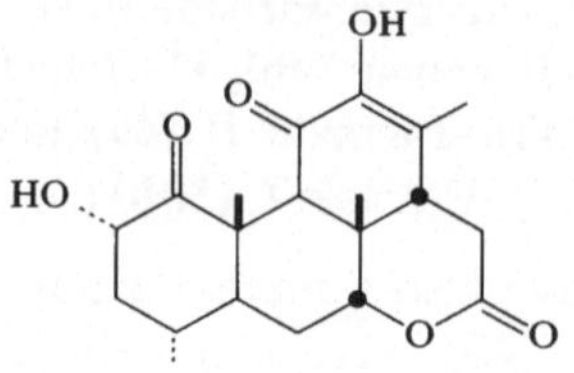

(**53**)

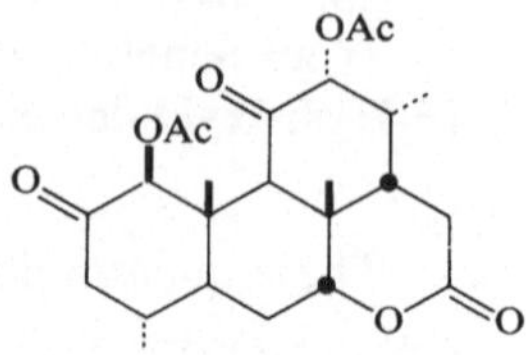

(**52**)

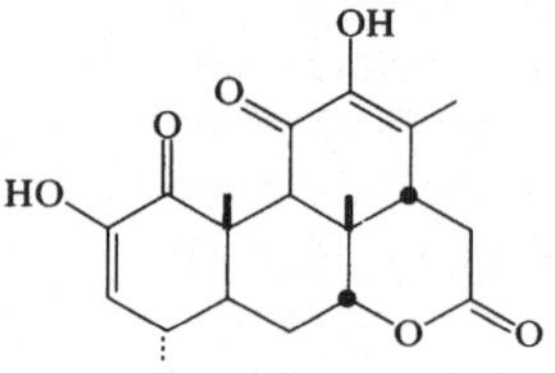

(**51**)

CH_2N_2 / BBr_3

(**1**)   Quassin

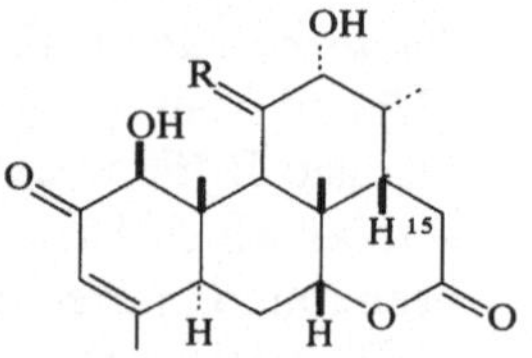

Klaineanone   (**46**)   R= H, OH
              (**47**)   R=O

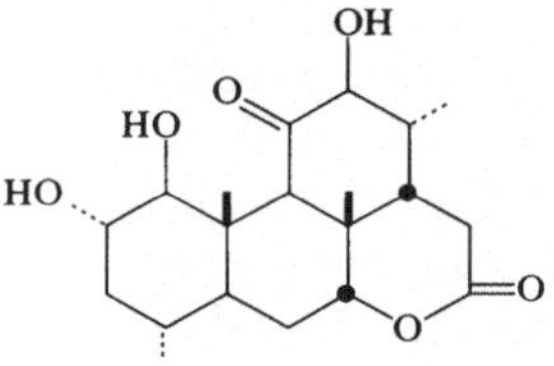

Chaparrolide   (**44**)

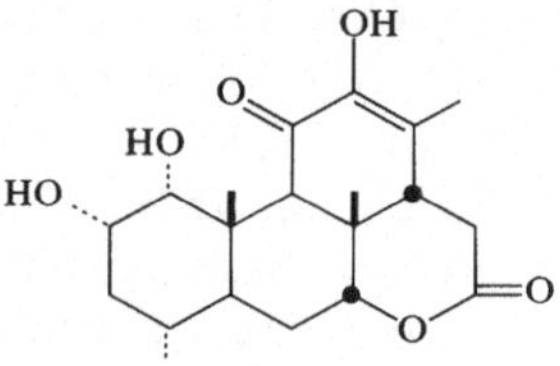

Castelanolide   (**45**)

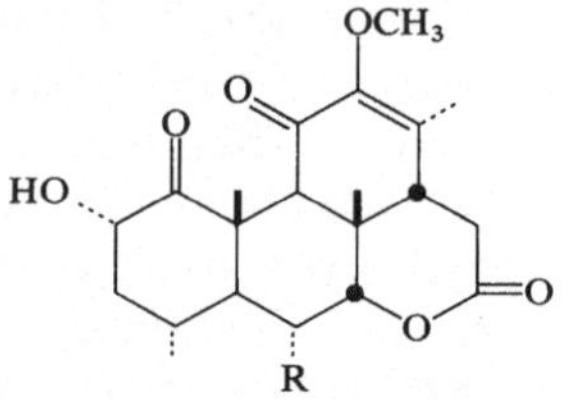

Picrasin B   (**49**)   R=H
6-Hydroxypicrasin B   (**50a**)   R=OH

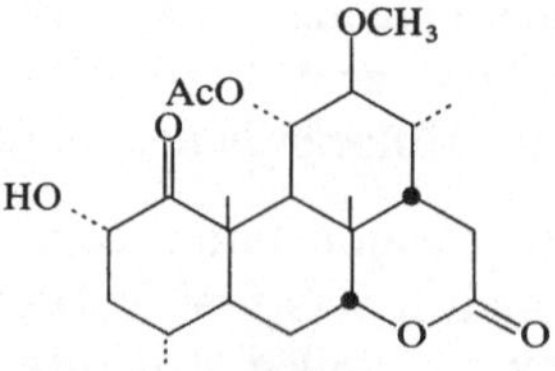

(**50b**)

Chart 8

References, pp. 147—150

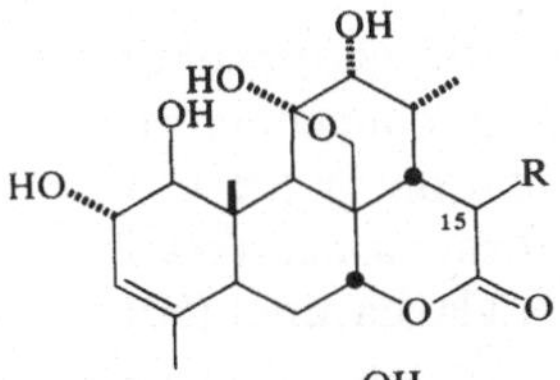

(54) $R = OCOC(OH)(CH_3)-C_2H_5$
Glaucarubin

(55) $R = OH$
Glaucarubol

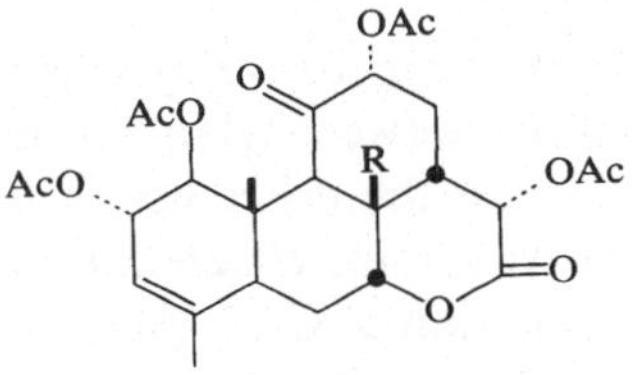

(56) $R = CH_2OAc$

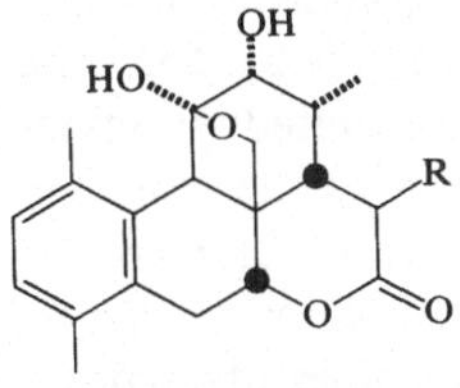

(57) $R = OH$
Glaucanol

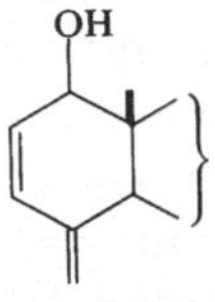

(58)

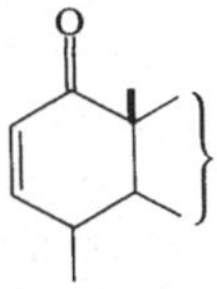

(59)

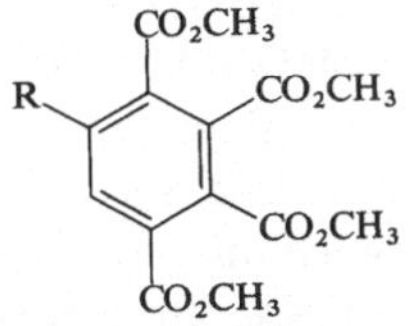

(60) a : $R = H$
b : $R = NO_2$

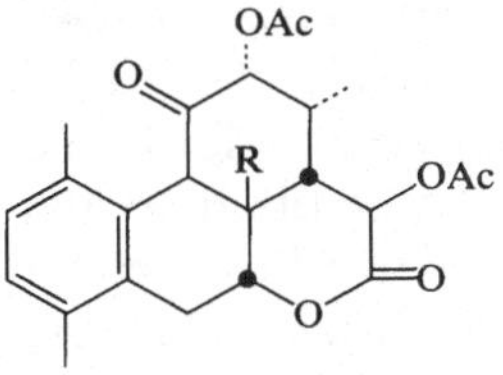

(61) $R = CH_2OAc$

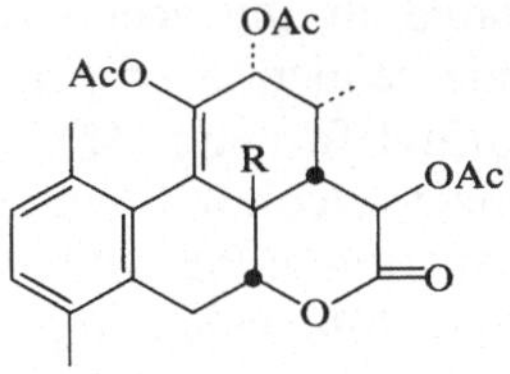

(62) $R = CH_2OAc$

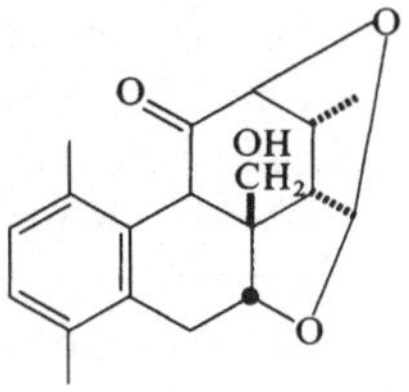

(63)

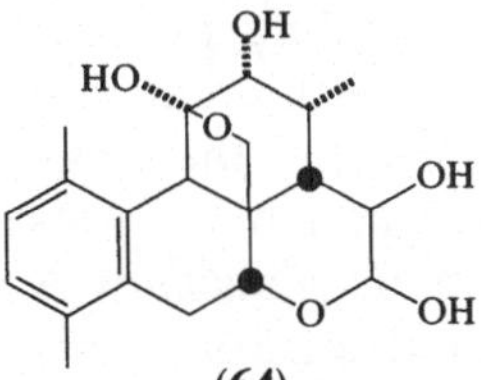

(64)

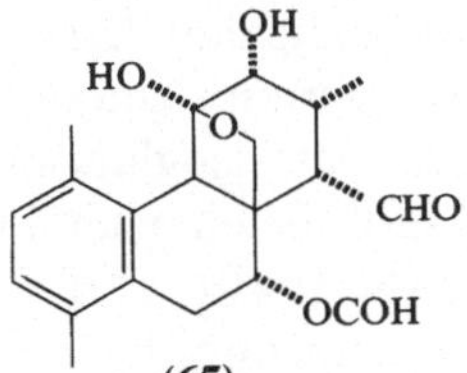

(65)

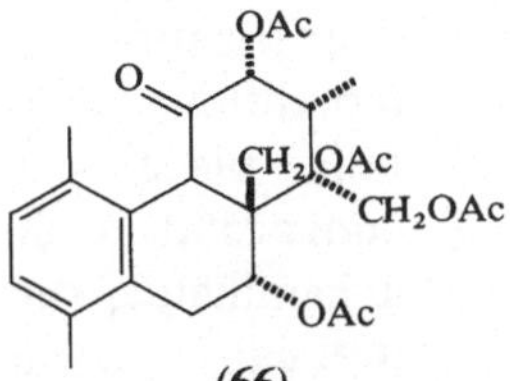

(66)

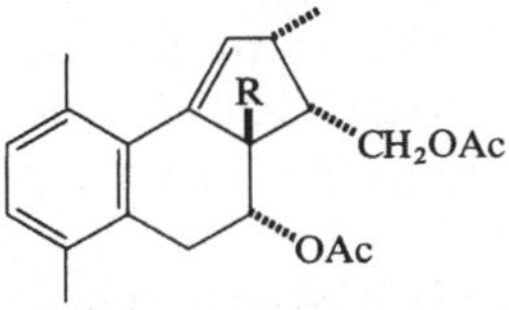

(67) $R = -CH_2OAc$

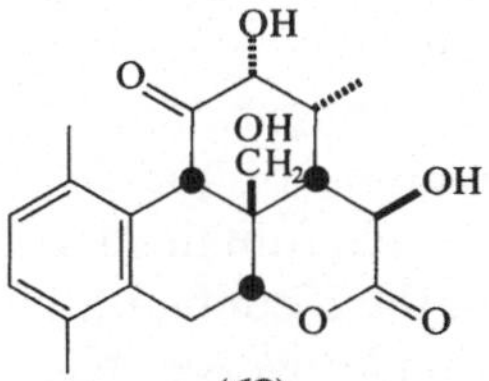

(68)

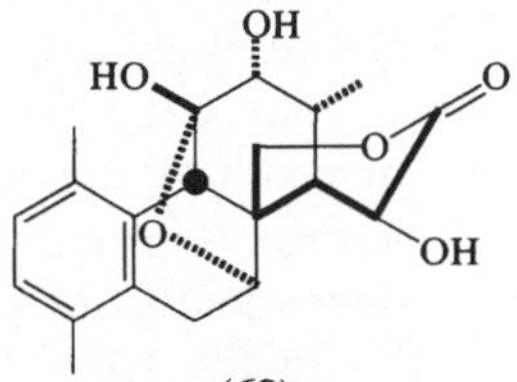

(69)

Chart 9

Glaucarubin is the 15-α-hydroxy-α-methylbutyrate of glaucarubol
(**55**). The major distinction between glaucarubin and its congeners on
the one hand and quassin on the other lies in the oxidation of the C-8
methyl to a hydroxymethyl group, which commonly forms a hemiketal
linkage with a carbonyl function at C-11. Acetylation of glaucarubol (**55**)
gives a penta-acetate (**56**) in which the carbonyl group at C-11 is
unmasked. A key reaction (*64*) in the study of glaucarubol (**55**) was
aromatization, under acidic conditions, of ring A to glaucanol (**57**), a
reaction characteristic of those members of the group which have a
a ring-A ene-diol system, as has already been indicated. The tetra-
substituted nature of the benzenoid ring-A of glaucanol (**57**) has been
proven by NMR evidence and by its oxidative degradation to 3,6-
dimethyl-phtalic anhydride of to the tetraester (**60a**) and (**60b**) (*64*). The
intermediate dienol (**58**) and enone (**59**) could also be isolated. Acetylation of glaucanol (**57**) under controlled conditions permits the isolation
of the triacetate (**61**) with an unmasked carbonyl group and of the
tetraacetate (**62**) which is also an enol acetate. The former has a strong
positive Cotton effect while the latter exhibits an ultraviolet absorption
consistent with conjugation of the enolic system with the aromatic ring.

The lactone function of glaucanol and the oxygen functions attached
to ring C are interrelated in the acetal (**63**). Dihydroglaucanol (**64**),
obtained by borohydride reduction of glaucanol, is readily cleaved by
periodic acid to the formyl formate (**65**) which with acids affords the
acetal (**63**) whose primary alcohol function could be oxidized to the
corresponding acid. Reduction of (**65**) with diborane followed by acetylation affords the tetraacetate (**66**) which on pyrolysis (*78*) furnishes the ring
C contracted product (**67**).

The stereochemistry assigned to glaucarubin rests heavily on the
isomerization products of glaucanol (**57**). Thus with boiling pyridine
(conditions which do not permit lactone opening) inversion at C-9
takes place, since the product, isoglaucanol C (**68**) forms the same
enol acetate as does glaucanol. Isoglaucanol C, however, no longer forms
a hemiketal; this is consistent with a B/C *cis* fusion. Isoglaucanol A
(**69**) can be obtained from glaucanol by reaction with alkali. Its formation must involve epimerization at C-9, which implies again that the B/C
fusion in glaucanol is *trans*.

The constitution and configuration of glaucarubin have been confirmed (*38*) by x-ray analysis of the 2-*p*-bromobenzoate. The (+)-α-
hydroxy-α-methylbutyric acid obtained from hydrolysis has been independently assigned (*56*) the S-configuration.

In the seeds of *Simarouba glauca*, glaucarubin (**54**) is accompanied
by three minor principles. These are glaucarubinone (**70a**), glaucarubolone (**70b**) (*22*) and $\Delta^{13,\,18}$-glaucarubin (*45*) (Chart. 10).

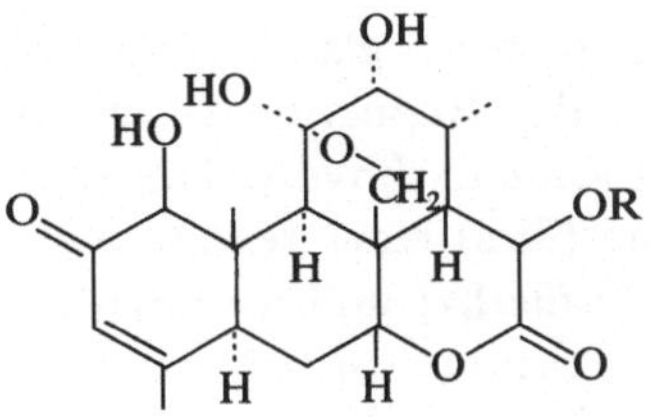

Glaucarubinone (70) a: $R = CO-\underset{\underset{OH}{|}}{\overset{\overset{CH_3}{|}}{C}}-C_2H_5$

Glaucarubolone b: $R = H$

c: $R = CO-\underset{}{\overset{\overset{CH_3}{|}}{CH}}-C_2H_5$

(71) $R = CO-\underset{\underset{OAc}{|}}{\overset{\overset{CH_3}{|}}{C}}-C_2H_5$

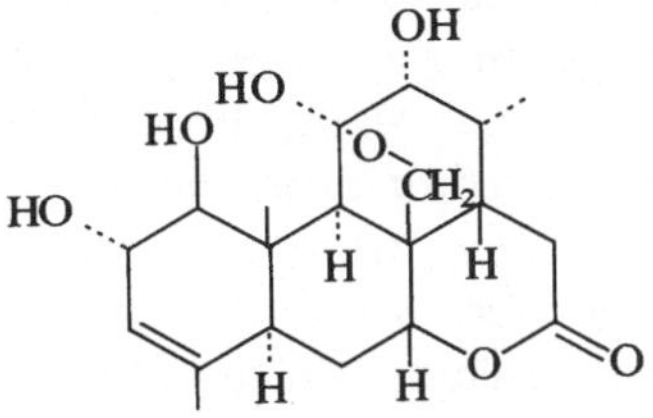

(72) Chaparrin

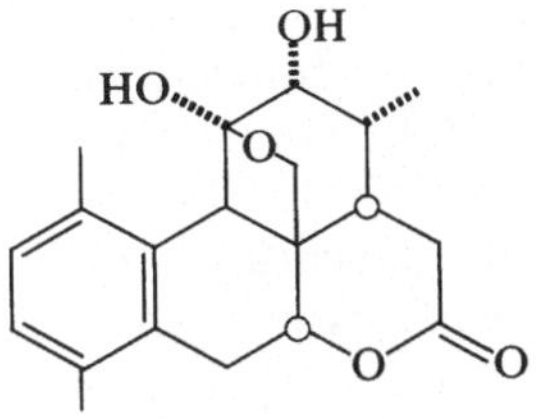

(75) Chaparrol

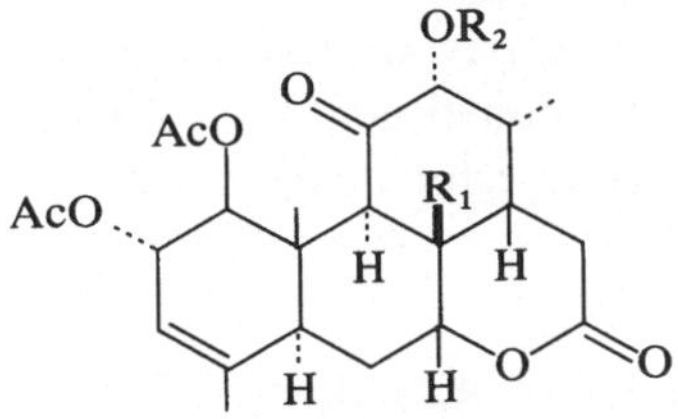

(73) $R_1 = CH_2OAc$; $R_2 = H$
(74) $R_1 = CH_2OAc$; $R_2 = Ac$

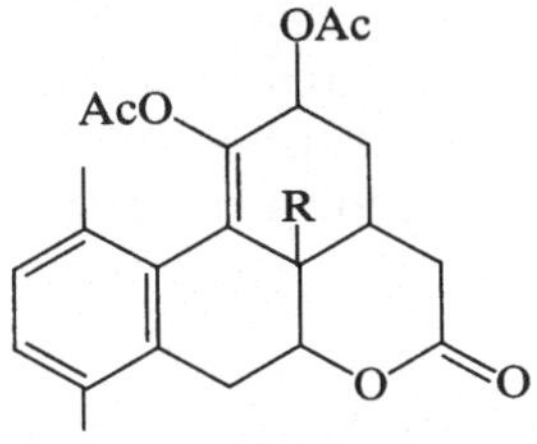

(76) $R = CH_2OAc$

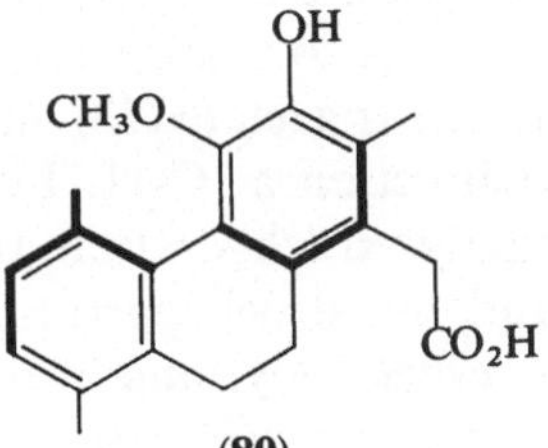

(80)

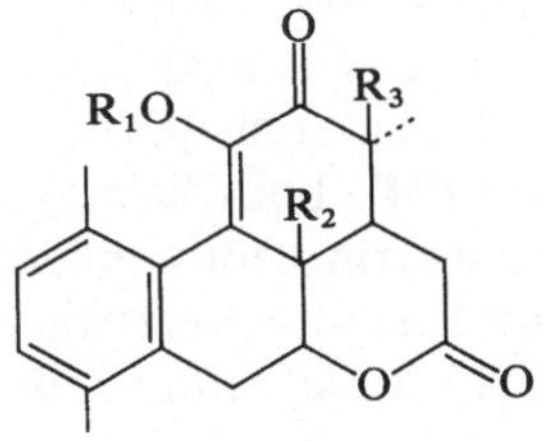

(77) $R_1 = Ac$; $R_2 = CH_2OAc$; $R_3 = H$
(78) $R_1 = CH_3$; $R_2 = CH_2OH$; $R_3 = H$
(79) $R_1 = Ac$; $R_2 = CH_2OAc$; $R_3 = Br$

Chart 10

Glaucarubinone (**70a**) has been correlated with glaucarubin by oxidation of the latter with manganese dioxide. Reduction of glaucarubolone (**70b**) followed by acidic treatment leads to dihydroglaucanol (**64**). Glaucarubinone (**70a**) or glaucarubolone (**70b**) react readily with diazomethane to give the corresponding 11-O-methyl and 1-O-methyl derivatives (*24*). The latter, which is the major product of the reaction, was shown to be a suitable starting material for the degradations required in the biosynthetic studies (*46, 49*).

Glaucarubin and glaucarubinone and their hydrolysis products, glaucarubol and glaucarubolone, occur also in simaroubaceous species other than *Simarouba glauca*. Thus, glaucarubolone (**70b**) has been found in the fruits of *Hannoa klaineana* (*59*) and in the wood of *Castela nicholsoni* (*44*). The latter species also contains glaucarubol and its 15-isovalerate (*44*). Glaucarubinone (**70a**) accompanied by a small amount of glaucarubin (**54**) has been isolated from the fruits of *Perriera madagascariensis* Courchet (*9*), a tree which is commonly called "Kirondro". Glaucarubinone (**70a**) occurs also in the trunk the bitter constituent of which has recently been characterized as (**71**), the α-methyl-α-acetoxybutyric ester of glaucarubolone (*82*).

Holocanthus emoryii A. Gray has also been found to contain glaucarubol (*71*) and 15-acetoxyglaucarubolone, trivially named holocanthone (*77*). Ailanthinone (**70c**), a minor constituent of *Ailanthus altissima* was shown to be the α-methylbutyric ester of glaucarubolone (*20*).

5. Chaparrin (72), Chaparrinone (81), and Ailanthone (87)

Chaparrin (**72**) which was isolated from the trunk of *Castela nicholsoni* Hook. was studied by GEISSMAN *et al.* (*26*) and by DE MAYO and co-workers (*14, 37*). Chaparrin is 15-desoxy-glaucarubol (**72**) and therefore its chemistry is analogous in many respects to that of glaucarubol (**55**). Acetylation of chaparrin (**72**) leads to the triacetate (**73**) and to the tetraacetate (**74**), both having a carbonyl function at C-11. Like glaucarubol, chaparrin readily undergoes acid-catalyzed dehydration, a reaction which leads first to anhydrochaparrin-A and anhydrochaparrin-B, which have the A-ring structure (**58**) and (**59**) respectively, and then to the partially aromatized chaparrol (**75**). This gives on acetylation the enol acetate (**76**).

Important information about the structure of chaparrin came from the study of the oxidation of chaparrol (**75**) with hexavalent bismuth. Treatment of chaparrol (**75**) with bismuth triacetate followed by acetylation or by methylation with diazomethane leads to the enol acetate (**77**) or the methyl ether (**78**). The ultraviolet absorption of these compounds and that of the triacetate (76) indicates conjugation of the enolic system with the aromatic ring. Bromination of the enol acetate (77) affords the monobromo derivative (79) in which the methyl doublet at C-13 has been replaced by a singlet.

Treatment of the methyl ether (78) with dilute alkali causes elimination of formaldehyde and gives the biphenyl (80). This reaction, which may be rationalized as involving a vinylogous reverse aldolisation followed by a vinylogous β-elimination, establishes the relative positions of the primary hydroxyl group and of the lactonic ethereal oxygen atom with respect to the enol system in compound (77). The biphenyl (80) was optically active and on the basis of its rotatory dispersion (37) could be assigned the S-configuration, thus establishing for chaparrin itself the absolute configuration shown in (72) (Chart 10). Chaparrol (75) gave with alkali two isomers (neochaparrol and isochaparrol) analogous to the formation of isoglaucanols A and C, as discussed for glaucarubin, this led to assignment of B/C *trans* and B/D *cis* fusions in chaparrol and therefore chaparrin.

Chaparrin has also been found, in very small amounts, in *Ailanthus atissima* (20).

Chaparrinone (81) is a major quassinoid found in the fruits of *Hannoa klaineana* Pierre and Engler (59). It was readily shown to be 2-dehydrochaparrin, *i.e.* 15-desoxyglaucarubolone, and has been correl-

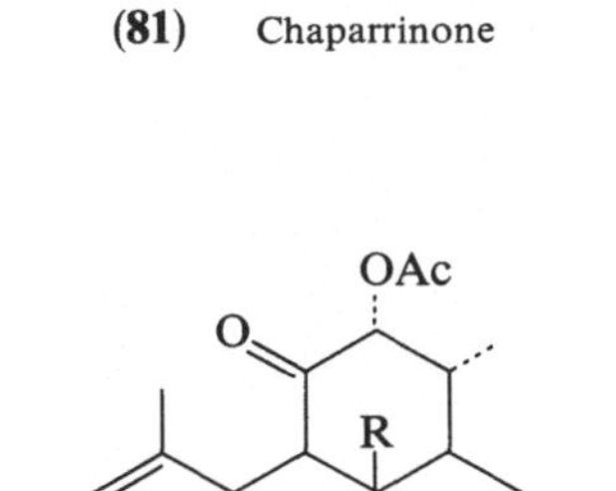

(81)   Chaparrinone

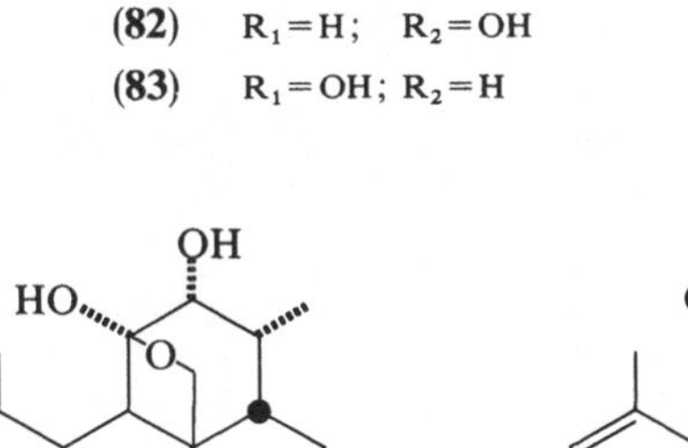

(82)   R₁=H;   R₂=OH

(83)   R₁=OH; R₂=H

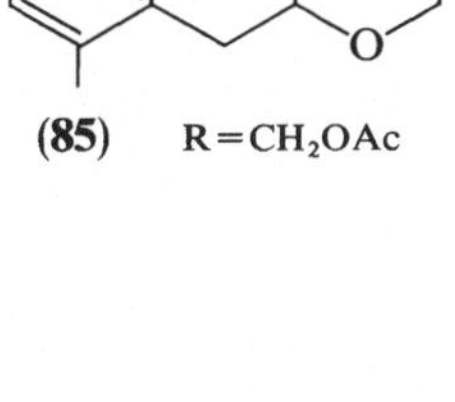

(85)   R=CH₂OAc

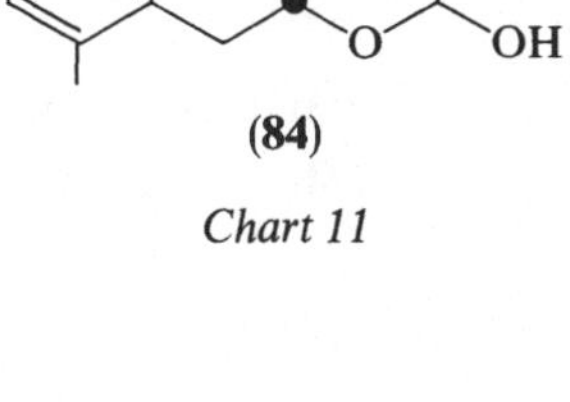

(84)

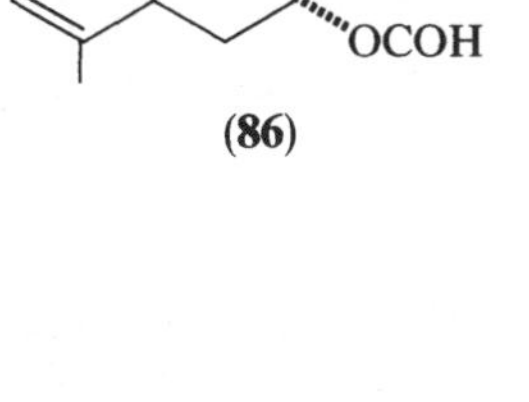

(86)

Chart 11

ated with both chaparrin (**72**) and glaucarubolone (**70b**). The former was converted to chaparrinone (**81**) by oxidation with manganese dioxide. Both compounds have also been, interrelated by the following reactions (*81*). Treatment of chaparrinone (**81**) with sodium borohydride in methanol leads to a mixture of products from which compounds (**82**) and (**83**) may be isolated. These are isomeric at C-2 and so their NMR spectra differ essentially only in the chemical shift of the angular methyl group. Under the same reduction conditions, chaparrin furnishes a methyl-derivative which was found to be identical with compound (**82**). It may also be noted that reduction od chaparrinone (**81**) followed by acidic treatment (*81*) leads to dihydrochaparrol (**84**).

Chaparrinone and glaucarubin (**54**) (and therefore their congeners) have been interrelated (*81*) through the glaucanol derivative (**86**). As in the case of neoquassin (**2**), treatment of dihydrochaparrol (**84**) with acetic anhydride leads to the cyclic enol-ether (**85**) which on ozonolysis affords compound (**86**). This was found to be identical with the acetylation product of the aldehyde formate (**65**) obtained by periodic acid oxydation of dihydroglaucanol (**64**) (Chart 11).

Chaparrinone also occurs as a minor constituent in the seeds of *Ailanthus altissima* (*20*).

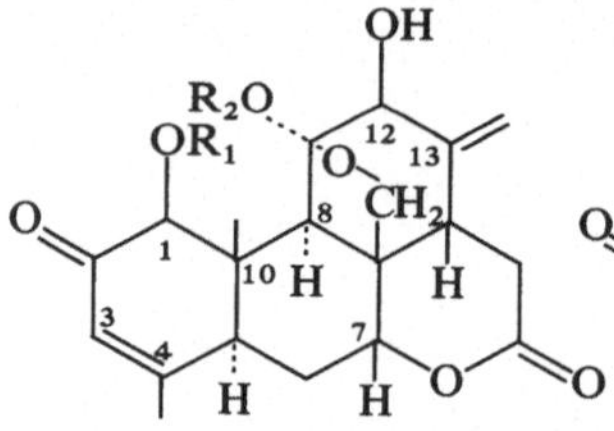

Ailanthone

(**87**) $R_1 = R_2 = H$

(**89**) $R_1 = CH_3$; $R_2 = H$

(**90**) $R_1 = H$; $R_2 = CH_3$

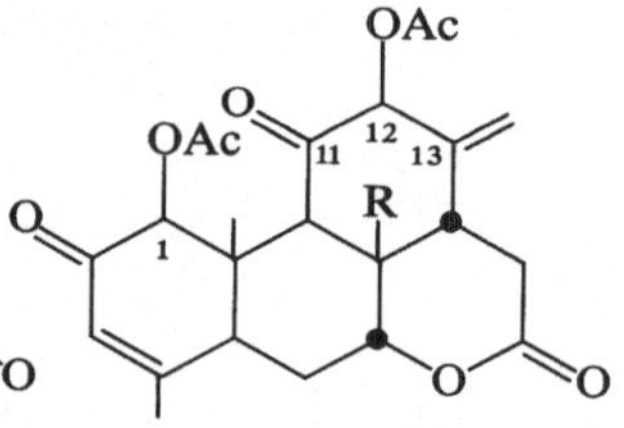

(**88**) $R = CH_2OAc$

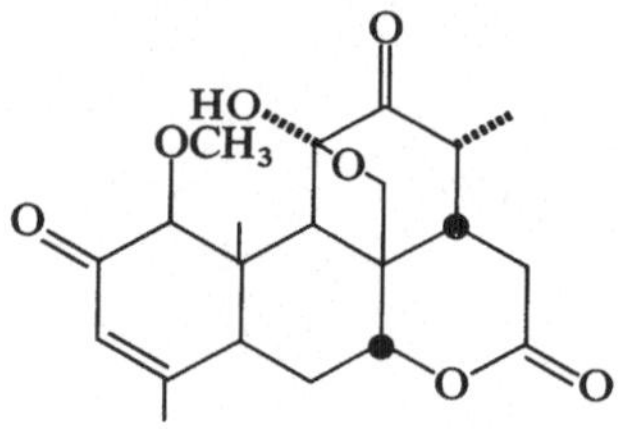

(**91**)

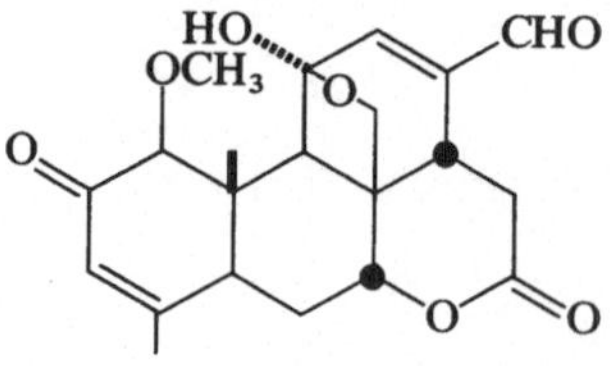

(**92**)

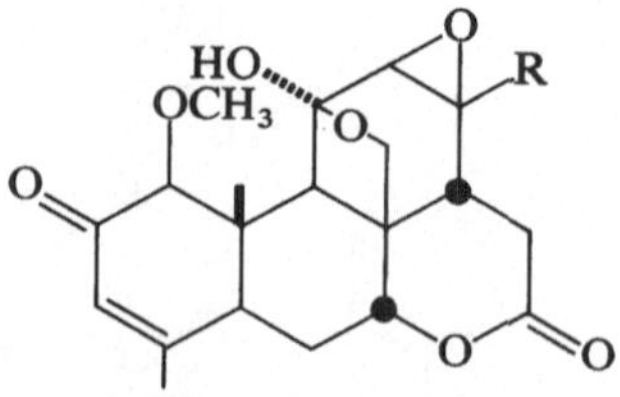

(**93**) $R = CH_2OH$

(**94**) $R = CO_2H$

Chart 12

Ailanthone (**87**) (Chart 12) is the chief bitter constituent of the seeds (*20, 66*) and bark (*11*) of *Ailanthus altissima* (Mill.). Swingle (= *Ailanthus glandulosa* Desf.) where it is accompanied by small amounts of several other quassinoids. This plant commonly called "Chinese tree of heaven", is one of the few Simaroubaceae to have been acclimatized to temperature areas.

Ailanthone (**87**) and chaparrinone (**81**) differ in the nature of the group at C-13 and in the configuration at C-12: the former has an exocyclic methylene function at C-13 and a β-equatorially oriented hydroxyl group at C-12. The structure of ailanthone (**87**) rests on detailed NMR investigations of its triacetate (**88**) and O-methyl-derivatives (**89**) and (**90**). The structure has been confirmed by correlation (*81*) with chaparrinone through a common derivative (**91**). The latter was obtained by selective catalytic hydrogenation of ailanthone (**87**) followed by treatment with diazomethane and by mild chromic acid oxidation and was found to be identical with the oxidation product of 1-O-methylchaparrinone.

Most of the chemistry of ailanthone (**87**) parallels that of chaparrinone (**81**), with the exception of reactions that are associated with the presence of the exocyclic methylene at C-13, such as the following (*20*). When treated with diluted alkali (*20*) ailanthone (**87**) readily isomerizes to isoailanthone which may be converted by reaction with diazomethane to (**91**). On the other hand, 11-O-methyl-ailanthone (**90**) is stable under these conditions. Another point of chemical difference between chaparrinone and ailanthone is the reaction with chromic acid. While oxidation of 1-O-methylchaparrinone with JONE's reagent (*20*) leads easily to the 12-keto compound (**91**), 1-O-methylailanthone gives a mixture of products, from which the unsaturated aldehyde (**92**), the oxidoalcohol (**93**) and the acid (**94**) may be isolated. The structures assigned to these compounds are based on their chemical and spectroscopic properties and on the following conversions. The oxidoalcohol (**93**) is converted to the acid (**94**) by Jone's oxidation and to the aldehyde (**92**) by acidic treatment.

It should be noted that most of the mass spectral studies dealing with quassinoids have been made on ailanthone and its numerous derivatives, particularly the 1-O- and 11-O-alkyl derivatives (*21*). The principal fragments in the mass spectrum of ailanthone (**87**) are found at m/e 135, 151 and 248 and correspond to ions (**95**) $[C_9H_{11}O]^+$, (**96**) $[C_9H_{11}O_2]^+$ and (**97**) $[C_{14}H_{18}O_4]^+$, respectively. In the mass spectra of 1-O-methyl-ailanthone (**89**) and 1-O-ethyl-ailanthone the positions of the latter two peaks are shifted by 14 and 28 mass units respectively, while the m/e 135 peak remains unshifted.

While ions of the type (**97**) are independent of the A-ring structure,

the oxidation level (or the substitution pattern) in ring A seems to play
a major role in the formation of ions (95) and (96).

(95)

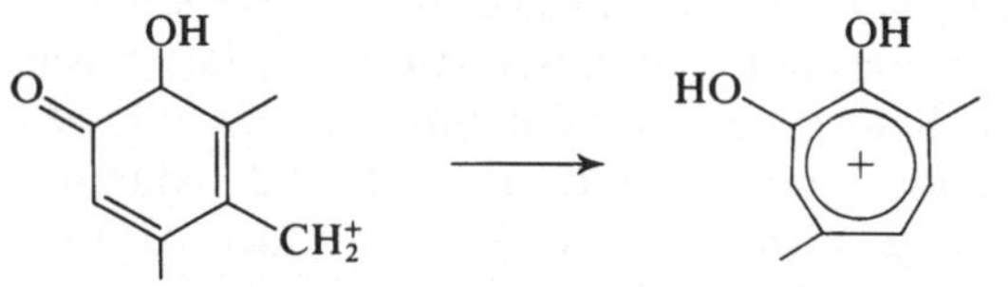

(96) 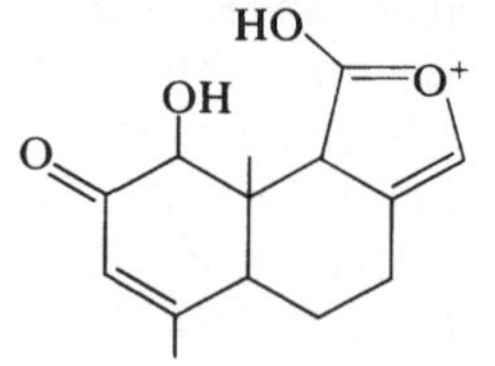 (97)

6. Bruceins A (98), B (99), C (100), D (105), E (106), F (107), G (110), Brusatol (101), and Simalikalactone D (111)

These quassinoids have the C-8 hydroxymethyl engaged in an ether
function which terminates at C-13. All except the three last have been isolated from the seeds of *Brucea amarissima* Merr. (*61—63*). Bruceine G (18)
and brusatol (*68*) have been found in the seeds of *Brucea sumatrana*
Roxb., which also contains bruceins D and E. Simalikalactone D has been
found in *Quassia africana* Baill. (*73*).

Bruceins A (98), B (99), C (100) and brusatol (101), the most oxygenated quassinoids so far known, differ in the nature of the ester
group at C-15. They are esters of isovaleric, acetic, 3,4-dimethyl-4-
hydroxy-2-pentenoic and senecioic acid, respectively. Mild alkaline
hydrolysis of each leads to the same tetrol named bruceolide (102) (*62*).

NMR studies show clearly that bruceolide, $C_{21}H_{26}O_{10}$, has the basic
skeleton of a C_{20}-quassinoid. The methoxyl present was found to be part
of a carbomethoxy grouping since basic or acidic hydrolysis gives the
corresponding acid (*62*); this, when treated with diazomethane regenerates
bruceolide and when treated with diazoethane gives the ethyl ester.
A diosphenol grouping was found to be present in bruceolide (102) as
shown by a positive ferric-chloride test and the ultraviolet absorption
at 280 nm which is shifted to 246 nm in the acetates. The location of the
diosphenol grouping in ring A and the absence of other oxygenated

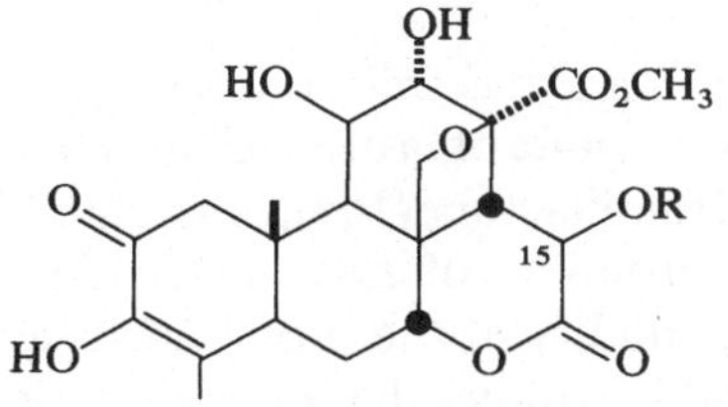

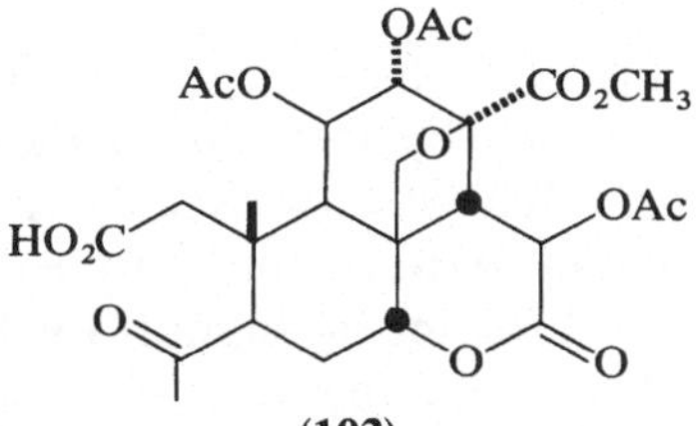

(103)

Brucein A   **(98)** $R = COCH_2 - CH(Me)_2$

Brucein B   **(99)** $R = Ac$

Brucein C   **(100)** $R = -COCH = \overset{Me}{\underset{OH}{C}} - C(Me)_2$

Brusatol   **(101)** $R = -CO - CH = C(Me)_2$

Bruceolide   **(102)** $R = H$

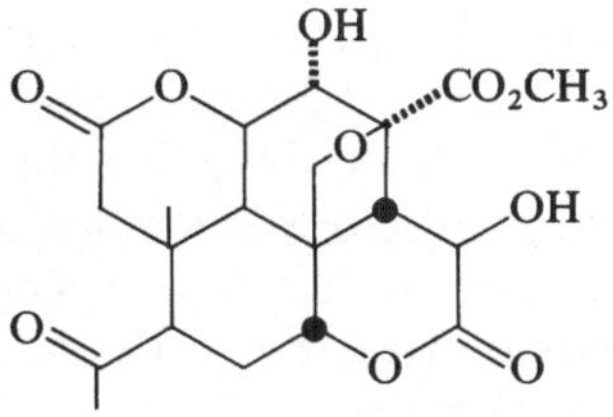

(104)

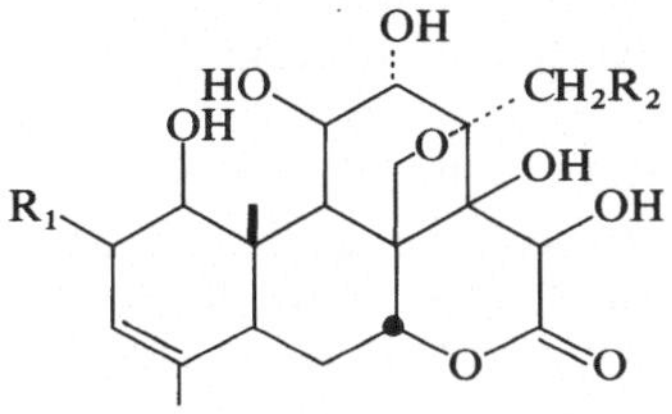

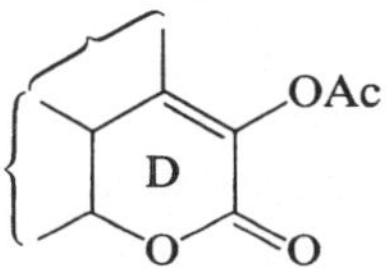

(108)

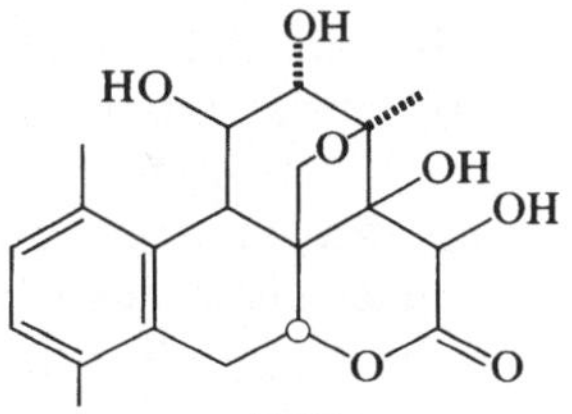

(109)

Brucein D   **(105)** $R_1 = O;$ $R_2 = H$

Brucein E   **(106)** $R_1 = \overset{H}{\underset{OH}{\diagup}}$; $R_2 = H$

Brucein F   **(107)** $R_1 = \overset{H}{\underset{OH}{\diagup}}$; $R_2 = OH$

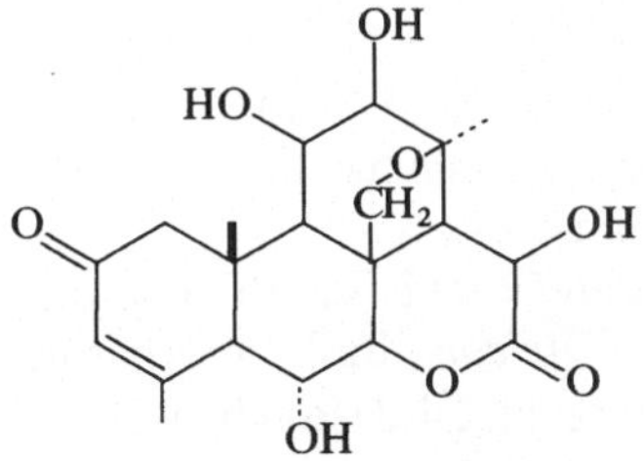

Brucein G   **(110)**

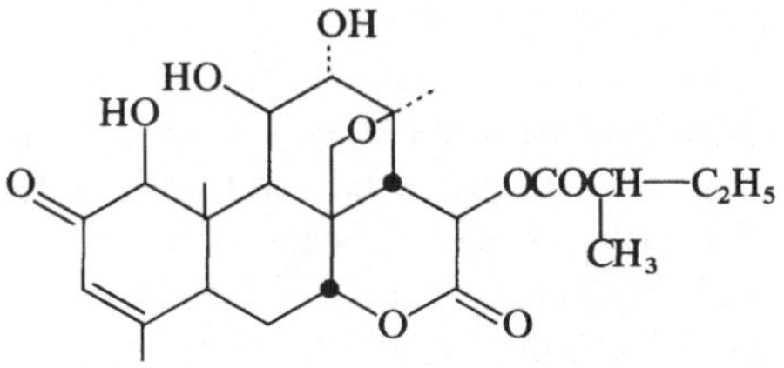

Simalikalactone D   **(111)**

Chart 13

functions from this ring follows from mass spectral arguments and has been confirmed by ozonolysis of the tetraacetate of bruceolide (*62*). This led to the acid (**103**) which with acid formed the dilactone (**104**), thus establishing the relation between the diosphenol grouping of ring A and the C-11 hydroxyl. Convincing support for the validity of the proposed structures for the bruceins comes from detailed NMR studies that involved extensive spin decoupling of several derivatives.

It is noteworthy that the acyl group at C-15 in bruceine C (**100**) does not seem to have been found in other naturally occurring compounds.

Bruceins A, B, C and brusatol are the only quassinoids so far known which have oxygenated functions at positions 2 and 3, and not at 1 and 2.

Bruceine D (**105**) (*61*), bruceine E (**106**) (*61*), bruceine F (**107**) (*63*) and simalikalactone B (**109**) (*73*) possess the more typical oxygenation pattern of ring A *i.e.* they lack the C-3 oxygen function and contain C-18 at a lower oxidation level (CH_3 or CH_2OH) than the other bruceins. In these bruceins the A-ring structure is similar to that found in most other quassinoids. The nature of ring A in brucein D (**105**) is clearly indicated by its spectroscopic properties (UV, NMR) and by its mass-spectroscopic fragmentation pattern.

Brucein E (**106**), which has also been described by the number Wst-63 (*69*), is 2-dihydrobrucein D and has been converted to brucein D by oxidation with manganese dioxide.

In accordance with the formulation of bruceins D and E, the NMR spectra of their derivatives show the C-15 proton as a downfield singlet. Treatment of the tetraacetate of brucein E (**106**) with thionyl chloride-pyridine leads to the anhydrotetraacetate (**108**) from which the tertiary hydroxyl has been eliminated (*69*). As expected, brucein E (**106**) undergoes aromatization with acids leading to brucinol (**109**) (*61*).

In Brucein F (**107**), the most non-lipophylic quassinoid so far known, the angular methyl group at C-13 is replaced by a primary alcohol grouping (*63*).

Bruceins D, E and F are the first quassinoids to have an α-glycol grouping at C-14/C-15. This may perhaps be derived from a 14,15-epoxide in a supposed precursor.

The co-occurrence in *Brucea* species of both the 1,2 and 2,3 oxygenated quassinoids may be of biosynthetic significance and suggests a common 3-oxygenated intermediate for both types.

Structure (**110**) has been proposed for brucein G (*18*). Simalikalactone D (**111**) has been recently isolated from *Quassia africana* Baill. originating from the Congo. It is the 15-α-methyl-butyrate of 14-desoxybrucein D (*73*).

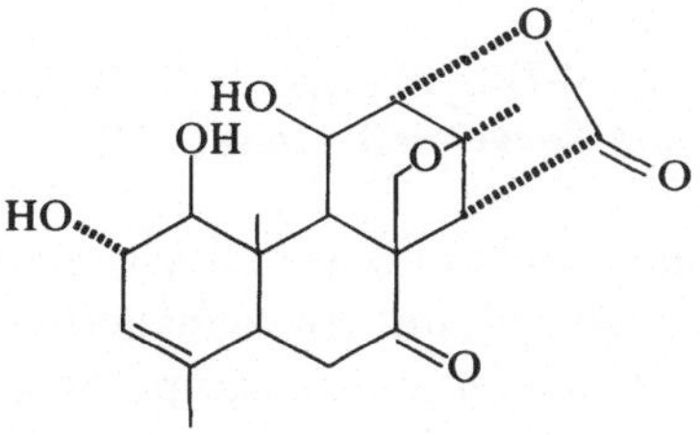

Samaderin C **(113)**

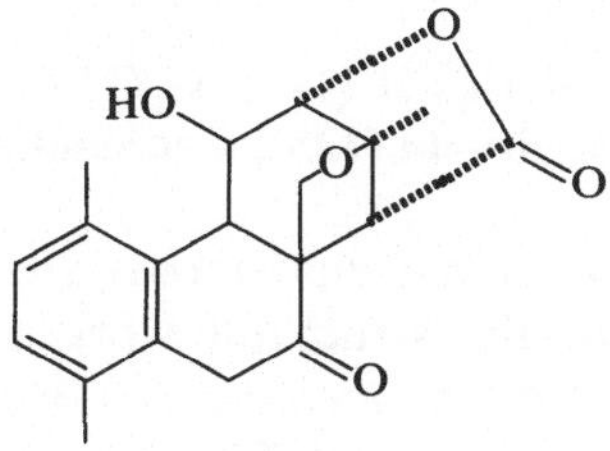

(114)

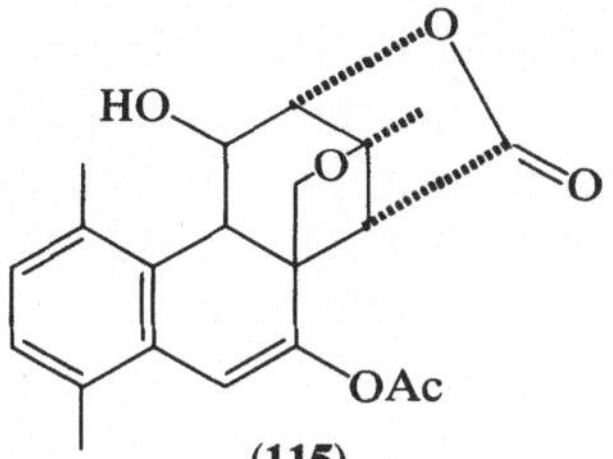

(115)

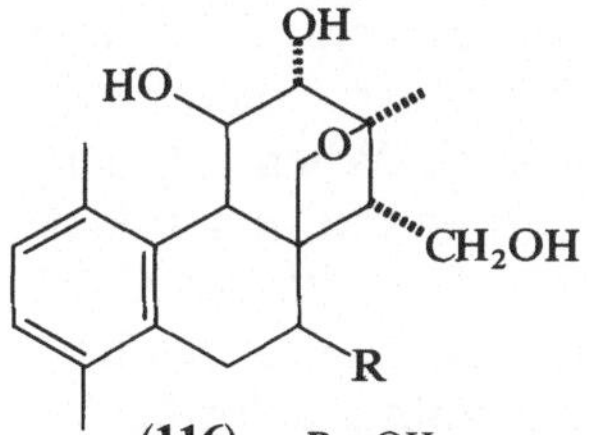

(116) R=OH
(117) R=H

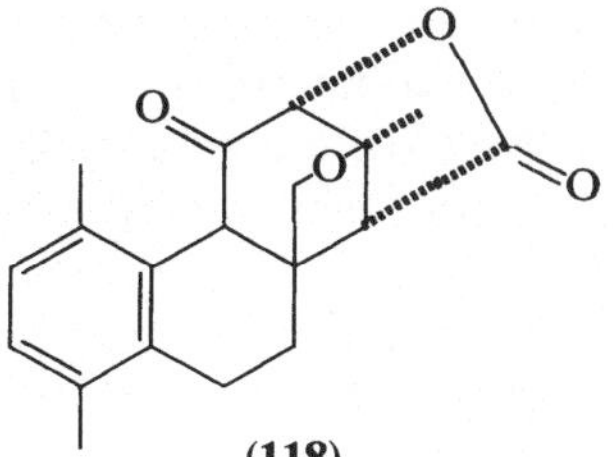

(118)

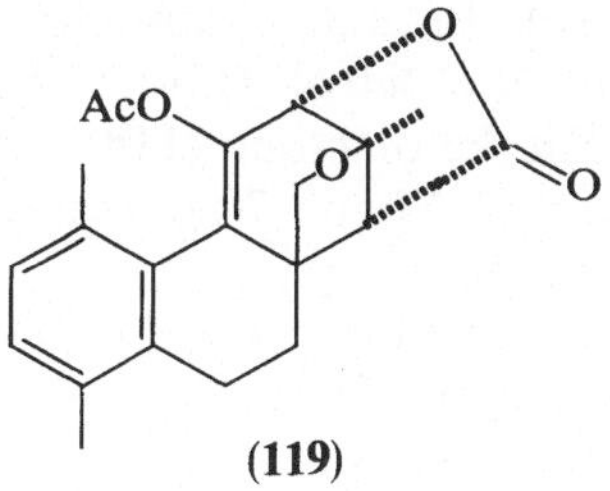

(119)

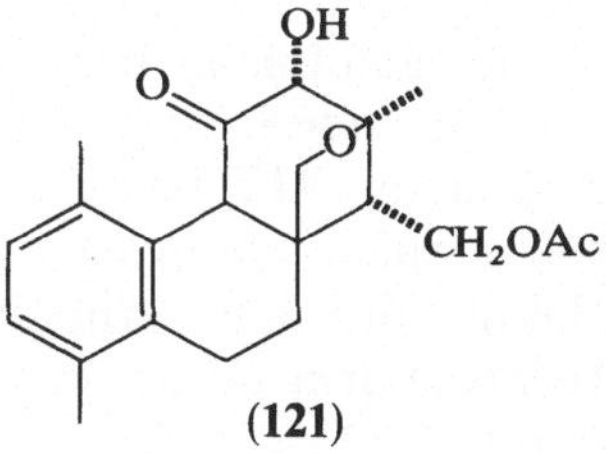

(121)

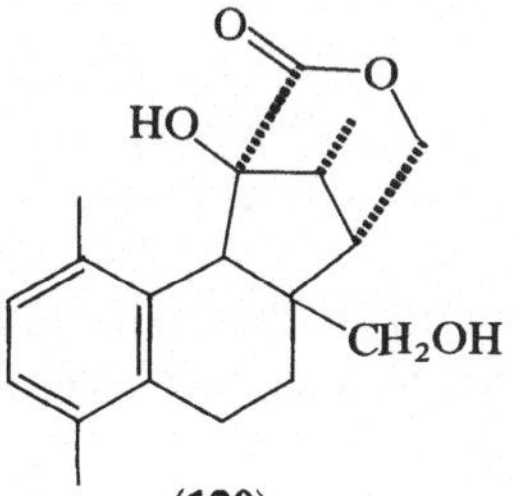

(120)

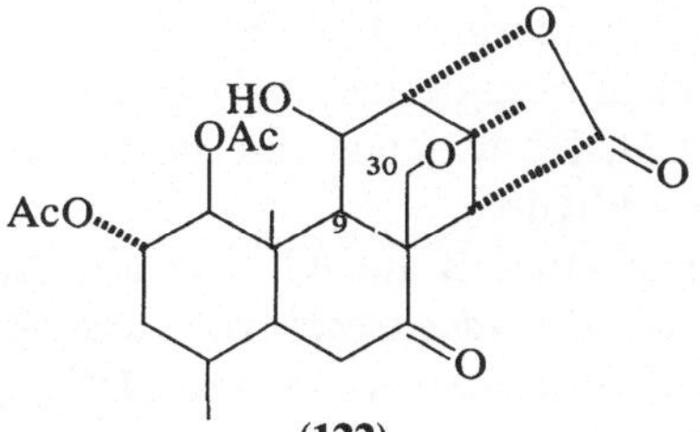

(122)

Chart 14

7. Samaderins B (112), C (113), D (124), 3,4-Dihydrosamaderin B (123), Cedronin (125), Cedronolin (126), and Eurycomalactone (127)

These seven quassinoids are C_{19}-compounds and are all γ-lactones. Most of the structural work has been carried out on samaderine C (**113**) from *Samadera indica* Gärtn., with which all others, except the last, have been correlated (*79, 80*).

The A-ring structure of samaderine C (**113**) is defined, above all, by its reaction with diluted mineral acids which leads to the partially aromatized compound (**114**), named samaderol. This was the starting point for further structural investigations into the remaining rings of samaderine C. When the acid-catalysed dehydration of samaderine C (**113**) is carried out under deuterating conditions (DCl, D_2O), the samaderol formed has its two C-6 hydrogens (and only these) exchanged by deuterium. Samaderol readily forms an enolacetate (**115**) whose NMR spectrum exhibits a lowfield one-proton singlet and whose ultraviolet absorption is consistent with conjungation of the enolic system with the aromatic ring. Clemmensen reduction of samaderol (**114**) gives 7-desoxo-samaderol. Lithium aluminium hydride reduction of samaderol (**114**) leads to the tetrol (**116**), while 7-desoxo-samaderol gives the triol (**117**). The ketolactone (**118**), obtainable by chromic acid oxidation of both triol (**117**) and 7-desoxosamaderol, again forms a conjugated enol acetate (**119**).

An interesting reaction which involves the oxygen functions attached to ring C is the formation (*79*) of the dihydroxy-δ-lactone (**120**) when a ketol (**121**) obtained from triol (**117**) by selective acetylation and mild chromic acid oxidation is treated with methanolic potassium hydroxide. The formation of (**120**) may be rationalized as involving cleavage of a β-ketoether followed by an intramolecular benzilic ester rearrangement.

The stereochemistry of samaderin C is based on the circular dichroism of the saturated ketone (**122**) and the long-range coupling of H-9 with one C-30 proton which requires a *trans* B/C fusion.

Samaderine B (**112**), 3,4-dihydro-samaderine B (**123**) and samaderine D (**124**) are minor constituents of *Samadera indica* Gärtn. They have been correlated with samaderine C (**113**) as follows. Oxidation of the latter with manganese dioxide gives samaderine B (**112**) which affords 3,4-dihydrosamaderine B (**123**) by selective catalytic hydrogenation. The structure of samaderine D (**124**), which has strong ultraviolet absorption, rests on NMR evidence and on its conversion to 3,4-dihydrocedronoline (**126**, 3,4-dihydro).

Samaderines B and C have also been found in the stem and leaves of *Samadera madagascariensis* Juss. (*43*). This species has recently been identified with *Samadera indica* (*43*).

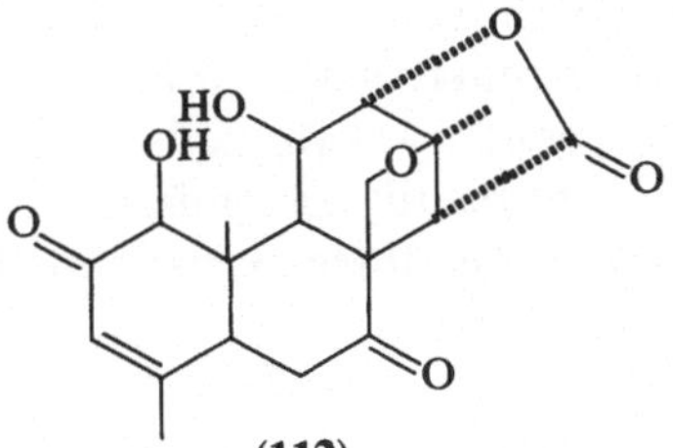

(112)

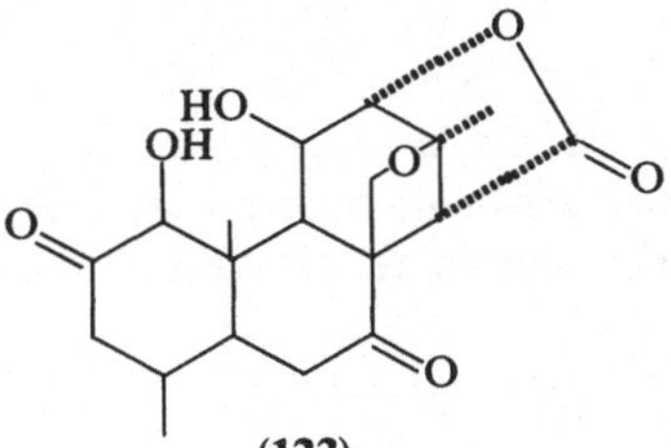

(123)

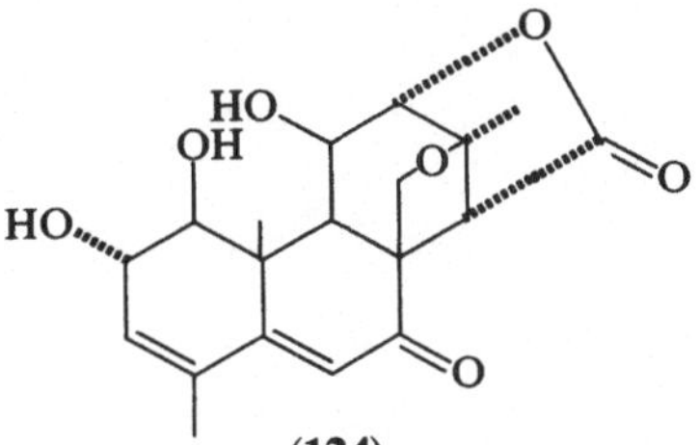

(124)

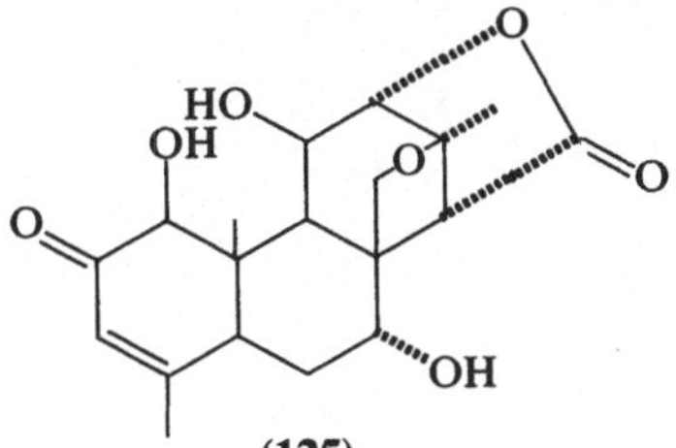

(125)

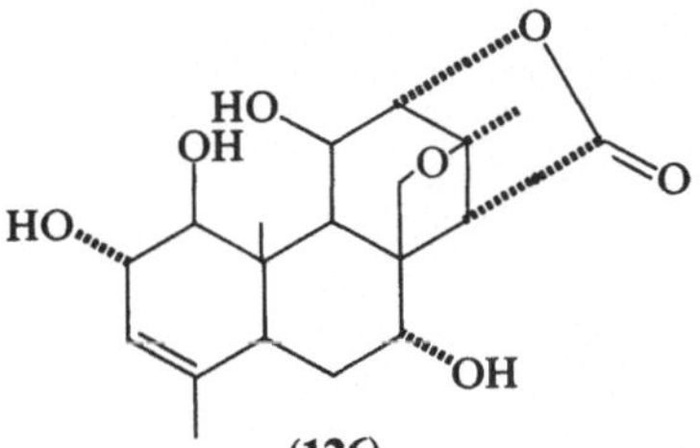

(126)

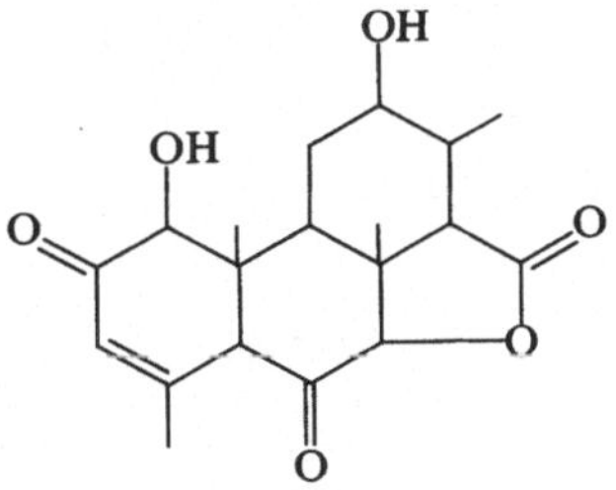

(127)

Chart 15

Cedronine (**125**) and cedronoline (**126**) have been isolated from the fruits of *Simaba cedron* Planch. (*80*). They differ from samaderines B and C in the nature of the function at C-7. The first two have a hydroxyl at this position as shown by hydrogenation experiments on the samaderines (*80*).

Structure (**127**) has been proposed for eurycomalactone, the bitter constituent of *Eurycoma longifolia* Jack. (*41*).

8. Simarolide (128) and Picrasin A (133)

Simarolide (**128**) has been isolated from the bark of *Simarouba amara* Aubl. (*57*). The constitution and absolute configuration of simarolide were determined by X-ray analysis (*7*) of its *m*-iodobenzoate and *p*-iodo-*m*-nitrobenzoate. Before these results became available, some chemical evi-

dence was obtained. These studies (57), which were limited by the un-
availability of *Simarouba amara*, may be summarized as follows. The
presence of an a acetoxy-group, one secondary hydroxyl, a keto-
group and a five membered lactone-ring were readily recognized. Simarolide was shown to contain one secondary and three tertiary methyl
groups.

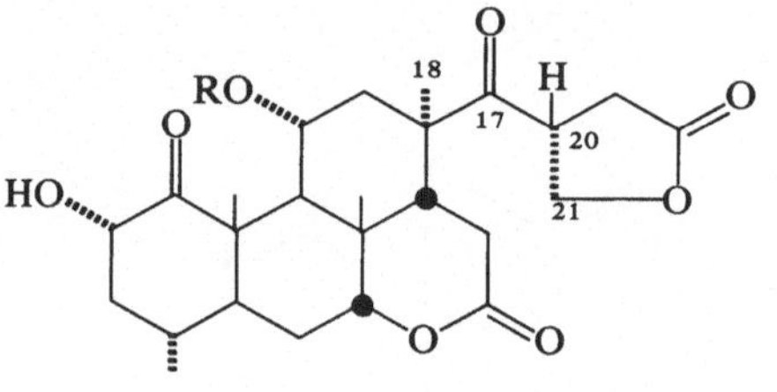

Simarolide   **(128)**   R = Ac
               (129)   R = H

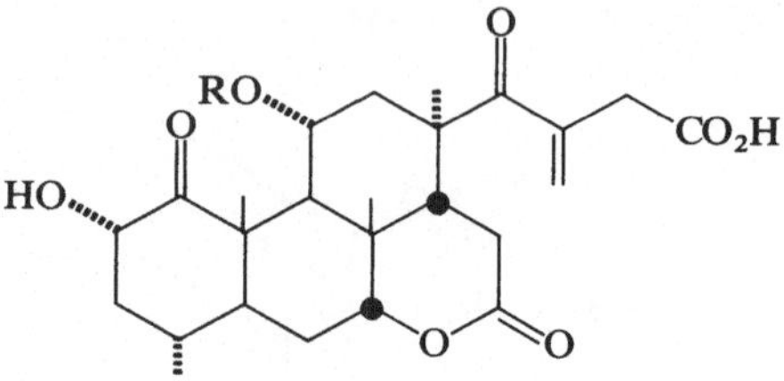

               (130)   R = Ac
               (131)   R = H

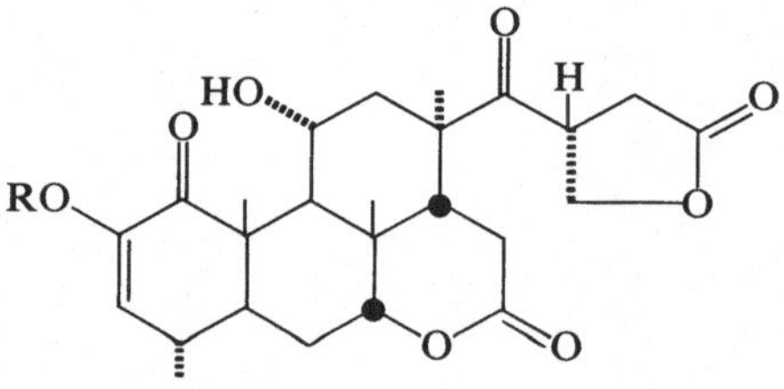

               (132)   R = H
Picrasin A   **(133)**   R = CH₃

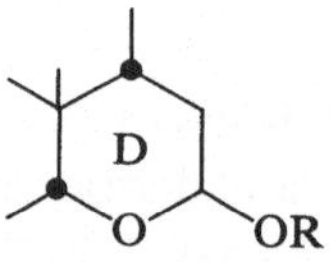

(134)   R = H
(135)   R = CH₃

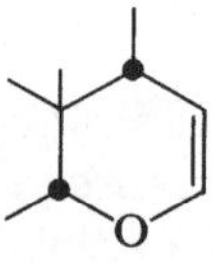

(136)

Chart 16

The structure of the γ-lactone ring and its position with respect to
the keto group were established by reaction with mild alkali, which
yields, besides desacetylsimarolide **(129)**, an acid **(130)** called simarenic
acid and its desacetyl derivative **(131)**. The presence of the side chain
in simarenic acid, as in **(130)**, is shown by mass spectral and NMR
evidence and by the formation of formaldehyde on ozonolysis. Furthermore, treatment of simarolide and simarenic acid with concentrated

phosphoric acid gives formaldehyde. A primary hydroxyl group must therefore be involved in the γ-lactone group and situated in the β-position with respect to a carbonyl group.

The secondary hydroxyl group is part of an α-ketol group which is sterically crowded by the 11-acetoxy group. Under similar conditions, (**129**) and (**131**), but not (**128**) and (**130**), readily consume one molecule of periodic acid reagent. Similarly, simarolide is recovered almost quantitatively after treatment with bismuth trioxide, but oxidation of deacetyl-simarolide (**129**) leads rapidly to a diosphenol (**132**).

A reaction involving the δ-lactone group is the reduction of simarolide with sodium borohydride leading to the hemiacetal (**134**) and to the corresponding methoxy-derivative (**135**). The former affords on treatment with acetic anhydride the cyclic enol ether (**136**).

Picrasin A [=nigakilactone G (*52*)], isolated from *Picrasma quassioides* Bennett (= *P. ailanthoides* Planchon), has been formulated as (**133**) (*33*). It has been correlated with simarolide (**128**) by methylation (*33*) of the diosphenol derivative (**132**).

Simarolide and picrasin A are the only quassinoids so far known which have a C_{25} skeleton. Their occurrence is of great biogenetic significance, as it will be seen later in this chapter.

IV. The Biogenetic Isoprene Rule and the Biosynthesis of the Quassinoids

In their paper on limonin (**3**), ARIGONI, BARTON, COREY, JEGER and their collaborators (*2*) first proposed a biogenetic scheme for the limonoids. According to this, the postulated precursor euphol (**137**), (a) loses four terminal side chain carbon atoms and (b) undergoes a skeletal rearrangement during which one methyl group migrates from C-14 to C-8. The last step could be triggered by oxidative attack on the nuclear double bond of a Δ^7-isomer of euphol (for instance *via* the epoxide; see p. 135).

By 1964 it had become apparent (*6, 15, 29, 64, 65*), that the group of substances related to quassin must represent a later stage along the same biosynthetic pathway. Although the side chain is completely missing in most of the quassinoids, the exceptions, the C_{25}-compounds [simarolide (**128**) and picrasin A (**133**)] have a γ-lactone group which is reminiscent of the furan ring present in the limonoids. Moreover the absolute configuration of simarolide (**128**) is that of the triterpenoids. The formation of the quassinoids from the proto-triterpene, apo-euphol or its C-20 epimer, apo-tirucallol (**138**) may be schematized as follows (*65*): one of the methyl groups at C-4 and four carbon atoms at the end of the side chain have been removed and carbons C_{20} to C_{23} have been converted to a γ-lactone ring (see the model conversion of melianodiol into an analogous C_{26} lactone, p. 140). As with the limonoids, ring D has been oxidatively expanded and the carboxyl group of C-16 has formed a lactone ring with a 7α-hydroxyl group. Further oxidation of the residual C-17 hydroxyl would then lead to the basic skeletons of simarolide and picrasin A.

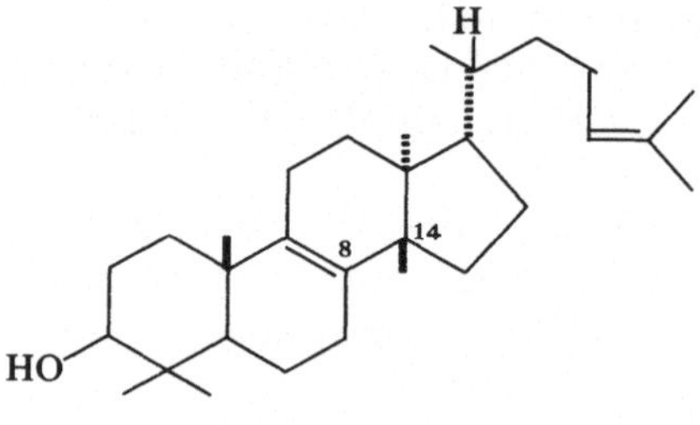

(137)   Euphol

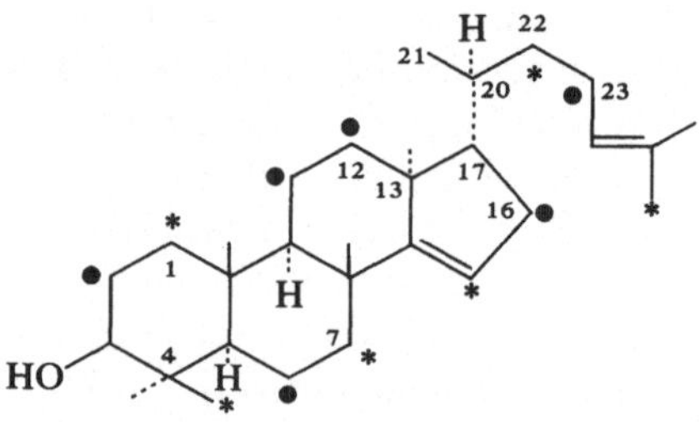

(138)   Apo-tirucallol

The C_{20}- and C_{19}-quassinoids would be formed by cleavage of the C-13/C-17 bond, formation of the C_{19}-compounds requiring the additional loss of carbon atom C-16. Simarolide and Picrasin A are the only Simaroubaceae bitter constituents which have no oxygenated function at C-12. This seems to indicate that in the other quassinoids the C-13/C-17 bond has been broken due to the presence of a β-dicarbonyl system (12,17-dione) or by retroaldol reaction of a 12-one -17-ol. But cleavage of the C-13/C-17 bond by the biological analog of a Bayer-Villiger oxidation of a 17-ketone should also be considered; the presence of an oxygenated group at C-13 in several quassinoids may be relevant in this context.

The triterpenoid biogenetic pathway for the quassinoids has been experimentally verified by MORON *et al.* (*46, 49*) using labelled mevalonate precursors (Chart 17).

Viable seeds of *Simaruba glauca* were chosen for these experiments and the incorporation of [2-^{14}C]- and [5-^{14}C]-mevalonic acid lactone **(139)** into the bitter constituents glaucarubinone **(140)** and glaucarubolone **(141)** was investigated. The expected labelling pattern in these compounds if formed by way of a triterpenoid intermediate is indicated in Formulae **138** and **140** by stars for [2-^{14}C]- or dots for [5-^{14}C]-mevalonate precursor. The prefixes [2-[14]C*MV] and [5-[14]C·MV] are added to the names of the products referring to the type of mevalonate used in each experiment.

(140)   R = —CO—C(CH₃)(OH)—C₂H₅

(141)   R = H

(139)

Hydrolysis of the radioactive glaucarubinone (**140**) isolated from both series of experiments showed that the α-hydroxy-α-methylbutyric acid residue was radio-inactive and that the label was confined to the C_{20}-component, glaucarubolone (**141**). It may be noted that isoleucine has been shown to be the biogenetic precursor of the hydroxy acid (*47*).

Before the oxidative degradation of the [2-^{14}C]-mevalonate-labelled glaucarubinone or glaucarubolone was studied it was important to know whether the labelled or the unlabelled methyl of the *gem*-dimethyl group of the hypothetical terpenoid precursor (**138**) was retained at position 4 in glaucarubolone. Acetic acid obtained upon Kuhn-Roth oxidation of [2-^{14}C*MV] glaucarubolone proved to be radioinactive. The three methyl groups and the carbon atoms C-4, C-10 and C-13 are therefore unlabelled. The methyl group at position 4 in glaucarubolone must then be derived from the methyl group of mevalonate and must correspond to the β axial methyl group at C-4 of the terpenoid precursor (**138**) (*1*). It is noteworthy that a similar situation was found in viridin (*27*) but not in fusidic acid (*28*) nor in the diterpenoid, rosenonolactone.

Comparison of the two samples of acetic obtained by Kuhn-Roth oxidation of [2-^{14}C*MV]-glaucarubolone and its partially aromatized product, dihydroglaucanol (**142**) showed that the label was specifically incorporated into C-1. Moreover, the tetra-esters (**143**) and (**144**) obtained by degradation of [2-^{14}C*]-dihydroglaucanol proved to carry one third of the original label. When obtained from [5-^{14}C·MV]-glaucarubolone, these esters had two fifths of the total radioactivity. Methylation of glaucarubinone to 1-O-methyl-glaucarubinone, followed by oxidation to the acid (**145**) and decarboxylation of the latter, permitted isolation of C-12. In this manner carbon atom C-12 was shown to be devoid of radioactivity in [2-^{14}C*MV]-glaucarubinone but to carry one fifth of the original label in [5-^{14}C·MV]-glaucarubinone. This result is consistent

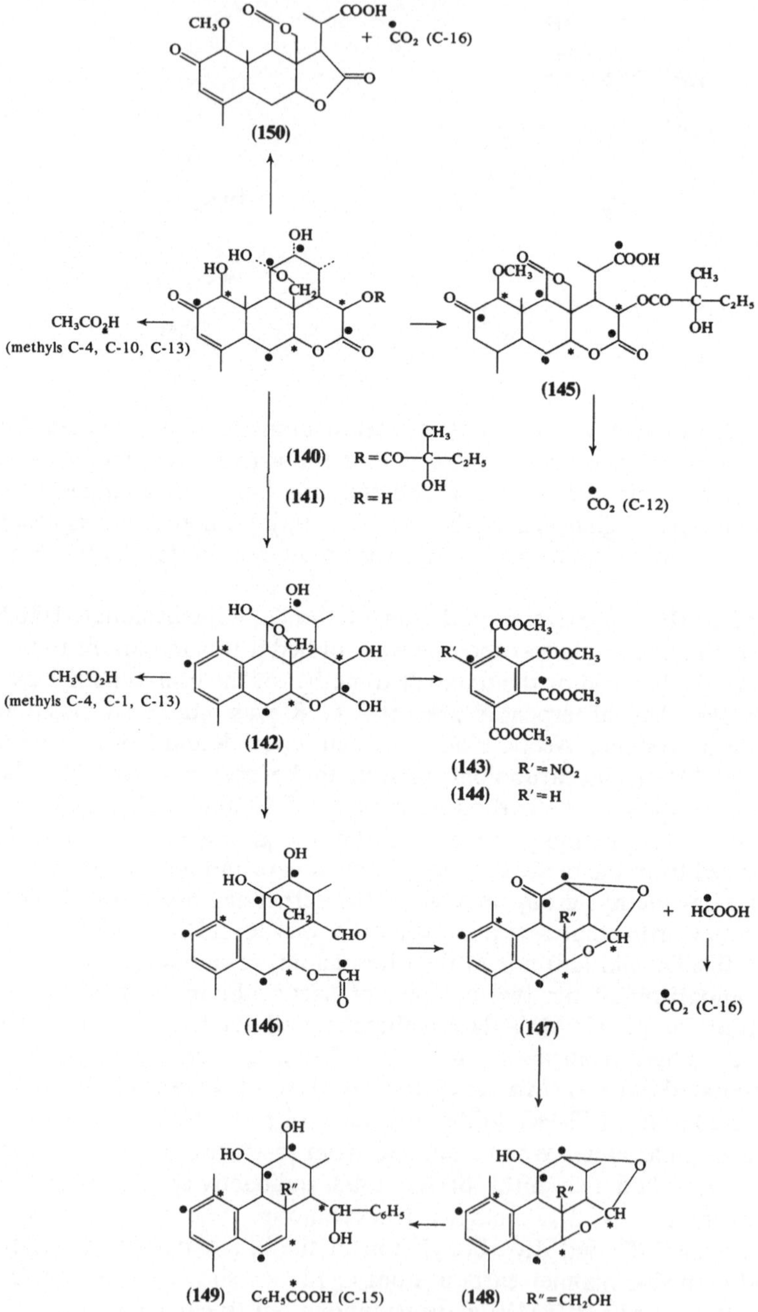

+ CO2 (C-16)
(150)
CH3CO2H
(methyls C-4, C-10, C-13)
(145)
(140) R=CO—C(CH3)(OH)—C2H5
(141) R=H
CO2 (C-12)
CH3CO2H
(methyls C-4, C-1, C-13)
(142)
(143) R′=NO2
(144) R′=H
(146)
(147)
+ HCOOH
CO2 (C-16)
(149) C6H5COOH (C-15)
(148) R″=CH2OH

Chart 17

with biogenesis of glaucarubinone by tail-to-tail condensation of two farnesyl residues to give a triterpenoid precursor. In contrast to C-12, carbon atom C-15 would be expected to be labelled in a $[2\text{-}^{14}C^*MV]$-glaucarubinone biosynthetised by way of a triterpenoid intermediate and unlabelled if a diterpenoid intermediate were involved. C-15 was carved out as benzoic acid by Kuhn-Roth oxidation of the phenylcarbinol (**149**), obtained by treatment of the $[2\text{-}^{14}C^*MV]$-ketal (**148**) with phenyl lithium and proved to carry about one third of the total label.

C-16 was isolated either by oxidation of 1-O-methyl-glaucarubolone to the 16-nor-acid (**150**) and CO_2 or by acidic hydrolysis of the formate (**146**) to the keto-ketal (**147**) and formic acid; it was found non-labelled in $[2\text{-}^{14}C^*]$ and labelled in $[5\text{-}^{14}C^\circ]$-glaucarubolone.

The labelling patterns observed in $[2\text{-}^{14}C^*MV]$- and $[5\text{-}^{14}C^\circ MV]$-glaucarubinone as well as the presence of five labelled carbon atoms in the latter prove unambigously the triterpene origin of these compounds and are consistent with a scheme involving a tetracyclic triterpene such apo-euphol or apo-tirucallol as a precursor of these bitter constituents of *Simaroubaceae*. Thus, the quassinoids are to be regarded as degraded triterpenoids.

As to the proto-triterpene, apo-tirucallol (**138**) or its 20-epimer, apo-euphol, it might be formed (Chart 18) either directly from the cationic intermediate (**151**, arrows) resulting from the cyclisation of squalene without the intervention of tirucallol (or euphol), or *via* tirucalla-7, 24-dien-3-ol (**152**) (or its 20-epimer) and subsequent black-migration of the C-14 methyl group to C-8. Evidence that such rearrangements of the C-14 methyl group can be brought about is provided by the oxidation of dihydrobutyrospermyl acetate (**153**) (dihydro-24,25-eupha-7-en-3β-acetyl) with chromic acid to 7-oxodihydro-24,25-apo-euphol (**154**) (*39*) and by the rearrangement of 7,8-α-epoxidotirucallol derivatives (**155**) to 7-α-hydroxy-compounds in the apo-tirucallol series (**156**).

The rearrangement of a 7,8-α-epoxide leading to a 7-α-hydroxy-apo-compound seems more likely than an oxidative rearrangement of a 7-ene-derivative to a 7-oxo-apo-compound. All the known quassinoids (and limonoids) have oxygen at C-7 and when it is present as a hydroxyl group (the latter being generally involved in a lactone ring in the quassinoids) the configuration is α. Moreover, the incorporation experiments with $[2\text{-}^3H]$ mevalonate into glaucarubinone (**140**) showed that the label at C-7 was retained (*48*).

Whether the limonoids derive from euphol (20βH) or tirucallol (20αH) is uncertain, since C-20 becomes trigonal in the course of the formation of (**151**).

It may be noted, that among the degraded triterpenes of known structure, only simarolide (**128**) and picrasin A (**133**) have an assymetric

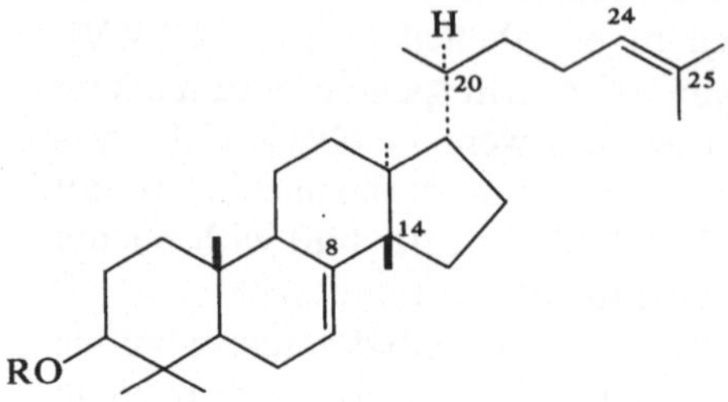

(152)

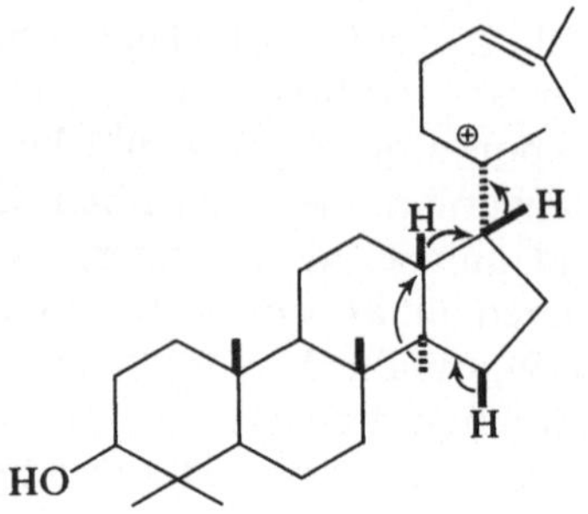

(151)

(153)   20-epimer
24,25-dihydro
R = Ac

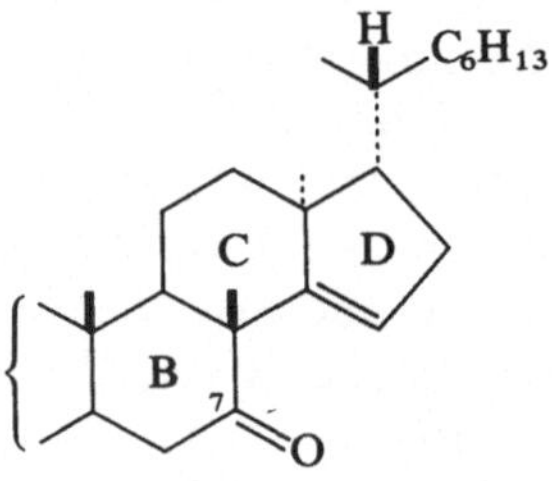

(154)

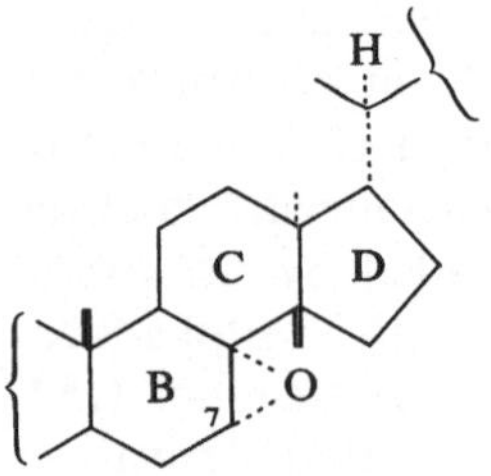

(155)

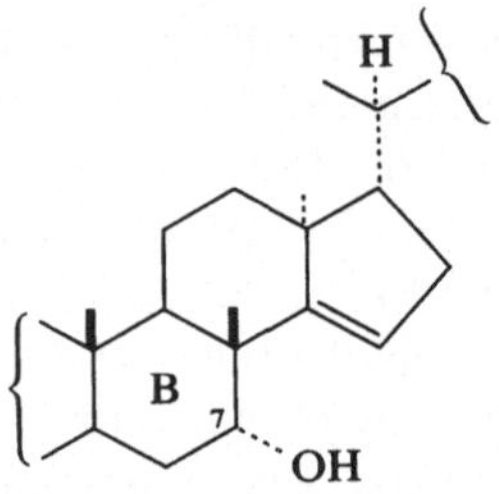

(156)

Chart 18

carbon atom at C-20, and that this carbon atom possesses the tirucallol configuration. However, because of the presence of the keto-group at C-17, the possibility of isomerisation at this centre during biosynthesis cannot be excluded.

It is not profitable at this stage to discuss the sequence of complex oxidations steps which introduce the oxygen functions of the quassinoids. However a comment on the way in which ring A of the quassinoids is modified may be in order. It was suggested (*50*) that this might arise by an eliminative decarboxylation of a 4β-carboxylate and a 3α-hydroxyl function as shown in formula (**157**) (where R may be a hydro-

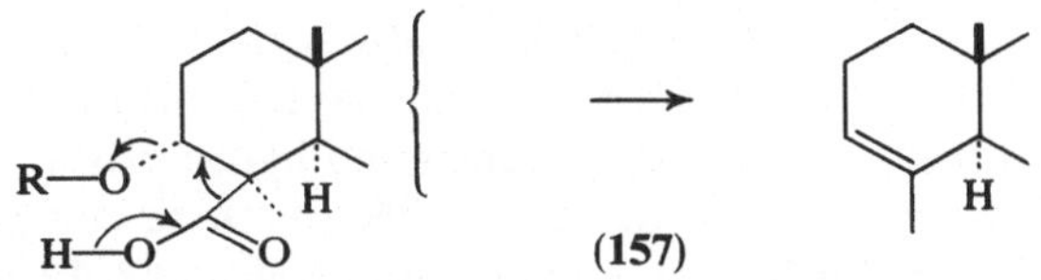

(157)

gen or a phosphate ester to give a better leaving group). At the time the suggestion was made no naturally occurring quassinoid with an oxygen function at C-3, was known nor were the results of the labelling experiments available. But the determination of the structures of bruceines A, B and C which have an oxygenated function at C-3, as well as the fact that it is the β-axial and not the α-equatorial methyl group which is retained in the quassinoids, seem to invalidate this hypothesis. It seems more plausible to assume that a 3-oxo-triterpenoid is an intermediate of all the quassinoids, and that one C-4 methyl group is eliminated by decarboxylation of a β-keto-acid.

In order to get some insight into the mode of formation of the triterpenic precursor of the quassinoids, mevalonic acid labelled at position 4R with tritium (158) was incorporated into glaucarubinone and glaucarubolone by viable seeds of *Simarouba glauca* (48).

As indicated, these quassinoids biosynthesed from mevalonic acid labelled with ^{14}C at position 5 possess five labelled carbon atoms (C-2,

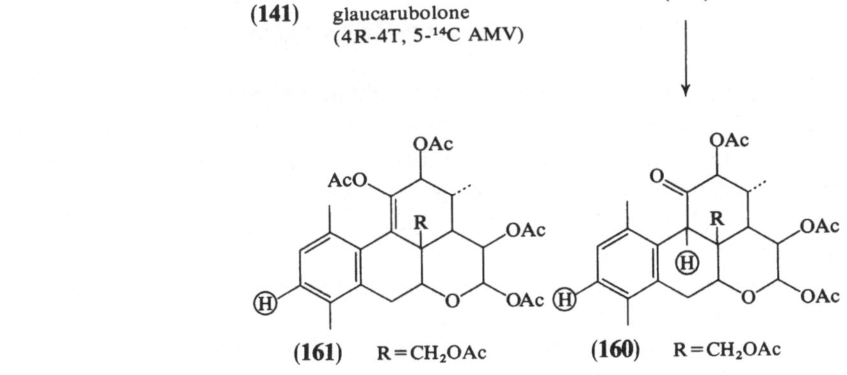

(158)

(141)   glaucarubolone
(4R-4T, 5-^{14}C AMV)

(159)

(161)   R = CH₂OAc

(160)   R = CH₂OAc

Chart 19

C-6, C-11, C-12 and C-16). When biosynthesized from mevalonic acid tritiated at position 4R (158) they should have, if all the tritium atoms were retained, three tritiums (see the encircled hydrogens in Chart 19), at position 3, 5 and 9. The atomic ratio $T/^{14}C$ in these compounds when biosynthesized from both of these mevalonic acids should be $3:5$.

The value of the atomic ratio of radioactive glaucarubinone and glaucarubolone, isolated from this experiment, indicates the presence of only two tritiums. They have been localized at position 5 and 9 as follows: dihydroglaucanol (159) obtained, as we have already mentionned, by reduction with sodium borohydride followed by aromatization, reveals a decrease in the atomic ratio $T/^{14}C$, in agreement with the elimination of proton 5. Acetylation of dihydroglaucanol under mild conditions yields the tetraacetate (160) which has the same atomic ratio $T/^{14}C$ as dihydroglaucanol. Under more drastic conditions of acetylation, dihydroglaucanol gives the enol acetate (161). The atomic ratio $T/^{14}C$ in this substance shows that the elimination of proton H_9 is accompanied by the nearly total loss of tritium in the molecule. These results show that two tritium atoms — at position 5 and 9 — are retained in glaucarubolone when it is biosynthesized from mevalonic acid tritiated at position 4R. The absence of tritium at position 3 supports the hypothesis that the loss of the α-equatorial methyl at C-4 proceeds *via* a compound having a carbonyl at position 3, for instance by decarboxylation of a β-keto-acid. The retention of tritium at position 9 excludes the involvement of a triterpenic precursor with a $\Delta^{8,9}$ double bond and makes that of 9,10-cyclopropane precursor unlikely. So, if a $\Delta^{7,8}$-euphol or $\Delta^{7,8}$-tirucallol is the precursor of the quassinoids, they should be formed by stabilisation of the intermediate carbocation at C-8 without the intervention of euphol or tirucallol.

V. Tetracyclic Triterpenes from Simaroubaceae

In recent years a number of side chain oxygenated derivatives of tirucalla-7,24-dien-3-ol have been found to occur together with limonoids. But it is only quite recently that two of these triterpenes have been isolated from a genus belonging to the Simaroubaceae. They are melanodiol (162), and its diacetate (163) which were isolated from *Samadera madagascariensis* (43).

Melanodiol (162) has been previously isolated from *Melia azadirachta* (Meliaceae) and has been obtained by acid-catalysed opening of the epoxide of melianone (164) by LAVIE *et al.* (40). The isolation of the diacetate (163) indicates that melianodiol (162) is not an artefact formed from melianone (164) during the extraction.

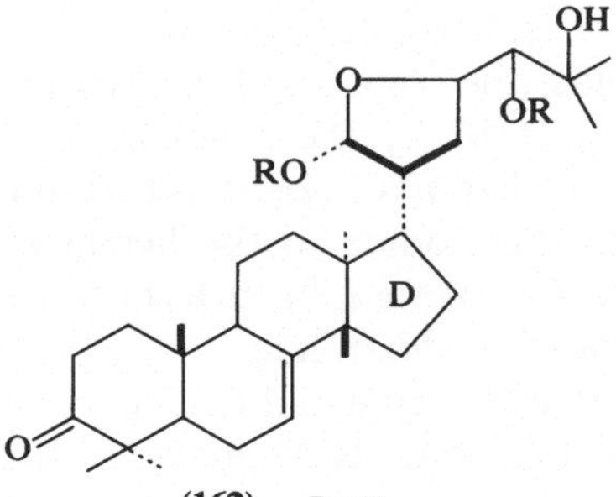

(162)   R=H
(163)   R=Ac

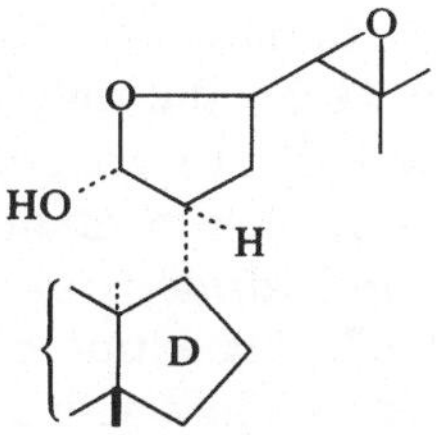

(164)

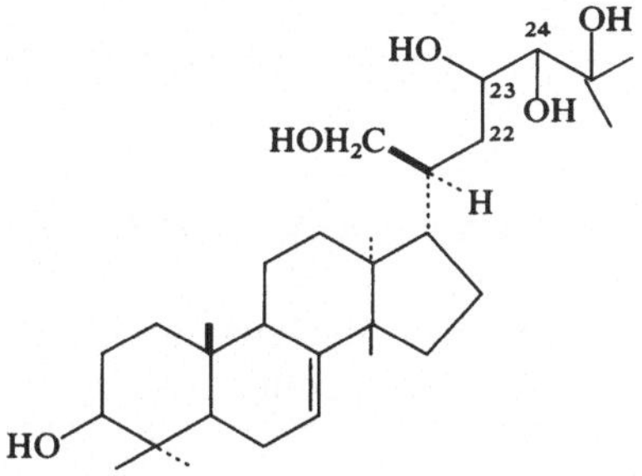

(166)

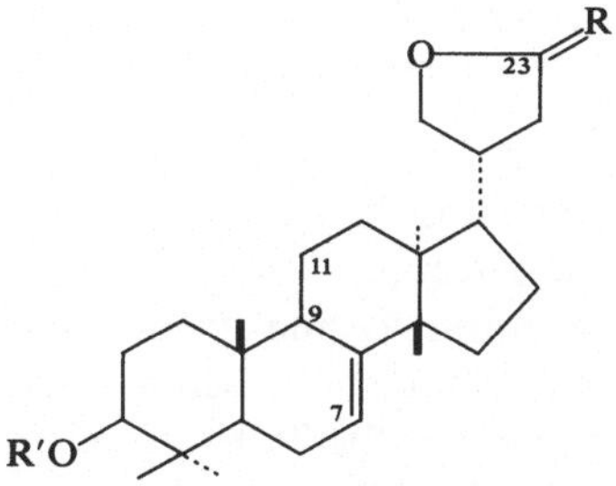

(167)   R=H, OH;   R′=H
(168)   R=O;   R′=H
(169)   R=O;   R′=Ac
(170)   R=O;   R′=Ac

7α,8α-expoxyde

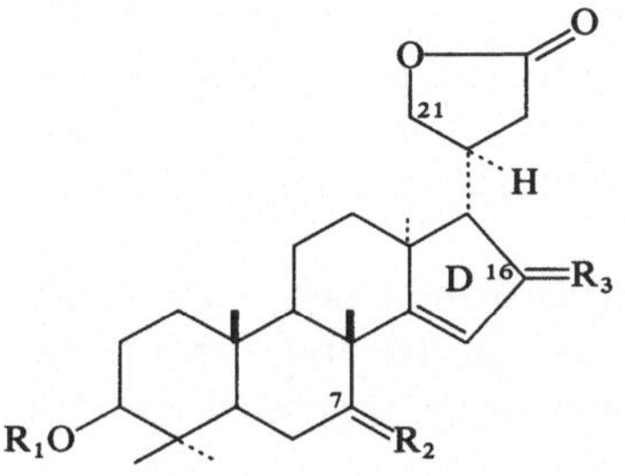

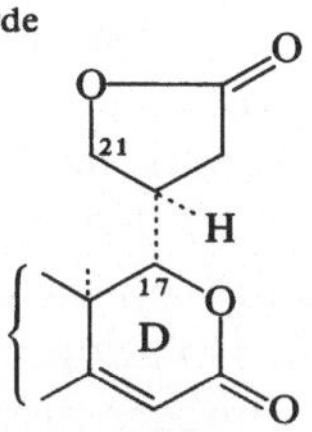

(175)

(171)   R₁=Ac; R₂=$\langle$ OH, H ; R₃=H₂
(172)   R₁=Ac; R₂=O;   R₃=H₂
(173)   R₁=Ac; R₂=$\langle$ OAc, H ; R₃=H₂
(174)   R₁=Ac; R₂=$\langle$ OAc, H ; R₃=O

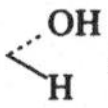

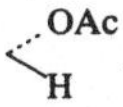

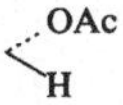

Chart 20

In model experiments directed towards simarolide (**128**) and Picrasin A (**133**), the only known C_{25}-quassinoids, melianodiol (**162**) was converted into an analogous C_{26} γ-lactone (**168**) by the following reactions: The pentaol (**166**) obtained from melianodiol (**162**) with sodium borohydride, was cleaved by sodium metaperiodate to the hemiacetal (**167**). Oxidation of the latter with silver carbonate (*19*) furnished compound (**168**) with a γ-lactone such as is found in simarolide.

It has been suggested previously (*3*) that the potential C-23 carboxyl group of simarolide might arise from a 23-keto-24,25-epoxide (**165**), as indicated:

$$R\text{—}\overset{23}{C}O\text{—}CH\text{—}C\overset{Me}{\underset{Me}{<}} \quad \longrightarrow \quad R\text{—}C\text{—}CH\text{—}C< \quad \longrightarrow \quad R\ COOH + OHCHC\overset{Me}{\underset{Me}{<}}$$

$$\text{(165)}$$

However, transformations similar to that shown in (**166**)→(**168**) may occur in nature.

According to one of the biogenetic hypotheses, the quassinoids as well as the limonoids, which are all oxygenated at C-7, could arise by rearrangement of the $7\alpha,8\alpha$-epoxide of a tirucall-7-ene to a 7α-hydroxy-Δ^{14}-apo-derivative. Such transformations have now been carried out on compound (**169**) following the procedure of BUCHANAN *et al.* (*8*). When (**169**) was treated with monoperphthalic acid in dry ether at 0° for 48 hour it yielded the $7\alpha,8\alpha$-epoxide (**170**) which rearranged with boron trifluoride etherate to give the apo-compound (**171**). Upon oxidation at 0° with Jones' reagent the latter compound gives the ketone (**172**) and upon acetylation it affords the diacetate (**173**), which, upon treatment with chromic acid in acetone is smoothly oxidised to the α,β-unsaturated ketone (**174**). When submitted to a Baeyer-Villiger oxidation the latter yields the α,β-unsaturated lactone (**175**) (*43*).

The isolation of side-chain oxygenated tirucall-7-ene derivatives from a simaroubaceous plant and the transformations reported above are clearly relevant to the biogenetic schemes that have been suggested for the quassinoids.

VI. Physiological Activity

A number of the simaroubaceous principles are well known in herbal medicine as effective antiamoebic agents (*25*). Thus, a preparation from *Castela nicholsoni*, from which chaparrin has been isolated, has been found very active although there still remain conflicting reports on the toxicity and harmful side effects. A medicinal preparation of this plant called "castamargina", is found in the Mexican market. *Brucea* species have yielded extracts revealing antiamoebic and other pharmacological properties. The total extract of *Simaba cedron* has been found to have anti-malaria activity and anti-inflammatory properties.

Physiological studies of the pure crystalline compounds have not yet been extensive. Glaucarubin has been studied carefully and was shown to be an effective amoebicide (*17*), less active than emetine *in vitro* but much more effective than carbasone in experimental amoebiasis in animals. It was also reported to be an effective tumor-necrotizing agent in mice. In France, glaucarubin is sold commercially as a medicinal preparation.

VII. Tables

Table 1. *Quassinoids of Type A*

Compound	Refs.	Mol. Formula	C=C	−OH	−CO	−CO₂H		M.p.	[α]
Quassin (1)	(74, 75) (55, 67)	$C_{22}H_{28}O_6$	2, 12	$2OCH_3$; $12OCH_3$	1, 11			221	+34
Neoquassin (2)	(74, 75 52)	$C_{22}H_{30}O_6$	2, 12	16; $2OCH_3$, $12OCH_3$	1, 11		7α, 16 oxide	231	+41
Glaucarubin (54)	(64)	$C_{25}H_{36}O_{10}$	3	1β, 2α, 11α, 12α, 15βOR₁			11β, 30 oxide	276	+45
Glaucarubinone (70a)	(22)	$C_{25}H_{34}O_{10}$	3	1β, 11α, 12α, 15βOR₁	2		11β, 30 oxide	230	+50
Δ¹³⁽¹⁸⁾-Glaucarubin	(45)	$C_{25}H_{34}O_{10}$	3, 13 (18)	1β, 2α, 11α, 12α, 15βOR₁			11β, 30 oxide	248	
Glaucarubol (55)	(44)	$C_{20}H_{28}O_8$	3	1β, 2α, 11α, 12α, 15β			11β, 30 oxide	288	+38
2′-Acetylglaucarubinone	(82)	$C_{27}H_{36}O_{11}$	3	1β, 11α, 12α; 15βOR₆	2		11β, 30 oxide	173	+74
Holocanthone	(77)	$C_{22}H_{28}O_9$	3	1β, 11α, 12α; 15βOAc	2		11β, 30 oxide	244	+70
Glaucarubolone (70b)	(22, 59, 44)	$C_{20}H_{26}O_8$	3	1β, 11α, 12α, 15β	2		11β, 30 oxide	258	−26
Δ¹³⁽¹⁸⁾-glaucarubolone	(45)	$C_{20}H_{24}O_8$	3, 13 (18)	1β, 11α, 12α, 15β	2		11β, 30 oxide	230	

Table 1 (continued)

Compound	Refs.	Mol. Formula	C=C	-OH	-CO	-CO$_2$H	M.p.	[α]
Chaparrin (**72**)	(*26, 14, 37*)	C$_{20}$H$_{28}$O$_7$	3	1β, 2α, 11α, 12α		11β, 30 oxide	300	
Chaparrinone (**81**)	(*59, 20*)	C$_{20}$H$_{26}$O$_7$	3	1β, 11α, 12α	2	11β, 30 oxide	242	−47
Ailanthone (**87**)	(*66, 11, 20*)	C$_{20}$H$_{24}$O$_7$	3, 13 (18)	1β, 11α, 12β	2	11β, 30 oxide	238	+12
Ailanthinone (**70c**)	(*20*)	C$_{25}$H$_{34}$O$_9$	3	1β, 11α, 12α; 15βOR$_2$	2	11β, 30 oxide	231	+88
Amarolide (**42**)	(*10, 20, 72*)	C$_{20}$H$_{28}$O$_6$		2α, 11α	1, 12		255	
11-Acetylamarolide	(*10, 20, 72*)	C$_{22}$H$_{30}$O$_7$		2α; 11αOAc	1, 12		265	
Klaineanone (**46**)	(*59, 81*)	C$_{20}$H$_{28}$O$_6$	3	1β, 11β, 12α	2		258	−52
11-Dehydroklaineanone (**47**)	(*59, 81*)	C$_{20}$H$_{26}$O$_6$	3	1β, 12α	2, 11		225	+8
Nigakilactone A (**33**)	(*55*)	C$_{21}$H$_{30}$O$_6$	2	11α, 12β, 2OCH$_3$	1		238°	+53°
Nigakilactone B (**34**) (=Simalikalactone A)	(*55, 73*)	C$_{22}$H$_{32}$O$_6$	2	11α, 2, 12βOCH$_3$	1		278°	+17
Nigakilactone C (**35**)	(*55, 54*)	C$_{24}$H$_{34}$O$_7$	2	11α, Ac, 2, 12βOCH$_3$	1		253°	+9
Nigakilactone E (**36**)	(*55, 54*)	C$_{24}$H$_{34}$O$_8$	2	13β, 11αAc, 2, 12β	1		280°	+36
Nigakilactone F (**37**)	(*55, 54*)	C$_{22}$H$_{32}$O$_7$	2	11α, 13β, 2, 12βOCH$_3$	1		265°	+46
Nigakilactone H (**38**)	(*52*)	C$_{22}$H$_{32}$O$_8$	2	11α, 14β, 2, 12βOCH$_3$	1		275°	+67

Table 1 (continued)

Compound	Refs.	Mol. Formula	C=C	$-OH$	$-CO$	$-CO_2H$	M.p. $[\alpha]$
Simalikalactone C (**39a**)	(73)	$C_{21}H_{28}O_6$	2	11α, $2OCH_3$	1, 12		$228°$ $+42$
Nigakihemiacetal A (**40**)	(52)	$C_{22}H_{34}O_7$		11α, 13β, 16, 2, $12\beta OCH_3$	1	7α, 16 oxide	$262°$ $+20$
Nigakihemiacetal C (**41**)	(53)	$C_{21}H_{32}O_6$		11α, 12β, 16, $12\beta OCH_3$	1	7α, 16 oxide	$205°$ $+19°$
Picrasin C (**50b**)	(36)	$C_{23}H_{34}O_7$		12α, $12\beta OCH_3$, $11\alpha Ac$	1		
Picrasin D (**39b**)	(34)	$C_{22}H_{30}O_6$	2	$11\alpha-CH_2-12\beta$, $2OCH_3$	1		
Picrasin E (**39c**)	(34)	$C_{22}H_{30}O_7$	2	14β, $11\alpha-CH_2-12\beta$, $2OCH_3$	1		
Picrasin F (**39d**)	(35)	$C_{22}H_{30}O_8$	2	13β, 14β, $11\alpha-CH_2-12\beta$, $2OCH_3$			
Chaparrolide (**44**)	(44)	$C_{20}H_{30}O_6$		1β, 2α, 12β	11		$130°$
Castelanolide (**45**)	(44)	$C_{20}H_{28}O_6$	12	1α, 2α, 12	11		$167°$
15-Hydroxy-Klaineanone (**48**)	(60)	$C_{20}H_{28}O_7$	3	1β, 11β, 12α, 15β	2		$224°$ $+70°$
Picrasin B (**49**) (=Simalikalactone B)	(32, 55, 73, 76)	$C_{21}H_{28}O_6$	12	12α, $12OCH_3$	1, 11		$254°$
6-Hydroxy-picrasin (**50a**)	(76)	$C_{21}H_{28}O_7$	12	2α, 6α, $12OCH_3$	1, 11		$288°$ $-48°$

Table 1 (continued)

Compound	Refs.	Mol. Formula	C=C	–OH	–CO	–CO₂H		M.p.	[α]
Brucein A (**98**)	*(62)*	$C_{26}H_{34}O_{11}$	3	3, 11β, 12α; 15βOR₃	2	18– CO₂CH₃	13β, 30 oxide	270	–86
Brusatol (**101**)	*(68)*	$C_{26}H_{32}O_{11}$	3	3, 11β, 12α; 15βOR₄	2	18– CO₂CH₃	13β, 30 oxide	276	
Brucein B (**99**)	*(62)*	$C_{23}H_{28}O_{11}$	3	3, 11β, 12α; 15βOAc	2	18– CO₂CH₃	13β, 30 oxide	266	–77
Brucein C (**100**)	*(62)*	$C_{28}H_{36}O_{12}$	3	3, 11β, 12α; 15βOR₅	2	18– CO₂CH₃	13β, 30 oxide	180°	–34°
Brucein D (**105**)	*(61)*	$C_{20}H_{26}O_9$	3	1β, 11β, 12α, 14β, 15β	2		13β, 30 oxide	290	–21
Brucein E (**106**)	*(61)*	$C_{20}H_{28}O_9$	3	1β, 2α, 11β, 12α, 14β, 15β			13β, 30 oxide	258	+25
Brucein F (**107**)	*(63)*	$C_{20}H_{28}O_{10}$	3	1β, 2α, 11β, 12α, 14β, 15β, 18			13β, 30 oxide	227	+22
Brucein G (**110**)	*(18)*	$C_{20}H_{26}O_8$	3	6β, 11β, 12α, 15β	2		13β, 30 oxide	258	+58
Simalikalactone D (**111**)	*(73)*	$C_{25}H_{34}O_9$	3	1, 11α, 12, 15R₂	2		13β, 30 oxide	230°	+53

R₁ = (S) 2-hydroxy-2-methylbutyrate
R₂ = 2-methylbutyrate
R₃ = isovalerate
R₄ = senecioate
R₅ = 3,4-dimethyl-1-hydroxy-4-penta-2-enoate
R₆ = 2-acetoxy-2-methylbutyrate

Table 2. *Quassinoids of Type B*

Compound	Refs.	Mol. Formula	C=C	−OH	−CO	−CO$_2$H		M.p.	[α]
Samaderine B (112)	(80, 79)	C$_{19}$H$_{22}$O$_7$	3	1β, 11β	2, 7		13β, 30 oxide	240	+67
3,4-Dihydrosamaderin B (123)	(79)	C$_{19}$H$_{24}$O$_7$		1β, 11β	2, 7		13β, 30 oxide	237	+70
Samaderine C (113)	(80, 79)	C$_{19}$H$_{24}$O$_7$	3	1β, 2α, 11β	7		13β, 30 oxide	268	+74
Samaderine D (124)	(79)	C$_{19}$H$_{22}$O$_7$	3, 5	1β, 2α, 11β	7		13β, 30 oxide	290	+88
Cedronine (125)	(80, 79)	C$_{19}$H$_{24}$O$_7$	3	1β, 7α, 11β	2		13β, 30 oxide	280	−12
Cedronoline (126)	(80, 79)	C$_{19}$H$_{26}$O$_7$	3	1β, 2α, 7α, 11β			13β, 30 oxide	267	+17
Eurycomalactone (127)	(41)	C$_{19}$H$_{24}$O$_6$	3	1, 12	2, 6		15,7 lactone	270	+10

Table 3. *Quassinoids of Type C*

Compound	Refs.	Mol. Formula	C=C	−OH	−CO	−CO$_2$H	M.p.	[α]
Simarolide (128) (=5)	(57, 7)	C$_{27}$H$_{36}$O$_9$		2α; 11αOAc	1, 17		270	+73
Picrasin A (133) (=Nigakilactone G)	(33, 52)	C$_{26}$H$_{34}$O$_8$	2	2OCH$_3$, 11αOH	1, 17		305°	+41

References

1. ARIGONI, D.: Biogenesis of terpenes in moulds and higher plants. In "Ciba Symposium on the Biosynthesis of Terpenes and Sterols", p. 231. London: J. and A. Churchill. 1959.
2. ARIGONI, D., D. H. R. BARTON, E. J. COREY, O. JEGER, L. CAGLIOTI, S. DEV, P. G. FERRINI, E. R. GLAZIER, A. MELERA, S. K. PRADHAN, K. SCHAFFNER, S. STERNHELL, J. F. TEMPLETON, and S. TOBINAGA: The constitution of limonin. Experienta **16**, 41 (1960).
3. BEVAN, G. W. L., D. E. U. EKONG, T. G. HALSALL, and P. TOFT: West African Timbers. Part XX. The structure of turreanthin, an oxygenated tetracyclic triterpene monoacetate. J. Chem. Soc. (London) C **1967**, 820.
4. BEERS, R. J. S., D. B. G. JAQUISS, A. ROBERTSON, and W. E. SAVIGE: Quassin and Neoquassin. II. J. Chem. Soc. (London) **1954**, 3672.
5. BEER, R. J. S., K. R. HANSON, and A. ROBERTSON: Quassin and Neoquassin. IV. J. Chem. Soc. (London) **1956**, 3280.
6. BREDENBERG, B-SON. J.: Biogenesis of the simarubaceae bitter compounds. Chem. Ind. (London) **1964**, 73.
7. BROWN, W. A. C., and G. A. SIM: The constitution and absolute stereochemistry of simarolide, the bitter principle of *Simarouba amara*. Proc. Chem. Soc. **1964**, 293.
8. BUCHANAN, J. G. ST. C., and T. G. HALSALL: The synthesis of the simplest meliacins (Limonoids) from tetranortirucallane. Triterpenoids containing a β-substituted furyl side-chain. Chem. Comm. **1969**, 242.
9. BOURGUIGNON, N., et J. POLONSKY: Sur les constituants amers des fruits du Kirondo *(Perriera madagascariensis)*. Bull. Soc. Chim. Biol. **46**, 1145 (1964).
10. CASINOVI, C. G., V. BELLAVITA, G. GRANDOLINI, and P. CECCHERELLI: Occurrence of bitter substances related to Quassin in *Ailanthus glandulosa*. Tetrahedron Lett. **1965**, 2273.
11. CASINOVI, C. G., P. CECCHERELLI, B. GRANDOLINI, and V. BELLAVITA: On the structure of ailanthone. Tetrahedron Lett. **1964**, 3991.
12. CASINOVI, C. G., P. CECCHERELLI e G. GRANDOLINI: La 18-ossiquassina nuovo amaroide isolato dalla *Quassia amara*. Ann. Ist. Super. Sanità **2**, 414 (1966).
13. CONNOLLY, J. D., K. H. OVERTON, and J. POLONSKY: The chemistry and biochemistry of the limonoids and quassinoids. In: "Progress in Phytochemistry (L. REINHOLD and Y. LIWSCHITZ, eds.). Vol. II, 385. London: Interscience Publishers. 1970.
14. DAVIDSON, T. A., T. R. HOLLANDS, P. DE MAYO, and M. NISBET: The structure of chaparrin. Canad. J. Chem. **43**, 2996 (1965).
15. DREYER, D. L.: A biogenetic proposal for the simaroubaceous bitter principles. Experientia **20**, 297 (1964).
16. — Limonoid bitter principles. In: Progress in the Chemistry of Organic Natural Products (L. ZECHMEISTER, ed.), Vol. 26, p. 190. Wien-New York: Springer. 1968.
17. DRUEY, J.: Amoebizide. Angew. Chem. **72**, 677 (1960).
18. DUNCAN, G. R., and G. B. HENDERSON: The structure of bruceine G. Experientia **24**, 768 (1968).
19. FETIZON, M., et M. GOLFIER: Oxydation sélective des alcools par le carbonate d'argent. C. R. Acad. Sc. Paris **267**, Série C, 900 (1968).
20. FOURREY, J.-L.: Determination de la structure des constituants d'*Ailanthus altissima*. Etude de l'Ailanthone, constituant principal, et des composés apparentés en spectrométrie de masse. Doctoral Thesis, Orsay, Paris 1968.
21. FOURREY, J.-L. C. D. DAS et J. POLONSKY: Etude des spectres de masse des constituants amers des simarubacées. Organ. Mass. Spectrometry **1968**, 819.
22. GAUDEMER, A., et J. POLONSKY: Structure de la glaucarubinone, nouveau principe amer isolé de *Simaruba glauca*. Phytochemistry **4**, 149 (1965).
23. GAUDEMER, A.: Etude des spectres de RMN des dérivés de la glaucarubine et de la

glaucarubinone. Structure de l'isoglaucanol A, un isomère du glaucanol. Bull. Soc. Chim. France **1967**, 406.

24. GAUDEMER, A., J.-L. FOURREY et J. POLONSKY: Etude de la méthylation par le diazométhane des hydroxyles de composés amers des simarubacées. Bull. Soc. Chim. France **5**, 1676 (1967).

25. GEISSMAN, T. A.: New substances of plant origin. Ann. Rev. Pharmacol. **4**, 305 (1964).

26. GEISSMAN, T. A., and G. A. ELLESTAD: The structure of chaparrin, and a note on glaucarubol. Tetrahedron Lett. **1962**, 1083.

27. GROVE, J. F.: Viridin, part VI. Evidence for a steroidal pathway in the biogenesis of viridin from mevalonic acid. J. Chem. Soc. (London) C. **1969**, 549.

28. GODTFREDSEN, W. O., W. VON DAEHNE, S. VANGEDAL, A. MARQUET, D. ARIGONI, and A. MELERA: The stereochemistry of fusidic acid. Tetrahedron **21**, 3505 (1965).

29. HALSALL, T. G., and APLIN, R. T.: A pattern of development in the chemistry of pentacyclic triterpenes. In: Progress in the Chemistry of Organic Natural Products (L. ZECHMEISTER, ed.). New York: Springer-Verlag **22**, 153 (1964).

30. HAM, E. A., H. M. SCHAFER, R. C. DENKEWALTER, and N. G. BRINK: Structural studies on Glaucarubin from *Simarouba glauca*. J. Am. Chem. Soc. **76**, 6066 (1954).

31. HANSON, K. R., D. B. G. JAQUISS, J. A. LAMBERTON, A. ROBERTSON, and W. E. SAVIGE: Quassin and Neoquassin. III. J. Chem. Soc. **1954**, 4238.

32. HIKINO, H., T. OHTA, and T. TAKEMOTO: Stereostructure of Picrasin B, Simaroubolide of *Picrasma quassioides*. Chem. Pharm. Bull. **18**, 219 (1970).

33. — — — Stereostructure of Picrasin A, Simaroubolide of *Picrasma quassioides*. Chem. Pharm. Bull. **18**, 1082 (1970).

34. — — — Stereostructure of picrasin D and E, simaroubolides of *Picrasma quassioides*. Chem. Pharm. Bull. **19**, 212 (1971).

35. — — — Stereostructure of picrasin F, simaroubolide of *Picrasma quassioides*. Chem. Pharm. Bull. **19**, 2203 (1971).

36. — — — Stereostructure of picrasin C, simaroubolide of *Picrasma quassioides*. Chem. Pharm. Bull. **19**, 2211 (1971).

37. HOLLANDS, T. R., P. DE MAYO, M. NISBET, and P. CRABBE: Terpenoids XIII. The stereochemistry of chaparrin, chaparrol and related compounds. Canad. J. Chem. **43**, 2996 (1965).

38. KARTHA, G., and D. J. HAAS: The direct determination of the crystal structure of p-bromobenzoate-glaucarubin. J. Am. Chem. Soc. **86**, 3630 (1964).

39. LAWRIE, W., W. HAMILTON, F. S. SPRING, and S. H. WATSON: Triterpenoids, part. LIII. The constitution and stereochemistry of butyrospermol. J. Chem. Soc. (London) **1956**, 3272.

40. LAVIE, D., M. K. JAIN, and S. R. SHPAN-GABRIELITH: A locust phagorepellent from two melia species. Chem. Comm. **1967**, 910.

41. LE-VAN-THOI and NGUYÊN-NGOC-SUONG: Constituents of *Eurycoma longifolia* Jack. J. Org. Chem. **35**, 1104 (1970).

42. LONDON, E., A. ROBERTSON, and H. WORTHINGTON: Quassin and Neoquassin I. J. Chem. Soc. (London) **1950**, 3431.

43. MERRIEN, M.-A., and J. POLONSKY: The natural occurrence of melianodiol and its diacetate in *Samadera madagascariensis* (simaroubaceae): Model experiments on melianodiol directed towards simarolide. Chem. Comm. **1971**, 261.

44. MITCHELL, R. E., W. STÖCKLIN, M. STEFANOVIĆ, and T. A. GEISSMAN: Chaparrolide and castelanolide, new bitter principles from *Castela nicholsoni*. Phytochemistry **10**, 411 (1971).

45. MORON, J.: Etude de la biosynthèse des principes amers des simarubacées. Doctoral Thesis, Orsay, Paris, 1968.

46. MORON, J., et J. POLONSKY: Sur l'origine triterpènique des constituants amers des simarubacées. Tetrahedron Lett. **1968**, 385.

47. MORON, J., et J. POLONSKY: Incorporation spécifique de la L-[U-^{14}C] isoleucine dans le groupement hydroxy-2-methyl-2-butyroyle de la glaucarubinone. European J. Biochem. **3**, 488 (1968).

48. MORON, J., M.-A. MERRIEN et J. POLONSKY: Sur la biosynthèse des quassinoides de *Simaruba glauca* (simarubaceae). Phytochemistry **10**, 585 (1971).

49. MORON, J., J. RONDEST, and J. POLONSKY: Sur la biosynthèse des constituants amers des simarubacées. Experientia **22**, 511 (1966).

50. MOSS, G. P.: Some aspects of triterpene bitter principle biosynthesis. Planta medica, Suppl. **1966**, 86.

51. MURAE, T., T. IKEDA, T. TSUYUKI, T. NISHIHAMA, and T. TAKAHASHI: The structures of nigakilactones E and F. Bull. Chem. Soc. Japan **43**, 969 (1970).

52. — — — — — Bitter principles of *Picrasma ailanthoides* Planchon. Nigakihemicetals A and B, and nigakilactones G and H. Bull. Chem. Soc. Japan **43**, 3021 (1970).

53. — — — — — Bitter principles of *Picrasma ailanthoides* Planchon. Nigakihemiacetal C. Chem. Pharm. Bull. **18**, 2590 (1970).

54. — — — — — Nigakilactones: Stereostructure and nuclear Overhauser effects. Tetrahedron Lett. **42**, 3897 (1971).

55. MURAE, T., T. TSUYUKI, T. IKEDA, T. NISHIHAMA, T. MASUDA, and T. TAKAHASHI: Bitter principles of *Picrasma ailanthoides* Planchon. Nigakilactones A, B, C, D, E and F. Tetrahedron **27**, 1545 (1971).

56. NYBURG, S. C., G. L. WALFORD, and P. YATES: The Configuration of Glaucarubin. Chem. Comm. **10**, 203 (1965).

57. POLONSKY, J.: The structure of simarolide, the bitter principle of *Simarouba amara*. Proc. Chem. Soc. **1964**, 292.

58. — Les principes amers des simarubacées. Pflanzliche Bitterstoffe. Planta medica, Suppl. **1966**, 107.

59. POLONSKY, J., et N. BOURGUIGNON-ZYLBER: Etude des constituants de *Hannoa klaineana* (simarubacée): Chaparrinone et klaineanone. Bull. Soc. Chim. France **1965**, 2793.

60. POLONSKY, J., and Z. BASKEVITCH: On the constituents of *Perriera orientalis*. Unpublished results.

61. POLONSKY, J., Z. BASKEVITCH, B. C. DAS et J. MULLER: Sur les constituants amers du *Brucea amarissima*: Structures des brucéines D et E. C. R. Acad. Sc. Paris **267**, Série C, 1346 (1968).

62. POLONSKY, J., Z. BASKEVITCH, A. GAUDEMER et B. C. DAS: Constituants amers de *Brucea amarissima*, structures des brucéines A, B et C. Experientia **23**, 424 (1967).

63. POLONSKY, J., Z. BASKEVITCH et J. MULLER: Constituants amers du *Brucea amarissima*: Structure de la brucéine F. C. R. Acad. Sc. Paris **268**, Série C, 1392 (1969).

64. POLONSKY, J., CL. FOUQUEY et A. GAUDEMER: Etude de la glaucarubine I — structure et stéréochimie du glaucanol. Bull. Soc. Chim. France **1964**, 1818.

65. — — — Etude de la glaucarubine II. — Structure du glaucarubol et de la glaucarubine. Bull. Soc. Chim. France **1964**, 1827.

66. POLONSKY, J., et J. L. FOURREY: Constituants des Graines d'*Ailanthus altissima* Swingle. Structure de l'Ailanthone. Tetrahedron Lett. **1964**, 3983.

67. POLONSKY, J., et E. LEDERER: Note sur l'isolement de la diméthoxy-2,6-benzoquinone des écores et du bois de quelques simarubacées et méliacées. Bull. Soc. Chim. France **1959**, 1157.

68. SIM, K. Y., J. SIMS, and T. A. GEISSMAN: Constituents of *Brucea sumatrana*. Roxb. I. Brusatol. J. Org. Chem. **33**, 429 (1968).

69. STÖCKLIN, W., and T. A. GEISSMAN: A new bitter principle from *Brucea sumatrana* Roxb. Tetrahedron Lett. **1968**, 6007.

70. STÖCKLIN, W., and T. A. GEISSMAN: Glaucarubol and glaucarubol-15-isovalerate from *Castela nicholsoni*. Phytochemistry **9**, 1887 (1970).

71. STÖCKLIN, W., L. B. DE SILVA, and T. A. GEISSMAN: Constituents of *Holocantha emoryi*. Phytochemistry **8**, 1565 (1969).

72. STÖCKLIN, W., M. STEFANOVIĆ, and T. A. GEISSMAN: Amarolide, isolation from *Castela nicholsoni* Hook and revision of its structure. Tetrahedron Lett. **27**, 2399 (1970).

73. TRESCA, J.-P., L. ALAIS et J. POLONSKY: Constituants amers du *Quassia africana* Baill. (simarubacée). Simalikalactones A, B, C, D et simalikahemiacétal A. C. R. Acad. Sc. Paris **273**, Série C, 601 (1971).

74. VALENTA, Z., S. PAPADOPOULOS, and C. PODEŠVA: Quassin and neoquassin. Tetrahedron **15**, 100 (1961).

75. VALENTA, Z., A. H. GRAY, D. E. ORR, S. PAPADOPOULOS, and C. PODEŠVA: The stereochemistry of quassin. Tetrahedron **18**, 1433 (1962).

76. VIALA, B., et J. POLONSKY: Etude des constituants amers de *Soulamea Pancheri* (simarubacée). C. R. Acad. Sc. Paris **271**, Série C, 410 (1970).

77. WALL, M. E., and M. WANI: The isolation and structure of holacanthone, a potent experimental anti-tumor agent. In: Abstracts Book, 7th Int. Symp. Chem. Nat. Prod., IUPAC, Riga, **1970**, 614.

78. ZYLBER, J.: Sur la pyrolyse d'α-acetoxy-cétones. Pyrolyse du tétraacétyl-7,12,15,32-nor-16-glaucanol. Experientia **25**, 240 (1969).

79. ZYLBER, J.: Etude des constituants de l'écoye de *Samadera indica;* structures des Samaderines, B, C et D. Doctoral Thesis, Orsay, Paris, **1967.**

80. ZYLBER, J., et J. POLONSKY: Sur les constituants de l'ecorce de *Samadera indica;* structure de la samadérine B et de la samadérine C. Bull. Soc. Chim. France **1964**, 2016.

81. ZYLBER, N.: Etude des constituants amers des fruits de *Perriera madagascariensis* et d'*Hannoa klaineana* (simarubacées). Doctoral Thesis, Orsay, Paris, 1968.

82. BOURGUIGNON-ZYLBER, N., et J. POLONSKY: Etudes des constituants des écorces et du bois de *Perriera madagascariensis*. Isolement et structure d'un nouvel alcaloide indolique et d'un nouveau quassinoide. Bull. Chim. Therap. **6**, 396 (1970).

(Received January 24, 1972)

Die Ergochrome
(Physiologie, Isolierung, Struktur und Biosynthese)

Von B. FRANCK und H. FLASCH, Münster

Mit 9 Abbildungen

Inhaltsübersicht

I. Einleitung

Mutterkorn, das Dauermycel des auf Roggenähren wuchernden Fadenpilzes *Claviceps purpurea*, enthält eine Gruppe hellgelber, schwach saurer, optisch aktiver Farbstoffe. Obwohl die ersten Präparate dieser Farbstoffe schon vor fast 100 Jahren von DRAGENDORFF und Mitarb. (*14*) isoliert wurden, konnte deren Strukturaufklärung im wesentlichen erst in den letzten Jahren durchgeführt werden. Dies ist darauf zurückzuführen, daß für die Reindarstellung der schwer trennbaren, teilweise diastereomeren Farbstoffe leistungsfähige chromatographische Methoden entwickelt werden mußten (*1, 27, 29, 47*). Als sich gezeigt hatte, daß diese Farbstoffe eine Naturstoffgruppe aus mehr als 10 Vertretern mit sehr ähnlichen Strukturen bilden, wurde für sie der Sammelname *Ergochrome* und eine einfache Nomenklatur vorgeschlagen (*19, 25, 29*).

Das Vorkommen der Ergochrome ist nicht auf den Mutterkornpilz beschränkt. Von verschiedenen Forschungsgruppen wurden sie kürzlich auch aus anderen Pilzen und Flechten isoliert. Dabei stellte sich heraus, daß es sich auch bei den toxischen Inhaltstoffen zahlreicher Reis- und Maisschimmelpilze um Ergochrome handelt (*69, 79*). Die Toxizität der Ergochrome scheint an die von strukturell verwandten Stoffen aus Nahrungsmittelpilzen heranzureichen, die wegen ihrer Leberschäden verursachenden Wirkung neuerdings Besorgnis hervorrufen (*18*). Die biologische Aktivität der Ergochrome ist jedoch geringer als die der Alkaloide aus Mutterkorn, die vor allem von STOLL und Mitarb. (*52, 70*) aufgeklärt wurden. In welchem Umfang die Ergochrome an den seit Jahrhunderten bekannten akuten Vergiftungen nach Verzehren von Mutterkorn in unreinem Getreide beteiligt sind, ist noch nicht geklärt.

Die Bedeutung der Ergochrome liegt neben ihren neuartigen Strukturen und ihrer Toxizität vor allem darin, daß sie zur Entdeckung eines bisher unbekannten Biosyntheseweges führten (*30, 49*). Sie entstehen durch oxidative Ringöffnung des Anthrachinons Emodin (**101**), ein aufschlußreiches Biosyntheseprinzip, das danach auch für weitere Naturstoffe nachgewiesen werden konnte (*44*).

Die Gewinnung sterisch einheitlicher Ergochrome, die Aufklärung ihrer Strukturen und ihrer Biosynthese ist das Ergebnis von Bemühungen mehrerer Arbeitskreise. Das Ergoflavin wurde von FREEBURN (*41*) rein dargestellt, die Secalonsäuren A, B und C, die Ergochrysine A und B sowie die Ergochrome AD, BD, CD und DD von FRANCK und Mitarb. (*23—25, 27, 29*), die Secalonsäuren B und C, die Ergochrysine A und B sowie das Ergoxanthin unabhängig von DE MAYO und Mitarb. (*1*) und die Secalonsäure D von STEYN (*69*). Die genannten Arbeitskreise sowie WHALLEY und Mitarb. (*3, 4*) haben sich eingehend mit der Strukturermittlung befaßt. Die Biosynthese der Ergochrome untersuchten FRANCK und

Mitarb. sowie GRÖGER und Mitarb. gemeinsam (*26, 30—32, 49*). Die bisher vorliegenden Ergebnisse werden im nachstehenden Aufsatz beschrieben. Dabei sind die älteren Untersuchungen an Ergochrom-Gemischen entsprechend ihrem Aussagewert berücksichtigt.

II. Allgemeine Informationen über Ergochrome

1. Strukturvariation und Nomenklatur der Ergochrome

Die bisher isolierten und aufgeklärten Ergochrome leiten sich sämtlich von der Grundstruktur (**1**) ab. So haben Secalonsäure A, das nach Menge und Verbreitung wichtigste Ergochrom, und Ergoflavin die Strukturen (**2**) und (**3**) [FRANCK und Mitarb. (*21, 29*), WHALLEY und Mitarb. (*4, 53, 59*)]. Es sind dimere Tetra- und Hexahydro-xanthone mit einer Verknüpfung der Molekülhälften in o-Stellung zur phenolischen Hydroxygruppe.

(1)

(2)

Secalonsäure A

(3)

Ergoflavin

Nur 6 strukturelle Variationen der Molekülhälften ergeben in Verbindung mit den durch die dimere Konstitution ermöglichten Kombinationen die Strukturen der 12 bekannten Ergochrome. Die 6 Strukturvariationen lassen sich durch die Formeln (**4**) bis (**9**) darstellen.

A 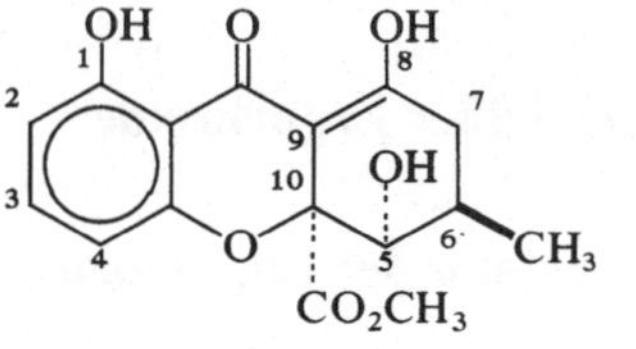

(**4**)

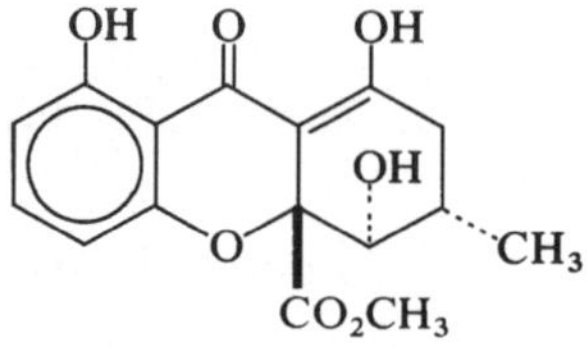 B

(**5**)

C 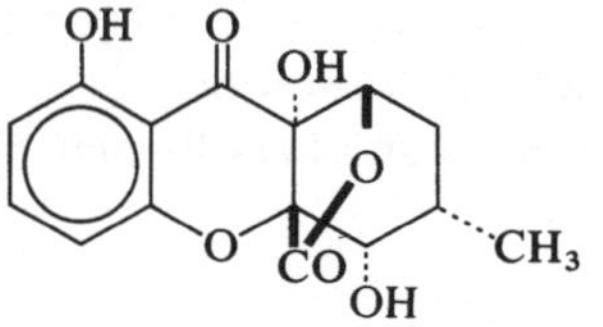

(**6**)

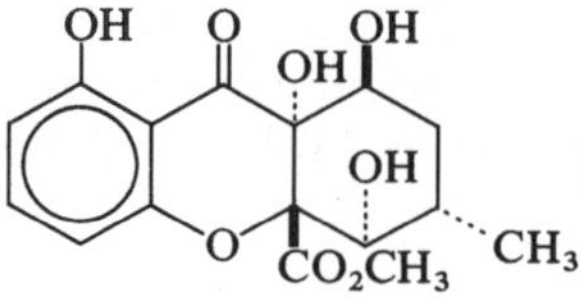 D

(**7**)

E 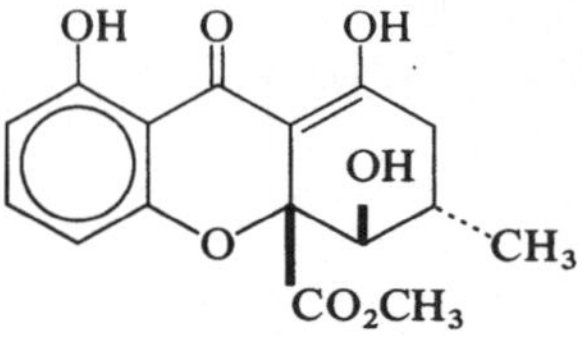

(**8**)

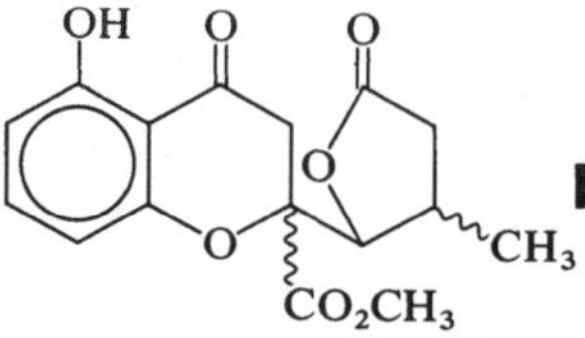 F

(**9**)

Von der Grundstruktur (**1**) leiten sie sich durch die folgenden formalen Umwandlungen ab:

a) Methylierung der Carboxylgruppe an C-10 und Variation der Konfigurationen an den C-Atomen 10, 5 und 6. Dies führt zu den Tetrahydroxanthonen (**4**), (**5**) und (**8**), den Molekülhälften der *Secalonsäuren* A, B, C und D. So bestehen die Secalonsäuren A (**2**), B und D aus einer Kombination von zwei gleichartigen Tetrahydro-xanthonen (**4**), (**5**) bzw. (**8**), während in Secalonsäure C eine unsymmetrische Kombination der Tetrahydro-xanthone (**4**) und (**5**) vorliegt.

b) Hydratisierung der Enoldoppelbindung und γ-Lactonringschluß liefert die Molekülhälfte (**6**) des symmetrischen Ergochroms *Ergoflavin*

Literaturverzeichnis: SS. 203—206

(3). Kombination von (6) mit den Molekülhälften der Secalonsäuren A und B (4, 5) ergibt die *Ergochrysine* A bzw. B.

c) Hydratisierung der Enoldoppelbindung und Methylierung der Carboxylgruppe an C-10. Das so abgeleitete Hexahydro-xanthon (7) ist die Molekülhälfte D der *Ergochrome* AD, BD, CD und DD, bei denen es sich um die vier möglichen dimeren Kombinationen von (7) mit (4), (5) und (6) handelt.

d) Säurespaltung der enolisierten β-Diketon-Gruppierung von Grundstruktur (1) und Lactonringschluß der so gebildeten γ-Hydroxysäure zu (9). Hierbei handelt es sich um eine Molekülhälfte des „seco-Ergochroms" *Ergoxanthin*. Sie ist im Ergoxanthin mit einem Hexahydro-xanthon von der Konstitution der Ergoflavinhälfte (6) kombiniert. In beiden Molekülhälften des Ergoxanthins sind jedoch die Konfigurationen der Chiralitätszentren noch ungeklärt.

Tabelle 1. *Bezeichnungen und Strukturen der Ergochrome*

Systematische Bezeichnung	Einzelname	Struktur
Ergochrom AA	Secalonsäure A	(4)—(4) = (2)
Ergochrom BB	Secalonsäure B	(5)—(5)
Ergochrom AB	Secalonsäure C	(4)—(5)
Ergochrom CC	Ergoflavin	(6)—(6) = (3)
Ergochrom AC	Ergochrysin A	(4)—(6)
Ergochrom BC	Ergochrysin B	(5)—(6)
Ergochrom AD	—	(4)—(7)
Ergochrom BD	—	(5)—(7)
Ergochrom CD	—	(6)—(7)
Ergochrom DD	—	(7)—(7)
Ergochrom EE	Secalonsäure D	(8)—(8)

Die Strukturen der Ergochrome lassen somit eine bemerkenswerte, biogenetisch begründete Systematik erkennen (Tabelle 1). Dies ermöglichte es, eine einfache Nomenklatur vorzuschlagen, durch welche die Struktur in leicht ersichtlicher Weise gekennzeichnet und die Einführung weiterer Einzelnamen für Naturstoffe dieses Strukturtyps vermieden werden kann (*27, 29*). Hiernach werden die Molekülhälften (4—9) der Verbindungen fortlaufend mit den Buchstaben A, B, C . . . und deren dimere Kombinationen als „Ergochrom AA", „Ergochrom AB" usw. bezeichnet. Voraussetzung dafür ist, daß die Strukturen einschließlich der relativen und absoluten Konfigurationen vollständig aufgeklärt sind. Wie Tabelle 2 zeigt, bilden die Ergochrome ein bemerkenswert vollständiges Sortiment dimerer Naturstoffe. Von den Molekülhälften A, B, C und D wurden sämtliche möglichen dimeren Kombinationen gefunden. Nur für die hier mit E bezeichnete Molekülhälfte (**8**) der Secalonsäure D wurden noch keine Kombinationen mit anderen Xanthon-Einheiten festgestellt.

Tabelle 2. *Bisher gefundene Kombinationen der Ergochromhälften*
*A, B, C, D und E (**4—8**)*

A	B	C	D	E	
AA	AB	AC	AD	—	A
	BB	BC	BD	—	B
		CC	CD	—	C
			DD	—	D
				EE	E

2. Isolierung und Trennung der Ergochrome

Bei der Gewinnung reiner Ergochrome bereitet nur die Auftrennung der Komponenten gewisse Schwierigkeiten. Die Isolierung der Ergochromgemische aus Mutterkorn und anderen Pilzmycelien ist dadurch sehr erleichtert, daß sie nur wenig polarer als die Pilzlipoide und Fettsäuren sind und daher gut mit organischen Lösungsmitteln extrahiert werden können. Andererseits ermöglicht ihr hohes Molekulargewicht und die dadurch bedingte geringe Löslichkeit eine einfache Abtrennung von den Lipoiden im Verdampfungsrückstand des Extraktionsmittels. Beim Aufnehmen des Rückstandes mit Petroläther bleiben nur die Ergochrome ungelöst zurück. Dieser bequeme Zugang zu den Rohfarbstoffpräparaten aus Mutterkorn kann als Ursache dafür angesehen werden, daß deren Isolierung schon frühzeitig [Dragendorff und Mitarb. (*14*)] und immer wieder [s. Zusammenstellung in (*21*)] nach Verfahren, die sich nur unwesentlich voneinander unterscheiden, beschrieben wurde. Einige dieser Präparate erhielten die Bezeichnungen *Sclererythrin* (*14*), *Sclerocristallin* (*14*), *Scleroxanthin* (*14*), *Scerojodin* (*14*), *Ergochrysin* (*55*), *Secalonsäure* (*58*), *Ergoxanthin* (*77*), *Ergoflavin* (*41*) und *Chrysergonsäure* (*72*). Beim Nacharbeiten aller Darstellungsverfahren und chromatographischer Untersuchung der Präparate stellte sich heraus, daß sie mit Ausnahme des Ergoflavins sämtlich Gemische aus bis zu 13 Komponenten darstellen [Gottschalk (*47*)]. Nur Ergoflavin konnte von Freeburn (*41*) schon 1912 ohne Anwendung der Chromatographie durch einfache Kristallisation rein dargestellt werden, da es nicht von Diastereomeren begleitet war. Auch einige der übrigen Namen wurden später zur Kennzeichnung sterisch einheitlicher Ergochrome beibehalten (*1, 3, 27, 29*).

a) Aus Mutterkorn

Der Ergochromgehalt von gesammeltem Wildmutterkorn sowie von Zuchtmutterkorn, das durch künstliche Beimpfung gewonnen wurde, schwankt je nach Herkunft zwischen 0,1 und 0,6%. Ähnlich liegen auch die Ausbeuten für Trockenmycel aus der Züchtung des Mutterkornpilzes auf künstlicher Nährlösung (Tabelle 3). Für die Isolierung des Ergochrom-Gemisches aus Mutterkorn und Trockenmycel von *Claviceps purpurea* hat sich nachstehendes Verfahren am besten bewährt [FRANCK und Mitarb. (29)].

Mit Petroläther entfettetes, gemahlenes Mutterkorn wird während 24 Stunden in 10-proz. Weinsäure suspendiert. Anschließend läßt man den Extrakt ablaufen, wäscht mehrmals mit Methanol und engt die vereinigten Extrakte weitgehend ein. Hierbei kristallisiert ein Teil der Ergochrome (etwa 20%) aus. Nachdem auf diese Weise die basischen und polaren Inhaltstoffe größtenteils aus dem Mutterkorn abgetrennt sind, wird dieses mehrfach mit Chloroform ausgekocht und die Chloroformauszüge eingedampft. Der zunächst noch ölige Rückstand wird nacheinander mit Petroläther und Eisessig ausgezogen und danach zusammen mit der ersten Fraktion aus Chloroform umkristallisiert. So erhält man ein hellgelbes, kristallisiertes Ergochrom-Gemisch, das anschließend durch Säulen- oder Schichtchromatographie aufgetrennt werden kann.

b) Aus Flechten

Den ersten Nachweis für das Auftreten von Ergochromen in anderen Mikroorganismen außer dem Mutterkornpilz erbrachten 1968 YOSIOKA und Mitarb. *(80)*. Sie konnten zeigen, daß es sich bei dem schon länger bekannten gelben Farbstoff *Entothein* aus der Flechte *Parmelia entotheiochroa Hue.* um Secalonsäure A handelt. Der Gehalt der getrockneten Flechte an Secalonsäure A ist mit 0,72% höher als im Mutterkorn. Außerdem isolierten die Autoren Secalonsäure A auch aus den drei verwandten Flechten *P. perisidians Nyl., P. aurulenta Tuck.* und *P. subaurulenta Nyl.* (Tabelle 3). Von zwei weiteren Arbeitskreisen [WACHTMEISTER *(76)*; OLLIS und Mitarb. *(62)*] wurden Ergochrome auch in den Flechten *Laurera purpurina* und *Buellia canescens* aufgefunden. Für die Gewinnung der Ergochrome aus Flechten eigneten sich ähnliche Extraktionsverfahren wie beim Mutterkorn.

c) Aus Schimmelpilzen

Von STEYN *(69)* wurde 1970 die Isolierung der Secalonsäure D aus *Penicillium oxalicum* und damit erstmalig das Vorkommen eines Ergochroms in einem Schimmelpilz beschrieben. Es handelt sich hierbei um einen auf Getreidenahrungsmitteln verbreiteten Pilz, der eingehend untersucht wurde, als sich herausstellte, daß seine Kulturen bei Ratten, Mäusen und jungen Enten akute Toxikosen, insbesondere Lebernekrosen, verursachen [Carlson und Mitarb. (10)]. Aus Reinkulturen von *Penicillium oxalicum* (Stamm M-555) auf sterilisiertem Maismehl wurde Secalonsäure D mit 0,3—0,5% Ausbeute (bez. auf getrocknetes verschimmeltes Maismehl) neben dem neuen Alkaloid Oxalin $C_{24}H_{25}N_5O_4$ gewonnen. Nach dem Ergebnis der biologischen Untersuchungen ist allein Secalonsäure D für die Giftigkeit des *Penicillium oxalicum* verantwortlich. Die Bildung von Secalonsäure D und Oxalin ist nicht an ein Wachstum des Pilzes auf Maismehl gebunden. Sie fand sich auch bei der Züchtung von 5 verschiedenen Stämmen dieses Organismus auf einem modifizierten Czapek-Medium mit 3% Maisquellextrakt (corn steep liquor) (69).

Die Isolierung von Secalonsäure A, des Antipoden der Secalonsäure D aus 57 Stämmen des Reisschimmelpilzes *Aspergillus ochraceus* gelang kürzlich Yamazaki und Mitarb. (79). Die Pilzstämme gewannen die Autoren aus Proben von verschimmeltem Reis, die im Bereich der Präfekturen Chiba und Miyagi in Japan gesammelt wurden. Aus *Aspergillus ochraceus* hatten van der Merwe und Mitarb. (73, 74) zuvor die Ochratoxine A, B und C (10)—(12) isoliert, deren Strukturen sich beträchtlich von der der Secalonsäure A (2) unterscheiden. Der Gehalt von Kulturen des *Aspergillus ochraceus* auf sterilisiertem Reis an Secalonsäure A wird mit 0,06% angegeben (79). Da dieser Wert auf feuchten Reis bezogen ist, entspricht er etwa dem für Secalonsäure D in *Penicillium oxalicum*. Im Gegensatz zum letztgenannten Pilz scheint *Aspergillus ochraceus* keine Ergochrome auf künstlicher Nährlösung zu bilden (26).

(10) R′=H, R=Cl
(11) R′=R=H
(12) R′=C_2H_5, R=Cl

Die Isolierung der Ergochrome aus den Schimmelpilzkulturen erfolgte in etwa ähnlicher Weise wie aus Mutterkorn. Die Kulturen wurden mit einem polaren Lösungsmittel ($CHCl_3$/CH_3OH bzw. Essigester) erschöpfend extrahiert, der Verdampfungsrückstand des Extraktes Verteilungen zwischen Chloroform und wäßrigen Phasen unterworfen und das auf diese Weise abgetrennte Ergochrom aus Chloroform umkristalliert.

d) Übersicht über das Vorkommen der Ergochrome

Die Entdeckung von Ergochromen in zahlreichen weiteren Mikroorganismen außer dem Mutterkornpilz *Claviceps purpurea* allein im Zeitraum von 1968 bis 1971 berechtigt zu der Annahme, daß Vertreter dieser neuen Naturstoffgruppe weit verbreitet sind. Dies ist auch deswegen nicht unwahrscheinlich, weil die Ergochrome bei ihrer Biosynthese aus dem in Pilzen sehr häufig auftretenden Emodin (**101**) hervorgehen (*49*). Die bisher veröffentlichten Angaben über die Isolierung eindeutig identifizierter Ergochrome aus Pilzkulturen und Flechten sind in Tabelle 3 zusammengefaßt.

Tabelle 3. *Vorkommen von Ergochromen in Pilzen und Flechten*

Organismus	Herkunft	Lit.	Ergochromgehalt (% bez. auf Trockensubst.)	Ergochrome (vgl. Tabelle 1)
Claviceps purpurea (Mutterkorn)	Österreich	(*29*)	0,28	AA, BB, AB
Claviceps purpurea (Mutterkorn)	Portugal	(*24*)	0,6	AA, BB, AB, AC, BC, CC, AD, BD, CD, DD
Claviceps purpurea (Mutterkorn)	Deutschland	(*24*)	0,18	AA, BB, AB, AC, BC, CC, AD, BD, CD, DD
Claviceps purpurea (Mutterkorn)	Kanada	(*29*)	0,42	AA, Spuren von BB und AB
Parmelia entotheiochroa	Japan	(*80*)	0,72	AA (Secalonsäure A)
Parmelia perisidians	Japan	(*80*)	0,36	AA (Secalonsäure A)
Parmelia aurulenta	Japan	(*80*)	0,005	AA (Secalonsäure A)
Parmelia subaurulenta	Japan	(*80*)	0,17	AA (Secalonsäure A)
Penicillium oxalicum (5 Stämme)	Südafrika	(*69*)	0,3—0,5	EE (Secalonsäure D)
Aspergillus ochraceus (57 Stämme)	Japan	(*79*)	etwa 0,3%	AA (Secalonsäure A)

e) Trennungsmethoden

Bei den aus Mutterkorn isolierten Ergochrom-Gemischen ließen deren gutes Kristallisationsvermögen, vergleichsweise scharfe Schmelzpunkte und ähnliches chemisches Verhalten der Komponenten lange Zeit keine Zweifel an der Einheitlichkeit dieser Präparate aufkommen. Das änderte sich erst, als von FRANCK und Mitarb. (*21, 27, 37*) chromatographische Methoden zur Untersuchung der Ergochrompräparate herangezogen wurden. Die zunächst verwendete Papierchromatographie erforderte wegen geringer Polarität und Löslichkeit der Ergochrome extreme

Lösungsmittelsysteme, wie n-Butanol/Triäthylamin/Pyridin/Wasser (4 : 4 : 1 : 6) auf Papier, das mit pH-6-Citratpuffer imprägniert ist oder Tetralin/90-proz. Eisessig (1 : 1) (*37*). Den entscheidenden Fortschritt ermöglichte bald nach ihrem Bekanntwerden die Schichtchromatographie. Es konnten die folgenden Trennungssysteme entwickelt werden, mit denen sich auch die komponentenreichsten Ergochrom-Präparate sowohl analytisch im Dünnschichtchromatogramm, als auch präparativ auf Schichtplatten oder in Säulen auftrennen lassen (*23, 27*):

1. Chloroform/Pentanon-2 (9 : 1) auf Kieselgel G, das mit Oxalsäure imprägniert ist,
2. Chloroform/Methanol (100 : 1) auf weinsaurem Kieselgel.

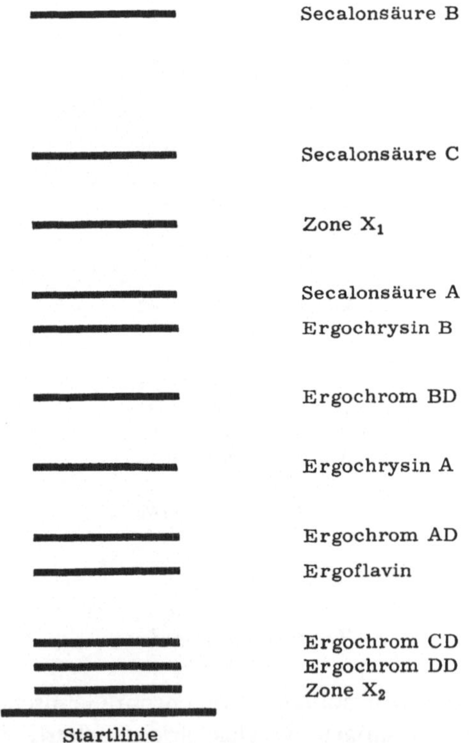

Abb. 1. Dünnschichtchromatogramm (schematisch) eines Ergochromgemisches (Chloroform/Methanol = 100 : 1 auf weinsaurem Kieselgel G) [Aus: Angew. Chem. *81*, 269 (1969)]

    Abb. 1 zeigt ein mit dem Trennungssystem 2. erhaltenes Dünnschichtchromatogramm der Ergochrome aus Mutterkorn. Die Reihenfolge der Ergochrome ist gut reproduzierbar und läßt eine systematische Ab-

Literaturverzeichnis: SS. 203—206

hängigkeit der Wanderungsgeschwindigkeit von der Struktur erkennen. Dies ermöglicht bequeme Identifizierungen neu isolierter Ergochrome durch Mitlaufenlassen von 2 Vergleichssubstanzen und Informationen über die Struktur von Umwandlungsprodukten. So nimmt die Wanderungsgeschwindigkeit der *symmetrischen* Ergochrome (vgl. Tabelle 1) in nachstehender Reihenfolge von oben nach unten ab:

Ergochrom BB (Secalonsäure B)
Ergochrom AA (Secalonsäure A)
Ergochrom CC (Ergoflavin)
Ergochrom DD.

Ganz entsprechend verhalten sich die durch Kombination verschiedener Molekülhälften gebildeten *unsymmetrischen* Ergochrome:

Ergochrom AB (Secalonsäure C)
Ergochrom BC (Ergochrysin B)
Ergochrom BD
Ergochrom AC (Ergochrysin A)
Ergochrom AD
Ergochrom CD.

3. Biologische Aktivität der Ergochrome

Nachdem die Ergochrome kürzlich als Toxine von Nahrungsmittelpilzen erkannt wurden, begann man mit einer eingehenden Untersuchung ihrer biologischen Aktivität. Die bisher vorliegenden Toxizitätsbestimmungen an Mäusen kennzeichnen sie als Pilztoxine von mittlerer Stärke.

Tabelle 4. *Toxizität der Ergochrome und anderer Pilztoxine*

Pilztoxin		Toxizität (LD_{50} an Mäusen in mg/kg)				Lit.
		peroral	intra-peritoneal	intravenös	subcutan	
Secalonsäure A	(2)	>250	<50			(79)
Secalonsäure D	(13)		42			(69)
Aflatoxin B_1	(14)	7,2*	6,0*			(9)
Sterigmatocystin	(15)	60	166			(63)
Luteoskyrin	(16)	221		6,6	147	(60)
Rubratoxin B	(17)	350	0,35			(63)
Citrinin	(18)				35	(63)

* Diese Werte wurden abweichend von den übrigen an Ratten bestimmt.

In Tabelle 4 sind die Toxizitäten an Mäusen bei verschiedener Applikationsform für die Secalonsäuren A (2) und D (13) mit denen einiger typischer Pilztoxine (14)—(18) verglichen. Die praktisch gleiche Toxizität der Antipoden Secalonsäure A und Secalonsäure D (**2, 13**) bei intraperitonealer Injektion läßt vermuten, daß deren Wirkung nicht durch die hydroaromatischen Ringe mit den Chiralitätszentren, sondern durch das aromatische Bichromonyl-System bestimmt ist. Dementsprechend zeigen die Secalonsäuren in diesen Molekülteilen gewisse Strukturähnlichkeit mit den Pilztoxinen Aflatoxin B_1 (**14**) und Sterigmatocystin (**15**). Spezifische Wirkungen reiner Ergochrome wurden bisher nicht geprüft, jedoch verursachte Maismehl, das von dem Secalonsäure-D-Bildner *Penicillium oxalicum* befallen war, bei Versuchstieren ausgedehnte Lebernekrosen (*69*).

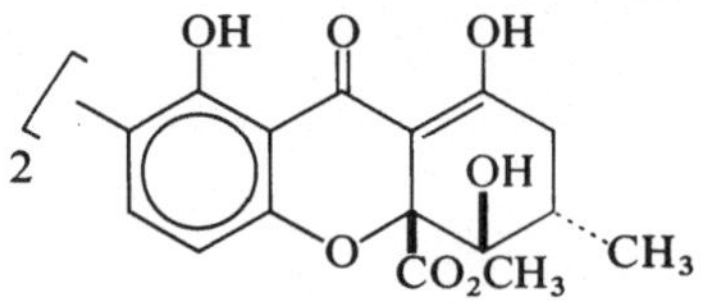

(**13**)

Secalonsäure D

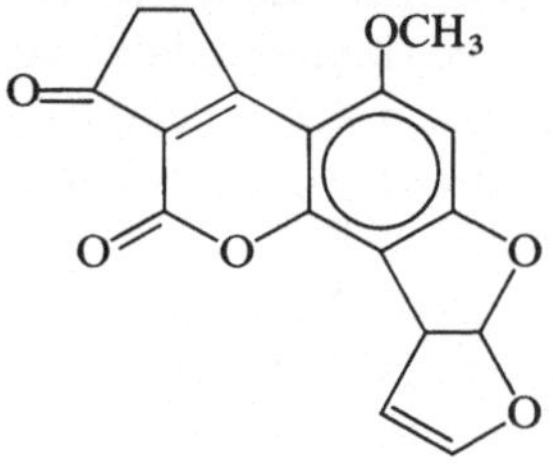

(**14**)

Aflatoxin B_1

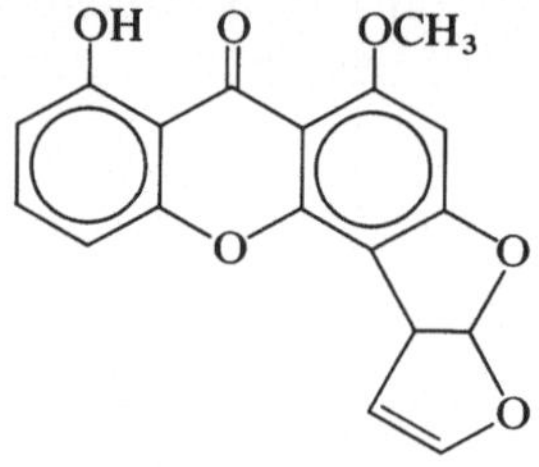

(**15**)

Sterigmatocystin

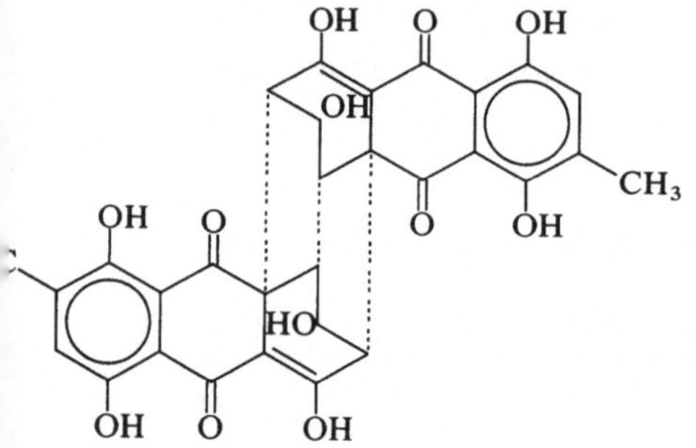

(**16**)

Luteoskyrin

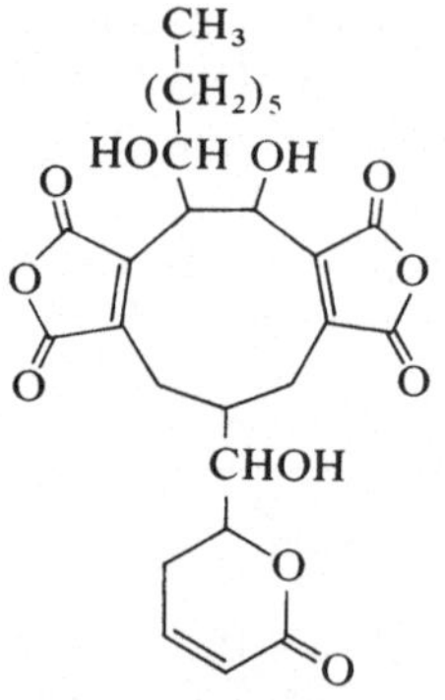

(**17**)

Rubratoxin

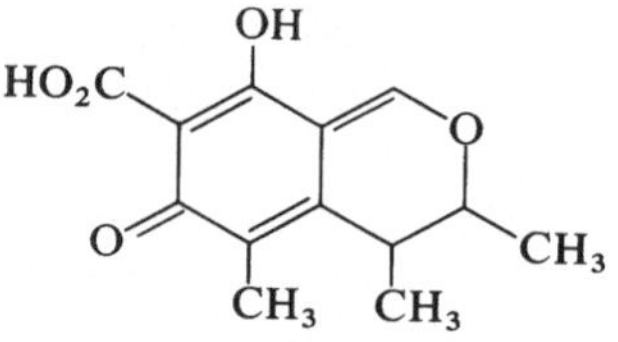

(**18**)

Citrinin

Eine antibiotische Wirkung gegenüber Staphylokokken in der Verdünnung von 1 : 40000 bis 1 : 100000 hatten Stoll und Mitarb. (*72*) für reine, aber nicht einheitliche Ergochrompräparate aus Mutterkorn beschrieben. Ähnlich schwache Wirkung beobachteten Yamazaki und

Mitarb. (*79*) im Scheibchentest mit *Bacillus subtilis* und *Piricularia oryzae* sowie FRANCK und Mitarb. (*33*) im tuberculostatischen und bacteriostatischen *(Staph. aureus)* Test an reinen Secalonsäuren.

III. Struktur und Konfiguration der Ergochrome

1. Methoden zur Strukturbestimmung

Die Ergochrome sind hellgelbe, optisch aktive Verbindungen mit massenspektroskopisch gut bestimmbaren Molekulargewichten zwischen 610 und 638. Alle bisher bekannten Vertreter dieser Naturstoffgruppe enthalten je 2 C-Methylgruppen, je 2 Carboxymethyl- oder γ-Lactongruppen, je 2 chelierte Arylketon-Gruppen und 4—8 acetylierbare Hydroxygruppen. Sie zeigen teilweise beträchtliche Unterschiede in der Acidität und ihrer Beständigkeit gegenüber milder Alkalieinwirkung. Bei allen Ergochromen besteht die Strukturbestimmung aus vier Teilabschnitten: a) Aromatisches Grundgerüst, b) Struktur von Ring C [vgl. (**1**)], c) Verknüpfung der Molekülhälften, d) relative und absolute Konfiguration der Chiralitätszentren von Ring C. Die für a)—c) verwendeten Methoden waren bei allen Ergochromen ähnlich und sollen daher im folgenden Abschnitt zusammenfassend behandelt werden. Die Bestimmung der Konfigurationen erforderte teilweise individuelle Untersuchungen, deren Beschreibung anschließend erfolgt.

a) Das aromatische Grundgerüst

Alle Ergochrome geben bei der Alkalischmelze oder beim Erhitzen mit $Ba(OH)_2$ im Einschlußrohr 2,4,2',4'-Tetrahydroxybiphenyl (**19**) und Methylbernsteinsäure. Dieser zunächst von BERGMANN (*7*) sowie von STOLL (*72*) für Ergochrom-Gemische beschriebene Abbau, der sich im 50-mg-Maßstab durchführen läßt, wurde später an reinen Ergochromen bestätigt (*16, 37, 47*). Das farblose (**19**) stellt nur einen Teil des konjugierten Systems der Ergochrome dar. Die Elektronenspektren der Ergochrome stimmen weitgehend mit denen der hellgelb gefärbten Modellverbindungen (**22**) bis (**24**) überein (*4, 29, 57*). Dabei ist das Spektrum der Secalonsäure A (**2**) und der übrigen drei Secalonsäuren dem des 5-Hydroxy-2-methylchromons (**22**) und das des Ergoflavins (**3**) dem des 5-Hydroxy-2-methylchromanons (**24**) etwas ähnlicher (Abb. 2 und 9). Für die Secalonsäuren kann ein Chromanon mit exoständiger Doppelbindung (**25**) in Betracht gezogen werden, da deren langwelliges Absorptionsmaximum (340 mμ) zwischen denen des 5-Hydroxychromons (**22**) und des 5-Hydroxychromanons (**24**) liegt.

OH

OH

OH

OH

(19)

OH O

(20)

oder

OH

(21)

OH O

R

CH₃

(22) R=H
(23) R=OH

OH O

CH₃

(24)

OH O

(25)

Hiernach enthalten alle Ergochrome als Grundgerüst ein dimeres 5-Hydroxychromanon, in dem die beiden Chromanon-Einheiten so angeordnet sind, daß daraus bei der Alkalischmelze 2,4,2′,4′-Tetrahydroxybiphenyl entstehen kann. Solche Anordnungen liegen im 8,8′- (20) und im 6,6′-Bichromanonyl (21) vor. Die Molekülhälften sind darin jeweils in p- bzw. o-Stellung zu den freien phenolischen Hydroxygruppen miteinander verknüpft. Die Frage, welches Bichromanonyl den Ergochromen entspricht bzw. wie deren Molekülhälften miteinander verbunden sind, erwies sich als recht schwierig, so daß sie für einige Ergochrome erst kürzlich endgültig geklärt werden konnte (40, 53). Diese Untersuchungen werden in einem späteren Abschnitt behandelt.

Aufschlußreich für die Struktur des aromatischen Grundgerüstes der Ergochrome sind weiterhin die NMR-Spektren (Abb. 3) (3, 29, 37). Im Falle der symmetrischen Ergochrome, wie z. B. Secalonsäure A (2) und B (64), zeigen sie im Aromatenbereich bei den δ-Werten 6,72 und 7,63 (Pyridin-d₅) Dubletts, die nach Intensität und Aufspaltung (J=8,5) zwei Paaren o-ständiger, aromatischer Protonen entsprechen. Dies zeigt, daß die Bichromanonyl-Grundgerüste (20, 21) der Ergochrome nur am Heterocyclus über die C-Atome 2 und 3 mit den restlichen Strukturteilen verbunden sein können. Die Resonanzabsorption der aromatischen Protonen gibt ferner Auskunft darüber, ob das Ergochrom aus zwei gleich-

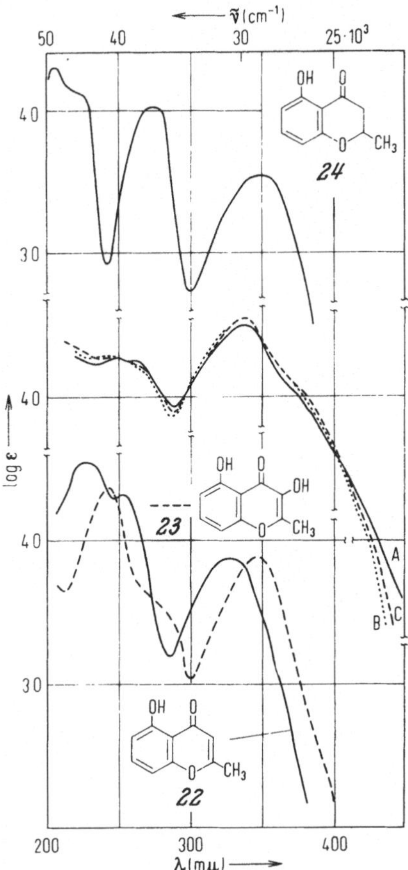

Abb. 2. Elektronenspektren der Secalonsäuren A (2), B (64) und C (26) sowie von 5-Hydroxy-2-methyl-chromon (22), 3,5-Dihydroxy-2-methylchromon (23) und 5-Hydroxy-2-methylchromanon (24) in Methanol. [Aus: Chem. Ber. 99, 3842 (1966)]

artigen Molekülhälften besteht, d. h. symmetrisch ist oder nicht. So kann ein unsymmetrisches Ergochrom daran erkannt werden, daß es im Aromatenbereich nicht zwei, sondern vier Dubletts von jeweils der halben Intensität aufweist. Es ist bemerkenswert, daß diese Signalaufspaltung sogar schon dann deutlich eintritt, wenn sich die beiden Molekülhälften, wie z. B. bei der Secalonsäure C (26), nur durch entgegengesetzte Konfiguration an einigen Chiralitätszentren von Ring C unterscheiden (Abb. 3). Diese Aussage der NMR-Spektren vereinfacht die Strukturermittlung symmetrischer Ergochrome außerordentlich, da bei ihnen so verfahren werden kann, als bestünden sie aus nur einer Molekülhälfte.

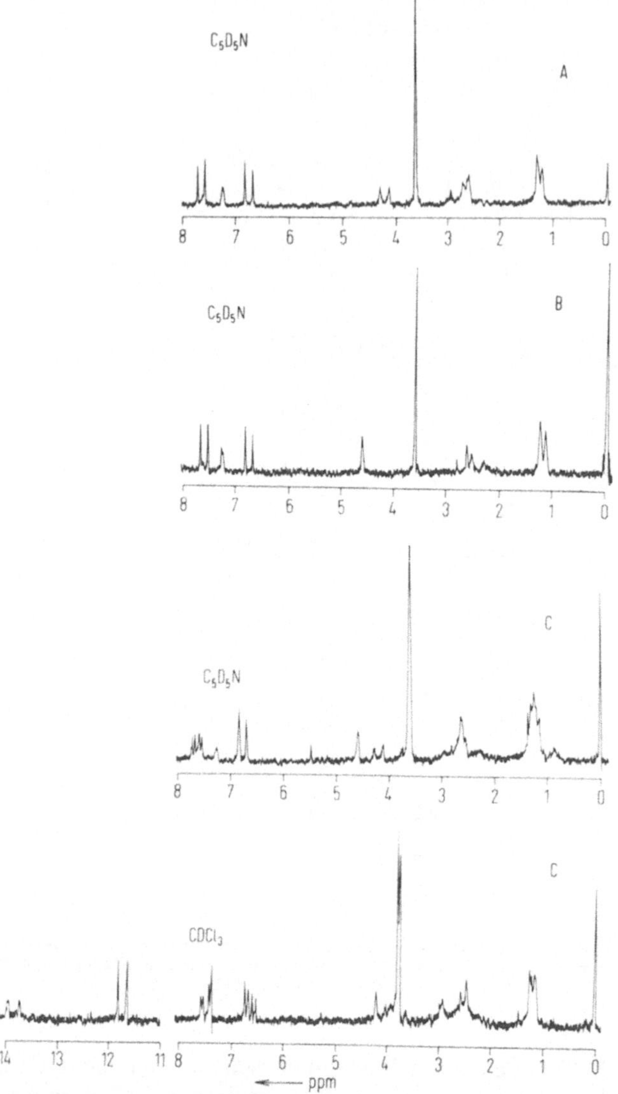

Abb. 3. NMR -Spektren der Secalonsäuren A (2), B (64) und C (26) in Deuteropyridin sowie von Secalon-säure C in Deuterochloroform. [Aus: Chem. Ber. *99*, 3842 (1966)]

b) *Struktur von Ring C*

Die weitere Strukturaufklärung der Ergochrome vollzog sich zunächst an Verbindungen, die nach Aussage der NMR-Spektren aus zwei gleich-artigen Molekülhälften aufgebaut sind (*4, 28, 29*). Es handelte sich

Literaturverzeichnis: SS. 203—206

$$(26)$$

Secalonsäure C

hierbei um Secalonsäure A (2) und um Ergoflavin (3). Aufgrund der durch Massenspektren und Elementaranalyse genau festgelegten Summenformeln und der nachgewiesenen funktionellen Gruppen ergeben sich in Verbindung mit dem Chromanon-Grundgerüst die Partialformeln (27) und (28) für die Molekülhälften dieser Ergochrome.

(27)

$C_7H_8O_4$

1 $-OH$

1 $=\overset{\overset{\textstyle OH}{|}}{C}-$

1 $\geq CH-CH_3$

1 $-CO_2CH_3$

(28)

$C_6H_6O_4$

2 $-OH$

1 $\geq CH-CH_3$

1 $-CO-O-$

Partialformeln der Molekülhälften von Secalonsäure A und Ergoflavin

Für Secalonsäure A (27) folgt aus der H-Atomzahl des restlichen Molekülteiles, daß dieser einen Ring und 2 Mehrfachbindungen enthalten muß. Bei den Mehrfachbindungen handelt es sich um die CO_2CH_3- und die Enol-Gruppe. Für den Ring stehen nach Abzug von CH_3 und CO_2CH_3 noch 4 C-Atome entsprechend einem angegliederten 6-Ring zur Verfügung. Die C-Methylgruppe befindet sich an einem tertiären C-Atom, da ihr Signal im NMR-Spektrum bei $\delta = 1{,}26$ als Dublett ($J = 6{,}0$ Hz) erscheint.

Der restliche Strukturteil der Ergoflavin-Molekülhälfte muß 2 Ringe und eine Mehrfachbindung enthalten. Die Mehrfachbindung und ein Ring können der γ-Lactongruppe zugeschrieben werden. Auch hier sind nach Abzug der funktionellen Gruppen vom restlichen Strukturteil noch 4 C-Atome für einen angegliederten 6-Ring übrig. Die C-Methylgruppe steht nach Aussage des NMR-Spektrums ebenfalls an einem tertiären C-Atom.

Aufschlußreich für die *Anordnung* der funktionellen Gruppen an Ring C ist der Befund, daß alle Ergochrome sowohl bei der Alkalischmelze (*16, 72*) als auch bei der Oxidation mit Kaliumpermanganat (*16, 23, 29*) Methylbernsteinsäure ergeben. Daraus ging hervor, daß die Ergochrome einen hydroaromatischen Ring mit dem Strukturteil (**29**) enthalten.

$$\begin{array}{ccccc} & O & CH_3 & & O \\ & \| & | & & \| \\ >C & - CH & - CH_2 & - & C< \end{array}$$

(**29**)

Die weitere Strukturbestimmung von Ring C erfolgte bei Secalonsäure A und Ergoflavin wegen ihres unterschiedlichen chemischen Verhaltens und der darauf basierenden Abbaumöglichkeiten auf getrennten Wegen,

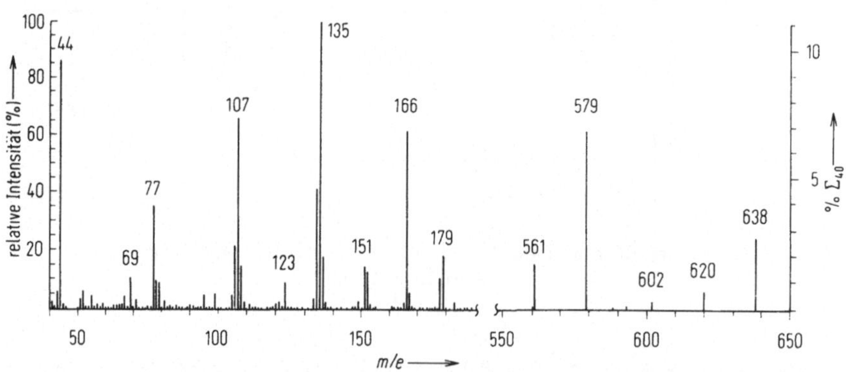

Abb. 4. Massenspektrum der Secalonsäure A (**2**) bei 35 eV. [Aus: Chem. Ber. **99**, 3842 (1966)]

die nacheinander mit ihren wesentlichen Zügen behandelt werden.

Secalonsäure A zeigt ebenso wie die anderen diastereomeren Secalonsäuren im Massenspektrum (Abb. 4) bei herabgesetzter Elektronenenergie (35 eV) drei intensive Fragmentionen bei m/e 166, 135 und 107 (*4*). Sie lassen sich mit der Abspaltung eines Kresotinsäure-methylesters

$$(2) \longrightarrow \text{[Ring]} \begin{cases} -CO_2CH_3 \\ -OH \\ -CH_3 \end{cases} \xrightarrow{-OCH_3} \text{[Ring]} \begin{cases} -CO \\ -OH \\ -CH_3 \end{cases} \xrightarrow{-CO} \text{[Ring]} \begin{cases} -OH \\ -CH_3 \end{cases}$$

m/e 166	m/e 135	m/e 107
(30)	**(31)**	**(32)**

(30) und dessen weiterer Fragmentierung erklären (*28, 29*). Hieraus geht hervor, daß bei der Secalonsäure A die Substituenten von Ring C so angeordnet sind, daß sie dessen leichte Abspaltung und Aromatisierung ermöglichen. Durch einen Ozonabbau in Essigester konnten dann die Positionen der Substituenten von Ring C festgelegt werden (*28, 29*). Neben R-(+)-Methylbernsteinsäure **(33)** entstand dabei in geringerer Menge α-Hydroxy-β-methyl-glutarsäure **(34)** in Form ihres γ-Lactons sowie α-Keto-β-methyl-glutarsäure **(35)**.

Bedenkt man, daß sich die Enol-OH-Gruppe der Secalonsäure A in β-Stellung zur Chromanon-Carbonylgruppe befinden muß, so können die C-Ketten der Ozonabbausäuren nur in der im Formelschema angeordneten Weise im Ring C vorgelegen haben. Für die Molekülhälfte der Secalonsäure A folgt hieraus die Struktur **(36)**. Eine alternative Anordnung der sekundären Hydroxygruppe an C-7 statt an C-5 könnte zwar ebenfalls die Bildung von **(34)** beim Ozonabbau erklären, doch stünde dies im Widerspruch zu dem Befund, daß die Secalonsäure keine α-Ketol-Reaktion mit Wismut(III)-acetat gibt (*37, 50, 64*). Bei dem durch Elektronenbeschuß im Massenspektrometer abgespaltenen Kresotinsäuremethylester handelt es sich daher um den Ester der 5-Hydroxy-3-methyl-benzoesäure **(37)**. **(37)** wurde auch bei der Alkalischmelze (*7*) und Pyrolyse (*3*) von Ergochrysin- bzw. Secalonsäuregemischen gewonnen. Außerdem ließ sich **(37)** auch aus Ergochrysin A **(61)** (*3*), Secalonsäure C **(26)** (*31*) und Ergochrysin B **(80)** (*31*) nach einem von WHALLEY und Mitarb. (*3*) beschriebenen zweistufigen Abbau mit Acetanhydrid/Pyridin und konz. Bromwasserstoffsäure erhalten. Hierbei wird zunächst unter den Bedingungen einer energischen Acetylierung Ring C aromatisiert und darauf das Benzophenon **(38)** nach Art einer Retro-Friedel-Crafts-Reaktion gespalten.

Beim *Ergoflavin* (*3*) basiert die endgültige Aufklärung der Struktur von Ring C auf einer Röntgenstrukturanalyse des Tetra-O-methylergoflavin-di-p-jodobenzoats **(39)** durch SIM und Mitarb. (*59*). Für die Molekülhälfte des Ergoflavins ergab sich daraus die Struktur **(41)**. Aus der Bildung des Tetra-O-methylergoflavins **(40)** bei der Methylierung mit Dimethylsulfat/ K_2CO_3 in Aceton (*16*) ging hervor, daß eine der beiden Hydroxygruppen von Ring C keine Acidität besitzt. Sie liegt im Tetramethyläther **(40)** frei

(33)   (34)   (35)

O_3

(36)

$e^{\ominus}$ (35 eV) Ac_2O/Pyr.

48% HBr

(37)   (38)

vor und muß neben der C-Methyl-Gruppe stehen, weil sie mit Jones-Reagens (*8*) zur Ketogruppe oxidiert werden kann und dies eine Verschiebung des NMR-Signals der Methylgruppe nach kleinerem Feld bewirkt (*4*). Hiernach konnte unabhängig von der Röntgenstrukturanalyse für Ergoflavin die Teilstruktur (**29**) zu (**42**) erweitert werden.

Die Molekülhälften (**5, 7, 8**) der übrigen Ergochrome mit Ausnahme der des Ergoxanthins (**85**) stimmen hinsichtlich Ring C bis auf abweichende Konfigurationen entweder mit der einer Secalonsäure überein (**5, 8**) oder sie können wie im Falle von (**7**) auf einfache Weise in die des Ergoflavins umgewandelt werden.

Literaturverzeichnis: SS. 203—206

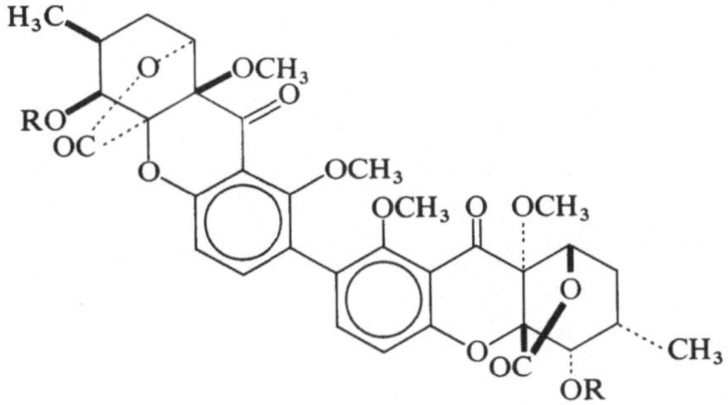

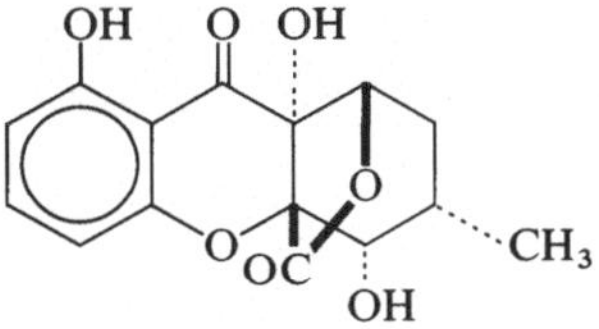

(39) R= —O—CO—⟨ ⟩—J

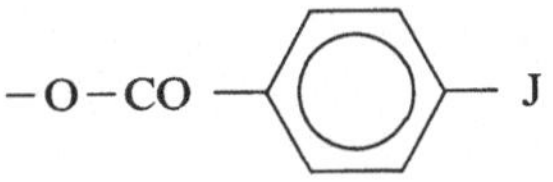

(40) R=H

(41)

$$-CHOH-\overset{\underset{\mid}{CH_3}}{CH}-CH_2-\overset{\overset{O}{\|}}{C}<$$

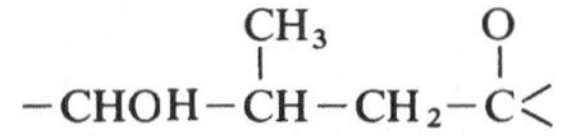

(42)

c) Verknüpfung der Molekülhälften

Da bei der Alkalischmelze der Ergochrome 2,4,2′,4′-Tetrahydroxy-biphenyl (19) entsteht, müssen deren Molekülhälften über die Ringe A in o- oder p-Stellung zur freien phenolischen Hydroxygruppe miteinander verknüpft sein. Das ist bei den Grundgerüsttypen (43)—(45) mit 2,2′-, 4,4′- und 2,4′-Verknüpfung der Fall. Um festzustellen, welcher Typ den Ergochromen zugrunde liegt, wurden von verschiedenen Arbeitskreisen folgende Verfahren benutzt:

a) Röntgenstrukturanalyse (59)
b) Abbau und NMR-Untersuchung der Methylierungsprodukte (53)
c) pKa-Messungen (23, 40).

Die vielfach (1, 3, 22, 37, 69) in derselben Absicht verwendete Farbreaktion nach GIBBS (46, 57) hat sich als unzuverlässig erwiesen (40, 53).

Durch die von SIM und Mitarb. (59) ausgeführte Röntgenstrukturanalyse des Tetra-O-methylergoflavin-di-p-jodobenzoats (39) ist für Ergoflavin die Verknüpfung der Molekülhälften an den C-Atomen 2 und 2′ bestimmt. Mit Hilfe dieser Kenntnis konnten WHALLEY und Mitarb. (53) den Verknüpfungstyp der Secalonsäure D (46) ermitteln, indem sie sowohl dieses Ergochrom als auch Ergoflavin (3) in dasselbe Abbauprodukt (51) überführten. Hierzu wurde zunächst 1,1′,9,9′-Tetra-O-methylergo-

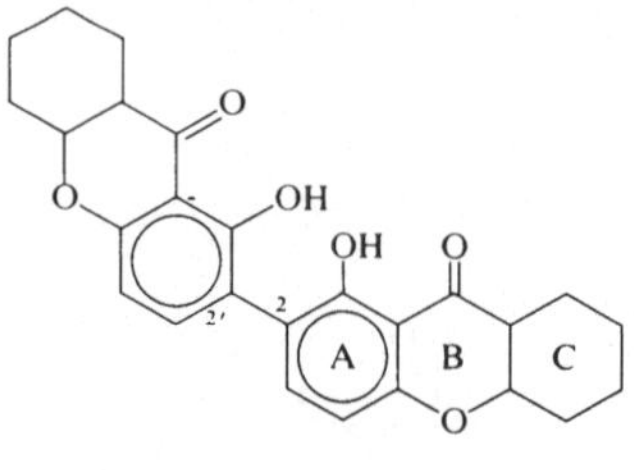

(43) 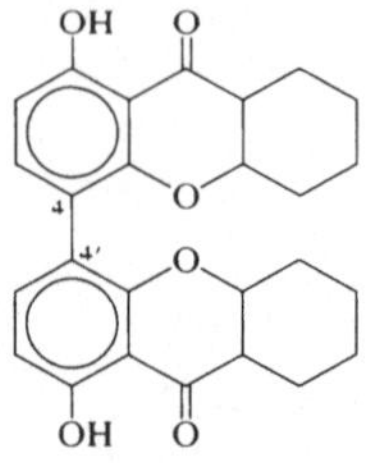   (44) 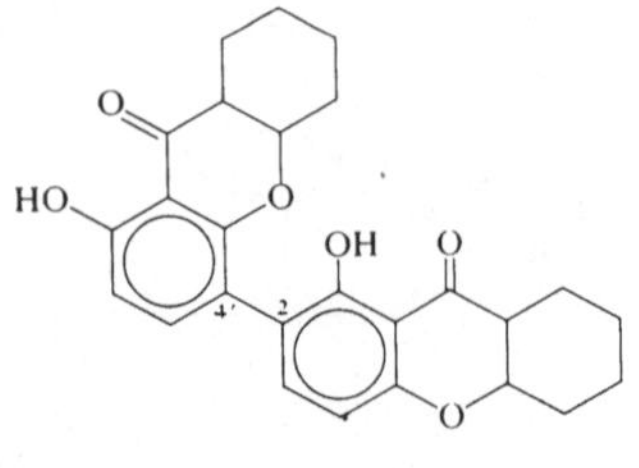   (45)

flavinon (**49**) durch Kochen mit verd. NaOH zum 3,3′-Diacetyl-4,4′-di-
hydroxy-2,2′-dimethoxybiphenyl (**50**) abgebaut und dieses anschließend
mit Diäthylsulfat/K_2CO_3 in Aceton zum 3,3′-Diacetyl-4,4′-diäthyl-2,2′-di-
methoxybiphenyl (**51**) veräthert. Für einen analogen Abbau der Secalonsäure D stellten die Autoren aus 8,8′-Di-O-methyl-secalonsäure D (**47**)
(*69*) durch Nachmethylierung mit Dimethylsulfat/K_2CO_3 in Aceton und
anschließender Hydrolyse der Enoläthergruppen an C-8,8′ zunächst
1,1′-Di-O-methylsecalonsäure (**48**) dar. Diese oxidierten sie zur entsprechenden *Secalononsäure* (**52**) und unterwarfen sie wie beim entsprechenden Ergoflavin-Derivat dem alkalischen Abbau. Das erhaltene

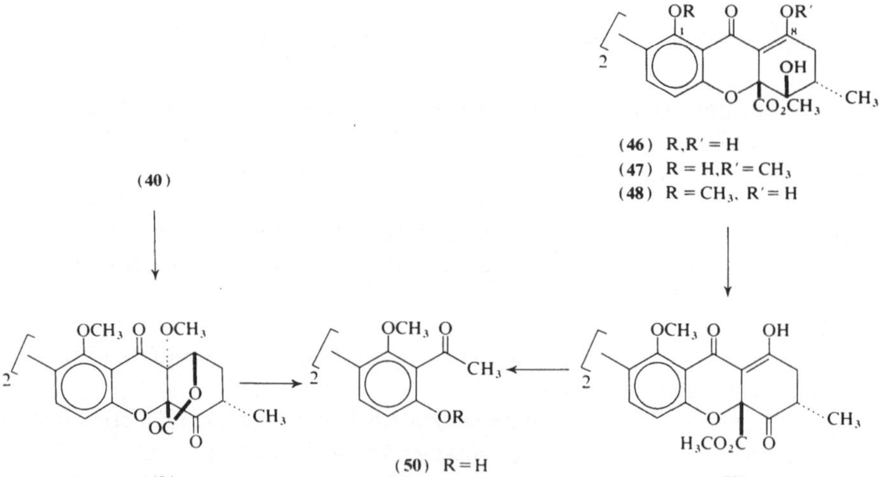

(40)

(46) R,R′ = H
(47) R = H,R′ = CH₃
(48) R = CH₃, R′ = H

(49)

(50) R = H
(51) R = C₂H₅

(52)

Biphenyl-Derivat (**50**) war nach Äthylierung mit dem Produkt (**51**) aus
Ergoflavin identisch.
Nach diesem Abbau haben Secalonsäure D (**46**) und damit auch ihr
Antipode, die Secalonsäure A (**2**), wie Ergoflavin (**3**) 2,2′-Verknüpfung
der Molekülhälften.

Literaturverzeichnis: SS. 203—206

Aufschlußreich für den Verknüpfungstyp sind ferner die NMR-Signale der aromatischen Methoxygruppen methylierter Ergochrome (*53*). So geben sich Methoxygruppen, die bei 3,3'-Diacetylbiphenylen (**50, 51, 53, 54**) und Bianthrachinonylen (**55**) (*15*) in o-Stellung zur Biphenylachse angeordnet sind, durch eine Verschiebung ihres NMR-Signals um etwa 0.4 ppm nach höherem Feld („Upfieldshift") zu erkennen (Tabelle 5). Dieselbe Verschiebung zeigen auch die aromatischen Methoxygruppen der Methylierungsprodukte von Ergoflavin (**3**), Ergochrysin A (**61**), Secalonsäure D (**46**), Secalonsäure A (**2**) und Secalonsäure C (**26**) (Tabelle 5), woraus ebenfalls hervorgeht, daß diese Ergochrome übereinstimmend dem 2,2'-Verknüpfungstyp (**43**) angehören.

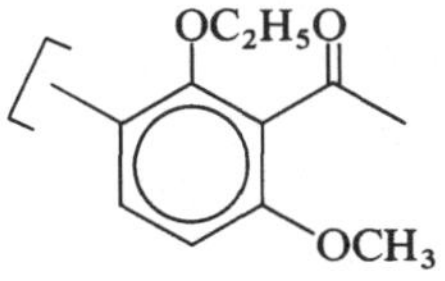

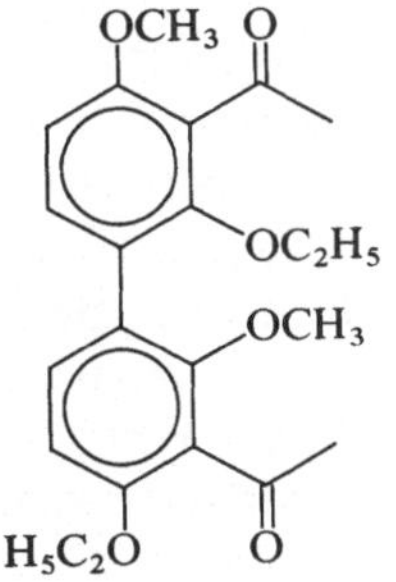

Tabelle 5. *Methoxyl-NMR-Signale (δ-Werte) von Biphenyl-Derivaten in [²H] Chloroform (15, 53)*

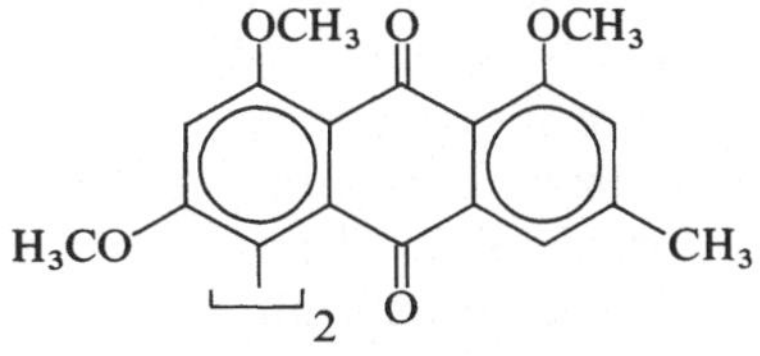

	„o-OCH₃"	„p-OCH₃"
3,3'-Diacetyl-4,4'-dihydroxy-2,2'-dimethoxybiphenyl (**50**)	3,55	—
3,3'-Diacetyl-4,4'-diäthoxy-2,2'-dimethoxybiphenyl (**51**)	3,47	—
3,3'-Diacetyl-2,2'-diäthoxy-4,4'-dimethoxybiphenyl (**53**)	—	3,85
3,3'-Diacetyl-2,4'-diäthoxy-2',4-dimethoxy-biphenyl (**54**)	3,46	3,87
Hexamethylskyrin (**55**)	3,69	4,05
1,1',9,9'-Tetra-O-methyl-ergoflavin (**40**)	3,58	—
1,1',9-Tri-O-methyl-ergochrysin A	3,57, 3,63	—
1,1'-Di-O-methylsecalonsäure D (**48**)	3,62	—
1,1'-Di-O-methylsecalonsäure A	3,62	—
1,1'-Di-O-methylsecalonsäure C	3,57, 3,60	—

Eine weitere Möglichkeit zur Bestimmung der Verknüpfungspositionen geben die pK-Werte der Ergochrome (*23, 40*). Biphenylderivate mit zwei Hydroxygruppen in o-Stellung zur Biphenylachse sind nämlich durch eine gegenüber anderen Phenolen beträchtlich erhöhte Acidität gekennzeichnet. Erklärt wird dies mit einer Stabilisierung des Monoanions durch eine starke Wasserstoffbrücke (**56**) (*61a*). Ergochrome mit einer

(**56**)

Verknüpfung der Molekülhälften in den o-Stellungen zu den phenolischen Hydroxygruppen (**43**) sollten daher bei der pK-Titration ebenfalls eine saure Gruppe zeigen. Das Ergebnis der pK-Messungen, die wegen begrenzter Löslichkeit der Ergochrome in 80-proz. wäßrigem Dimethylformamid durchgeführt wurden, ist in Tabelle 6 zusammengestellt. Die in diesem Lösungsmittel bestimmten scheinbaren pK-Werte liegen um etwa 2 pK-Einheiten höher als in Wasser. Ihr Aussagewert für Vergleichszwecke ist dadurch nicht beeinträchtigt. Wie aus dem Vergleich von 2,2'-Dihydroxybiphenyl (**57**), 4,4'-Dihydroxybiphenyl (**58**) und 2,2',4,4'-Tetrahydroxybiphenyl (**59**) hervorgeht, übertrifft die Acidität einer o,o'-

Tabelle 6. *pK-Werte von Ergochromen und Vergleichsverbindungen in 80-proz. wäßrigem Dimethylformamid (23, 40)*

Substanz	pK_{DMF}-Werte (Anzahl der titrierten sauren Gruppen eingeklammert)	
	o,o'-Dihydroxybiphenyl-Gruppierung	Andere saure Gruppen (Enol u. a.)
2,2'-Dihydroxybiphenyl (**57**)	9,4 (1)	
4,4'-Dihydroxybiphenyl (**58**)		>14
2,2',4,4'-Tetrahydroxybiphenyl (**59**)	9,8 (1)	>14
Benzoesäure		7,8 (1)
5-Hydroxy-2-methylchromon (**22**)		>14
Ergoflavin (**3**)	9,4 (1)	
Ergochrysin A (**61**)	10,9 (1)	8,3 (1)
Ergochrysin B (**80**)	11,1 (1)	8,8 (1)
Secalonsäure A (**2**)		8,7 (2)
Secalonsäure B (**64**)		9,0 (2)
Secalonsäure C (**26**)		8,7 (2)
Secalonen A (**60**)	12,3 (1)	
Secalonen B	12,7 (1)	
Secalonen C	12,5 (1)	

Literaturverzeichnis: SS. 203—206

Dihydroxybiphenyl-Gruppierung die der anderen phenolischen Hydroxygruppen um mehr als 4 pK_{DMF}-Einheiten. Die pK-Werte der nicht acidifizierten Phenolgruppen sowie der chelierten, wie z. B. von (**22**), liegen in diesem Lösungsmittel oberhalb der Erfassungsgrenze bei pK 14.

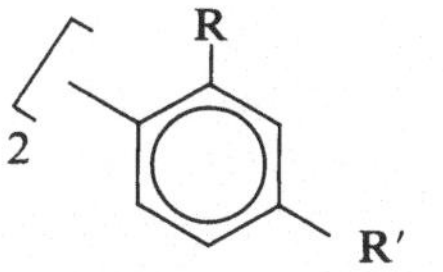

(**57**) R = OH, R' = H
(**58**) R = H, R' = OH
(**59**) R, R' = OH

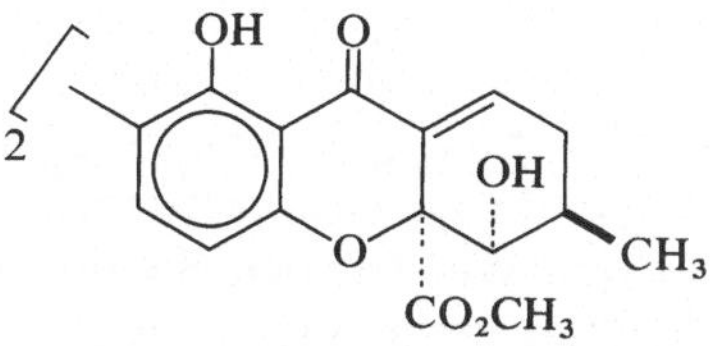

(**60**)

Secalonen A

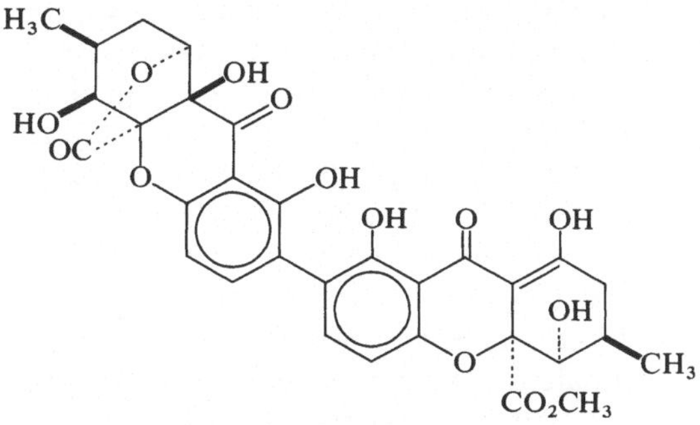

(**61**)

Ergochrysin A

Erwartungsgemäß hat Ergoflavin (**3**) eine saure Gruppe vom gleichen pK-Wert wie 2,2'-Dihydroxybiphenyl (**57**), woraus hervorgeht, daß sich die zur Ausbildung der Wasserstoffbrücke im Monoanion erforderliche Konformation [vgl. (**56**)] auch in den größeren Ergochrom-Molekülen einstellen kann. Beim Ergochrysin A (**61**) und Ergochrysin B (**80**) stellt die Enolgruppe der Secalonsäure-Molekülhälften die am stärksten saure Gruppe mit pK_{DMF}-Werten von 8,3 bzw. 8,8 dar. Beide Ergochrysine haben zwei weitere pK_{DMF}-Werte bei 10,9 und 11,1, die das Vorliegen einer o,o'-Dihydroxybiphenyl-Gruppierung anzeigen (*23*).

Bei der Titration der Secalonsäuren A, B und C (**2, 64, 26**) in wäßrigem Dimethylformamid können nur die pK-Werte der Enolprotonen bei 8,7—9,0 und darüber hinaus keine sauren Gruppen bestimmt werden. Dies ist einer der Befunde, die FRANCK und Mitarb. (*29*) zu der später korrigierten (*40, 53*) Annahme führten, daß die Secalonsäuren A, B und C dem 4,4'Verknüpfungstyp (**44**) entsprächen. Im Falle einer 4,4'-Verknüpfung wäre bei den Secalonsäuren keine saure o,o'-Dihydroxybiphenyl-Gruppierung vorhanden. Zweifel an dieser Schlußfolgerung aus den pK-Titrationen der Secalonsäuren ergaben sich aus einer Diskussion der pK-Werte der Ergochrysine. Bei diesen Ergochromen ist die Acidität der o,o′-Dihydroxybiphenyl-Gruppierung gegenüber der des Ergoflavins um mehr als 1,5 Einheiten vermindert. Dies kann damit erklärt werden, daß der Dissoziation der o-Hydroxygruppe die der stärker sauren Enol-Gruppierung vorausgeht. Wenn dies zutrifft, dann sollte die Acidität einer o-Hydroxygruppe, der Secalonsäuren, noch wesentlich stärker herabgesetzt sein, da deren Proton unter Überwindung erheblicher elektrostatischer Wechselwirkungen erst dann abgespalten werden kann, wenn bereits zwei Enolatanion-Gruppen vorliegen. Der pK_{DMF}-Wert einer o,o′-Dihydroxy-biphenyl-Gruppierung der Secalonsäuren könnte deshalb so hoch liegen, daß er in dem verwendeten Titrationssystem nicht mehr bestimmt wird. Die Richtigkeit dieser Annahme und damit der 2,2′-Verknüpfungstyp (**43**) für die Secalonsäure konnte Franckowiak (*40*) durch Darstellung und Titration von Secalonsäure-Derivaten, denen die Enol-Gruppe fehlt, beweisen. Solche Derivate, die „Secalonene", wie z. B. (**60**) (s. u.), können in drei Reaktionsschritten aus den Secalonsäuren dargestellt werden (*6, 40*). Die Secalonene A (**60**), B und C enthalten jeweils nur eine saure Gruppe mit den pK_{DMF}-Werten 12,3, 12,7 und 12,5 (Tabelle 6), die o,o′-Dihydroxybiphenyl-Gruppierungen zugeschrieben werden muß. Die gegenüber Ergoflavin bei den Secalonenen um 3 pK-Einheiten herabgesetzte Acidität ist darauf zurückzuführen, daß die o-Hydroxygruppen auch eine starke Wasserstoffbrücke zum benachbarten Chromoncarbonyl bilden. Anders als beim Ergoflavin (**3**) ist die Akzeptoreigenschaft der Carbonylgruppe von (**60**) nicht durch eine zweite Hydroxygruppe vermindert.

Durch drei voneinander unabhängige Methoden ist somit gesichert, daß in allen bisher bekannten Ergochromen die Molekülhälften in 2,2′-Stellung entsprechend (**43**) miteinander verknüpft sind.

2. Methoden zur Konfigurationsbestimmung und Struktur der Secalonsäuren A, B und D sowie des Ergoflavins

Die Ergochrome enthalten jeweils sechs, acht und zehn Chiralitätszentren, die sich in deren Ringen C auf die C-Atome 5, 6, 8, 9 und 10 verteilen (**62**). Bei allen Ergochromen mit Ausnahme des Ergoxanthins (**85**) konnten die relativen und absoluten Konfigurationen aufgeklärt werden [Franck und Mitarb. (*6, 21, 23, 24, 28, 29*), Sim und Mitarb. (*59*), Steyn (*69*) sowie Whalley und Mitarb. (*4*)]. Dies erfolgte im Falle des Ergoflavins durch eine Röntgenstrukturanalyse und bei den übrigen Ergochromen mit Hilfe von chemischen und spektroskopischen Methoden. Obwohl es sich bei diesen Untersuchungen um die Festlegung der Konfigurationen von insgesamt 86 Chiralitätszentren handelte, konnten sie

(62)

mit vergleichsweise geringem Aufwand durchgeführt werden. Das ist vor allem dadurch ermöglicht, daß die 11 bisher vollständig aufgeklärten Ergochrome aus nur 5 verschiedenen Molekülhälften aufgebaut sind und diese zusammen nicht mehr als 19 Chiralitätszentren aufweisen. Für 5 davon, nämlich für die des Ergoflavins (3) ergab sich die Konfiguration aus der Röntgenstrukturanalyse. Für weitere 5 Chiralitätszentren in den Molekülhälften der D-Ergochrome (7) ließ sich die Konfiguration durch eine einfache chemische Überführung in Ergoflavin ermitteln (s. u.). Somit verblieb als Kern der stereochemischen Untersuchungen mit chemischen und spektroskopischen Methoden die Konfigurationsbestimmung der Molekülhälften der vier Secalonsäuren A, B, C und D an den C-Atomen 6, 5 und 10. Über die hierzu verwendeten Methoden wird nachstehend berichtet.

a) Konfiguration der C-Methylgruppe an C-6

Das Chiralitätszentrum C-6, das die C-Methylgruppe trägt, kann bei allen Ergochromen unter Erhaltung seiner Konfiguration durch $KMnO_4$-Oxidation oder Ozonolyse als Methylbernsteinsäure (33) abgespalten werden (3, 4, 23, 29, 69). Dabei beträgt die Ausbeute 15—20%. Bei der durch Alkalischmelze von Ergochromen erhaltenen Methylbernsteinsäure wurde keine optische Aktivität festgestellt (16, 72). Nach oxidativem Abbau der einzelnen Ergochrome wurde sowohl R-(+)-, S-(−)- als auch race-

Tabelle 7. *Konfigurationen der beim oxidativen Abbau von Ergochromen erhaltenen Methylbernsteinsäure* (33)

Ergochrom	Konfiguration der Methylbernsteinsäure	Lit.
Secalonsäure A (2)	R − (+)	(29)
Secalonsäure B (64)	S − (−)	(29)
Secalonsäure D (65)	S − (−)	(69)
Ergoflavin (3)	S − (−)	(4)
Ergochrysin A (61)	R, S − (±)	(3, 23)
Ergochrysin B (80)	S − (−)	(23)

178   B. Franck und H. Flasch:

mische Methylbernsteinsäure erhalten (Tabelle 7). Es treten somit sämtliche möglichen Kombinationen der Konfiguration an C-6 bei den Ergochromen auf.

b) Konfiguration der sekundären Hydroxygruppe an C-5

Zur Bestimmung der absoluten Konfiguration an C-5 der Ergochrome dienten — abgesehen von der beim Ergoflavin durchgeführten Röntgenstrukturanalyse — NMR-Spektren und Ozonabbau (*21, 23, 28, 29, 69*). Dabei ergab sich die absolute Konfiguration der Hydroxylgruppe aus deren relativer Anordnung zum benachbarten Chiralitätszentrum an C-6. Zur Vereinfachung wurden diese stereochemischen Untersuchungen zu-

(63)

(64)

(65)

(66)

Literaturverzeichnis: SS. 203—206

nächst an Ergochromen vorgenommen, bei denen es sich nach den NMR-Spektren um symmetrische Moleküle handelt (s. o.), deren Molekülhälften also auch hinsichtlich der Konfiguration identisch sind. Diese Ergochrome können so untersucht werden, als bestünden sie nur aus einer Molekülhälfte.

Im Falle der Secalonsäuren A (63) und D (65) ließ sich die Konfiguration der Chiralitätszentren an C-5 folgendermaßen festlegen (29, 69). Die in [^{2}H$_5$]Pyridin gemessenen NMR-Spektren (vgl. Abb. 3) zeigen für das C-5-Proton Dubletts bei 4,19 ppm (J = 10,5 Hz) (Tabelle 8). Aus der Kopplungskonstanten folgt, daß die Protonen an C-5 und C-6 trans-diaxial stehen und dementsprechend die sekundäre Hydroxygruppe an C-5 trans-diäquatorial zur benachbarten C-Methylgruppe angeordnet ist. Von den beiden möglichen Konformationen (63a) und (63b) für Ring C von Secalonsäure A liegt somit die zweite vor. Da Secalonsäure D am C-Methylzentrum entgegengesetzte Konfiguration hat, kommt ihr die spiegelbildisomere Konformationsformel (65a) zu (29) [die Spiegelbildisomerie von (63b) und (65a) wird deutlich, sobald man eine der Formeln in der Papierebene um 180° dreht].

Tabelle 8. NMR-Signale (ppm) und Kopplungskonstanten (Hz) der
C-5-Protonen einiger Ergochrome in [^{2}H$_5$] Pyridin

Ergochrom	C-5	$J_{5,6}$	C-5′	$J_{5,6}$
Secalonsäure A (63)	4,19 (d)	10,5	4,19 (d)	10,5
Secalonsäure B (64)	4,59 (s)	—	4,59 (s)	—
Secalonsäure C (26)	4,20 (d)	10,5	4,60 (s)	—
Secalonsäure D (65)	4,19 (d)	10,5	4,19 (d)	10,5
Ergoflavin (66)	4,60 (s)	—	4,60 (d)	—
Ergochrysin A (61)	4,19 (d)	10,5	4,60 (s)	—
Ergochrysin B (80)	4,58 (s)	—	4,58 (s)	—

Bei der *Secalonsäure B* (64) zeigt sich das C-5-Proton mit einem Singulett bei 4,59 ppm (Abb. 3, Tabelle 8). Wegen des Fehlens einer erkennbaren Spinaufspaltung (J < 2 Hz) kommen für die Protonen an C-5 und C-6 nur die Anordnungen *diäquatorial, axial-äquatorial* und *äquatorial-axial* mit Diederwinkeln von jeweils 60° [KARPLUS (56)] in Betracht. Von diesen drei Möglichkeiten entfällt die erste, da sie wegen der sterisch ungünstigen Lage der Substituenten an den C-Atomen 5 und 6 in das Konformere mit diaxialen Protonen umklappen würde. Somit verbleiben für den Ring C von Secalonsäure B die Konformationsformeln (64a) und (64b) mit *cis*-Konfiguration der C-5-Hydroxygruppe zur Methylgruppe an C-6. Von diesen beiden energetisch nahezu gleichwertigen Konformationen dürfte (**64a**) überwiegen, da das C-5-Proton wegen Übereinstimmung der NMR-Signale (Tabelle 8) wahrscheinlich ebenso wie im Ergoflavin (**66a**) äquatorial steht (*29*).

Die *Ergochrysine A* und *B* zeigen für das C-5-Proton NMR-Signale wie sie zu erwarten sind, wenn diese beiden Ergochrome eine Kombination einer Ergoflavin-Hälfte (**66**) mit jeweils einer Molekülhälfte von Secalonsäure A (**63**) bzw. Secalonsäure B (**64**) darstellen (*23*) (Tabelle 8).

Für *Secalonsäure A* (**63**) konnte die Konfiguration der Hydroxygruppe an C-5 zusätzlich durch Ozonabbau bestimmt werden (*29, 54*). Wie bereits im Zusammenhang mit der Strukturbestimmung von Ring C der Ergochrome erwähnt, ließ sich nach Ozonolyse von Secalonsäure A (**63**) u. a. α-Hydroxy-β-methylglutarsäure (**34**) in Form ihres γ-Lactons (**67**) isolieren. Konfigurationsänderung an C-5, kenntlich am Auftreten von Diastereomeren des Lactons (**67**), erfolgte dabei nicht. Infolgedessen müßte bei *trans*-Stellung der Substituenten an den C-Atomen 5 und 6

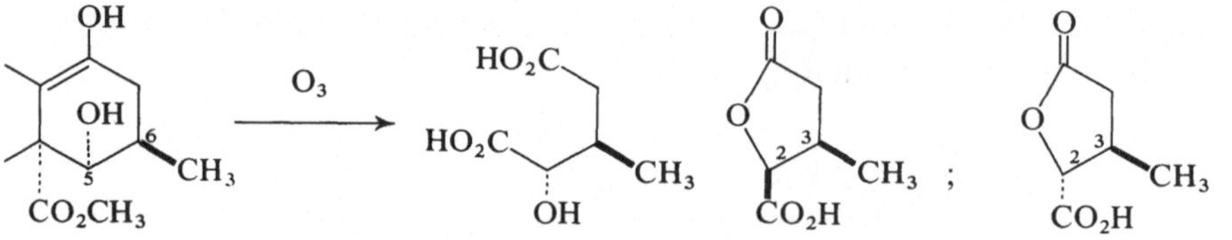

der Secalonsäure A (**63**) das Abbaulacton die Konfigurationsformel (**67**) mit Carboxyl- und C-Methylgruppe in *cis*-Anordnung haben. Das ist, wie aus dem NMR-Spektrum des Abbaulactons hervorgeht, der Fall. So gibt dessen C-2-Proton ($\delta = 4{,}97$ ppm) mit dem C-3-Proton eine Spinkopplung von $J = 7{,}6$ Hz in guter Übereinstimmung mit dem für den Diederwinkel von 0° bei *cis*-Anordnung berechneten Wert ($J = 8{,}2$ Hz) [Karplus (*56*)]. Diese Konfigurationsbestimmung des Abbaulactons mit Hilfe der NMR-Spektroskopie wurde auch durch Synthese und vergleichende Untersuchung der diastereomeren Lactone (**67**) und (**68**) bestätigt (*54*).

c) Konfiguration am Ringverknüpfungszentrum C-10

Die Ermittlung der absoluten Konfiguration an C-10 der Ergochrome ist durch zwei günstige Umstände erleichtert. Einmal bestimmt die Carboxylgruppe an C-10 als angulärer Substituent an einem polycyclischen Keton bei Rotationsdispersion und Circulardichroismus das Vorzeichen des Cotton-Effektes [Djerassi (*12, 13*)]. Zum zweiten zeigen alle Ergochrome mit einer oder zwei Secalonsäure-Molekülhälften im Massenspektrum ein Fragment, das durch Abspaltung der CO_2CH_3-Gruppe gebildet wird und dessen Intensität in charakteristischer Weise von der sterischen Hinderung dieser Gruppe abhängt. Dies ermöglichte die Festlegung der absoluten Konfiguration an C-10 der Ergochrome mit Hilfe von Rotationsdispersion, Circulardichroismus und Massenspektrometrie (*21, 23, 29, 69*).

Tabelle 9. *Vorzeichen der Cotton-Effekte einiger Ergochrome im Bereich des längstwelligen Absorptionsmaximums*

Ergochrom	$\lambda_{max}(\mu m)$	ORD (Chloroform)	CD (Dioxan)
Secalonsäure A (**63**)	340	−	−
Secalonsäure B (**64**)	339	+	+
Secalonsäure C (**26**)	339		±
Secalonsäure D (**65**)	337	+	+
Ergoflavin (**66**)	381	+	+
Ergochrysin A (**61**)	336		(−)
Ergochrysin B (**80**)	336		+

In Tabelle 9 sind die Vorzeichen der Cotton-Effekte der ORD- und CD-Kurven für 7 Ergochrome im Bereich der längstwelligen Absorptionsmaxima zusammengestellt. Hiernach lassen sich die *symmetrisch* aufgebauten Ergochrome in zwei Gruppen einteilen. Einen negativen Cotton-Effekt zeigt Secalonsäure A (**63**), einen positiven die Secalonsäuren B (**64**) und D (**65**) sowie Ergoflavin (**66**). Die ORD-Kurven der Secalonsäuren A und D haben, wie für optische Antipoden zu erwarten, einen spiegelbildlichen Verlauf (Abb. 5 und 6). Demgegenüber ist die ORD-Kurve von Secalonsäure A (**63**) zu denen der diastereomeren Secalonsäure B (**64**) und Ergoflavin (**66**) nur im Bereich der längstwelligen Absorptionsmaxima weitgehend spiegelbildlich (Abb. 5). Nach umfangreichen Untersuchungen, die vor allem von DJERASSI (*12*) an optisch aktiven, α,β-ungesättigten, polycyclischen Ketonen durchgeführt wurden, kann hieraus abgeleitet werden, daß sich die beiden genannten Gruppen von Ergochromen in der Konfiguration an einer Ringverknüpfung unterscheiden (*29*). Zwei typische, den Ergochromen strukturell verwandte Beispiele für die Gültigkeit dieser Strukturabhängigkeit des Cotton-Effektes bilden die isomeren Steroidketone (**69**) und (**70**) mit nahezu spiegelbildlichen ORD-Kurven (*13*).

(**69**)   (**70**)

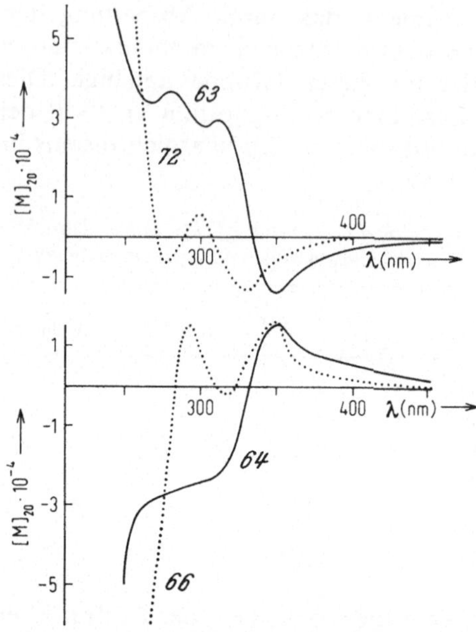

Abb. 5. ORD -Kurven von Secalonsäure A (**63**), Secalonsäure B (**64**), Secalon A (**72**) und Ergoflavin (**66**) in Chloroform. [Aus: Angew. Chem. *81*, 274 (1969)]

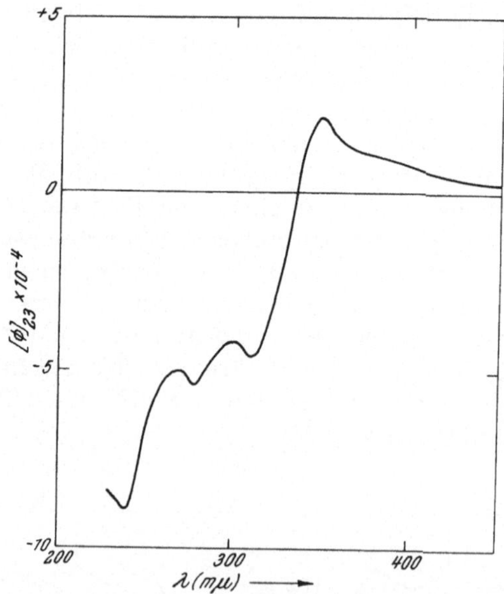

Abb. 6. ORD-Kurve der Secalonsäure D (**65**) in Chloroform. [Aus: Tetrahedron *26*, 51 (1970)]

Literaturverzeichnis: SS. 203—206

Auch die *absolute Konfiguration* am Ringverknüpfungszentrum C-10 der Ergochrome kann mit einiger Sicherheit bereits aus den Cotton-Effekten ihrer ORD-Kurven abgeleitet werden (*29*). Da für Ergoflavin die Konfiguration an C-10 entsprechend der Stereoformel (**66**) durch eine Röntgenstrukturanalyse festgelegt ist (*59*), sollten die Ergochrome mit gleichem Cotton-Effekt in der Stellung der Carboxygruppe (oberhalb der Schreibebene) übereinstimmen. Dies gilt für die Secalonsäuren B (**64**) und D (**65**). Dagegen müßte Secalonsäure A (**63**) hiernach an C-10 entgegengesetzte Konfiguration aufweisen.

Ein sicherer Konfigurationsvergleich mit Ergoflavin aufgrund der Rotationsdispersion ist möglich, wenn eine der Secalonsäuren auf chemischem Wege unter Erhaltung der Stereochemie in ein Derivat mit dem Chromophor des Ergoflavins übergeführt werden kann. Dies ließ sich für Secalonsäure A (**63**) mit der nachstehenden Reaktionsfolge erreichen [FRANCK und Mitarb. (*6, 21*)]. Secalonsäure A (**63**) wurde unter milden Bedingungen in ihr Hexaacetat übergeführt und dessen Enolacetat-Doppelbindung katalytisch zu (**71**) hydriert. Anschließende Entacetylierung durch Umesterung bewirkte zugleich Dehydratisierung des entstehenden Aldols zum α,β-ungesättigten Keton, dem Secalonen A (**60**), das erneut katalytisch hydriert wurde. Die entstandene Verbindung (**72**) erhielt die Bezeichnung *Secalon A,* da sie sich von Secalonsäure A durch das Fehlen der sauren Enolgruppierung unterscheidet. Secalon (**72**) stimmt hinsichtlich seines konjugierten π-Elektronensystems mit Ergoflavin (**66**) überein und sollte daher für einen Vergleich der Rotationsdispersion gut geeignet sein. Tatsächlich zeigen die ORD-Kurven von Secalon A (**72**) und Ergoflavin (**66**) im ganzen Wellenlängenbereich einen sehr ähnlichen, aber spiegelbildlichen Verlauf (Abb. 5). Danach müssen sich diese beiden Verbindungen durch entgegengesetzte Konfiguration am Ringverknüpfungszentrum C-10 unterscheiden.

(**71**)

(**72**)

Secalon A

Die absolute Konfiguration der Ergochrome an C-10 konnte außer durch den Vergleich mit Ergoflavin auch durch Bestimmung der Konfiguration relativ zum benachbarten Chiralitätszentrum an C-5 ermittelt werden. Für C-5 wurde die absolute Konfiguration, wie schon beschrieben, durch NMR-Korrelation zum Chiralitätszentrum an C-6 und dessen

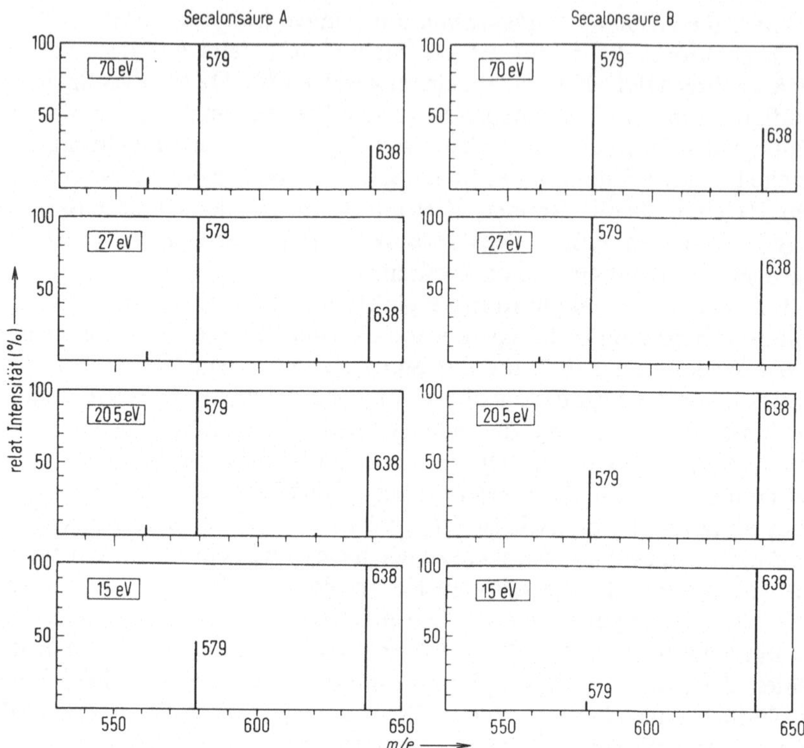

Abb. 7. Massenspektren der Secalonsäuren A (63) und B(64) bei 70, 27, 20,5 und 15 eV im m/e-Bereich 530—650. [Aus: Chem. Ber. *99*, 3842 (1966)]

Überführung in optisch aktive Methylbernsteinsäure festgelegt. Die relative Anordnung der Substituenten an C-10 und C-5 ergab sich aus den Massenspektren (*29*). So lassen z. B. die Massenspektren erkennen, daß die CO_2CH_3-Gruppe der Secalonsäure A (**63**) beim Elektronenbeschuß leichter abgespalten wird, als die von Secalonsäure B (**64**) (Abb. 7). Bei dieser Fragmentierung entsteht ein intensives Ion der Masse 579. Bezogen auf die Intensität dieses Fragmentions ist die des Molekülions bei Secalonsäure A kleiner als bei Secalonsäure B. Demnach hat in Secalonsäure A (**63**) die CO_2CH_3-Gruppe an C-10 die sterisch stärker gehinderte, weniger stabile *cis*-Anordnung zur benachbarten Hydroxygruppe an C-5, aus der sie leichter herausgespalten wird. Dieser Stabilitätsunterschied von Secalonsäure A (**63**) und Secalonsäure B (**64**) kommt erwartungsgemäß in Massenspektren mit herabgesetzter Elektronenenergie noch stärker zum Ausdruck (Abb. 7). Die leichter erfolgende Abspaltung der CO_2CH_3-Gruppe kann bei der Secalonsäure A nicht etwa auf einer sterischen Wechselwirkung mit der Methylgruppe an C-6 beruhen, denn

Literaturverzeichnis: SS. 203—206

hinsichtlich ihrer Konfigurationen an C-10 und C-6 verhalten sich die Secalonsäuren A und B wie Antipoden.

Um abzusichern, daß die geringere Stabilität der Secalonsäure A gegenüber Secalonsäure B beim Elektronenbeschuß auf sterischer Behinderung der CO_2CH_3-Gruppe durch die benachbarte Hydroxygruppe beruht, wurde auch das massenspektroskopische Verhalten von Umesterungsprodukten der Secalonsäuren untersucht. Durch längeres Kochen der Secalonsäuren mit Alkoholen in Gegenwart katalytischer Mengen konz. Schwefelsäure ließen sich mit guten Ausbeuten die Derivate (73)—(79) gewinnen (75). Deren Massenspektren zeigen eindeutig (vgl. Abb. 8), daß Einführung größerer Alkylreste in die Alkoxycarbonylgruppe den Stabilitätsunterschied zwischen Secalonsäure A und B beim Elektronenbeschuß beträchtlich erhöht (75).

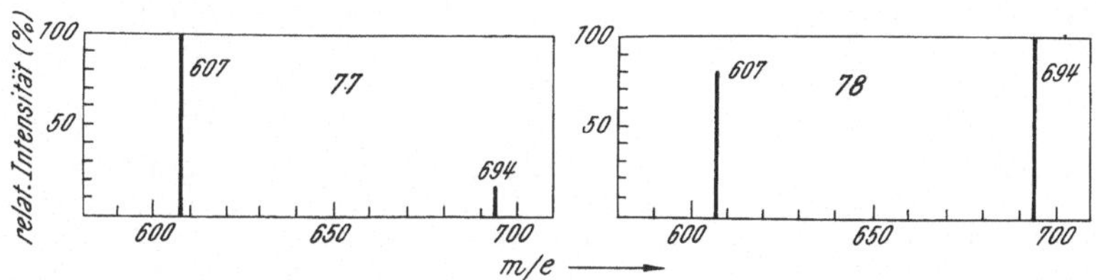

Abb. 8. Massenspektren der Diisopropyl-secalonsäuren A (77) und B (78) bei 70 eV im m/e-Bereich 580—710 (75)

Ebenso wie Secalonsäure A (63) verhält sich erwartungsgemäß auch die enantiomere Secalonsäure D (65) bei der Massenspektrometrie (69).

(63) R = CH_3
(73) R = C_2H_5
(75) R = n–C_3H_7
(77) R = i–C_3H_7
(79) R = n–C_8H_{17}

(64) R = CH_3
(74) R = C_2H_5
(76) R = n–C_3H_7
(78) R = i–C_3H_7

Umesterungsprodukte der Secalonsäuren A (63) und B (64)

3. Struktur der Secalonsäure C sowie der Ergochrysine A und B

Durch die in den vorhergehenden Abschnitten beschriebenen Struktur-
und Konfigurationsuntersuchungen sind die *symmetrischen* Ergochrome
Secalonsäure A, Secalonsäure B, Secalonsäure D und Ergoflavin vollständig aufgeklärt. Sie haben die Strukturen (**63**), (**64**), (**65**) und (**66**).
Soweit dabei auch die Secalonsäure C sowie die Ergochrysine A und B
behandelt wurden, ließen die Untersuchungsergebnisse bereits erkennen,
daß diese Ergochrome *unsymmetrische* Kombinationen von Molekülhälften der Secalonsäuren A und B und des Ergoflavins darstellen (*1, 3,
23, 29*). Das zeigten pK-Messungen (Tabelle 6), NMR-Spektren (Tabelle
8), Rotationsdispersion und Circulardichroismus (Tabelle 9) sowie die
Konfigurationsbestimmung der Methylgruppe an C-6 (Tabelle 7) dieser
unsymmetrischen Ergochrome. Ebenso verhielten sich diese Ergochrome
auch bei allen sonstigen Untersuchungen wie Mischungen aus jeweils zwei
symmetrischen Ergochromen, entsprechend *Schema 1* (*23, 29*).

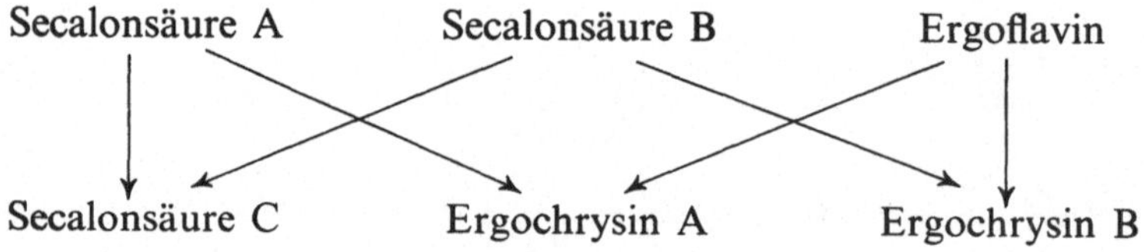

Secalonsäure A　　　　　Secalonsäure B　　　　　Ergoflavin

Secalonsäure C　　　　　Ergochrysin A　　　　　Ergochrysin B

Schema 1. Kombination von Eigenschaften und Molekülhälften der symmetrischen Ergochrome (obere Reihe) in den unsymmetrischen Ergochromen (untere Reihe)

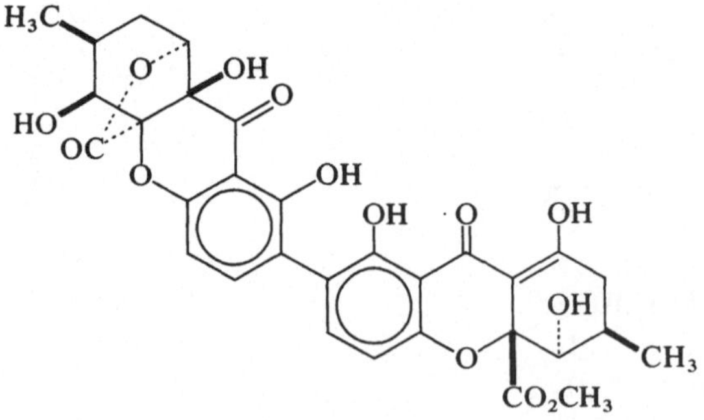

(**80**)

Ergochrysin B

Literaturverzeichnis: SS. 203—206

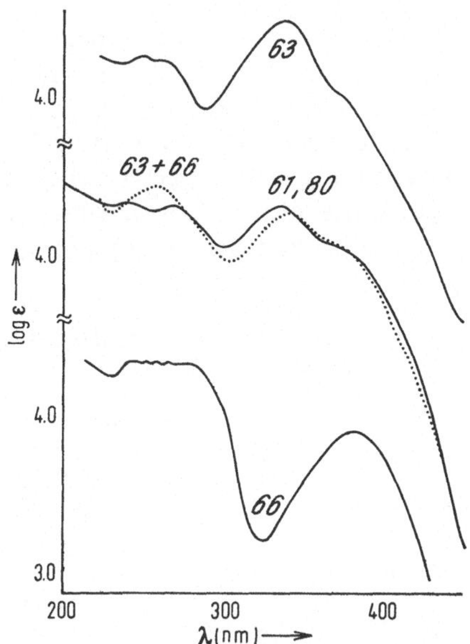

Abb. 9. Elektronenspektren von Ergochrysin A (**61**) und B (**80**), Secalonsäure A (**63**), Ergoflavin (**66**) sowie der äquimolaren Mischung von (**63**) und (**66**) in Methanol. [Aus: Chem. Ber. *99*, 3863 (1966)]

Um dies an einem Beispiel zu demonstrieren, sind in Abb. 9 die Elektronenspektren der diastereomeren Ergochrysine A (**61**) und B (**80**) mit denen von Secalonsäure A (**63**) und Ergoflavin (**66**) sowie einer äquinormalen Mischung aus (**63**) und (**66**) zusammengestellt. Somit haben Secalonsäure C, Ergochrysin A und Ergochrysin B die Strukturen (**26**), (**61**) und (**80**).

4. Strukturen der Ergochrome AD, BD, CD und DD

Im Rahmen einer systematischen Untersuchung über das Vorkommen von Ergochromen isolierten FRANCK und Mitarb. (*24, 25*) aus deutschem, portugiesischem und österreichischem Mutterkorn vier Ergochrome, die sich durch Empfindlichkeit gegenüber säurehaltigen Lösungsmitteln auszeichneten. Wie die nachstehend beschriebene Strukturermittlung ergab, handelt es sich bei diesen Ergochromen um dimere Kombinationen der Molekülhälften **A**, **B** und **C** (**4**)—(**6**) mit einem weiteren Xanthon-Derivat **D** (**7**). Nach der anfangs mitgeteilten Nomenklatur der Ergochrome erhielten sie daher die Bezeichnungen, AD, BD, CD und DD. Obwohl sich von diesen 4 Ergochromen nur Mengen zwischen 2 und 38 mg isolieren ließen, konnten deren Strukturen praktisch vollständig aufgeklärt werden (*21, 24, 25*). Dies ist dem Umstand zu danken, daß sich die Ergochrome AD, BD, CD und DD bei kurzem Erwärmen in Eisessig in eines der schon vorher bekannten Ergochrome AC (Ergochrysin A) (**61**), BC (Ergochrysin B) (**80**) und CC (Ergoflavin) (**66**) umwandeln. Diese Umwandlung tritt in gewissem Umfang schon bei der Chromatographie mit säurehaltigen Lösungsmittelsystemen ein und erklärt die erwähnte Empfindlichkeit der D-Ergochrome.

Tabelle 10. *Einige Eigenschaften der Ergochrome AD (**83**), BD (**84**),
CD (**82**) und DD (**81**) (24)*

Eigenschaften	Ergochrome			
	AD	BD	CD	DD
Mol.-Gew. (massen-spektrometr.)	656	656	642	674
CO_2CH_3 (cm^{-1})	2 (1730)	2 (1735)	1 (1735)	2 (1735)
γ-Lacton (cm^{-1})	—	—	1 (1795)	—
λ_{max} (CH$_3$OH) (mμ)	338, 246	339, 249	364, 241	361, 247
R_4-Wert*	0,37	0,74	0,14	0,09
FeCl$_3$-Reaktion	rot	rot	grün	grün
Ergochrom, das nach Erhitzen in Eisessig entsteht	AC (Ergochrysin A)	BC (Ergochrysin B)	CC (Ergoflavin)	CC (Ergoflavin)

* Laufstrecke dividiert durch die von Secalonsäure A (**63**) im System Chloroform/ Methanol (99 : 1) auf Kieselgel G, das mit 7,5% Weinsäure imprägniert ist.

In Tabelle 10 sind die für die Strukturaufklärung wichtigsten Eigenschaften der vier D-Ergochrome zusammengefaßt. Besonders aufschlußreich ist der Befund, daß bei deren Umwandlung in das entsprechende C-Ergochrom das Molekulargewicht um 32 Masseneinheiten, entsprechend einer Abspaltung von Methanol, abnimmt. Gleichzeitig verschwindet nach Aussage des IR-Spektrums eine Estercarbonylbande, und es tritt eine zusätzliche γ-Lactonbande bei 1780 cm^{-1} auf. Lediglich das Ergochrom DD weicht dadurch von diesem Verhalten ab, daß offensichtlich 2 Mol Methanol abgespalten werden und zwei γ-Lactongruppen entstehen. Hiernach ist anzunehmen, daß sich die Molekülhälfte **D** der neuen Ergochrome von der des Ergoflavins durch Methanolyse der γ-Lactongruppe unterscheidet. Beim Erwärmen mit Eisessig kann daraus die Ergoflavinhälfte **C** leicht gebildet werden. Die Richtigkeit dieser Annahme ließ sich durch eine Partialsynthese des Ergochroms DD (**81**) aus Ergoflavin (**66**) und anschließende Rückverwandlung bestätigen (*24*). So erwies sich Ergochrom DD (**81**) als identisch mit Ergoflavinsäure-dimethylester, der schon durch alkalische Hydrolyse des Ergoflavins und Veresterung der erhaltenen

Ergoflavinsäure mit Diazomethan erhalten worden war (*16*). Durch Erhitzen in Eisessig während 3 Stunden ließ sich (**81**) in Ergoflavin (**66**) zurückverwandeln (*24*). Demgegenüber gab kurzzeitiges Erwärmen in Eisessig unter partieller Lactonisierung das Ergochrom CD (**82**) neben Ergoflavin. Nach diesen Befunden kommen für die Ergochrome AD, BD,

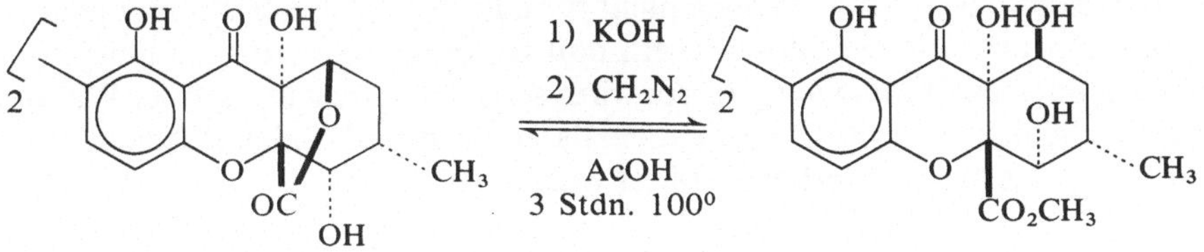

1) KOH

2) CH_2N_2

AcOH

3 Stdn. 100⁰

(**66**)

Ergoflavin

(**81**)

Ergochrom DD

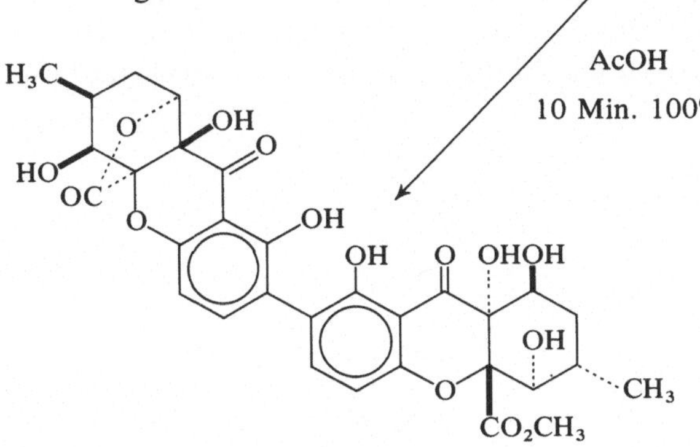

AcOH

10 Min. 100⁰

(**82**)

Ergochrom CD

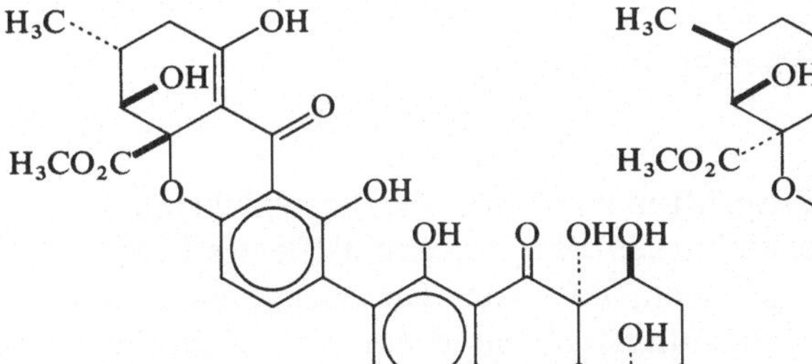

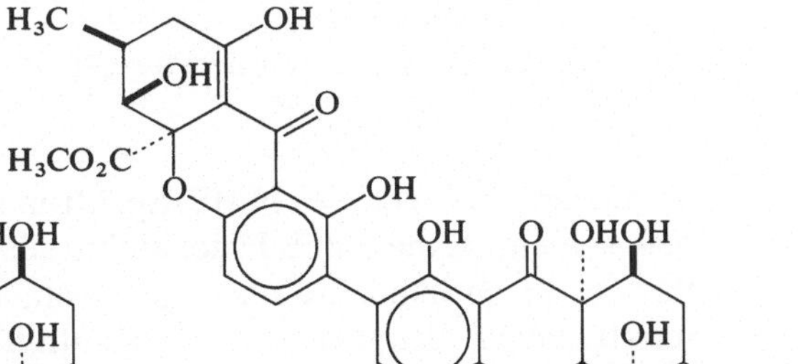

(**83**)

Ergochrom AD

(**84**)

Ergochrom BD

CD und DD nur die Strukturen (**83**), (**84**), (**82**) und (**81**) in Betracht. Allerdings war hierfür noch auszuschließen, daß es sich nicht um die Antipoden dieser Strukturformeln handelt. Für die Ergochrome AD (**83**) und BD (**84**) ging dies eindeutig daraus hervor, daß die Molrotationen ihrer Säureumwandlungsprodukte mit denen von Ergochrysin A (**61**) bzw. Ergochrysin B (**80**) übereinstimmen (*24*). Bei den Ergochromen CD (**82**) und DD (**81**) reichte die Substanzmenge nicht für eine zusätzliche Bestimmung der Molrotation aus. Die Möglichkeit, daß bei deren Säureumwandlung ein Antipode des Ergoflavins gebildet wird, scheidet jedoch mit großer Wahrscheinlichkeit aus, weil bisher niemals ein Antipode oder Racemat des Ergoflavins aus Mutterkorn isoliert wurde.

5. Struktur des Ergoxanthins

Aus portugiesischem Mutterkorn konnten DE Mayo und Mitarb. (*1, 2*) ein seco-Ergochrom, das *Ergoxanthin,* isolieren. Seine Struktur wurde bis auf die Konfiguration der Chiralitätszentren entsprechend (**85**) aufgeklärt (*1, 2, 53*).

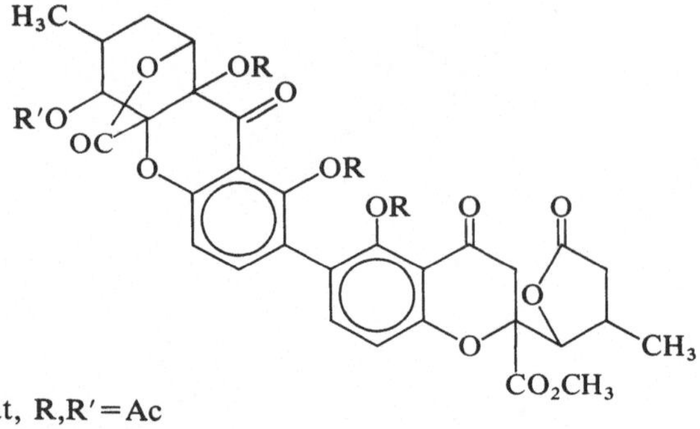

(**85**) Ergoxanthin, R,R′=H
(**86**) Ergoxanthin-tetraacetat, R,R′=Ac
(**87**) Ergoxanthin-trimethyläther, R=CH₃; R′=H

Einem Vergleich des NMR-Spektrums mit dem des Ergoflavins (**66**) ließ sich entnehmen, daß Ergoxanthin (**85**) eine Molekülhälfte des Ergoflavins (**6**) enthält. Daraus ergab sich, daß von den nachgewiesenen funktionellen Gruppen des Ergoxanthins folgende auf die „Xanthin-Hälfte" (**9**) entfallen: 1 γ-Lactongruppe, 1 CO$_2$CH$_3$-Gruppe, 1 Hydroxygruppe, die sich acetylieren und mit Diazomethan methylieren läßt, sowie 1 C-Methylgruppe. Aus der Ähnlichkeit von UV-, IR-Spektrum und FeCl$_3$-Reaktion ging hervor, daß das Ergoxanthin in beiden Molekülhälften den Chromophor des Ergoflavins (**66**) enthält. Die weitere Untersuchung zeigte, daß sich die „Xanthin-Hälfte" aus den Strukturteilen (**88**) und (**90**) zusammensetzt (*2*). So konnte Ergoxanthin-tetraacetat (**86**) durch Erhitzen mit Acetanhydrid/Pyridin in ein Pentaacetat übergeführt werden, das gegenüber (**86**) eine zusätzliche, leicht verseifbare Enolacetat-Gruppe enthält. Für letztere ist die Anordnung (**89**) anzunehmen, da mit der Überführung des Tetraacetats (**86**) in das Pentaacetat im NMR-Spektrum ein 2-Protonen-Singulett bei 3,03 ppm (—CH_2—CO—) verschwindet und ein 1-Protonen-Singulett bei 5,45 ppm (=CH—) neu erscheint. Das NMR-Spektrum des Ergoxanthins ließ die benachbarte Anordnung von zwei tertiären Protonen entsprechend (**90**) erkennen. Da außer den Strukturteilen (**88**) und (**90**) sowie der CO_2CH_3-Gruppe keine weiteren C-Atome verfügbar sind, wurde für die „Xanthin-Hälfte" aufgrund dieser Befunde die Struktur (**92**) angegeben. Ebenso wie in allen anderen Ergochromen sind die Molekülhälften des Ergoxanthins in den o-Stellungen zu den phenolischen Hydroxygruppen miteinander verbunden (2,2'-Verknüpfung). Dies geht daraus hervor (s. o.), daß im NMR-Spektrum des 1,1',9-Tri-O-methylergoxanthins (**87**) die Signale der beiden aromatischen Methoxylgruppen (6-Protonen-Singulett bei 3,56 ppm) einen „Upfield-shift" zeigen und mit denen des 1,1',9,9'-Tetra-O-methylergoflavins (**40**) übereinstimmen (*53*). Für Ergoxanthin ergibt sich somit die Struktur (**85**).

(88) (89) (90)

1) 0,1 n NaOH

2) Kochen m. AcOH

3) CH_2N_2

(91) (92)

Die für Ergoxanthin ermittelte Struktur (**85**) gab Anlaß zu Betrachtungen über die Bildungsweise dieses seco-Ergochroms (*2*). Es erschien möglich, daß dieses bei der Biosynthese oder auch bei der Isolierung aus einem Ergochrysin (**61, 80**) durch Ringöffnung von dessen Secalonsäure-Hälfte (**91**) nach dem Mechanismus der Säurespaltung von β-Diketonen und nachfolgende Lactonisierung zu (**92**) entsteht. Daher untersuchten DE Mayo und Mitarb. (*2*), ob sich Ergoxanthin durch Alkalieinwirkung aus Ergochrysin A oder B gewinnen läßt. Nach Erhitzen von Ergochrysin A (**61**) mit verd. Natronlauge, Lactonisierung durch Kochen in Eisessig, Remethylierung der Carboxylgruppe mit Diazomethan und schicht-chromatographischer Auftrennung des Reaktionsgemisches wurde entsprechend der Umwandlung (**91**)→(**92**) ein „ψ-Ergoxanthin" erhalten, das nach IR-Spektrum und $FeCl_3$-Reaktion mit (**85**) nahezu überein-stimmt, sich jedoch durch etwas kleineren R_F-Wert unterschied. Das entsprechende Umwandlungsprodukt aus Ergochrysin B (**80**) differierte in seinen Eigenschaften noch stärker von (**85**). Hiernach wurde ange-nommen, daß das Ergoxanthin nicht aus einem der beiden Ergochrysine entsteht (*2*).

IV. Biosynthese der Ergochrome

1. Hypothesen

Aufschlußreich für die Biosynthese der Ergochrome war die Isolierung der Anthrachinon-carbonsäuren *Endocrocin* (**93**) und *Clavorubin* (**94**) aus Mutterkorn durch Franck und Mitarb. (*36, 39*). Von diesen ist das zuerst von Asahina und Mitarb. (*5*) in der Flechte *Nephromopsis endocrocea* auf-gefundene Endocrocin auch in anderen Pilzen verbreitet (*43, 65, 68*). Die Ähnlichkeit der Substituenten-Anordnung des Endocrocins (**93**) mit der der Ergochrome, wie z. B. Secalonsäure A (**63**), legte die Vermutung nahe, daß die Ergochrome durch eine oxidative Ringöffnung aus einem Anthrachinon hervorgehen (*3, 19, 20*). Im einzelnen könnte nach dieser Hypothese die Ergochrom-Biosynthese über die im *Schema 2* beschriebe-nen Teilschritte a)—f) verlaufen. Wesentliches Prinzip dieses Biosynthese-weges ist die Bildung eines Hydroxanthons (**97**) durch oxidative Ring-öffnung eines Anthrachinons zu einer Benzophenoncarbonsäure (**96**) und deren oxidative Cyclisierung nach dem Mechanismus der Phenoloxida-tion. Alternativ zu diesem Schema wurde von Whalley und Mitarb. (*3*) sowie von Hassall und Mitarb. (*11*) die Bildung der Benzophenoncarbon-säure (**96**) durch Kondensation von zwei Benzolderivaten in Betracht gezogen.

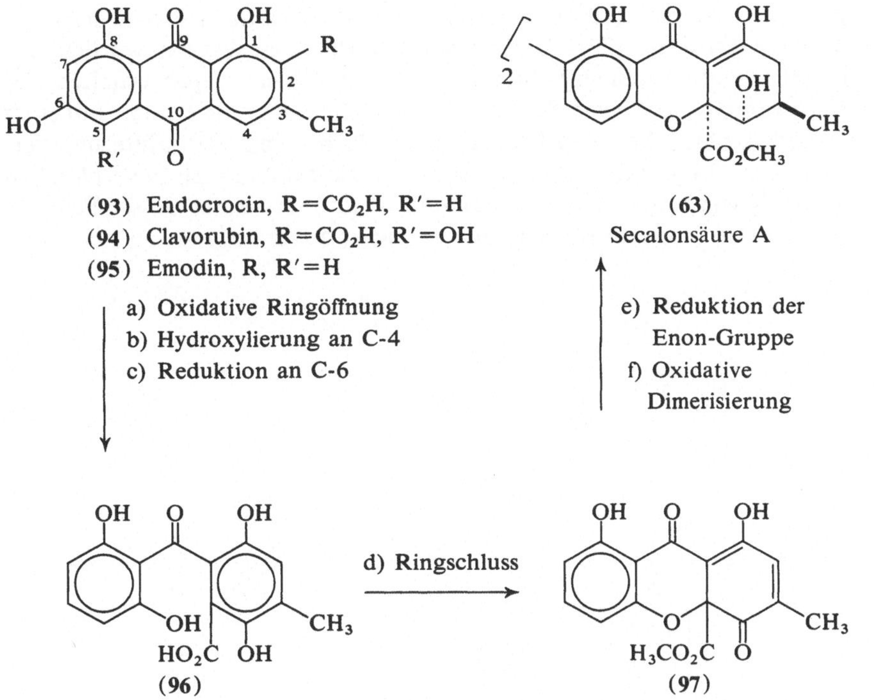

(93) Endocrocin, $R = CO_2H$, $R' = H$
(94) Clavorubin, $R = CO_2H$, $R' = OH$
(95) Emodin, R, $R' = H$

a) Oxidative Ringöffnung
b) Hydroxylierung an C-4
c) Reduktion an C-6

(63)
Secalonsäure A

e) Reduktion der
   Enon-Gruppe
f) Oxidative
   Dimerisierung

(96)

d) Ringschluss

(97)

Schema 2. Mögliche Teilschritte der Ergochrom-Biosynthese

2. Fütterungsversuche

Gemeinsam von FRANCK und Mitarb. sowie von GRÖGER und Mitarb. durchgeführte Fütterungsversuche mit radioaktiv markierten Biosynthese-Vorstufen, die nun beschrieben werden sollen, zeigten, daß die Ergochrome entsprechend *Schema 2* aus einem Anthrachinon gebildet werden (*30, 31, 49*). Dies war zugleich der erste Nachweis einer oxidativen Ring-öffnung von Anthrachinonen in der Naturstoff-Biosynthese. Später wurde dieses Prinzip auch bei der Biosynthese anderer Naturstoffe fest-gestellt (*44*).

Die Fütterungsversuche wurden wesentlich vereinfacht durch die Feststellung, daß sich die Ergochrome auch aus Oberflächen-Kulturen des Mutterkornpilzes *Claviceps purpurea* auf synthetischen Nährlösungen gewinnen lassen (*31*). Hierdurch ist zugleich bewiesen, daß die Bildung der Mutterkorn-Farbstoffe unabhängig vom Stoffwechsel der Roggen-pflanze erfolgt. Im Gegensatz hierzu unterscheiden sich die aus Mutterkorn und aus Flüssigkeitskulturen gewonnenen Mutterkorn-*Alkaloide* meistens in ihrer Struktur *(71)*. Die besten Ergochromausbeuten wurden aus Oberflächenkulturen eines Stammes 467 von *Claviceps spec.* erhalten, der sich in Japan auf *Elymus Mollis* fand *(31)*. Die Züchtung erfolgte auf einer Mannit-Rohrzucker-Nährlösung *(48)*. Als bei späteren Fütterungsversuchen die Ergochrom-Bildung durch diesen Stamm mehrmals plötzlich ausblieb, wurde auch *Penicillium oxalicum (69)* verwendet, dessen Produktion von Secalonsäure D sich als besonders zuverlässig erwies.

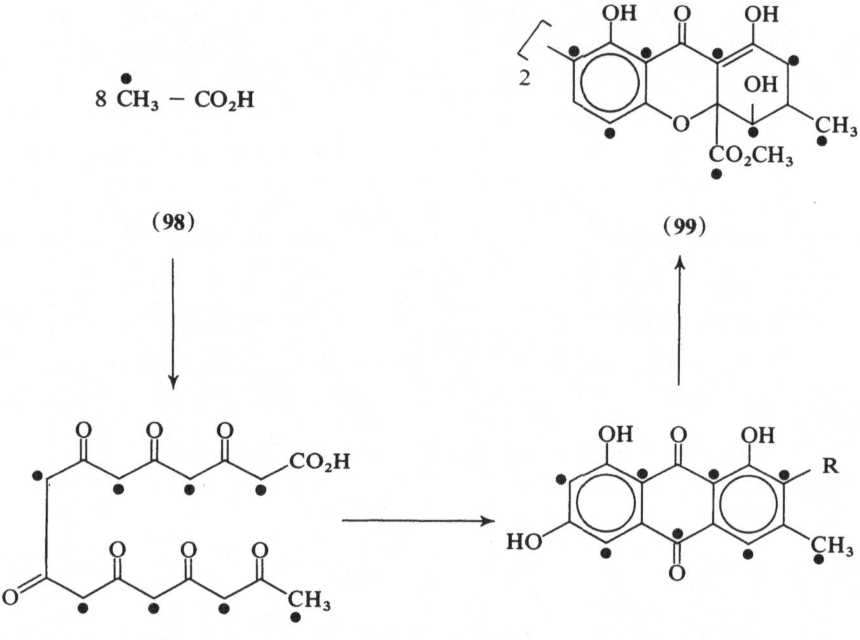

$$8 \; \overset{\bullet}{C}H_3 - CO_2H$$

(98)

(99)

(100)

(101) R=H

(102) R=CO₂H

Zunächst wurde in drei getrennten Ansätzen an Oberflächenkulturen des Mutterkornpilzes Natriumacetat verfüttert, das in der Carboxylgruppe mit ^{14}C sowie in der Methylgruppe mit ^{14}C bzw. ^3H markiert war *(30, 31)*. Emodin (**101**) und damit auch Endocrocin (**102**) entsteht nach Gatenbeck *(42)* bei der Biosynthese durch Pilze *(Penicillium islandicum)* aus 8 Molekülen Essigsäure über Acetyl-, Malonyl-Coenzym A und möglicherweise über eine Heptaketopalmitinsäure (**100**). Wenn die Ergochrome (**99**) unter oxidativer Ringöffnung aus den Anthrachinonen (**101**) oder (**102**) hervorgehen, so würden sich z. B. die Methyl-C-Atome der Essigsäure (•) in charakteristischer Weise (**99**) über das Molekül verteilen. Alle übrigen

C-Atome müßten aus der Carboxyl-Gruppe stammen. Zur Auswertung der Fütterungsversuche mit radioaktiv markierten Natriumacetaten und weiteren markierten Vorstufen war daher ein spezifischer Abbau der Ergochrome erforderlich, mit dem sich möglichst viele ihrer C-Atome zur Radioaktivitätsmessung einzeln isolieren lassen. Die verwendeten Abbaureaktionen (*19*) sind in *Schema 3* für Secalonsäure C (**103**) nach Verfütterung von [2-^{14}C]-Natriumacetat zusammengefaßt. Davon läßt sich der in Anlehnung an WHALLEY und Mitarb. (*3*) entwickelte Abbau mit Acetanhydrid/Pyridin und konz. Bromwasserstoffsäure nur auf Secalonsäuren und die entsprechende Molekülhälfte der Ergochrysine anwenden. Hierbei wird die Secalonsäure C (**103**) in das Tetrahydroxybiphenyl (**c**), 5-Hydroxy-3-methyl-benzoesäure (**e**) und zwei Moleküle CO_2 (**d**) gespalten, die dem ursprünglichen Xanthon-Carbonyl entstammen. Abspaltung der Carboxylgruppe der Kresotinsäure (**e**) durch Schmidt-Abbau ermöglicht die Isolierung des C-Atoms der Secalonsäure-Carboxylgruppe bzw.

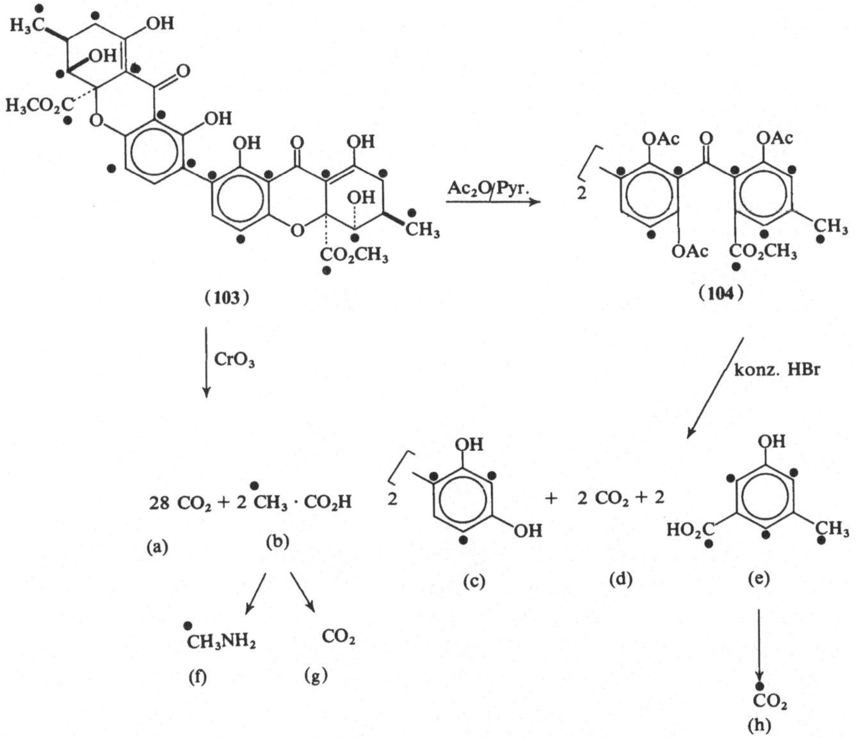

Schema 3. Spezifischer Abbau der nach Verfütterung von [2-^{14}C]-Natriumacetat erhaltenen Secalonsäure C (**103**) zur Bestimmung der Radioaktivitätsverteilung (*31*)

des Chinon-Carbonyls C-10 der Anthrachinon-Vorstufe als CO_2 (h). Die C-Atome der durch Kuhn-Roth-Oxidation mit CrO_3 abgespaltenen Essigsäure (b) wurden nach Schmidt-Abbau als Methylamin (f) und CO_2 (g) einzeln isoliert.

Die Radioaktivitätsverteilung in der Secalonsäure C (**103**) (Tabelle 11) zeigte eindeutig, daß deren biogenetisch besonders aufschlußreiche C-Atome der Xanthon-Carbonyl- (**d**) und der Carboxylgruppe (**h**) aus der Carboxyl- bzw. Methylgruppe des verfütterten Acetats hervorgehen. Ebenso entsprach auch die Radioaktivität der übrigen C-Atome von Secalonsäure C sowie von weiteren Ergochromen, die nach Verfütterung markierter Essigsäuren isoliert wurden (*31*), dem Biosynthese-Schema.

Tabelle 11. *Radioaktive C-Atome der nach Verfütterung von [2-^{14}C]- und [1-^{14}C]- Natriumacetat an den Mutterkornpilz (Claviceps spec.) erhaltenen Secalonsäuren C (**103**) und ihrer Abbauprodukte (s. Schema 3) (31)*

| Substanz | Radioaktive C-Atome nach Verfütterung von | | | |
| | $^{14}CH_3CO_2Na$ | | $CH_3{}^{14}CO_2Na$ | |
	Ber.	Gef.*	Ber.	Gef.*
Secalonsäure C (**103**)	16	16,6	14	15,4
CO_2 (**a**) (CrO_3-Oxidation)	0,5	0,55	14/32 =0,44	0.42
CH_3CO_2H(**b**) (CrO_3-Oxidation)	1	1	1	1
CH_3NH_2 (**f**) aus (**b**)	1	1	0	—
CO_2 (**g**) aus (**b**)	0	0,01	1	1
CO_2 (**d**) (HBr-Abbau)	0	0,09	1	0,58
5-Hydroxy-3-methyl-benzoesäure (**e**)	5	4,6	3	2,3
CO_2 (**h**) aus (**e**)	1	0,64	0	0,07

* Zur Ermittlung der *gefundenen* Anzahl radioaktiver C-Atome wurden die Radioaktivitäten aller Substanzen durch die der Essigsäure (**b**) von der Kuhn-Roth-Oxidation dividiert.

Einen direkten Beweis der Ergochrom-Biosynthese aus einem Anthrachinon sowie Einblick in die biogenetische Sequenz der Ergochrome brachten weitere Fütterungsversuche (*26, 32, 49*) mit speziellen radioaktiven Biosynthese-Vorstufen (Tabelle 12). Von den radioaktiven Biosynthese-Vorstufen wurden [U-^{14}C]-Emodin (**101**) (*49*), [U-^{14}C]-Shikimisäure (**105**), [U-^{14}C]-Secalonsäure B (**64**) sowie [U-^{14}C]-Ergochrysin B (**80**) biosynthetisch (*32*), [U-^{3}H]-Emodin (**101**) durch Wilzbach-Tritiierung (*17, 78*) mit 10 Ci Tritiumgas und [10-^{14}C]-Endocrocin (**102**) durch Totalsynthese (*34*) gewonnen. Die Fütterungsversuche geben mit fol-

Tabelle 12. *Einbauraten radioaktiv markierter Biosynthese-Vorstufen in Ergochrome nach Verfütterung an den Mutterkornpilz (Claviceps spec.) und Penicillium oxalicum (26, 32, 49)*

Verfütterte Vorstufe	Organismus	Radioaktive Biosynthese-produkte	Spez. Einbaurate (%)
[U-^{14}C]-Emodin (**101**)*	*Claviceps spec.*	Secalonsäure B (**64**)	0,38
		Ergochrysin B (**80**)	0,60
[U-^{14}C]-Emodin (**101**)	*Penicillium oxalicum*	Secalonsäure D (**65**)	0,53
[U-^{3}H]-Emodin (**101**)	*Claviceps spec.*	Ergochrysin B (**80**)	1,5
		Ergochrysin A (**61**)	1,3
[10-^{14}C]-Endocrocin (**102**)	*Claviceps spec.*	Secalonsäure B (**64**)	0,0005
		Ergochrysin B (**80**)	0,005
[10-^{14}C]-Endocrocin (**102**)	*Penicillium oxalicum*	Secalonsäure D (**65**)	0,005
[U-^{14}C]-Shikimisäure (**105**)	*Claviceps spec.*	—	—
[U-^{14}C]-Secalonsäure B (**64**)	*Claviceps spec.*	Secalonsäure B (**64**)	1,3
		Ergochrom BD (**84**)	0,90
		Ergochrysin B (**80**)	0,34
		Ergoflavin (**66**)	0,12
[U-^{14}C]-Ergochrysin B (**80**)	*Claviceps spec.*	Ergochrysin B (**80**)	0,19
		Secalonsäure B (**64**)	0,002
		Ergoflavin (**66**)	0,025

* Die Bezeichnung „U" (uniform) dient hier, wie auch sonst üblich (*66*), zur Kennzeichnung einer Radioaktivitätsverteilung, die sich über eine größere Anzahl der betreffenden Atome des Moleküls erstreckt.

genden Ergebnissen (Tabelle 12) weitreichende Auskunft über den Verlauf der Ergochrom-Biosynthese:

1. Emodin (**101**) wird unter Ringöffnung in Ergochrome übergeführt. Die für 5 Ergochrome gefundenen spez. Einbauraten liegen mit 0,38 bis 1,5% vergleichsweise recht hoch.

2. Die spez. Einbauraten für Endocrocin (**102**) in Ergochrome sind kleiner als 0,005% und damit nicht signifikant für einen Einbau.

3. Shikimisäure (**105**) wird nicht in Ergochrome eingebaut.

(**105**)

4. Verfütterung radioaktiver Secalonsäure B (**64**) an den Mutterkornpilz ergab ein Gemisch aus vier Ergochromen, deren Radioaktivitäten
den nachstehenden spez. Einbauraten entsprechen:

            Ergochrom BB (Secalonsäure B)   1,3%
            Ergochrom BD   0,90%
            Ergochrom BC (Ergochrysin B)   0,34%
            Ergochrom CC (Ergoflavin)   0,12%.

Die Abnahme der spez. Einbauraten in der angegebenen Reihenfolge erlaubt die Annahme, daß die Ergochrom-Hälfte B (**5**) nacheinander in D
(**7**) und C (**6**) umgewandelt wird (s. *Schema 4*). Die Verringerung der spez.
Radioaktivität der wiedergewonnenen Secalonsäure B (scheinbare Einbau-
rate 1,3%) gegenüber der eingesetzten Substanz ist durch Verdünnung mit
inaktiver Secalonsäure verursacht, die zum Zeitpunkt der Verfütterung
bereits gebildet war. Die Ergochrom-Hälften B, D und C haben übereinstimmende Konfigurationen am Ringverknüpfungszentrum C-10. Die
Ergochrom-Hälfte A (**4**) mit entgegengesetzter Konfiguration an C-10
scheint hiernach kein Folgeprodukt von B zu sein. Die Einbauraten der
nach Verfütterung von radioaktivem Ergochrysin B (**80**) isolierten
Ergochrome stehen ebenfalls mit der biogenetischen Sequenz B→D→C in
Einklang (Tabelle 12).

Aus diesen Ergebnissen folgt der in *Schema 4* zusammengefaßte Verlauf der Ergochrom-Biosynthese. Von allgemeiner Bedeutung ist der
Befund, daß Emodin (**101**) und nicht das im Mutterkorn als Begleiter der
Ergochrome festgestellte Endocrocin (**102**) als Biosynthese-Vorstufe dient.
Die bisherige Annahme, daß Endocrocin der biogenetische Vorläufer des
Emodins und zahlreicher weiterer Anthrachinon-Naturstoffe ist (*45*),
scheint somit nicht richtig zu sein. Vielmehr muß angenommen werden,
daß bei der Anthrachinon-Biosynthese aus Acetat-Einheiten die Abspaltung der endständigen Carboxylgruppe bereits während der Kondensation der hypothetischen Heptaketopalmitinsäure (**100**) und nicht erst
nach der Bildung des aromatischen Endproduktes eintritt. Endocrocin
scheint somit auf einem Nebenweg der Anthrachinon-Biosynthese zu
stehen. In Übereinstimmung hiermit fanden Steglich und Mitarb. (*67*) bei
Untersuchungen über die Biosynthese von Anthrachinon-Pigmenten aus
Arten des Pilzes *Dermocybe,* daß Endocrocin nicht in Emodin und verwandte Anthrachinone übergeführt wird.

Es erscheint möglich, daß Emodin auch Biosynthese-Vorstufe einer
ganzen Reihe weiterer Naturstoffe ist, die daraus durch oxidative Ringöffnung an verschiedenen Stellen hervorgehen. So konnte kürzlich Gatenbeck (*44*) für das Benzophenon-Derivat Sulochrin (**107**) die Biosynthese aus Emodin (**101**) nachweisen. Im Gegensatz zur Bildung der
Benzophenon-Zwischenstufe (**96**) der Ergochrom-Biosynthese muß hierbei
oxidative Spaltung der Bindung zwischen Chinoncarbonyl an C-10 und

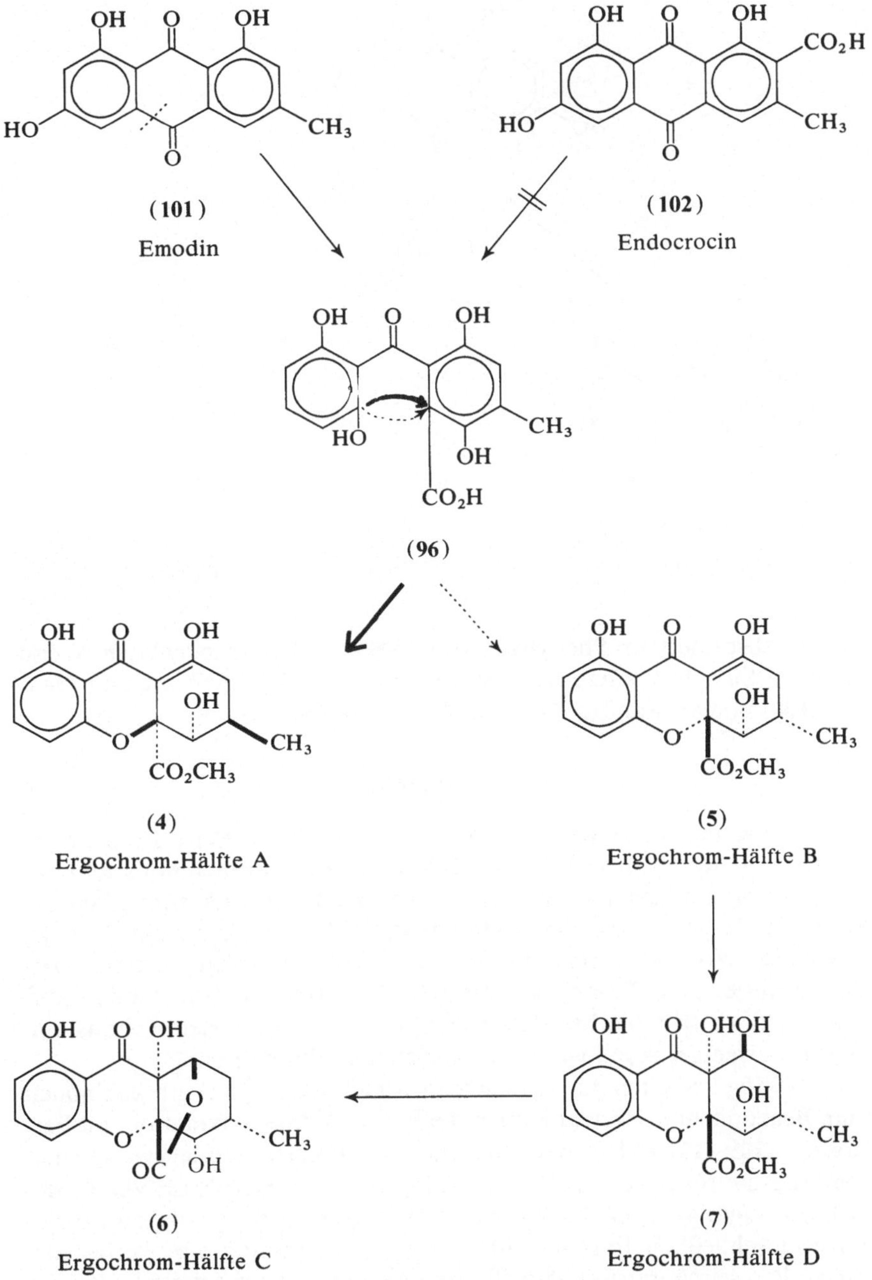

(101)

Emodin

(102)

Endocrocin

(96)

(4)

Ergochrom-Hälfte A

(5)

Ergochrom-Hälfte B

(6)

Ergochrom-Hälfte C

(7)

Ergochrom-Hälfte D

Schema 4. Biosynthese der Ergochrom-Hälften nach dem Ergebnis der Fütterungsversuche (*26, 31, 32, 49*)

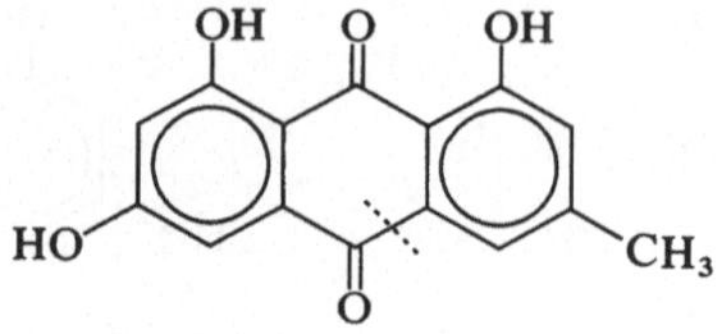

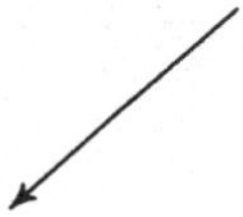

(101)

Emodin

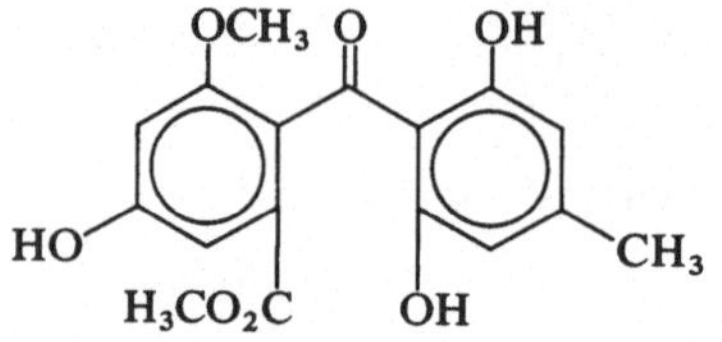

(107)

Sulochrin

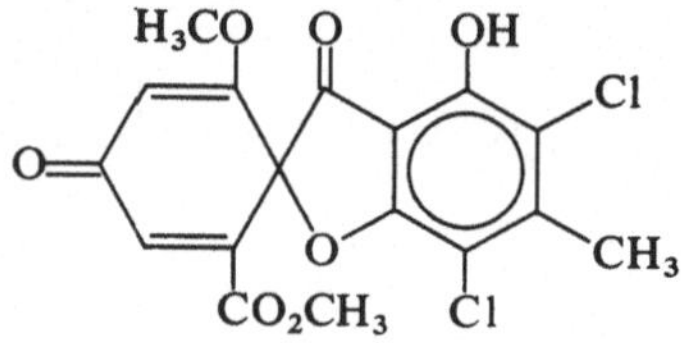

(108)

Geodin

*methyl*substituiertem Phenylrest von (101) in der angedeuteten Weise erfolgen. Sulochrin käme seinerseits als Vorstufe des Antibioticums Geodin (108) und verwandter Naturstoffe in Betracht.

3. Modellreaktionen

Für die mit Hilfe der Isotopentechnik nachgewiesene Ergochrom-Biosynthese durch oxidative Ringöffnung eines Anthrachinons und anschließende Cyclisierung des gebildeten Benzophenons zum Xanthon *(Schema 4)* gab es in der umfangreichen Anthrachinonchemie keine Analogie. Wohl läßt sich durch energische Einwirkung starker Oxidationsmittel oder Basen der chinoide Ring von Anthrachinonen aufspalten, doch tritt hierbei stets weiterer Abbau des Benzophenons ein, da dieses weniger stabil ist als das Ausgangsanthrachinon.

Als Möglichkeiten zur schonenden *Ringöffnung* von Anthrachinonen zum Benzophenon-Derivat kamen die Baeyer-Villiger-Oxidation mit Persäuren (109→111→113) und die säurekatalysierte Umlagerung eines Anthron-hydroperoxids (110→112→114) in Frage. Der erste Weg erwies sich als ungeeignet, da 19 systematisch ausgewählte Hydroxyanthrachinone einschließlich Emodin (101) bei der Einwirkung von Perbenzoesäure, Peressigsäure und Pertrifluoressigsäure nicht reagierten *(35)*. Im Gegensatz dazu wird Benzophenon unter diesen Bedingungen mit hoher Ausbeute gespalten *(51)*. Die oxidative Ringöffnung über ein Anthron-

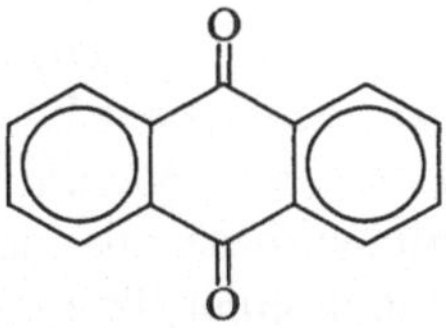

(109) 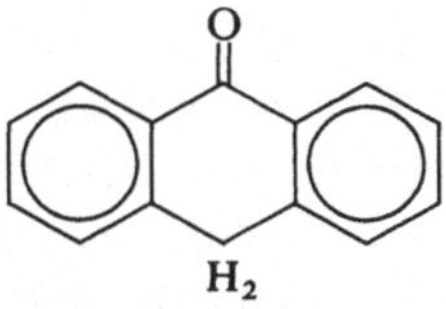   (110)

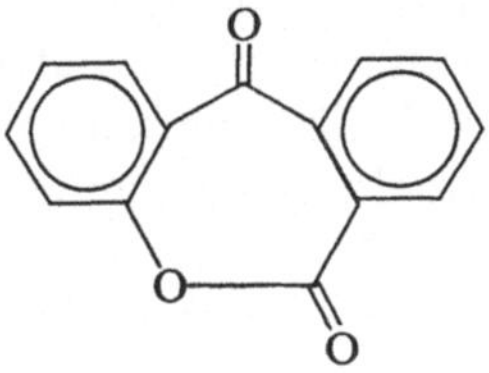

(111) 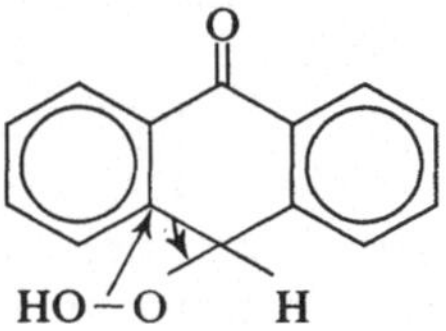   (112)

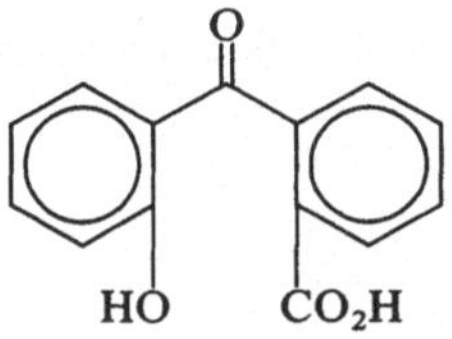

(113) 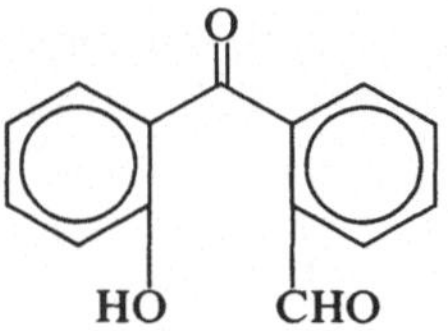   (114)

hydroperoxid wurde schon von MONEY (*61*) im Zusammenhang mit biogenetischen Hypothesen diskutiert. Anthrone sind Zwischenstufen der
Anthrachinon-Biosynthese aus Acetateinheiten. Auch können sie leicht
durch Reduktion von Anthrachinonen gebildet werden. Wegen der Un-
beständigkeit des Anthron-hydroperoxids wurde zur Untersuchung der
säurekatalysierten Umlagerung das 10-Methylanthron-hydroperoxid (**115**)
dargestellt (*35*). Einwirkung von Eisessig/4 *n* H_2SO_4 auf (**115**) bei 40°
ergab mit 4,5% Ausbeute das gesuchte Produkt (**116**). Diese erstmalig
durchgeführte Ringöffnung eines Anthrons über das 10-Methylanthron-
hydroperoxid kann als Modellreaktion für die Bildung eines Benzophenons (**96**) aus Emodin (**101**) bei der Ergochrom-Biosynthese angesehen
werden (*21*).

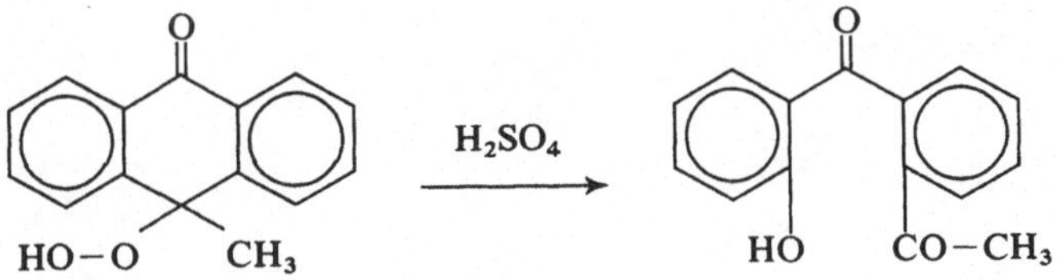

(115)   (116)

An die Ringöffnung des Emodins schließt sich bei der Ergochrom-Biosynthese die Cyclisierung des entstandenen Benzophenons an. Die Nachahmung dieses Reaktionsschrittes in Modellversuchen ist sowohl für das Verständnis des Reaktionsmechanismus der Biosynthese als auch zum Aufbau des Ergochrom-Grundgerüstes im Rahmen einer Totalsynthese von Interesse. Hierfür bot sich eine intramolekulare Phenoloxidation an. So konnten FRANCK und Mitarb. (*38*) durch Synthese und oxidative Kondensation des Benzophenons (**117**) das Xanthon-Derivat (**118**) darstellen. Dabei führt die Oxidation zunächst zum p-Chinon von (**117**), das bei anschließender milder Alkalieinwirkung intramolekular kondensiert. Reduktion von (**118**) mit Natriumborhydrid ergab ein Gemisch von zwei diastereomeren Dienolen, aus dem (**120**) schichtchromatographisch abgetrennt wurde. Dieses ließ sich katalytisch zu (**119**) hydrieren.

In enger Anlehnung an die Biosynthese gelang hiermit die Totalsynthese des racemischen Xanthons (**119**), dessen relative Konfiguration an allen drei Chiralitätszentren einer Molekülhälfte der Secalonsäure A (**63**) entspricht. Es unterscheidet sich davon nur durch eine Methylgruppe anstatt der CO_2CH_3-Gruppe an C-10 und stellt somit eine *10-Methyl-hemi-secalonsäure A* (**119**) dar.

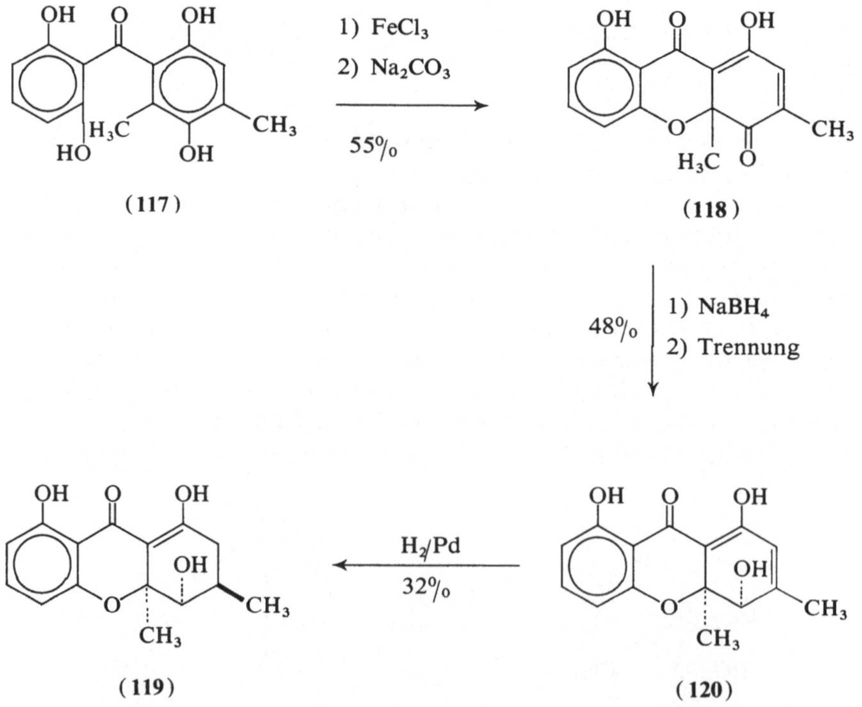

Literaturverzeichnis: SS. 203—206

Literaturverzeichnis

1. ABERHART, D. J., Y. S. CHEN, P. DE MAYO, and J. B. STOTHERS: Mould Metabolites IV. The Isolation and Constitution of some Ergot Pigments. Tetrahedron (London) **21,** 1417 (1965).

2. ABERHART, D. J., and P. DE MAYO: Mould Metabolites V. The Constitution of Ergoxanthin. Tetrahedron (London) **22,** 2359 (1966).

3. APSIMON, J. W., J. A. CORRAN, N. G. CREASEY, W. MARLOW, W. B. WHALLEY, and K. Y. SIM: The Chemistry of Fungi. Part XLVII. The Constitution of Ergochrysin A, Secalonic Acid A, and Secalonic Acid B. J. chem. Soc. (London) **1965,** 4144.

4. APSIMON, J. W., J. A. CORRAN, N. G. CREASEY, K. Y. SIM, and W. B. WHALLEY: The Chemistry of Fungi. Part XLVI. The Constitution of Ergoflavin. J. chem. Soc. (London) **1965,** 4130.

5. ASAHINA, Y., und F. FUZIKAWA: Untersuchungen über Flechtenstoffe, LV. Mitteil.: Über Endocrocin, ein neues Oxyanthrachinon-Derivat. Ber. dtsch. chem. Ges. **68,** 1558 (1935).

6. BEINERT, W.: Dissertation, Universität Kiel, 1970.

7. BERGMANN, W.: Über die gelben Farbstoffe des Mutterkorns, II. Mitteil.: Das Ergochrysin. Ber. dtsch. chem. Ges. **65,** 1489 (1932).

8. BOWERS, A., T. G. HALSALL, E. R. H. JONES, and A. J. LEMIN: The Chemistry of the Triterpenes and Related Compounds. Part XVIII. Elucidation of the Structure of Polyporenic Acid C. J. chem. Soc. (London) **1953,** 2548.

9. BUTLER, W. H.: Acute Toxicity of Aflatoxin B_1 in Rats. Brit. J. Cancer **18,** 756 (1964).

10. CARLSON, W. W., J. TUITE, and P. MISLIVEC: Investigations of the Toxic Effects in Mice of Certain Species of Penicillium. Toxicol. Appl. Pharmacol. **13,** 372 (1968).

11. DAVIES, J. S., V. H. DAVIES, and C. H. HASSALL: The Biosynthesis of Phenols. Part XX. Synthesis of Anthrachinones through Carbanions of *ortho*-Substituted Benzophenones. J. chem. Soc. (London) C **1969,** 1873.

12. DJERASSI, C.: Optical Rotatory Dispersion, S. 64. New York: McGraw-Hill. 1960.

13. DJERASSI, C., R. RINIKER, and B. RINIKER: Optical Rotatory Dispersion Studies. VII. Application to Problems of Absolute Configurations. J. Amer. chem. Soc. **78,** 6362 (1956).

14. DRAGENDORFF, G., und V. PODWYSSOTSKI: Über die wirksamen und einige andere Bestandteile des Mutterkorns. Naunyn-Schmiedebergs Arch. exp. Pathol. Pharmakol. **6,** 153 (1877).

15. ECKERT, H.-G.: Diplomarbeit, Universität Münster, 1971.

16. EGLINTON, G., F. E. KING, G. LLOYD, J. W. LODER, J. R. MARSHALL, A. ROBERTSON, and W. B. WHALLEY: The Chemistry of Fungi. Part XXXV. A Preliminary Investigation of Ergoflavin. J. chem. Soc. (London) **1958,** 1833.

17. EVANS, E. A.: Tritium and its Compounds. London: Butterworths. 1966.

18. FEUELL, A. J.: Types of Mycotoxins in Foods and Feeds. In L. A. GOLDBLATT: Aflatoxin, p. 187. New York: Academic Press. 1969.

19. FRANCK, B.: Mutterkorn-Farbstoffe. Angew. Chem. **76,** 864 (1964); Angew. Chem. internat. Edit. **3,** 763 (1964).

20. — Struktur der Mutterkorn-Farbstoffe. Festschrift Prof. Dr. K. Mothes, S. 153. Jena: Fischer. 1965.

21. — Struktur und Biosynthese der Mutterkorn-Farbstoffe. Angew. Chem. **81,** 269 (1969); Angew. Chem. internat. Edit. **8,** 251 (1969).

22. FRANCK, B., und G. BAUMANN: Synthese und Wessely-Moser-Umlagerung von 8,8'- und 6,6'-Bichromonylen. Chem. Ber. **96,** 3209 (1963).

23. — — Isolierung, Struktur und absolute Konfiguration der Ergochrysine A und B. Chem. Ber. **99,** 3863 (1966).

24. Franck, B., und G. Baumann: Isolierung, Struktur und absolute Konfiguration der Ergochrome AD, BD, CD und DD. Chem. Ber. **99**, 3875 (1966).
25. Franck, B., G. Baumann und U. Ohnsorge: Ergochrome, eine ungewöhnlich vollständige Gruppe dimerer Farbstoffe aus *Claviceps purpurea*. Tetrahedron Letters **1965**, 2031.
26. Franck, B., und H. Flasch: Versuche 1971.
27. Franck, B., und E. M. Gottschalk: Diastereomere Secalonsäuren. Angew. Chem. **76**, 438 (1964); Angew. Chem. internat. Edit. **3**, 441 (1964).
28. Franck, B., E. M. Gottschalk, U. Ohnsorge und G. Baumann: Struktur der Secalonsäuren A und B. Angew. Chem. **76**, 438 (1964); Angew. Chem. internat. Edit. **3**, 441 (1964).
29. Franck, B., E. M. Gottschalk, U. Ohnsorge und F. Hüper: Trennung, Struktur und absolute Konfiguration der diastereomeren Secalonsäuren A, B und C. Chem. Ber. **99**, 3842 (1966).
30. Franck, B., F. Hüper, D. Gröger und D. Erge: Seco-Anthrachinone nach einem neuen Biosyntheseprinzip. Angew. Chem. **78**, 752 (1966); Angew. Chem. internat. Edit. **5**, 728 (1966).
31. — — — — Biosynthese der Ergochrome. Chem. Ber. **101**, 1954 (1968).
32. — — — — Unveröffentlicht.
33. Franck, B., und U. Ohnsorge: Versuche 1964.
34. Franck, B., U. Ohnsorge und H. Flasch: Einfache Totalsynthese von [10-^{14}C]-Endocrocin. Tetrahedron Letters **1970**, 3773.
35. Franck, B., V. Radtke und U. Zeidler: Modellreaktionen zur biologischen Ringöffnung der Anthrachinone. Angew. Chem. **79**, 935 (1967); Angew. Chem. internat. Edit. **6**, 952 (1967).
36. Franck, B., und T. Reschke: Isolierung der Hydroxy-anthrachinoncarbonsäuren Endocrocin und Clavorubin aus Roggenmutterkorn. Chem. Ber. **93**, 347 (1960).
37. Franck, B., O. W. Thiele und T. Reschke: Zur Konstitution der Secalonsäure. Chem. Ber. **95**, 1328 (1962).
38. *a)* Franck, B., und U. Zeidler: Oxidative Cyclisierung von Hydroxy-benzophenonen. Chem. Ber. 1972, im Druck;
b) Franck, B., J. Stöckigt und U. Zeidler: Stereospezifische Totalsynthese der 10-Methyl-hemisecalonsäure A. Chem. Ber. 1972, im Druck.
39. Franck, B., und I. Zimmer: Konstitution und Synthese des Clavorubins. Chem. Ber. **98**, 1514 (1965).
40. Franckowiak, G.: Diplomarbeit, Universität Münster, 1970.
41. Freeburn, A.: Untersuchungen über eine gelb gefärbte Verbindung aus Mutterkorn. Pharmac. J. **34**, 568 (1912); Chem. Zentralbl. **83**, 39 (1912, II).
42. Gatenbeck, S.: Incorporation of Labelled Acetate in Emodin in *Penicillium islandicum*. Acta chem. scand. **12**, 1211 (1958).
43. — Biosynthesis of Anthraquinones in Lower Fungi. Svensk kem. Tidskr. **72**, 188 (1960).
44. Gatenbeck, S., and L. Malmström: On the Biosynthesis of Sulochrin. Acta chem. scand. **23**, 3493 (1969).
45. Geissman, T. A.: The Biosynthesis of Phenolic Plants Products. In P. Bernfeld: Biogenesis of Natural Compounds, 2. Ed., p. 769. Oxford: Pergamon Press. 1967.
46. Gibbs, H. D.: Phenol Tests. III. The Indophenol Test. J. biol. Chemistry **73**, 649 (1927).
47. Gottschalk, E. M.: Dissertation, Universität Göttingen, 1964.
48. Gröger, D.: Über die Bildung von Clavin-Alkaloiden in Submerskultur. Arch. Pharmaz. **292**, 389 (1959).
49. Gröger, D., D. Erge, B. Franck, U. Ohnsorge, H. Flasch und F. Hüper: Emodin als Biosynthesevorstufe der Ergochrome. Chem. Ber. **101**, 1970 (1968).
50. Guider, J. M., T. H. Simpson, and D. B. Thomas: Anthoxanthins. Part II. Derivatives of Katuranin and Kaempferol. J. chem. Soc. (London) **1955**, 171.

51. HASSALL, C. H.: The Baeyer-Villiger Oxidation of Aldehydes and Ketones. Organic Reactions **9,** 73 (1957).

52. HOFMANN, A.: Die Mutterkorn-Alkaloide. Stuttgart: Enke. 1964.

53. HOOPER, J. W., W. MARLOW, W. B. WHALLEY, A. D. BORTHWICK, and R. BOWDEN: The Chemistry of Fungi. Part LXV. The Structures of Ergochrysin A, Isoergochrysin A, and Ergoxanthin, and of Secalonic Acids A, B, C, and D. J. chem. Soc. (London) **1971,** 3580.

54. HÜPER, F.: Diplomarbeit, Universität Kiel, 1965.

55. JACOBY, C.: Das Sphacelo-Toxin, der spezifisch wirksame Bestandteil des Mutterkorns. Naunyn-Schmiedebergs Arch. exp. Pathol. Pharmakol. **39,** 85 (1897).

56. KARPLUS, M.: Vicinal Proton Coupling in Nuclear Magnetic Resonance. J. Amer. chem. Soc. **85,** 2870 (1963).

57. KING, F. E., T. J. KING, and L. C. MANNING: An Investigation of the Gibbs Reaction and its Bearing on the Constitution of Jacareubin. J. chem. Soc. (London) **1957,** 563.

58. KRAFT, F.: Über das Mutterkorn. Arch. Pharmaz. Ber. dtsch. pharmaz. Ges. **244,** 336 (1906).

59. McPHAIL, A. T., G. A. SIM, J. D. M. ASHER, J. M. ROBERTSON, and J. V. SILVERTON: Fungal Metabolites. Part IV. The Structure of Ergoflavin: X-Ray Analysis of Tetra- O-Methylergoflavin-Di-p-jodobenzoate. J. chem. Soc. (London) **B 1966,** 18.

60. MIYAKE, M., and M. SAITO: Liver Injury Tumors induced by Toxins of *Penicillium islandicum* growing on yellowed Rice. In G. N. WOGAN: Mycotoxins in Foodstuffs, p. 133. Cambridge, Massachusetts: M. I. T. Press. 1965.

61. MONEY, T.: A Postulated Biosynthesis of Some "Anomalous" Natural Phenolic Compounds. Nature **199,** 592 (1963).

61a. MUSSO, H., und H.-G. MATTHIES: Acidität und Wasserstoffbrücken bei Hydroxy-biphenylen und Hydroxy-biphenylchinonen. Chem. Ber. **94,** 356 (1961).

62. OLLIS, W. D.: Persönl. Mitteilung.

63. REISS, J.: Gefährliche Stoffwechselprodukte von Pilzen auf Futter- und Nahrungsmitteln. Naturwiss. Rundschau **22,** 203 (1969).

64. RIGBY, W.: A New Reagent for the Oxidation of Acyloins to Diketones. J. chem. Soc. (London) **1951,** 793.

65. SHIBATA, S., and S. NATORI: Metabolic Products of Fungi. II. Metabolic Products of *Aspergillus amstelodami.* Pharmac. Bull. Japan **1,** 160 (1953).

66. SCHÜTTE, H.-R.: Radioaktive Isotope in der organischen Chemie und Biochemie. Weinheim: Verlag Chemie. 1966.

67. STEGLICH, W., R. ARNOLD, W. LÖSEL, and W. REININGER: Biosynthesis of Anthraquinone Pigments in *Dermocybe.* Chem. Commun. (London) **1972,** 102.

68. STEGLICH, W., W. LÖSEL und V. AUSTEL: Anthrachinon-Pigmente aus *Dermocybe sanguinea* (Wulf. ex. Fr.) Wünsche und *D. semisanguinea* (Fr.). Chem. Ber. **102,** 4104 (1969).

69. STEYN, P. S.: The Isolation, Structure and Absolute Configuration of Secalonic Acid D, the Toxic Metabolite of *Penicillium oxalicum.* Tetrahedron **26,** 51 (1970).

70. STOLL, A.: Recent Investigations on Ergot Alkaloids. Fortschr. Chem. org. Naturstoffe **9,** 114 (1952).

71. STOLL, A., A. BRACK, H. KOBEL, H. HOFFMANN und R. BRUNNER: Die Alkaloide eines Mutterkornpilzes von *Pennisetum typhoideum* Rich. und deren Bildung in saprophytischer Kultur. Helv. chim. Acta **37,** 1815 (1954).

72. STOLL, A., J. RENZ und A. BRACK: Über gelbe Farbstoffe im Mutterkorn. Helv. chim. Acta **35,** 2022 (1952).

73. VAN DER MERWE, K. J., P. S. STEYN, and L. FOURIE: Mycotoxins. II. The Constitution of Ochratoxins A, B, and C, Metabolites of *Aspergillus ochraceus.* J. chem. Soc. (London) **1965,** 7083.

74. VAN DER MERWE, K. J., P. S. STEYN, L. FOURIE, DE B. SCOTT, and J. J. THERON: Ochratoxin A, a Toxic Metabolite produced by *Aspergillus ochraceus*. Nature **205**, 1112 (1965).

75. Von der Heyde, O.: Diplomarbeit, Universität Kiel, 1966.
76. Wachtmeister, C. A.: Persönl. Mitteilung.
77. Wenzell, W. T.: Ergoxanthin, ein neuer wirksamer Bestandteil des Mutterkorns. Amer. J. Pharmacy **82**, 410 (1910); Chem. Zentralbl. **81**, 1390 (1910, II).
78. Wilzbach, K. E.: Tritium-Labeling by Exposure of Organic Compounds to Tritium-Gas. J. Amer. chem. Soc. **79**, 1013 (1957).
79. Yamazaki, M., Y. Maebayashi, and K. Miyaki: The Isolation of Secalonic Acid A from *Aspergillus ochraceus* cultured on Rice. Chem. Pharm. Bull. (Japan) **19**, 199 (1971).
80. Yosioka, I., T. Nakanishi, S. Izumi, and I. Kitagawa: Structure of a Lichen Pigment Entothein and its Identity with Secalonic Acid A, a Major Ergot Pigment. Chem. Pharm. Bull. (Japan) **16**, 2090 (1968).

(Eingelaufen am 21. April 1972)

The Chemistry of Biflavanoid Compounds

By H. D. LOCKSLEY, Salford, England

With 4 Figures

Contents

Acknowledgement: I wish to record my appreciation of the help given to me by Professor Nobusuke Kawano, Faculty of Pharmaceutical Sciences, Nagasaki University.

I. Introduction

As constituents of plants, flavanoids are amongst the most frequently encountered natural products. Since interest in flavanoids as a class of compounds can be traced as far back as the period of Robert Boyle preoccupation with them can justifiably be claimed to have played an important part in the development of chemistry. Until recently only mono-meric forms of flavanoids were thought to exist*. This is curious since the dimeric species, or biflavanoids, occur quite abundantly in some plants making their isolation not especially difficult. The first biflavanoid, a biflavone, was isolated in 1929 by Furukawa who extracted the autumnal leaves of the maidenhair tree, *Ginkgo biloba L.* and obtained a yellow pig-ment, "compound B", which later acquired the name ginkgetin. There can be little doubt that much of the impetus for this investigation was generated by the curious botanical position of this ancient tree, for it is the only species of the order Ginkgoales to survive to the present day. Fossilised remains of the plant establish that it was in existence some 250 million years ago (during the Cretaceous or even the Jurassic period of the earth's evolution) (*124*).

Although several reviews on biflavanoids already exist (*4, 55, 91*), there has been such an upsurge of work since publication of the last in 1962 that a re-examination of the literature is now warranted.

At the time of the last review only six members of the biflavone family were known and these belonged to two distinct structural types, *viz.* the

---

* Biflavanoids in which the C-4 carbonyl groups of rings I-C and II-C appear in reduced form are not included in this review. For a recent discussion of these compounds see K. Weinges, W. Bähr, W. Ebert, K. Göritz, and H.-D. Marx, „Konstitution, Entstehung und Bedeutung der Flavonoid-Gerbstoffe" in "Progress in the Chemistry of Organic Natural Products", ed. L. Zechmeister, Springer-Verlag, Vienna, Vol. XXVII, 1969, pp. 158—260.

amentoflavone **(1)** type (with apigenin units linked C-3'/C-8), and the hinokiflavone* **(2)** type (with apigenin units linked C-4'/C-6 through ether oxygen). Since that time biflavones formed from units of apigenin and belonging to two new types have been isolated, *viz.* the cupressuflavone **(3)** type (with a symmetrical C-8/C-8 interflavone linkage) and the agathisflavone **(4)** type (with an unsymmetrical C-6/C-8 interflavone linkage).

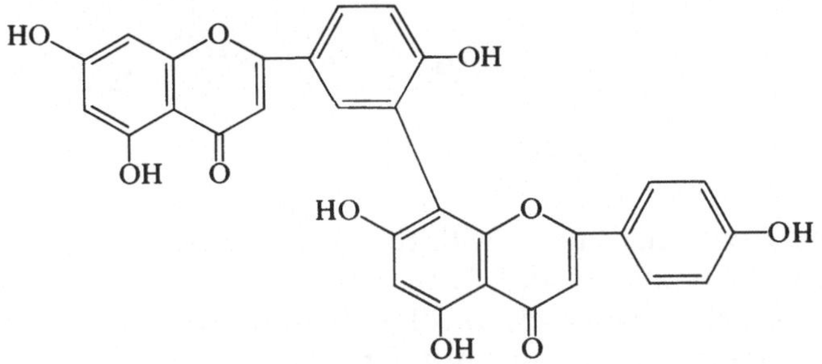

(1) Amentoflavone

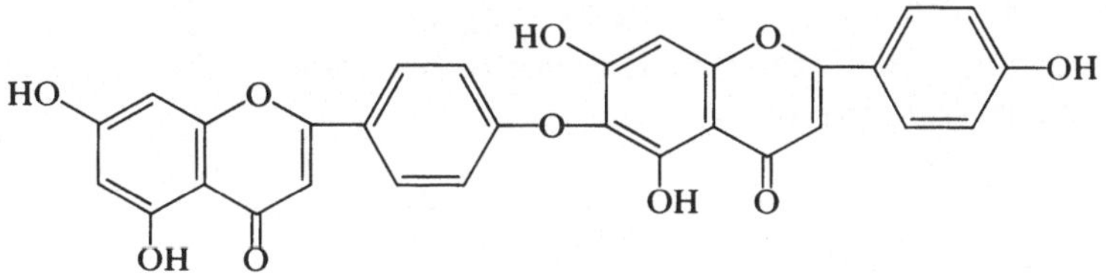

(2) Hinokiflavone

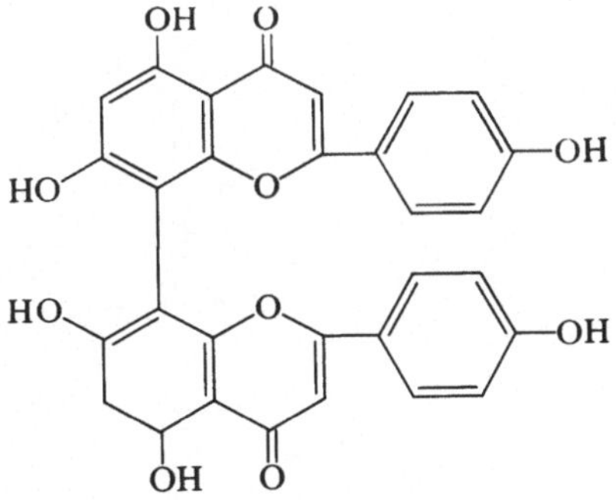

(3) Cupressuflavone

---

* At the time of the last comprehensive review in 1961, hinokiflavone **(2)** was thought to have the C-4'-*O*-C-8 linkage. Later work has shown this to be incorrect: the linkage is C-4'-*O*-C-6.

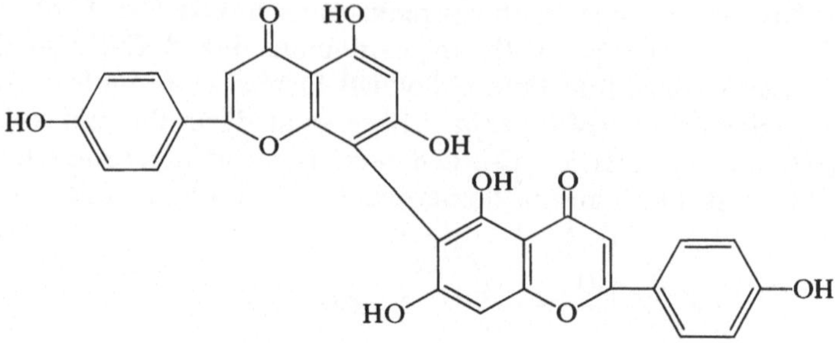

(**4**) Agathisflavone

Recently two biflavanoids with the carbon skeletons (**5**) and (**6**), i.e. with one flavone ring reduced, have been added to this list, namely I-2, I-3-dihydroamentoflavone (**5**) and I-2, I-3-dihydrohinokiflavone (**6**). In addition to these, a completely new family of C-3/C-8 linked flavanone-flavones has been discovered, *viz.* volkensiflavone (**7**) and morelloflavone (**8**). Related to these is another new series of biflavanones GB-1 (**9**), GB-1a (**10**), GB-2 (**11**), and GB-2a (**12**), possessing a C-3/C-8 linkage between units of naringenin and, respectively, aromadendrin, naringenin, taxifolin, and eriodictyol.

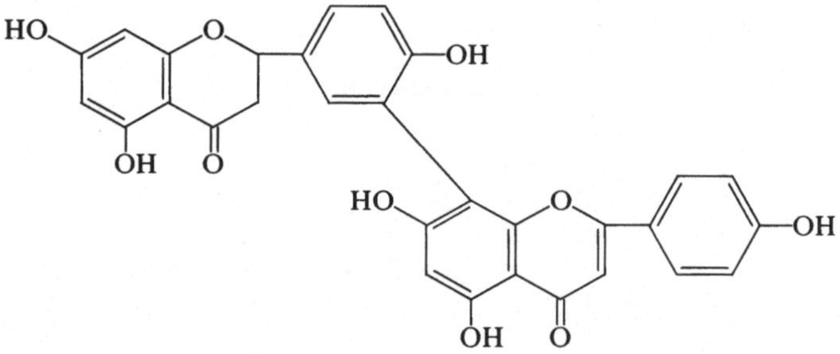

(**5**) Dihydroamentoflavone

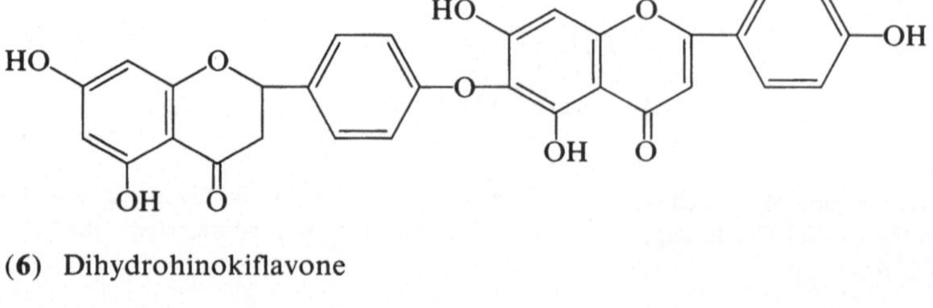

(**6**) Dihydrohinokiflavone

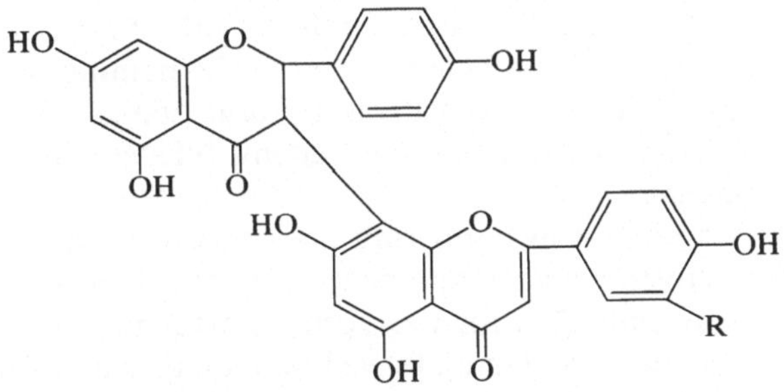

(**7**) Volkensiflavone (R = H)
(**8**) Morelloflavone (R = OH)

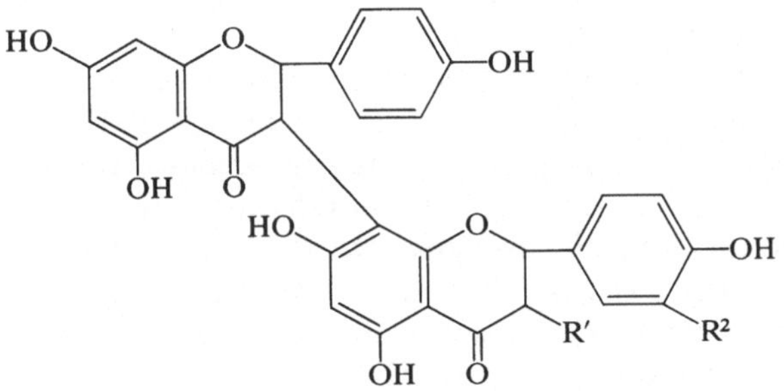

(**9**) GB-1   (R′ = OH, R² = H)
(**10**) GB-1a (R′ = R² = H)
(**11**) GB-2   (R′ = R² = OH)
(**12**) GB-2a (R′ = H, R² = OH)

II. Nomenclature for Biflavanoids

The rapid growth in the literature has led to the introduction of several systems of nomenclature in addition to the widespread use of trivial names to describe biflavanoids. In order to rationalize the nomenclature, we at Salford sought the views of Professors W. D. OLLIS, N. KAWANO, T. R. SESHADRI, A. PELTER, and Drs. R. S. CAHN and L. C. CROSS. From these exchanges of views the following system of nomenclature evolved (*36*).

The generic name biflavanoid is advocated in place of biflavonyl *etc.*, to describe the family of flavanoid dimers. The ending "oid" may then be modified to cover specific types of flavanoid dimers such as biflavanone, biflavone, biflavan, *etc.*, and for mixed systems, flavanone-flavone. This system follows general I.U.P.A.C. policy. The unmodified monomer nomenclature is utilized as a generic term in the naming of dimeric, trimeric, and tetrameric *etc.*, derivatives by insertion of the appropriate Greek prefices, bi-, ter-, quater-, *etc.,* giving biflavanoid, terflavanoid, quaterflavanoid, *etc.*

To identify specific rings and ring positions in flavanoids and their polymeric derivatives, we have retained the present long accepted system [exemplified in formula (71) for naringenin], extending it in the case of polymeric flavanoids by assigning to each monomer unit a Roman numeral, I, II, III, etc., running in sequence from one end of the molecule to the other. The points of linkage between neighbouring flavanoid units are identified by a combination of a Roman numeral (to identify the flavanoid unit) and an Arabic numeral (to identify the position of the interflavanoid link), the two numerals being coupled with a hyphen and enclosed within square brackets.

A few examples of this nomenclature will serve best to illustrate its use. GB-1 (9) would have the systematic name II-3, I-4′, II-4′, I-5, II-5, I-7, II-7-heptahydroxy [I-3, II-8] biflavanone, and amentoflavone (1) would be named I-4′, II-4′, I-5, II-5, I-7, II-7-hexahydroxy [I-3′, II-8] biflavone. Volkensiflavone (7), which possesses a flavanone and a flavone unit, would be named I-4′, II-4′, I-5, II-5, I-7, II-7- hexahydroxyflavanone [I-3, II-8] flavone. Hinokiflavone (2), whose flavone units are linked through oxygen, would show this feature in the name II-4′, I-5, II-5, I-7, II-7- pentahydroxy [I-4′-*O*-II-6] biflavone.

In this nomenclature the generic term biflav*a*noid has been adopted in preference to biflav*o*noid since in general the saturated system is regarded as the parent for the purposes of nomenclature. It should be noted that this concept has already been adopted in the name "neoflav*a*noid" rather than "neoflav*o*noid".

III. Chemistry of Biflavones of the Amentoflavone (1),
Family-Structures of Ginkgetin (13),
Isoginkgetin (23), and Sciadopitysin (14)

In the period prior to the impact of n.m.r. and mass spectrometry on natural products chemistry (i.e. the period roughly before 1960), chemists were forced to rely heavily upon structural evidence built up by established techniques of classical degradation. The use of these methods almost always demanded an abundant supply of the raw material, and where large complex molecules were involved, frequently left the final structure in doubt, due to a variety of limitations in the method (e.g. failure to collect and identify enough fragment molecules to enable the original to be reconstructed decisively; failure to recognize the occurrence of rearrangements in the original carbon skeleton during degradation, etc.). The early structural chemistry of ginkgetin (13) and sciadopitysin (14), a related biflavanoid pigment from the "umbrella pine" *Sciadopitys verticella* Sieb. & Zucc., provides good illustration of the formidable difficulties encountered by chemists of the pre-1960 period.

(13) Ginkgetin

(14) Sciadopitysin

The outline of the structural work on ginkgetin (13) and sciadopitysin (14) is included here merely to illustrate the manner in which classical degradative techniques were utilized. More comprehensive reviews of this structural chemistry already exist (*4, 55, 91*).

In the early studies carried out on the biflavones ginkgetin (13) and sciadopitysin (14) much use was made of alkali degradation reactions since the outcome of these reactions could reasonably be expected to follow the pattern already established for monoflavones under such conditions. Such degradation reactions have been reviewed elsewhere (*127*) but since apigenin (15) is the key unit of the four biflavones (1—4) it is of interest to examine the manner in which it decomposes under the

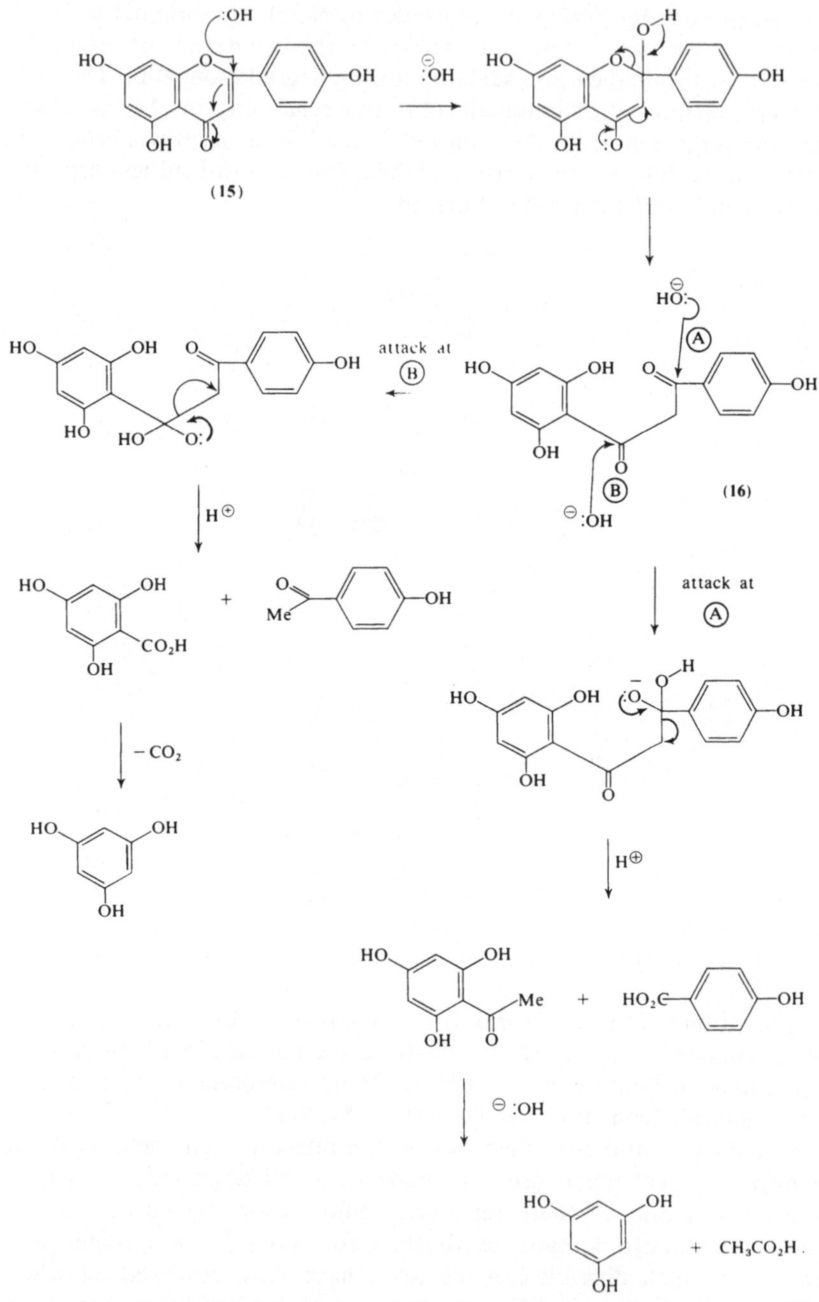

H. D. LOCKSLEY:

Chart 1. Products from alkali degradation of apigenin (15)

References, pp. 305—311

influence of alkali. The mechanism of decomposition shown in Chart 1, leads initially to the formation of β-diketone (16) which further decomposes through attack by hydroxyl ion at either carbonyl group and gives ultimately some or all of the characteristic degradation products shown. The vigour of the alkali treatment tends to determine the number of products formed: fusion with alkali at 220° — possibly the most vigorous treatment — tends to give fewer degradation products than, for example, treatment with hot methanolic barium hydroxide or hot aqueous potassium hydroxide. Use of the fusion reaction in cases where the flavone nucleus carries methoxy groups is considered inadvisable, for demethylation is frequently observed to accompany degradation; however, methoxy groups survive under most other alkaline conditions. Alkaline hydrogen peroxide is a much favoured alternative means of effecting controlled oxidative degradation and has been utilized with mono and biflavanoids alike (3, 4). Under these conditions apigenin trimethyl ether for example furnishes p-methoxybenzoic and 2-hydroxy-4,6-dimethoxybenzoic acids in good yield (3).

The work of NAKAZAWA (90—92, 94) showed ginkgetin (13) to have a molecular formula $C_{30}H_{16}O_8(OMe)_2$ and to contain four phenolic hydroxy groups through its formation of tetraacetyl, tetramethyl, and tetraethyl derivatives. Formation of a dimethyl ether derivative using milder methylation conditions demonstrated that two of the four phenolic hydroxy groups displayed greater resistence to methylation than the other two. When fused with alkali, ginkgetin gave acetic acid, phloroglucinol, p-hydroxybenzoic acid, and an acid $C_9H_{10}O_5$ of unknown constitution (Chart 2). Hydrolytic decomposition with 20% aqueous potassium hydroxide gave different products, namely p-hydroxyacetophenone, phloroglucinol monomethyl ether and a ketoflavone, $C_{24}H_{18}O_7$, of unknown structure (Chart 2). Though definitive evidence was lacking, NAKAZAWA believed the acid $C_9H_{10}O_5$ to be a derivative of phenylacetic acid, and the ketoflavone $C_{24}H_{18}O_7$ to be a deoxybenzoin with the constitution (17) or (18). Consideration of these degradation products led NAKAZAWA to postulate erroneously structure (19) or (20) for ginkgetin in which two units of genkwanin (5,4'-dihydroxy-7-methoxyflavone) were linked either between C-3 and C-8 or C-3 and C-6, respectively. The work outlined above illustrates well the difficulties and the limitations of the classical approach to structure elucidation.

Much of the structural work on ginkgetin which followed was concentrated on attempts to synthesize the ketoflavones (17) and (18). Though this was successfully achieved (95—98), the efforts were unrewarding for neither synthetic material proved to be identical with the degradation product, $C_{24}H_{18}O_7$, from ginkgetin. Consequently, formulae (19) and (20) were abandoned as possible structures for ginkgetin.

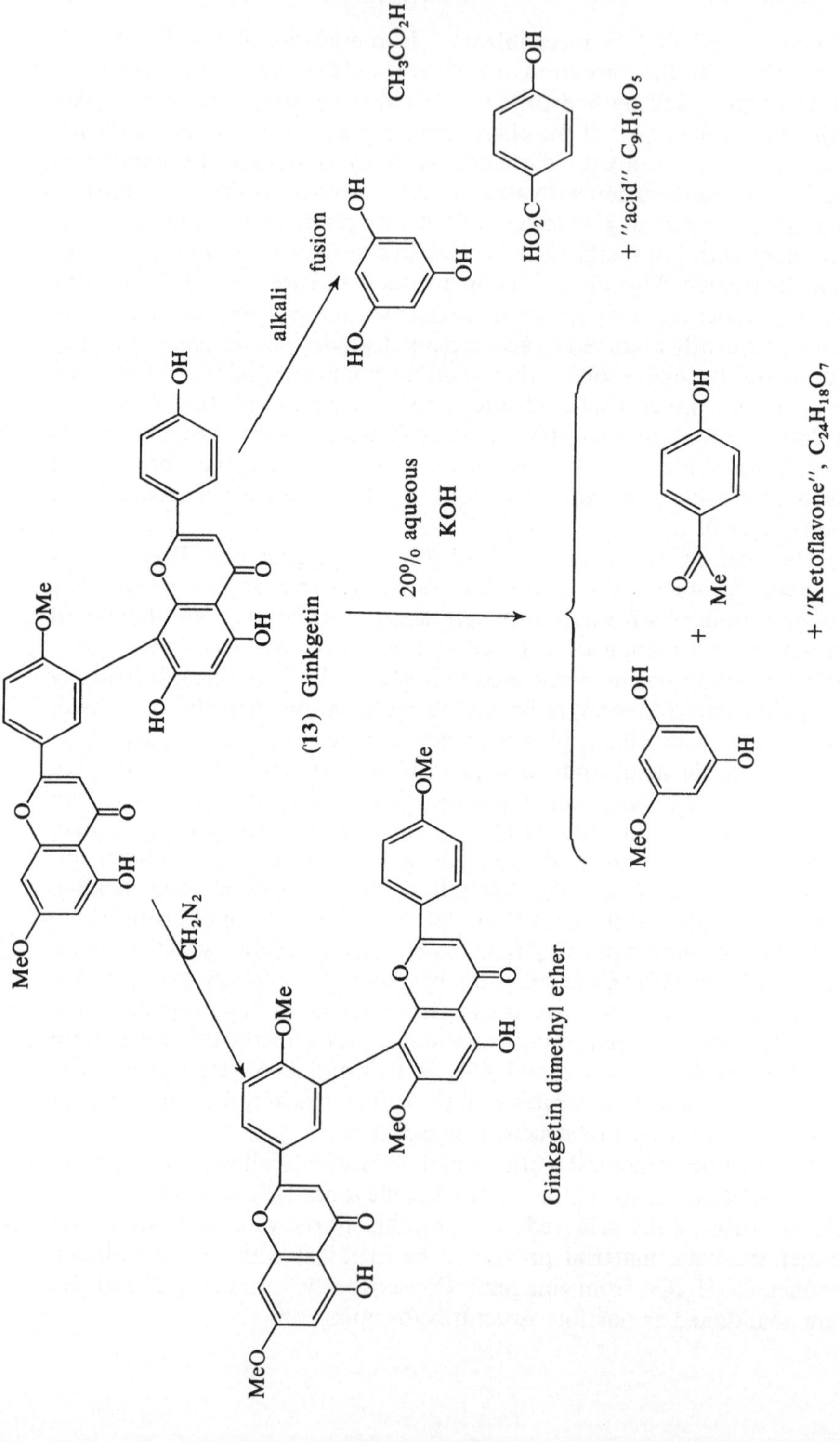

Chart 2. Alkali degradation reactions of ginkgetin (13)

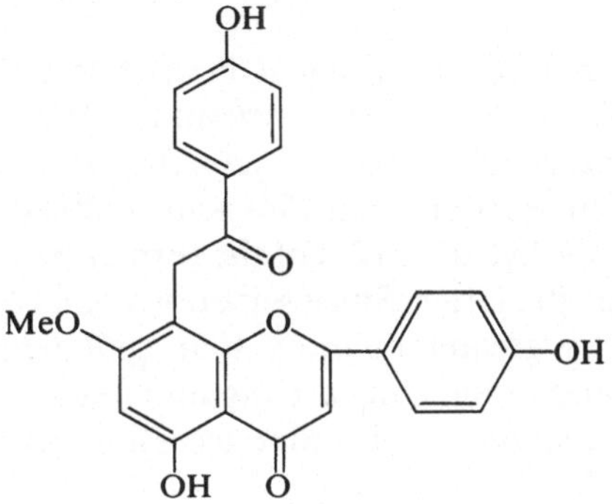

(17)

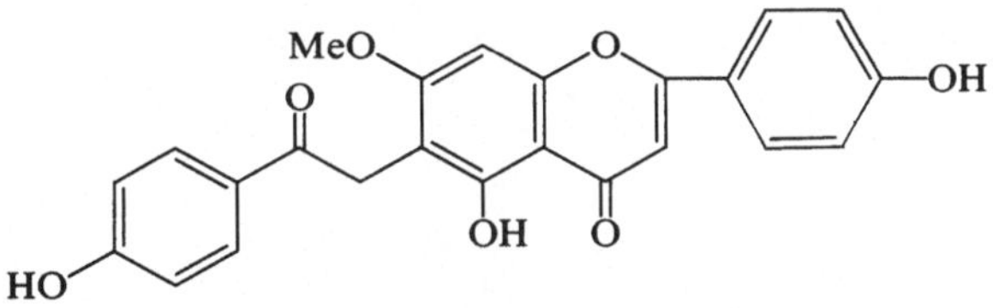

(18)

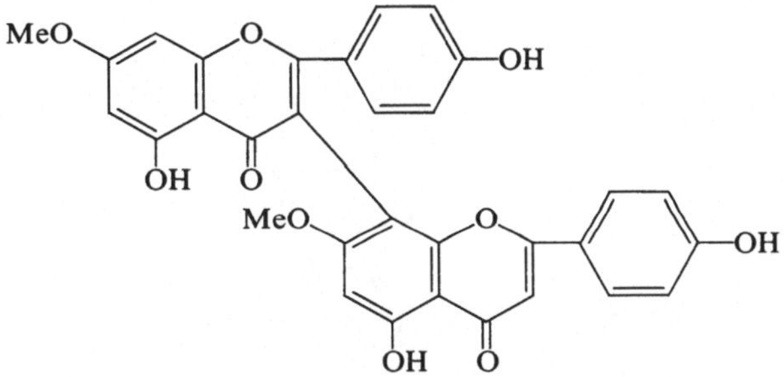

(19)

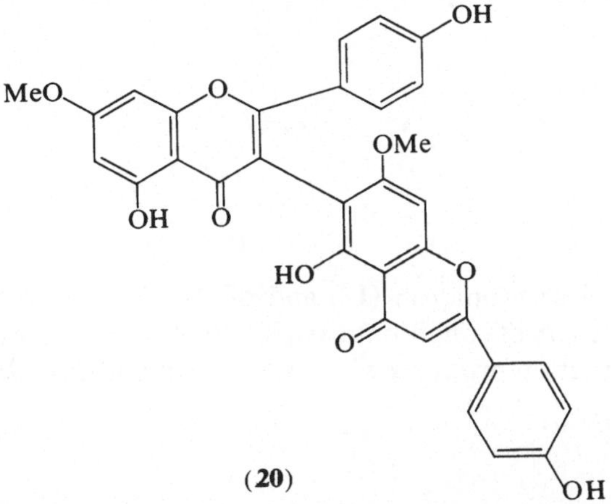

(20)

In 1956, 15 years after the work of Nakazawa, Kariyone and Kawano published (*42—46, 51—53*) an account of their structural work on the related biflavone, sciadopitysin, $C_{30}H_{12}O_4(OH)_3(OMe)_3$. A structural relationship between sciadopitysin (**14**) and ginkgetin (**13**) was quickly established; the trimethyl ether (**21**) of the former being identical with the tetramethyl ether of the latter. Since sciadopitysin (**14**) could be obtained more readily and in greater quantity than ginkgetin, the attention of chemists in Japan and in the United Kingdom tended to be diverted away from ginkgetin to a study of this more accessible natural product.

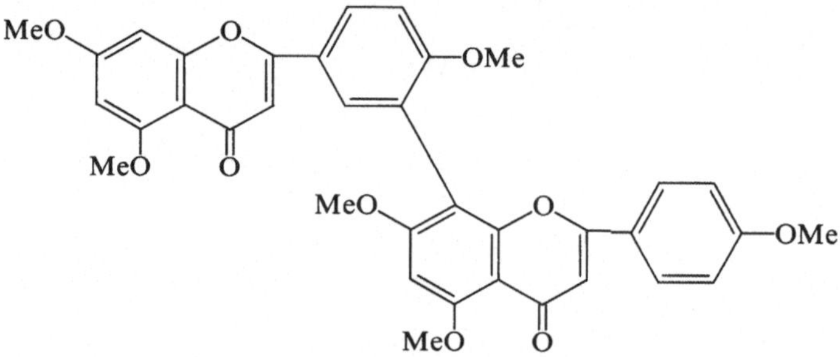

(**21**) Sciadopitysin trimethyl ether ($\equiv$ Gingketin tetramethyl ether)

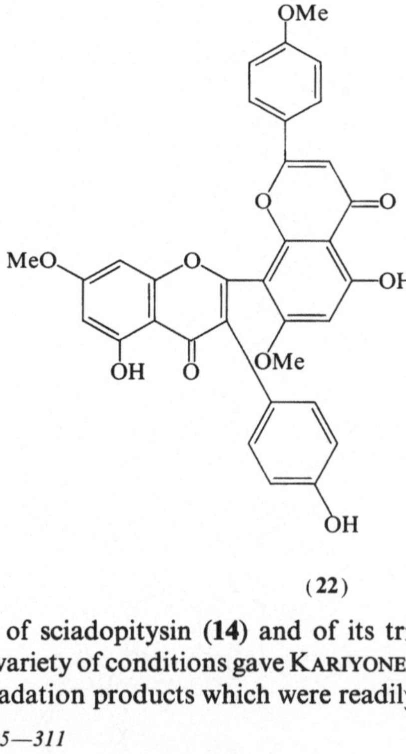

(**22**)

Treatment of sciadopitysin (**14**) and of its trimethyl ether (**21**) with alkali under a variety of conditions gave Kariyone and Kawano a number of simple degradation products which were readily identifiable (see Chart

3); however, other fragments proved too large to be identified completely, consequently only a tentative structure **(22)** for sciadopitysin emerged from this work. Through the structural relationship already established, it followed that ginkgetin must also have this carbon skeleton.

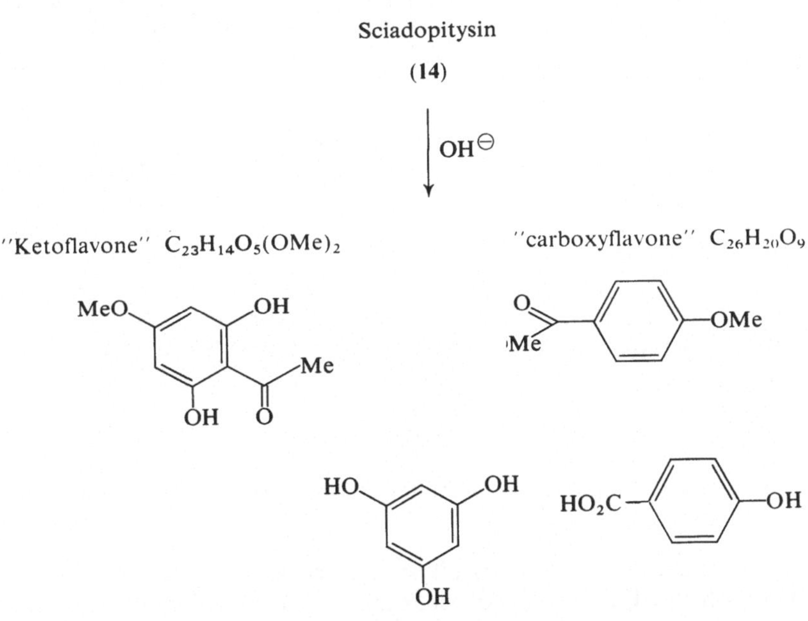

(23) Isoginkgetin

WILSON BAKER and his co-workers at the University of Bristol, realizing that there were inadequacies in the evidence for the isoflavone-flavone structure **(22)** deduced for sciadopitysin, decided to reinvestigate the whole problem afresh (*3*). Ginkgetin **(13)**, and a new metabolite, iso-ginkgetin **(23)**, were isolated from *Ginkgo biloba L.* Ginkgetin tetramethyl

Chart 3a. Alkali degradation of sciadopitysin **(14)**

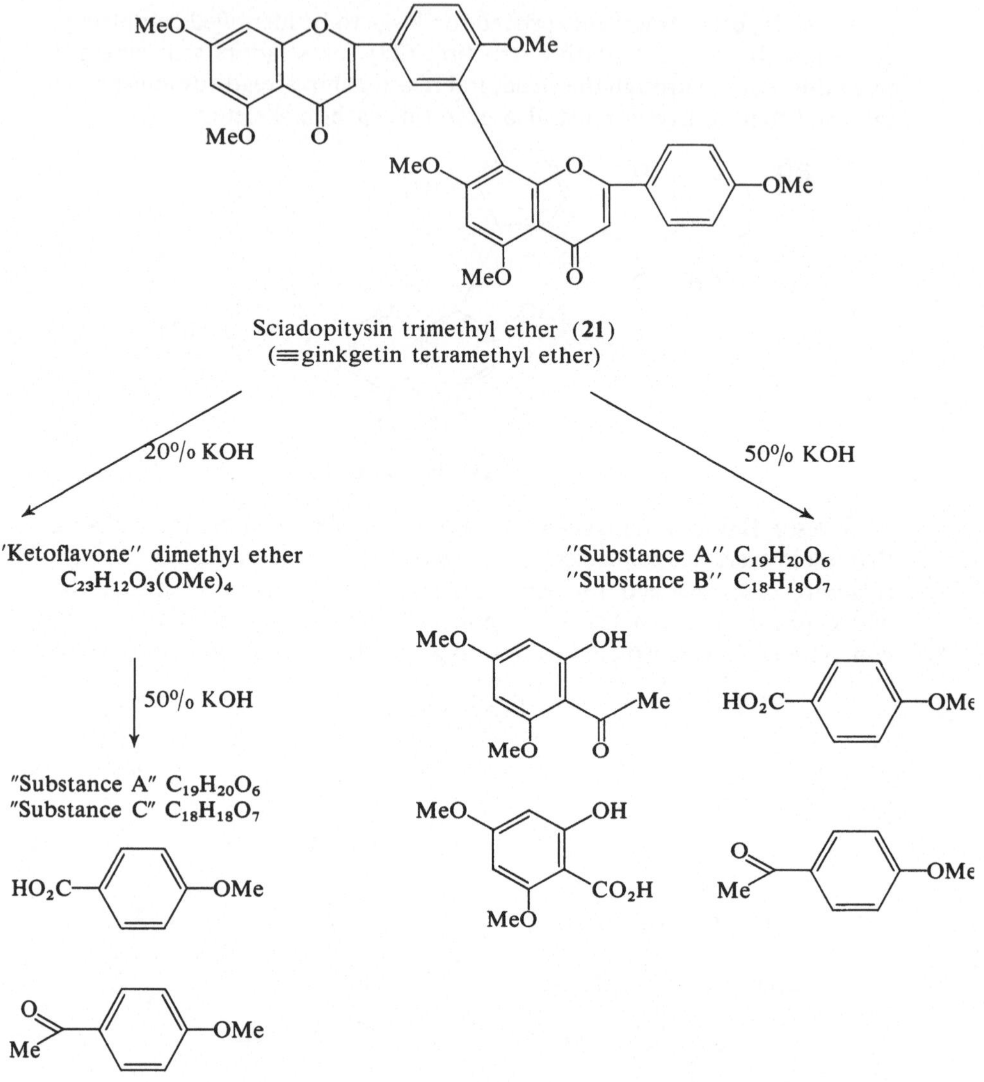

Sciadopitysin trimethyl ether (**21**)
(≡ginkgetin tetramethyl ether)

20% KOH

50% KOH

"Ketoflavone" dimethyl ether
$C_{23}H_{12}O_3(OMe)_4$

"Substance A" $C_{19}H_{20}O_6$
"Substance B" $C_{18}H_{18}O_7$

50% KOH

"Substance A" $C_{19}H_{20}O_6$
"Substance C" $C_{18}H_{18}O_7$

Chart 3b. Products of alkali degradation of sciadopitysin trimethyl ether (**21**)

ether (**21**) (identical with sciadopitysin trimethyl ether) was cleanly degraded with alkaline hydrogen peroxide to give *p*-methoxybenzoic (**24**) and 2-hydroxy-4,6-dimethoxybenzoic (**25**) acids and a dicarboxylic acid, $C_{17}H_{16}O_8$, with partial structure $C_{12}H_4(OMe)_3(OH)(CO_2H)_2$ (Chart 4).

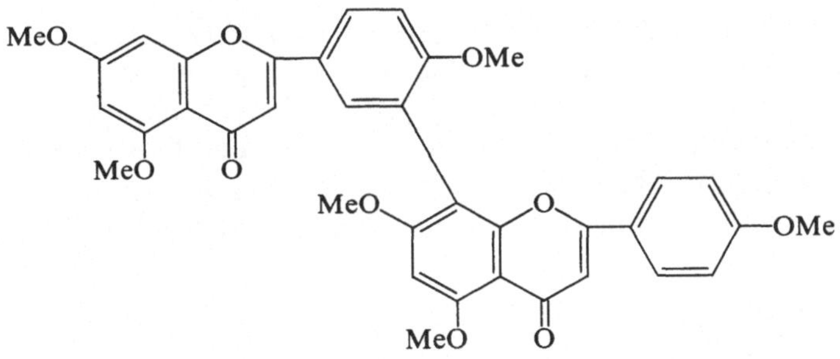

(**21**) Ginkgetin tetramethyl ether $\equiv$ sciadopitysin trimethyl ether

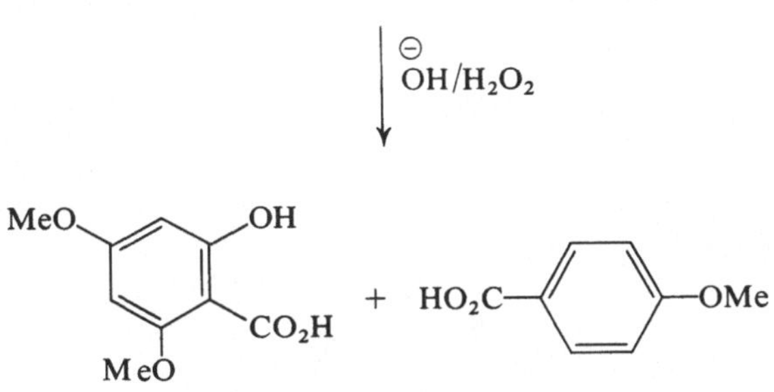

$$\overset{\ominus}{O}H/H_2O_2$$

(**25**)   (**24**)

+ "Dicarboxylic acid" $C_{17}H_{16}O_8$
or $C_{12}H_4(OMe)_3(OH)(CO_2H)_2$

Chart 4. Degradation of ginkgetin tetramethyl ether (**21**) with alkaline
hydrogen peroxide

Ultraviolet spectroscopy was utilized with conspicuous success to show
that this dicarboxylic acid was in fact a derivative of biphenyl possessing
ortho substituents of sufficient bulk to cause steric inhibition of resonance
between the aryl nuclei. Virtually the same ultraviolet spectrum was produced by an equimolar mixture of *p*-methoxybenzoic (**24**) and 2-hydroxy-
4,6-dimethoxybenzoic (**25**) acids which, if suitably combined through a
diaryl linkage were capable of providing a biphenyl with the same molecular composition and functional groups as those present in the degradation
product.

Speculation about the biogenetic origin of biflavones such as ginkgetin
(**13**) also favoured a diaryl linkage since this could be brought about
readily through phenol oxidative coupling of two molecules of apigenin.
The theory (*6, 16, 122, 123*), furthermore, demands that the ring carbons
involved in the diaryl linkage be situated *ortho* or *para* to a phenolic

222 H. D. Locksley:

group. Consequently, the two fragments (24) and (25) (Chart 4) could be
linked in one of two ways only (26) and (27), giving to ginkgetin tetramethyl ether either structure (28) or (21). Of these, the latter was preferred
since no difficulty was encountered in alkylating all four hydroxy-groups
of ginkgetin (13). The alternative structure (28) possesses a crowded C-5
hydroxy-group in ring II-A which would be expected to alkylate with
difficulty.

(26)

(27)

(28)

To reinforce the spectroscopic and biogenetic evidence, Baker and
his co-workers (3) carried out degradative experiments using the parent
biflavones, ginkgetin (13), isoginkgetin (23), and sciadopitysin (14). In
each case alkaline hydrogen peroxide produced 4-methoxy-isophthalic
acid (29) (Chart 5) showing conclusively that each biflavone contained a
linkage at C-3' of ring I-B. Other acidic products from the degradation were
p-methoxybenzoic acid (24) from isoginkgetin (23) and sciadopitysin (14),
and p-hydroxybenzoic acid (30) from ginkgetin (13): these latter acids

arise from ring II-B in each case (see Chart 5). From this and earlier
(*51—53, 94—98, 42—46*) evidence (Chart 2) it follows that ginkgetin must
have either structure (13) or (31): no ambiguity exists in defining the structures of sciadopitysin and isoginkgetin (all methoxy groups appear in the
fragments), which must have structures (14) and (23), respectively.

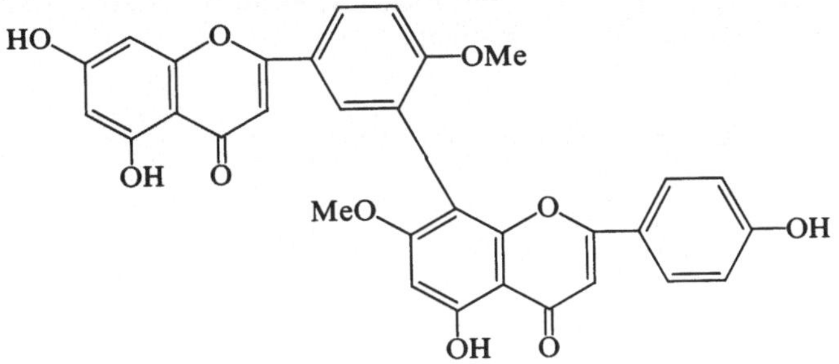

(31)

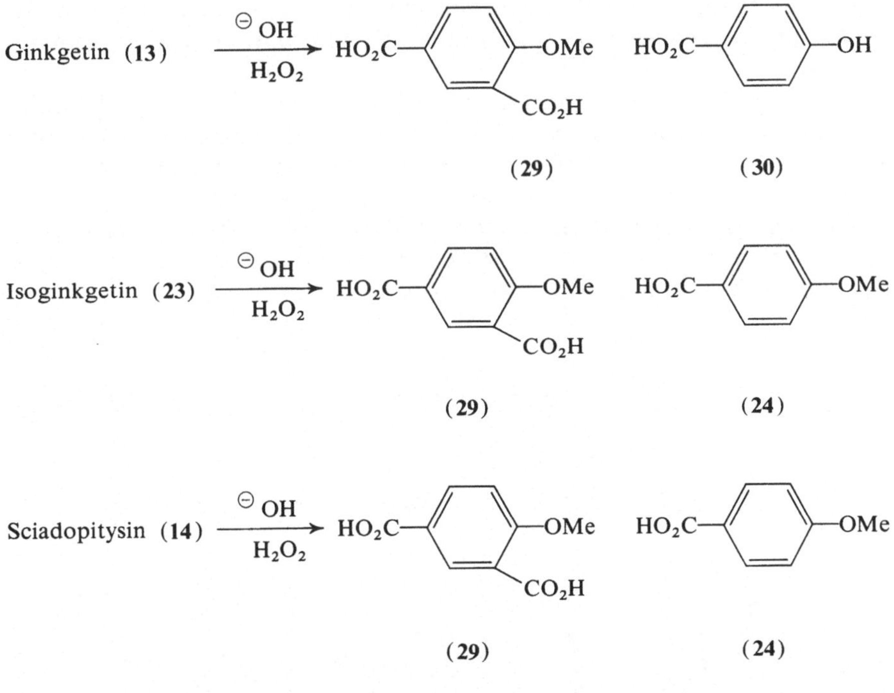

Ginkgetin (13) $\xrightarrow[\text{H}_2\text{O}_2]{^{\ominus}\text{OH}}$

(29)   (30)

Isoginkgetin (23) $\xrightarrow[\text{H}_2\text{O}_2]{^{\ominus}\text{OH}}$

(29)   (24)

Sciadopitysin (14) $\xrightarrow[\text{H}_2\text{O}_2]{^{\ominus}\text{OH}}$

(29)   (24)

Chart 5. Degradation products from ginkgetin (13), isoginkgetin (23) and
sciadopitysin (14) with alkaline hydrogen peroxide

An ingenious use of ultraviolet spectroscopy enabled a choice to be made between structures (13) and (31) for ginkgetin. The effect of added base on the ultraviolet spectra of simple flavones such as apigenin (15) and its methylated derivatives had been the subject of earlier studies (73). In neutral ethanol apigenin (15) and related flavones show two bands I and II, at *ca.* 270 nm and *ca.* 330—340 nm. The addition of strong base (e. g. NaOH) produces the following effects,

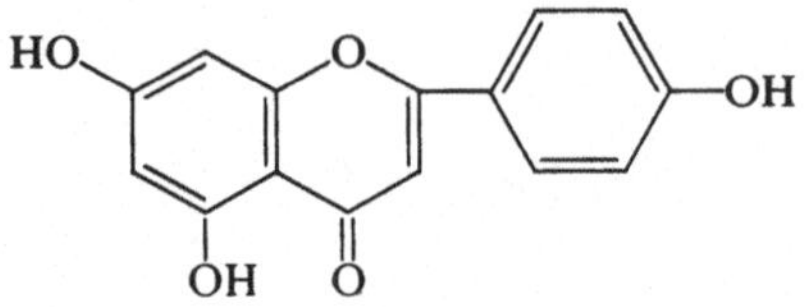

(15)

(a) when a 7-hydroxy group is present,
   -band I shifts from 270 to 280 nm with a great increase in intensity, band II shifts from 330 to 370 with a reduction in intensity.
(b) when a 4'-hydroxy group is present,
   -band I shifts from 270 to 280 nm with a reduction in intensity, band II shifts very considerably, from 330 to 400 nm, with an increase in intensity.

Electronically these bathochromic shifts are the result of charge delocalisation in the anion as exemplified in (32) and (33). When hydroxy groups are present at C-7 and C-4' in the *same* molecule, it is the 7-hydroxy group which ionises more readily in base since it is the more acidic. In practice, however, an equilibrium is established in which ionisation occurs at both hydroxy functions, that at the C-7 group being the more predominant at low concentrations of base.

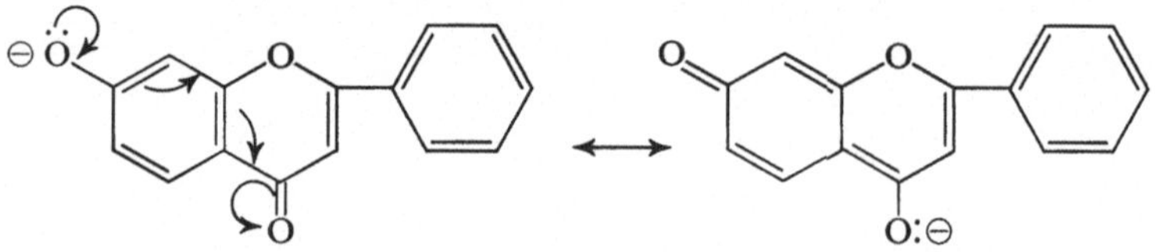

(32)

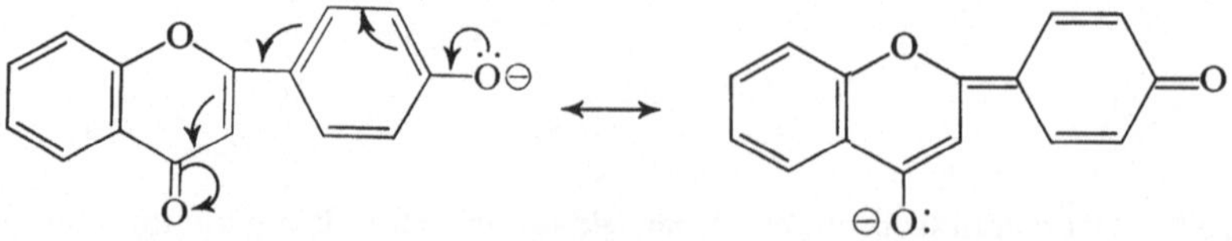

(33)

When the effect of added base on the biflavones (**23**), (**14**), and ginkgetin (**13**) or (**31**) was studied (*3*) the following results emerged.
Isoginkgetin (**23**): band I, shifted from 271.5 to 280 nm with a marked *increase* in intensity; band II, shifted from 330 to 376.5 nm with a considerable *decrease* in intensity.
Sciadopitysin (**14**): band I shifted from 271.5 to 287 nm with a marked *increase* in intensity; band II shifted from 330 to 378 nm with a very considerable *decrease* in intensity.
Ginkgetin (**13**) or (**31**): band I shifted from 271.5 to 284 nm with a moderate *increase* in intensity; band II shifted from 335 to 397 nm with a *decrease* in intensity appreciably less than that for isoginkgetin (**23**) and for sciadopitysin (**14**).

These shifts clearly showed isoginkgetin (**23**) to be a flavone of type (a) with two 7-hydroxy groups but lacking free hydroxyls at positions I-4' or II-4' in accordance with the structure (**23**) proposed on degradative evidence. The spectral changes for sciadopitysin (**14**) were also of type (a) in accord with the proposed structure except that much more concentrated base (*N*/50) was required to effect complete ionisation of the 7-hydroxy group, the presence of the hydroxy-group at the sterically crowded C-7 position of ring II-A being cited as the reason for this abnormally low acidity. Analysis of the spectral changes for ginkgetin showed that it had some of the features of a type (b) spectral shift thus indicating the presence of a 4'-hydroxy group, except that a *decrease* rather than an *increase* in intensity was observed. This apparently anomalous behaviour was considered to be in accordance with the presence of a hydroxy group at C-4' of ring II-B, the intensity reduction being caused by the presence of a 7-hydroxyl in the same flavone unit with greatly reduced acidity. Location of these two hydroxyls in flavone unit II rationalises the behaviour and leads unambiguously to structure (**13**) for ginkgetin.
The sophisticated use of ultraviolet spectroscopy outlined above marked the beginning of a new era in biflavanoid chemistry, for although classical degradation procedures still continued to be utilized for the determination of structure their importance began to decline with the advent of modern spectroscopic methods. A great number of biflavones of the amentoflavone family (**1**) have been isolated since the pioneering work on ginkgetin (**13**), isoginkgetin (**23**) and sciadopitysin (**14**): 49 plant species and varieties have yielded amentoflavone (**1**) and its derivatives to date. A full list of these is given in Tables 1 and 5 at the end of this chapter.

IV. Biflavones of the Hinokiflavone (2) Family

The biflavones so far discussed are all derivatives of the parent 3',8-linked biflavone, amentoflavone (1). However, biflavones belonging to three other families are now known to exist and frequently these co-occur with members of the amentoflavone family as well as with other types of biflavanoids.

Hinokiflavone (2), initially isolated in 1958 by KARIYONE and SAWADA from the leaves of *Chamaecyparis obtusa* Endl. (*49*), was at first thought to contain six hydroxyls like amentoflavone (1). In the following year, however, FUKUI and KAWANO showed (*17, 18, 19, 41, 56*) conclusively that hinokiflavone contained five hydroxy groups. Although the ultraviolet spectrum of hinokiflavone differed significantly from that of ginkgetin (13), that of the pentamethyl ether (34) showed greater similarity, so a

(34) Hinokiflavone pentamethyl ether

(35)

biflavone was immediately suspected. Degradation sequences carried out on hinokiflavone (2) and on its pentamethyl ether (34) (see Charts 6 and 7) provided sufficient evidence for structure (35) to be advanced for the parent molecule with structure (2) a less likely alternative. However, the favoured structure (35) ultimately proved to be an incorrect choice. Though the rational Ullmann synthesis (*68, 99*) (Chart 8) seemed at first to substantiate structure (35), it later became apparent through the elegant synthetic work of NAKAZAWA (*93*) (Chart 9) that a Wesseley-Moser rearrangement had

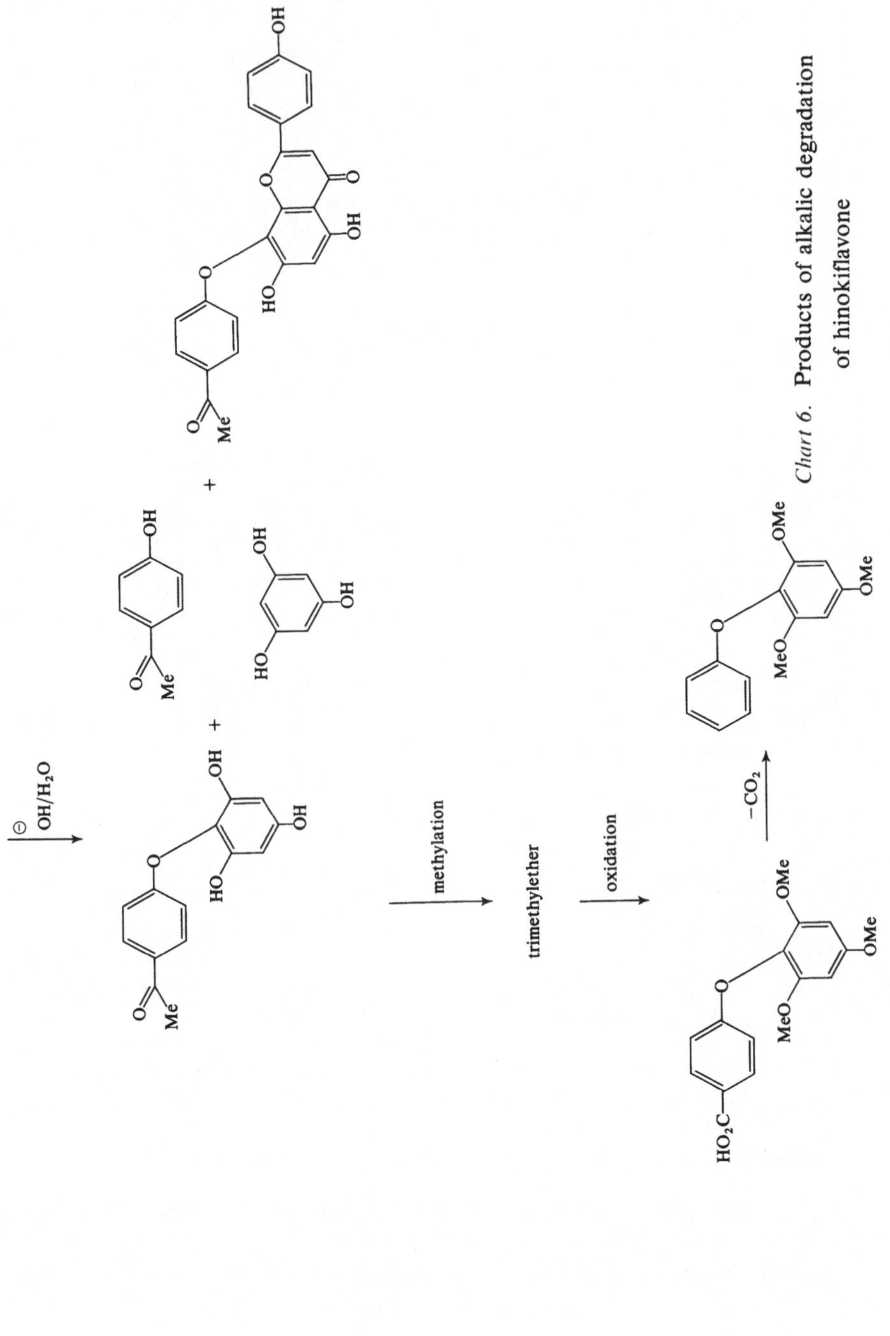

Chart 6. Products of alkalic degradation of hinokiflavone

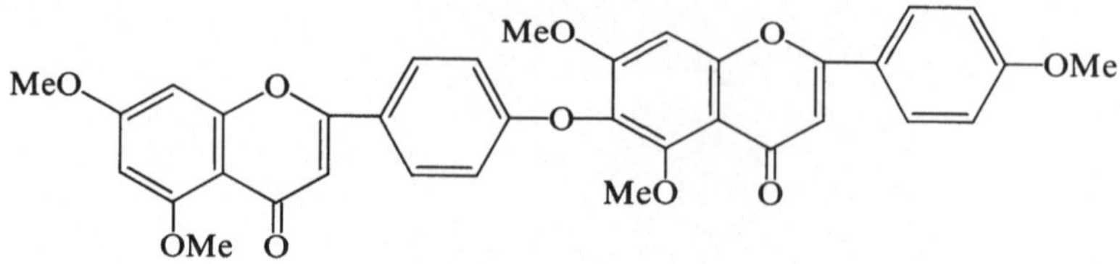

(34) hinokiflavone pentamethyl ether
[or pentamethyl ether of (35)]

$\ominus$
OH/H$_2$O$_2$

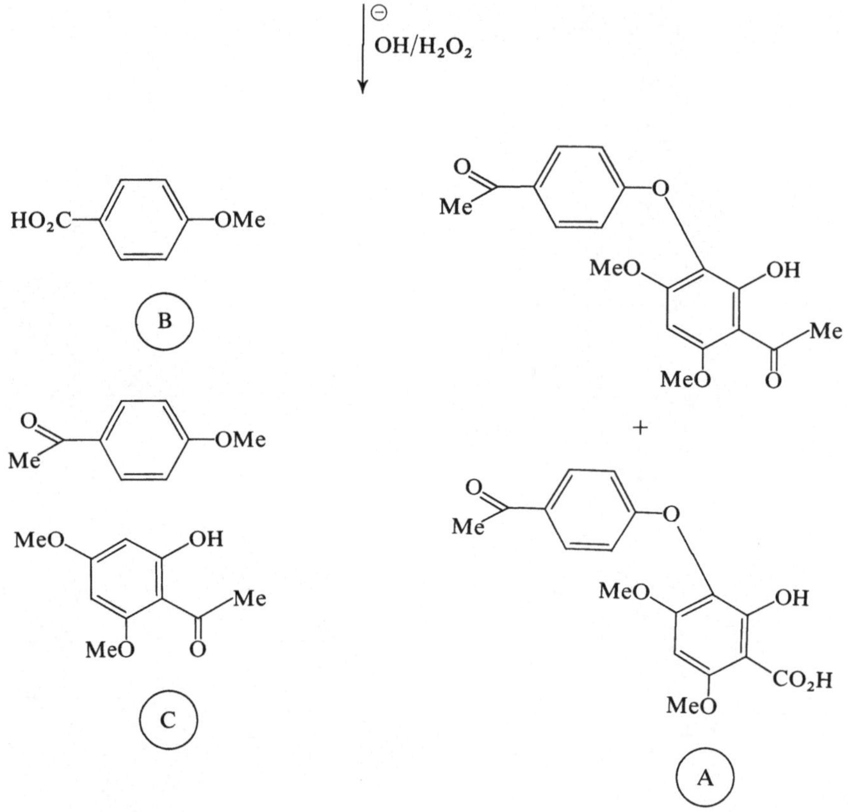

Products (A), (B), (C) were also obtained in good yield when

(34) [or the pentamethyl ether of (35)] was treated with Ba(OH)$_2$ in methanol

Chart 7. Products of degradation of hinokiflavone pentamethyl ether with
alkaline hydrogen peroxide

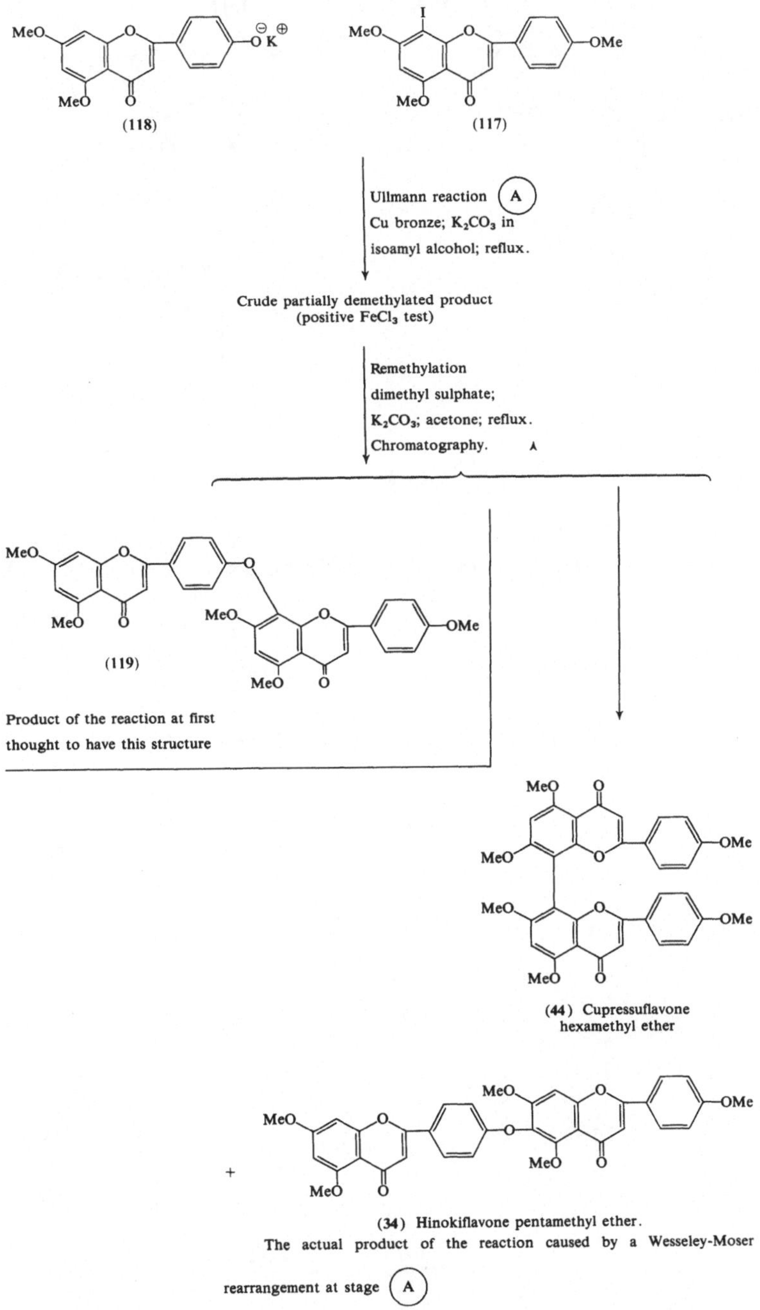

(118) (117)

Ullmann reaction Ⓐ
Cu bronze; K_2CO_3 in
isoamyl alcohol; reflux.

Crude partially demethylated product
(positive $FeCl_3$ test)

Remethylation
dimethyl sulphate;
K_2CO_3; acetone; reflux.
Chromatography. ⅄

(119)

Product of the reaction at first
thought to have this structure

(44) Cupressuflavone
hexamethyl ether

+

(34) Hinokiflavone pentamethyl ether.
The actual product of the reaction caused by a Wesseley-Moser

rearrangement at stage Ⓐ

Chart 8. Seshadri's synthesis of hinokiflavone pentamethyl ether

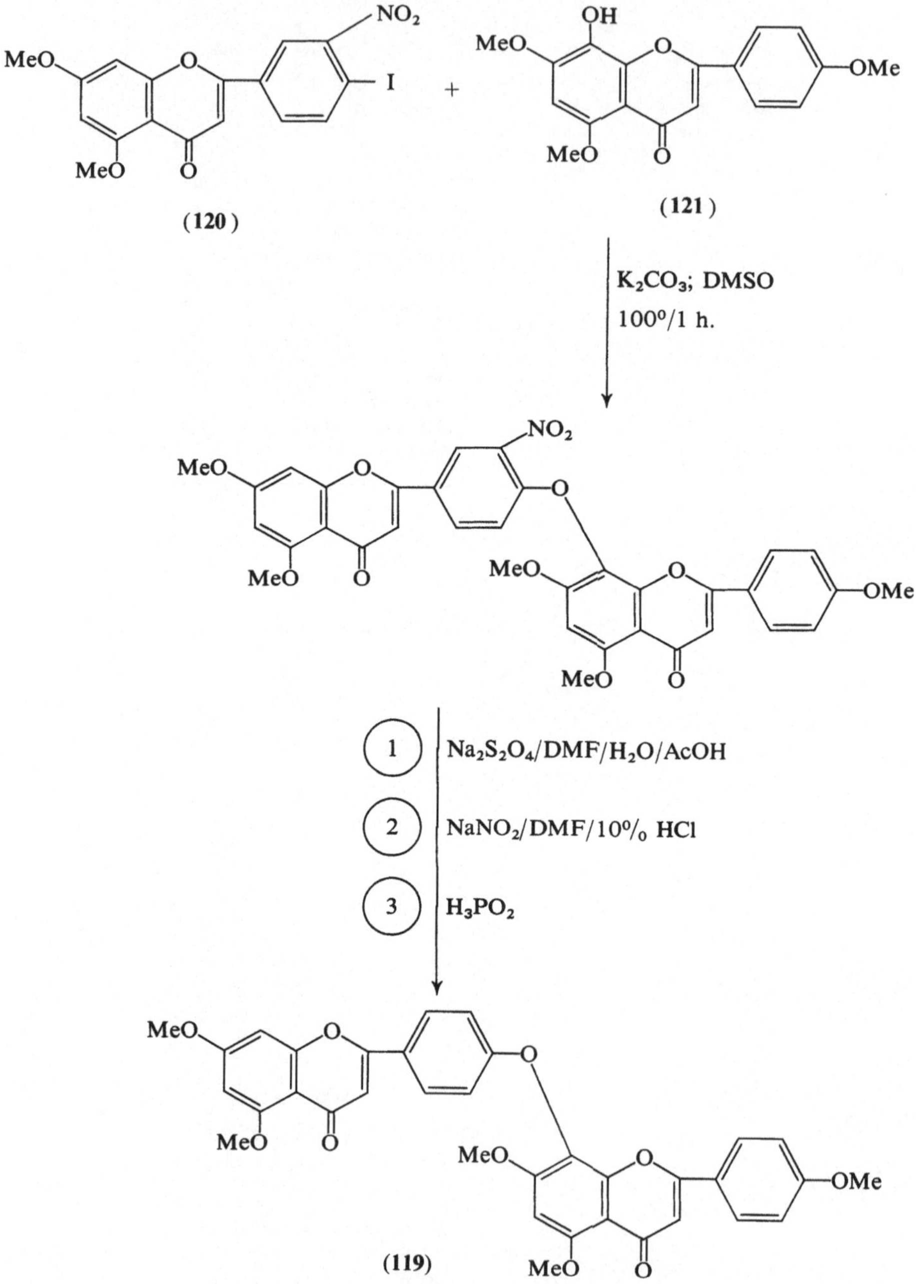

(120)   +   (121)

K_2CO_3; DMSO
$100°/1$ h.

1 | $Na_2S_2O_4/DMF/H_2O/AcOH$

2 | $NaNO_2/DMF/10\%$ HCl

3 | H_3PO_2

(119)

Chart 9 (Part 1). Nakazawa's synthesis of hinokiflavone
pentamethyl ether [incorrect structure (119)]

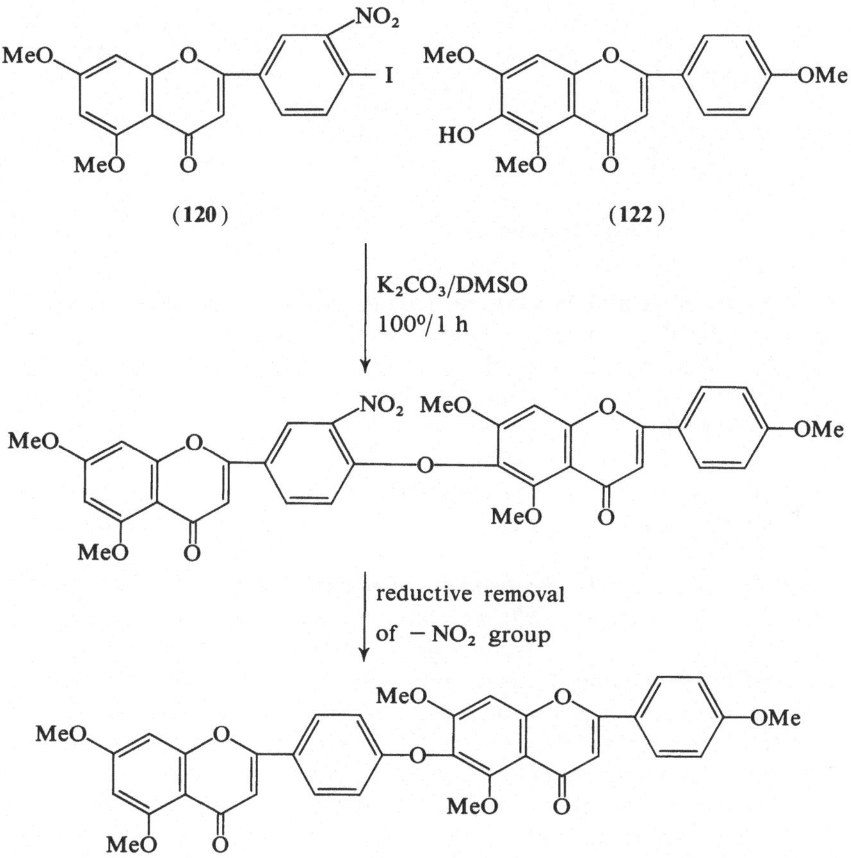

(**120**)   (**122**)

K$_2$CO$_3$/DMSO

100°/1 h

reductive removal

of $-$ NO$_2$ group

(**34**) Hinokiflavone pentamethyl ether

Chart 9 (Part 2). Nakazawa's synthesis of hinokiflavone pentamethyl ether
[correct structure (**34**)]

occurred during the Ullmann reaction in SESHADRI's synthesis affording
not the anticipated 4'-*O*-8 but the 4'-*O*-6 isomer. The discovery of this
rearrangement required that the structure for hinokiflavone be revised
(*102*) also to the 4'-*O*-6 biflavone ether (**2**), and this assignment has been
very adequately substantiated (*102, 116*) recently by use of n.m.r. solvent
shifts which are discussed more fully later.

Several partially methylated derivatives of hinokiflavone occur in nature
(Table 2 at end of chapter). In general their structures have been determined by the use of methods already outlined. For example, the structures

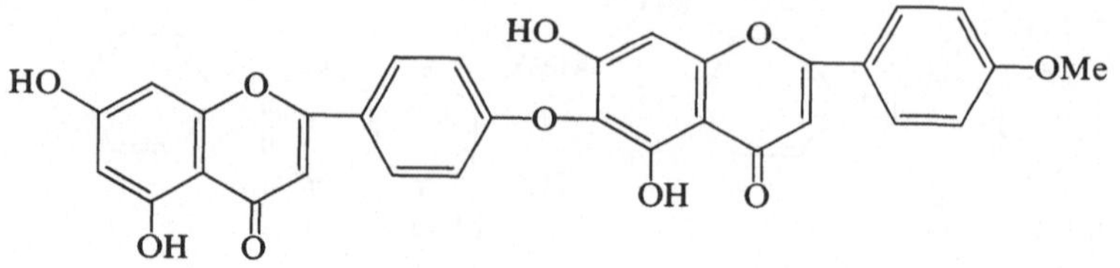

(**36**) Cryptomerin A

of cryptomerin A and B, (**36**) and (**37**) (Chart 11) were determined (*84*)
by examining the influence of added alkali upon the ultraviolet spectra
and by alkali degradation (methanolic barium hydroxide of the tetraethyl
derivatives) [Chart 10 summarizes the degradation of cryptomerin B
triethyl ether (**38**)].

Compound (**39**) was recognized as a diaryl ether derivative through
the similarity of its ultraviolet spectrum to that of (**40**) obtained from
hinokiflavone pentamethyl ether (**34**) when it was similarly degraded
(Chart 10).

Isocryptomerin (**41**), the product of partial demethylation of cryptomerin B (**37**) (Chart 11) (*80*) was later identified as a natural product
present in the leaves of *Chamaecyparis obtusa* Endl. (*19*), the original
source of hinokiflavone (**2**), base induced ultraviolet shifts being utilized
once again to provide proof for the site of the single methoxy group.

A similar technique of synthesis by demethylation (*83, 85*) [in this case
with hinokiflavone pentamethyl ether (**34**)] has been used to prepare and
provide structural proof for chamaecyparin (*81*) (**42**) and neocryptomerin
(*83, 85*) (**43**), two partial methyl ether derivatives of hinokiflavone isolated
from natural sources.

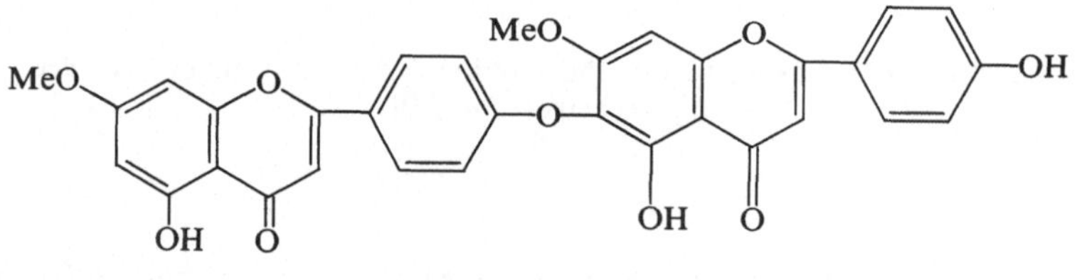

(**42**) Chamaecyparin

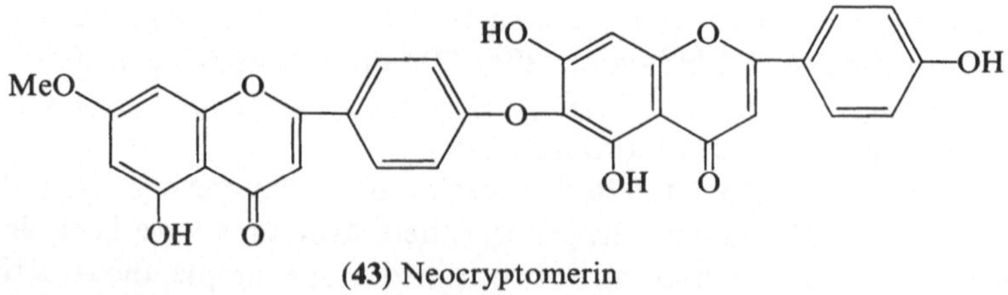

(**43**) Neocryptomerin

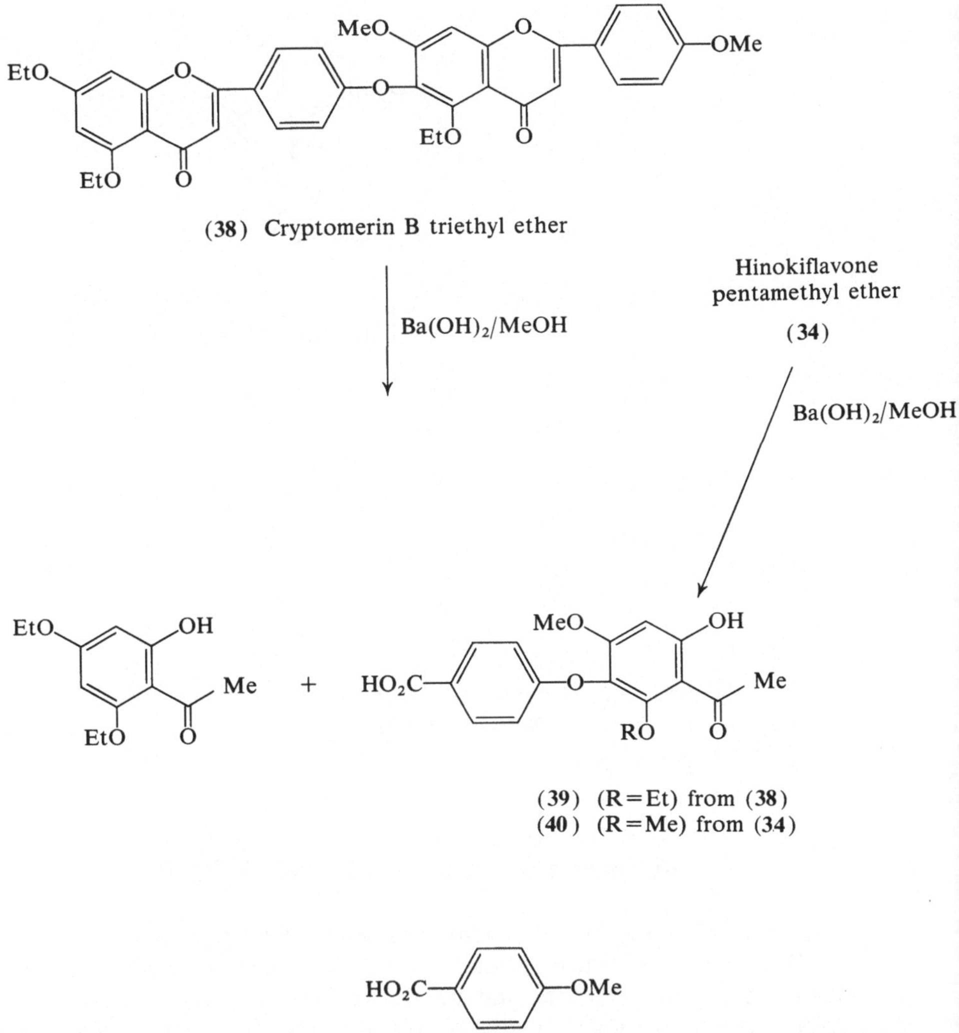

(**38**) Cryptomerin B triethyl ether

Hinokiflavone
pentamethyl ether
(**34**)

Ba(OH)$_2$/MeOH

Ba(OH)$_2$/MeOH

(**39**) (R = Et) from (**38**)
(**40**) (R = Me) from (**34**)

Note: A C—4′—O—C—6 linkage is shown for cryptomerin B triethylether (**38**)
[and for cryptomerin A (**36**)]. Originally these two natural products were thought
to have C—4′—O—C—8 linkages: the degradative scheme above has been
re-interpreted using the correct structures.

Chart 10. Alkali degradation reactions of cryptomerin B triethyl ether (**38**)
and hinokiflavone pentamethyl ether (**34**)

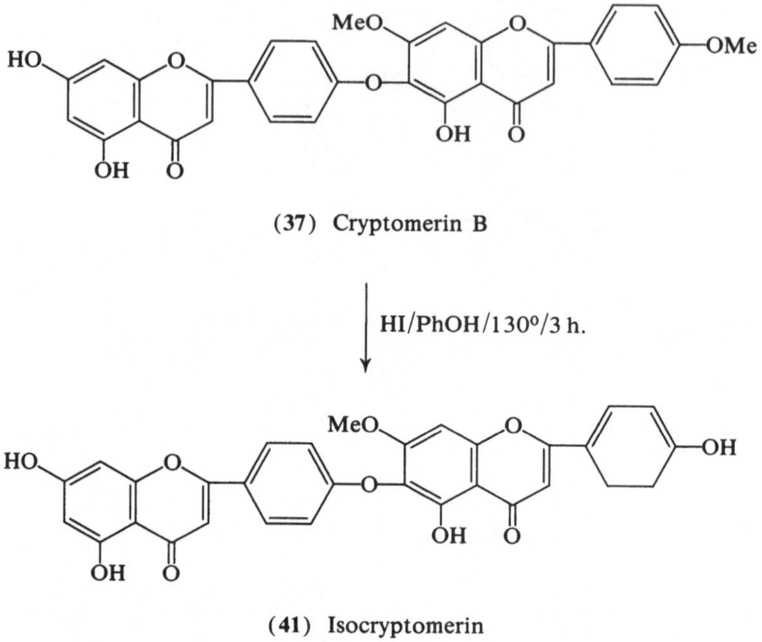

(37) Cryptomerin B

HI/PhOH/130°/3 h.

(41) Isocryptomerin

Chart 11. Synthesis of isocryptomerin (41) from cryptomerin B (37) by partial demethylation

V. Biflavones of the Cupressuflavone (3) Family

Cupressuflavone (3), the parent member of the symmetrically linked [I-8, II-8]-biflavones, was first isolated from *Cupressus torulosa* (the Himalayan cyprus) by SESHADRI and his co-workers (*88, 89*) in 1964. The molecular formular $C_{30}H_{12}O_4(OH)_6$, the ultraviolet and infrared spectra, and various diagnostic colour reactions suggested a biflavone structure possibly composed of two apigenin (15) units as in the case of amentoflavone (1) and hinokiflavone (2). A striking degree of symmetry in the molecule was clearly evident from even a cursory examination of the n.m.r. spectrum of the hexamethyl ether derivative (44) which, being remarkably similar to that of apigenin trimethyl ether (45) and showing only *two* equivalent high field phloroglucinol ring protons, at once suggested either a C-6/C-6 or a C-8/C-8 interflavone linkage. The latter was chosen on somewhat slender evidence: it is observed that for apigenin (15) and its trimethyl ether (45) the C-8 proton resonates at slightly lower

(44) Cupressuflavone hexamethyl ether

(45) Apigenin trimethyl ether

field than that at C-6, the difference being, however, only *ca.* 0.2 p.p.m. The relevant two proton signal in cupressuflavone hexamethyl ether **(44)** resonates at τ 3.40 which corresponds more closely to the value for a C-6 proton than that for a C-8 making the linkage C-8/C-8. Though this very simple comparison takes no account of the possibility of shielding or deshielding influences by the second apigenin unit as a substituent of the first, the original choice of a C-8/C-8 linkage has withstood the test of more reliable evidence gained from n.m.r. solvent shift studies (*30, 114*) which are discussed later.

Quite fortuitously, cupressuflavone hexamethyl ether **(44)** had already been synthesized by NAKAZAWA (*92, 94*) in 1962, it being an unwanted by-product of his Ullmann synthesis of ginkgetin **(13)** and its derivatives (see Chart 12): an exchange of samples (*89*) established identity. The work of SESHADRI and his colleagues is notable in that for the first time it makes use of n.m.r. spectroscopy as an aid to the structural determination of biflavanoids.

In contrast with the amento- **(1)** and hinokiflavone **(2)** families, naturally occurring methyl ether derivatives of cupressuflavone **(3)** are as yet rare: to date only three such derivatives have been positively identified (see Table 3 at end of chapter) though the existence of others is suspected (*30, 61, 62, 75, 118*).

The increasing degree of sophistication in chromatography has enabled more complex mixtures of biflavones to be resolved for the first time, but since these methods tend to afford only small quantities of material, an ever pressing need has arisen to solve structures by means other than those of classical degradation. Clearly spectral methods offer great opportunities in this connection, and it is the emergence of n.m.r. spectroscopy and mass spectrometry in particular which has had the greatest influence on bi-flavanoid structural work.

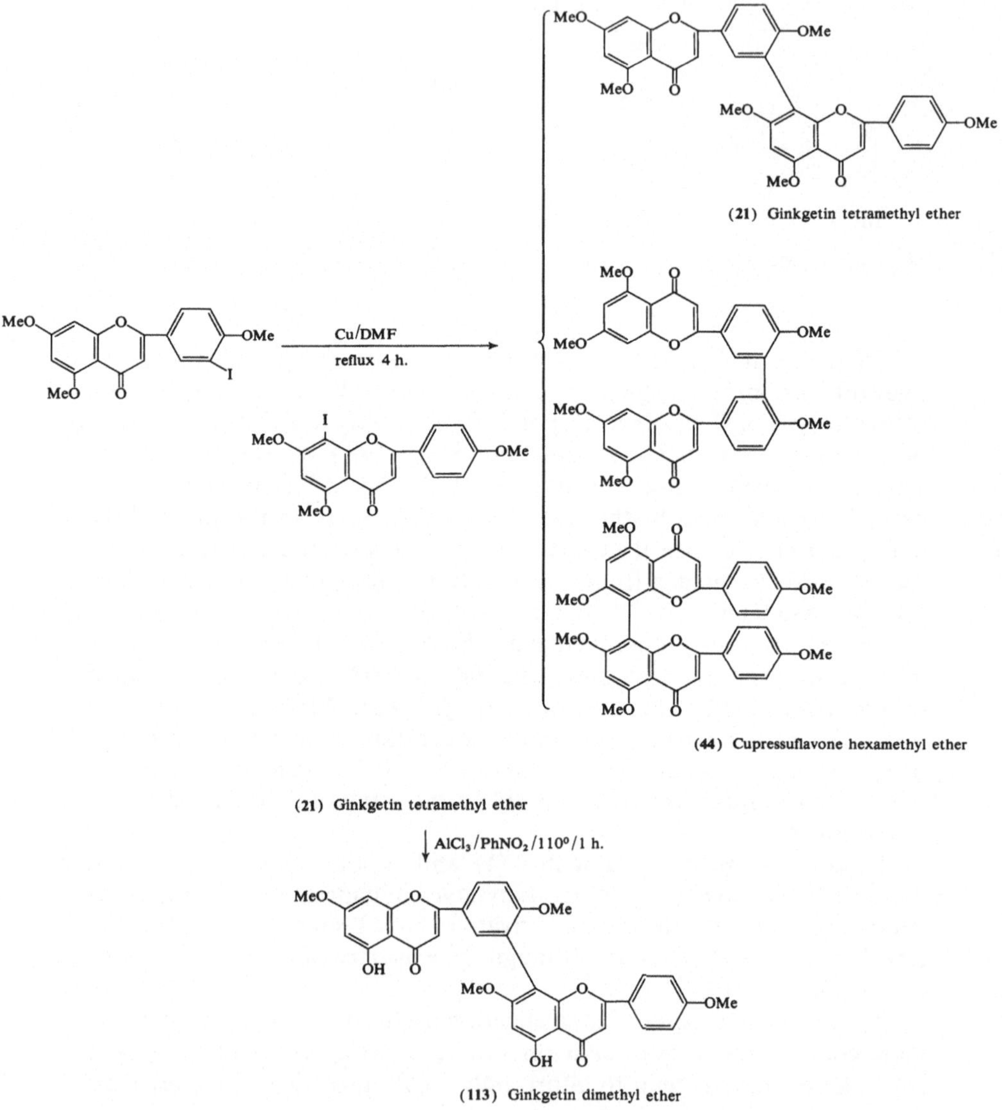

H. D. LOCKSLEY:

Cu/DMF
reflux 4 h.

(21) Ginkgetin tetramethyl ether

(44) Cupressuflavone hexamethyl ether

(21) Ginkgetin tetramethyl ether

$AlCl_3/PhNO_2/110°/1$ h.

(113) Ginkgetin dimethyl ether

Chart 12. Syntheses by Nakazawa of ginkgetin tetramethyl ether (**21**), ginkgetin dimethyl ether (**113**), cupressuflavone hexamethyl ether (**44**), etc.

VI. Biflavones of the Agathisflavone (4) Family

It is convenient to consider the fourth class of natural biflavone, agathisflavone (**4**), at this stage since natural *O*-methyl derivatives of this biflavone (Table 4 at end of chapter) and those of cupressuflavone (**3**) (Table 3) often co-occur and tend to have been investigated by the use of similar methods.

The first account of biflavones belonging to the agathisflavone family appeared in 1969 when from the plant *Agathis palmerstonii* two biflavones WA[I] and WA[VII] (**46**) and (**47**) were isolated (*117*) which clearly possessed a new type of biflavone nucleus. Complete methylation of WA[I] and WA[VII] gave one and the same derivative (**48**) (agathisflavone hexamethyl ether) whose n.m.r. spectrum in displaying a greater degree of complexity than that of cupressuflavone hexamethyl ether (**44**), indicated an unsymmetrical linkage between flavone units. Ultraviolet and n.m.r. spectral characteristics suggested that these units were *O*-methyl derivatives of apigenin (**15**) as before. The n.m.r. spectrum showed only two phloroglucinol ring protons to be present but since these had dissimilar chemical shifts (τ 3.4—3.5) and were uncoupled, an unsymmetrical C-6/C-8 linkage between units of apigenin trimethyl ether (**45**) provided the only possible explanation for this and other features in the spectrum of the permethyl derivative (**48**).

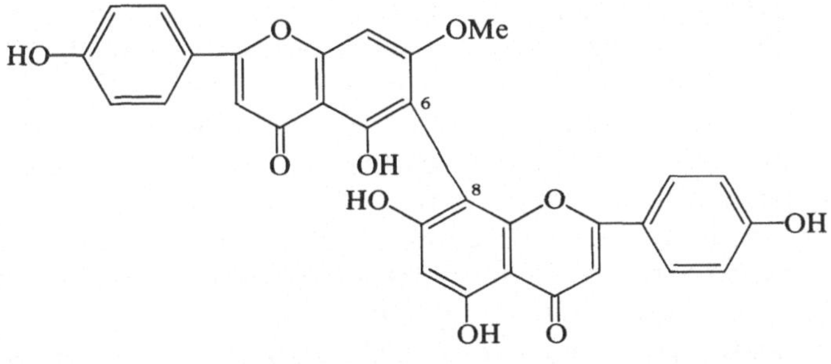

(**46**) WA[I]

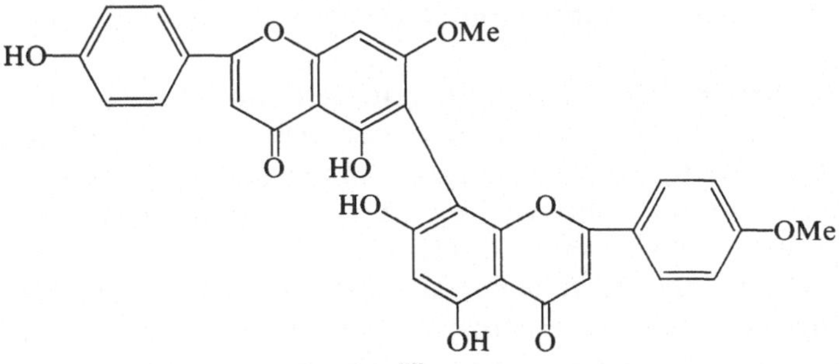

(**47**) WA[VII]

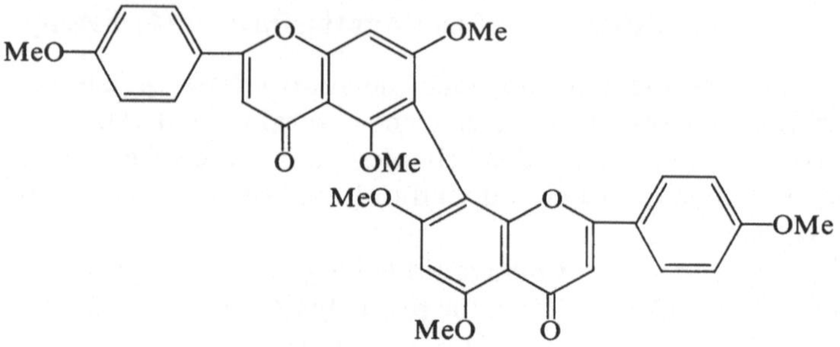

(**48**) Agathisflavone hexamethyl ether

This conclusion was very amply supported by independent evidence provided by solvent shift studies (*9, 23, 32, 70, 129*).

It is now well established (*9, 23, 32, 70, 129*) that in compounds containing a methoxy group attached to an aromatic nucleus which is flanked by *two ortho*-substituents (each belonging to a well-defined group) (*108*) the methoxy protons experience either no marked change in chemical shift or a slight negative (downfield) shift (solvent shift $\Delta = \tau_{C_6D_6} - \tau_{CDCl_3}$) when the solvent is gradually changed from pure d_1-chloroform (or carbon tetrachloride) to pure d_6-benzene. This behaviour is in contrast to the situation which prevails when there is either only *one* or *no* substituent at all *ortho* to the methoxy group. In the latter case a marked positive (upfield) shift in the order of $\Delta + 0.4$ p.p.m. is observed for the methoxy protons. The solvent shift technique has found wide application as an aid to structure determination; it is particularly useful in defining unambiguously positional isomers, as in the following cases.

Every methoxy group in cupressuflavone hexamethyl ether (**44**) and amentoflavone hexamethyl ether (**21**) is flanked by at least one unsubstituted *ortho*-position. However, this would not be the case if the interflavone linkage were C-6/C-6 or C-6/C-3′, respectively, for in these cases the C-5 methoxyls of rings I-A and II-A, or I-A respectively, would be flanked by two *ortho* substituents (*viz.* an aryl group and a carbonyl group).

Independent experimental proof that these two biflavones do possess the C-8/C-8 and C-8/C-3′ linkages, respectively, originally advanced for them, is provided by the observation that on increasing the concentration of d_6-benzene in d_1-chloroform every methoxy group signal experiences a marked *positive* solvent shift (*30, 115—117*). To further test the reliability of this criterion, hinokiflavone pentamethyl ether (**34**) and agathisflavone hexamethyl ether (**48**) were similarly examined (*98, 102, 113, 114, 116, 117*) since these do contain linkages involving the C-6 position of ring

II-A. That these structures are the correct ones is again amply substantiated for in both these cases one of the methoxy group signals experiences a quite marked negative solvent shift, in contrast to the remainder which show normal positive shifts (see Figures 1—3).

Location of the O-methyl groups in the agathisflavone derivatives WA^I and WA^{VII} (**46**) and (**47**) (*117*) was achieved using the acetate shift technique discussed in the next section.

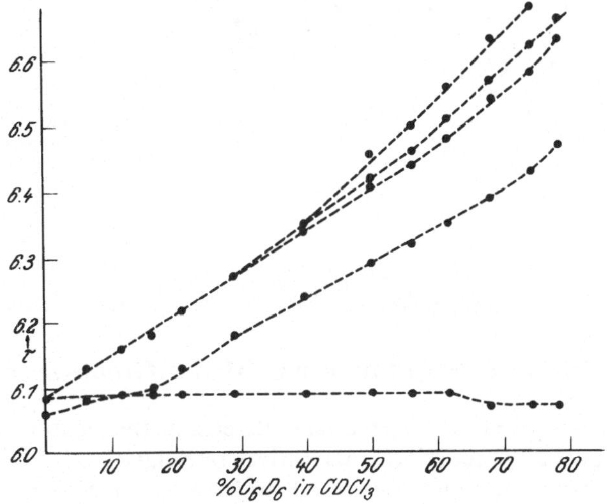

Fig. 1. Solvent shifts of methoxy-group protons in hexa-O-methylhinokiflavone (**34**) for solutions in d$_6$-benzene/d-chloroform. Reproduced by Tetrahedron Letters (*116*) by kind permission of the authors and the publisher, Pergamon Press Ltd., Oxford, U.K.

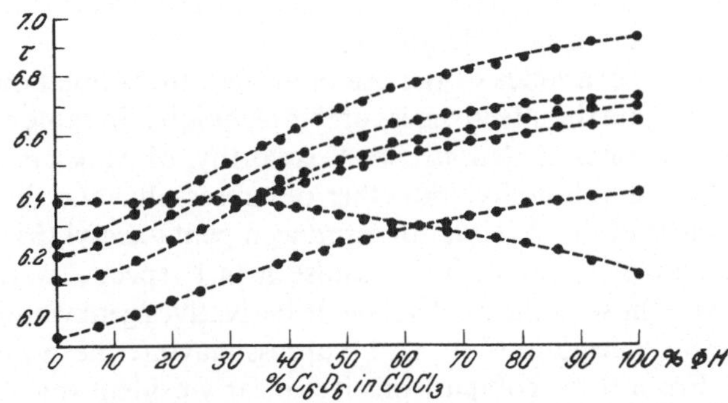

Fig. 2. Solvent shifts of methoxy-group protons in hexa-O-methylagathisflavone (**48**) for solutions in d$_6$-benzene/d-chloroform. Reproduced from Experientia (*117*) by kind permission of the authors and the publisher, Birkhäuser Verlag, Basel, Switzerland.

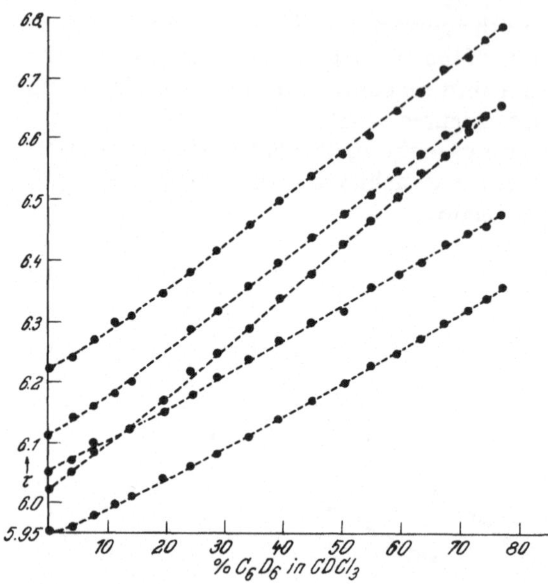

Fig. 3. Solvent shifts of methoxy-group protons in II-4′, I-5, I-7, II-7-pentamethoxy[I-4′-*O*-II-8] bi-flavone (penta-*O*-methyl*iso*hinoki-flavone) [(**119**) Chart 9] for solutions in d₆-benzene/d-chloroform. Reproduced from Tetrahedron Letters (*116*) by kind permission of the authors and the publisher, Pergamon Press Ltd., Oxford, U. K.

VII. Methods for the Location of O-Methyl Groups in Biflavones

Three techniques have been used to determine the location of *O*-methyl groups in partially methylated derivatives of biflavones.

(i)   Determining the influence of added base on ultraviolet spectra,

(ii) determining the effects produced by acetylation of free hydroxyl groups on the n.m.r. chemical shifts of ring protons, and using other n.m.r. techniques such as solvent shifts,

(iii) detailed analysis of the mass spectral fragmentation patterns of partially and fully methylated or acetylated biflavones.

The first of these, method (i), and the n.m.r. solvent shift technique in (ii), have already been fully discussed.

Some specific examples of the use of acetate shifts [method (ii)] serve best to demonstrate the effectiveness of this technique. In some recent work (*75*) on the biflavones of *Agathis alba* Foxworthy, Mashima *et al.* isolated a mono- (**49**) and a dimethyl (**50**) ether of cupressuflavone: both were of unknown constitution. Acetylation produced penta-acetyl (**51**) and tetra-acetyl (**52**) derivates, respectively, whose n.m.r. spectra were carefully compared with those of the model flavone derivatives genkwanin diacetate (**53**), acacetin diacetate (**54**), and cupressuflavone hexa-acetate (**55**) (Chart 13). From these comparisons it is clearly evident that the natural monomethyl derivative of cupressuflavone has its *O*-methyl group at C-7 rather than at C-4′ since the chemical shifts for the ring protons correspond more closely to those of genkwanin diacetate (**53**) than to

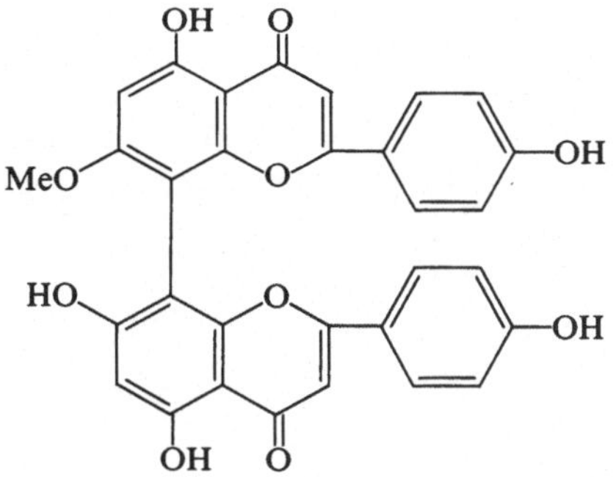

(49) 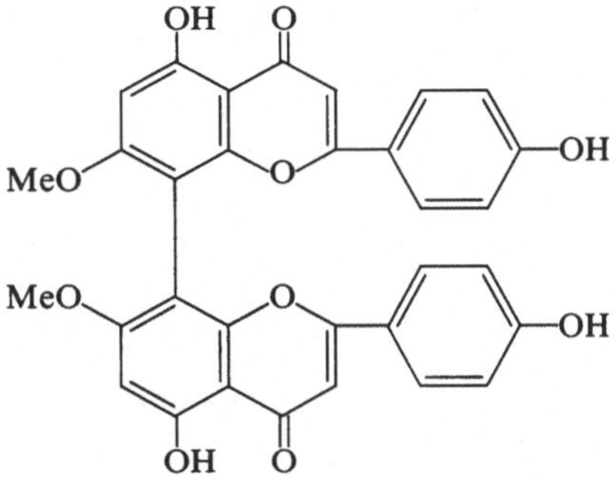   (50)

those of acacetin diacetate (**54**). Signals for the flavone unit devoid of
O-methyl groups compare favourably with those in cupressuflavone
hexaacetate (**55**). Similar comparisons in the case of the acetyl derivative
(**52**) of the natural di-*O*-methyl cuppressuflavone (**50**) lead to the conclusion that the *O*-methyl groups are symmetrically placed at the two C-7
positions of rings I-A and II-A, as in (**50**), since again the model flavone
genkwanin diacetate (**53**) provides the better correlation in chemical shift
values (Chart 13).

Similar n. m. r. acetate comparative studies have been utilized to assign
the structure (**56**) to a dimethyl ether derivative (W 11) of cupressuflavone
found in *Araucaria cunninghamii* and *A. cookii* (*30*). In this case the two
O-methyl groups are assigned the two C-4′ of rings I-B and II-B*, since the
ring protons in the tetra-acetyl derivative (**57**) correspond best with those
of acacetin diacetate (**54**) (Chart 13). Other instances were the acetate
shift technique has been used to determine the methylation pattern in derivatives of cupressuflavone (**3**) and agathisflavone (**4**) can be found in
references *13, 60, 62, 63,* and *86.*

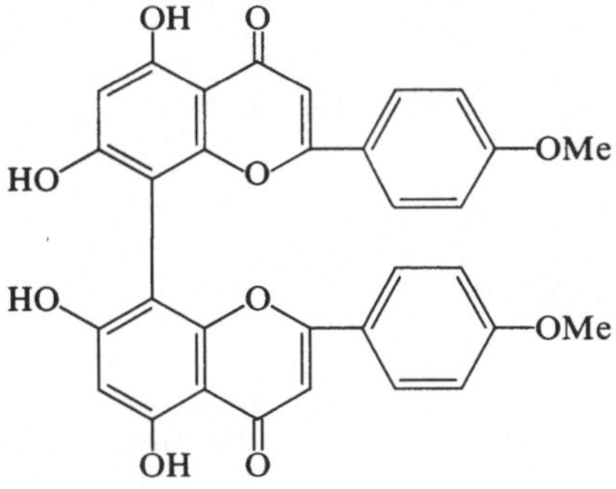

(56) W 11

* A revision of this structure has appeared. The new structure is the C-8/C-8 bigenkwanin
structure (**50**), identical with that obtained from *Agathis alba* Foxworthy.

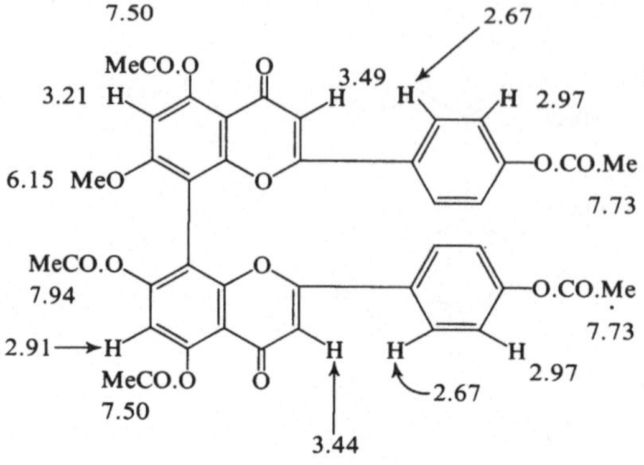

(51)

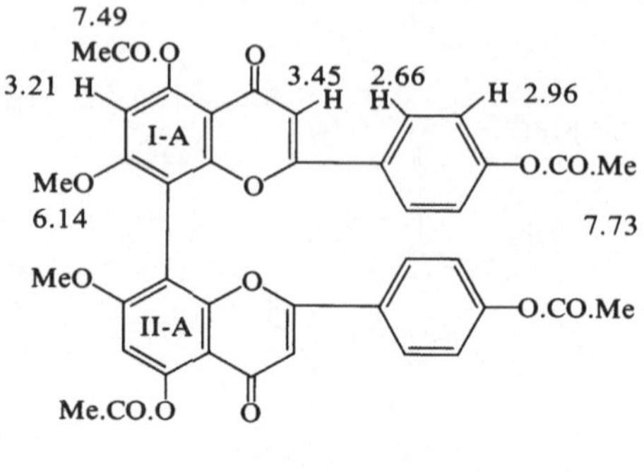

(52)

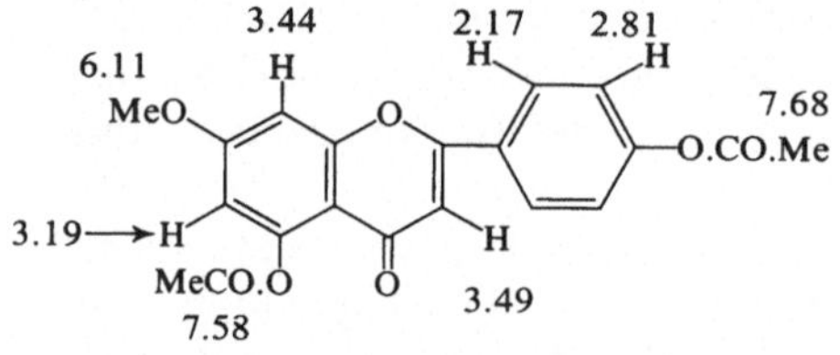

(53) Genkwanin diacetate

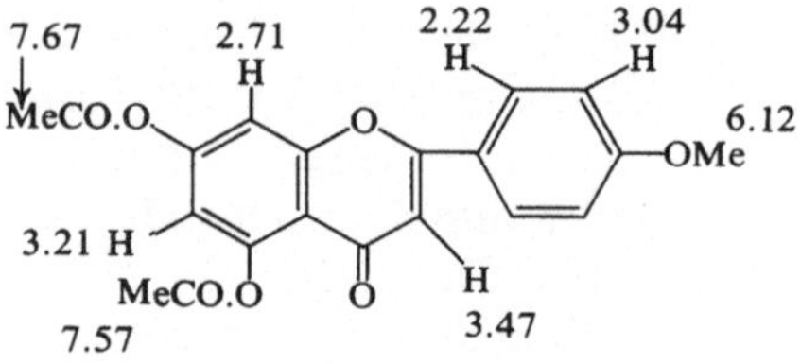

(54) Acacetin diacetate

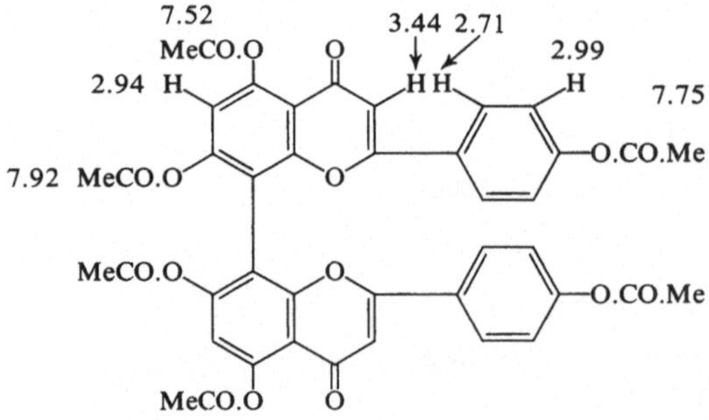

(55) Cupressuflavone hexaacetate

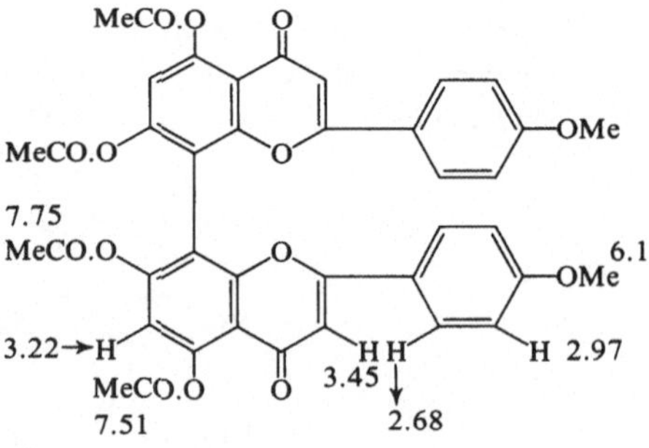

(57) W11 tetra-acetate.

[N.B. This structure has since been revised. The product W11 is identical with (50)].

Chart 13. Chemical shifts of protons (in τ units) for various acetylated flavones and biflavones

The effectiveness of this technique rests on the availability of empirical chemical shift data collated from various methylated and acetylated derivatives of apigenin (15), and of biflavones whose constitutions have already been reliably assigned by other methods. In this connection the pioneering n. m. r. acetate shift work of MASSICOT, MARTHE, and HEITZ (76) on monoflavanoids has proved invaluable.

The third method (iii), mass spectrometry, has not been used widely as yet but it offers great potential. Several papers have appeared on the subject of the mass spectral fragmentation of flavanoid compounds (2, 5, 8, 10, 31, 35, 109, 110, 119, 120), but only one specifically examines the spectra of biflavones (100). KINGSTON (64) has recently summarized the manner in which monoflavones fragment as follows:

(a) Flavones with fewer than four hydroxy groups do not readily fragment, a consequence of the stability of their molecular ion.
(b) Flavones with fewer than four hydroxy groups tend to undergo decomposition predominantly by way of the retro Diels-Alder process. This and other common fragmentation processes are shown in Chart 14 using apigenin (15) as a typical example.
(c) An M-1$^+$ ion is often found in the mass spectra of flavones, its origin is, however, obscure.
(d) The presence of ion C (Chart 14), frequently more intense when a 3-hydroxy group is present, is attributed to the alternative mode of retro Diels-Alder fragmentation also depicted in Chart 14.
(e) Doubly charged ions are frequently present.
(f) When heavily substituted with hydroxyls and methoxyls, the flavone tends to fragment in a less predictable manner, the retro Diels-Alder process becomes insignificant and the spectrum is dominated by the molecular ion and ions at M-15, M-28, and M-43.

A more specific study of the mass spectral fragmentation of the permethyl ether derivatives (21), (34), and (44) of amentoflavone (1), hinokiflavone (2) and cupressuflavone (3) has been made by SESHADRI and his co-workers (100). The following fragmentation schemes (Charts 15, 16 and 17) have been proposed to explain the appearance of some of the ions observed.

A note of caution needs to be sounded in connection with the appearance of the ions with m/e 311 in the spectra of cupressuflavone and amentoflavone hexamethyl ethers (44) and (21), for these ions can be attributed either to fission at the diaryl linkage giving two fragments with identical mass to charge ratios, or to the formation of a doubly charged molecular ion, M^{2+}, or contributions from both, since there is no way in which these contributions can be quantified. In large aromatic molecules like biflavones, ions bearing two positive charges are favoured so it is potentially dangerous to claim without reservation that the appearance of an ion

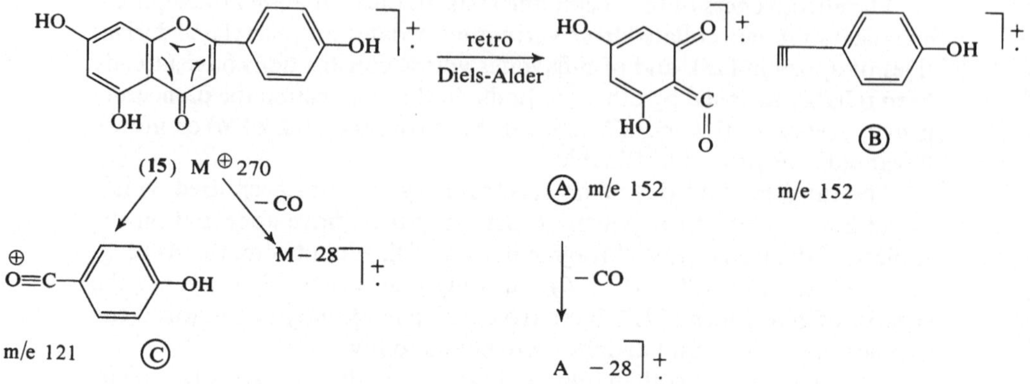

retro Diels-Alder

(15) M⊕270

−CO

M−28⟧⁺·

O≡C⊕—⟨⟩—OH

m/e 121   Ⓒ

Ⓐ m/e 152

Ⓑ m/e 152

−CO

A −28⟧⁺·

Mass spectral fragmentation of apigenin (15).

retro Diels-Alder

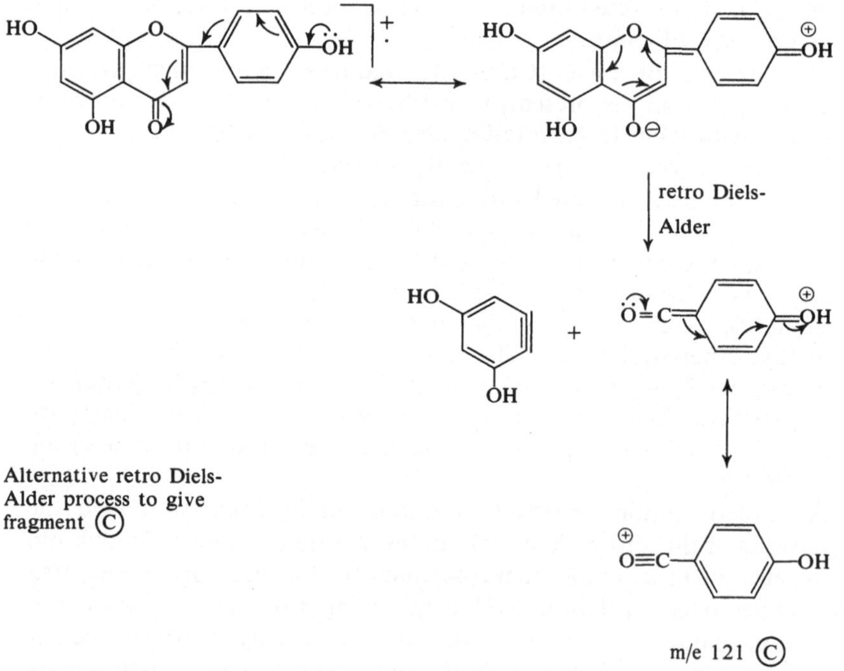

Alternative retro Diels-Alder process to give fragment Ⓒ

m/e 121 Ⓒ

Chart 14. Mass spectral fragmentation of apigenin (15)

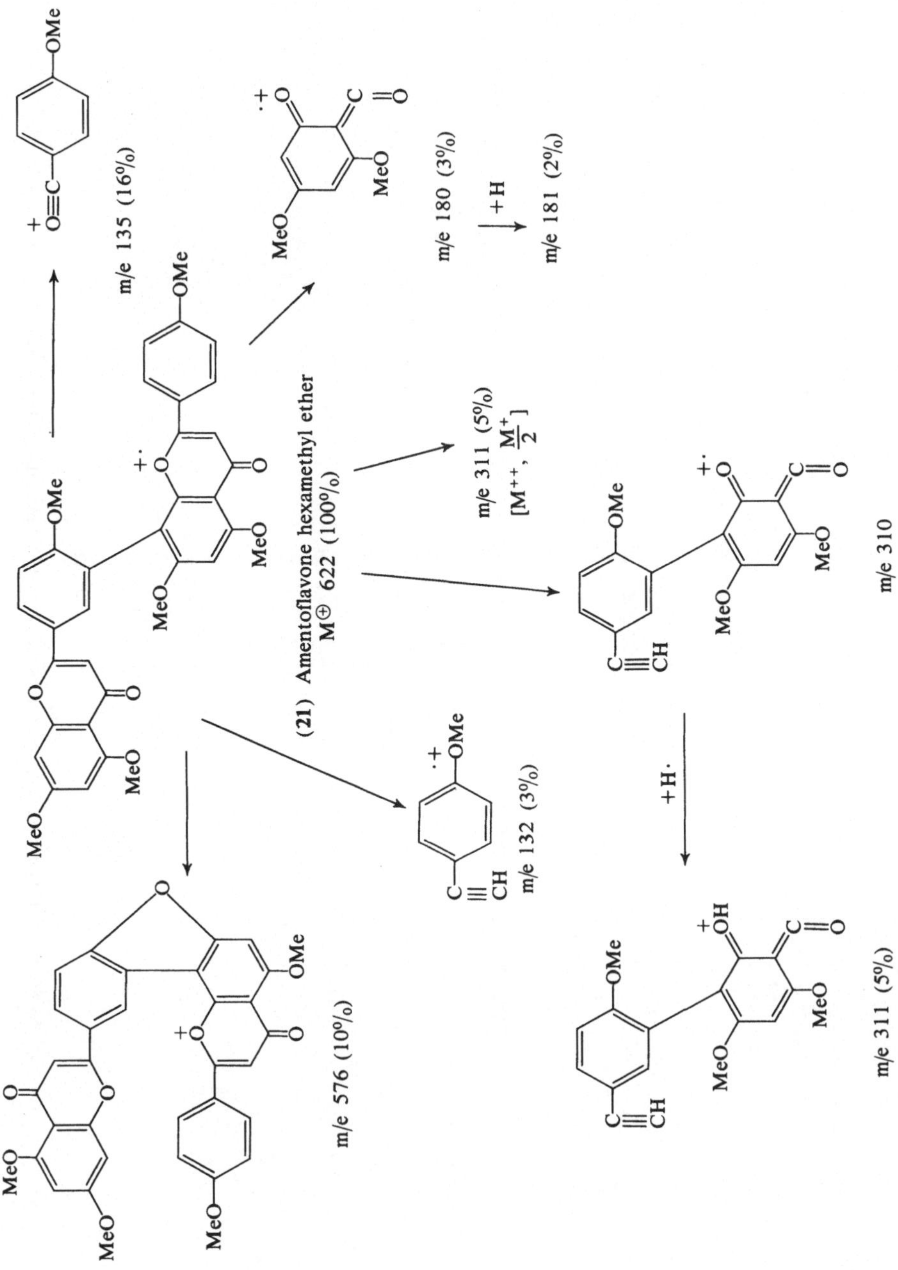

m/e 135 (16%)

MeO

OMe

m/e 180 (3%) $\xrightarrow{+H}$ m/e 181 (2%)

(21) Amentoflavone hexamethyl ether
M$^{\oplus}$ 622 (100%)

m/e 311 (5%)
[M^{++}, $\frac{M^+}{2}$]

m/e 310

m/e 132 (3%)

+H·

m/e 311 (5%)

m/e 576 (10%)

Chart 15. Mass spectral fragmentation of amentoflavone hexamethyl ether (21)

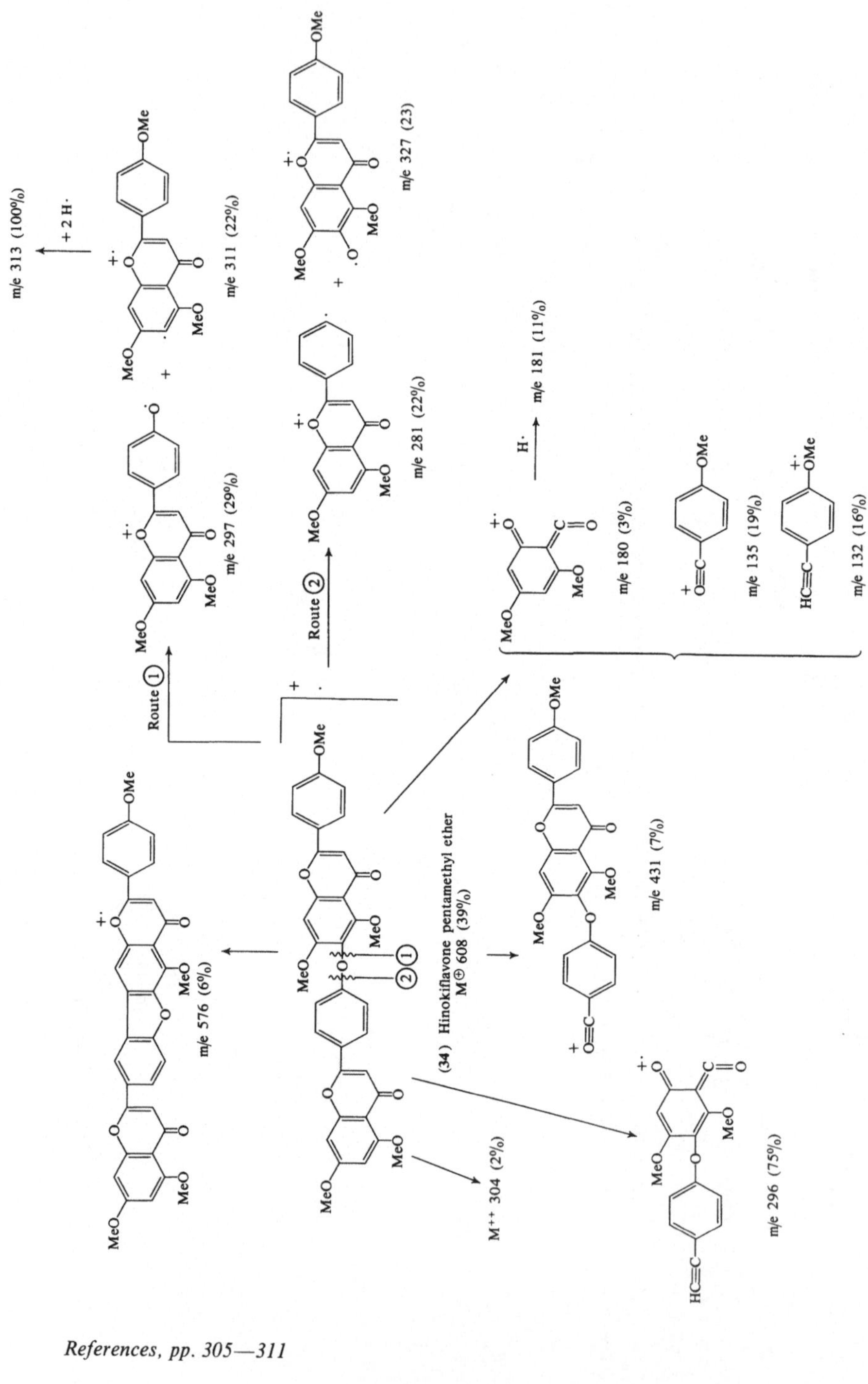

H. D. LOCKSLEY:

m/e 313 (100%)

+ 2 H·

m/e 311 (22%)

m/e 327 (23)

m/e 297 (29%)

m/e 281 (22%)

Route ①

Route ②

m/e 181 (11%)

H·

m/e 180 (3%)

m/e 135 (19%)

m/e 132 (16%)

m/e 576 (6%)

(34) Hinokiflavone pentamethyl ether
M⊕ 608 (39%)

m/e 431 (7%)

M⁺⁺ 304 (2%)

m/e 296 (75%)

Chart 16. Mass spectral fragmentation of hinokiflavone pentamethyl ether (34)

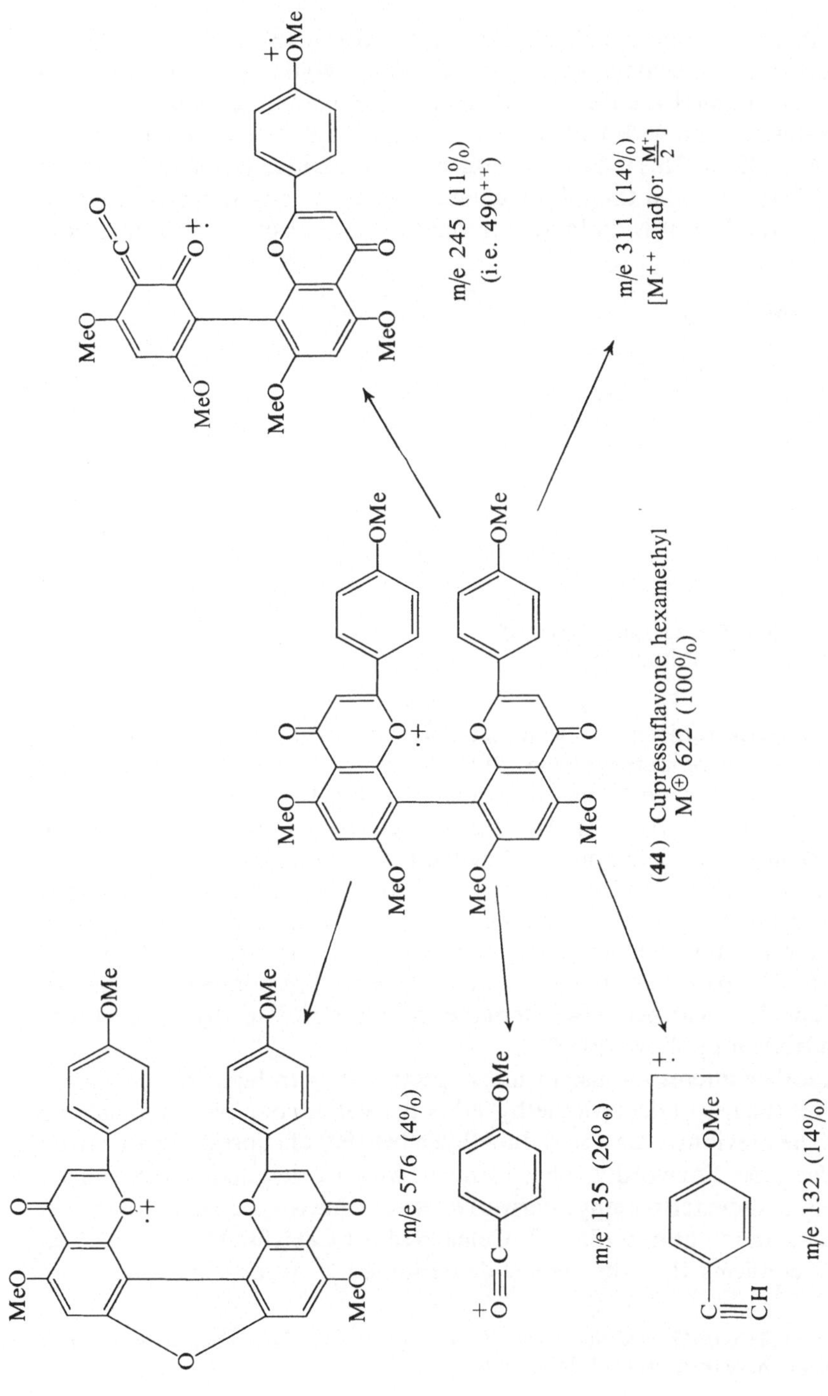

Chart 17. Mass spectral fragmentation of cupressuflavone hexamethyl ether (**44**)

with this m/e value is indicative of symmetry in the molecule. Support for this is clearly evident when one examines (*100*) the spectrum of hinokiflavone pentamethyl ether (**34**) which, in spite of an absence of symmetry, still shows an ion with half the mass (m/e 304; 2%) of the molecular ion (M$^+$ 608; 39%). This ion can be reasonably attributed to the formation of an M^{2+} species and though its abundance is low in relation to the base peak at m/e 313, it is quite high in relation to that of the molecular ion.

(**58**) Podocarpusflavone B
(≡ Putraflavone)

Mass spectrometry has been used (*20*) successfully to clarify the methylation pattern of putraflavone (**58**)*. Mass spectral and other evidence shows this biflavone to be an isomer of ginkgetin (**13**), *i.e.* a dimethyl ether of amentoflavone (**1**). The n.m.r. spectrum of putraflavone has low field signals at τ − 3.15 and − 2.9 indicative of hydrogen bonded hydroxy groups at the C-5 positions of rings I-A and II-A. Since these positions can be eliminated as possible sites of *O*-methylation, it follows that the two ether groups must be located at positions 7-or 4'- on rings I-A, II-A, I-B, or II-B. The pattern of mass spectral fragmentation outlined in Chart 18 provides clear evidence in favour of the unsymmetrical methylation pattern for putraflavone shown in (**58**).

Another interesting use of mass spectrometry in biflavone chemistry involves the use of deuteriomethyl ether derivatives of a natural biflavone (*75*). The previously discussed dimethyl ether (**50**) of cupressuflavone from *Agathis alba* Foxworthy, on treatment with deuteriated diazomethane affords a dideuteriomethyl derivative whose mass spectrum (Chart 19) contains an ion at m/e 135.077 (Calculated for $C_9H_5D_3O^+$, m/e 135.076) which confirms that the two *O*-deuteriomethyl groups introduced are

---

* Putraflavone (**58**) is identical with podocarpusflavone B. The Indian authors (*20*) do not seem to have been aware of this identity.

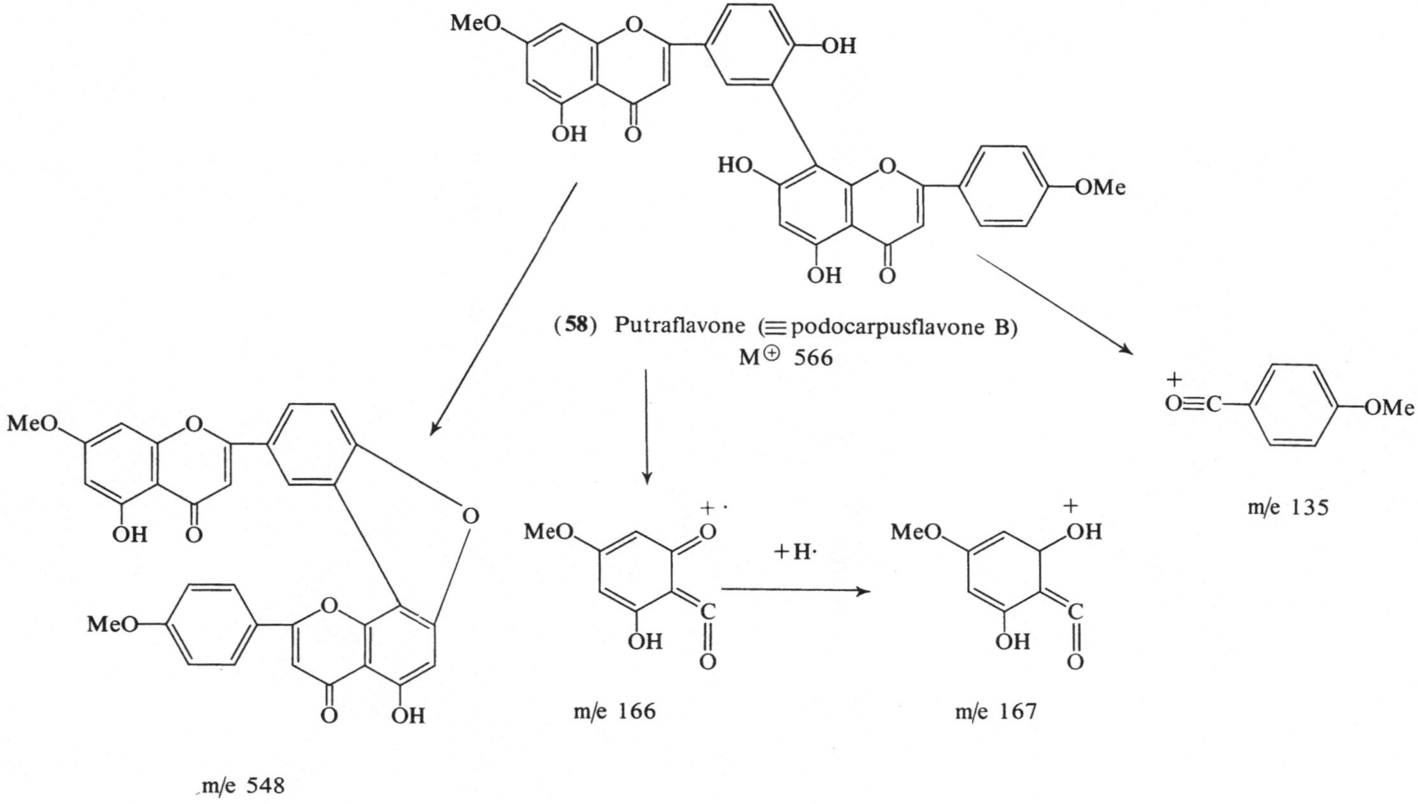

m/e 135

(58) Putraflavone (≡podocarpusflavone B)
M$^{\oplus}$ 566

+H·

m/e 167

m/e 166

m/e 548

Chart 18. Mass spectral fragmentation of putraflavone (58) (≡podocarpusflavone B)

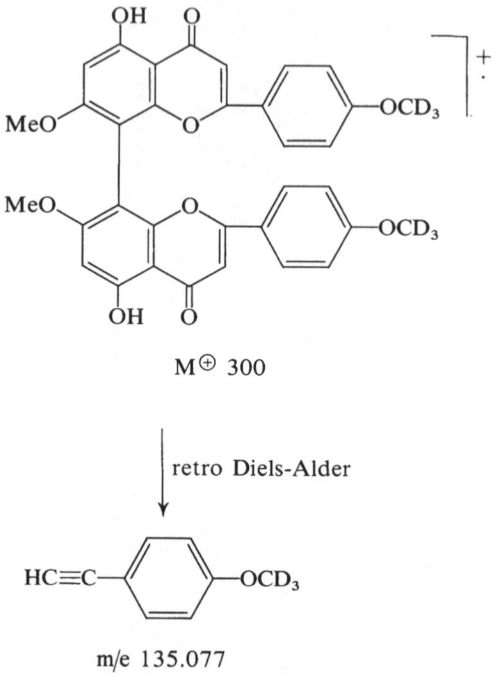

$$M^{\oplus}\ 300$$

retro Diels-Alder

m/e 135.077

Chart 19. Mass spectral fragmentation of a dideuteriomethyl derivative of a natural dimethyl ether derivative of cupressuflavone

located at the C-4′ positions of rings I-B and II-B of the biflavone nucleus. This evidence supports the conclusion reached independently that the *O*-methyl groups of the parent biflavones are located at the two C-7 positions of rings I-A and II-A as in (**50**).

Combined use of n.m.r. spectroscopy and mass spectrometry has been effective (*72*) in determining the structure of heveaflavone (**59**), a trimethyl ether of amentoflavone (**1**) isolated from the leaves of the commeri-

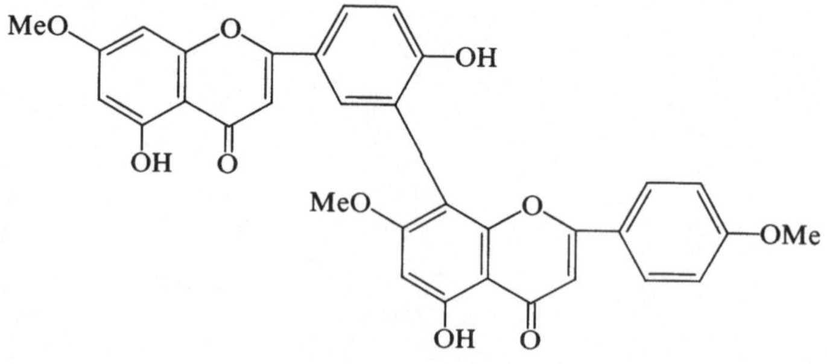

(**59**) Heveaflavone

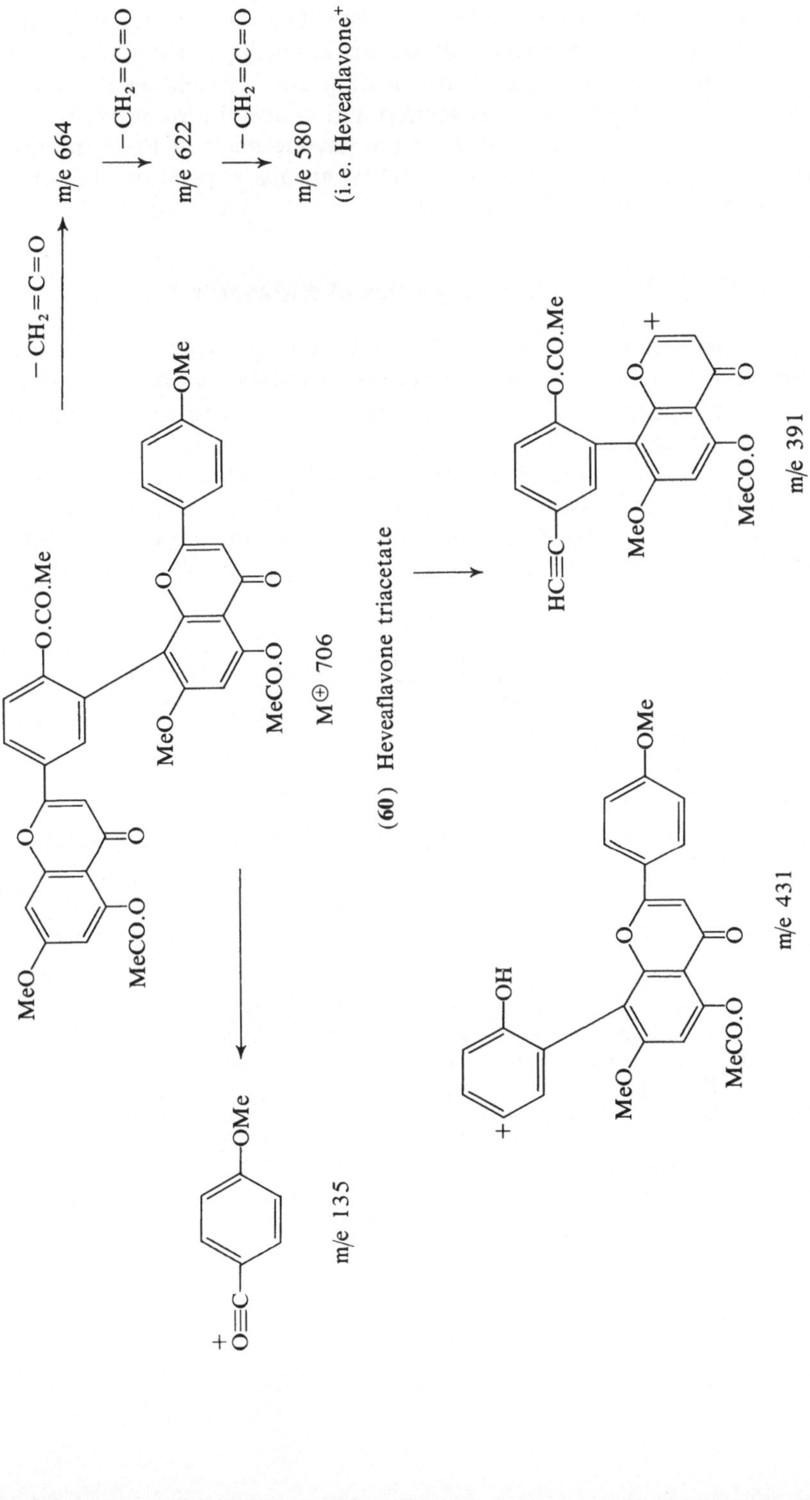

Chart 20. Mass spectral fragmentation of heveaflavone triacetate (60)

252 H. D. LOCKSLEY:

cally important rubber tree, *Hevea brasiliensis*. The mass spectral fragmentation of the triacetyl derivative (**60**) (Chart 20) neatly consolidates more
tenuous n.m.r. evidence based on shielding and deshielding influences
experienced by the protons in *O*-methyl and *O*-acetyl groups when they
are located at positions 4',5- or 7- of the flavone nucleus. Both spectral
methods point to the presence of *O*-methyl groups at positions I-7, II-7,
and II-4' in the parent biflavone (**59**).

VIII. Chiral Properties of Biflavones

Early predictions (*89*) that biflavones such as amentoflavone (**1**),
cupressuflavone (**3**), and their naturally occurring ether derivatives
should show optical activity through steric inhibition to rotation about
the diaryl linkage were not realized until 1968 when a joint Anglo-Indian
group (*30, 112, 115, 117, 13, 61, 62*) reported the presence of optically
active methyl ether derivatives (**61**), (**56**), (**62**), (**46**), and (**47**) of cupressuflavone (**3**), amentoflavone (**1**), and agathisflavone (**4**). Later the parent

(**61**)

(**62**)

biflavones, amentoflavone (1) and cupressuflavone (3) were isolated (*113, 114*) in optically active form from several plant species. Failure to detect optical activity in biflavones prior to this date cannot easily be explained: theories which claim that lack of optical activity is the result of racemisation during isolation of the biflavones seem unconvincing when PELTER *et al.* (*114*) have shown that cupressuflavone (3) is optically stable to treatment with boiling acetic anhydride and that methylation proceeds normally to give optically active cupressuflavone hexamethyl ether (44). Racemisation does occur *via* the Wessely-Moser rearrangement when (+)-cupressuflavone (3) is heated at 130—140° in a mixture of hydriodic acid and acetic anhydride.

The detection of optical activity in biflavones is of immense biogenetic significance for if coupling of the precursor molecule apigenin (15) occurs through phenol oxidative coupling (*6, 16, 122, 123*), as is commonly believed, the process must be under stereospecific control. On the other hand, if further experimentation is able to positively exclude poor isolation technique as a cause of racemisation, the continued appearance of racemic biflavones would again be of importance since it would suggest then that phenol oxidative coupling is not necessarily exclusively stereospecific.

IX. Reduced Forms of Amentoflavone and Hinokiflavone

Only four representatives of this new type of reduced biflavone have so far been isolated. Leaves of the plant *Metasequoia glyptostroboides* Hu and Cheng were first examined in 1958 for the presence of biflavones by two Japanese groups (*50, 121*): both reported the presence of hinokiflavone (2). Reinvestigating the same plant ten years later, BECKMANN and his co-workers (*7*) confirmed the presence of hinokiflavone (2) and found in addition amentoflavone (1), their methyl ether derivatives (41) (see Chart 11)

(63) Sotetsuflavone

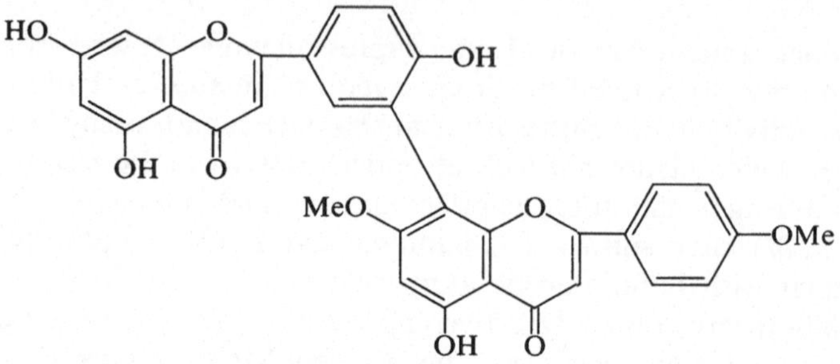

(**64**) Amentoflavone-II-7,II-4′-dimethyl ether

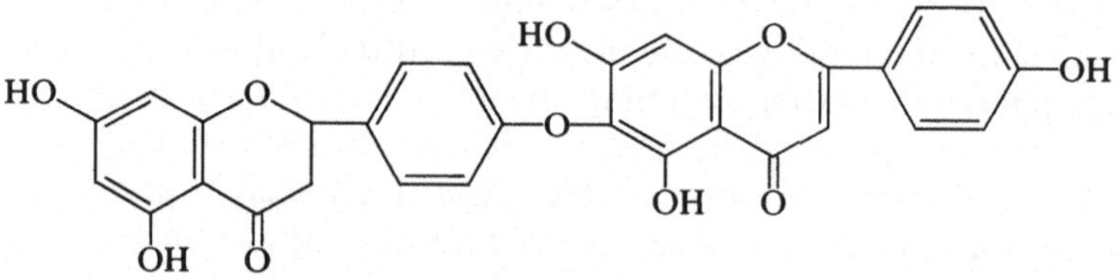

(**65**) Dihydrohinokiflavone

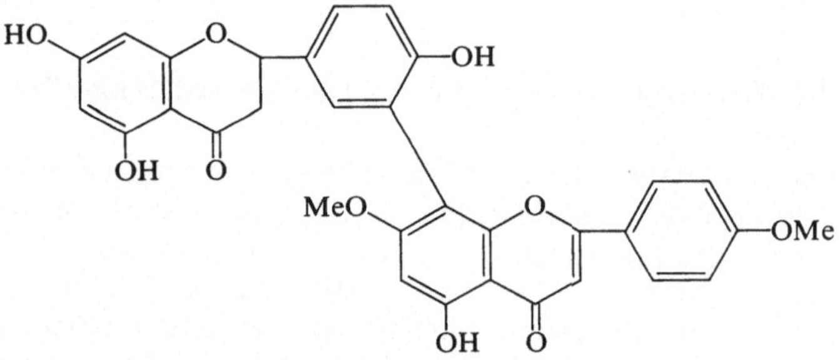

(**66**) Dihydroamentoflavone-II-7,II-4′-dimethyl ether

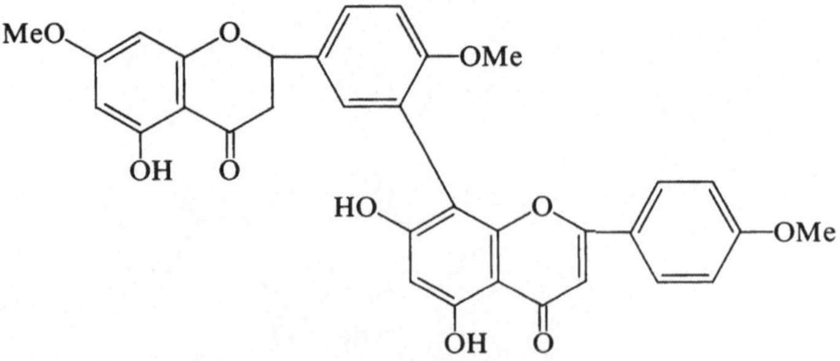

(**67**) Dihydrosciadopitysin

(**68**) Dihydroamentoflavone

(**63**), (**64**), and (**14**), as well as apigenin (**15**), and three hitherto unknown reduced biflavones, 2,3-dihydrohinokiflavone (**65**), the 2,3-dihydroamento-flavone dimethyl ether (**66**), and 2,3-dihydrosciadopitysin (**67**). The same group (*21*) has also reexamined the leaves of *Cycas revoluta* Thunb. (*48, 59, 121*) and found 2,3-dihydroamentoflavone (**68**) and 2,3-dihydrohinoki-flavone (**65**) as well as amentoflavone (**1**), hinokiflavone (**2**) and sotetsufla-vone (**63**).

Signals in the n.m.r. spectra of the penta- and tetramethyl ether derivatives (**69**) and (**70**), prepared from (**65**) and (**66**), respectively, leave little doubt that the parents of these reduced biflavones are constructed from a unit each of apigenin (**15**) and naringenin (**71**) or their O-methyl derivatives. In particular the AXY pattern of signals so characteristic of the protons at C-2 and C-3 of ring C in naringenin (**71**) appear in the spectra of the dihydrobiflavone derivatives (**69**) and (**70**) at τ 4.7 (1H) (quartet) and 7.2 (2H) (multiplet) in good agreement with those for naringenin (*36*) (**71**) H$_A$ τ 4.65 q (1H) (J=12.0 and 3.5 Hz); H$_x$ τ 6.90 q (1H) (J=12.0 and 17.0 Hz); H$_y$ τ 7.36 q (1H) (J=3.5 and 17.0 Hz).

(**69**) Dihydrohinokiflavone pentamethyl ether

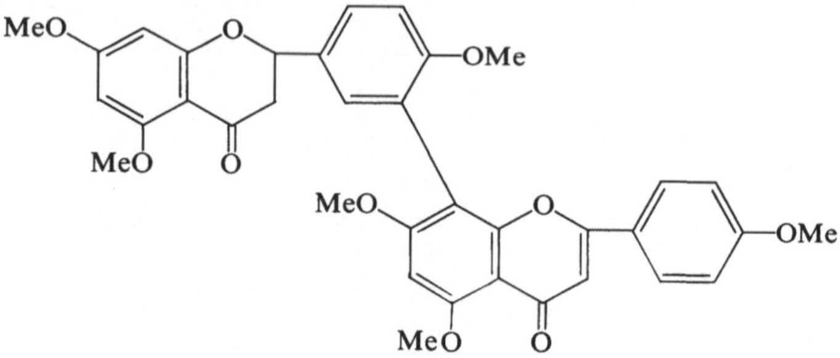

                     H. D. Locksley:

(70)

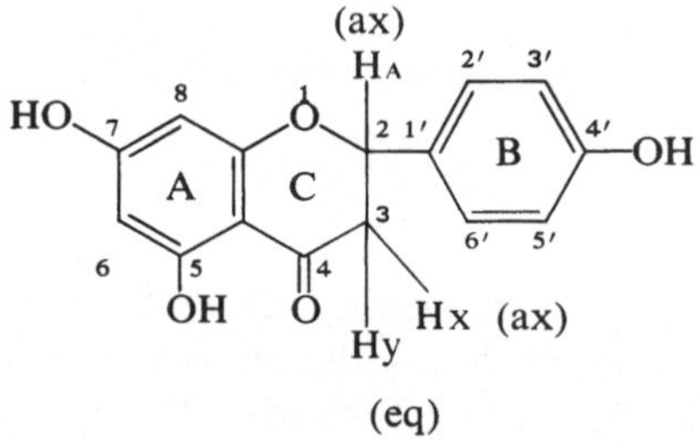

(71) Naringenin

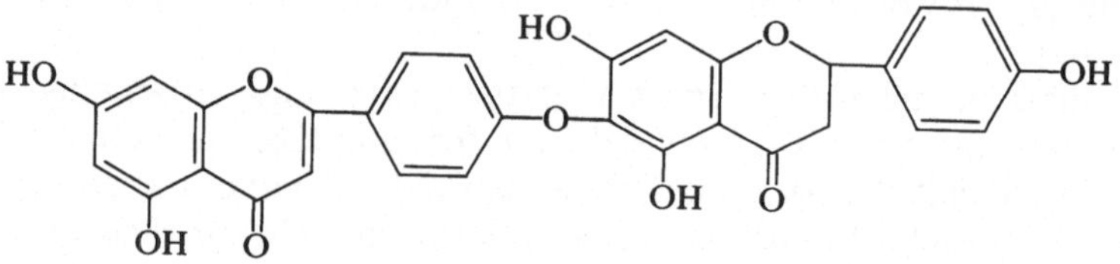

(72)

To identify the carbon skeleton of each dihydrobiflavone, the permethyl ethers (69) and (70) derived from the parent molecules were dehydrogenated using chloranil to give hinokiflavone pentamethyl ether (34) [from 69] and amentoflavone hexamethyl ether (21) [from 70] (Chart 21).

There are two ways in which the apigenin-naringenin link might be forged in the natural dihydrobiflavones. For example, in the case of dihydrohinokiflavone these would be (65) and (72), both of which comply with the dehydrogenation evidence. To differentiate between these two possibilities and those pairs of structures which can be similarly contrived for dihydroamentoflavone and its derivatives, BECKMANN and his co-workers (7) carried out degradation experiments with alkali. Dihydrohinokiflavone (65) gave p-hydroxyacetophenone, a typical flavone degradation product, thus providing clear evidence in favour of structure (65) since only with this mode of coupling is the apigenin (15) unit free to release this fragment, in contrast to the alternative (72).

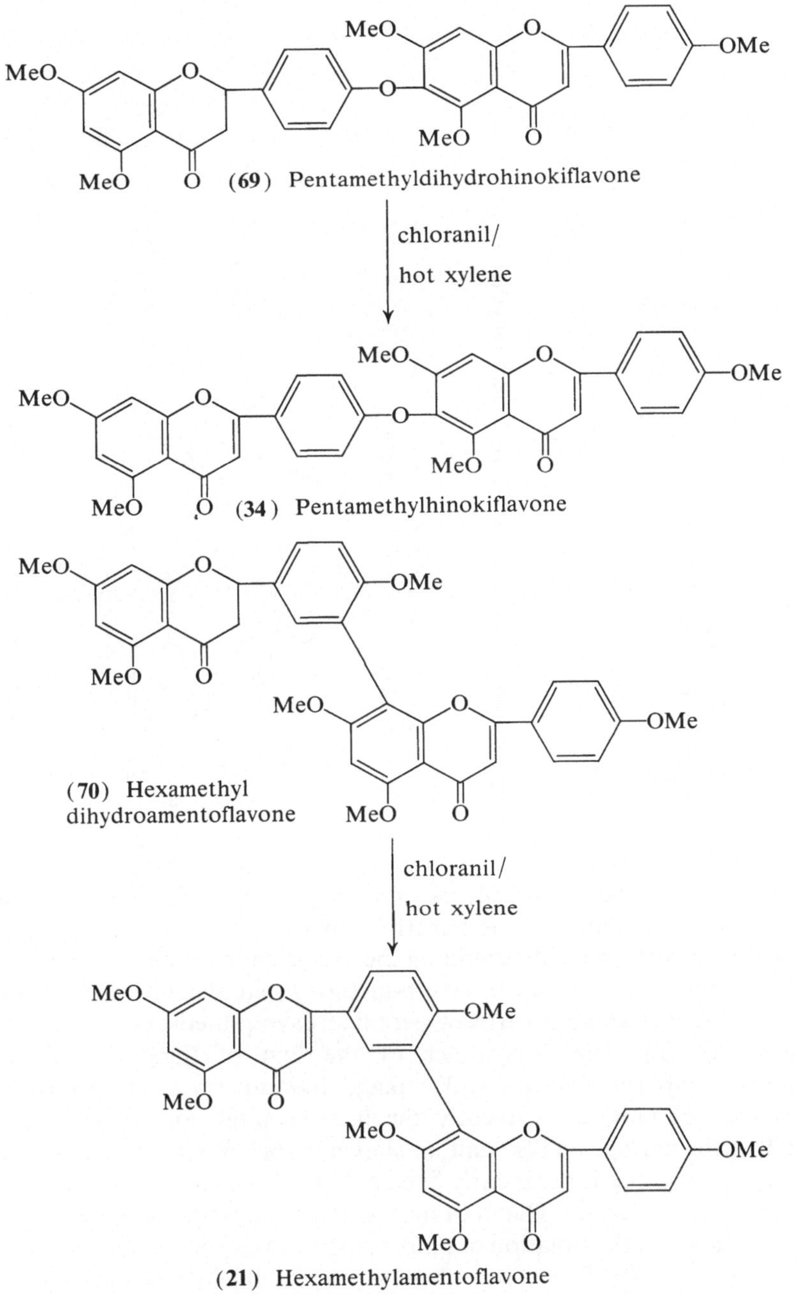

(**69**) Pentamethyldihydrohinokiflavone

chloranil/
hot xylene

(**34**) Pentamethylhinokiflavone

(**70**) Hexamethyl
dihydroamentoflavone

chloranil/
hot xylene

(**21**) Hexamethylamentoflavone

Chart 21. Dehydrogenation reactions of pentamethyldihydrohinokiflavone
(**69**) and hexamethyldihydroamentoflavone (**70**)

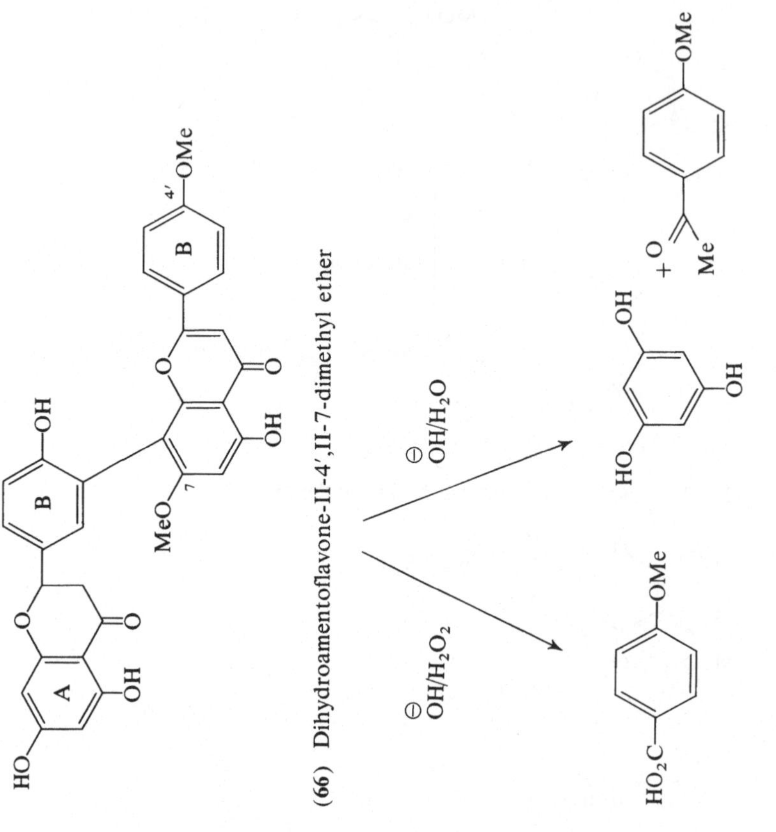

(66) Dihydroamentoflavone-II-4',II-7-dimethyl ether

Chart 22. Alkali degradation reactions of dihydro-amentoflavone-II-4',II-7-dimethyl ether (**66**)

Similar degradative procedures were adopted to confirm structure **(66)** for the dihydroamentoflavone dimethyl ether where the linkage problem is combined with one of determining the distribution of the two *O*-methyl groups. Hot alkali treatment gave *p*-methoxyacetophenone and phloroglucinol while alkaline hydrogen peroxide gave p-methoxybenzoic acid (see Chart 22). The appearance of the first of these degradation products establishes that the C-3' linkage does not involve ring B of the apigenin **(15)** unit: consequently the interflavanoid link must involve ring B of the naringenin **(71)** unit, as shown in **(66)**. Whilst the position of one methoxy group is necessarily fixed at C-4' of the flavone unit, that for the second is less certain. That it cannot be placed on ring A of the flavanone unit is clear from the isolation of phloroglucinol as a degradation product. The authors prefer C-7 of the flavone ring as the site for the second methoxy-group though there is insufficient evidence presented to exclude the two remaining possible sites C-4' (flavanone ring B) and C-5 (flavone ring A).

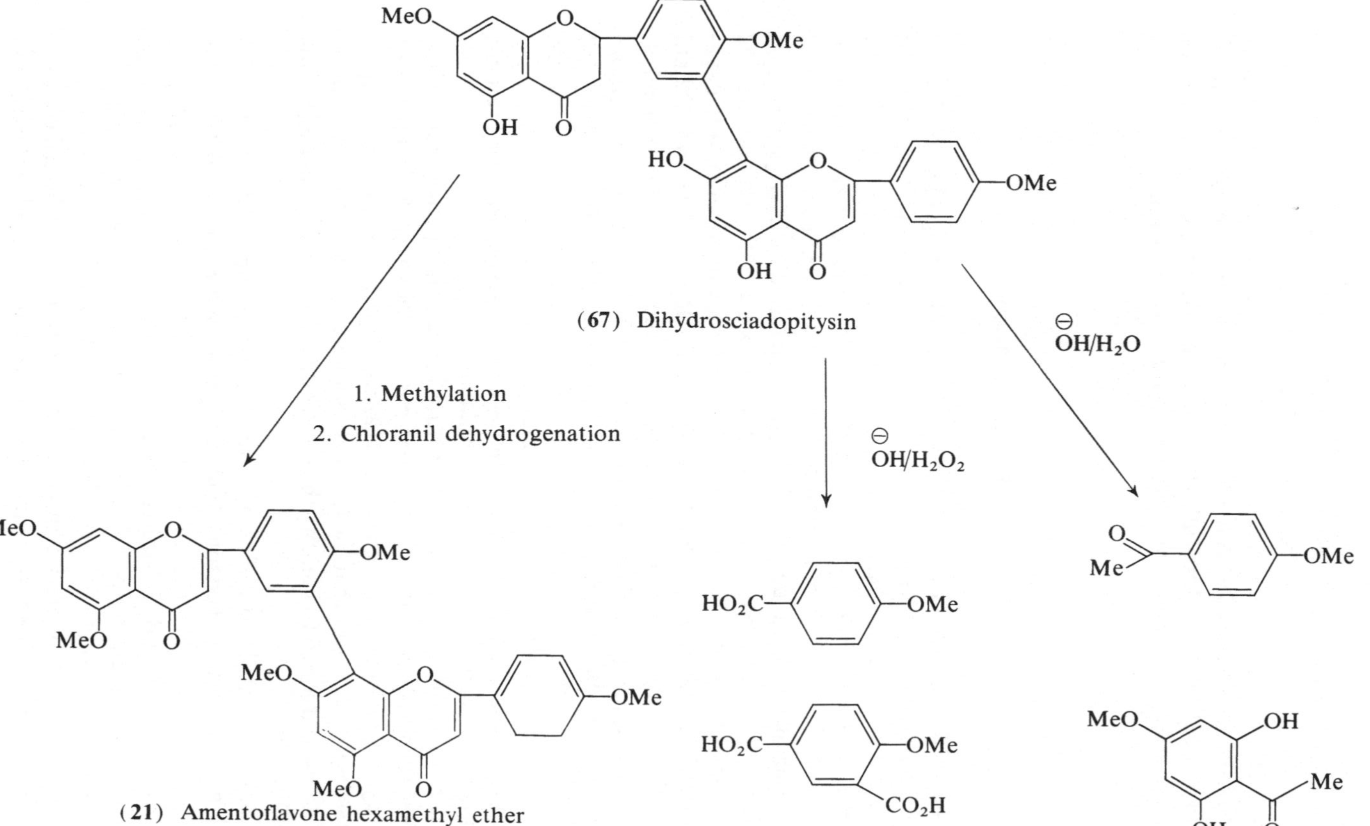

Chart 23. Some reactions of dihydrosciadopitysin (67)

17*

Degradative evidence summarized in Chart 23 provides unambiguous evidence in favour of structure (**67**) for dihydrosciadopitysin. Dihydro-amentoflavone (**68**) from *Cycas revoluta* Thunb. (*21*) yielded a hexamethyl ether derivative identical with the permethyl ether (**70**) derived from natural dihydroamentoflavone dimethyl ether (**66**), showing that the naringenin-apigenin linkage in these two natural products is the same.

X. The C-3/C-8 Biflavanones

The "GB" biflavanones, GB-1 (**9**), GB-1a (**10**), GB-2 (**11**), and GB-2a (**12**), (Chart 24), so-called because they are major constituents of the heartwood of *Garcinia buchananii* Baker, were first isolated in 1967 by JACKSON *et al.* (*34*). Subsequent work showed that they are also present in *G. eugeniifolia* Wall (*33, 36*). High resolution mass spectrometry provided the molecular formulae $C_{30}H_{22}O_{11}$, $C_{30}H_{22}O_{10}$, $C_{30}H_{22}O_{12}$, and $C_{30}H_{22}O_{11}$, respectively; their ultraviolet spectra with maxima at 292 (log ε 28.0→31.4) and 329 nm correlated well with those of simple flava-nones such as naringenin which has λ_{max} 288 (log ε, 17.8) and 330 nm (4.86). The molar extinction coefficients for the more intense band at 292 nm are approximately twice those of simple flavanones; this feature and the size of the molecules (i.e. C_{30}) shown by mass spectrometry led inevitably to the conclusion that the GB metabolites are flavanone dimers, i.e. biflavanones. Addition of sodium acetate produced in each case a bathochromic shift in the ultraviolet spectrum suggesting the presence of an acidic hydroxyl at C-7 in each biflavanone (*38*).

Each biflavanone contains in its infrared spectrum intense absorptions at 3300 (broad band) and 1650—1655 cm^{-1} indicating the presence of hydroxy-groups and hydrogen bonded carbonyl groups, respectively. With these strong indications that the GB metabolites consisted of two flavanone units it only remained to identify the constituent flavanones and their man-ner of linkage.

Classical degradation of a mixture of GB-1 (**9**) and GB-1a (**10**) with alkali gave *p*-hydroxybenzoic acid and phloroglucinol, whilst similar treatment of GB-2 (**11**) and a mixture of GB-2 (**11**) and GB-2a (**12**) gave *p*-hydroxybenzoic and 3,4-dihydroxybenzoic acids as well as phloro-glucinol. These degradation fragments are typical of those expected from simple flavanones such as naringenin (**71**) and taxifolin (**73**) etc. (Chart 25), but unfortunately, no larger fragment(s) retaining the interflavanone linkage could be isolated, so other means were sought to solve the linkage problem.

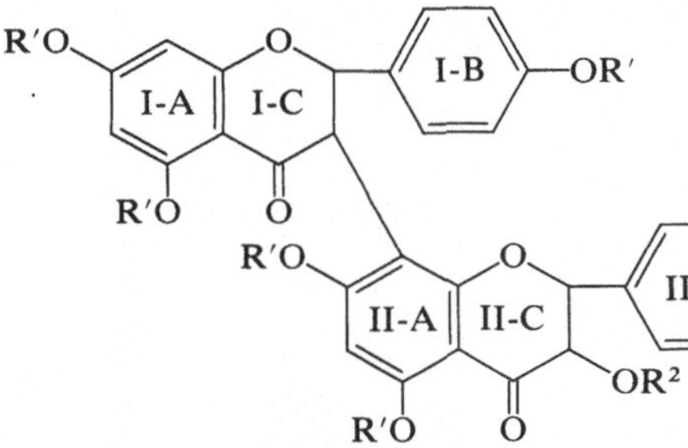

(9) GB-I ($R^1 = R^2 = H$)

(74) GB-I hexamethyl ether
      ($R^1 = Me$, $R^2 = H$)

(75) GB-I heptamethyl ether
      ($R^1 = R^2 = Me$)

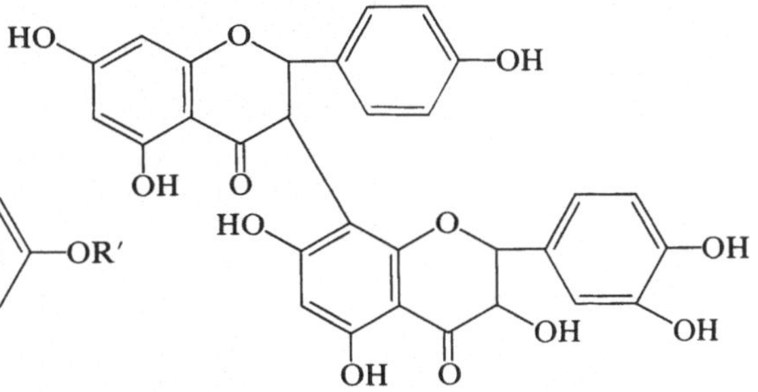

(11) GB-2

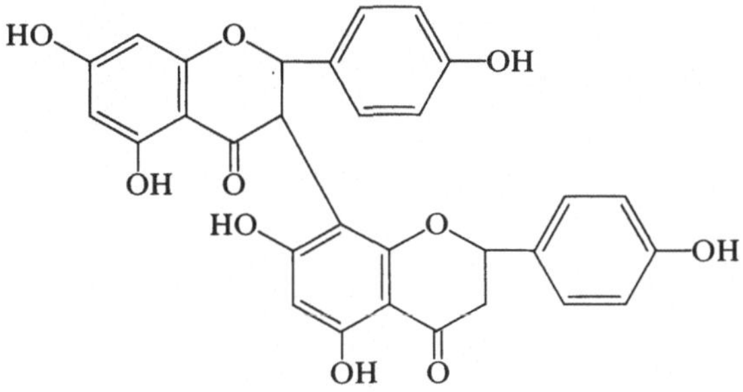

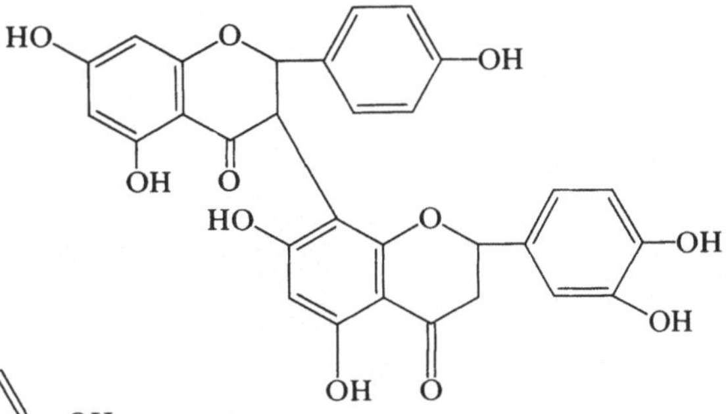

(12) GB-2a

(10) GB-Ia

Chart 24. Biflavones GB-I (9), GB-Ia (10), GB-2 (11), and GB-2a (12)

GB-1 (9), the most abundant and most easily purified of the four GB metabolites was selected for more detailed study. Methylation with dimethyl sulphate and potassium carbonate in acetone gave a mixture of products from which the hexamethyl ether [(74), Chart 24] was isolated. Its i.r. (absorption at 3350 cm^{-1}), n.m.r. (6 methoxyls at τ 6.00—6.30), and mass spectra [M, 642.2175, $C_{30}H_{15}O_4$ (OMe)$_6$(OH) requires M, 642.2101] clearly showed that one hydroxy group in the original molecule was resistent to methylation and so had unique properties. Treatment either of the hexamethyl ether (74) or alternatively GB-1 (9) itself under Purdie methylation conditions (methyl iodide, silver oxide in dimethylformamide at room temperature) effected complete methylation giving the heptamethyl ether [(75), Chart 24] the n.m.r. spectrum of which

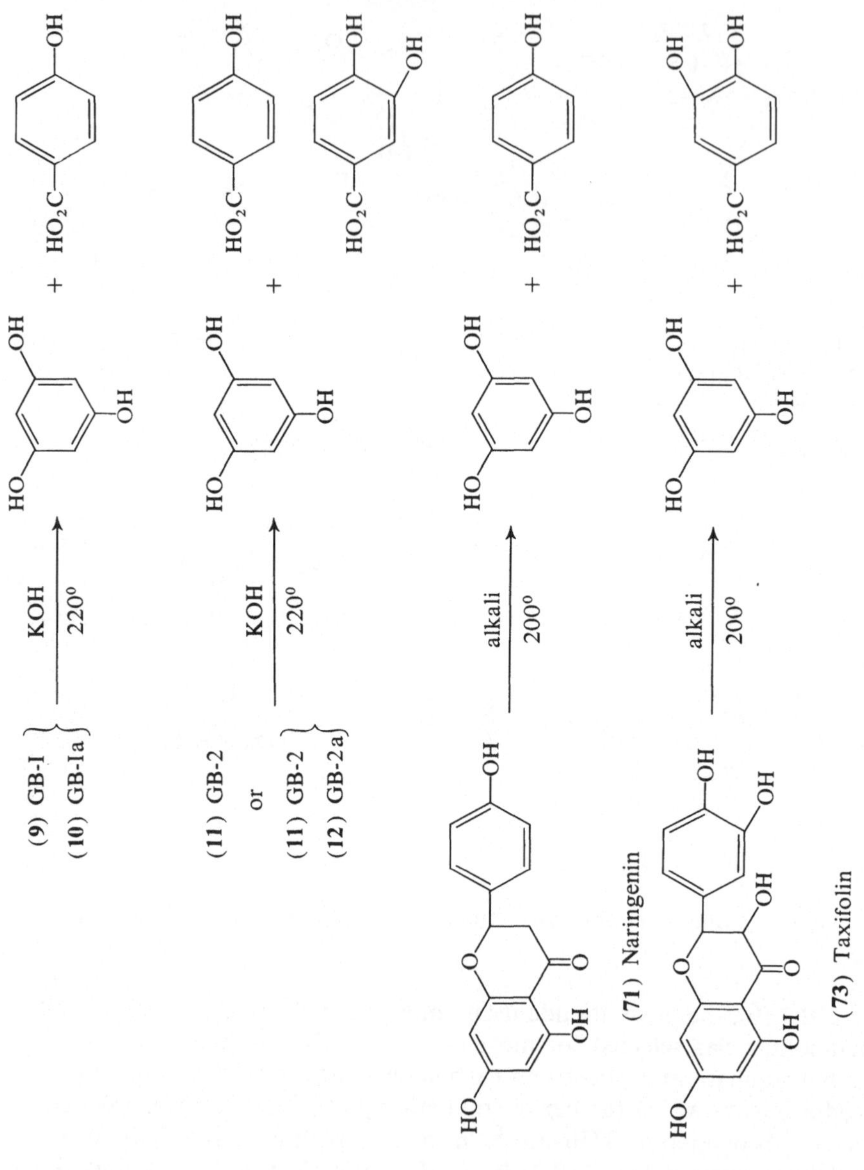

H. D. Locksley:

Chart 25. Alkali degradation reactions of biflavanones (9)—(12) and flavanones (71) and (73)

possessed a new methoxy-signal at τ 6.70 in addition to the six occurring at lower field (τ 6.11—6.29) and seen already in the spectrum of the hexamethyl ether (74). The appearance of a methoxy-group signal at such a high field position is typical of an *O*-methyl group attached to saturated carbon, making the hydroxy group from which it is derived alcoholic; this conclusion also affords an acceptable explanation for the marked reluctance of the hydroxy group to methylate. Since flavanones bearing an alcoholic hydroxy group at C-3 occur commonly in nature (*125*), it was considered likely that GB-1 (**9**) might contain a 3-hydroxyflavanone unit as one of its two flavanone components.

The n.m.r. spectra of the parent GB biflavanones were recorded at room temperature in several solvents but only poorly resolved spectra resulted until it was found that solutions in d_6-dimethylsulphoxide at 100° (with hexamethyldisiloxane as the internal calibration standard) gave very well defined spectra (*36*) (see Table 6 at end of chapter). In this medium the spectrum of GB-1 (**9**) shows sharp signals at τ −1.57 and −2.03 consistent with the presence of hydrogen bonded hydroxy groups at the two C-5 positions, other observable hydroxy-groups appearing as a broad signal centred at τ 0.90; all disappear on addition of deuterium oxide. The aromatic protons of ring I-B and II-B appear as slightly overlapping *ortho*-split doublets at τ 2.92 and 3.06 coupled to their accompanying doublets at τ 3.33 and 3.40 (*J* 9.0 Hz). Such AA′XX′ arrangements indicate the presence in each ring of a C-4′ hydroxy-group.

Three more aromatic signals appear at the high field position of τ 4.22 where they correlate well with those for C-6 and C-8 protons of the phloroglucinol ring A in naringenin (τ 4.30) (*36*). The appearance of only *three* such protons and not *four* establishes that one end of the interflavanone linkage involves either the C-6 or the C-8 position of a flavanone ring A.

In the middle region of the spectrum there are four one-proton doublets, designated *a*, *b*, *c*, and *d*, respectively, with centres at τ 4.56, 5.07, 5.56, and 5.97 (in each case *J* = 12.0 Hz). Spin decoupling experiments showed doublet *a* to be coupled to doublet *c*, and likewise *b* to *d*. Doublets *b* and *d* were assigned to the C-2 and C-3 protons, respectively, of the 3-hydroxyflavanone unit II, since the same two protons in the model 3-hydroxyflavanones, aromadendrin tetramethyl ether (**76**) (in d_1-chloroform) (14) and taxifolin-5,7,3′,4′-tetramethyl ether (**77**) (in d_1-

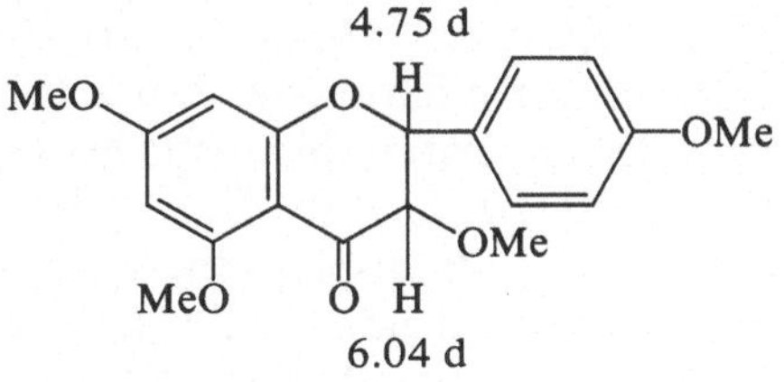

(**76**) Aromadendrin
    tetramethyl ether

(77) Taxifolin tetramethyl ether

chloroform) (14) are at τ 4.75 d, 6.04, and 5.03 d, 5.55 d, respectively. Doublets *a* and *c* were therefore assigned to the ring C protons of the flavanone unit I. All the evidence points to this being a unit of naringenin, but since *two* and not *three* aliphatic protons are present in its ring C, the interflavanone link must involve C-3 as in (9) so as to leave two vicinally placed protons at C-2 and C-3. The chemical shifts and coupling constants of the two protons *a* and *c* are in keeping with this explanation. The interflavanone link is therefore between C-3 of ring I-C and C-8 (or C-6) of ring II-A. Any point of linkage other than C-3 of ring I-C would produce a different coupling pattern for the four aliphatic protons, as would any other location of the hydroxy group at C-3 of ring II-C, whilst any diaryl linkage would require *five* aliphatic protons to be present having more complex spin-spin interactions.

The n.m.r. spectrum of GB-1a [(10), Chart 24] (Table 1), shows no significant differences from that of GB-1 (9) in the aromatic region; the differences lie in the aliphatic region, where a much more complex splitting pattern representing the spin-spin interactions of *five* aliphatic protons is present. Two of the five protons have chemical shifts (τ 4.54 d and 5.55 d) and coupling constants (*J* 12 Hz) which identify them as belonging to ring C of flavanone unit I as in GB-1 (9), the remaining three constitute an AXY pattern typical of that shown by the ring C protons of naringenin (71). Consequently, GB-1a lacks the 3-hydroxy-group present in ring II-C of GB-1 (9) and is thus constructed of two units of naringenin (71) linked C-3/C-8 (or C-6) (see Chart 24 and Table 6).

The n.m.r. spectra of the biflavanones GB-2 (11) and GB-2a (12) (Chart 24) clearly show that a relationship exists between these and GB-1 (9) and GB-1a (10), respectively. That for GB-2 (11), which contains one more oxygen atom than GB-1 (9), displays the same pattern of aliphatic signals but differs in its aromatic region (Table 6), there being *seven* rather than eight low field aromatic protons. This points to the additional oxygen atom being located either on ring I-B or ring II-B. Solubility of GB-2 (11) in aqueous sodium tetraborate and the isolation of 3,4-dihydroxybenzoic and *p*-hydroxybenzoic acids as alkali degradation products

(**78**)

(**79**) Aromadendrin

(Chart 25) show that rings I-B and II-B consist of 4′-hydroxyphenyl and 3′,4′-dihydroxyphenyl systems, or *vice versa,* giving GB-2 either structure (**11**) or (**78**). A choice between these two structures is provided by a detailed study of the manner in which GB-2 and the other biflavanones fragment in the mass spectrometer (Chart 26).

Monoflavanones such as naringenin (**71**) and aromadendrin (**79**) fragment predominantly through the retro-Diels-Alder process (Chart 26) and by the release of the *p*-hydroxybenzyl ion (or its equivalent hydroxy-tropylium ion) (*35*). When the mass spectrometer is operated such that the temperature of the ionisation chamber is *ca.* 320°, naringenin (**71**) is observed to lose phloroglucinol, but as this loss can be induced by purely thermal means outside of the mass spectrometer, its formation may be due only in part to electron impact (*35*). No loss of phloroglucinol occurs at the normal operating temperature of the mass spectrometer (*ca.* 200° in the ionisation chamber) (*35*). GB-2 (**11**), by analogy (*36*), loses phloroglucinol initially (Chart 26) giving a peak at (M-126)$^{+}$ which then goes on to fragment *via* the retro-Diels-Adler process to give an ion with m/e 296. In addition, GB-2 (**11**) fragments by way of two successive retro-Diels-Alder sequences around ring I-C and II-C to give the ion at m/e 270. To fit these observed final fragments, ring B, which is retained in both these fragments, must carry *one* hydroxy group only. If it were to carry two

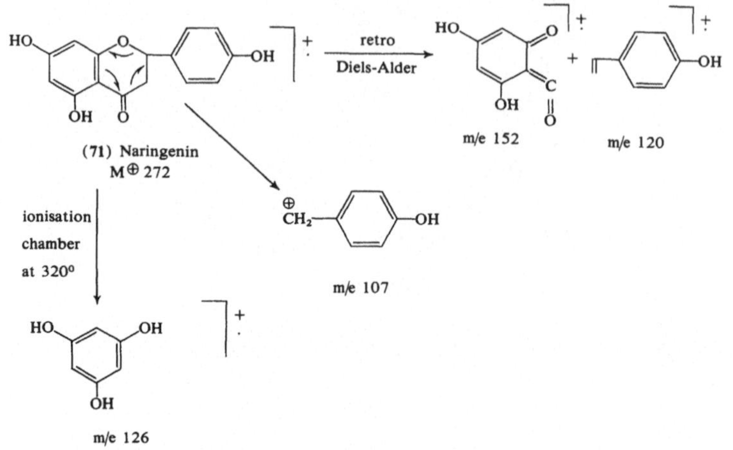

Mass spectral fragmentation of naringenin (71)

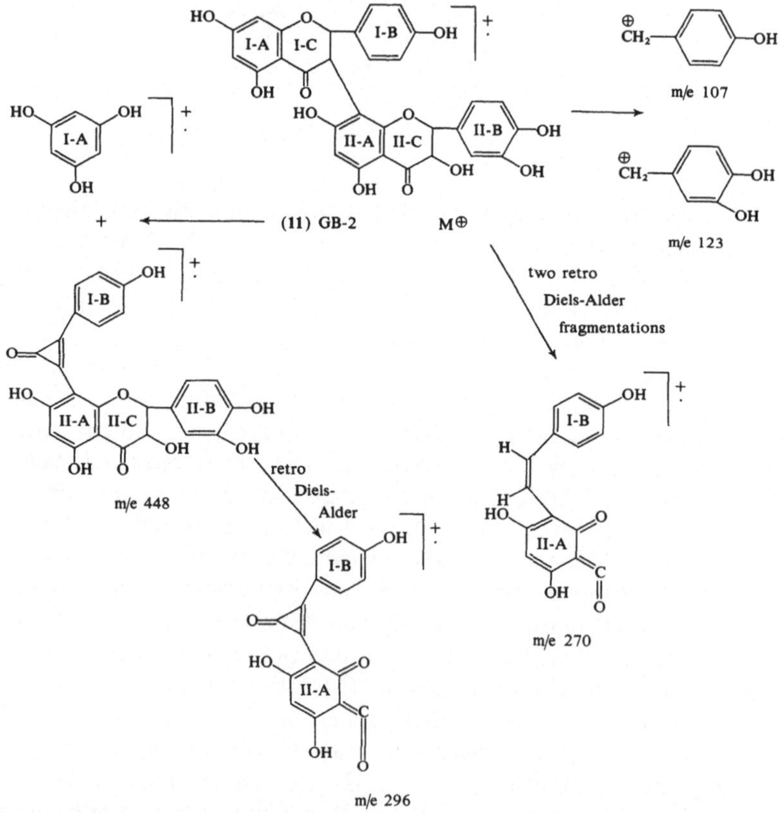

Chart 26. Mass spectral fragmentations of naringenin (**71**) and GB-2 (**11**)

References, pp. 305—311

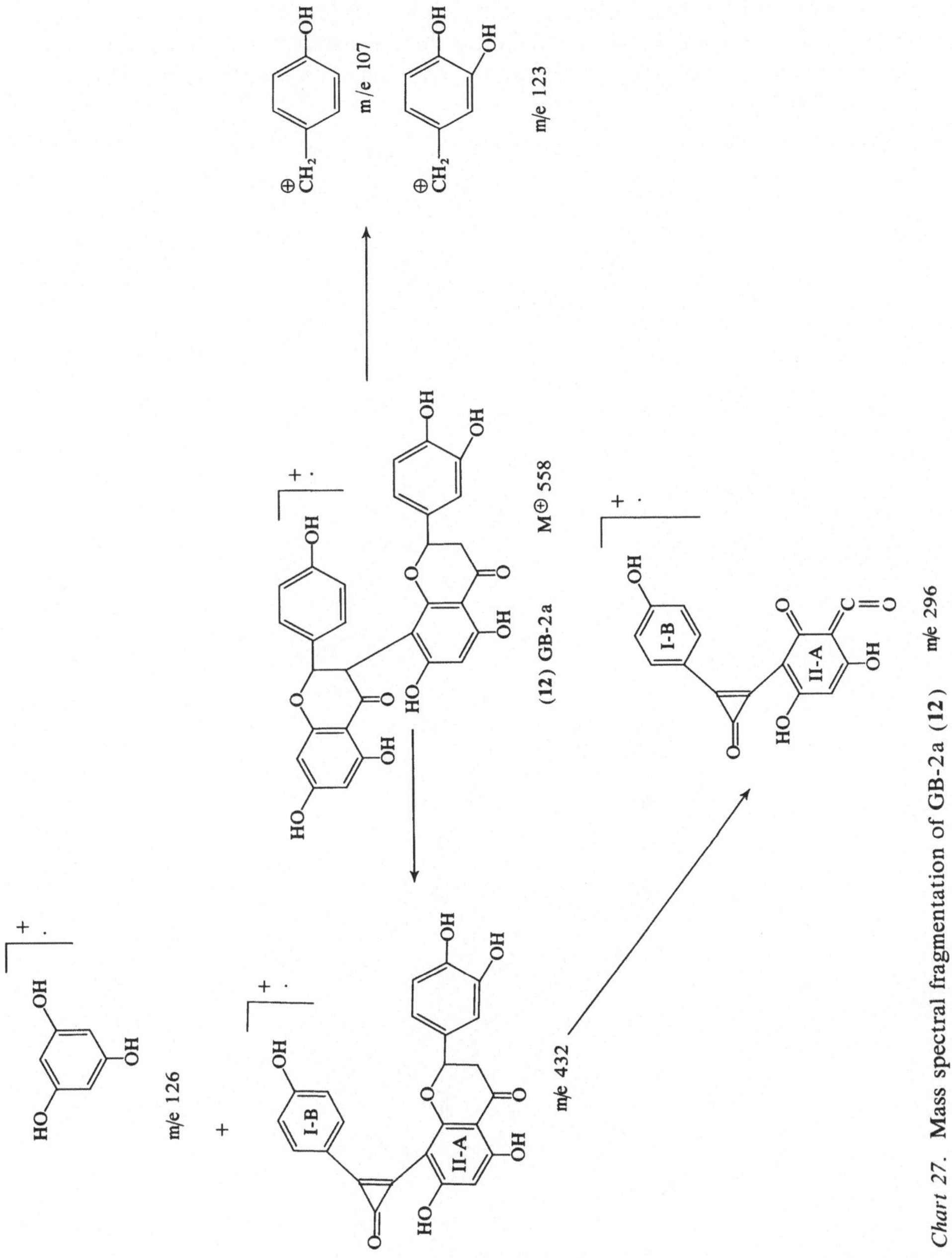

Chart 27. Mass spectral fragmentation of GB-2a (**12**)

 H. D. LOCKSLEY:

hydroxy groups (making it a 3,4-dihydroxyphenyl group) these ions would need to have m/e values of 312 and 286, respectively. Consequently, it can be confidently concluded that GB-2 has the constitution (11) and not (78).

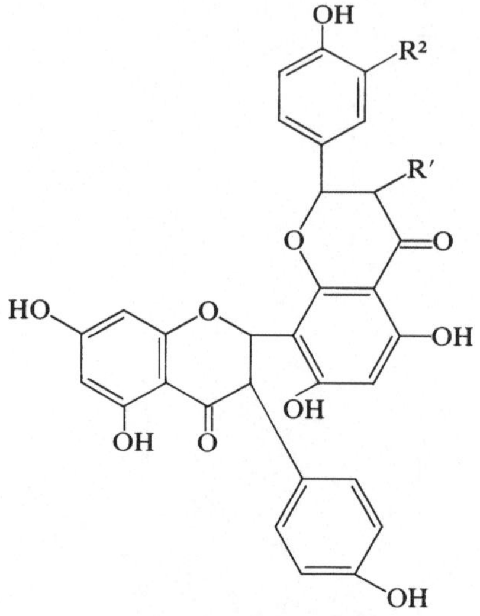

(80) (R′=OH, R²=H)
(81) (R′=R²=H)
(82) (R′=R²=OH)
(83) (R′=H, R²=OH)

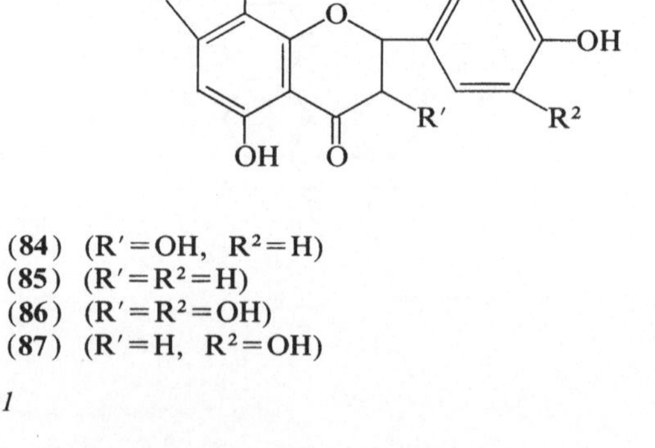

(84) (R′=OH, R²=H)
(85) (R′=R²=H)
(86) (R′=R²=OH)
(87) (R′=H, R²=OH)

References, pp. 305—311

From its n.m.r. spectrum it is clear that GB-2a differs from GB-2 in lacking the C-3 hydroxy group of ring II-C (Table 6). Similar mass spectral evidence (Chart 27) to that outlined for GB-2 (**11**) establishes that the 4′-hydroxyphenyl and 3′,4′-dihydroxyphenyl systems constitute rings I-B and II-B, respectively, as shown in formula (**12**).

At an early stage structures (**80**)—(**83**) and (**84**)—(**87**) were considered to be likely alternatives to those already discussed (**9**)—(**12**), since in a broad sense most of the evidence outlined is compatible with them also.

To remove completely any chance of error, further structural evidence was obtained through the use of classical degradation techniques. When methylated under the Purdie conditions, GB-1a (**10**) (Chart 24) gave a heptamethyl ether whose spectral properties were in accord with structure (**88**). In particular, the replacement of the three naringenin-like AXY protons of ring II-C by two new signals at lower field indicated that ring II-C had opened to produce a chalcone. Furthermore, the ultraviolet

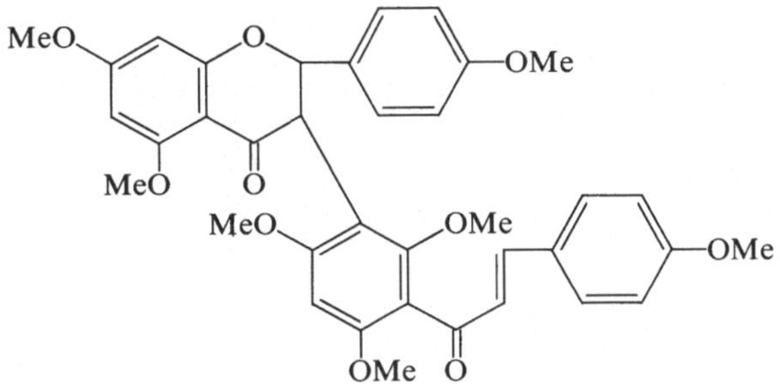

(**88**) GB-Ia heptamethyl ether

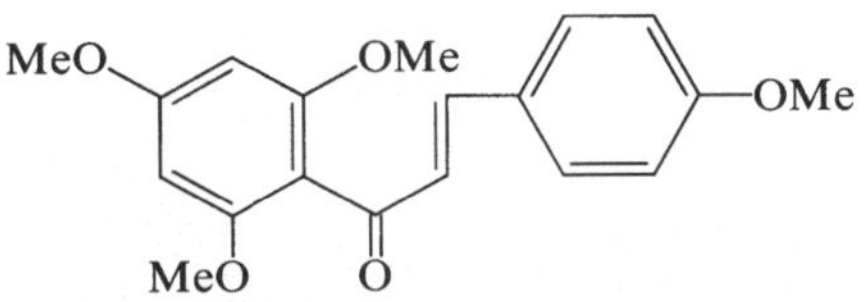

(**89**) 2′,4,4′,6-tetramethoxychalcone

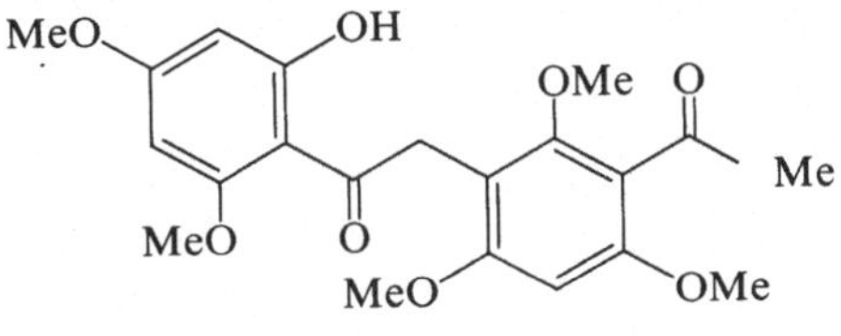

(**90**)

spectrum of the heptamethyl ether was typical of that expected for a combination of flavanone and chalcone spectra [e. g. naringenin (71) and 2′,4,4′,6′-tetramethoxychalcone (89)]. Alkali degradation of the ether (88) gave as a major product the deoxybenzoin (90) whose structure was established through the rational synthesis shown in Chart 28. The alternative structures (91) and (92) for GB-1a heptamethyl ether derived from the parents (81) and (85), respectively, would have been expected to give as comparable alkali degradation products the deoxybenzoins (93) and

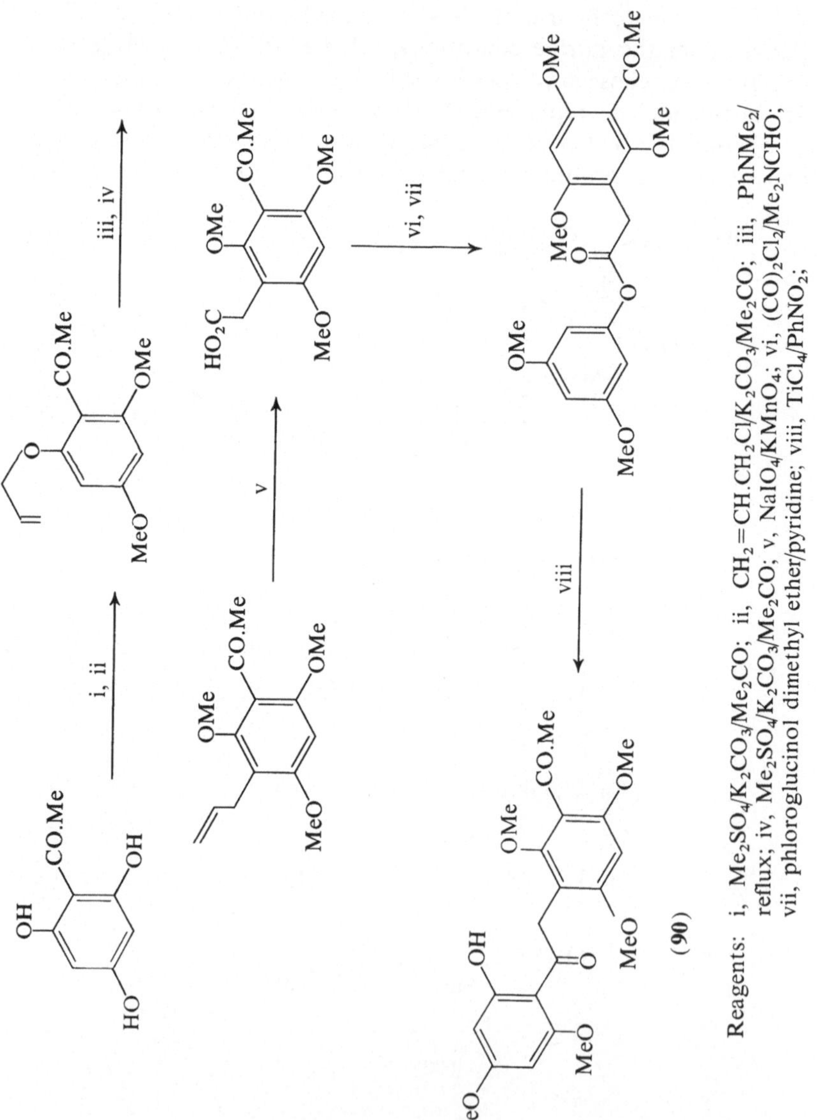

Chart 28. Synthesis of deoxybenzoin (90)

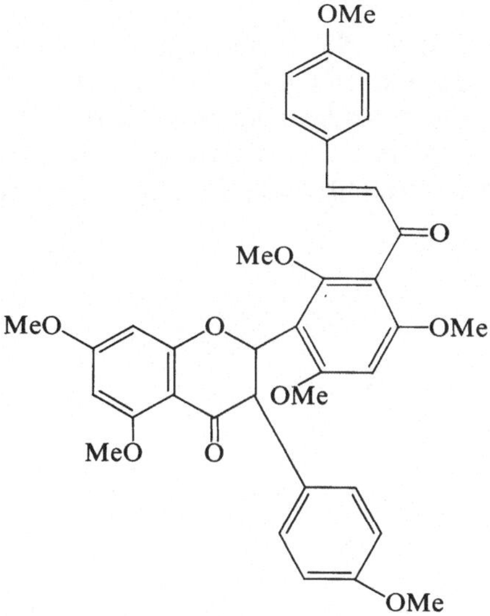
OMe
O
MeO
MeO
O
OMe
OMe
MeO
O
OMe

(**91**)

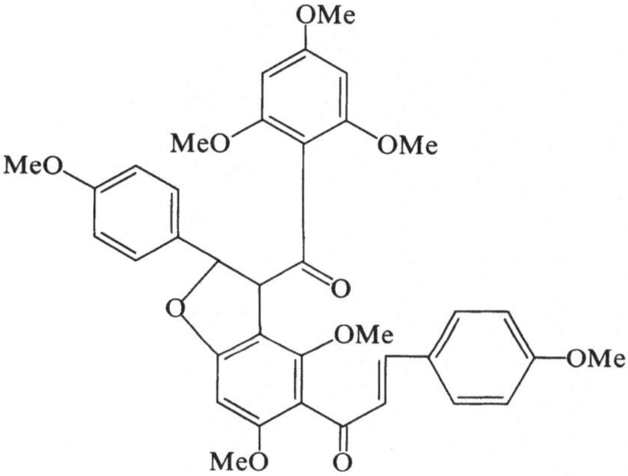
OMe
MeO
OMe
MeO
O
O
OMe
OMe
MeO
O

(**92**)

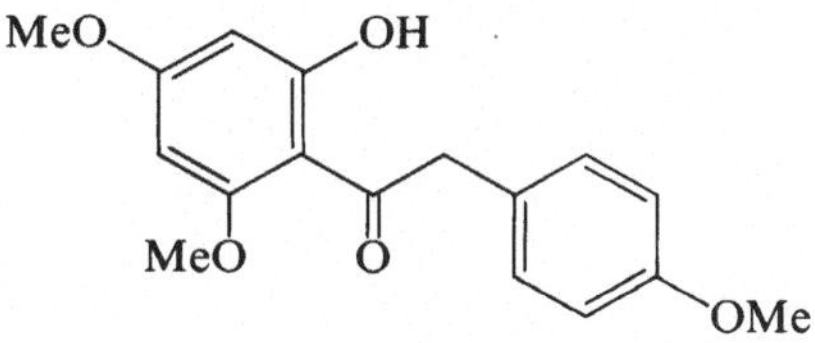
MeO
OH
MeO
O
OMe

(**93**)

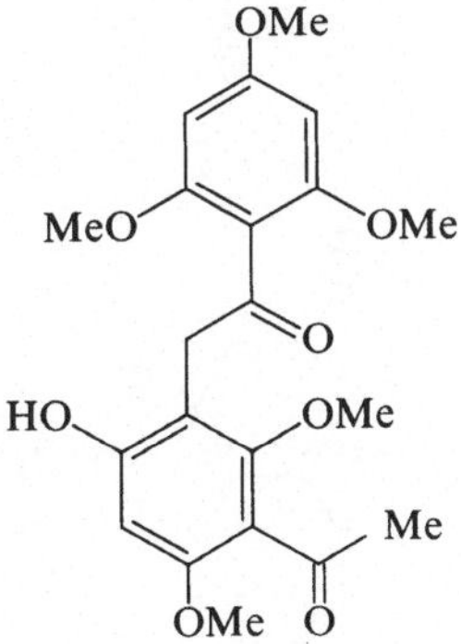
OMe
MeO
OMe
O
HO
OMe
Me
OMe
O

(**94**)

(94), respectively. The first of these (93) was synthesized (Chart 29) and shown to be absent from the alkali degradation mixture. The second (94), though isomeric with the deoxybenzoin isolated (90), possesses a quite different methylation pattern. This evidence clearly provides unambiguous proof that structures (10) and (88) are the correct ones for

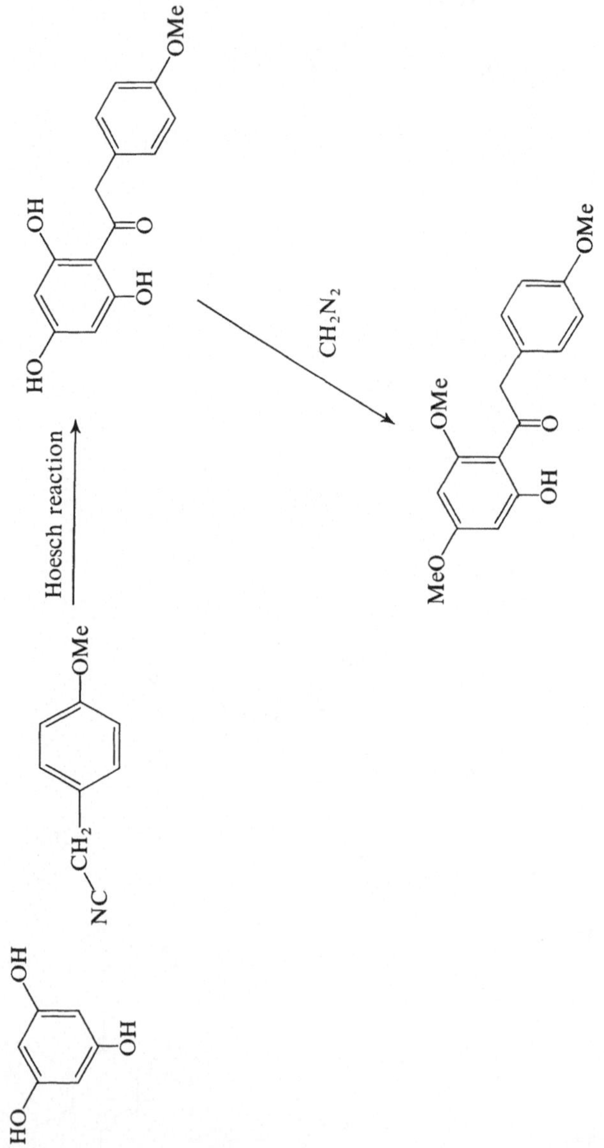

Chart 29. Synthesis of deoxybenzoin (93)

GB-1 a and its heptamethyl ether, respectively. From this result, and from the very close similarity of their spectra (including the optical rotatory dispersion curves, see Table 7 at end of chapter) it is assumed that all four biflavanones possess the same carbon skeleton giving them the constitutions (9)—(12).

However, two aspects of the biflavanone structures (9)—(12) still required investigation, namely,

(i) the position of the interflavanone linkage at ring II-A: C-3/C-8 or C-3/C-6,

(ii) the absolute stereochemistry of the biflavanones.

In the case of (i) a C-3/C-8 interflavanone linkage was preferred since no extreme difficulty was experienced in methylating the C-5 hydroxy group of ring II-A in GB-1 (9) and GB-1 a (10). A C-3/C-6 linkage would leave the C-5 hydroxy group in a very crowded environment rendering its methylation difficult. However, although this argument has been used by ourselves and others (*3, 37, 130*) as a means of distinguishing positional isomers of this type, it is quite erroneous since the facts do not support it: hinokiflavone (2) and agathisflavone (4) and their naturally occurring ether derivatives readily give permethyl ether derivatives in spite of their possessing interflavone linkages at C-6 of ring II-A. Consequently, a more reliable criterion for the linkage question was required: use of the solvent shift technique of PELTER and his co-workers (*30*) provided this criterion. Figure 4 shows graphically the changes in chemical shifts which occur in the signals of the seven methoxy groups of GB-1 heptamethyl ether [(75), Chart 24] on increasing the concentration of d_6-benzene in d_1-chloroform. All methoxy signals show positive (upfield) shifts indicating (with the exception of the C-3 methoxy group on ring II-C) that each has at least one free *ortho*-position. This evidence supports the C-3/C-8 linkage for GB-1 heptamethyl ether. In the absence of any rearrangement occurring

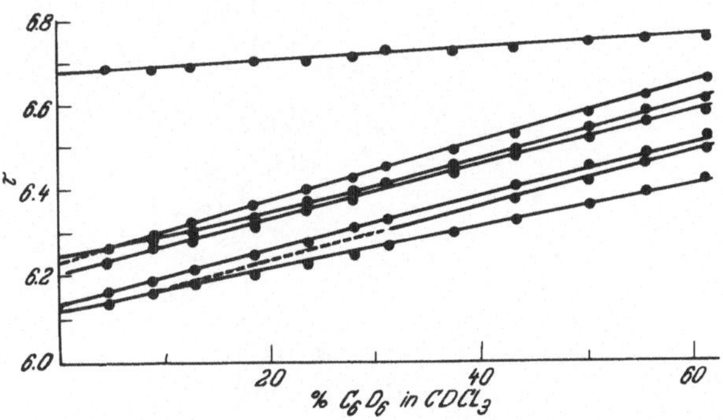

Fig. 4. Solvent shifts of methoxy-group protons in GB-1 heptamethyl ether [(75) Chart 24] for solutions d_6-benzene/d-chloroform. Reproduced from J. Chem. Soc. (C) (*36*) by kind permission of the authors and the publishers, The Chemical Society, London, U.K.

during methylation, it is assumed that GB-1 (**9**), and the remaining three biflavanones, GB-1a (**10**), GB-2 (**11**), and GB-2a (**12**) also possess the C-3/C-8 linkage.

The question of the absolute stereochemistry i.e. (ii) above, of the biflavanones remains unsolved. The large coupling constant ($J = 12.0\,\text{Hz}$) observed for the C-2 and C-3 protons of ring C is consistent with the *trans*-diaxial disposition of hydrogen atoms. Thus in GB-1 (**9**) and GB-2 (**11**) the four substituents on rings I-C and II-C are equatorially bonded. The absolute stereochemistry of these groups is, however, unknown. GB-1 (**9**) and GB-2 (**11**), which have the maximum number of chiral

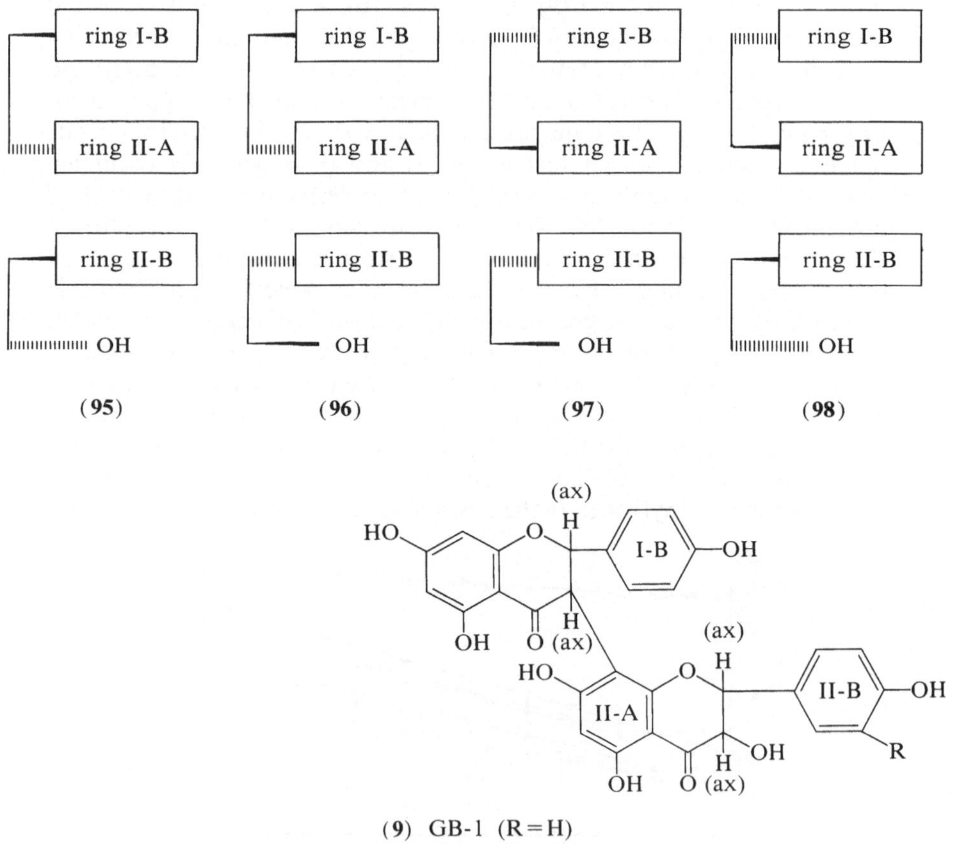

(**95**)   (**96**)   (**97**)   (**98**)

(**9**) GB-1 (R=H)

(**11**) GB-2 (R=OH)

Chart 30. Diagrammatic representation of the stereoisomeric forms of GB-1 (**9**) and GB-2 (**11**)

centres, could have any one of the following configurations [(**95**)—(**98**), Chart 30], assuming that all the chiral centres exhibit optical activity. Should this not be so, then GB-1 (**9**) would be represented by appropriate combinations of stereoisomers (**95**) to (**98**). In addition to this form of chirality there is the added possibility that steric inhibition to rotation around the interflavanone linkage could double the number of stereoisomers. Unfortunately, X-ray crystallography cannot be used to solve the stereochemical question as neither the parents nor their derivatives are crystalline. Some information may be gained from a detailed study of the optical rotatory dispersion (see Table 7) and circular dichroism curves of the parent biflavanone molecules in relation to those of model mono-flavanones.

XI. Flavanone-flavones Possessing a C-3/C-8 Interflavanoid Linkage

Only two members of this family have been isolated to date, namely morelloflavone (**8**) and volkensiflavone (**7**) (Chart 31). Morelloflavone (**8**), the first to be discovered, was isolated trom the heartwood of *Garcinia morella* by VENKATARAMAN and his co-workers in 1967 (*40*). It gave colour tests typical for a flavanoid and showed absorption maxima in the ultra-violet region at 255 sh, 275, 288, and 345 nm, and in the infrared at 1645 cm$^-$1 (hydrogen bonded carbonyl). Complete methylation with dimethyl sulphate provided the heptamethyl ether (**99**) (Chart 31) whose mass spectrum suggested the molecular formula $C_{37}H_{34}O_{11}$, giving for morelloflavone (**8**) itself the formula $C_{30}H_{20}O_{11}$, consistent with micro-analytical data. The infrared spectrum of the derivative (**99**) possessed two absorptions at 1645 and 1670 cm^{-1}. Such spectral changes on methylation are reminiscent of the behaviour of flavone and flavanone systems, respectively, bearing hydroxy-groups at C-5 (*15*).

Boiling ethanolic potassium hydroxide degraded the ether (**99**) giving veratric acid (**100**), acetoveratrone (**101**), and a deoxybenzoin whose properties (colour tests, n.m.r., and mass spectral characteristics) were consistent with structure (**102**)* (Chart 32).

---

* This same deoxybenzoin (**102**) has since been synthesized (*36*). However, there is a wide and inexplicable discrepancy in the m.p.s. reported (viz., m.p. 116°, KARANJGAOKAR *et al.* (*40*)) and 153—156° (JACKSON *et al.* (*36*)). JOSHI *et al.* (*37*), who have also isolated the compound as a degradation product of morelloflavone give m.p. 166°. The n.m.r. spectral data are broadly in agreement.

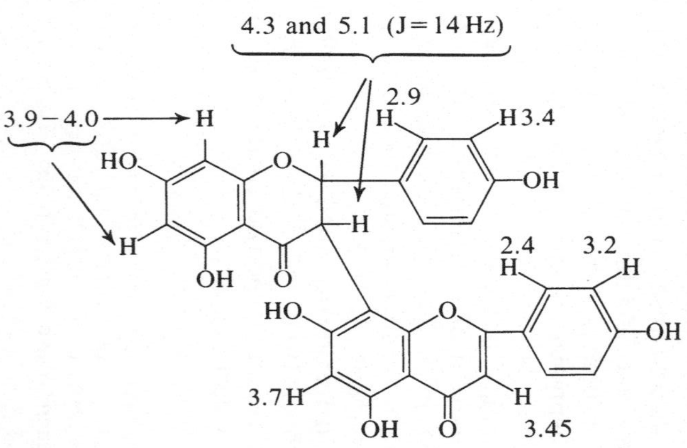

H. D. LOCKSLEY:

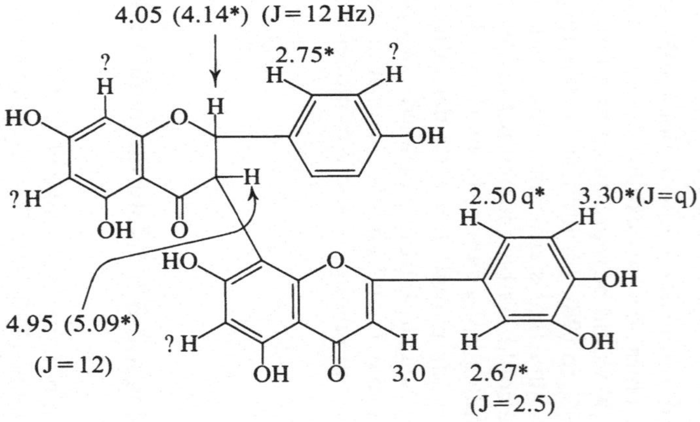

(8) Morelloflavone

* These chemical shift values are those for the hepta-methyl ether (**99**) (Me in place of H in −OH) of morelloflavone.

(Hydrogen bonded hydroxyls at positions I-5 and II-5 appear as singlets at $\tau -2.66$ and -3.50)

(7) Volkensiflavone

(hydrogen bonded hydroxyls at positions I-5 and II-5 appear as singlets at $\tau -2.7$ and -2.1)

Chart 31. N.m.r. chemical shifts for protons in volkensiflavone (**7**), morelloflavone (**8**) and its heptamethyl ether (**99**)

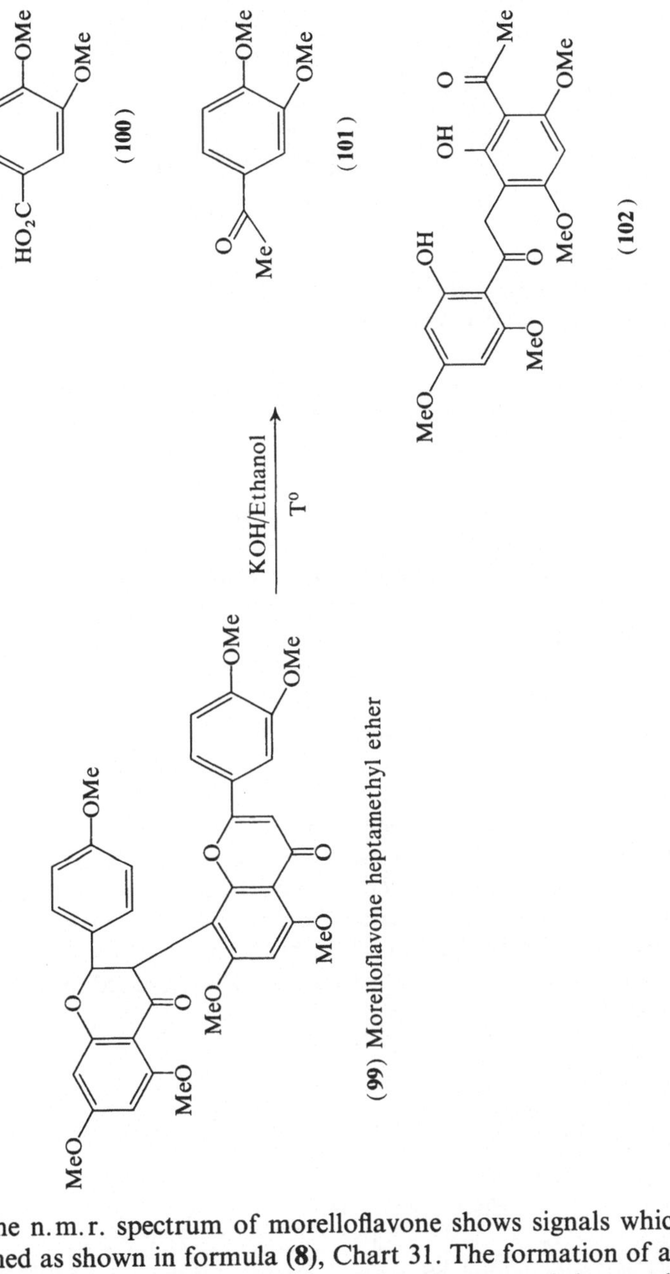

(100)

(101)

(102)

(99) Morelloflavone heptamethyl ether

KOH/Ethanol
T°

Chart 32. Alkali degradation of morelloflavone heptamethyl ether (99)

The n.m.r. spectrum of morelloflavone shows signals which can be assigned as shown in formula (8), Chart 31. The formation of acetoveratrone [(101), Chart 32] establishes that the flavone unit in morelloflavone (8) contains as its ring B the 3,4-dihydroxyphenyl system, and isolation of a deoxybenzoin with the constitution (102), [Chart 32] rather than (103)

(103)

fixes the interflavanoid linkage as C-3/C-8 rather than the alternative C-3/C-6 (assuming that no Wesseley-Moser rearrangement occurs during methylation).

Detailed analysis of the mass spectrum of morelloflavone heptamethyl ether (99) leads to the interpretation shown in Chart 33 through which further evidence is provided in support of structure (8) for morello-flavone.

The closely related flavanone-flavone, volkensiflavone (7), (Chart 31), which is composed of C-3/C-8 linked units of naringenin (71) and apigenin (15), was isolated in 1970 from the heartwood of *Garcinia volkensii* Engl. (*27*), along with morelloflavone (8) and the known biflavanones GB-1a (10) and GB-2a (12). Its ultraviolet and infrared spectral charac-teristics resemble those of morelloflavone (8), while its n. m. r. spectrum run in d_6-dimethylsulphoxide at 100° contained signals which could be readily assigned as shown in formula (7), Chart 31.

The interpretation of the mass spectrum of volkensiflavone shown in Chart 34 also lends considerable support for structure (7).

To confirm this structural assignment GB-1a (10), whose structure had already been proved (*34, 36*), was dehydrogenated using iodine and potassium acetate in acetic acid. The monodehydrogenation product (7) proved to be identical with volkensiflavone. The bisdehydrogenation product (104), which accompanies it is of historical interest as it possesses the carbon skeleton of the erroneous structure (19) proposed for ginkgetin by NAKAZAWA (*90*).

(104)

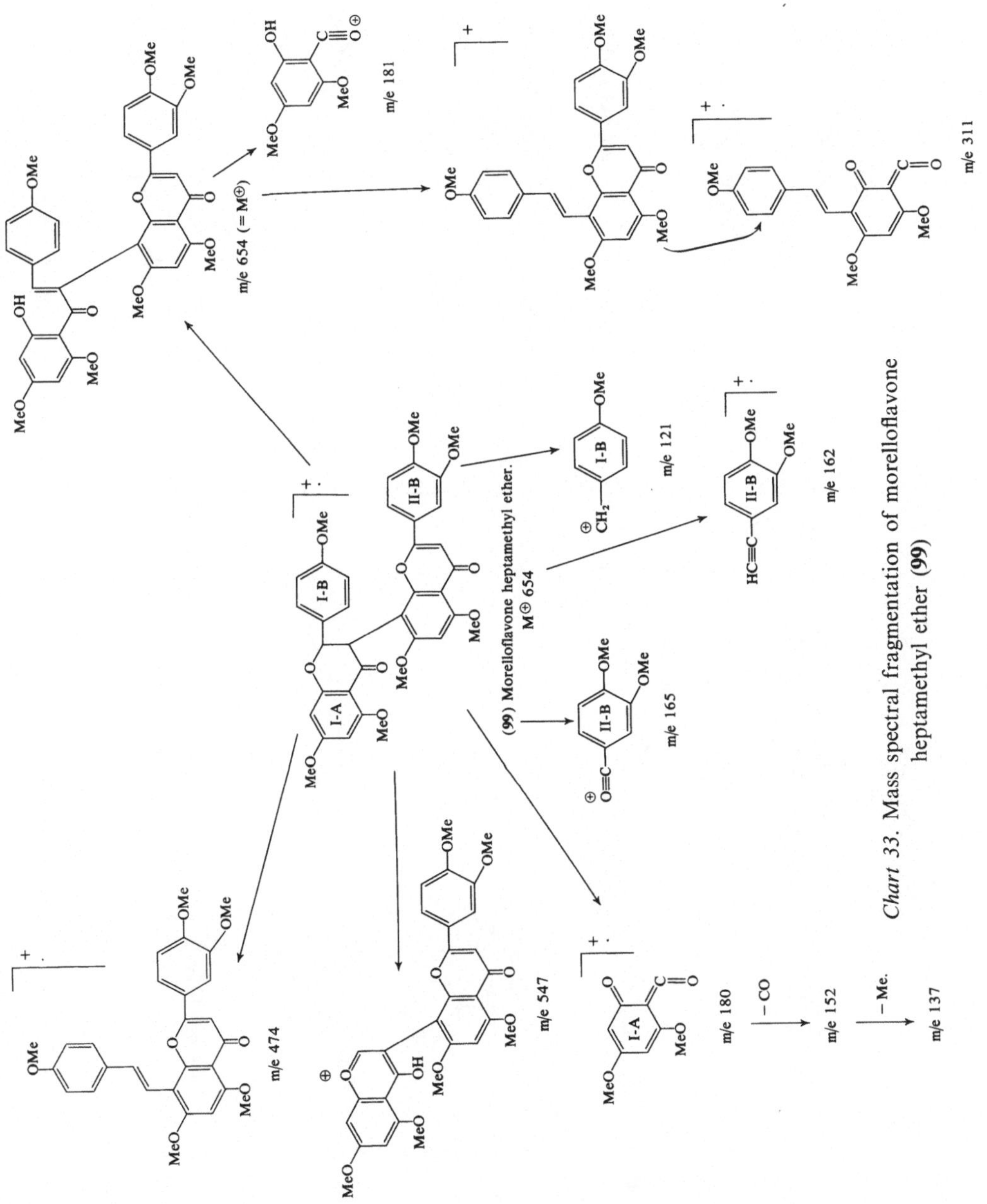

Chart 33. Mass spectral fragmentation of morelloflavone heptamethyl ether (**99**)

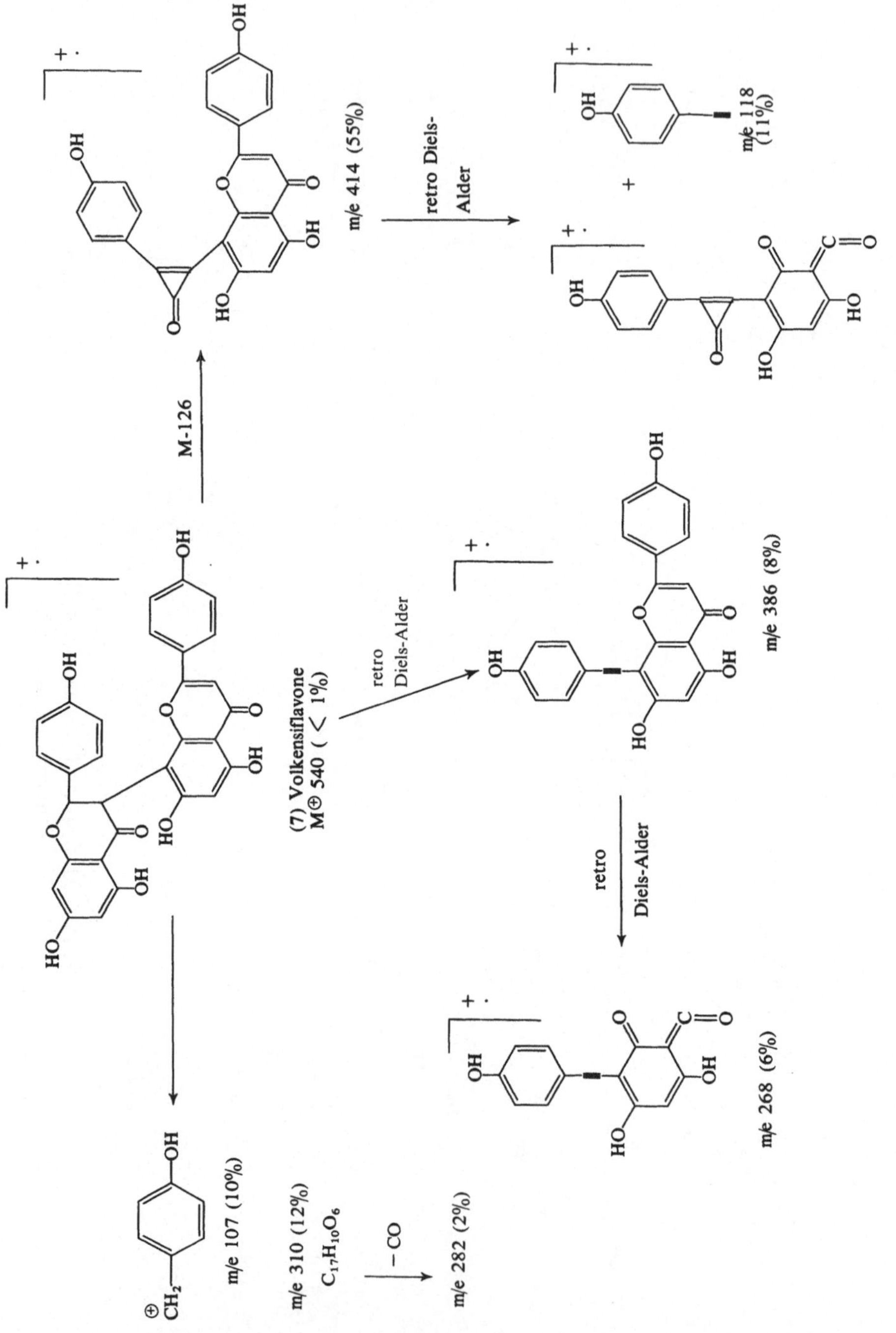

Chart 34. Mass spectral fragmentation of volkensiflavone (7)

Very soon after the work on volkensiflavone (**7**) was published the results of an independent investigation of *Garcinia talboti* appeared (*37*). The roots were shown to contain volkensiflavone (**7**) (the authors refer (*37*) to it as talbotaflavone) and morelloflavone (**8**), (Chart 31). This elegant work confirms very adequately structure (**7**) for volkensiflavone but additionally, the use of solvent shift studies on the permethyl ethers of volkensiflavone and morelloflavone provides unambiguous evidence that they each have the C-3/C-8 linkage rather than the alternative C-3/C-6 one.

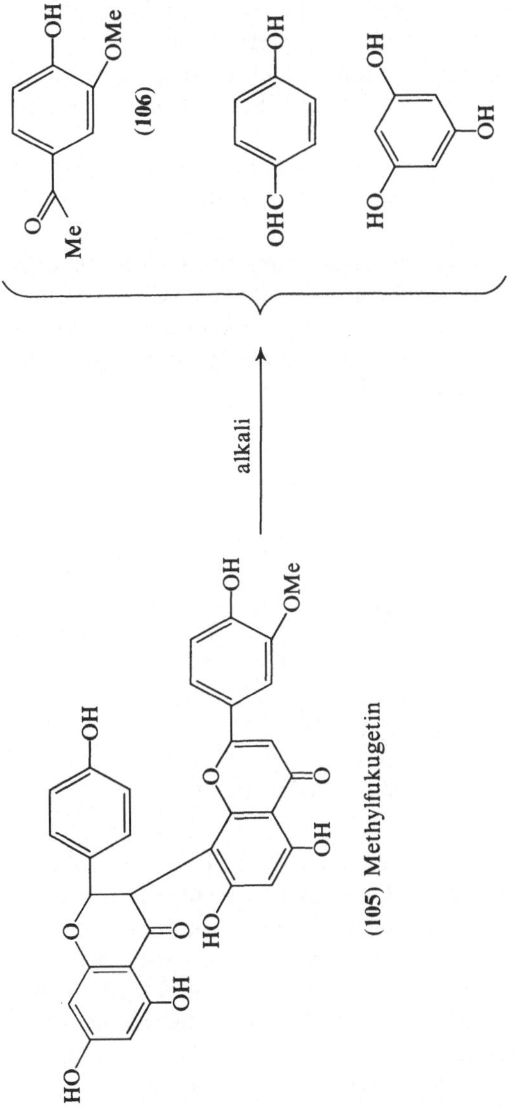

Chart 35. Alkali degradation of methylfukugetin (**105**)

Other *Garcinia* species (*G. spicata* Hook f., *G. xanthochymus* Hook f., *G. multiflora* Champ., *G. linii* Chang, and *G. livingstonii*) have now been examined and from the bark of *G. spicata,* KONOSHIMA *et al.* (*65—67*), have isolated both racemic and optically active forms of morelloflavone (**8**), (±)-volkensiflavone (**7**), (Chart 31), GB-1a (**10**) and GB-2a (**12**), (Chart 24). Independent structural evidence for morelloflavone (to which the authors refer as fukugetin) supports that provided earlier by VENKATARAMAN and his co-workers (*40*).

Also present in *G. spicata* is a natural racemic *O*-methyl ether of morelloflavone (Chart 35) referred to as methylfukugetin (**105**). Its n.m.r. spectrum (singlet at τ 6.08 in d_6-acetone) showed that it contained one methoxy-group whose location at C-3' of ring II-B was established through alkali degradation and the identification of acetoguaiacol (**106**), (Chart 35).

The bark of *G. multiflora* and of *G. linii* has been similarly investigated (*66*). The former contains (±)-morelloflavone (**8**), (+)-morelloflavone (**8**), volkensiflavone (**7**), GB-1 (**9**), GB-1a (**10**), GB-2 (**11**), and GB-2a (**12**) (Charts 31 and 24), while the latter contains (+)-morelloflavone (**8**), GB-1 (**9**), GB-1a (**10**), GB-2 (**11**), and GB-2a (**12**) (Charts 31 and 24).

Garcinia livingstonii has yielded an interesting collection of biflavanoids. From the heartwood and bark morelloflavone (**8**) and volkensiflavone (**7**) have been isolated, while the leaves have yielded the biflavones

(**107**) Podocarpusflavone A

amentoflavone (**1**) and podocarpusflavone A (**107**). This is the first time that biflavones and flavanone-flavones possessing quite different interflavanoid linkages have been found together in the same plant. This co-occurrence represents an interesting chemotaxonomic phenomenon since *Garcinia* species were not formerly thought likely to produce biflavones derived from apigenin units linked in the manner characteristic of the amentoflavone series.

The heartwood of *Allanblackia floribunda* Oliver contains volkensiflavone (7) and morelloflavone (8) in addition to benzophenones and xanthones (*71*). The genera *Allanblackia* and *Garcinia* are both members of the subfamily Clusioideae of the family Guttiferae.

XII. Natural Glycosides of Biflavanoids

On reinvestigation, the fresh bark of *G. spicata* was found to contain fukugiside (108), the first natural *O*-glycoside derivative of a biflavanoid to be isolated (*65*). Hydrolysis with dilute acid, acid ion exchange resin, or emulsin afforded glucose and (±)-morelloflavone (8) (the parent glucoside is optically active), the last reaction showing that the sugar and aglycone are β-linked. Hydrolysis of the permethyl ether derivative (109) of fukugiside (108), formed through use of the Purdie methylation conditions in dimethylformamide, gave a hexamethyl ether of morelloflavone whose ultraviolet spectrum exhibited a marked bathochromic shift on addition of sodium acetate. Thus C-7 is implicated as the site of the free hydroxy-group; however, though the authors prefer the hydroxy group to be sited on the flavone unit, the evidence presented in this preliminary account of their work is minimal. C-7 of the flavanone unit remains a plausible alternative siting for the hydroxyl and, in the original molecule, the glucose unit. Alkaline degradation of the hexamethyl ether (110) might offer the least ambiguous means of settling the structure of fukugiside.

(108) Fukugiside (R′=H, R²=glucosyl)
(109) Fukugiside permethyl ether (R′=Me, R²=pentamethyl glucosyl)
(110) (R′=Me, R²=H)

Spicataside (111) and xanthochymusside (112) are two new biflavanoid glycosides isolated recently by KONOSHIMA *et al.* (*66*). The first of these comes from the fresh bark of *G. spicata* and is a β-*O*-glucosyl derivative of (±)-volkensiflavone. Methylation studies suggest that the sugar is attached at C-7 of the flavone unit as with fukugiside (**108**), though here, as before, full evidence of this is awaited.

(**111**) Spicataside (R = glucosyl)

(**112**) Xanthochymusside (R = glucosyl)
(**11**) R = H (≡ GB−2a)

Xanthochymusside (**112**), isolated (*66*) from the wood and leaves of *G. xanthochymus*, is the first *O*-glycosyl derivative of a biflavanone to be isolated. Hydrolysis with dilute sulphuric acid or with emulsin generates GB-2a (**11**), identical with an authentic sample, as the aglycone and glucose, the glucosidic linkage being β. The site of the *O*-glucosyl group was determined by partial dehydrogenation (iodine, sodium acetate in ethanol) of xanthochymusside (**112**) which furnished fukugiside (**108**) the sugar-aglycone link is therefore probably at C-7 of ring II-A in xanthochymusside (**112**) as it is believed to be in fukugiside (**108**).

XIII. Synthesis of Biflavanoids

The synthesis of biflavanoids has attracted almost as much attention as the associated work of isolation and structure elucidation, but as yet this activity has been restricted almost totally* to the synthesis of members of the biflavone families (1)—(4). Synthesis of the reduced biflavanoids (5)—(12) introduces problems of stereochemistry which can be expected to offer a greater challenge to the chemist of the future. Broadly speaking the synthetic approaches to the biflavones (1)—(4) fall into five distinct categories.

1. Coupling of two flavone nuclei by the Ullmann reaction.
2. Ullmann synthesis of suitably substituted biphenyls followed by their heteroannulation to biflavones.
3. Wesseley-Moser rearrangement of existing biflavones [interconversion of cupressuflavone (3) and agathisflavone (4)].
4. Partial methylation and demethylation of natural or synthetic biflavones.
5. Phenol oxidative coupling of flavones.

1. Ullmann Coupling of Flavones

As a means of combining units of flavones to give biflavones of the amentoflavone (1) and hinokiflavone (2) families, the Ullmann reaction seems to have been the natural choice of NAKAZAWA (*92—94*) who pioneered this method in what was historically the earliest attempt to synthesize biflavones of this type. This old reaction in its original form was by nature vigorous (the reaction temperature is normally in the region 260—300° and no solvent is used) and yields are rarely very high. To circumvent these shortcomings NAKAZAWA (*92*) carried out trial reactions in solvents for the first time, for which the relatively high boiling dimethylformamide and dimethylsulphoxide proved ideal. In these media resinification and oxidation are minimal and yields of coupled products in the order of 30% are commonly achieved.

The dimethyl (113) and tetramethyl (21) ethers of ginkgetin (13), prize targets for synthesis over a number of years, were successfully prepared by NAKAZAWA (*92*) in 1962 using the modified Ullmann reaction shown in Chart 12. Low yields of C-3'/C-3' and C-8/C-8 coupled biflavones accompanied ginkgetin tetramethyl ether (21) but these self-coupled biflavones could be obtained in superior yield when the coupling reaction was performed in the absence of the second component. In this way NAKAZAWA

---

* A partial synthesis of GB-1a (10) has been effected (*36*).

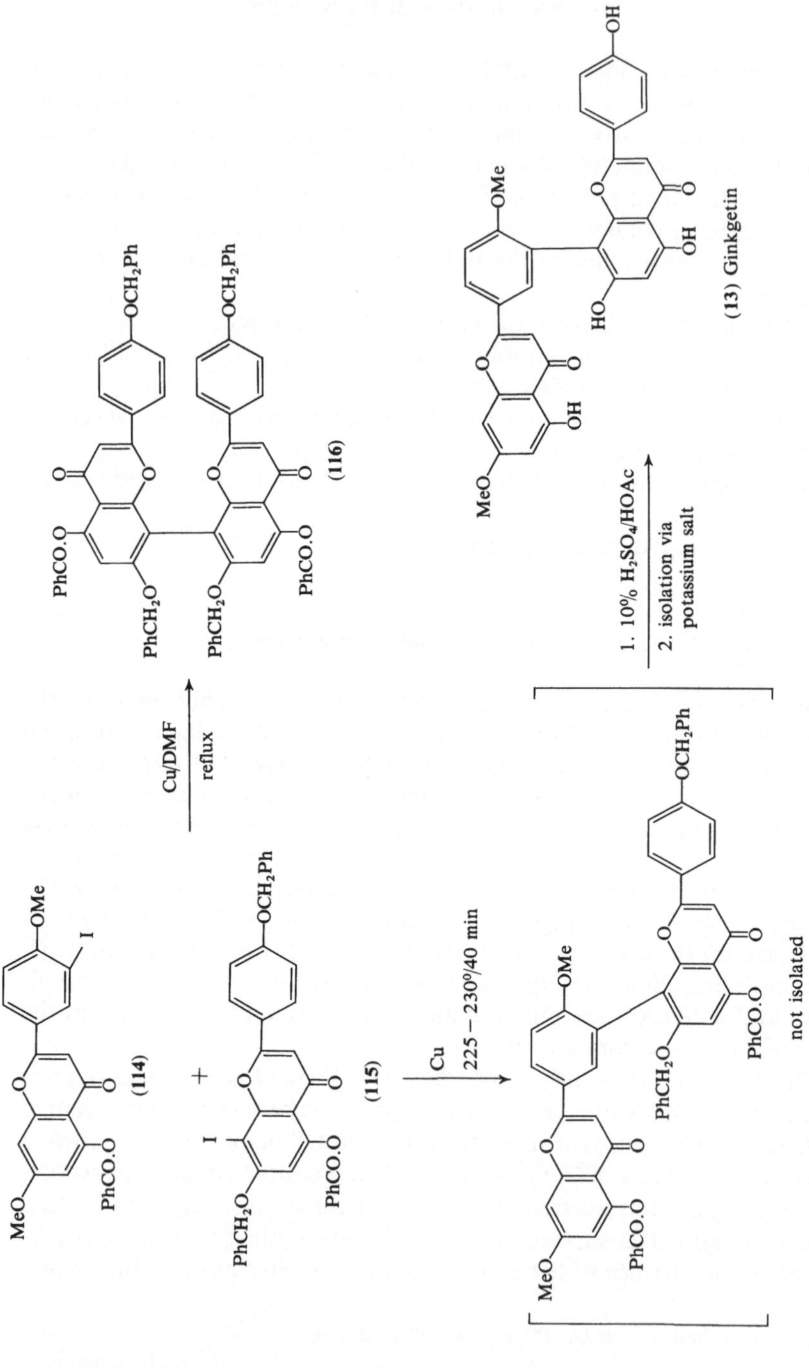

Cu/DMF
reflux

(114)

+

(115)

Cu
225 — 230°/40 min

(116)

not isolated

1. 10% H_2SO_4/HOAc
2. isolation via
   potassium salt

(13) Ginkgetin

Chart 36. Nakazawa's synthesis of ginkgetin (13)

prepared in 33% yield cupressuflavone hexamethyl ether (**44**) two years before the discovery of the parent biflavone (**3**) by SESHADRI (*88*).

With these syntheses successfully completed NAKAZAWA (*94*) attempted to prepare ginkgetin (**13**) itself by use of the modified Ullmann reaction. Condensation of the two iodinated flavones (**114**) and (**115**) (Chart 36) unexpectedly gave only the C-8/C-8 coupled product (**116**): omission of the solvent completely changed the course of the reaction and the required C-3'/C-8 coupled product formed, which on hydrolytic work-up afforded ginkgetin (**13**) in 21% overall yield (Chart 36).

Work on the first synthesis of hinokiflavone pentamethyl ether commenced at a time when the biflavone was believed to possess the C-4'-*O*-C-8 linkage as in (**35**). The preliminary report of this work by SESHADRI and his co-workers (*68*), appeared in 1966; the full account in 1968 (*99*). The Ullmann diaryl ether synthesis was used to link the precursor flavones, 8-iodo-4',5,7-trimethoxyflavone (**117**) and the potassium salt of 5,7-dimethoxy-4'-hydroxyflavone (**118**) (Chart 8), but the product gave a positive ferric chloride test showing that demethylation had taken place during condensation. Remethylation of the crude reaction mixture followed by purification gave cupressuflavone hexamethyl ether (**44**) [the self-coupled product of (**117**)] and a product identical with hinokiflavone pentamethyl ether, the former, however, being the major product. It was thus concluded from the nature of the starting materials that hinokiflavone must possess the C-4'-*O*-C-8 linkage as in (**119**), and this conclusion remained unchallenged for two years when NAKAZAWA published (*93*) details of his independent work on the synthesis of hinokiflavone. Curiously, his early attempt to achieve the synthesis of hinokiflavone followed almost exactly the method used by SESHADRI (*99*), but in his hands the synthesis was unsuccessful. The failure, which was attributed to lack of reactivity of the halogen atom in the electron-rich phloroglucinol ring of (**117**) (Chart 8), proved fortuitously beneficial for it motivated NAKAZAWA (**93**) to develop an alternative approach to the synthesis.

To enhance the reactivity of the halogen for Ullmann condensation, NAKAZAWA (*93*) introduced an *ortho*-nitro group whereupon in the model reaction shown in Chart 37 the diaryl ether was obtained in 65% yield. Applying the method to the synthesis of hinokiflavone the two components (**120**) and (**121**) (Chart 9) were successfully coupled in 85% yield with dimethylsulphoxide as the high boiling solvent. Subsequent reductive removal of the nitro group by standard methods produced the C-4'-*O*-C-8 coupled product (**119**) which however was *not identical* with hinokiflavone pentamethyl ether (**34**) derived from the natural biflavone.

The same synthetic route was employed to prepare unambiguously the isomeric C-4'-*O*-C-6 pentamethyl ether (**34**) from the reagents (**120**) and (**122**) (Chart 9). This product was identical with the pentamethyl ether (**34**)

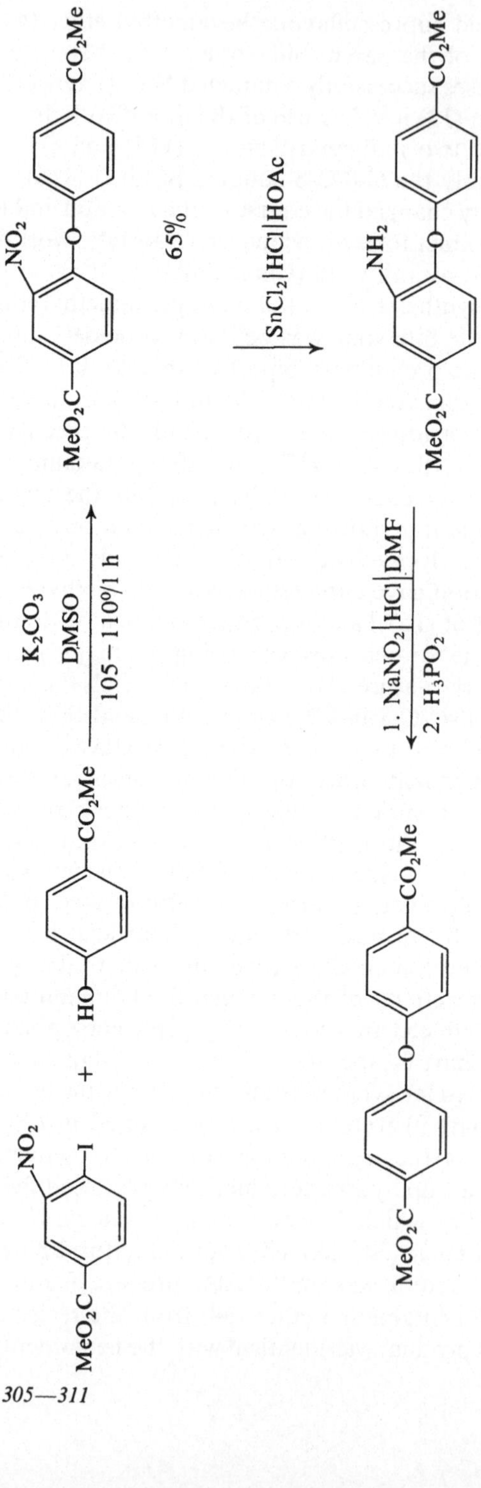

Chart 37. Nakazawa's model reaction sequence for synthesis of hinokiflavone penta-methyl ether (**34**)

of natural hinokiflavone, thus proving conclusively that an error had occurred earlier when the C-4'-*O*-C-8 structure was assigned to hinokiflavone on the basis of degradative evidence.

NAKAZAWA (*93*) showed that hydriodic acid in acetic anhydride was able to convert the C-4'-*O*-C-8 isomer (**119**) of hinokiflavone pentamethyl ether into the natural C-4'-*O*-C-6 hinokiflavone (**34**) by way of the Wesseley-Moser rearrangement, and later work of SESHADRI and his co-workers (*102*) established that the same rearrangement had occurred during the course of their Ullmann reaction (Chart 8). Benzene induced n.m.r. solvent shifts carried out independently by two research groups (*102, 116*) confirm that the linkage in the naturally occurring hinokiflavone is indeed C-4'-*O*-C-6.

Further extension of NAKAZAWA's activated halogen procedure has enabled CHEN *et al.* (*11, 12*), to synthesize the symmetrical and unsymmetrical biflavone ethers (**123**) and (**124**).

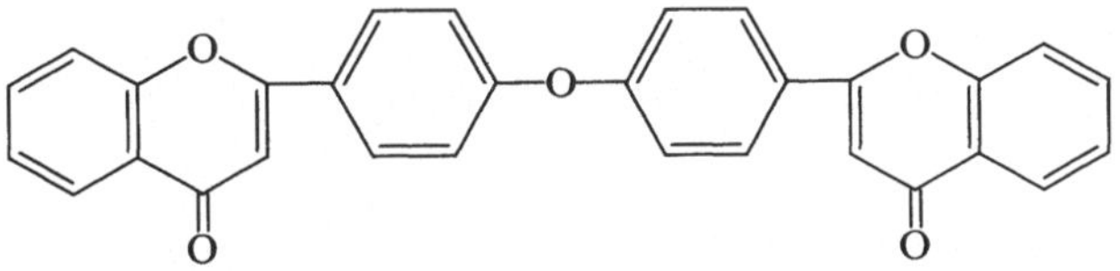

(**123**)

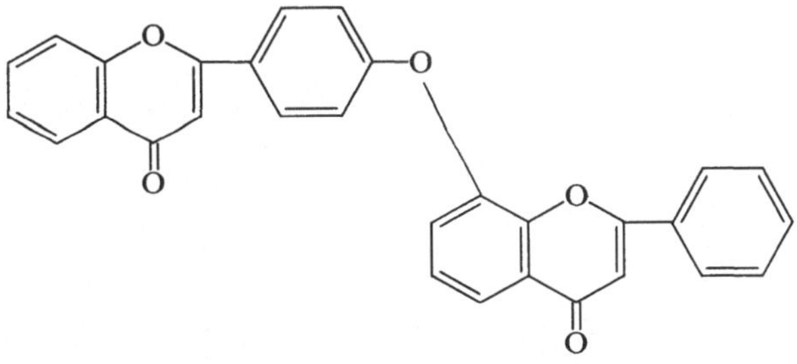

(**124**)

2. Ullmann Synthesis of Biflavones via Biphenyl Precursors

MATHAI and co-workers (*39, 77, 78*) first introduced this approach to the synthesis of biflavones in 1964. Two typical examples of the method are shown in Chart 38. However, none of the biflavones prepared by this Indian group occurs naturally: AHMAD and RAZAQ (*1*) appear to have

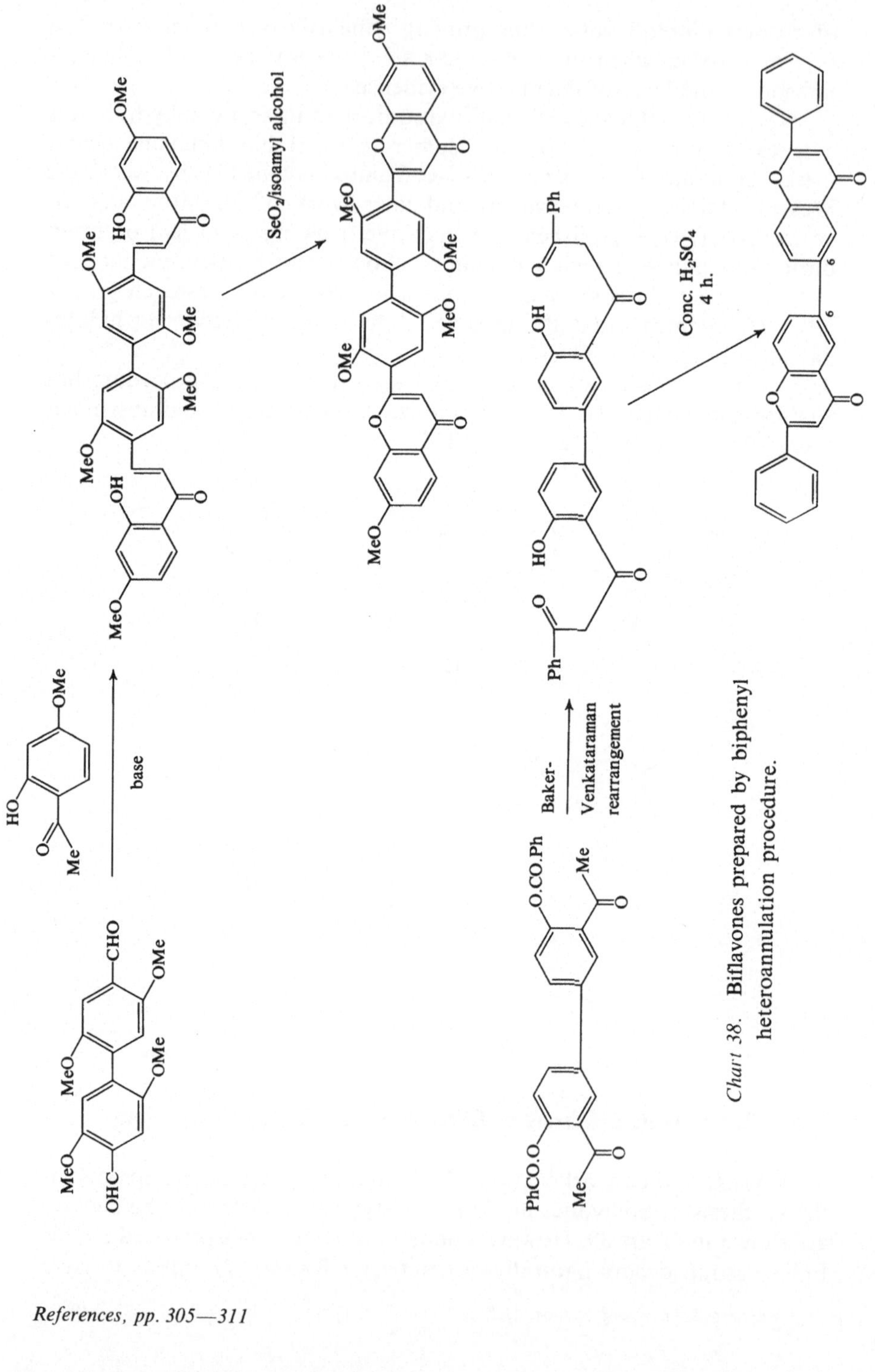

Chart 38. Biflavones prepared by biphenyl heteroannulation procedure.

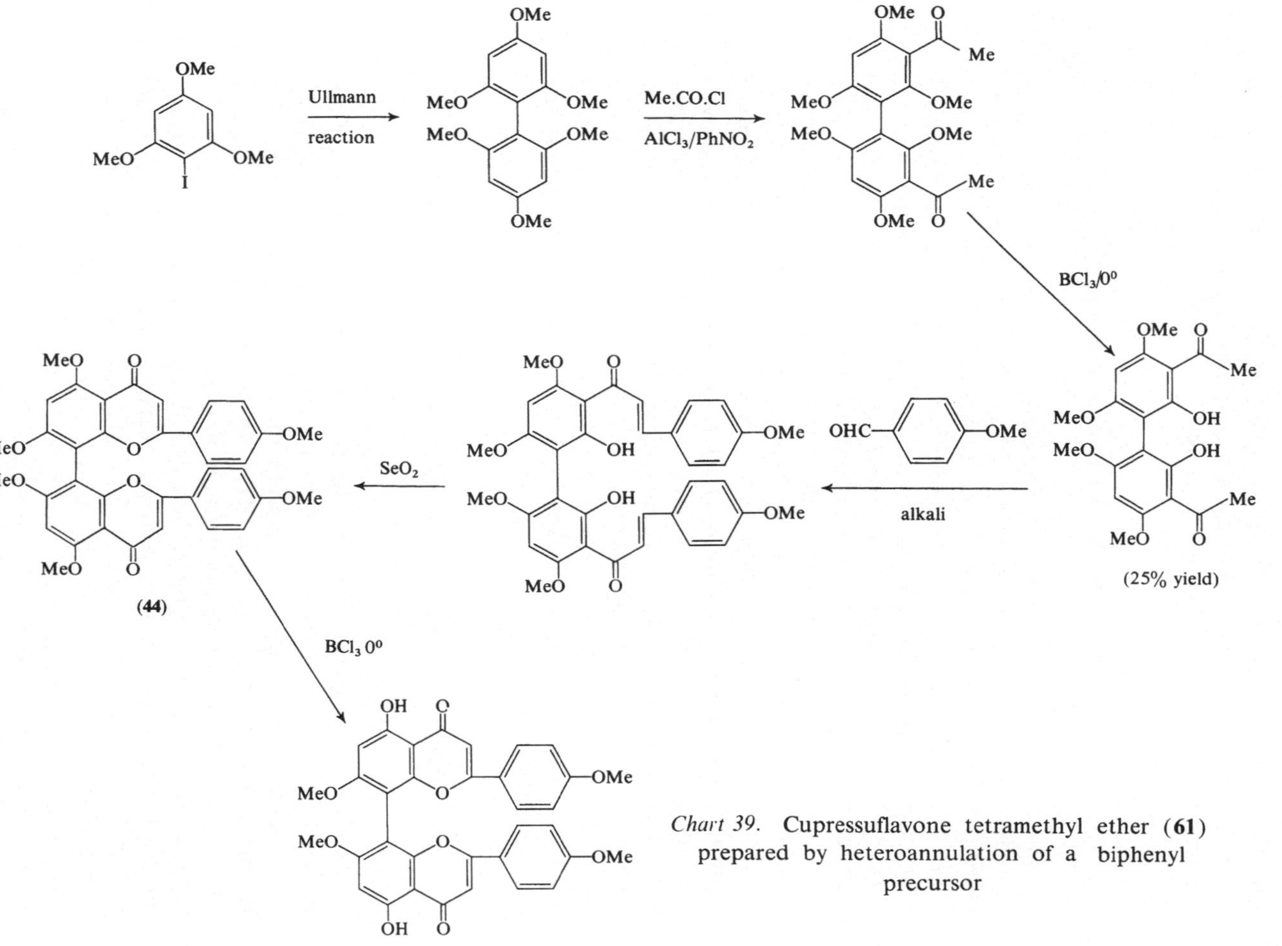

Chart 39. Cupressuflavone tetramethyl ether (**61**) prepared by heteroannulation of a biphenyl precursor

quickly realized the potential of the method, however, and have successfully synthesized the hexamethyl (**44**) and tetramethyl (**61**) ethers of cupressuflavone (**3**) as shown in Chart 39.

3. Wesseley-Moser Rearrangements

The unexpected and undesired Wesseley-Moser rearrangement which attended Seshadri's synthesis (*68, 99, 102*) of hinokiflavone pentamethyl ether (**34**) [see Chart 8] has been put to beneficial use by Pelter et al. (*114*). Seshadri and his co-workers (*89*) had earlier noted the possibility that cupressuflavone (**3**) and its derivatives might conceivably undergo Wessely-Moser rearrangement to produce C-6/C-6 or C-6/C-8 linked biflavones, but no successful conversions were achieved in their preliminary experiments. Though initially discouraged by these failures, Pelter et al. (*114*), nevertheless treated (+)-cupressuflavone hexamethyl ether (**44**) with hydriodic acid in acetic anhydride at 130—140° for 8 hours (typical Wesseley-Moser conditions) after which time the reaction was worked up and remethylated to give a mixture of ($\pm$)-agathisflavone hexamethyl ether (**48**) and ($\pm$)-cupressuflavone hexamethyl ether (**44**) in the ratio 3 : 2 (w/w): the conversion constituted the first preparation of a member of the agathisflavone family (**4**). Benzene induced n.m.r.. solvent shifts were used to verify the linkage positions in agathisflavone (**4**) and cupressuflavone (**3**).

4. Partial Demethylation and Methylation of Natural and Synthetic Biflavones

As a means of preparation, the partial demethylation of more fully methylated natural or synthetic biflavones was first explored in 1967 by Miura and Kawano (*80*) who treated the naturally occurring hinokiflavone derivative cryptomerin B (**37**) with hydriodic acid in phenol at 130° for 3 hours to obtain isocryptomerin (**41**) (Chart 11) (*79*). As one might expect from the foregoing, there is in these reactions a considerable danger of inducing Wesseley-Moser rearrangement in addition to demethylation, especially when a C-8 interflavone linkage is known to be present. However, the above authors have excluded this possibility in their synthesis by continuing the reaction to the stage of complete demethylation to give hinokiflavone (**2**), identical with the natural product. The position of the methoxy group in isocryptomerin (**41**) was established by use of the technique of alkali-induced ultraviolet shifts. Neocryptomerin (**43**) and chamaecyparin (**42**) have been synthesized (*81*) from hinokiflavone pentamethyl ether (**34**) using a similar partial demethylation technique (*83, 85*).

In the amentoflavone series where Wesseley-Moser rearrangements might be anticipated to occur readily, sequoiaflavone (**125**) has been successfully synthesized (*82*) from sciadopitysin (**14**) by partial demethylation without any aparent rearrangement occurring.

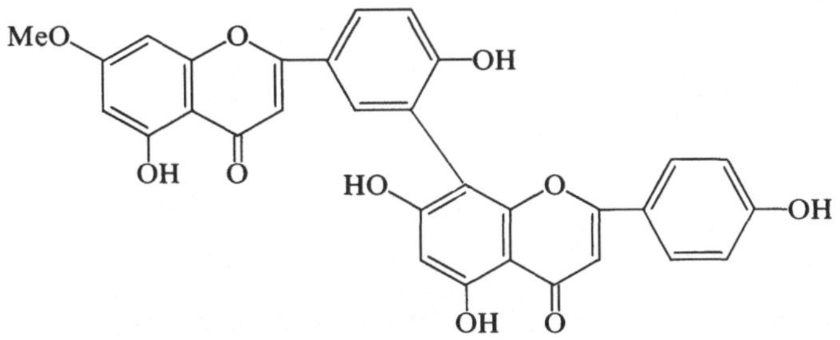

(**125**) Sequoiaflavone

A large number of partial methyl ethers of cupressuflavone (**3**) and agathisflavone (**4**) has been prepared by the controlled demethylation of cupressuflavone hexamethyl ether (**44**). In the initial study NATARAJAN, MURTI, and SESHADRI (*101*) obtained only the dimethyl (**50**) and tetramethyl (**61**) ethers of cupressuflavone. Repetition of the reaction by RAHMAN and his co-workers showed (*26*) that there is an added complication, for the Wesseley-Moser rearrangement accompanies demethylation yielding derivatives of both agathisflavone (**4**) and cupressuflavone (**3**), the full list of identified products is shown in Chart 40. The *O*-methyl derivatives of agathisflavone (**4**) have not been fully characterized but on remethylation they all give agathisflavone hexamethyl ether (**48**). Agathisflavone hexamethyl ether (**48**) subjected to the same demethylation conditions gives an identical mixture of *O*-methyl derivatives. From this study it is concluded that demethylation follows the sequence C-5, then C-4′, and finally C-7; precisely the same sequence is observed for apigenin trimethyl ether (**45**) (*24*).

Partial methylation of cupressuflavone (**3**) with 5.5 equivalents of dimethyl sulphate has been used by SESHADRI and his co-workers (*101*) to obtain the tetra- (**61**) and penta- (**126**) -methyl ethers of cupressuflavone.

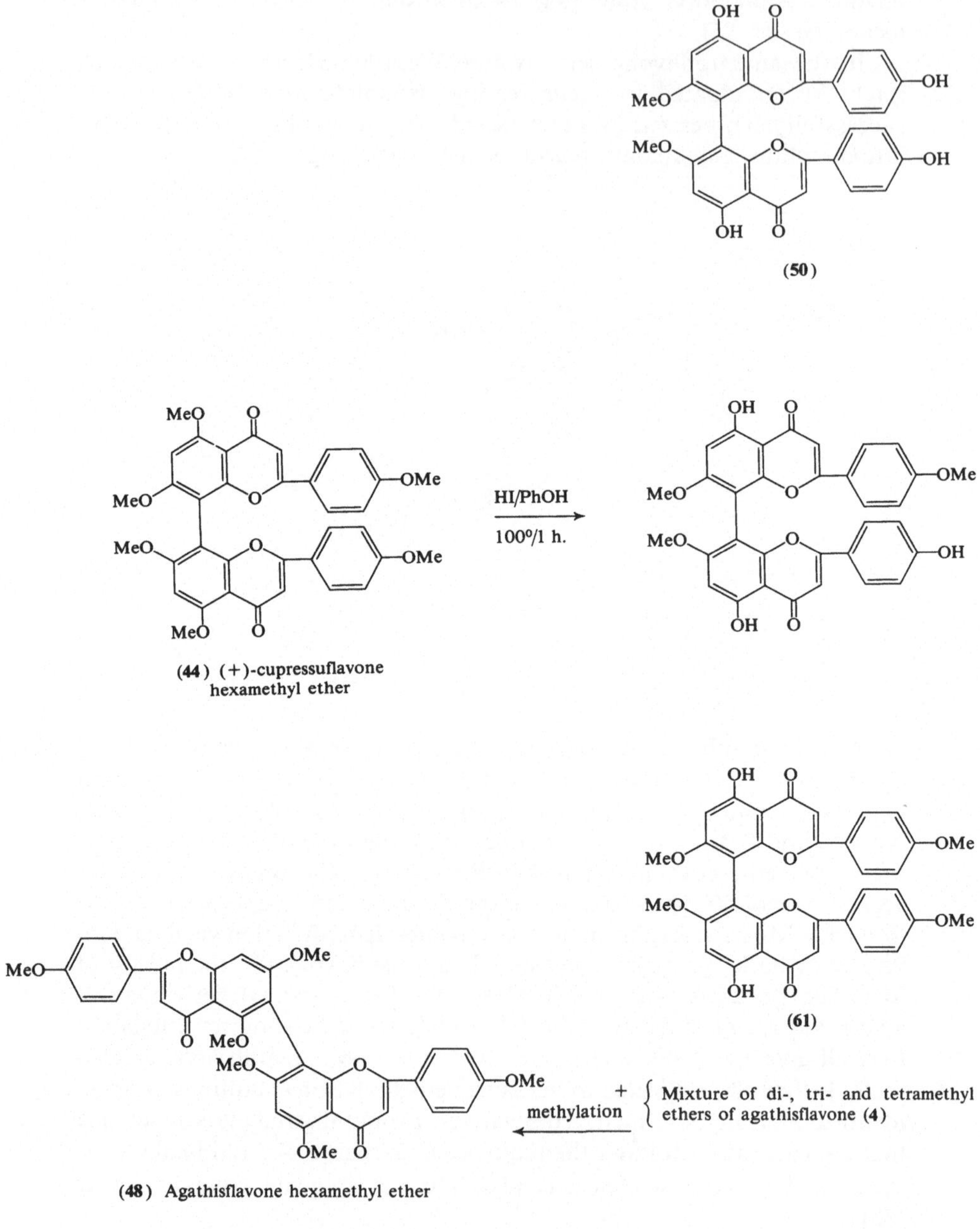

(50)

(44) (+)-cupressuflavone
hexamethyl ether

HI/PhOH

100°/1 h.

(61)

(48) Agathisflavone hexamethyl ether

+ { Mixture of di-, tri- and tetramethyl
methylation ethers of agathisflavone **(4)**

Chart 40. Selective demethylation of cupressuflavone hexamethyl ether (**44**); generation
of agathisflavone derivatives via Wesseley-Moser rearrangement

References, pp. 305—311

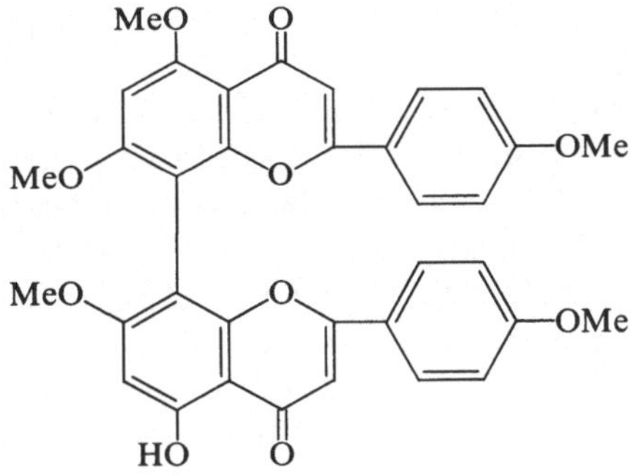

(126) Cupressuflavone pentamethyl
ether

5. Phenol Oxidative Coupling of Flavones as a Route to Biflavones

Of all the methods discussed, the dimerisation of apigenin (15) and its derivatives by oxidative coupling offers the most stimulating and aesthetically pleasing route to the biflavones since it most closely follows the process which is believed to occur in nature.

BAKER, OLLIS, and their co-workers (3) were the forerunners of several who have suggested that the concept of phenol oxidative coupling might offer an explanation for the biogenesis of biflavones. WAISS and his co-workers (87) investigated the oxidative coupling of apigenin (15) using alkaline potassium ferricyanide and isolated two biflavones with C-3'/C-3 and C-3/C-3 interflavone linkages which arise presumably by appropriate spin-pairing of the mesomeric radical (127)—(129) (Chart 41). Though these forms of interflavone linkage have not as yet been encountered in nature, there seems no reason to doubt that they will in all probability be found in the future if the theory of radical coupling is itself valid.

Unfortunately, self-coupling of the mesomeric radical (127)—(129) shown in Chart 41, cannot easily explain the formation of the natural biflavones which, without exception, possess an interflavone linkage at C-6 or C-8 of ring A. Electron spin resonance studies show (69) that delocalisation of an unpaired electron initially generated at the C-4' hydroxy-group in apigenin (15) occurs only in rings B and C (Chart 41): no hyperfine splitting has been observed at the carbon atoms of ring A. Thus, in order to achieve an interflavone linkage to ring A, WAISS and his co-workers (87) believe that a radical initially generated at C-4' in apigenin (15) and delocalized (Chart 41), attacks electrophilically the electron rich C-6 or C-8 positions of the phloroglucinol ring of an intact apigenin molecule: i. e. radical substitutions occurs in preference to radical pairing.

However, if a radical were to be generated at either of the hydroxy-groups in ring A of apigenin, delocalisation of the electron in this ring would enable pairing between the independently generated radicals (**128**) and (**130**) to occur as shown in Chart 42 to give for example amento-flavone (**1**). Alternatively, self-coupling of (**130**) or coupling with (**131**) would offer an explanation of how cupressuflavone (**3**) and agathisflavone (**4**) are formed. Radical substitution fails to offer an explanation for the formation of these latter two types of biflavone, if it is accepted that no free radical can exist on ring A.

Recent experiments of NATARAJAN, MURTI, and SESHADRI (*104*) suggest that radicals can be generated at ring A of an apigenin derivative provided that the C-4′ and C-7 hydroxyls are first blocked by methylation. Thus oxidative dimerisation of apigenin-4′, 7-dimethyl ether (**132**) with ferric chloride in boiling dioxan gives a product in 6% yield whose properties suggest that it is the C-6/C-6 coupled biflavone (**133**). This symmetrically coupled analogue of cupressuflavone (**3**), whose properties it most resembles, has not been encountered in nature as yet but once again theory implies that it could be found in the future.

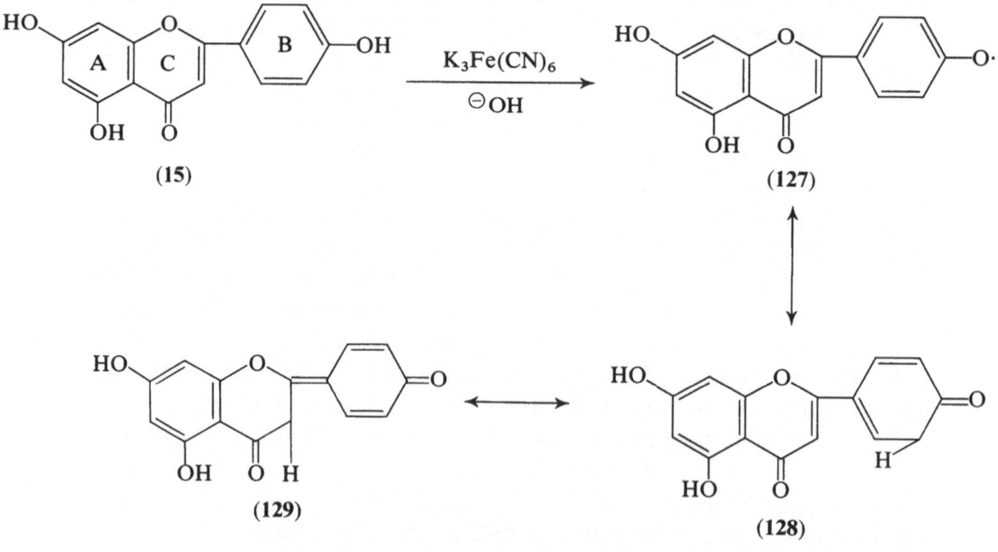

Chart 41.   Generation of an apigeninyl radical and its delocalisation into rings B and C

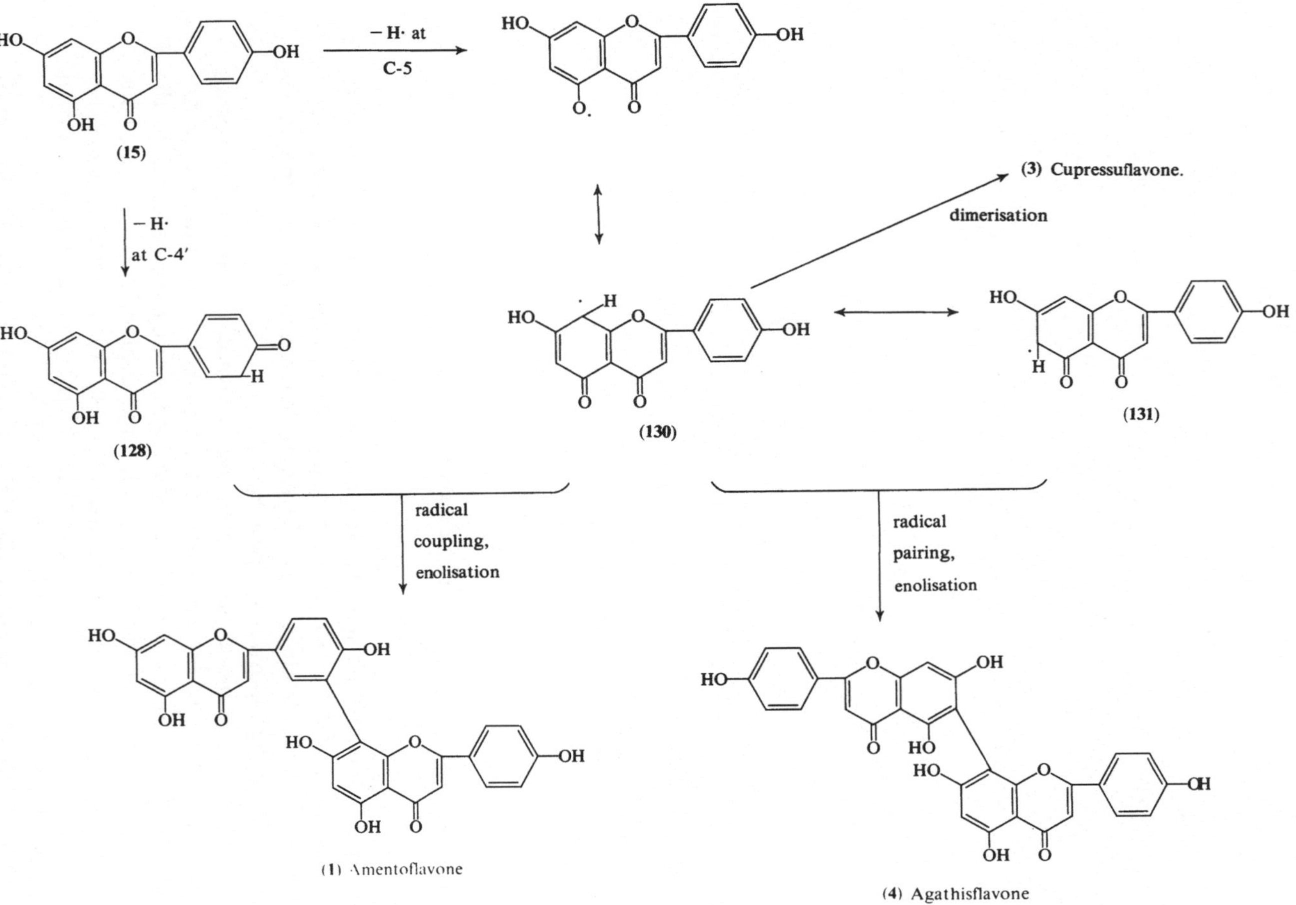

Chart 42. Scheme for the proposed formation of amentoflavone (1), cupressuflavone (3) and agathisflavone (4) from apigenin (15)

(132)

(133)

XIV. Pharmacology

An interesting medical use for the extract of the leaves of *Ginkgo biloba* L., has been reported recently by Seshadri and co-workers (*106*). The extract, marketed in West Germany under the name "Tebonin" is said to have the following properties.

"(a) Possesses blood-flow increasing and vascular dilating properties and is recommended for use in hormonal, vascular, blood-flow and nutrional disorders, (b) the use of Tebonin over a long period leads to the healing of chronic vasotropic ulcers, (c) cerebral blood-flow deficiency due to cerebrosclerotic vascular processes also responds favourably, and (d) in cases of high blood pressure Tebonin has benefical effects while in normal or low blood pressures no effect is noticed."

The result of tests designed to assess peripheral vasodilation, anti-bradykinin and antispasmogenic activity of various flavones and biflavones are recorded.

XV. Tables

Table 1. *Biflavones of the Amentoflavone Series*

	R	R′	R″	R‴	Biflavone	Designation
1.	H	H	H	H	amentoflavone	A_0
2.	Me	H	H	H	sequoiaflavone	A_{1se}
3.	H	Me	H	H	bilobetin	A_{1b}
4.	H	H	Me	H	sotetsuflavone	A_{1so}
5.	H	H	H	Me	podocarpusflavone A	A_{1p}
6.	Me	Me	H	H	ginkgetin	A_{2g}
7.	H	Me	H	Me	isoginkgetin	A_{2i}
8.	Me	H	H	Me	{ podocarpusflavone B (≡ putraflavone)	A_{2p}
9.	Me	Me	H	Me	sciadopitysin	A_{3s}
10.	H	Me	Me	Me	kayaflavone	A_{3k}
11.	Me	H	Me	Me	heveaflavone	A_{3h}
12.	Me	Me	Me	Me	{ amentoflavone tetramethyl ether	A_4

Table 2. *Biflavones of the Hinokiflavone Series*

Table 2 (continued)

	R	R′	R″	Biflavone	Designation
1.	H	H	H	hinokiflavone	B_0
2.	Me	H	H	neocryptomerin	B_{1n}
3.	H	Me	H	isocryptomerin	B_{1i}
4.	H	H	Me	cryptomerin A	B_{1c}
5.	H	Me	Me	cryptomerin B	B_{2cr}
6.	Me	Me	H	chamaecyparin	B_{2ch}
7.	Me	H	Me	{ hinokiflavone- I-7,II-4′-dimethyl ether	B_{2a}
8.	Me	Me	Me	{ hinokiflavone trimethyl ether	B_3

Table 3. *Biflavones of the Cupressuflavone Series*

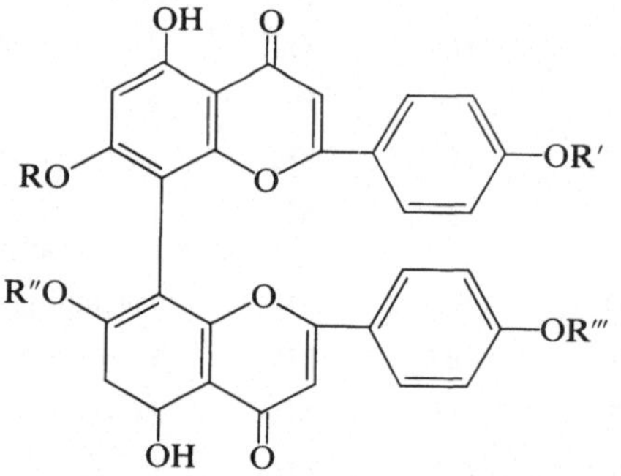

	R	R′	R″	R″	Biflavone	Designation
1.	H	H	H	H	cupressuflavone	C_0
2.	Me	H	H	H	{ I-7-*O*-methyl- cupressuflavone	C_{1a}
3.	Me	H	Me	H	{ I-7,II-7-di-*O*-methyl- cupressuflavone	C_{2a}
4.	Me	Me	Me	Me	{ cupressuflavone — tetramethyl ether	C_4

Table 4. *Biflavones of the Agathisflavone Series*

	R	R'	R''	R'''	Biflavone	Designation
1.	H	H	H	H	agathisflavone	D_0
2.	Me	H	H	H	I-7-*O*-methylagathis-flavone	D_{1a}
3.	Me	H	H	Me	I-7,II-4'-di-*O*-methyl-agathisflavone	D_{2a}
4.	Me	H	Me	H	I-7,II-7-di-*O*-methyl-agathisflavone	D_{2b}

Table 5. *Natural Distribution of Biflavanoids*[a]

Araucariaceae

Agathis alba (75, 61) — A_0 $A_{1?}$ $A_{2?}$ $A_{3?}$ B_0 C_0 C_{1a} C_{2a} C_3 D_0 D_{1a} D_{2a}

A. palmerstonii (117, 61) — A_0 $A_{1?}$ $A_{2?}$ $A_{3?}$ B_0 $B_{1?}$ C_0 $C_{1?}$ $C_{2?}$ $C_{3?}$ D_0 D_{1a} D_{2a}

Araucaria bidwillii (62, 61) — A_0 A_{1b} $A_{2?}$ $A_{3?}$ B_0 C_0 C_{1a} C_{2a} $C_{3?}$ D_{1a} D_{2a} D_{2b}

A. cookii (30, 61, 115) — $A_{1?}$ $A_{2?}$ A_{3k} A_4 B_0 $B_{1?}$ $C_{1?}$ C_{2a} $C_{3?}$ C_4 D_{2a}

A. cunninghamii (30, 61, 75, 118) — $A_{1?}$ $A_{2?}$ A_{3k} A_4 B_0 $B_{1?}$ $C_{1?}$ C_{2a} $C_{2?}$ $C_{3?}$ C_4 D_1 $D_{2?}$

[a] The key to abbreviations A_0, B_0, C_0, D_0, *etc.*, can be found by consulting Tables 1—4. Symbols which do not appear in these Tables are given below,

E_0 = volkensiflavone (**7**); F_0 = morelloflavone (**8**); G_0 = GB-1 (**9**); H_0 = GB-1a (**10**);

J_0 = GB-2 (**11**); K_0 = GB-2a (**12**); E_{1gl} = volkensiflavone-7-*O*-β-glucosyl (spicataside) (**111**);

F_{1gl} = morelloflavone, 7-*O*-β-glucosyl (fukugiside) (**108**); E_{1f} = methylfukugetin (**105**)

K_{1gl} = xanthochymusside (**112**); HA_0 = dihydroamentoflavone (**68**);

HB_0 = dihydrohinokiflavone (**65**);

HA_{2d} = dihydroamentoflavone-II-4',II-7-dimethyl ether (**66**);

HA_{3s} = dihydrosciadopitysin (**67**); A_{2d} = amentoflavone-II-4',II-7-dimethyl ether (**64**).

Caprifoliaceae

Viburnum opulus (29) — A_0
V. prunifolium (29) — A_0

Casuarinaceae

Casuarina cunninghamiana (105) — C_0
C. equisetifolia (105) — No biflavanoids present
C. junghuhniana (105) — B_0
C. littoralis (105) — No biflavanoids present
C. montana (105) — No biflavanoids present
C. rigida (105) — No biflavanoids present
C. stricta (121) — B_0
C. suberosa (105) — B_0

Cephalotaxaceae

Cephalotaxus drupacea (63, 121) — $A_0\ A_{1se}\ A_{2g}\ A_{3k}\ A_{3s}$
C. nana (63, 121) — $A_0\ A_{1se}\ A_{2g}\ A_{3k}\ A_{3s}$

Cupressaceae

Biota orientalis (81, 103, 121) — $B_0\ B_{1?}\ B_{2?}$
B. orientalis var. *mexican* (103) — B_0
Callitris glauca (121) — B_0
Chamaecyparis obtusa (79, 80, 121) — $B_0\ B_{1i}\ B_{2?}$
C. obtusa var. *breviamea* (81, 121) — $B_0\ B_{1i}\ B_{2ch}$
C. pisifera (81, 121) — $B_0\ B_{1i}\ B_{2?}$
C. pisifera var *filifera* (121) — B_0
C. pisifera var. *squarrosa* (81, 121) — $B_0\ B_{1i}\ B_{2ch}$
Cupressus arizonica (81, 121) — $B_0\ C_0$
C. funebris (103, 114, 121) — $A_0\ B_0\ B_{1i}\ C_0$
C. glauca (103) — $A_0\ B_0\ C_0$
C. goveniana (81, 103) — $A_0\ B_0\ C_0$
C. sempervirens (88, 89, 103, 114) — $A_0\ C_0$
C. sempervirens var. *stricta* (103) — $A_0\ C_0$
C. torulosa (88, 89, 103) — $A_0\ B_0\ C_0$
Juniperus chinensis (112, 116) — $A_0\ B_0\ B_{1?}$
J. conferta (121) — $A_{3k}\ B_0$
J. rigida (74) — $A_0\ A_{1p}$
J. utilis (121) — $A_{3k}\ B_0$
Libocedrus decurrens (121) — B_0
L. formosana (121) — B_0
Sabina chinensis (121) — $A_{3k}\ B_0$
S. procumbens (121) — $A_{3k}\ B_0$
S. sargentii (121) — $A_{3k}\ B_0$
S. sargentii var. *kaizuka* (121) — $A_{3k}\ B_0$
S. virginiana (121) — $A_{3k}\ B_0$
Thuja occidentalis (121) — B_0
T. orientalis (113, 116) — $A_0\ B_0$
T. standishii (81, 121) — $B_0\ B_{1?}\ B_{2?}$
Thujopsis dolabrata (81, 121) — $A_{1so}\ A_{3s}\ B_0\ B_{1i}\ B_{2ch}\ B_{3?}$

Cycadaceae

Cycas revoluta (21, 48, 59, 121) — $A_0\ B_0\ A_{1so}\ HA_0\ HB_0$
Zamia angustifolia (25) — $A_0\ A_{1b}\ A_{1se}\ A_{2g}\ A_{3s}\ A_4$

References, pp. 305—311

Euphorbeaceae

Hevea brasiliensis (72) A_{3h}

Ginkgoaceae

Ginkgo biloba (3, 4, 25, 86) A_{1b} A_{2g} A_{2i} A_{3s}

Guttiferae

Allanblackia floribunda (71) E_0 F_0
Garcinia buchananii (33, 34, 36) G_0 H_0 J_0 K_0
G. eugeniifolia (33, 34, 36) G_0 H_0 J_0 K_0
G. linii (66) F_0 G_0 H_0 J_0 K_0
Garcinia livingstonii (111) E_0 F_0 A_0 A_{1p}
G. morella (40) F_0
G. multiflora (66) E_0 F_0 G_0 H_0 J_0 K_0
G. spicata (65, 66, 67) E_0 E_{1g1} F_0 F_{1g1} F_{1f} H_0 K_0
G. talboti (37) E_0 F_0
G. volkensii (27) E_0 F_0 H_0 K_0
G. xanthochymus (66) K_{1g1}

Podocarpaceae

Dacrydium cupressinum (28) A_4
Podocarpus chinensis (121) A_{3k}
P. gracilior (13) A_0 A_{1p} A_{2i} A_{3k}
P. macrophylla (86, 121) A_0 A_{1p} A_{2p} A_{3k} A_{3s} B_0 B_{1n}
P. nagi (86, 121) A_0 A_{1p} A_{2i} A_{3k}

Selaginellaceae

Selaginella tamariscina (107) A_0 B_0 B_{1i}
S. nipponica (107) A_0 $A_{1?}$ $A_{2?}$
S. pachystachys (107) A_0 $A_{1?}$ $A_{2?}$

Taxaceae

Taxus cuspidata (121) $A_{1?}$ $A_{2?}$ A_{3s}
T. floriana (121) A_{3s}
Torreya nucifera (47, 121) $A_{2?}$ A_{3k}

Taxodiaceae

Cryptomeria japonica (57, 84, 121) A_{3k} A_{3s} B_0 B_{1c} B_{2cr}
C. japonica var. araucarioides (121) A_{1so} A_{3k} A_{3s}
C. japonica var. zindai (81) B_0 B_{1c} B_{2cr}
Cunninghamia lanceolata (121) A_{1so} A_{3k} B_0
C. lanceolata var. Konishii (82, 121) A_{1se} A_{3k} B_0
Glyptostrobus pensilis (81, 121) B_0 B_{1c} B_{1n} $B_{2?}$
Metasequioia glyptostroboides (7, 50, 121) B_0 B_{1i} A_0 A_{1so} A_{2d} A_{3s} HB_0 HA_{2d} HA_{3s}
Sciadopitys verticillata(42—46,51—53,55,121) A_{2i} A_{3s}
Sequoia sempervirens (82, 121) A_0 A_{1se} $A_{2?}$ A_{3k} B_0 $B_{1?}$
Taiwania cryptomerioides (121) B_0
Taxodium distichum (121) B_0
T. mucronatum (116) B_0

Table 6. ^1H *NMR Spectral Data for Biflavanones* (**9**)—(**12**) (τ values; J in Hz; spectra run at 100 MHz) (*36*)

Biflavanone	C-5 OH, rings I-A, II-A	Other OH's	C-6 rings I-A & II-A, C-8 rings II-A	C-2/C-6 rings I-B & II-B	C-3-/C-5 rings I-B & II-B	C-2/C-3 ring I-C		C-2 ring II-C	C-3 ring II-C	Solvent
GB-1 (**9**)	−2.03 s (1 H) −1.57 s (1 H)	0.90 (2 H) (2 H)	4.14 s (1 H) 4.18 d (1 H) 4.27 d (1 H) ($J = 3.0$)	2.92 d (2 H) 3.06 d (2 H) ($J = 9.0$)	3.33 d (2 H) 3.40 d (2 H) ($J = 9.0$)	4.56 d (1 H) ($J = 12.0$)	5.56 d (1 H) ($J = 12.0$)	5.07 d (1 H) ($J = 12.0$)	5.97 d (1 H) ($J = 12.0$)	a
GB-1a (**10**)	c	c	4.27 c (3 H)	2.98 d (2 H) 3.05 d (2 H) ($J = 9.0$)	3.39 d (2 H) 3.48 d (2 H) ($J = 9.0$)	4.54 d (1 H) ($J = 12.0$)	5.55 d (1 H) ($J = 12.0$)	4.75 q (1 H) ($J = 11.0$ & 5.5)	7.10 c (1 H) 7.39 c (1 H)	b
GB-2 (**11**)	c	c	4.18 s 4.19 d 4.24 d ($J = 2.0$)	2.99 d (2 H) 3.20—3.50 c (5 H)		4.50 d (1 H) ($J = 12.0$)	5.51 d (1 H) ($J = 12.0$)	5.16 d (1 H) ($J = 12.0$)	5.99 d (1 H) ($J = 12.0$)	a
GB-2a (**12**)	−2.31 s (1 H) −2.14 s (1 H)	1.44 br s (4 H)	4.10 c (3 H)	2.82 d (2 H) 3.08—3.20 c (5 H)		4.71 br d (1 H)	5.32 br d (1 H)	*ca.* 4.3 c (1 H)	6.91 c (1 H) 7.35 c (1 H)	b

a d_6-dimethylsulphoxide at 100°; hexamethyldisiloxane as internal standard.
b d_6-acetone at room temperature; tetramethylsilane as internal standard.
c not observed; protons undergoing rapid exchange.

Table 7. *Optical Rotatory Dispersion Data for Biflavanones* **(9)—(12)**

Metabolite	Molecular rotation (°)	λ nm	Molecular amplitude (°)
GB-1 **(9)**	− 19,800 tr	306	− 501
	+ 30,300 pk	285	
	− 6,150 pk	240	+ 478
	− 54,000 tr	225	
GB-1a **(10)**	− 20,300 tr	303	− 387
	+ 18,400 pk	284	
	− 6,430 pk	240	+ 456
	− 52,000 tr	229	
GB-2 **(11)**	− 16,050 tr	306	− 487
	+ 32,600 pk	288	
	− 9,750 pk	241	+ 262
	− 36,000 tr	227	
GB-2a **(12)**	− 18,300 tr	303	− 264
	+ 8,070 pk	283	
	− 7,280 pk	239	+ 204
	− 22,720 tr	221	

Notes: pk = peak, tr = trough.

References

1. AHMAD, S., and S. RAZAQ: A New Approach to the Synthesis of Symmetrical Biflavones. Tetrahedron Letters **1971**, 4633.

2. AUDIER, H.: Étude des Composés flavoniques par Spectrométrie de Masse. Bull. Soc. chim. France **1966**, 2892.

3. BAKER, W., A. C. M. FINCH, W. D. OLLIS, and K. W. ROBINSON: The Structures of the Naturally Occurring Biflavonyls. J. Chem. Soc. **1963**, 1477.

4. BAKER, W., and W. D. OLLIS: Biflavonyls, in W. D. OLLIS Recent Developments in the Chemistry of Natural Phenolic Compounds, p. 152. Oxford: Pergamon. 1961.

5. BARNES, C. S., and J. L. OCCOLOWITZ: The Mass Spectra of Some Naturally Occurring Oxygen Heterocycles and Related Compounds. Austral. J. Chem. **17**, 975 (1964).

6. BARTON, D. H., and T. COHEN: Festschrift A. Stoll, p. 117. Basel: Birkhäuser. 1957.

7. BECKMANN, S., H. GEIGER, and W. DE G. PFLEIDERER: Biflavone und 2,3-Dihydrobi-flavone aus *Metasequoia glyptostroboides*. Phytochemistry **10**, 2465 (1971).

8. BOWIE, J. H., and D. W. CAMERON: Electron Impact Studies, Part 2. Mass Spectra of Quercetagetin Derivates. Austral. J. Chem. **19**, 1627 (1966).

9. BOWIE, J. H., J. RONAYNE, and D. H. WILLIAMS: Solvent Effects in Nuclear Resonance Spectroscopy, Part 10. Solvent Shifts induced by Benzene in *ortho-* and *meta-* Substituted Methoxybenzenes. J. Chem. Soc. (B) **1967**, 535.

10. BOWIE, J. H., and P. Y. WHITE: Electron Impact Studies, Part 39. Proximity Effects in the Mass Spectra of Aromatic Carbonyl Compounds Containing Adjacent Methoxy-substituents. J. Chem. Soc. (B) **1969**, 89.

11. Chen, F. C., T. Ueng, C. Y. Chen, and P. W. Chang: Synthesis of Biflavonyl Ethers. J. Chinese Chem. Soc. **17**, 251 (1970).

12. Chen, F. C., T. Weng, and Y. S. Chen: Synthesis of Diflavonyl Ethers, Part 3. Tai-Wan K'o Hsueh **23**, 46 (1969) [Chem. Abstr. **73**, 109630 P (1970)].

13. Chexal, K. K., B. K. Handa, W. Rahman, and N. Kawano: Some Optically Active Biflavones from *Podocarpus gracilior*. Chem. and Ind. (London) **1970**, 28.

14. Clark-Lewis, J. W.: Flavan Derivatives, Part 21. Nuclear Magnetic Resonance Spectra, Configuration, and Conformation of Flavan Derivatives. Austral. J. Chem. **21**, 2059 (1968).

15. Dean, F. M.: Naturally Occurring Oxygen Ring Compounds, pp. 280 and 333. London: Butterworths. 1963.

16. Erdtmann, H., and C. A. Wachtmeister: Festschrift A. Stoll, p. 144. Basel: Birkhäuser. 1957.

17. Fukui, Y.: Studies on the Chemical Constituents of the Plants of Coniferae and Allied Orders, Part 40. Studies on the Structure of Hinokiflavone, a Flavonoid from the Leaves of *Chamaecyparis obtusa* Endlicher (4). Degradation of Hinokiflavone Pentamethyl Ether in Ethanolic Potassium Hydroxide Solution and the Structures of Substances Y and Z. Yakugaku Zasshi **80**, 752 (1960).

18. — Studies on the Chemical Constituents of the Plants of Coniferae and Allied Orders, Part 41. Studies on the Structure of Hinokiflavone, a Flavonoid from the Leaves of *Chamaecyparis obtusa* Endlicher (5). Degradation of Hinokiflavone Pentamethyl Ether in Methanolic Barium Hydroxide Solution. Yakugaku Zasshi **80**, 756 (1960).

19. Fukui, Y., and N. Kawano: The Structure of Hinokiflavone, A New Type Bisflavonoid. J. Amer. Chem. Soc **81**, 6331 (1959).

20. Garg, H. S., and C. R. Mitra: Putraflavone, A New Biflavonoid from *Putranjiva roxburghii*. Phytochemistry **10**, 2787 (1971).

21. Geiger, H., and W. de G. Pfleiderer: Über 2,3-Dihydrobiflavone in *Cycas revoluta* (Cycadaceae). Phytochemistry **10**, 1936 (1971).

22. Geissman, T. A.: ed. The Chemistry of Flavonoid Compounds, Preface p. vii. Oxford: Pergamon Press. 1962.

23. Grigg, R. J. A. Knight, and P. Roffey: NMR Solvent Shifts and Structure Elucidation in Coumarins. Tetrahedron **22**, 3301 (1966).

24. Gripenberg, J.: Flavones, in: The Chemistry of Flavonoid Compounds, ed. T. A. Geissman, p. 406. Oxford: Pergamon Press. 1962.

25. Handa, B. K., K. K. Chexal, W. Rahman, M. Okigawa, and N. Kawano: Occurrence of Bisflavones in *Zamia* (Cycadaceae). Phytochemistry **10**, 436 (1971).

26. Handa, B. K., K. K. Chexal, (Miss) Talat Mah, and W. Rahman: Some Observations on Partial Demethylation of Biflavonyls. J. Indian Chem. Soc. **48**, 177 (1971).

27. Herbin, G. A., B. Jackson, H. D. Locksley, F. Scheinmann, and W. A. Wolstenholme: The Biflavonoids of *Garcinia volkensii* Engl. (Guttiferae). Phytochemistry **9**, 221 (1970).

28. Hodges, R.: 5,5'-Dihydoxy-7,4',7'',4'''-tetramethoxy-8,3'''-biflavone from *Dacrydium cupressinum* Lamb. Austral. J. Chem. **18**, 1491 (1965).

29. Hörhammer, L., H. Wagner, and H. Reinhardt: Isolierung des Bis-(5,7,4'-Trihydroxy-)-flavons „Amentoflavone" aus der Rinde von *Viburnum prunifolium* L. (Amerikan. Schneeball). Naturwiss. **52**, 161 (1965).

30. Ilyas, M., J. N. Usmani, S. P. Bhatnagar, M. Ilyas, W. Rahman, and A. Pelter: WBl and Wll, the First Optically Active Biflavones. Tetrahedron Letters **1968**, 5515.

31. Itagaki, Y., T. Kurokawa, S. Sasaki, C. T. Chang, and F.-C. Chen: The Mass Spectra of Chalcones, Flavones, and Isoflavones. Bull. Chem. Soc., Japan **39**, 538 (1966).

32. Jackson, B., H. D. Locksley, and F. Scheinmann: Extractives from Guttiferae, Part 7. The Isolation and Structure of Seven Xanthones from *Calophyllum scriblitifolium* Henderson and Wyatt-Smith. J. Chem. Soc. (*C*) **1967**, 2500.

33. JACKSON, B., H. D. LOCKSLEY, and F. SCHEINMANN: The Structure of the "GB" Biflava-nones from *Garcinia buchananii* Baker and *G. eugeniifolia* Wall. Chem. Comm. **1968**, 1125.

34. JACKSON, B., H. D. LOCKSLEY, F. SCHEINMANN, and W. A. WOLSTENHOLME: The Isolation of a New Series of Biflavanones from the Heartwood of *Garcinia buchananii* Baker. Tetrahedron Letters **1967**, 787.

35. — — — — Comments on the Thermal and Mass Spectral Fragmentation of Flava-nones and Biflavanones. Tetrahedron Letters **1967**, 3049.

36. — — — — Extractives from Guttiferae, Part 22. The Isolation and Structure of Four Novel Biflavanones from the Heartwoods of *Garcinia buchananii* Baker and *G. eugenii-folia* Wall. J. Chem. Soc. (*C*) **1971**, 3791.

37. JOSHI, B. S., V. N. KAMAT, and N. VISWANATHAN: The Isolation and Structure of Two Biflavones from *Garcinia talboti* Raiz (Guttiferae). Phytochemistry **9**, 881 (1970).

38. JURD, L.: Spectral Properties of Flavonoid Compounds, in: The Chemistry of Flavonoid Compounds, ed. T. A. GEISSMAN, p. 107. Oxford: Pergamon Press. 1962.

39. KANAKLAKSHMI, B.: Synthesis of some 3′,3′′′; 4′,4′′′ and 6,6′′-Biflavonyls. J. Indian Chem. Soc. **46**, 279 (1969).

40. KARANJGAOKAR, C. G., P. V. RADHAKRISHNAN, and K. VENKATARAMAN: Morello-flavone, a 3-(8-)-Flavonylflavanone from the Heartwood of *Garcinia morella*. Tetra-hedron Letters **1967**, 3195.

41. KARIYONE, T., and Y. FUKUI: Studies on the Chemical Constituents of the Plants of Coniferae and Allied Orders, Part 38. Studies on the Structure of Hinokiflavone, a Flavonoid from the Leaves of *Chamaecyparis obtusa* Endlicher (2). Composition of Hinokiflavone and its Degradation in KOH Solution. Yakugaku Zasshi **80**, 746 (1960).

42. KARIYONE, T., and N. KAWANO: Studies on the Structure of Sciadopitysin, a Flavonoid from the Leaves of *Sciadopitys verticillata* Sieb et Zucc., Part 1. Yakugaku Zasshi **76**, 448 (1956).

43. — — Studies on the Structure of Sciadopitysin, a Flavonoid from the Leaves of *Sciadopitys verticillata* Sieb et Zucc., Part 2. Degradation of Sciadopitysin. Yakugaku Zasshi **76**, 451 (1956).

44. — — Studies on the Structure of Sciadopitysin, a Flavonoid from the Leaves of *Sciadopitys verticillata* Sieb et Zucc., Part 3. Structure of Ketoflavone and Carboxy-flavone. Yakugaku Zasshi **76**, 453 (1956).

45. KARIYONE, T., N. KAWANO, and M. MAEYAMA: Chemical Constituents of the Plants of Coniferae and Allied Orders, Part 31. Studies on the Structure of Sciadopitysin, a Flavonoid from the Leaves of *Sciadopitys verticillata* Sieb et Zucc., Part 5. Degradation of Sciadopitysin Trimethyl Ether in Ethanolic Potassium Hydroxide Solution, Part 2. Yakugaku Zasshi **79**, 382 (1959).

46. KARIYONE, T., N. KAWANO, and H. MIURA: Chemical Constituents of the Plants of Coniferae and Allied Orders, Part 35. Studies on the Structure of Sciadopitysin, a Flavonoid from the Leaves of *Sciadopitys verticillata* Sieb et Zucc., Part 6. Degra-dation of Sciadopitysin Trimethyl Ether in Methanolic Barium Hydroxide Solution. Yakugaku Zasshi **79**, 1182 (1959).

47. KARIYONE, T., and T. SAWADA: Studies on Flavonoids of the Leaves of Coniferae and Allied Plants, Part 1. On the Flavonoid from the Leaves of *Torreya nucifera* Sieb et Zucc. Yakugaku Zasshi **78**, 1010 (1958).

48. — — Studies on Flavonoids in the Leaves of Coniferae and Allied Plants, Part 2. On the Flavonoids from the Leaves of *Cycas revoluta* Thunb., and *Cryptomeria japonica* D. Don var *araucarioides* Hort. Yakugaku Zasshi **78**, 1013 (1958).

49. — — Studies on Flavonoids in the Leaves of Coniferae and Allied Plants, Part 4. On the Flavonoid from the Leaves of *Chamaecyparis obtusa* Endl. Yakugaku Zasshi **78**, 1020 (1958).

50. KARIYONE, T., M. TAKAHASHI, K. ISOI, and M. YOSHIKURA: Chemical Constituents of the Plants of Coniferae and Allied Orders, Part 20. Studies on the Components of the Leaves of *Metasequoia glyptostroboides* Hu et Cheng., Part 1. Yakugaku Zasshi **78**, 801 (1958).

51. KAWANO, N.: Studies on the Structure of Sciadopitysin, a Flavonoid from the Leaves of *Sciadopitys verticillata* Sieb et Zucc., Part 4. Degradation of Sciadopitysin Trimethyl Ether in Ethanolic Potassium Hydroxide Solution. Yakugaku Zasshi **76**, 457 (1956).

52. — Studies on the Structure of Sciadopitysin, a Flavonoid from the Leaves of *Sciadopitys verticillata* Sieb et Zucc., Part 7. Chem. Pharm. Bull., Japan **7**, 698 (1959).

53. — Studies on the Structure of Sciadopitysin, a Flavonoid from the Leaves of *Sciadopitys verticillata* Sieb et Zucc., Part 8. Chem. Pharm. Bull., Japan **7**, 821 (1959).

54. — The Structure of Kayaflavone. Chem. Pharm. Bull., Japan **9**, 358 (1961).

55. — Recent Advances in the Chemistry of Biflavones, in: Chemistry of Natural and Synthetic Colouring Matters and Related Fields, eds. T. S. GORE, B. S. JOSHI, S. V. SUNTHANKAR, and B. D. TILAK, p. 177. New York and London: Academic Press. 1962.

56. KAWANO, N., and Y. FUKUI: Studies on the Chemical Constituents of Coniferae and Allied Orders, Part 39. Studies on the Structure of Hinokiflavone, a Flavonoid from the Leaves of *Chamaecyparis obtusa* Endlicher, Part 3. The Structures of Substance X and Ketoflavone. Yakugaku Zasshi **80**, 749 (1960).

57. KAWANO, N., H. MIURA, and A. C. WAISS: Naturally Occurring Hinokiflavone Methyl Ethers. Chem. and Ind. (London) **1964**, 2020.

58. KAWANO, N., and M. YAMADA: The Structure of Sotetsuflavone. J. Amer. Chem. Soc. **82**, 1505 (1960).

59. — — Chemical Constituents of the Plants of Coniferae and Allied Orders, Part. 45. The Structure of Sotetsuflavone. Yakugaku Zasshi **80**, 1576 (1960).

60. KHAN, N. U., W. H. ANSARI, W. RAHMAN, M. OKIGAWA, and N. KAWANO: Two New Biflavonyls from *Araucaria cunninghamii*. Chem. Pharm. Bull., Japan **19**, 1500 (1971).

61. KHAN, N. U., W. H. ANSARI, J. N. USMANI, M. ILYAS, and W. RAHMAN: Biflavonyls of the Araucariales. Phytochemistry **10**, 2129 (1971).

62. KHAN, N. U., M. ILYAS, W. RAHMAN, M. OKIGAWA, and N. KAWANO: On the Bisflavones in the Leaves of *Araucaria bidwillii* Hooker. Tetrahedron Letters **1970**, 2941.

63. — — — — — Occurrence of Biflavonyls in *Cephalotaxus*. Phytochemistry **10**, 2541 (1971).

64. KINGSTON, D. G. I.: Mass Spectrometry of Organic Compounds, Part 6. Electron-Impact Spectra of Flavonoid Compounds. Tetrahedron **27**, 2691 (1971).

65. KONOSHIMA, M., and Y. IKESHIRO: Fukugiside, the First Biflavonoid Glycoside from *Garcinia spicata* Hook f. Tetrahedron Letters **1970**, 1717.

66. KONOSHIMA, M., Y. IKESHIRO, S. MIYAHARA, and KUN-YING YEN: The Constitution of Biflavonoids from *Garcinia* Plants. Tetrahedron Letters **1970**, 4203.

67. KONOSHIMA, M., Y. IKESHIRO, A. NISHINAGA, T. MATSUURA, T. KUBOTA and H. SAKAMOTO: The Constitution of the Flavanoids from *Garcinia spicata* Hook f. Tetrahedron Letters **1969**, 121.

68. KRISHNAN, S. K., V. V. S. MURTI, and T. R. SESHADRI: Synthesis of Hinokiflavone Pentamethyl Ether. Current Science **35**, 64 (1966).

69. KUHNLE, J. A., J. J. WINDLE, and A. C. WAISS: Electron Paramagnetic Resonance Spectra of Flavonoid Anion-Radicals. J. Chem. Soc. (B) **1969**, 613.

70. LOCKSLEY, H. D., and I. G. MURRAY: Extractives from Guttiferae, Part 16. Biogenetic-type Synthesis of Xanthones from their Benzophenone Precursors. J. Chem. Soc. (C) **1970**, 392.

71. — — Extractives from Guttiferae, Part 19. The Isolation and Structure of Two Benzophenones, Six Xanthones, and Two Biflavonoids from the Heartwood of *Allanblackia floribunda* Oliver. J. Chem. Soc. (C) **1971**, 1332.

72. MADHAV, R.: Heveaflavone — a New Biflavonoid from *Hevea brasiliensis*. Tetrahedron Letters **1969**, 2017.

73. Mansfield, G. H., T. Swain, and C. G. Nardstrom: Identification of Flavones by the Ultra-violet Absorption Spectra of Their Ions. Nature 172, 23 (1953).

74. Mashima, T., M. Okigawa, and N. Kawano: On the Bisflavones in the Leaves of *Juniperus rigida* Sieb et Zucc. Yakugaku Zasshi 90, 512 (1970).

75. Mashima, T., M. Okigawa, N. Kawano, N. U. Khan, M. Ilyas, and W. Rahman: Tetrahedron Letters 1970, 2937.

76. Massicot, J., J.-P. Marthe, and S. Heitz: Résonance magnétique nucléaire de Produits naturels, Part 7. Nouvelles données sur les Dérivés flavoniques. Bull. Soc. chim., France 1963, 2712.

77. Mathai, K. P., B. Kanakalakshmi, and S. Sethna: Studies in Chromones and Flavones, Part 7. Synthesis of some 3′,3‴- and 6,6′,6″-Biflavonyls. J. Indian Chem. Soc. 44, 148 (1967).

78. Mathai, K. P., and S. Sethna: Chromones and Flavones, Part 5. Synthesis of Some 3′,3‴- and 6,6″-Biflavonyls. J. Indian Chem. Soc. 41, 347 (1964).

79. Miura, H.: The Isolation of Isocryptomerin from the Leaves of *Chamaecyparis obtusa* Endlicher. Yakugaku Zasshi 87, 871 (1967).

80. Miura, H., and N. Kawano: The Partial Demethylation of Flavones, Part 2. Formation of Isocryptomerin. Chem. Pharm. Bull., Japan 15, 232 (1967).

81. —— On the Distribution of Bisflavones in the Leaves of Taxodiaceae and Cupressaceae Plants. Yakugaku Zasshi 88, 1459 (1968).

82. — — Sequoiaflavone in the Leaves of *Sequoia sempervirens* and *Cunninghamia lanceolata* var *Konishii* and its Formation by Partial Demethylation. Yakugaku Zasshi 88, 1489 (1968).

83. —— The Partial Demethylation of Flavones, Part 4. Formation of New Bisflavones Hinokiflavone-7,7″-dimethyl Ether and Neocryptomerin. Chem. Pharm. Bull., Japan 16, 1838 (1968).

84. Miura, H., N. Kawano, and A. C. Waiss: Cryptomerin A and B, Hinokiflavone Methyl Ethers from the Leaves of *Cryptomeria japonica*. Chem. Pharm. Bull., Japan 14, 1404 (1966).

85. Miura, H., T. Kihara, and N. Kawano: New Bisflavones from *Podocarpus* and *Chamaecyparis* Plants. Tetrahedron Letters 1968, 2339.

86. ——— Studies on Bisflavones in the Leaves of *Podocarpus macrophylla* and *P. nagi*. Chem. Pharm. Bull., Japan 17, 150 (1969).

87. Molyneux, R. J., A. C. Waiss, and W. F. Haddon: Oxidative Coupling of Apigenin. Tetrahedron 26, 1409 (1970).

88. Murti, V. V. S., P. V. Raman, and T. R. Seshadri: Cupressuflavone, a New Member of the Biflavonyl Group. Tetrahedron Letters 1964, 2995.

89. ——— Cupressuflavone, a New Biflavonyl Pigment. Tetrahedron 23, 397 (1967).

90. Nakazawa, K.: Yakugaku Zasshi 61, 90, 174, 228 (1941).

91. — The Chemical Structure of Ginkgetin, in: Gifu Yakka Daigako Kiyo, 1962, No. 12, 1 [Chem. Abstr. 59, 2759 d (1963)].

92. — Syntheses of Nuclear-substituted Flavonoids and Allied Compounds, Part 9. Syntheses of Tetramethyl Ether and Dimethyl Ether of Ginkgetin. Chem. Pharm. Bull., Japan 10, 1032 (1962).

93. — Syntheses of Ring-substituted Flavonoids and Allied Compounds, Part 11. Synthesis of Hinokiflavone. Chem. Pharm. Bull., Japan 16, 2503 (1968).

94. Nakazawa, K., and M. Ito: Syntheses of Ring-substituted Flavonoids and Allied Compounds, Part 10. Synthesis of Ginkgetin. Chem. Pharm. Bull., Japan 11, 283 (1963).

95. Nakazawa, K., and S. Matsuura: Syntheses of Nuclearsubstituted Flavonoids and Allied Compounds, Part 4. Synthesis of 8-(*p*-Hydroxyphenacyl)-5,7,4′-trihydroxyflavone Methyl Ethers. On the Structure of the Degradation Flavone formed from Ginkgetin. Yakugaku Zasshi 74, 40 (1954).

96. — — Syntheses of Nuclearsubstituted Flavonoids and Allied Compounds, Part 6.

New Syntheses of 8-Methylacacetin-5,7-dimethyl Ether and 8-(β-Carboxyethyl)-5,7,4'-trimethoxyflavone. Yakugaku Zasshi **75,** 467 (1955).

97. NAKAZAWA, K., and S. MATSUURA: Syntheses of Nuclearsubstituted Flavonoids and Allied Compounds, Part 5. On the Structure of the Flavone formed by Degradation from Ginkgetin (2). Synthesis of 8-(β-Anisoylethyl)-5,7,4'-trihydroxyflavone Methyl Ethers. Yakugaku Zasshi **75,** 68 (1955).

98. NAKAZAWA, K., and S. TSUBOUCHI: Syntheses of Nuclearsubstituted Flavonoids and Allied Compounds, Part 7. On the Structure of the Flavone formed by Degradation from Ginkgetin (3). Synthesis of 6-(p-Hydroxyphenacyl)-5,7,4'-trihydroxyflavone Methyl Ether. Yakugaku Zasshi **75,** 716 (1955).

99. NATARAJAN, S., V. V. S. MURTI, and T. R. SESHADRI: Studies in the Biflavonyl Series, Part 2. Synthesis of 5,7,5'',7'',4''''-Pentamethoxy-4',8''-biflavonyl Ether. Indian J. Chem. **6,** 549 (1968).

100. ——— Biflavonyls, Part 3. Mass Spectrometry of Biflavones. Indian J. Chem. **7,** 751 (1969).

101. ——— Biflavonoids, Part 5. Synthesis of Two Partial Methyl Ethers of Cupressuflavone and of 8,8''-Bichrysinyl Tetramethyl Ether. Indian J. Chem. **8,** 113 (1970).

102. ——— Biflavonoids, Part 6. Some Observations on the Structure and Synthesis of Hinokiflavone. Indian J. Chem. **8,** 116 (1970).

103. ——— Biflavones of Some Cupressaceae Plants. Phytochemistry **9,** 575 (1970).

104. ——— Oxidative Dimerization of Apigenin-7,4'-dimethyl Ether. Indian J. Chem. **9,** 383 (1971).

105. ——— Chemotaxonomical Studies of Some *Casuarina* species. Phytochemistry **10,** 1083 (1971).

106. NATARAJAN, S., V. V. S. MURTI, T. R. SESHADRI, and A. S. RAMASWAMY: Some New Pharmacological Properties of Flavonoids and Biflavonoids. Current Science **39,** 533 (1970).

107. OKIGAWA, M., CHIANG WU HWA, N. KAWANO, and W. RAHMAN: Biflavones in *Selaginella* Species. Phytochemistry **10,** 3286 (1971).

108. PELTER, A., and P. I. AMENECHI: Isoflavonoid and Pterocarpinoid Extractives of *Lonchocarpus laxiflorus*. J. Chem. Soc. (C) **1969,** 887.

109. PELTER, A., and P. STAINTON: The Mass Spectra of Oxygen Heterocycles, Part 5. The Mass Spectra of 2'-Hydroxyflavonoids. J. Chem. Soc. (C) **1967,** 1933.

110. PELTER, A., P. STAINTON, and M. BARBER: The Mass Spectra of Oxygen Heterocycles, Part 2. The Mass Spectra of Some Flavonoids. J. Heterocyclic Chem. **2,** 262 (1965).

111. PELTER, A., R. WARREN, K. K. CHEXAL, B. K. HANDA, and W. RAHMAN: Biflavonyls from Guttiferae. *Garcinia livingstonii*. Tetrahedron **27,** 1625, 3724 (1971).

112. PELTER, A., R. WARREN, N. HAMECD, M. ILYAS, and W. RAHMAN: Biflavonyls Pigments from *Juniperus chinensis* (Cupressaceae). J. Indian Chem. Soc. **48,** 204 (1971).

113. PELTER, A., R. WARREN, N. HAMEED, N. U. KHAN, M. ILYAS, and W. RAHMAN: Biflavonyl Pigments from *Thuja orientalis* (Cupressaceae). Phytochemistry **9,** 1897 (1970).

114. PELTER, A., R. WARREN, B. HANDA, K. K. CHEXAL, and W. RAHMAN: On the Occurrence, Isomerization, and Racemisation of Some Optically Active Biflavonyls. Indian J. Chem. **9,** 98 (1971).

115. PELTER, A., R. WARREN, M. ILYAS, J. N. USMANI, S. P. BHATNAGAR, R. H. RIZVI, M. ILYAS, and W. RAHMAN: The Structure of W 13, the First Optically Active Biflavone of the Amentoflavone Series. Experientia **25,** 350 (1969).

116. PELTER, A., R. WARREN, J. N. USMANI, M. ILYAS, and W. RAHMAN: The Use of Solvent Induced Methoxy Shifts as a Guide to Hinokiflavone Structure. Tetrahedron Letters **1969,** 4259.

117. PELTER, A., R. WARREN, J. N. USMANI, R. H. RIZVI, M. ILYAS, and W. RAHMAN: The Isolation and Characterization of Two Members of a New Series of Naturally Occurring Biflavones. Experientia **25,** 351 (1969).

118. RAHMAN, W., and S. P. BHATNAGAR: A New Biflavonyl AC_3 from *Araucaria cunninghamii.* Tetrahedron Letters **1968,** 675.

119. REED, R. I., and J. M. WILSON: Electron Impact and Molecular Dissociation, Part 12. The Cracking Patterns of Some Rotenoids and Flavones. J. Chem. Soc. (C) **1963,** 5949.

120. SASAKI, S., ITAGAKI, KUROKAWA, T. WATANABE, and T. AOYAMA, Shitsuryo Bunseki **1966,** 14, 82.

121. SAWADA, T.: Studies on Flavonoids in the Leaves of Coniferae and Allied Plants, Part 5. Relation between the Distribution of Bisflavonoids and Taxonomical Position in the Plants. Yakugaku Zasshi **78,** 1023 (1958).

122. SCOTT, A. I.: Oxidative Coupling of Phenolic Compounds. Quart. Rev. **19,** 1 (1965).

123. — Some Natural Products Derived by Phenol Oxidation, in: Oxidative Coupling of Phenols, eds. W. I. TAYLOR and A. R. BATTERSBY, Marcel Dekker, p. 95. New York, 1967.

124. SEWARD, A. C.: The Story of the Maidenhair Tree. Science Progress **1938,** No. 127, 420.

125. SHIMOKORIYAMA, M.: Flavanones, Chalcones and Aurones, in: The Chemistry of Flavonoid Compounds, ed. T. A. GEISSMANN, p. 286. Oxford: Pergamon Press. 1962.

126. USMANI, J. N., N. U. KHAN, and W. RAHMAN: Flavonoids from the Leaves of *Casuarina equisetifolia* Linn (Casuarinaceae). J. Indian Chem. Soc. **47,** 179 (1970).

127. VENKATARAMAN, K.: Methods for Determining the Structures of Flavonoid Compounds, in: The Chemistry of Flavonoid Compounds, ed. T. A. GEISSMAN, p. 70. Oxford: Pergamon Press. 1962.

128. VOIRIN, B., and P. LEBRETON: Recherches chimiotaxonomiques sur les Plantes vasculaires. Sur la presence de Biflavones chez *Psilotum triquetrum* Swartz. Comptes rendus (D) **262,** 707 (1966).

129. WILSON, R. G., J. H. BOWIE, and D. H. WILLIAMS: Solvent Effects in NMR Spectroscopy. Solvent Shifts of Methoxyl Resonances in Flavones induced by Benzene; an Aid to Structure Elucidation. Tetrahedron **24,** 1407 (1968).

130. YATES, P., and G. H. STOUT: The Structure of Mangostin. J. Amer. Chem. Soc. **80,** 1961 (1958).

Addendum

Since April 1972, several important additions to the literature on biflavanoids have occurred. These include references concerned with the isolation of biflavanoids from Araucariaceae (*135*), Cycadaceae (*133*), Euphorbiaceae (*131*) and Taxodiaceae (*134*), a publication dealing with benzene-induced shifts in flavanoids (*136*), a reference to mass spectrometry of flavanoids (*137*) and a publication on the mechanism of the Ullman and related reactions (*132*).

131. CHANDRAMOULI, N., S. NATARAJAN, V. V. S. MURTI, and T. R. SESHADRI: Structure of Heveaflavone. Indian J. Chem. **9,** 895 (1971).

132. COHEN, T., and T. POETH: Copper-Induced Coupling of Vinyl Halides. Stereochemistry of the Ullmann Reaction. J. Amer. Chem. Soc. **94,** 4363 (1972).

133. DOSSAJI, S. F., E. A. BELL, and J. W. WALLACE: Biflavones of *Dioon.* Phytochemistry **12,** 371 (1973).

134. GEIGER, H., and W. DE GROOT-PFLEIDERER: Die Biflavone von *Taxodium distichum* Rich. Phytochemistry **12,** 465 (1973).

135. KHAN, N. U., M. ILYAS, W. RAHMAN, T. MASHIMA, M. OKIGAWA, and N. KAWANO: Biflavones from the Leaves of *Araucaria bidwillii* Hooker and *Agathis alba* Foxworthy (Araucariaceae). Tetrahedron **28,** 5689 (1972).

136. NATARAJAN, S., S. C. DATTA, V. V. S. MURTI, and T. R. SESHADRI: Role of Steric Factors on Benzene-induced Shift of Methoxyl Resonances in Flavonoids. Indian J. Chem. **9,** 813 (1971).

137. VAN DE SANDE, C., J. W. SERUM, and M. VANDEWALLE: Studies in Organic Mass Spectrometry — XII. Mass Spectra of Chalcones and Flavanones. The Isomerisation of 2'-Hydroxychalcone and Flavanone. Organic Mass Spectrometry **6,** 1333 (1972).

(Received, April 10, 1972)

Chemie der Makrolid-Antibiotica

Von W. KELLER-SCHIERLEIN, Zürich

Mit 1 Abbildung

Inhaltsübersicht

I. Einleitung

Unter dem Namen Makrolid-Antibiotica (*329*) wurden ursprünglich, im Anschluß an die Konstitutionsaufklärungen des Erythromycins, Methymycins und Carbomycins, einige lipophile basische Antibiotica mit hoher Wirkung vorwiegend gegen Grampositive Bakterien zusammengefaßt, die ausschließlich von Actinomyceten produziert wurden (*152*). Kreuzresistenz-Versuche deuteten auf einen gemeinsamen Wirkungsmechanismus hin. Bei allen damals bekannten Vertretern der Gruppe konnte der basische Charakter auf die Anwesenheit von miteinander nahe verwandten Aminozuckern zurückgeführt werden.

Je mehr sich die Konstitutionsaufklärungen neuer oder schon länger bekannter Antibiotica häuften, desto heterogener wurde die Familie der Makrolide, die heute praktisch nur noch ein gemeinsames Merkmal, einen makrozyklischen Lactonring, besitzt. Denn die chemische Vielfalt findet ihr Gegenstück in einer großen Mannigfaltigkeit hinsichtlich der biologischen Wirkung. Zu den antibakteriellen „klassischen" Makroliden sind im Laufe der Jahre Verbindungen mit vorwiegend antifungischer Wirkung, die Polyene und die Venturicidine, ferner das antivirale und antibakterielle Borrelidin, gestoßen. Und nicht einmal hinsichtlich ihrer biologischen Herkunft ist die Einheitlichkeit der Gruppe gewahrt geblieben; denn mehrere Stoffwechselprodukte niederer Pilze mit vorwiegend antifungischer und cytostatischer Wirkung müssen auf Grund ihrer Struktur und Biogenese zu den Makroliden gezählt werden.

Die große Zahl an Verbindungen zwingt uns, bei der Darstellung der Chemie der Makrolide im Rahmen dieses Referates eine Auswahl zu treffen. Wir ließen uns dabei nicht in erster Linie von der medizinischen oder biologischen Bedeutung leiten. Die bisherigen Arbeiten über Makrolid-Chemie, etwa die beiden letzten Jahrzehnte umfassend, fielen in eine Epoche rascher Entwicklung der analytischen organischen Chemie, die nur in wenigen Gebieten der Naturstoff-Chemie so klar zum Ausdruck kommt, wie gerade hier. Die ersten Untersuchungen zur Konstitution der Makrolide (z. B. Erythromycin, Methymycin, Narbomycin, Carbomycin) könnten etwa noch dem ausgehenden „klassischen" Zeitalter zugerechnet werden, in dem fast ausschließlich chemische Umwandlungen und der Abbau zu einfacheren Verbindungen für konstitutionelle Schlußfolgerungen maßgebend waren. In einer Zwischenphase (z. B. Lankamycin) wurden zwar moderne instrumentalanalytische Methoden in starkem Maße angewendet, im wesentlichen aber zur Stützung bereits chemisch erarbeiteter Strukturelemente benützt. In neueren Arbeiten tritt dann der systematische Einsatz der NMR.-Spektroskopie (z. B. Leucomycin A_3, Borrelidin) und der Massenspektrometrie (Borrelidin, Polyene) stark in den Vordergrund, während man versucht, die eigent-

lichen chemischen Methoden auf das unumgänglich notwendige zu beschränken. Und schließlich kommt der Integration der Röntgen-Strukturanalyse in die organische Chemie (z. B. Venturicidin A) eine stets wach-
sende Bedeutung zu.

Die neuere Entwicklung auf dem Gebiet des Erythromycins zeigt aber
auch, eine wie stark verfeinerte Information wir durch die modernen
Methoden erhalten können. Unsere Kenntnisse beschränken sich keineswegs mehr nur auf die Konstitution und Konfiguration der Antibiotica,
sondern lassen uns tiefere Einblicke in die Konformation im festen und
flüssigen Zustand gewinnen, die für ein Verständnis der Wirkungsweise
von besonderer Bedeutung ist.

Dem Bestreben, diese Entwicklung besonders herauszustreichen,
mußte die Darstellung von an sich bedeutungsvollen älteren Arbeiten wie
etwa der Konstitutionsaufklärung des Oleandomycins durch CELMER
und Mitarb. (*59, 100, 140*) geopfert werden. Aber auch von den neueren
Arbeiten können manche nur am Rand erwähnt werden. Schließlich kann
auf einige Gruppen von Antibiotica, die eine Verwandtschaft zu den Makroliden vermuten lassen, wie etwa die Oligomycine, Scopamycin und
seine Verwandten, die Bundline und Ansamycine, überhaupt nicht eingegangen werden.

Vor allem aber können im Rahmen dieser Zusammenfassung die
bedeutungsvollen Arbeiten zur Aufklärung der Wirkungsweise der
Makrolide nicht behandelt werden. Der Leser muß daher auf andere
Referate verwiesen werden (z. B. *208, 302, 324*). Die wichtigsten Ergebnisse der Untersuchungen über die Biosynthese der Makrolide sind bereits
früher zusammengefaßt worden (*121*).

II. Antibakterielle Makrolid-Glykoside aus Actinomyceten

Zu dieser ältesten Gruppe der Familie gehören einige in der Humanmedizin häufig gebrauchte Antibiotica, darunter an erster Stelle das
Erythromycin. Ihre Vertreter sind ausnahmslos Glykoside, und meistens
ist mindestens einer der Zuckerbausteine ein Aminozucker, der dem Antibioticum einen basischen Charakter verleiht. In neuerer Zeit sind aber
auch mehrere neutrale Antibiotica mit ausschließlich stickstofffreien
Zuckerbausteinen aufgeklärt worden, die ihrem Aufbau und ihrer biologischen Wirkung nach eindeutig dieser Gruppe zugerechnet werden
müssen. Sie läßt sich in zwei Untergruppen aufteilen, die sich ihrer Struktur nach deutlich voneinander unterscheiden. Wahrscheinlich sind auch
Unterschiede hinsichtlich des Wirkungsmechanismus vorhanden; denn
volle gekreuzte Resistenz findet man oft nur innerhalb der Untergrup-

pen, während zwischen den Untergruppen meist nur partielle Kreuz-
resistenz auftritt (*152*). Der Angriffsort ist aber für beide Untergruppen
derselbe: das Ribosom, an dem sich die Proteinsynthese abspielt.

1. Die Erythromycin-Gruppe

Diese Untergruppe umfaßt zwei Makrolide mit einem zwölfgliedrigen
Lactonring (Methymycin und Neomethymycin) sowie zahlreiche Vertreter mit einem vierzehngliedrigen Lactonring.

a) Die Konstitution des Erythromycins

Das Erythromycin war das erste Makrolid-Antibioticum, dessen
Konstitution vollständig aufgeklärt wurde. Seine Entdeckung geht auf
das Jahr 1952 zurück (*197*), wo es unter dem Fabriknamen „Ilotycin"
zum erstenmal erwähnt wird. In einer ausgedehnten biologischen Studie
erwies es sich als hoch aktiv, insbesondere gegen gram-positive Bakterien, bei geringer Toxicität (*253*).

Die ersten Abbauversuche führten zur Isolierung von zwei neuartigen Zuckern (*64, 104, 129*), dem basischen Desosamin und einem
neutralen Zucker mit einer verzweigten Kohlenstoffkette, der Cladinose,
deren Chemie in einem besonderen Abschnitt (S. 413) beschrieben werden
soll.

Die chemische Untersuchung des Aglykonteils gestaltete sich aus zwei
Gründen als besonders schwierig. Einerseits erwies sich das Aglykon
unter den Bedingungen der Glykosidspaltung als unstabil. Erst nachdem
die Ketogruppe des Erythromycins (**1**), deren Anwesenheit aus einem
niedrigen UV.-Absorptionsmaximum bei 280 nm (*197, 104*) und einem
IR.-Absorptionsmaximum bei 1700 cm^{-1} hervorging, mittels Natriumborhydrid reduziert worden war, konnte durch stufenweisen Abbau ein
Derivat des Aglykons, das Dihydro-erythronolid (**4**), in reiner Form
gefaßt werden (*276, 319*; s. Schema 1).

Andererseits führte der weitere Abbau des Dihydro-erythronolids
zunächst zu Gemischen, die sich mit den damals verfügbaren Methoden
nicht trennen ließen. Wenn man sich vergegenwärtigt, daß sich die Papierchromatographie und die Gegenstromverteilung damals in den ersten
Anfängen befanden, die Dünnschichtchromatographie noch nicht erfun-
den und auch die Säulenchromatographie noch wenig entwickelt waren,
stellt der erfolgreiche Abschluß dieser Pionierarbeit der Konstitutionsaufklärung des Erythromycins innerhalb von wenigen Jahren K. Gerzon
und seinen Mitarbeitern ein hervorragendes Zeugnis für experimentelles
Geschick und Unternehmungsgeist aus. Der Erfolg war oft nur dadurch
möglich, daß konsequent Gemische über mehrere Stufen weiterverarbei-
tet wurden, bis sich durch Umwandlungen von funktionellen Gruppen

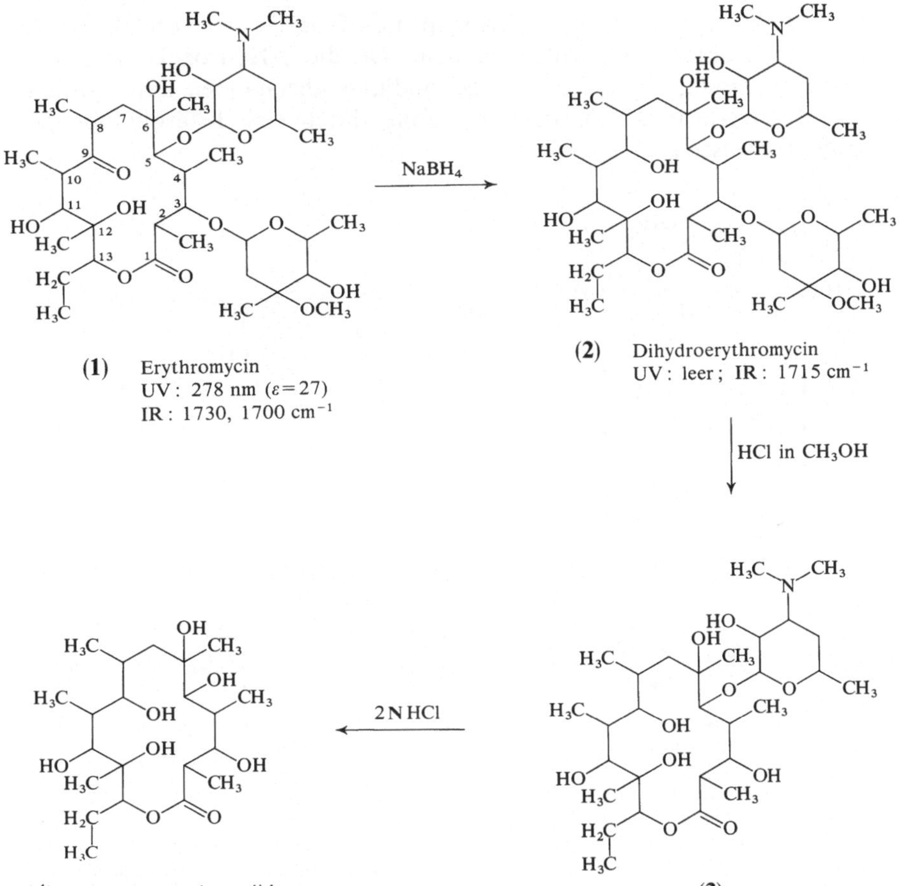

(1) Erythromycin
UV: 278 nm ($\varepsilon = 27$)
IR: 1730, 1700 cm^{-1}

NaBH$_4$

(2) Dihydroerythromycin
UV: leer; IR: 1715 cm^{-1}

HCl in CH$_3$OH

(4) Dihydro-erythronolid
UV: leer; IR: 1710 cm^{-1}

2 N HCl

(3)

Schema 1. Gewinnung von Dihydro-erythronolid

oder weiteren Abbau leichter isolierbare Produkte gebildet hatten. Ein Glück war allerdings, daß wegen der bald einsetzenden Großproduktion reichliche Mengen an Erythromycin für wissenschaftliche Untersuchungen zur Verfügung standen.

Dihydro-erythronolid, $C_{21}H_{40}O_8$ **(4)**, verbraucht bei der Oxydation mit Natriumperjodat zwei Mol des Reagens, was für die Anwesenheit von entweder zwei 1,2-Diolgruppierungen oder einer 1,2,3-Triolgruppierung spricht (*319*). Der präparative Abbau mit dem gleichen Reagens (*276, 319*) führte zu einem Gemisch, aus dem sich nach alkalischer Verseifung dank ihrer Leichtflüchtigkeit eine Verbindung $C_5H_{10}O_2$ rein gewinnen ließ (Schema 2). Auf Grund der Oxydation zu 2,3-Pentandion

(6) muß es sich um das 3-Hydroxy-pentan-2-on (5) oder sein Isomeres mit vertauschten Funktionen handeln. Da die α-Hydroxyketongruppe der Verbindung 5 die Perjodat-Behandlung überstanden hatte, mußte die Hydroxylgruppe vor der Verseifung durch eine Esterbindung geschützt gewesen sein.

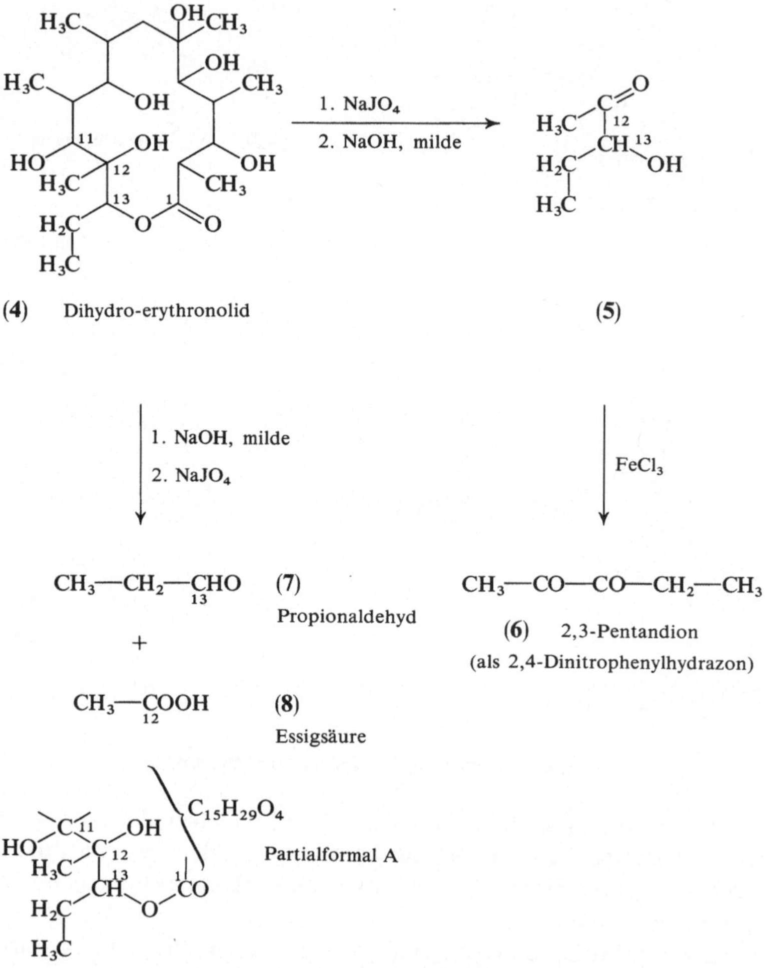

(4)   Dihydro-erythronolid

1. NaJO$_4$
2. NaOH, milde

(5)

1. NaOH, milde
2. NaJO$_4$

FeCl$_3$

CH_3—CH_2—$\underset{13}{C}HO$ (7)
Propionaldehyd

+

CH_3—$\underset{12}{C}OOH$ (8)
Essigsäure

CH_3—CO—CO—CH_2—CH_3

(6)   2,3-Pentandion
(als 2,4-Dinitrophenylhydrazon)

$C_{15}H_{29}O_4$

Partialformal A

Schema 2. Oxidativer Abbau von Dihydro-erythronolid (1. Teil)

Dementsprechend wurden bei einer Umkehrung der Reihenfolge der beiden Reaktionen — zuerst alkalische Verseifung und anschließend Oxydation — drei Mol Perjodat verbraucht. Die beiden leicht flüchtigen

Produkte, je etwa 1 Mol Propionaldehyd und Essigsäure, mußten aus demselben Molekülbereich entstanden sein, wie vorher das Abbauprodukt 5. Demnach ergibt sich für den Bereich der Ringkohlenstoffatome C-11

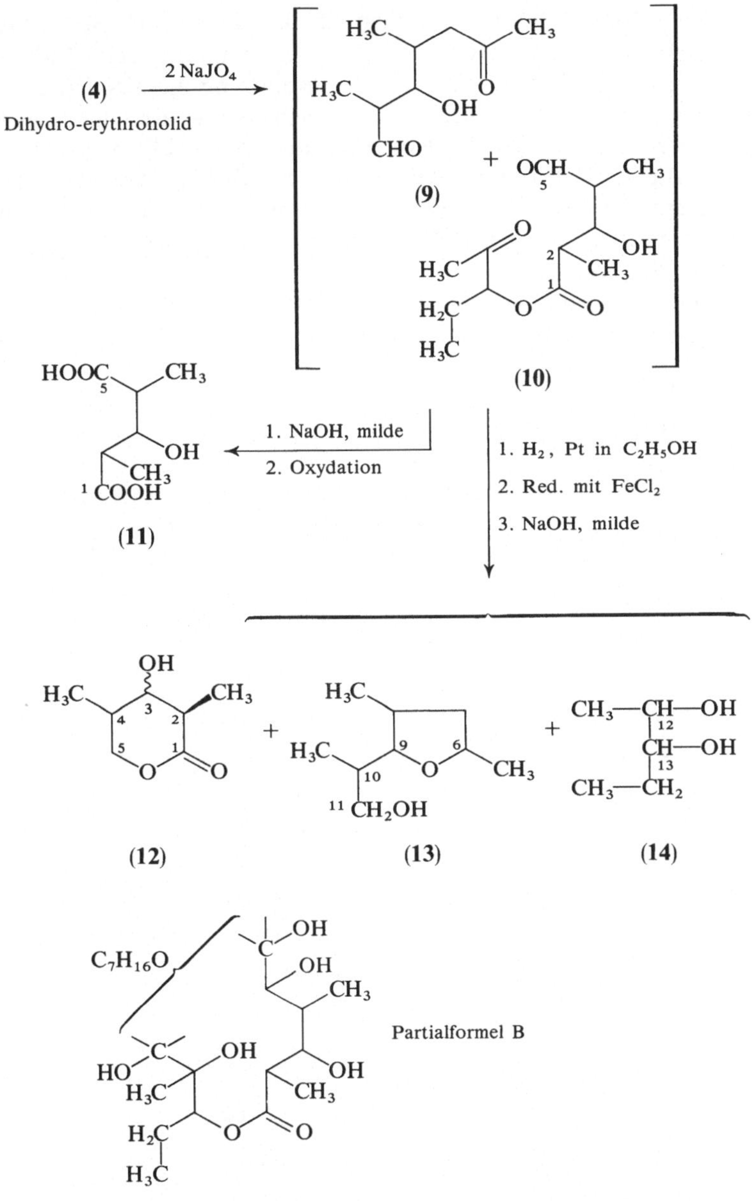

Schema 3. Oxidativer Abbau von Dihydro-erythronolid (2. Teil)

bis C-13 des Erythromycins die Partialformel A. Da außer für die Lacton-gruppe keine Anhaltspunkte für weitere Estergruppen vorlagen, mußte die Carbonylgruppe der Partialformel A dem C-Atom 1 des Lactonringes entsprechen.

Aus den schwerflüchtigen Anteilen der Oxidation mit Perjodsäure und anschließender Hydrolyse konnte erst nach Oxidation einer Aldehyd-gruppe in geringer Menge eine reine Verbindung herauskristallisiert werden. Sie erwies sich beim direkten Vergleich als identisch mit einer der diastereomeren α, α′-Dimethyl-β-hydroxyglutarsäuren **(11)** *(320)*. Aus der Beziehung dieser Verbindung mit später zu diskutierenden Abbau-produkten ergab sich, daß sie den C-Atomen 1 bis 5 des Lactonrings entsprechen mußte, wodurch die Partialformel A zur Teilformel B erweitert werden konnte.

Abbauprodukte, die gesamthaft alle 21 Kohlenstoffatome des Agly-kons enthielten, wurden dagegen erst erhalten, als das Gemisch der pri-

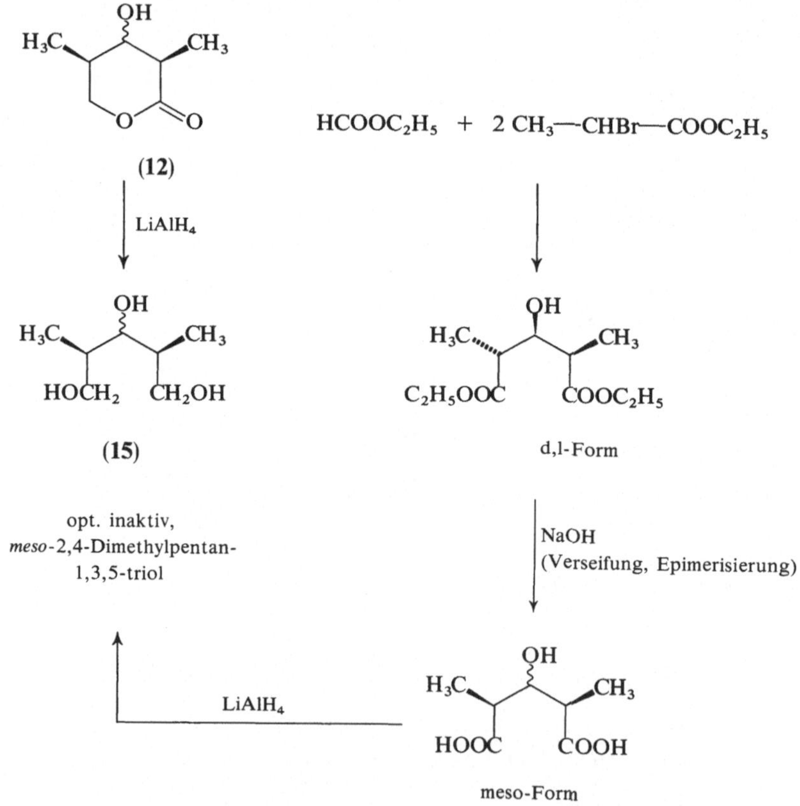

$$HCOOC_2H_5 \;\; + \;\; 2\,CH_3\text{—}CHBr\text{—}COOC_2H_5$$

Schema 4. Konstitutionsaufklärung von Abbauprodukten

Literaturverzeichnis: SS. 445—460

mären Oxidationsprodukte (**9** und **10** im Formelschema 3) zunächst katalytisch, anschließend mittels Eisen(2)-chlorid reduziert und endlich alkalisch hydrolysiert wurde (*115*). Das optisch aktive Lacton **12** gab bei der Reduktion mit Lithiumaluminiumhydrid ein optisch inaktives 2,4-**Dimethylpentan-1,3,5-triol, also eine der beiden möglichen Meso-Ver**bindungen der Formel **15,** deren Identität durch Vergleich mit einem gemäß Formelschema 4 hergestellten synthetischen Präparat gesichert wurde.

Die Konstitution des Pentan-2,3-diols **14,** eines zweiten Produktes aus dem Abbaugemisch, ergab sich aus der Oxidation zum bereits erwähnten Diketon **6** und aus der Bildung von Propionaldehyd beim Abbau mit Natriumperjodat.

Die Eigenschaften des cyclischen Aethers **13,** $C_9H_{18}O_2$, mit drei nach KUHN-ROTH erfaßbaren C-Methylgruppen und einer freien Hydroxylgruppe passen auf die angegebene Struktur. Seine vollständige Konstitution ergibt sich aus weiteren Abbau-Versuchen mit dem Oxidationsgemisch (**9**+**10**) (Schema 5). Nach der Oxidation des rohen Abbaugemisches mit Trifluorperessigsäure und anschließenden Hydrolyse mit Natronlauge ließen sich 2 Produkte isolieren. Die α,α'Dimethyl-β-hydroxyglutarsäure war identisch mit der bereits früher (Schema 3) isolierten Mesoform **11** und stammt demnach aus dem gleichen Molekülteil, den C-Atomen 1 bis 5. Das neutrale Produkt der Reaktionsfolge, das rein isoliert wurde, erwies sich als ein Diastereomeres (**16**) des früher beschriebenen Lactons **12.** Wegen der verschiedenen Stereochemie konnte diese Verbindung aber nicht aus dem gleichen Molekülteil stammen (C-1 bis C-5); ihre Entstehung erklärt sich vielmehr durch eine Bayer-Villiger-Oxidation und gleichzeitige Oxidation einer Aldehydgruppe des C_9-Zwischenproduktes **9,** dessen Konstitution sich aus diesen Abbaureaktionen und aus der Isolierung des Hydroxyäthers **13** (Schema 3) ergibt.

Daß von den beiden Carboxylgruppen der Dicarbonsäure **11** eine durch Oxidation einer Aldehydgruppe entstanden ist, während die andere der ursprünglichen Lactongruppierung entstammt, ergibt sich aus der Isolierung des α,β-ungesättigten Aldehyds **17** bei der milden alkalischen Verseifung des Gemisches (**9**+**10**). Seine Entstehung ist aus dem Zwischenprodukt **10** leicht interpretierbar über die nicht isolierbaren Zwischenprodukte **18** und **19,** welch letztere als „vinyloge β-Oxosäure" leicht decarboxylieren kann.

Schließlich konnte, wenn auch in unreiner Form, der Ketoaldehyd **9** selber aus dem Hydrolysegemisch isoliert werden. Die positive Jodoformreaktion des Produktes ist nicht unbedingt zwingend für die Anwesenheit einer Methylketongruppierung, da die Bildung von Jodoform auch **durch eine Verunreinigung mit dem Hydroxyketon 5 (Schema 2) erklär**bar ist. Die Konstitution des Oxidationsproduktes **9** wird aber, abgesehen

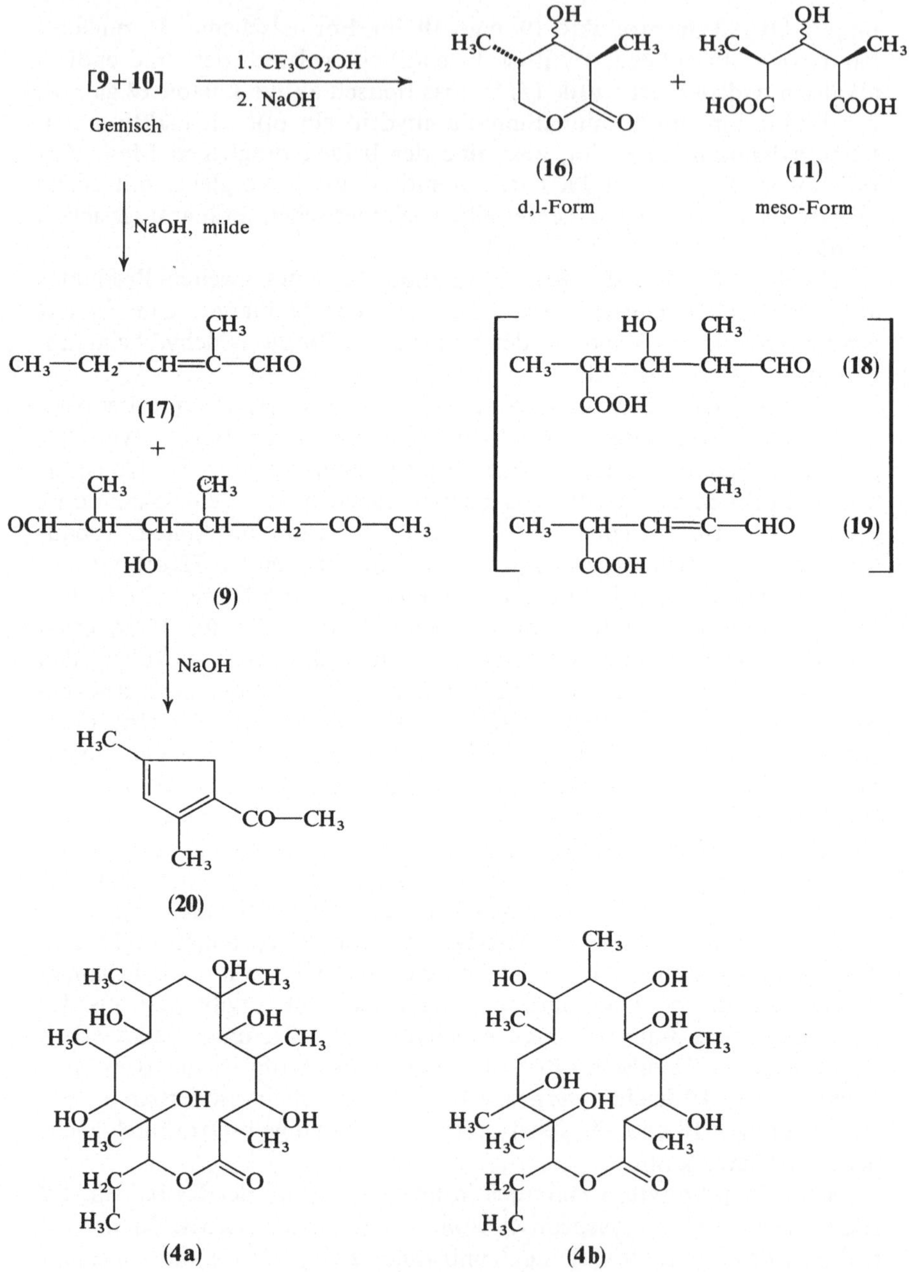

 W. Keller-Schierlein:

[9+10]
Gemisch

1. CF_3CO_2OH
2. NaOH

(16) d,l-Form

(11) meso-Form

NaOH, milde

$CH_3-CH_2-CH=C(CH_3)-CHO$

(17)

+

(9)

NaOH

(18)

(19)

(20)

(4a)

(4b)

Alternativformeln für das Dihydro-erythronolid

Schema 5. Weitere Umwandlungen von Abbauprodukten

Literaturverzeichnis: SS. 445—460

von den bereits erwähnten Beziehungen, bestätigt durch die cyclisierende **Aldolkondensation unter gleichzeitiger Wasserabspaltung zum Dienon 20**, das durch ein kristallines 2,4-Dinitrophenylhydrazon mit einem Absorptionsmaximum bei 402 nm (log ε 4,88) charakterisiert wurde.

Durch diese Reaktionen werden die Konstitutionen der Verbindungen **9** und **10**, die zusammen alle Kohlenstoffatome des Aglykons enthalten, festgelegt. Sie lassen sich auf zwei Arten zum Makrolidring zusammenfügen: **4a** und **4b**. Aus biogenetischen Überlegungen wurde schon bei diesem Stand der Untersuchungen der Struktur **4a** mit regelmäßigem Aufbau des Kohlenstoffgerüstes der Vorzug gegeben. Endgültig ausschließen ließ sich die Formel **4b** erst auf Grund von Abbauergebnissen, die mit dem intakten Glykosid durchgeführt wurden (*321*).

Abbauversuche, die sich mit dem Bereich der C-Atome 11 bis 13 des Erythromycins (**1**) befassen und sich eng an bereits beim Dihydroerythronolid besprochene Reaktionen anlehnten, ergaben zunächst, daß der Lactonring auch beim Antibioticum zum C-Atom 13 geschlossen ist, und daß die Hydroxylgruppen an C-11 und C-12 frei sind. Ein wichtiges Umwandlungsprodukt des Erythromycins, das durch milde Säureeinwirkung erhalten wird, ist das Spiroketal **21**, Erythralosamin ($C_{29}H_{49}NO_8$). Durch Oxidation dieses Amins mit Chromsäure in Eisessig-Wasser wurde ein nicht näher charakterisiertes Produkt **22** erhalten, dessen Analysen auf eine Bruttoformel $C_{29}H_{51}NO_{11}$ hinweisen. Es sind demnach noch sämtliche C-Atome des Erythralosamins darin enthalten. Ein kleineres Abbauprodukt wurde erst durch energische saure Hydrolyse der Verbindung **22** erhalten. Das leicht isolierbare neutrale Produkt **23** besitzt eine γ-Lactongruppierung (IR.-Absorptionsmaximum 1763 cm^{-1}), einen α,β-ungesättigten δ-Lactonring (IR: 1715 cm^{-1}; UV: 209 nm, log ε 4,11) sowie vier nach KUHN-ROTH nachweisbare C-Methylgruppen. Nach der katalytischen Hydrierung wurde daraus ein gesättigtes Dilacton erhalten (UV leer), das jetzt nur noch eine breite Carbonylabsorption bei 1730 bis 1765 cm^{-1} im IR.-Absorptionsspektrum aufweist.

Die Entstehung des Abbauproduktes **23** aus Erythralosamin (**21**) über die beiden Reaktionsstufen ließ sich nur auf eine Weise interpretieren: Elimination der β-ständigen Sauerstoffunktion an C-3, Abspaltung des Aminozuckers an C-5 unter „Umlactonisierung", oxidative Spaltung im Bereich der α,β-ungesättigten Ketonfunktion, die im Erythralosamin als Ketalgruppe vorliegt. Das Abbauprodukt **23** muß demnach noch die C-Atome 1 bis 9 des ursprünglichen Makrolidringes enthalten.

Die Konstitution der Dilactone **23** und **24** wurde durch weiteren Abbau vollständig aufgeklärt. Zunächst wurde das gesättigte Lacton **24** mit Lithiumaluminiumhydrid zum Tetrol **25** reduziert, das im IR.-Absorptionsspektrum keine Carbonylbande mehr zeigte. Eine 1,2-Diol-Gruppierung in diesem Tetrol wurde durch eine Spaltung mit Perjodsäure

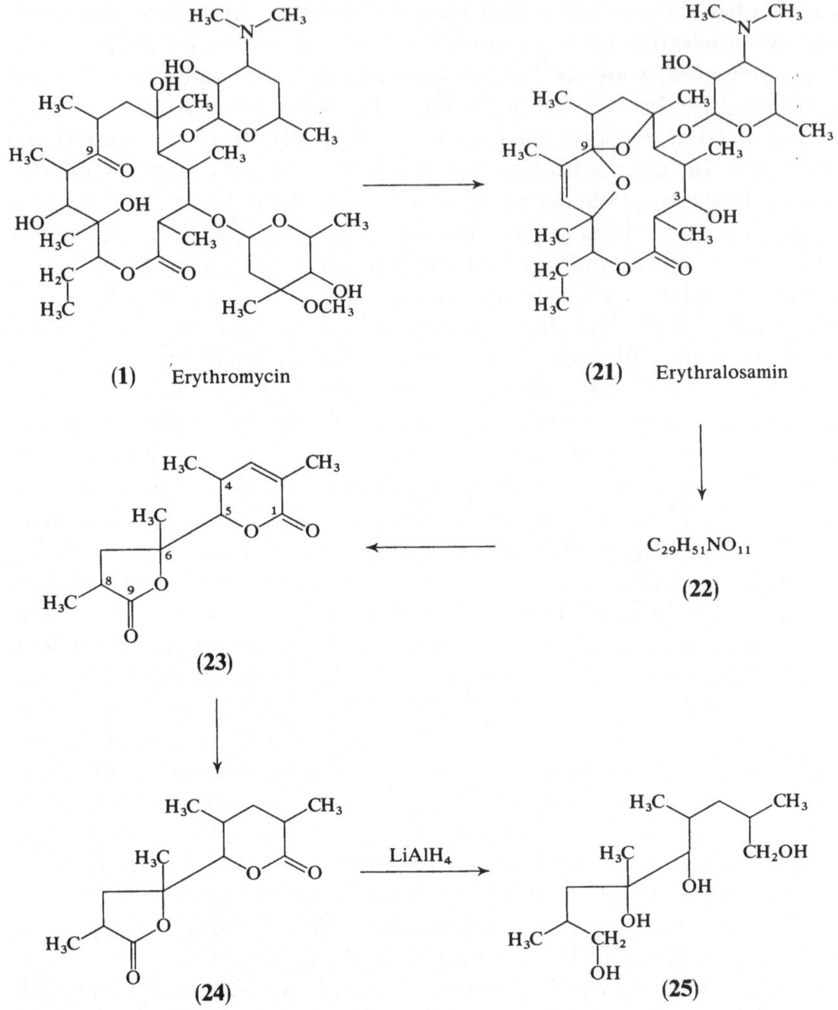

(1) Erythromycin

(21) Erythralosamin

$C_{29}H_{51}NO_{11}$

(22)

(23)

(24) $\xrightarrow{\text{LiAlH}_4}$ **(25)**

Schema 6. Herstellung und Abbau des Erythralosamins

nachgewiesen. Allerdings konnte nur das eine der zu erwartenden Abbauprodukte, das 2,4-Dimethylpentan-1,5-diol (**26**) als Bis-3,5-dinitrobenzoat
rein gefaßt und mit einem synthetischen Präparat identifiziert werden,
während das aus den C-Atomen 6 bis 9 zu erwartende Diol bei der Aufarbeitung verlorenging. Identifizierbare Abbauprodukte, die zusammen
alle 13 C-Atome des Dilactons **24** enthielten, konnten dagegen gefaßt
werden, wenn das Lacton zunächst mit Natronlauge zum Dinatriumsalz
der entsprechenden Dihydroxydicarbonsäure hydrolisiert und dieses in

alkalischer Lösung mit Natriumperjodat oxydiert wurde. Die α-Methyl-
lävulinsäure (27) wurde in weitgehend racemisierter Form erhalten, und
auch die α,γ-Dimethylglutaraldehydsäure (28) war erwartungsgemäß
stereochemisch nicht völlig einheitlich. Trotzdem ließen sich beide
Produkte als 2,4-Dinitrophenylhydrazone mit entsprechenden synthetischen Derivaten eindeutig identifizieren. Die beiden Abbauprodukte beweisen streng die Strukturen der Dilactone 23 und 24, die ihrerseits nur
mit Struktur 4a, nicht aber mit der Formel 4b für das Dihydroerythronolid vereinbar sind.

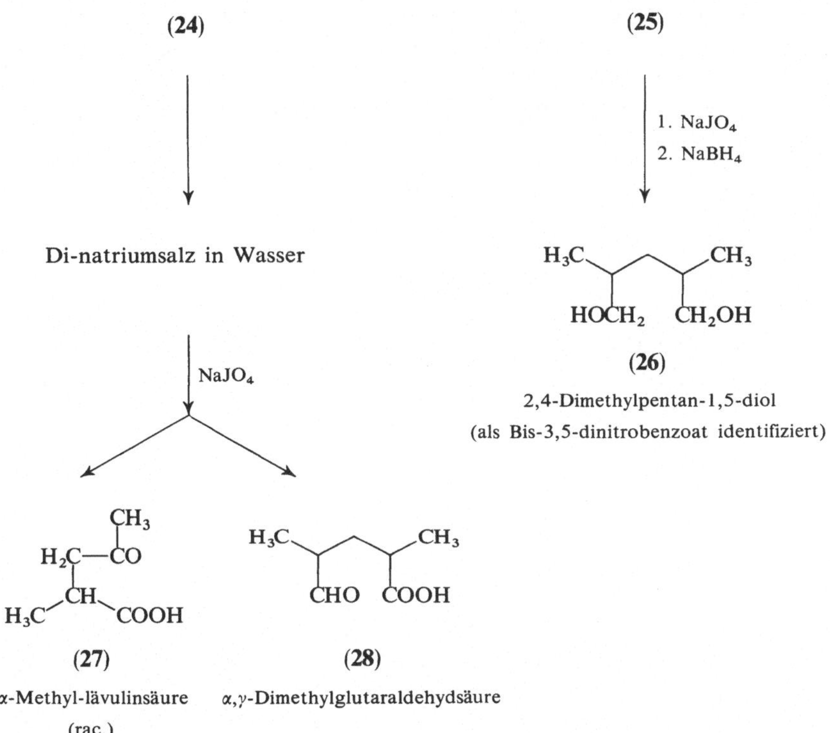

Schema 7. Konstitutionsaufklärung des Tetrols (25) und des Dilactons (24)

Gleichzeitig wird durch diesen Abbau auch die Lage der Ketogruppe
des Erythromycins festgelegt. Schon auf Grund früher erwähnter Abbauergebnisse konnten die Sauerstoffatome an C-11, C-12 und C-13 als
Carbonylsauerstoffe ausgeschlossen werden. Die Bildung des Lactons 23 zeigt, daß auch die O-Atome an C-3, C-5 und C-6 dafür
nicht in Betracht kommen. Es bleibt als einzige Sauerstoffunktion diejenige
an C-9 übrig.

Auch die möglichen Stellungen der beiden Zuckerreste werden durch die erwähnten Abbauversuche bereits eingeschränkt: Die Stellung 13 trägt das Lactonring-Sauerstoffatom; O-C-9 ist das Keton-Sauerstoffatom; die C-Atome 11 und 12 tragen auch im Erythromycin freie Hydroxylgruppen, denn die Bindung zwischen diesen beiden C-Atomen ist mit Perjodsäure spaltbar. Es bleiben somit die C-Atome 3, 5 und 6 als mögliche Träger der beiden Zuckerreste.

Das O-Desosaminyl-dihydroerythronolid **3** (Schema 1), das neben Methylcladinosid bei der milden Methanolyse von Dihydroerythromycin entsteht, wird von Perjodsäure nur zwischen den C-Atomen 11 und 12, nicht aber zwischen C-5 und C-6 angegriffen. Somit muß der Desosaminrest entweder in Stellung 5 oder 6 stehen. Entscheidend für die Fixierung der Lage beider Zuckerreste war eine Verbindung **29**, die aus Erythromycin in wässeriger Lösung bei pH 2,5 innerhalb weniger Minuten entsteht. Die Verbindung **29**, $C_{37}H_{65}NO_{12}$, unterscheidet sich von Erythro-

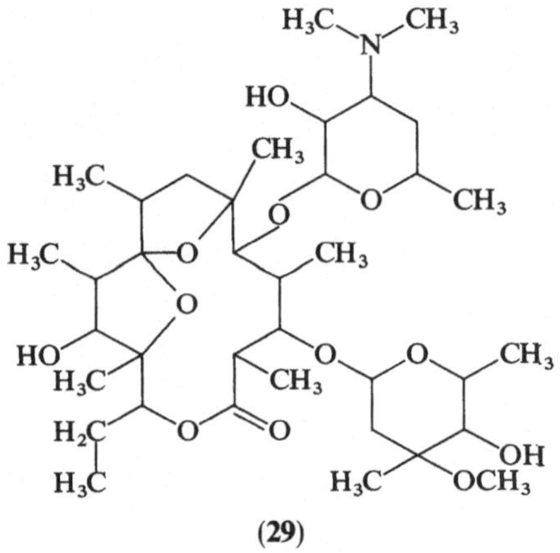

(29)

Anhydro-erythromycin

(Spiro-Ketal)

mycin durch den Verlust eines Moleküls Wasser und wurde daher Anhydroerythromycin genannt. Im Infrarot-Absorptionsspektrum ist die Bande bei 1700 cm^{-1}, im UV-Spektrum die schwache Bande bei 280 bis 300 nm nicht mehr vorhanden, die Verbindung ist demnach kein Keton. Andererseits zeigt sie im IR.-Absorptionsspektrum keine Anhaltspunkte für eine C=C-Doppelbindung. Dagegen ist eine starke IR.-Absorptionsbande bei 905 cm^{-1} mit der Formulierung als Spiroketal bestens verträglich, die auch dadurch gestützt wird, daß das N-Oxid der Verbindung **29** nicht mit Perjodat reagiert, und die Verbindung selber mit verdünnter Säure leicht in Erythralosamin (**21**, Schema 6) übergeht. In günstiger Position zur Ketogruppe an C-9 für eine Spiroketalbindung befinden sich einerseits das Sauerstoffatom an C-12 (Beteiligung von O-C-11 würde zu einem gespannten Vierring führen), andererseits eine Hydroxylgruppe an C-5 oder C-6. Damit bleibt für die Lage der Cladinose nur noch C-3 übrig.

Für die Lage des Desosaminrestes an C-5 konnte durch Abbau kein strikter Beweis erbracht werden, doch ließen zahlreiche Beobachtungen die Stellung 5 gegenüber 6 bevorzugen. Insbesondere schien seine Lage an einem tertiären Sauerstoffatom von Anfang an unwahrscheinlich, da sonst beim Dihydroerythromycin eine Abspaltung des Desosaminrestes durch Elimination wahrscheinlicher wäre als durch Hydrolyse. Schließlich wurden für das Methymycin, das Pikromycin und das Narbomycin unabhängig die 5-Stellung des Desosaminrestes bewiesen, so daß aus Analogiegründen auch beim Erythromycin die Position 5 den Vorzug verdient. Eine Bestätigung für diese Annahme brachte viel später eine Röntgenstrukturanalyse (s. unten).

b) Konfiguration und Konformation des Erythromycins

Schon die im vorangehenden Abschnitt erwähnten Abbauprodukte sowie weitere, hier nicht näher besprochene Untersuchungen (*88*), ließen Schlußfolgerungen über gewisse stereochemische Beziehungen zu. Im Ganzen aber blieben unsere Kenntnisse über die Stereochemie des Erythromycins lange Zeit fragmentarisch. Die vollständige Konfiguration wurde mit Hilfe der Röntgen-Strukturanalyse abgeleitet (*128*). Als kristallines Derivat mit einem Schweratom diente das Erythromycinhydrojodid, das mit zwei Molekülen Wasser orthorhombische Kristalle der Raumgruppe $P2_1 2_1 2_1$ bildet. Da einerseits nur 1516 unabhängige Reflexionen benützbar waren und andererseits die Jodatome eine ungünstige Verteilung im Raumgitter zeigten, bot die Analyse zunächst große Schwierigkeiten. Erst als die anomale Dispersion an den Jodatomen systematisch ausgenützt wurde, gelang die Ableitung der Struktur im Sinne der Formel **30**. Darnach besitzt das Aglykon die 2-(*R*), 3-(*S*), 4-(*S*), 5-(*R*), 6-(*R*), 8-(*R*), 10-(*R*), 11-(*R*), 12-(*S*), 13-(*R*)-Chiralität, in Übereinstimmung mit Teilbefunden aus Abbauergebnissen. Interessanterweise ist die **L**-Cladinose α-glykosidisch, das **D**-Desosamin β-glykosidisch an das Aglykon gebunden. Die Stereochemie des Erythromycins stimmt, mutatis mutandis, völlig überein mit derjenigen des Oleandomycins, die CELMER (*59*) unabhängig mit chemischen und spektroskopischen Methoden ableiten konnte, was ihn zur Aufstellung eines fruchtbaren generellen Modells der Makrolid-Stereochemie anregte (*61*).

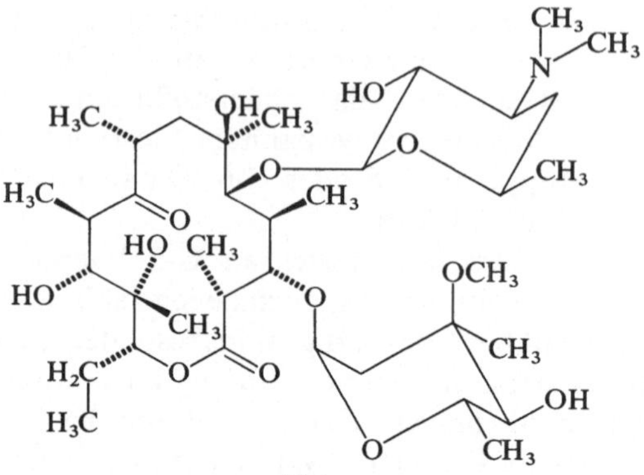

(30)

Konfiguration des Erythromycins

Obwohl die Kristallstrukturanalyse neben der Konfiguration auch ein vollständiges Bild der Konformation des Moleküls im Kristallgitter lieferte, gaben sich verschiedene amerikanische Forscher mit diesem Ergebnis nicht zufrieden. Sie stützten sich auf die plausible Annahme, daß das Antibioticum nicht aus der kristallinen Phase, sondern aus einer Lösung heraus mit den Ribosomen in Wechselwirkung tritt, und daß daher die Konformation in flüssiger Phase für das Verständnis der antibitotischen Wirkung bedeutungsvoller sei als die Kristall-Konformation. Ein erstes konformatives Modell von CELMER (*58*) stützt sich vor allem auf theoretische Überlegungen. Mit großem experimentellem Aufwand haben dagegen PERUN, EGAN und Mitarb. (*248, 249, 250*), MITSCHER und Mitarb. (*207, 249*) sowie DEMARCO (*76, 77*) eine Lösung des Problems angestrebt. Ein großes Tatsachenmaterial stand dabei vor allem der Gruppe von PERUN zur Verfügung, die über 50 Derivate des Erythromycins und seines Aglykons vergleichend untersuchen konnten, die teilweise durch chemische Umwandlungen aus Erythromycin oder Erythromycin B (s. unten) hergestellt worden waren (vgl. z. B. *179, 247*), teils durch Fermentierung mittels künstlicher Mutanten von *Streptomyces erythreus* mit blockierter Erythromycinsynthese zugänglich waren (*192, 193, 194*). Die letzteren Produkte wurden ihrerseits wieder mannigfaltig chemisch umgewandelt.

Den Ausgangspunkt der Betrachtungen bildete ein Modell, das einem ungestörten Diamantmuster entsprach. Dieses erwies sich aber als energetisch ungünstig, da vor allem die pseudoaxiale Anordnung der Methylgruppen an den C-Atomen 2, 4 und 6 zu starken 1,3-Wechselwirkungen Anlaß gaben. Eine erste Verfeinerung des Modells erfolgte daher durch eine derartige Verdrillung, daß diese starken Wechselwirkungen minimalisiert wurden. Eine weitere Verfeinerung ergab sich aus einer eingehenden Analyse der NMR.-Spektren zahlreicher Derivate. Die erste Aufgabe bestand darin, sämtliche Signale den zahlreichen Protonen exakt zuzuordnen. Wegen der zahlreichen gleichartigen chemischen Gruppierungen war dies bei Spektren, die bei 60 oder 100 MHz aufgenommen worden waren, nicht möglich, da viele Signale zu unentwirrbaren Signalhaufen überlappten. Erst bei 220 MHz gemessene Spektren gaben eine genügende Auflösung. Bei der Zuordnung der Signale spielte die Doppelresonanz-Spektrometrie eine wesentliche Rolle, die schließlich auch den zweiten Schritt ermöglichte: die Bestimmung der Kopplungskonstanten zwischen allen benachbarten Ringprotonen, die unmittelbar mit der Konformation zusammenhängen.

Der erste wesentliche Befund war die Feststellung einer weitgehenden Invarianz der Kopplungskonstanten vergleichbarer Protonenpaare in den verschiedenen Verbindungen, woraus auf eine weitgehende Unabhängigkeit der Konformation von kleinen chemischen Veränderungen geschlossen werden konnte (*248, 249, 250*). Das Vorhandensein oder Fehlen einzelner Hydroxylgruppen, die Acetylierung von Hydroxylgruppen und selbst Glykosidreste vermögen die Konformation nur geringfügig zu verändern. Und sogar die Reduktion der Ketogruppe an C-9 zu zwei epimeren Carbinolgruppen ist ohne entscheidenden Einfluß auf die Konformation des Ringgerüstes. Die Aufnahme von Spektren in Lösungsmitteln verschiedener Polarität gaben ebenfalls ein recht einheitliches Bild. Das bedeutet, daß die Konformation auch von Wechselwirkungen mit Lösungsmittelmolekülen weitgehend unabhängig ist. Die geringe Temperaturabhängigkeit der Spektren bewies ferner, daß kein Gleichgewicht verschiedener Konformationen ähnlicher Stabilität vorliegt, sondern daß eine einzige, relativ starre Konformation vorherrscht. Dieser Befund überrascht, da Molekülmodelle von Makroliden recht flexibel sind. Offensichtlich sorgt ein Zusammenwirken von zahlreichen nichtklassischen Spannungen und möglicherweise Dipol-Dipol-Wechselwirkungen dafür, daß *eine* Konformation gegenüber allen anderen energetisch deutlich bevorzugt ist.

(31)

6-Desoxy-erythronolid B

Für eine exakte Konformationsanalyse eignet sich das 6-Desoxy-erythronolid B (**31**) besonders gut, da hier mit Ausnahme des C-Atoms 9 alle Ringkohlenstoffatome mindestens ein Wasserstoffatom tragen, während die Kette vicinaler Protonen etwa im Erythromycin (**30**) mehrfach unterbrochen ist. Die Verbindung **31** ist eine Zwischenstufe der Erythromycin-Biosynthese und wird von einer Mutante von *Streptomyces erythreus* angereichert (*194*). In der Tabelle 1 sind die Kopplungskonstanten aller Ringprotonen registriert, die zur Verfeinerung des Modells herangezogen wurden (*249*), während Abb. 1 eine mit den experimentellen Befunden bestens vereinbare Konformation wiedergibt. Verschiedene andere gelegentlich diskutierte Konformationen stehen zu den experimentellen Daten im Wiederspruch.

Tabelle 1 (*25*). *NMR.-Daten für 6-Desoxyerythronolid B in CDCl₃*

Protonen-Kopplung	Kopplungs-konstanten	Protonenbeziehung
$J_{2,\,3}$	10,5	a, a
$J_{3,\,4}$	<1	a, e
$J_{4,\,5}$	2,5	e, a
$J_{5,\,6}$	4,7	a, e
$J_{6,\,7a}$	4,7	e, a
$J_{6,\,7e}$	10,2	e, e gegenst.
$J_{7a,\,7e}$	15	geminal
$J_{7a,\,8}$	13	a, a
$J_{7e,\,8}$	4,0	e, a
$J_{10,\,11}$	2,0	a, e
$J_{11,\,12}$	10,2	e, e gegenst.
$J_{12,\,13}$	1,5	e, a

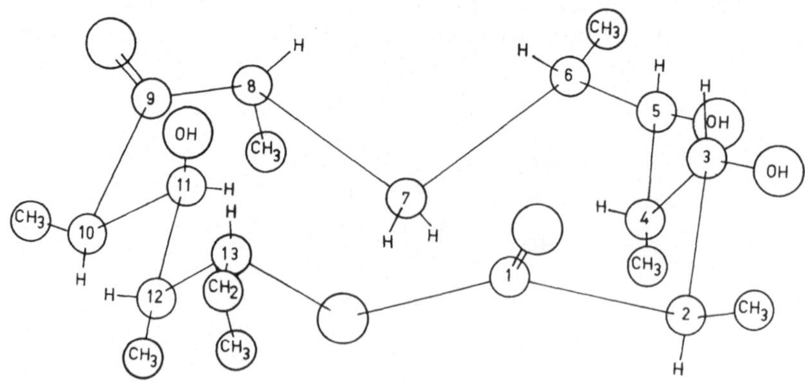

Abb. 1

Bei der Beurteilung der Konformation aus den Kopplungskonstanten gemäß der Beziehung von KARPLUS (*158*) ist im Falle eines 14gliedrigen Ringes zu beachten, daß im Gegensatz zum Cyclohexanring auch zwei benachbarte *äquatoriale* (e) Wasserstoffatome in antiperiplanarer Anordnung zueinander stehen können (e, e gegenständig in Tabelle 1), was z. B. bei den H-Atomen an C-11 und C-12 (Abb. 1) schön zum Ausdruck kommt. Die Tabelle 1 zeigt, daß anti-periplanare Beziehungen zwischen den Wasserstoffpaaren 2—3, 6—7e, 7a—8, 11—12 bestehen müssen (J über 10), was im Modell (Abb. 1) berücksichtigt ist. Besonders kleine Kopplungskonstanten (unter 2 Hz.) finden wir zwischen den Wasserstoffpaaren 3—4, 10—11 und 12—13, was durch C-H-Torsionswinkel um 90° herum im Modell zum Ausdruck kommt.

Die einzige Abweichung von der Konstanz der Kopplungskonstanten, die die Protonen an C-4 und C-5 betrifft, findet eine zwanglose Erklärung ebenfalls aus der angegebenen Konformation. Während die Kopplungskonstante $J_{4,5}$ bei Aglykonderivaten (freie Hydroxyle an C-3 und C-5) etwa 2,5 Hz. beträgt, steigt sie bei Diglykosiden (Cladinoserest an C-3, Desosaminrest an C-5) bis gegen 7 Hz. an. In O-Acetylderivaten (OAc an C-3 und C-5) ist die Abweichung etwas geringer: $J_{4,5}$ etwa 5,5 Hz. Bei Monoglykosiden (OH an C-3, Desosaminylrest an C-5) kommt die Kopplungskonstante dem Normalwert von 2,5 Hz. wieder nahe. Cyclische Derivate, bei denen die Sauerstoffatome Glieder eines Fünfrings sind (z. B. Phenylboronate, s. unten) besitzen eine Kopplungskonstante von etwa 2,5 Hz.

Eine plausible Erklärung für diese Beobachtungen besteht darin, daß die sperrigen Glykosidreste an C-3 und C-5 eine geringfügige Verdrillung um die C-4—C-5-Bindung bewirken, ohne die übrige Konformation in erheblichem Maße zu stören. Acetylreste verursachen eine ähnliche Verdrillung, aber in geringerem Ausmaß.

Mit der aus NMR.-Spektren abgeleiteten Konformationen (Abb. 1) sind Studien über den Cirkulardichroismus von MITSCHER (*207, 249*) in bestem Einklang. Der konformativen Einheitlichkeit im Bereich der erythromycinartigen Makrolide entspricht eine Uniformität im Verlauf der Kurven aller untersuchten Vertreter der Gruppe mit negativen Cotton-Effekten bei etwa 290 nm (Keton-Chromophor) und 225 nm (Lacton-Carbonylgruppe). Diese Werte sind auf Grund einer erweiterten Octantregel und der Lactonregel gut verträglich mit der angenommenen Konformation; und geringfügige Unterschiede in den Amplituden sind zwanglos interpretierbar.

Zur Stützung der Konformation (Abb. 1) konnten weiter zahlreiche Beobachtungen über chemische Reaktivitäten angeführt werden. Von den zahlreichen Beispielen sei hier nur ein besonders instruktives herausgegriffen: Das 6-Desoxy-erythronolid B (**31**) wird von Natriumborhydrid zu den beiden epimeren 9-Dihydroverbindungen (**32** und **33**) reduziert, die sich durch Chromatographie trennen lassen. Während das 9-(*S*)-Dihydro-6-desoxyerythronolid (**32**) mit Phenylboronsäure glatt ein Bis-phenylboronat (**34**) bildet, kann unter den gleichen Reaktionsbedingungen aus dem 9-(*R*)-Epimeren (**33**) nur ein Mono-phenylboronat (**35**) erhalten werden. Die unterschiedliche Reaktivität ist aus einer stabilen Konformation **35** leicht verständlich.

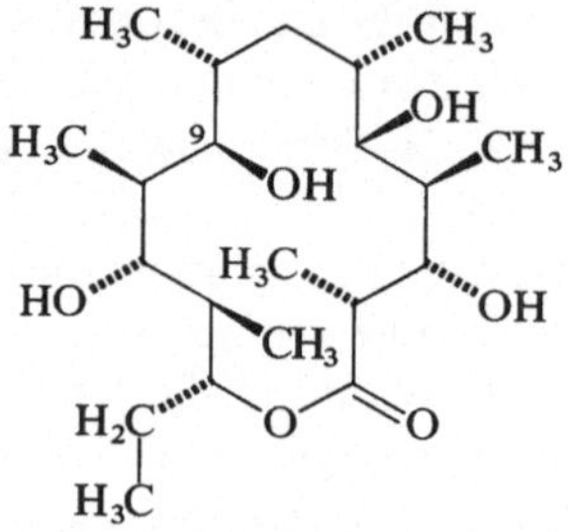

(**32**)

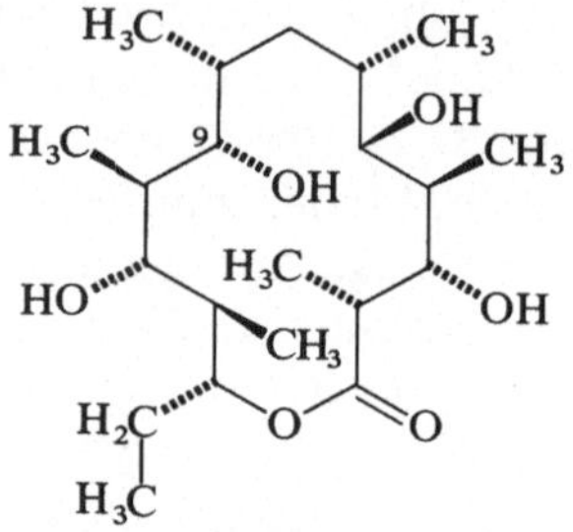

(**33**)

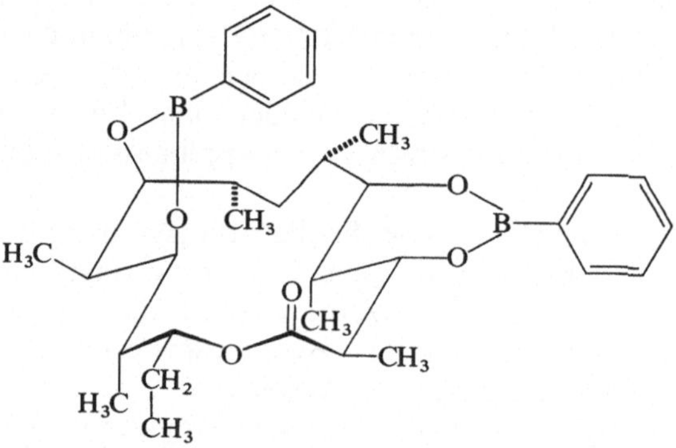

(**34**)

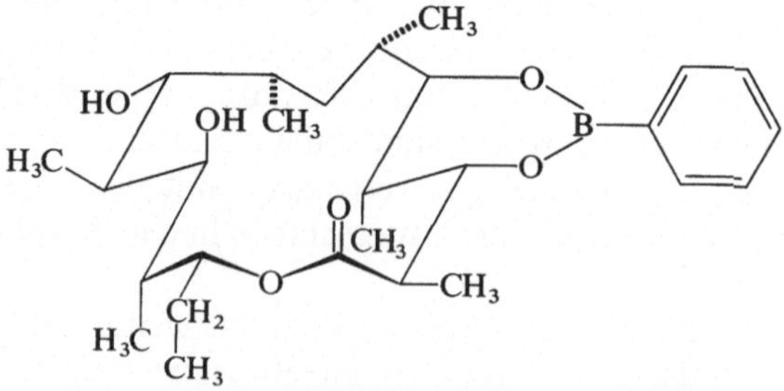

(**35**)

Schema 8. Bildung cyclischer Phenylboronate

Eine ähnliche Konformation hat, ebenfalls vorwiegend aus der Analyse von NMR.-Spektren DEMARCO (*76, 77*) abgeleitet. An der Konformation (Abb. 1) fällt auf, daß sie von derjenigen im Kristallgitter (*128*) nur wenig abweicht. Wenn die Konformation in flüssiger Phase tatsächlich so starr ist, wie sich den Untersuchungen von PERUN und Mitarb. entnehmen läßt, ist dieser Befund aber kaum verwunderlich. Ein weiterer Befund am Modell in Abb. 1 wird möglicherweise für das Verständnis der biologischen Wirkung bedeutungsvoll werden. Wenn man durch den Makrolidring eine „beste Ebene" legt, die das Molekül in eine obere und eine untere Hälfte trennt, dann fällt auf, daß alle polaren Gruppen (OH, C=O) in die obere Hälfte zu liegen kommen, während die untere Hälfte weitgehend unpolar ist. Diese Anordnung birgt möglicherweise den Schlüssel für die Aufdeckung der Wechselwirkung zwischen Antibioticum und Ribosom in sich. In diesem Zusammenhang ist es vielleicht kein Zufall, daß das C-Atom 7 des Erythromycins und der meisten seiner Verwandten keine Sauerstoffunktion trägt, obwohl es sich biogenetisch von einer Carboxylgruppe der Propionsäure ableitet. Eine Hydroxylgruppe an C-7 würde nämlich in die untere (unpolare) Molekülhälfte zu liegen kommen.

c) Weitere Antibiotica der Erythromycin-Gruppe

Das Erythromycin (oft Erythromycin A genannt) wird von *Streptomyces erythreus* zusammen mit zwei Begleit-Antibiotica gebildet, die aus den Mutterlaugen durch Chromatographie in reiner Form isoliert werden konnten (*251, 318*). Ihre Konstitutionsaufklärung durch chemischen Abbau erfolgte in engem Zusammenhang mit derjenigen des Hauptantibioticums. Das Erythromycin B (**36**) unterscheidet sich chemisch von ihm durch das Fehlen der Hydroxylgruppe an C-12 (*116, 322*). Das Erythromycin C (**37**) besitzt das gleiche Aglykon wie das Erythromycin A, hingegen ist der Cladinoserest durch den Mycaroserest (ohne die Methylgruppe an O-C-3′) ersetzt (*145, 318*). Die Konfiguration und Konformation dieser beiden Nebenantibiotica stimmt gemäß den oben ausführlich geschilderten Untersuchungen von PERUN, EGAN und Mitarb. völlig mit der des Erythromycins A überein.

Unter allen Verwandten des Erythromycins nehmen das Methymycin (**38a**) und sein Begleiter Neomethymycin (**38b**) eine Sonderstellung ein, indem sie nur einen 12gliedrigen Lactonring aufweisen. Die Konstitutionsaufklärung der Methymycine gehört mit zu den Pionierarbeiten der Makrolidchemie (*85, 86, 87, 91*). Die vollständige Stereochemie wurde aber erst in neuerer Zeit durch eine Kombination von Abbaureaktionen und NMR.-Spektren abgeleitet (*189, 263*). Die Chiralität an vergleichbaren Zentren stimmt mit derjenigen der Erythromycine überein.

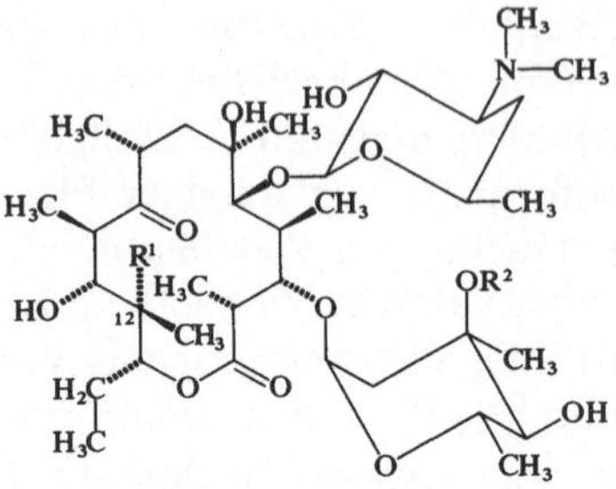

Schema 9. Mit Erythromycin verwandte Antibiotica

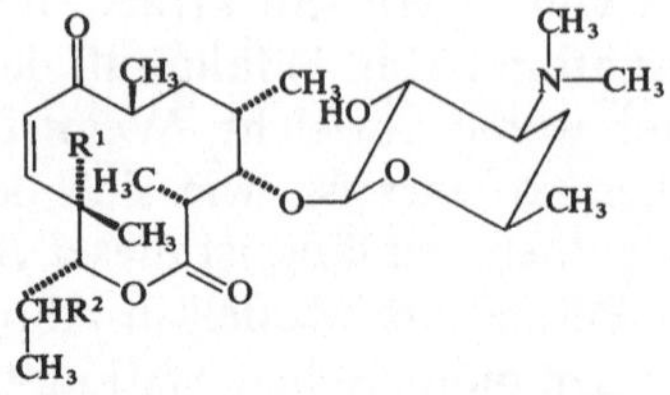

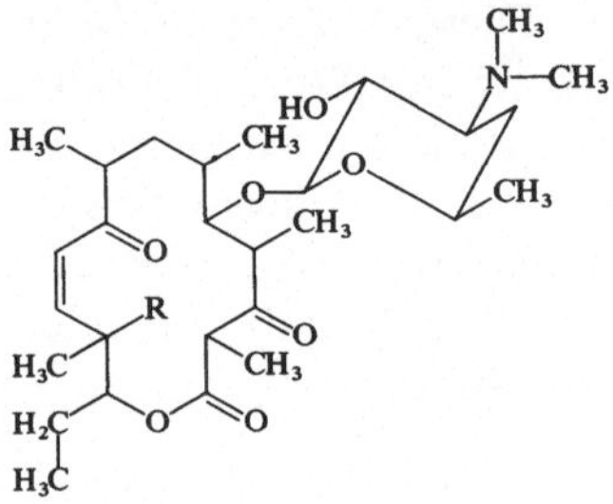

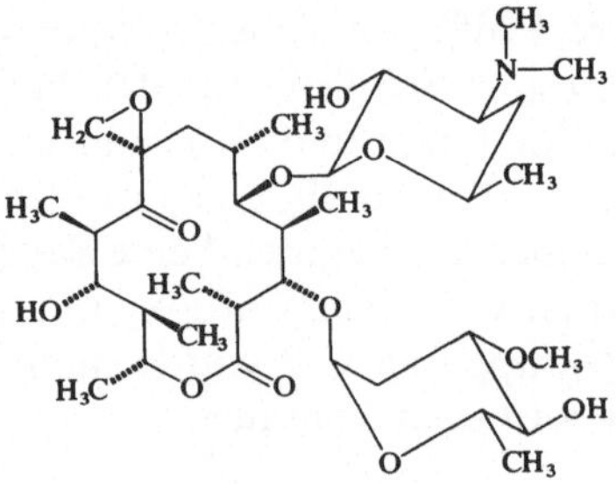

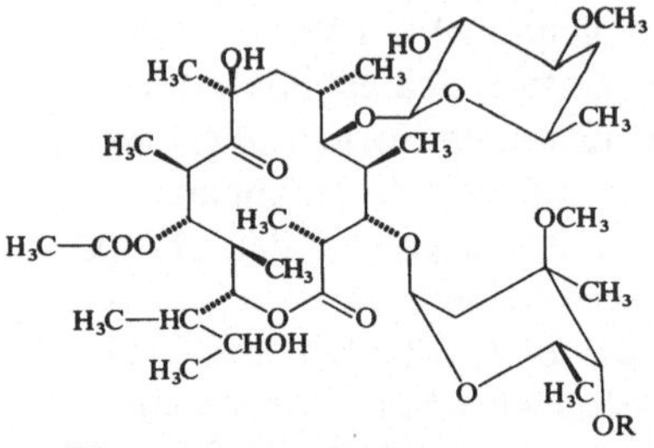

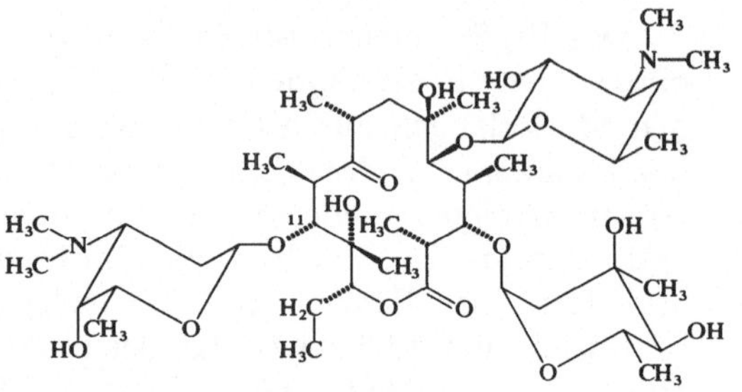

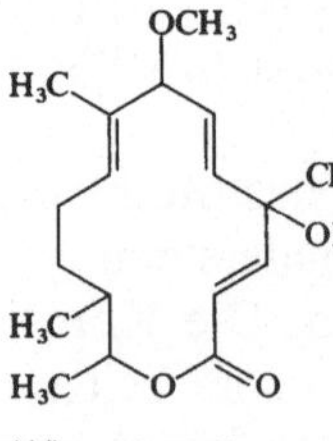

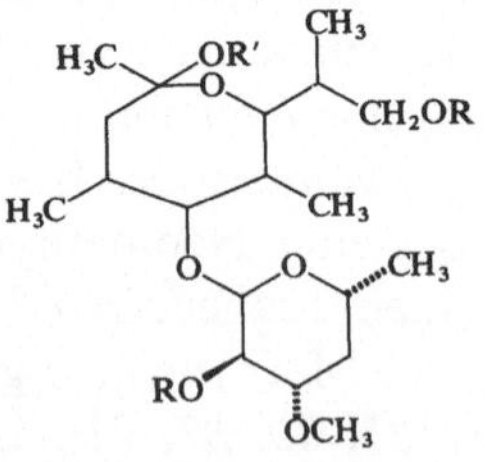

Literaturverzeichnis: SS. 445—460

Das Narbomycin (**39a**) ist das am einfachsten gebaute Glykosidmakrolid mit einem vierzehngliedrigen Ring, indem es nur einen Zuckerbaustein (Desosamin) enthält (*71, 254*).

Die Konstitutionsaufklärung des Pikromycins (**39b**) geht im wesentlichen auf zahlreiche Abbaureaktionen von BROCKMANN und Mitarb. (*38, 39, 41, 42*) sowie ANLIKER und Mitarb. (*8, 9, 10*) zurück. Auf Grund der damals noch ungenauen Methoden zur Molekulargewichtsbestimmung wurde eine Bruttoformel $C_{25}H_{43}NO_6$ und eine Struktur isomer zu derjenigen des Methymycins abgeleitet. Der Fehler wurde auch viel später von OGURA und Mitarb. (*224*) nicht aufgedeckt, obwohl sie bei der Konstitutionsaufklärung des Amaromycins (=Pikromycin) die NMR.-Spektroskopie eingehend anwendeten. Erst zwei unabhängige massenspektrometrische Untersuchungen zeigten, daß dem Pikromycin tatsächlich die Bruttoformel $C_{28}H_{47}NO_7$ mit dem Mol-Gewicht 525 zukommt (*216, 264*). Durch Neuinterpretation der früheren Abbauergebnisse und mit Hilfe von NMR.-Spektren war es dann einfach, unzweideutig die Struktur **39b** zu beweisen.

Die Stereochemie von Narbomycin und Pikromycin ist erst in Fragmenten bekannt (*88, 263*), doch dürfte sie auf Grund biogenetischer Überlegungen mit derjenigen der Erythromycine übereinstimmen (*61*).

Von einiger chemotherapeutischer Bedeutung ist, neben dem Erythromycin, in dieser Gruppe auch das Oleandomycin (**40**). Auffallende Merkmale sind der Epoxidring und der relativ einfache Zucker, L-Oleandrose, der anstelle der Cladinose steht (*62, 100, 140*). Die Konfigurationsaufklärung (*59*), die völlig unabhängig von der des Erythromycins und mit anderen Methoden durchgeführt wurde, ergab völlige Übereinstimmung an allen Chiralitätszentren.

Die Erkenntnis, daß das Lankamycin (*144*) zu den Makroliden der Erythromycingruppe gehört, bildete seinerzeit eine kleine Überraschung, weil es als neutrale Verbindung zunächst gar nicht ins damalige Bild der Makrolide als einer Gruppe von stickstoffhaltigen Basen paßte. Das Lankamycin enthält nämlich anstelle des Desosamins den stickstofffreien Zucker **D**-Lankavose (*152*). Der Zuckerbaustein an C-3 ist die 4-O-Acetyl-L-arcanose, die mit der Cladinose nahe verwandt ist. Die ursprünglich abgeleitete Struktur (*166*) erfuhr kürzlich auf Grund einer eingehenden NMR.-spektroskopischen Analyse mit Spinentkopplung eine geringfügige Korrektur zur Formel **42**, indem die Stellung der beiden Zucker vertauscht werden mußte (*97*).

Als Folge dieser Revision müssen auch zahlreiche Abbauprodukte der ursprünglichen Arbeit (*166*) neu formuliert werden. Die einschneidendsten Veränderungen erfahren die cyclischen Produkte XXXI bis XXXIII (*166*), die jetzt einen siebengliedrigen statt eines fünfgliedrigen Cycloketalringes gemäß Formel (**45**) besitzen.

Das gründliche Studium von hochaufgelösten NMR.-Spektren mit Hilfe der Doppelresonanztechnik erlaubte auch, durch Vergleich der Kopplungskonstanten mit denen der Erythromycinderivate die Konfiguration des Lankamycins mit Ausnahme der beiden Chiralitätszentren in der Seitenkette abzuleiten. Und schließlich zeigten diese Untersuchungen von EGAN und MARTIN (*97*), daß auch die Konformation in Lösung derjenigen der Erythromycine entspricht (vgl. Abschnitt I/1/b).

Das Kujimycin A, das in einem Falle als Begleiter des Lankamycins isoliert wurde (*234*), erwies sich als ein 4'-O-Desacetyllankamycin (**41**). Auf Grund der Untersuchungen von EGAN sind auch in der publizierten Formel des Kujimycins A (*231*) die beiden Zucker zu vertauschen.

Der am kompliziertesten gebaute Vertreter der Erythromycin-Gruppe ist das Megalomycin A, das zusammen mit mehreren noch nicht näher untersuchten verwandten Verbindungen von einem *Micromonospora*-Stamm gebildet wird (*190, 315*). (*Micromonospora* ist eine mit *Streptomyces* verwandte Gattung der Familie Actinomycetaceae). Dieses Antibioticum (**43**) ist ein Derivat des Erythromycins, das in Stellung 11 einen zusätzlichen Aminozucker, D-Rhodosamin, trägt (*186, 187*).

Ein strukturell der Erythromycingruppe nahestehendes Antibioticum mit einem 14gliedrigen Lactonring aber ohne Zuckerreste ist das Albocyclin (*217*), für das die Struktur **44** abgeleitet wurde (*218, 219, 220, 256*). Mit dem Albocyclin verwandt ist das Cineromycin B (Des-O-methylalbocyclin).

2. Die Carbomycin-Gruppe

Diese Gruppe umfaßt eine Reihe von Glykosiden 16gliedriger Makrolactone. Die antibiotischen Wirkungsspektren ähneln denen der Erythromycin-Gruppe auffallend mit vorwiegender Aktivität gegen Gram-positive und säurefeste Bakterien. Vielversprechend schien zunächst auch eine Beobachtung über eine Aktivität gegenüber Rickettsien (*94*), doch haben sich diese Antibiotica anscheinend als spezifische Mittel gegen Rickettsiosen nicht durchgesetzt.

Innerhalb dieser Untergruppe besteht ein besonders enger struktureller Zusammenhang zwischen den Carbomycinen A und B (=Magnamycin A und B) (*94, 141, 290, 309*), den Spiramycinen I, II und III (=Foromacidine A, B und C) (*72, 252*) und den Leucomycinen A_1 bis A_9 (*130, 131, 132, 226*). In engem Zusammenhang stehen daher auch die Strukturaufklärungen dieser drei Typen von Antibiotica. Ursprünglich wurden zwar die Arbeiten getrennt in Angriff genommen. Vollständige Klarheit über die Strukturen aller Glieder dieser Familie (s. Formelschema 10) wurde aber erst erhalten, als es gelang, sie durch chemische Reaktionen zueinander in Beziehung zu bringen.

Literaturverzeichnis: SS. 445—460

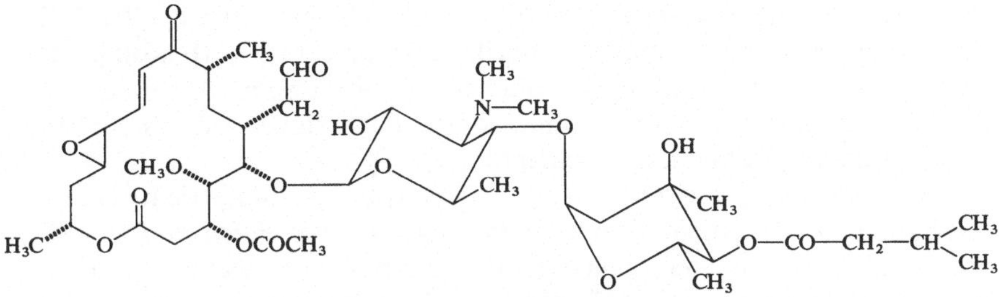

(**46**)   Carbomycin A (Magnamycin)

KJ oder
CrCl$_2$ in CH$_3$COOH

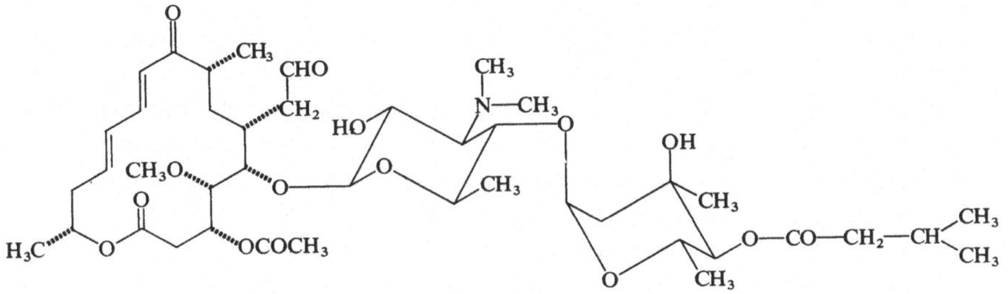

(**47**)   Carbomycin B (Magnamycin B)

MnO$_2$ in Chlf.

(**49**)   Leucomycin A$_3$

Schema 10. Die Strukturen der Carbomycine, Leucomycine und Spiramycine
Verknüpfung der Carbomycine und der Spiramycine mit den Leucomycinen
(Fortsetzung s. S. 339 und 340)

a) Carbomycine, Leucomycine und Spiramycine

Es wäre für die Übersichtlichkeit dieses Referates nicht von Nutzen,
dem historischen Gang der Ereignisse zu folgen und alle Wege und Irr-
wege der Strukturermittlung im einzelnen zu schildern. Es sei mir daher
erlaubt, die Geschichte am Ende zu beginnen, mit der chemischen Ver-

knüpfung der verschiedenen Antibiotica der Gruppe, anschließend die Strukturermittlung der zuletzt eingehend erforschten Verbindung, des Leucomycins A_3, auszuführen, und zuletzt die älteren Arbeiten über die Carbomycine und Spiramycine soweit darzustellen, als es für den Zusammenhang notwendig erscheint.

Die Beziehung zwischen dem Carbomycin A (**46**), dem Hauptantibioticum aus einem Stamm von *Streptomyces halstedii* (*94, 290, 309*), und einem aus dem gleichen Stamm erhaltenen zweiten Antibioticum, dem Carbomycin B ((**47**) (*141*), wurde schon früh erkannt, indem es gelang, das Carbomycin A mit einer α,β-ungesättigten Ketogruppe (UV.-Absorptionsmaxima 238 [log ε 4,22] und 327 nm [log ε 1,88]) und einem Epoxid-Sauerstoffatom mittels Kaliumjodid zum Carbomycin B mit einer konjugierten Dienongruppe (λ_{max} 278 nm, log ε 4,39) zu reduzieren (*329*). Später konnte die gleiche Reduktion auch mit Chromdichlorid in essigsaurer Lösung durchgeführt werden (*175*). Mit dem Carbomycin B war anderseits das Oxydationsprodukt identisch, das aus Leucomycin A_3 durch Erhitzen mit aktivem Mangandioxid in Chloroform erhalten wurde (*131, 230*). Der Zusammenhang zwischen den Leucomycinen A_1 bis A_9 (**48** bis **55**) ergab sich, abgesehen von weitgehenden spektroskopischen Übereinstimmungen, durch teilweise identische Acetylierungsprodukte (A_1 und A_3; A_4 und A_5; A_6 und A_7; A_8 und A_9) und teilweise identische Produkte der partiellen Methanolyse (A_1, A_5, A_7 und A_9 einerseits; A_3, A_4, A_6 und A_8 anderseits).

Die Überführung des Spiramycins II (**57**) und des Leucomycins A_3 (**49**) in ein gemeinsames Umwandlungsprodukt gestaltete sich schwieriger, weil unter den Bedingungen, die für die hydrolytische Abspaltung des Aminozuckers notwendig waren, das Aglykon unstabil ist und eine Allylumlagerung erleidet (s. unten). Eine einwandfreie Verknüpfung gelang erst, als man für die Partialhydrolyse anstelle des Spiramycins II (**57**) sein Tetrahydroderivat (**57a**), das durch Hydrierung mit einem Palladiumkatalysator zugänglich ist, einsetzte (*233*). Neben dem Aminozucker Forosamin (*60*) und der Mycarose (*61*) erhielt man eine lipophile Base, das Tetrahydroforocidin B (**59**), das sich noch aus dem Tetrahydroaglykon und dem Aminozucker Mycaminose zusammensetzt. Das Tetrahydroforocidin B erwies sich als identisch mit dem entsprechenden Hydrolyseprodukt des Tetrahydroleucomycins A_3 (**62**). Im letzteren Fall wurde als zweites Spaltprodukt die 4-O-Isovaleroylmycarose erhalten.

Das Leucomycin wurde beinahe zur gleichen Zeit zum erstenmal isoliert wie das Carbomycin (*132*). Bald wurde auch seine Verwandtschaft mit dem Erythromycin und dem Carbomycin erkannt und gleichzeitig die Verschiedenheit von den damals bekannten Makroliden nachgewiesen (*271, 273*). Eine eingehende chemische Untersuchung wurde aber dadurch verzögert, daß das Leucomycin ein schwer trennbares Gemisch

Schema 10 (1. Fortsetzung)

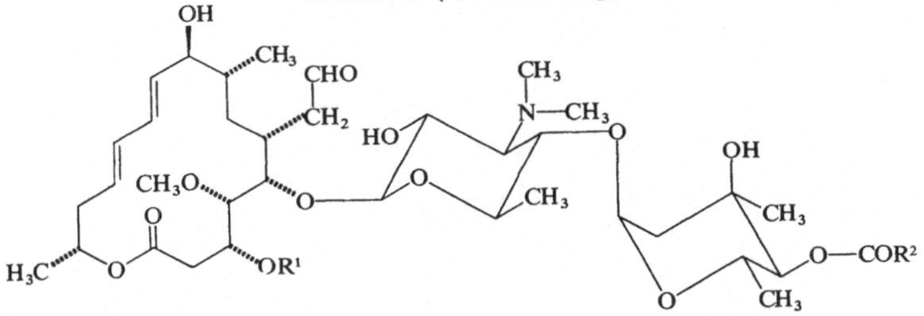

		R^1	R^2
(48)	Leucomycin A_1	H	$-CH_2-CH\langle{CH_3 \atop CH_3}$
(49)	Leucomycin A_3	$-COCH_3$	—do—
(50)	Leucomycin A_5	H	$-CH_2-CH_2-CH_3$
(51)	Leucomycin A_4	$-COCH_3$	—do—
(52)	Leucomycin A_7	H	$-CH_2-CH_3$
(53)	Leucomycin A_6	$-COCH_3$	—do—
(54)	Leucomycin A_9	H	$-CH_3$
(55)	Leucomycin A_8	$-COCH_3$	—do—

(56)	Spiramycin I (Foromacidin A)	$R=H$
(57)	Spiramycin II (Foromacidin B)	$R= -COCH_3$
(58)	Spiramycin III (Foromacidin C)	$R= -CO-CH_2-CH_3$

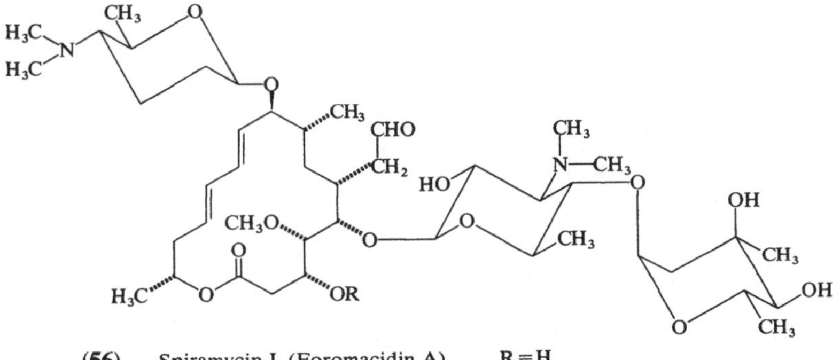

zahlreicher miteinander nahe verwandter Verbindungen darstellt (*272*). Die Untersuchungsergebnisse blieben daher lange Zeit unklar und fragmentarisch (vgl. z. B. *311, 312, 313*).

Die Auftrennung des Gemisches in acht chemisch einheitliche Verbindungen durch Chromatographie an Kieselgel gelang erst HATA, OMURA und Mitarb. (*130, 131, 226*), denen wir auch die rasch anschließende Konstitutionsaufklärung der Hauptkomponente, A_3, verdanken (*228, 229, 230*). Mit Leucomycin A_3 identisch ist das Antibioticum Josamycin (*225, 236*).

Schema 10 (2. Fortsetzung)

(**47**)   Spiramycin II

H_2, Pd

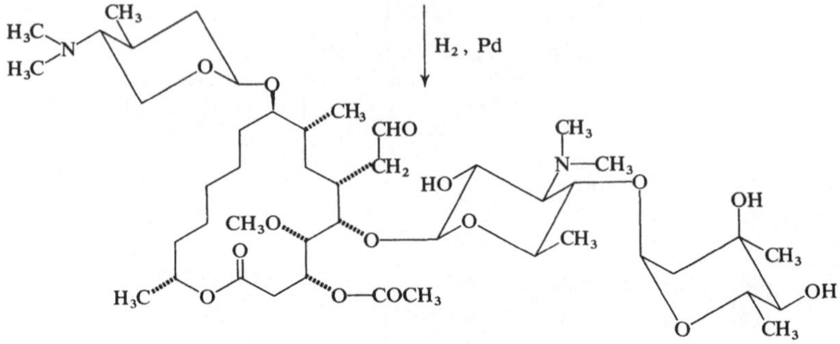

(**57a**)   Tetrahydrospiramycin II

0,3 NHCl, 60°

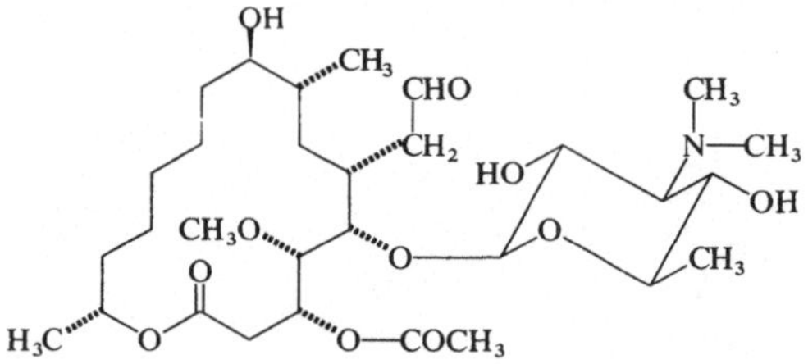

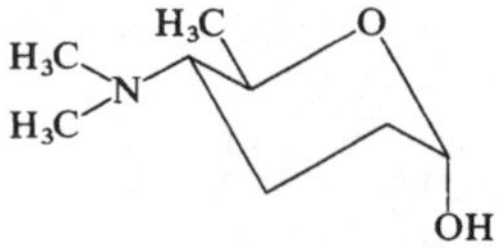

(**59**)   Tetrahydroforocidin B

(**60**)   Forosamin

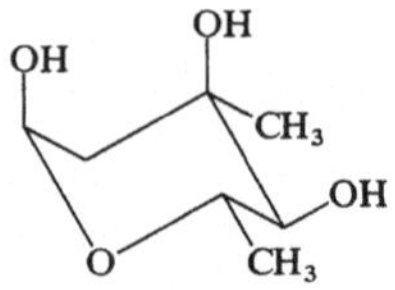

(**61**)   Mycarose

0,2 N HCl

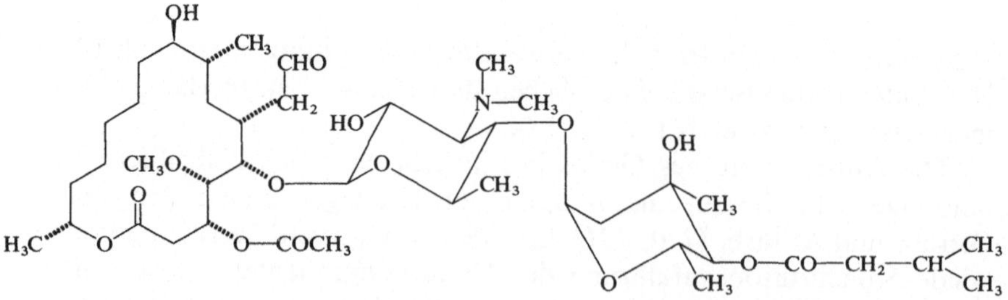

(**62**)   Tetrahydroleucomycin A_3

Die Analysen der farblosen, lipophilen Base mit Smp. 120 bis 121° ließen, wie das im Bereich der Makrolid-Antibiotica und Verbindungen ähnlicher Zusammensetzung die Regel ist (*329*), nicht sofort eine eindeutige Bruttoformel errechnen. Diese, $C_{42}H_{69}NO_{15}$ war daher erst gesichert, als nach eingehenden Abbauversuchen eine Struktur wenigstens in groben Zügen abgeleitet werden konnte. Hingegen ergab eine Methoxylbestimmung nach ZEISEL die Anwesenheit einer O-Methylgruppe. Die alkalische Hydrolyse, die drei Mol Natronlauge verbrauchte, ergab beinahe zwei Mol flüchtige Säuren, die durch Papierchromatographie und die NMR.-Spektren der Natriumsalze als Essigsäure und Isovaleriansäure identifiziert wurden.

Aufschlußreich über zahlreiche Atomgruppierungen waren die spektroskopischen Eigenschaften des Leucomycins (*228*). Das IR.-Absorptionsspektrum zeigt mit Banden bei 1728 und 1746 cm^{-1} mindestens zwei Carbonylgruppen an, während ein starkes Absorptionsmaximum bei 1230 cm^{-1} bedeutet, daß der analytisch nachweisbare Acetylrest in einer Estergruppe anzutreffen ist. Eine schwache Bande bei 2725 cm^{-1} ist charakteristisch für Aldehyde, eine solche bei 1661 cm^{-1} für eine oder mehrere Doppelbindungen. Daß das Doppelbindungssystem in einer konjugierten Diengruppe vorliegt, geht aus dem UV.-Absorptionsspektrum mit einem Maximum bei 231,5 nm (log ε 4,46) hervor.

Im NMR.-Spektrum werden die Methoxylgruppe durch ein Singulett (Integral 3 H) bei 3,47 ppm und die Acetylgruppe durch ein solches bei 2,22 ppm bestätigt. Weiter ergibt sich die Anwesenheit von zwei N-Methylgruppen durch ein Singulett bei 2,49 ppm (6 H). Die beiden N-Methyle müssen wegen der Anwesenheit von nur einem Stickstoffatom in einer Dimethylaminogruppe enthalten sein. Das Aldehyd-Wasserstoffatom gibt ein Singulett bei 9,56 ppm. Im Thiosemicarbazon des Leucomycins A$_3$ ist das Signal des entsprechenden Protons (—CH=N—) in ein Triplett mit J=5 Hz. aufgespalten, was beweist, daß in Nachbarschaft dazu eine Methylengruppe liegt: —CH$_2$-CHO. Im Falle des Carbomycins ist eine analoge Aufspaltung beim Oxim, nicht aber beim freien Aldehyd beobachtet worden (*175*). Die Signale von vier Vinylwasserstoffatomen im Bereich von 5,3 bis 6,7 ppm werden später eingehend zu diskutieren sein.

Im Leucomycin A$_3$ lassen sich zwei Hydroxylgruppen leicht acetylieren. Da sich eine davon später als Bestandteil eines Zuckerbausteins erweisen wird, muß die andere dem Aglykon angehören (*228*). Die Beziehung der letzteren zum Diensystem ergibt sich aus der Oxydation mit aktiviertem Mangandioxid, wobei ein Dienon mit einem hohen UV.-Absorptionsmaximum bei 278 nm (wie Carbomycin B, s. oben) entsteht. Das Di-O-acetyl-leucomycin A$_3$ zeigt im IR.-Absorptionsspektrum noch eine deutliche Hydroxyl-Absorption. Die nicht acetylierbare Hydroxylgruppe ist offensichtlich das tertiäre Hydroxyl im Mycaroseteil des Moleküls (s. unten).

Die beiden Doppelbindungen lassen sich in Gegenwart eines Palladiumkatalysators leicht hydrieren. Das Tetrahydroleucomycin A_3 (62) und sein Diacetylderivat besitzen weder im UV.- noch im IR.-Absorptionsspektrum die charakteristischen Banden für Doppelbindungen. Mit einem Platinkatalysator in Eisessig wird zusätzlich die Aldehydgruppe zu einer primären Alkoholgruppe reduziert; es entsteht das Hexahydroleucomycin A_3.

Eine Methanolyse unter milden Bedingungen spaltet selektiv die Mycarosid-Bindung. Das kleinere Spaltstück ist das α-Methyl-4-O-isovaleroyl-mycarosid (64) (neben geringen Mengen des β-Anomeren), das schon als Baustein des Carbomycins gut bekannt war (s. oben). Das basische Spaltprodukt, $C_{33}H_{57}NO_{12}$ (228), eine lipophile Base, enthält noch den Aminozucker. Der Aglykonteil bleibt bei dieser Reaktion nicht unverändert: zur ursprünglichen O-Methylgruppe sind drei neue hinzugekommen, während die Aldehydgruppe und die sekundäre Hydroxylgruppe nicht mehr vorhanden sind. Demnach wurde der Aldehyd unter den Methanolysebedingungen in das Dimethylacetal (63) übergeführt. Schwerer zu erklären ist die scheinbare Methylierung des sekundären Alkohols. Offensichtlich erfolgt sie unter Allylumlagerung, so daß dem basischen Methanolyseprodukt wahrscheinlich die Konstitution 63 (Formelschema 11) zuzuschreiben ist. Allylumlagerungen der Diencarbinyl-Gruppierung sind bei ähnlichen Reaktionen eingehend studiert worden (s. unten). Bei der energischen sauren Hydrolyse der Base 63 wurde die Mycaminose (65) als krist. Hydrochlorid gebildet, während sich das Aglykon zersetzte. Die Mycaminose war in diesem Zeitpunkt ebenfalls bereits als Baustein anderer Antibiotica bekannt (s. Abschnitt VI, 4).

Gewisse Einzelheiten des NMR.-Spektrums treten beim Abbauprodukt 63 deutlicher in Erscheinung als beim intakten Antibioticum. Insbesondere ist wegen des Fehlens der vier C-Methylgruppen des Isovaleroylmycarose-Teils die Gegend um 1 ppm jetzt eindeutig interpretierbar. Sie enthält drei Dublette von (CH)-CH$_3$-Gruppen. Da eine davon im Mycaminose-Teil enthalten ist, muß das Aglykon deren zwei besitzen.

Schließlich ließ sich eine Lactongruppierung durch eine milde alkalische Hydrolyse des Abbauproduktes 63 nachweisen, wobei sich eine amorphe amphothere Substanz bildete. Die spektralen Eigenschaften lassen noch alle Bauelemente der Verbindung 63 erkennen. Offenbar ist außer der Öffnung eines Lactonringes keine andere Veränderung eingetreten.

Die bisher dargestellten Eigenschaften und einfachen Reaktionen lassen sich in der Partialformel A zusammenfassen (228). Über die Lage

(49) Leucomycin A_3

CH_3OH, HCl

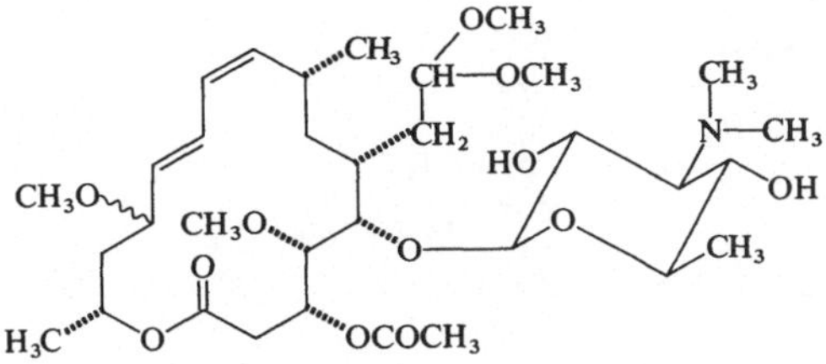

(63) Demycarosyl-O-methyl-iso-leucomycin A_3-dimethylacetal

$+$

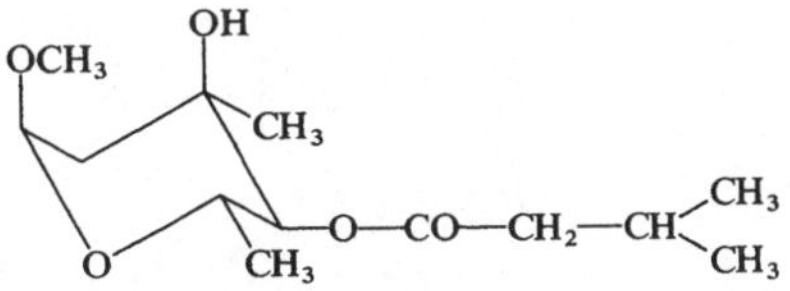

(64) α-Methyl-4-O-isovaleroyl-mycarosid

HCl
in Wasser

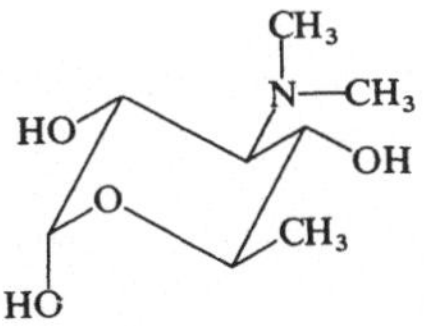

(65) Mycaminose

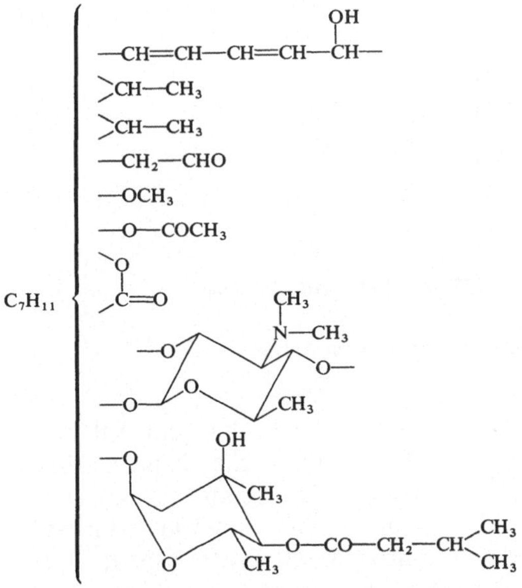

C_7H_{11}

Partialformel A des Leucomycins A_3

Schema 11. Solvolytischer Abbau von Leucomycin A_3

(63) Demycarosyl-O-methyl-iso-leucomycin A_3-dimethylacetal

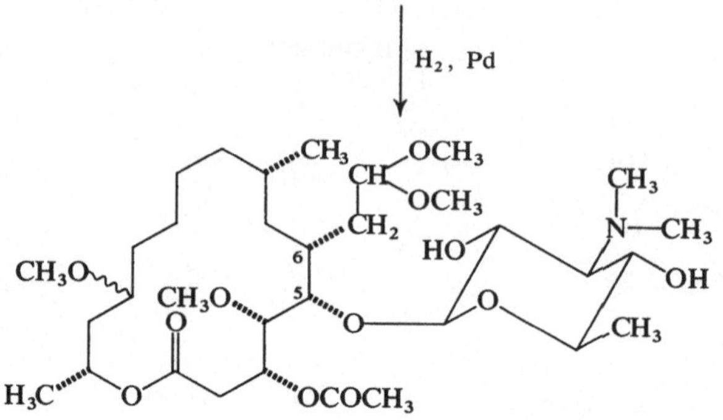

H_2, Pd

(66)

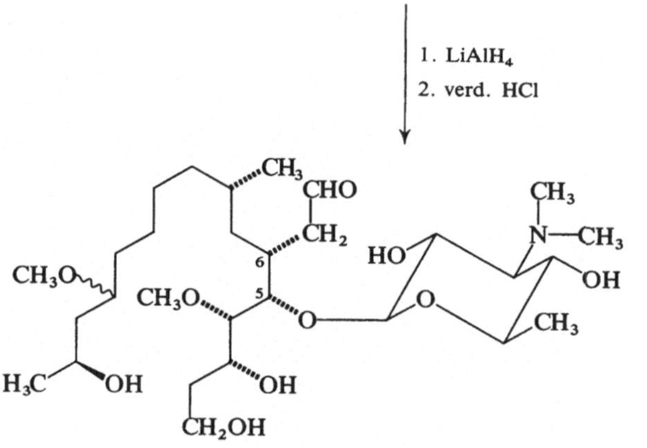

1. $LiAlH_4$
2. verd. HCl

(67) $C_{29}H_{57}NO_{10}$

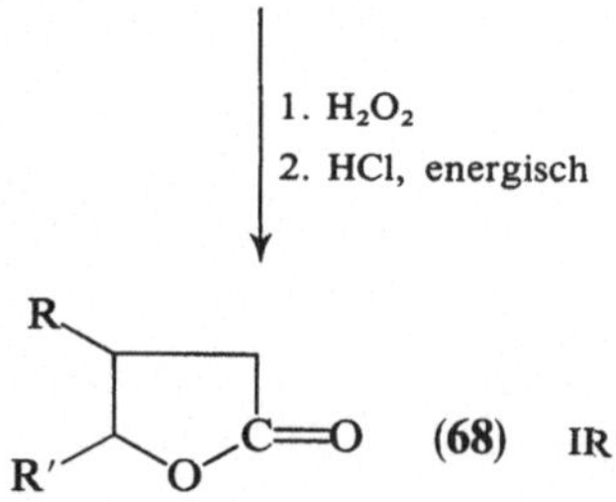

1. H_2O_2
2. HCl, energisch

(68)   IR: 1775 cm^{-1}

Schema 12. Lage der Mycaminose am Aglykon

des Aminozuckers in bezug auf die nächste Nachbarschaft am Aglykon
gab eine Versuchsreihe Auskunft, die sich schon vorher beim Carbomycin bewährt hatte. Das Demycarosyl-O-methyl-tetrahydro-isoleuco-
mycin A_3 (66) wurde zunächst mit Lithiumaluminiumhydrid reduziert.
Bei der milden sauren Aufarbeitung wurde durch Hydrolyse der Di-
methylacetalgruppe die Aldehydgruppe regeneriert. Dem amorphen
Reaktionsprodukt dürfte auf Grund der Elementarzusammensetzung,

$C_{29}H_{57}NO_{10}$, und der spektroskopischen Eigenschaften die Struktur **67** zukommen. Nach der Oxidation der Aldehydgruppe mittels Wasserstoffsuperoxid erfolgte zunächst keine Lactonbildung. Der Ringschluß erfolgt aber spontan nach der hydrolytischen Abspaltung des Aminozuckers. Das Lacton ist auf Grund des IR.-Absorptionsmaximums bei 1775 cm^{-1} ein γ-Lacton (**68**) und bestimmt damit die Lage des Aminozuckers am Sauerstoffatom in γ-Stellung zur Aldehydgruppe.

Die Lage des Isovaleroylmycarose-Restes an einem der beiden Hydroxylsauerstoffe der Mycaminose ergab sich zuerst aus einem Vergleich von pK-Werten des Leucomycins A_3 und einiger seiner Derivate (Tab. 2).

Tabelle 2. *pK-Werte von Leucomycin A_3 und einiger seiner Derivate**

Struktur		pK'a L 50% EtOH	chem. Verschiebung von $N\text{-}CH_3$ ($CDCl_3$)
(**69**)	Leucomycin A_3	6,70	2,49 ppm
(**70**)	Diacetyl-leucomycin A_3	5,69	2,40 ppm
(**71**)	Demycarosyl-leucomycin A_3	7,81	2,60 ppm
(**72**)	Triacetyl-demycarosyl-leucomycin A_3	5,40	2,28 ppm

* Vgl. Formelschema 13.

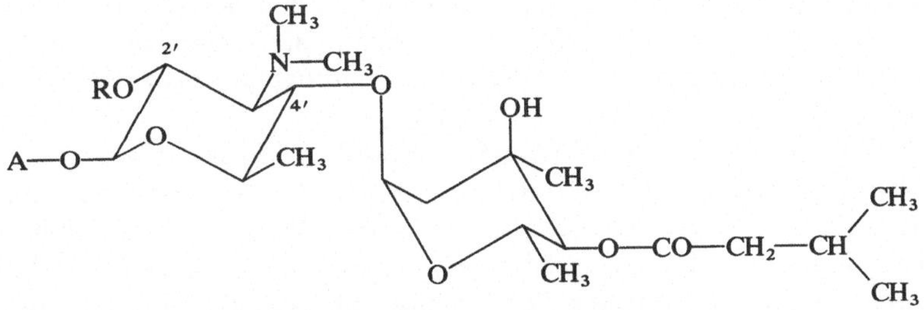

(**69**) R=H (**70**) R=CH_3CO A=Aglykon-Rest

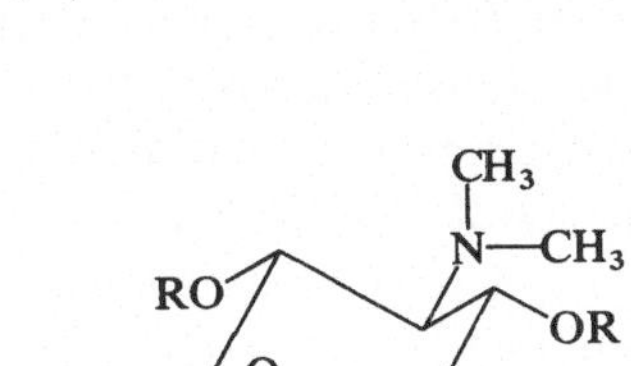

(**71**) R=H (**72**) R=CH_3CO

Schema 13. Partialstrukturen zur Tabelle 2

Die stark unterschiedlichen Basizitäten der vier Verbindungen der Tabelle 2 — über die Herstellung der Abbauprodukte **71** und **72** wird später näher zu berichten sein — lassen sich folgendermaßen erklären: **Die basischste Verbindung der Reihe ist das Demycarosylleucomycin A$_3$ (71)** mit zwei freien Hydroxylgruppen in Nachbarschaft zur Dimethylaminogruppe. Acetylierung der beiden Hydroxylgruppen bewirkt eine Verringerung der Basizität um mehr als zwei pK-Einheiten (**72**), da wegen der elektronenanziehenden Wirkung der Acetylgruppen die Elektronendichte am Stickstoffatom erheblich verringert wird. Aus dem gleichen Grund erfolgt parallel dazu eine drastische Verschiebung des N-CH$_3$-Signals im NMR.-Spektrum nach höherem Feld. Eine etwas geringere Elektronenverschiebung in der gleichen Richtung bewirkt die Glykosidbindung in den Verbindungen **69** und **70**. Diese nehmen daher, in plausibler Reihenfolge, Zwischenstellungen zwischen den beiden Extremen sowohl hinsichtlich der pK-Werte wie der chemischen Verschiebungen der N-Methylsignale ein.

Ähnliche pK-Verschiebungen wurden auch in der Carbomycinreihe beobachtet (*329*). Es erscheint allerdings fraglich, ob dermaßen große Effekte ausschließlich auf induktive Wechselwirkungen zurückgeführt werden dürfen. Dieser Einwand ändert aber nichts an der Reihenfolge der Effekte und deren Interpretierbarkeit durch die Strukturen im Schema 13.

Dadurch wird die Lage des Mycaroserestes an O-C-2′ oder O-C-4′ der Mycaminose klar, ein Entscheid zwischen den beiden Möglichkeiten ergibt sich aus diesen Betrachtungen jedoch nicht. Dieser wurde auf klassische Weise durch totale Methylierung und Abbau erbracht (Formelschema 14).

Bei der Methylierung mittels Methyljodid und Silberoxid werden zwei Hydroxylgruppen veräthert und gleichzeitig die Aminogruppe quaternisiert (**73**). Die energische Hydrolyse des Methiodid-dimethyläthers liefert u. a. das Mycaminosederivat **74** mit einer einzigen O-Methylgruppe. Der Abbau mittels Perjodsäure gab Acetaldehyd, der als 2,4-Dinitro-phenylhydrazon identifiziert wurde. Dieser Abbau beweist, daß die Hydroxylgruppe an C-4′ im Zuckerderivat **74** frei ist und damit diese Stellung als Träger des zweiten Zuckers im Antibioticum feststeht.

Die Frage der Konfiguration an den C-Atomen 1′ und 1″ der beiden Zuckerreste konnte nur für den Mycaminoserest mit Hilfe von NMR.-Spektren gelöst werden. Das Signal von H-C-1′ (Anomeriezentrum der Mycaminose) erscheint im NMR.-Spektrum des Leucomycins A$_3$ sowie zahlreicher Derivate isoliert bei 4,30 ppm als Dublett mit einer Kopplungskonstante von 7,4 Hz., die charakteristisch ist für trans-diaxiale Anordnung in sechsgliedrigen Ringen mit Sessel-Konformation. Damit steht die axiale Lage von H-C-1′ und die β-Konfiguration des Mycaminose-

(49)   Leucomycin A_3

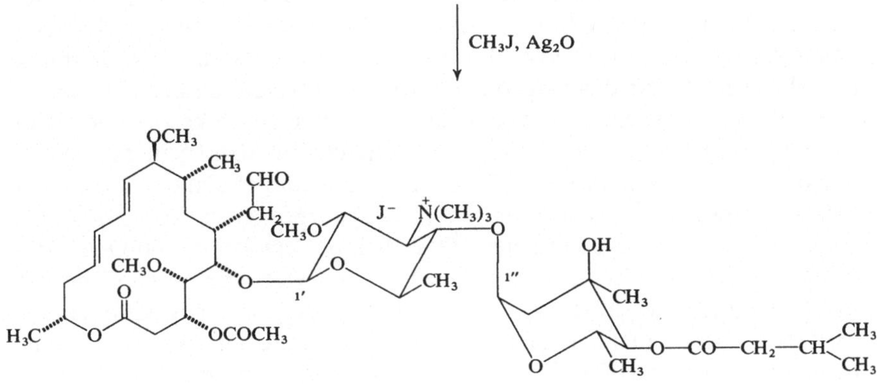

CH_3J, Ag_2O

(73) O,O'-Dimethyl-leucomycin A_3-methiodid

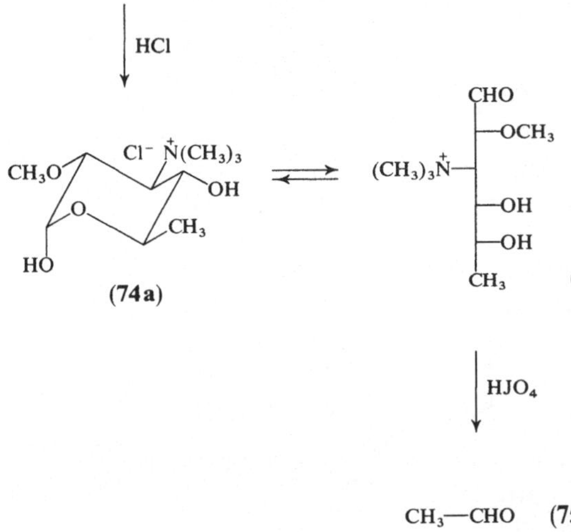

HCl

(74a)   **(74b)**

HJO_4

CH_3-CHO **(75)**

Acetaldehyd
(als 2,4-Dinitrophenylhydrazon identifiziert)

Schema 14. Methylierung und Abbau von Leucomycin A_3

restes fest. Das Signal von H-C-1″ der Mycarose ist in den Spektren des Leucomycins überlagert von anderen Signalen und daher nicht im Detail interpretierbar.

Für die Konfiguration an H-C-1″ ergibt sich dagegen eine gute Abklärungsmöglichkeit mit Hilfe der IR.-Spektren. Ein Vergleich der beiden Spektren der beiden anomeren Methylglykoside der 4-O-Isovaleroylmycarose in Tetrachlorkohlenstoff bei hoher Verdünnung (0,004 M) zeigte für das α-Anomere (**76**) eine OH-Bande bei 3530 cm^{-1}, was auf eine schwache intramolekulare Wasserstoffbrücke zwischen dem axialen tertiären Hydroxyl und dem axialen Glykosid-Sauerstoffatom hinweist. Beim β-Glykosid (**77**) ist eine intramolekulare Wasserstoffbrücke unmöglich, weshalb die Hydroxylabsorption bei 3615 cm^{-1} entsprechend einer ungebundenen OH-Gruppe erscheint. Di-O-Acetylleucomycin A$_3$ mit der einzigen freien Hydroxylgruppe an C-3″ besitzt die OH-Bande bei 3515 cm^{-1}. Leucomycin A$_3$ ist demgemäß ein α-Mycarosid mit einer intramolekularen Wasserstoffbrücke (s. Partialformel B). Mit der α-Konfiguration an C-1″ sind auch Drehungsverschie-

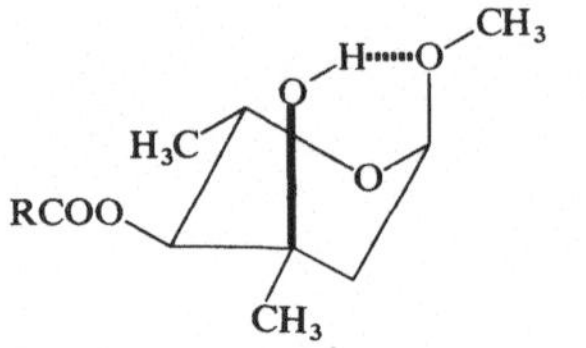

(**76**) 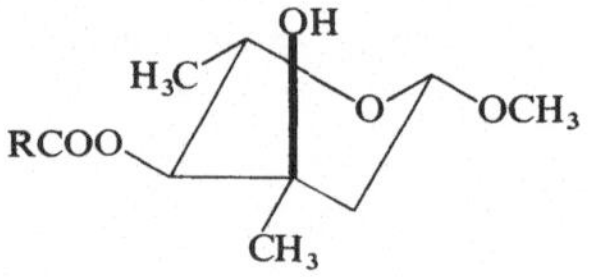   (**77**)

α-Methyl-4-O-isovaleroyl-mycarosid   β-Methyl-4-O-isovaleroyl-mycarosid
ν(OH) = 3530 cm^{-1} ν(OH) = 3615 cm^{-1}

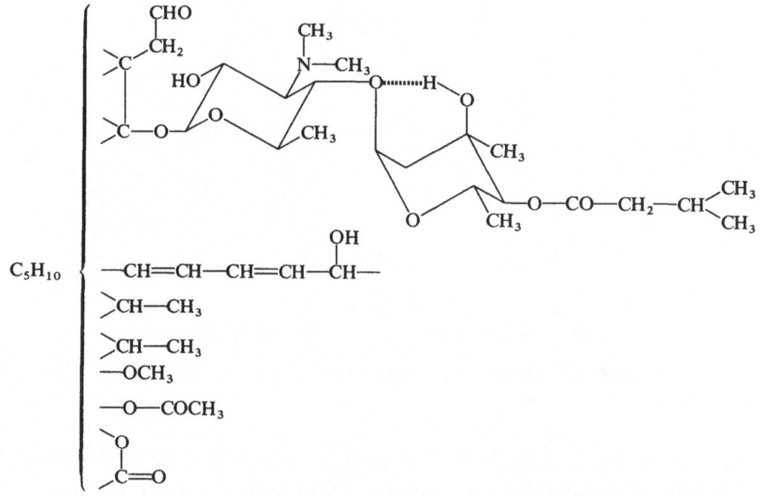

Partialformel B von Leucomycin A$_3$

Literaturverzeichnis: SS. 445—460

bungen zwischen Leucomycin A_3 ($[M]_D = -458°$ in Chlf.) und dem Demycarosyl-leucomycin A_3 ($[M]_D = -124°$ in Chlf.) bestens vereinbar (*229*). Auf Grund dieser Befunde läßt sich die Partialformel A (Seite 343) zu einer Partialformel B erweitern

Über den Zusammenhang von Atomgruppen über einen größeren Bereich des Aglykons gab ein Abbau mit Ozon Aufschluß. Das Ozonid wurde in Methanol bei $-60°$ bereitet und mit Wasserstoffsuperoxid in saurer Lösung gespalten (Formelschema 15). Nach der alkalischen Hydrolyse noch vorhandener Esterbindungen wurde das rohe Säuregemisch mit Diazomethan verestert und die destillierbaren Ester gaschromatographisch getrennt. Auf Grund des Verhaltens bei der Gaschromatographie und mittels der NMR.-Spektren konnten die beiden Komponenten als Isovaleriansäuremethylester (**78**), der dem Isovaleroylmycaroseteil entstammt, und als β-Hydroxybuttersäuremethylester (**79**) identifiziert werden. Die Carboxylgruppe in der Vorstufe zu **79** muß durch oxidativen Angriff auf das Doppelbindungssystem entstanden sein. Die Hydroxylgruppe kann nicht mit der einzigen sekundären Hydroxylgruppe des Aglykons, die ja allylständig ist, identisch sein und entstammt demnach der Lactongruppe. Damit ergibt sich für das Aglykon die Partialformel C, die auf Grund von NMR.-spektroskopischen Befunden sogleich zur Partialformel D erweitert werden konnte. Dem Wasserstoffatom an C-9 ließ sich ein Signal bei 4,05 ppm im NMR.-Spektrum des Leucomycins A_3 dadurch zuordnen, daß es beim Carbomycin B (**47**) fehlt und im Di-O-acetyl-leucomycin A um mehr als eine δ-Einheit nach tieferem Feld verschoben erscheint. Das Signal ist ein Doppeldublett mit $J_{9,10} = 9$ Hz. und $J_{9,8} = 4,2$ Hz., was durch Doppelresonanz-Studien leicht zu bestätigen war. Damit ergibt sich, daß C-8 ein und nur ein Wasserstoff, und wegen der chemischen Verschiebung von H-C-8 (das Signal befindet sich in einem Haufen bei etwa 2 ppm) kein Sauerstoffatom trägt. C-Atom 8 ist demnach eine Verzweigungsstelle. Schließlich ergab sich ebenfalls aus Doppelresonanz-Versuchen noch die *trans*-Geometrie an beiden Doppelbindungen. Die Wasserstoffatome an C-11 (δ 6,60 ppm) und C-12 (6,05 ppm) ließen sich leicht zuordnen. Ihre Signale sind Doppeldublette mit $J_{11,10} = 15,4$ Hz., $J_{11,12} = 10$ Hz., $J_{12,11} = 10$ Hz. und $J_{12,13} = 15,2$ Hz. Die beiden großen Kopplungskonstanten von etwa 15 Hz. beweisen die *trans*-Anordnung an beiden Doppelbindungen.

Die Struktur im Bereich der Kohlenstoffatome 1 bis 8 wurde am Leucomycin A_3 selber nur unvollständig studiert. Sie ergibt sich aus der oben beschriebenen Beziehung zum Carbomycin (Formelschema 9). Es sind daher im Folgenden noch diejenigen Abbaureaktionen des Carbomycins aus der Arbeit von WOODWARD zu behandeln, die sich speziell mit diesem Teil der Molekel befassen. Diese Reaktionen waren zwar schon 1957 (*329*) bekannt, sind aber 1965 (*330*) auf Grund ergänzender Stu-

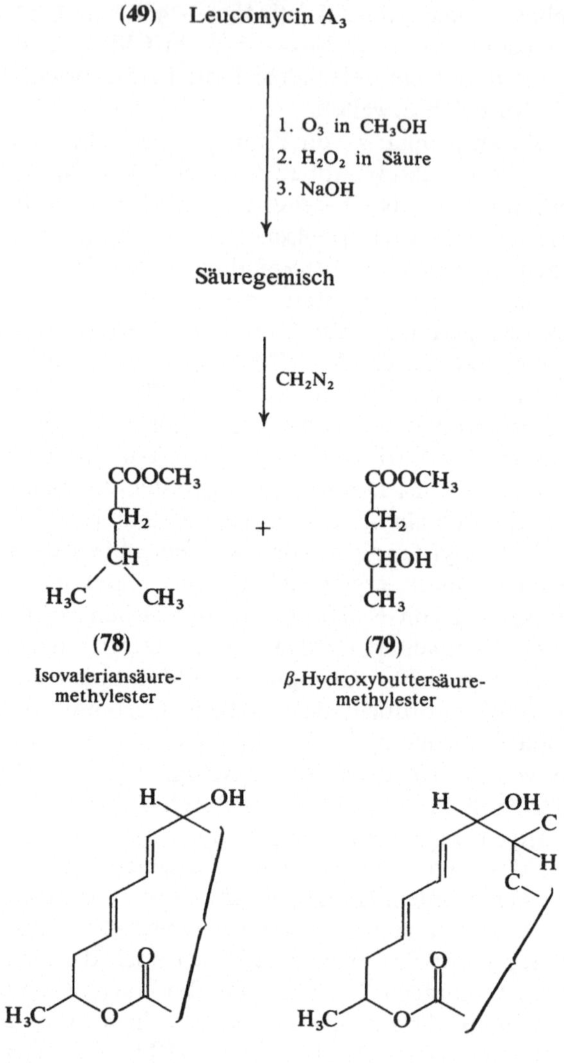

(49)   Leucomycin A$_3$

1. O$_3$ in CH$_3$OH
2. H$_2$O$_2$ in Säure
3. NaOH

↓

Säuregemisch

↓ CH$_2$N$_2$

Erweiterte Partialformeln von Leucomycin A$_3$

Schema 15. Abbau von Leucomycin A$_3$ mit Ozon

dien, insbesondere NMR.-Spektren, neu interpretiert worden, was zu einer Revision der ursprünglichen Carbomycinformel Anlaß gab (vgl. auch *175*).

Carbomycin (**46**) wurde zunächst einer Mischoxidation mit Perjodsäure und Kaliumpermanganat unterworfen und das rohe Oxidationspro-

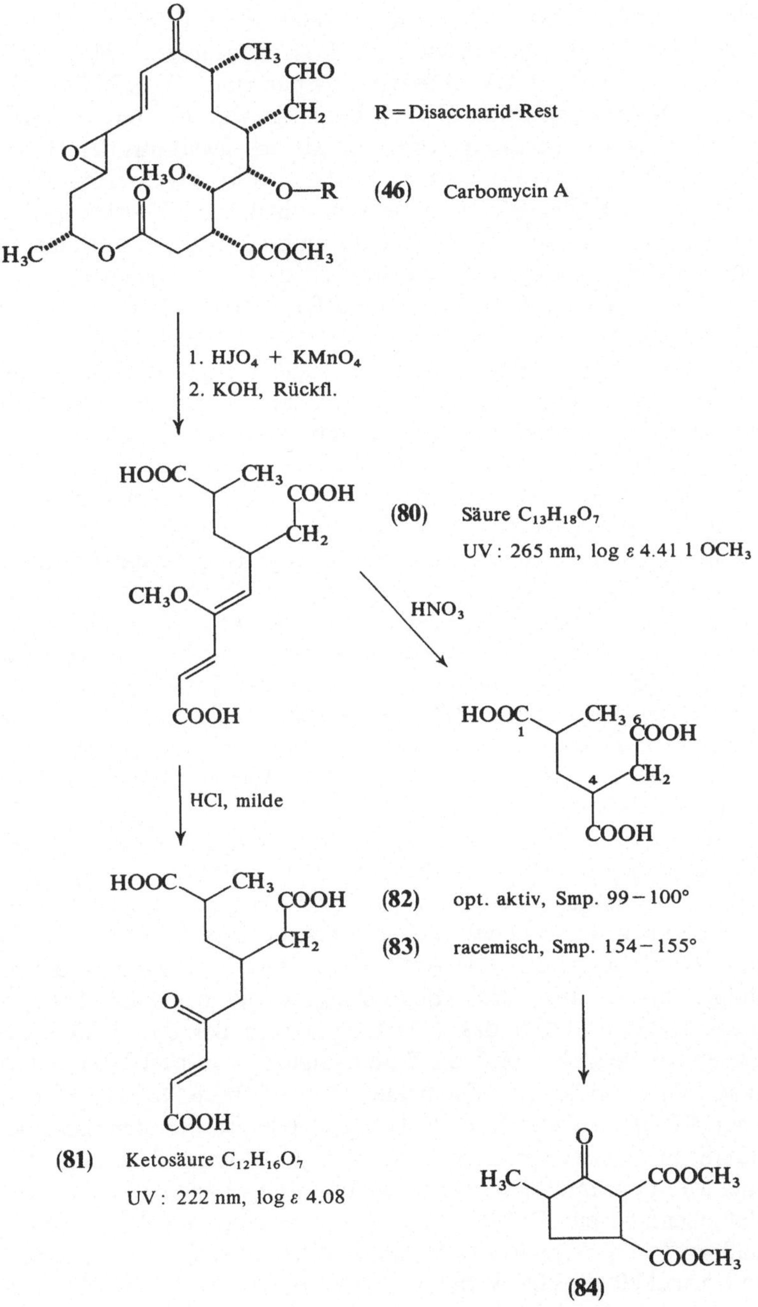

R = Disaccharid-Rest

(46) Carbomycin A

1. HJO_4 + $KMnO_4$
2. KOH, Rückfl.

(80) Säure $C_{13}H_{18}O_7$

UV: 265 nm, log ε 4.41 1 OCH_3

HNO_3

HCl, milde

(82) opt. aktiv, Smp. 99–100°

(83) racemisch, Smp. 154–155°

(81) Ketosäure $C_{12}H_{16}O_7$

UV: 222 nm, log ε 4.08

(84)

Schema 16. Oxidativer Abbau von Carbomycin

dukt energisch mit Kalilauge behandelt. Nach der Aufarbeitung kristallisierte aus dem Reaktionsgemisch eine Tricarbonsäure $C_{13}H_{18}O_7$ (**80**) aus, die auf Grund ihres UV.-Absorptionsspektrums (λ_{max} 265 nm, log ε 4,41) eine $\alpha,\beta,\gamma,\delta$-ungesättigte Diencarbonsäure war. Die Lage der nach Zeisel nachweisbaren Methoxylgruppe ergab sich leicht aus der Hydrolyse, die mit verdünnter Säure unter milden Bedingungen erfolgte und eine Säure $C_{12}H_{16}O_7$ (**81**) lieferte. Diese ist gemäß ihrem UV.-Absorptionsspektrum (λ_{max} 222 nm, log ε 4,08) eine α,β-ungesättigte γ-Ketosäure; die Ketogruppe fixiert die Lage der ursprünglichen O-Methylgruppe.

Die Struktur im ungesättigten Teil der Säure **81** wird erhärtet durch die Spaltung, die sie beim Erhitzen mit Kalilauge erleidet. Unter Verlust von zwei C-Atomen bildet sich eine gesättigte Ketodicarbonsäure (**85**, Formelschema 17), die auf Grund des Abbaus mit Hypojodit zu Jodoform und einer Tricarbonsäure $C_9H_{14}O_6$ (**86**) ein Methylketon sein muß. Der gesättigte Teil des Abbauproduktes **80** ergab sich aus dem weiteren oxidativen Abbau mit Salpetersäure (Schema 16). Dieser gab zunächst eine gesättigte, optisch aktive Tricarbonsäure $C_8H_{12}O_6$ (Smp. 99—100°). Diese stellt offenbar ein thermodynamisch ungünstiges Diastereomeres der Konstitution **82** dar, denn ihr Methylester wird beim Erhitzen mit Natriummethylat zu einer optisch inaktiven Form isomerisiert, die bei der alkalischen Hydrolyse die zu **82** stereoisomere, optisch inaktive Säure **83** (Smp. 154—155°) gibt. Die letztere war identisch mit einer synthetischen Säure der Konstitution **83**. Interessant ist der Befund, daß der Trimethylester der isomerisierten Säure **83** nach Dieckmann leicht ein Cyclopentanonderivat **84** lieferte, während der Trimethylester der Säure **82** diese Reaktion nicht eingeht.

Der nächste Schritt zur Konstitutionsaufklärung galt der Frage, welche der drei Carboxylgruppen der Säure **82** durch Angriff an der Doppelbindung neu entstanden war. Durch Einfügen des Bauelements **82** in die den Partialformeln B und D des Leucomycins A_3 entsprechenden Partialformeln des Carbomycins ist die Antwort bereits gegeben und damit die vollständige Konstitution des Abbauproduktes **80** bewiesen. Erhärtet wird diese Struktur durch das NMR.-Spektrum der Säure **80** (*330*). Dieses zeigt im Bereich von 5 bis 7 ppm Signale von drei Wasserstoffatomen an Doppelbindungen. Die beiden Dublette bei 6,88 und 5,85 ppm bilden ein AB-System mit $J_{AB} = 15$ Hz. und entsprechen demnach den *trans*-ständigen Wasserstoffatomen an den C-Atomen 2 und 3. Das Signal bei 5,12 ppm muß daher dem H-Atom an C-5 zugeordnet werden. Es bildet ebenfalls ein Dublett ($J = 10$ Hz), woraus abgeleitet werden muß, daß an C-6 ein einziges Wasserstoffatom sitzt. Dieser Forderung wird die Formel **80** gerecht, während eine ursprünglich in Betracht gezogene Alternativstruktur (**80 a**) mit zwei Wasserstoffatomen an C-6 (*329*) ausgeschlossen werden konnte.

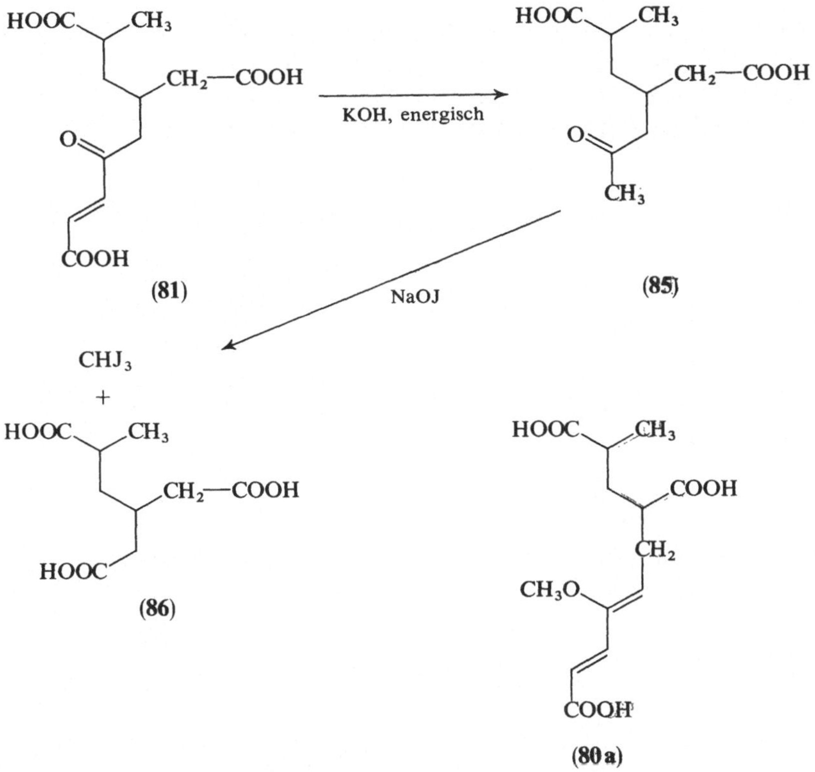

$$HOOC-CH(CH_3)-CH_2-CH(CH_2-COOH)-CO-CH=CH-COOH \quad (81)$$

$$\xrightarrow{\text{KOH, energisch}}$$

$$HOOC-CH(CH_3)-CH_2-CH(CH_2-COOH)-CO-CH_3 \quad (85)$$

$$\big\downarrow \text{NaOJ}$$

$$CHJ_3$$
$$+$$

(86)

(80a)

Schema 17. Konstitutionsaufklärung von Abbauprodukten des Carbomycins

Aus den Partialformeln B und D des Leucomycins und dem Abbauprodukt **80** des Carbomycins ergibt sich das vollständige Lactongerüst
der ganzen Antibiotica-Familie. Die leichte Bildung des Diens **80** bei
der Behandlung mit Alkali durch Elimination erfordert die Anwesenheit
von Sauerstoff-Funktionen in β- und δ-Stellung zum Lactoncarbonyl
(C-3 und C-5). Davon ist gemäß Partialformel B die Stellung 5 durch
die Glykosidbindung belegt. Für die Stellung 3 bleibt als letzte noch nicht
plazierte Gruppe die Acetoxygruppe übrig, womit die Strukturen der
Carbomycine und damit auch diejenige der Leucomycine und Spiramycine vervollständigt werden. Mit den neuen Formeln der Spiramycine
(233) sind ältere Abbauversuche von PAUL und TCHELITCHEFF (244, 245,
246), auf die hier nicht näher eingegangen werden soll, aufs Beste
interpretierbar.

Die Konfiguration der Leucomycine (und damit der Carbomycine
und Spiramycine) beruht wiederum auf einer Röntgenstrukturanalyse
(139). Dabei war das Ergebnis zuerst eher verwirrend. Es stellte sich
nämlich erst später eindeutig heraus, daß das Derivat, das für die

Strukturanalyse eingesetzt wurde, nicht in der einfachen Beziehung zum Leucomycin A_3 stand, wie man ursprünglich angenommen hatte (*227, 232*). Je nach den Bedingungen wurden bei der sauren Hydrolyse zwei verschiedene Demycarosylderivate des Leucomycins erhalten (**87** und **88**).

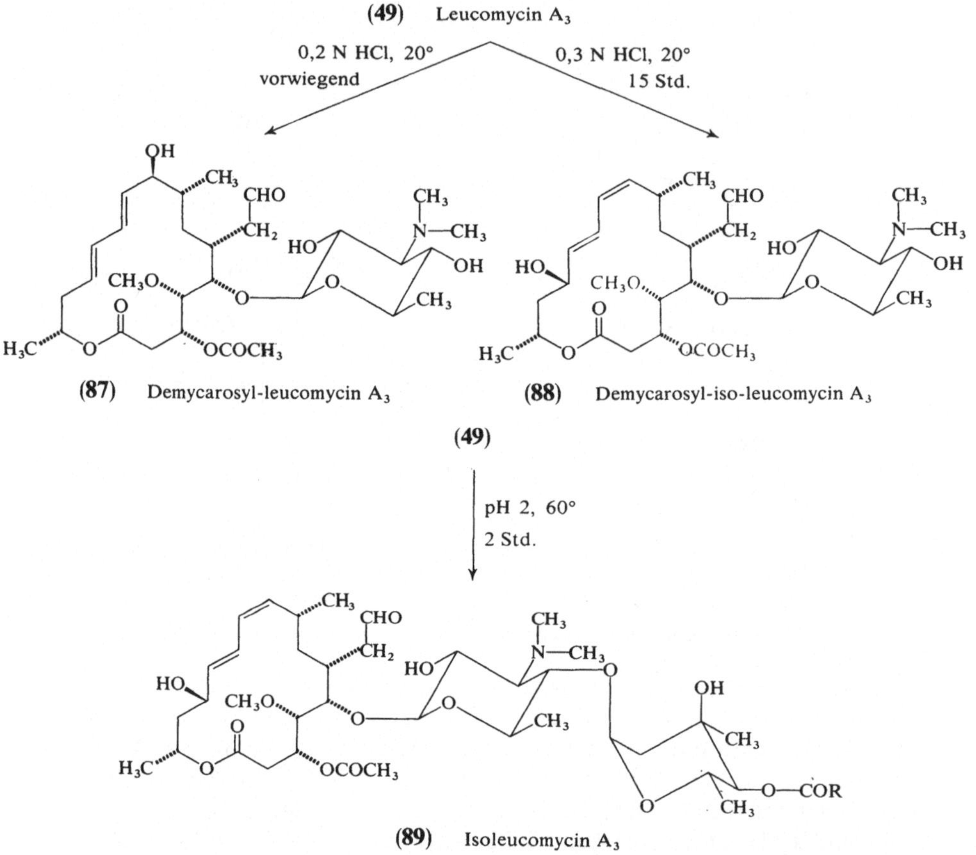

(**49**)   Leucomycin A_3

0,2 N HCl, 20°   0,3 N HCl, 20°

vorwiegend   15 Std.

(**87**)   Demycarosyl-leucomycin A_3 (**88**)   Demycarosyl-iso-leucomycin A_3

(**49**)

pH 2, 60°

2 Std.

(**89**)   Isoleucomycin A_3

Schema 18. Allylumlagerung von Leucomycin A_3

Obwohl anzunehmen war, daß durch die erhöhte Säurekonzentration (Schema 18, rechts) eine Umlagerung eher begünstigt wurde, gelang eine befriedigende Zuordnung auf spektroskopischem Wege erst, als es gelang, die Allylumlagerung auch ohne Glykosidspaltung zu bewirken, und das Isoleucomycin A_3, das in Benzol viel schwerer löslich ist als das ursprüngliche Antibioticum, rein zu isolieren. NMR.-spektroskopische Übereinstimmungen bzw. Verschiedenheiten erlaubten, die Verbindung **87** zum unveränderten Leucomycin A_3, das Produkt **88** zum Isoleucomycin A_3 (**89**) in Beziehung zu setzen.

Da es sich um eine recht heikle Entscheidung handelte, wurde die Lage der Hydroxylgruppe im Demycarosylleucomycin A_3 unabhängig durch einen Abbau sichergestellt. Der im Schema 19 dargestellte Abbau nach Bayer-Villiger zur 7-Hydroxyoctansäure (**92**) bedarf wohl keines weiteren Kommentars.

(**87**)   Demycarosyl-leucomycin A_3

MnO_2 in Chlf.

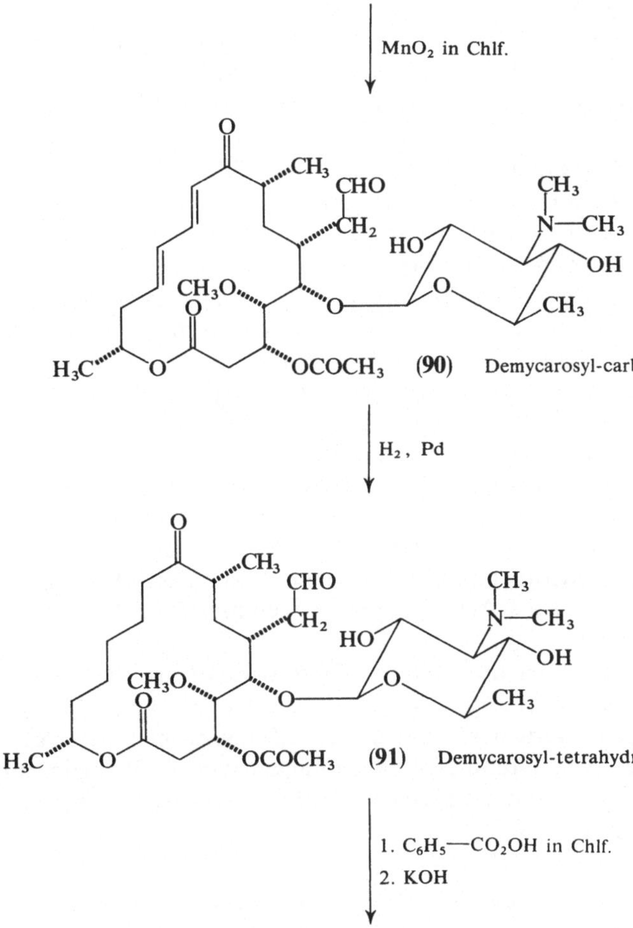

(**90**)   Demycarosyl-carbomycin B

H_2, Pd

(**91**)   Demycarosyl-tetrahydrocarbomycin B

1. C_6H_5—CO_2OH in Chlf.
2. KOH

$$CH_3\text{—}CHOH\text{—}(CH_2)_5\text{—}COOH$$

(**92**)   7-Hydroxy-octansäure

(identifiziert durch Gaschromatographie des Methylesters)

Schema 19. Abbau des Demycarosyl-Leucomycins A_3

23*

Als die Röntgenstrukturanalyse (*139*) begonnen wurde, war von den beiden Demycarosylderivaten erst das eine (**88**) bekannt. Das Hydrobromid gab aus Äthylalkohol orthorhombische Kristalle der Raumgruppe $P2_12_12_1$ mit nicht stöchiometrischen Mengen an Kristallalkohol. Die Einheitszelle enthielt vier Moleküle. Mit CuK_α-Strahlung konnten 2450 unabhängige Reflexionen gemessen und ausgewertet werden. Die Verfeinerung wurde bis zu einem R-Faktor von 17,7% vorangetrieben. Die absolute Konfiguration wurde nach der Methode von Bijvoet aus 23 Reflexionen von Weissenberg-Aufnahmen bestimmt. Das Ergebnis ist in der Formel **88** wiedergegeben und in allen früheren Formeln bereits berücksichtigt. Insbesondere konnten die β-Konfiguration der Mycaminosylbindung und die *trans*-Geometrie an beiden Doppelbindungen bestätigt werden.

Wegen der Allylumlagerung bei der Herstellung des Derivates **88** gibt dessen Konfiguration keine Auskunft über die räumliche Lage der Hydroxylgruppe an C-9 des Leucomycins A_3. Diese mußte daher gesondert untersucht werden, wobei sich die Anwendung der Benzoat-Regel (*35*) als nützlich erwies (*232*). Diese benützt Drehungsverschiebungen, die von relativ geringfügigen aber voraussehbaren Konformationsänderungen durch den Eintritt großer Substituenten verursacht werden, zur Bestimmung des Chiralitätssinnes an chiralen Carbinolgruppen.

Für eine zweifelsfreie Anwendung dieser Beziehung war zunächst ein Derivat des Leucomycins herzustellen, das als einzige acylierbare Hydroxylgruppe diejenige an C-9 besaß. Durch eine Acetylierung unter milden Bedingungen in verdünnter Lösung (Aceton) konnte ein Monoacetyl-leucomycin A_3 (**93**) erhalten werden. Die starke Veränderung des pK-Wertes (6,70 bei Leucomycin A_3; 5,48 beim Monoacetat **93**, vgl. auch Tabelle 2) bewies, daß die Acetylgruppe in Stellung 2′ des Mycaminoserestes eingetreten war.

Für den Vergleich der molekularen Drehungen ($[M]_D$-Werte) wurden einerseits das Monoacetat **93** und dessen 3,5-Dinitrobenzoat (**94**), andererseits das Tetrahydroderivat **95** und dessen 3,5-Dinitrobenzoat (**96**) verwendet. Die negativen Drehungsverschiebungen von $-308°$ beim ungesättigten und $-116°$ beim gesättigten Alkohol (s. Tabelle 3) weisen auf die *R*-Chiralität an C-9 hin.

Tabelle 3. *Molekulare Drehungen von Derivaten des Leucomycins A_3 in Methanol*

Substanz	Formel	M_D (MeOH).
Monoacetyl-leucomycin A_3	(**93**)	$-622°$
3,5-Dinitrobenzoat	(**94**)	$-930°$
Tetrahydro-monoacetylleucomycin A_3	(**95**)	$-818°$
3,5-Dinitrobenzoat	(**96**)	$-934°$

(**49**)   Leucomycin A$_3$ (pK' 6,70 in 50% Äthanol)

Ac$_2$O in Aceton

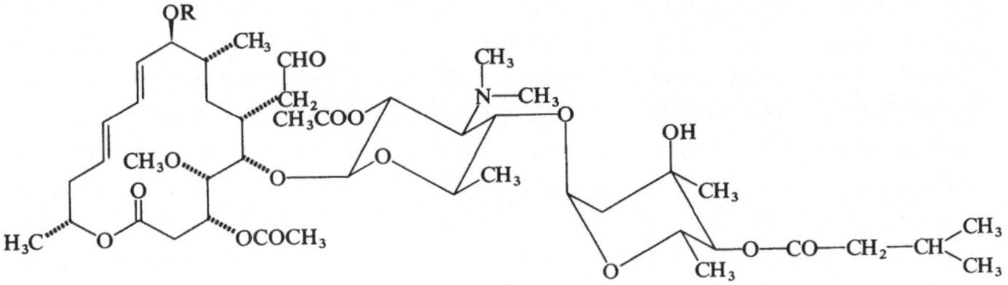

(**93**)   R=H
pK$_a'$ 5,48 (50% Äthanol)

(**94**)   R=—CO— [3,5-dinitrophenyl]

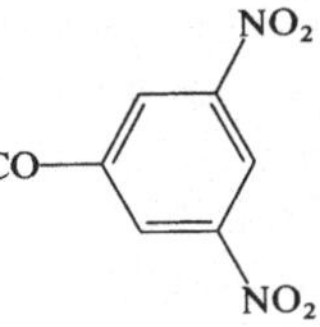

H$_2$, Pd

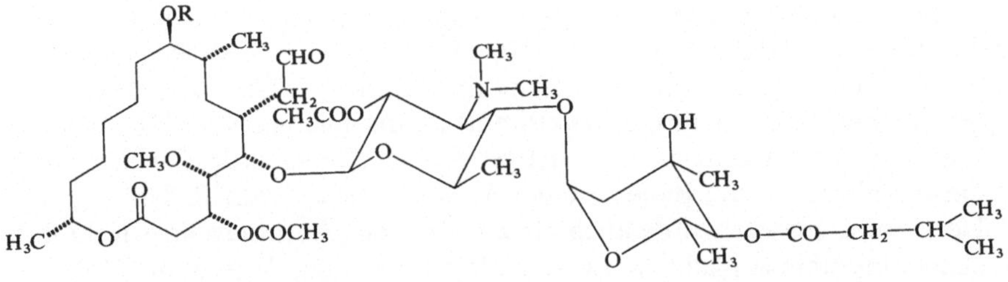

(**95**)   R=H

(**96**)   R=—CO— [3,5-dinitrophenyl]

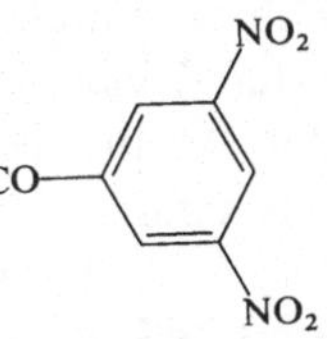

Schema 20. Herstellung der Substanzen der Tabelle 3

b) Weitere Antibiotica der Carbomycingruppe

Wie in der Erythromycingruppe finden wir auch hier eine Reihe von Substanzen mit verwandter Konstitution und ähnlicher biologischer Wirkung wie die Carbomycine, Leucomycine und Spiramycine. Besonders ähnlich den Leucomycinen sind die beiden Antibiotica YL-704 A und B (**97** und **98**), die als 3-O-Propionylanaloge der Leucomycine A_3 und A_4 erkannt wurden und damit auch dem Spiramycin III sehr nahe stehen (*285*). Ihre Konstitutionsaufklärung beruht zum Teil auf einer eingehenden spektroskopischen Untersuchung, wobei besonders der Vergleich des massenspektrometrischen Zerfalls mit dem des Leucomycins A_3 eine große Rolle spielt. Schließlich gelang auch eine direkte Verknüpfung mit der Leucomycinreihe: Das Tri-O-propionylderivat (**99**) des Leucomycins A_1 (**48**) erwies sich als identisch mit dem Propionylierungsprodukt von YL-704 A. Da andererseits die Antibiotica YL-704 A und B bei der milden Hydrolyse das gleiche Demycarosylderivat geben, dürfen die Konstitutionen und Konfigurationen beider Verbindungen als gesichert angesehen werden.

Zwei nahe miteinander verwandte basische Makrolide sind das Tylosin (*124*), ein Aldehyd, und der entsprechende primäre Alkohol, Relomycin (*316*). Tylosin läßt sich mit Natriumborhydrid zu Relomycin reduzieren. Im Vergleich mit Carbomycin besitzt das Aglykon in Stellung 4 eine Methylgruppe statt des Methoxyls sowie zusätzliche Verzweigungen an den C-Atomen 12 und 14 (vgl. Formeln **100** und **101**). Mit den neutralen Makroliden Chalcomycin und Neutramycin (s. unten) hat das Tylosin den neuen Zucker Mycinose in Stellung 14′ gemeinsam.

Das Acumycin (**102**) wurde 1962 von Bickel und Mitarb. (*24*) entdeckt und kurz darauf unter dem Namen Cirramycin B als Begleiter des Cirramycins A (**103**) von Koshiyama und Mitarb. (*173, 174, 294*) erneut beschrieben. Die endgültige Strukturaufklärung gelang erst 1970 (*286*) und führte zur Entdeckung eines neuartigen 4-Oxo-Derivates der Zuckerreihe, die vor allem für die Biogenese der Makrolidzucker von Bedeutung ist. Der gleiche Zucker wurde auch als Baustein des Anthracyclin-Antibioticums Cinerubin A nachgewiesen und dort Cinerulose A genannt (*164*).

Schließlich kommen auch in der Carbomycin-Reihe zwei neutrale Antibiotica, das Chalcomycin (**104**) und das Neutramycin (**105**) vor (*206, 326*). Der den Aminozucker ersetzende neutrale Zuckerbaustein ist, wie beim entsprechenden Vertreter der Erythromycinereihe (Lankamycin) die Lankavose (= Chalcose). Die Aglykone der zuletzt erwähnten Verbindungen **100** bis **105** zeichnen sich gegenüber dem Carbomycin und seinen nächsten Verwandten durch eine erhöhte Anzahl von C_1-Seitenketten aus, während die Lage der funktionellen Gruppen am Aglykon grosso modo dieselbe bleibt.

Literaturverzeichnis: SS. 445—460

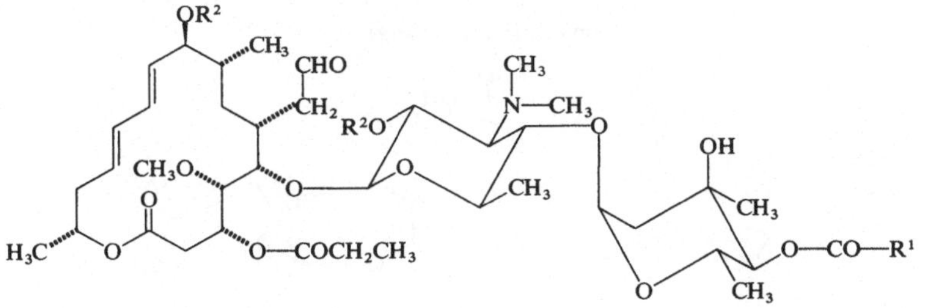

R¹ and R² table:

	R¹	R²	
(97)	—CH₂—CH(CH₃)₂	H	Antibioticum YL-704 A
(98)	—CH₂—CH₃	H	Antibioticum YL-704 B
(99)	—CH₂—CH(CH₃)₂	—CO—CH₂—CH₃	Tripropionyl-leucomycin A₁

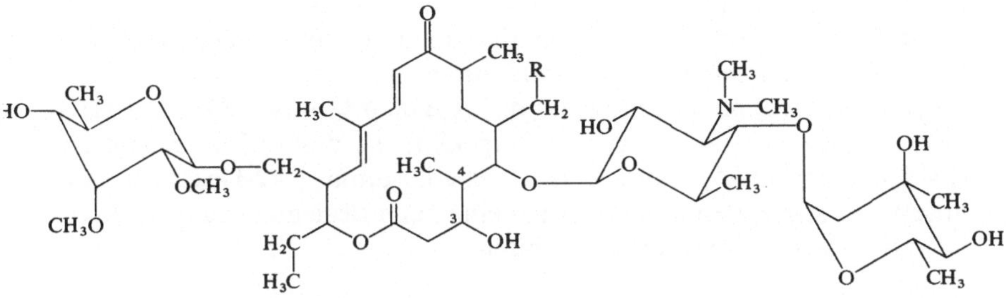

(100) R = —CHO Tylosin   **(101)** R = —CH₂OH Relomycin

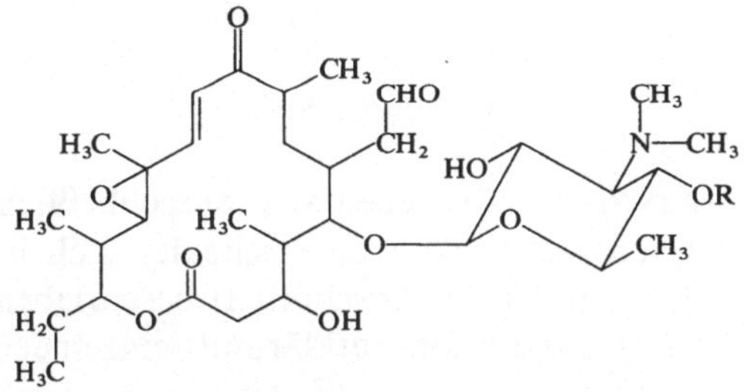

(102) R = (Cinerulose A-Rest)   **(103)** R = H

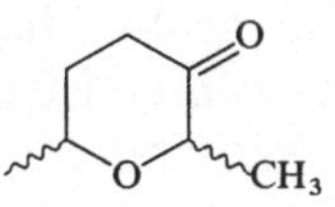

Acumycin   Cirramycin A₁
(Cirramycin B, Antibioticum B-5894)

Schema 21. Weitere Makrolide vom Carbomycin-Typus (s. auch S. 360)

Schema 21 (Fortsetzung)

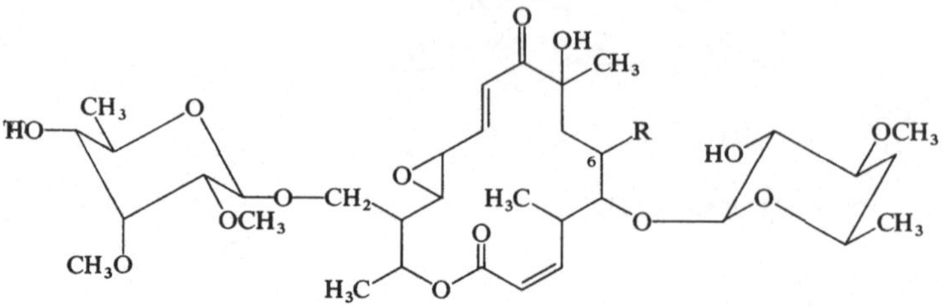

(104) R=CH$_3$: Chalcomycin (Aldgamycin D)
(105) R=H: Neutramycin

Die Stereochemie dieser Verbindungen ist erst fragmentarisch oder gar nicht untersucht worden. Die Erfahrungen aus der Erythromycin-Untergruppe lassen aber erwarten, daß auch hier an vergleichbaren Zentren übereinstimmende Chiralitäten vorliegen.

In diese Gruppe scheint ferner das Angolamycin (*47, 73*) zu gehören, dessen Aglykon zwar noch nicht aufgeklärt ist, dessen Zuckerbausteine (L-Mycarose, D-Mycinose und D-Angolosamin [=2-Desoxymycaminose]) und spektralen Eigenschaften eine nahe Beziehung zu den Makroliden der Carbomycingruppe andeuten.

III. Polyen-Makrolide

1. Allgemeines

Die Polyen-Makrolide, die ebenfalls ausschließlich Stoffwechselprodukte von Actinomyceten sind, unterscheiden sich in ihrem ganzen Charakter dermaßen von den im Abschnitt II beschriebenen Verbindungen, daß vor der Konstitutionsaufklärung einzelner Vertreter der Gruppe kaum jemand einen näheren Zusammenhang mit den Erythromycin- und Carbomycin-Antibiotica geahnt hatte. Die Polyene sind hochwirksam gegen Pilze und Hefen, während Bakterien kaum gehemmt werden. Für einige Polyene sind Aktivitäten gegen Protozoen (z. B. Trichomycin) und Tumorzellen (Trienin) beschrieben worden (*13*).

Im Gegensatz zu den antibakteriellen Makrolid-Glykosiden sind die Polyene in den meisten organischen Lösungsmitteln sehr schwer löslich. Auch sie sind häufig Gemische mehrerer verwandter Verbindungen. Ihre Trennung ist aber viel schwieriger als die der erythromycinartigen Antibiotica. Sie erfolgt meist durch mehrfache Gegenstromverteilungen über viele Stufen wie etwa bei Trichomycin-Komplex (*134, 221*), beim Levorin A (*34*) und beim Nystatin (*275*), oder durch mehrfache Verteilungschromatographie an Diatomeenerde oder an gepuffertem Kieselgel wie beim Filipin (*21*). Da diese Methoden recht aufwendig sind, und vor allem wegen der geringen Löslichkeit dieser Verbindungen in den dafür geeigneten Lösungsmitteln, können in der Regel auch bei großem experimentellen Aufwand nur geringe Mengen an Antibiotica-Gemisch getrennt werden, so daß nur in seltenen Fällen größere Mengen einheitlicher Substanzen für eine eingehende chemische Untersuchung zur Verfügung standen. Abbauversuche in größerem Maßstab wurden daher gelegentlich mit Gemischen durchgeführt (z. B. Filipin, s. unten).

Eine weitere Schwierigkeit besteht darin, daß die isolierten Verbindungen nur schwer charakterisierbar sind. Nur wenige Vertreter der Gruppe sind als gut kristallisierbare Substanzen beschrieben. Häufiger fallen sie in schlecht ausgebildeten lösungsmittelhaltigen Kristallen oder gar nur als amorphe Pulver an. Charakteristische Schmelzpunkte lassen sich kaum angeben, da Zersetzungsprozesse in der Regel schon weit unterhalb der Schmelztemperatur einsetzen.

Aber auch die spektroskopische Charakterisierung ist oft recht unsicher. Die einzelnen Komponenten einer Gruppe unterscheiden sich strukturell manchmal so geringfügig voneinander (vgl. z. B. Fungichromin und Filipin), daß sich die Unterschiede weder im IR.- noch im NMR.-Spektrum klar manifestieren. Erst kürzlich ist die Massenspektrometrie für die Analytik dieser relativ hochmolekularen Substanzen zugänglich gemacht worden. Am besten scheint sich dabei die Methode von PANDEY und RINEHART (*238*) zu bewähren, die durch Ionisierung bei niedriger Spannung Molekülionen von peracetylierten Polyen-Antibiotica beobachten konnten. Nicht ganz so eindeutige, aber doch meist brauchbare Resultate haben GOLDING, RICKARDS und BARBER (*118*) mit Massenspektren von Trimethylsilylderivaten erhalten. Eine generelle Anwendung der Massenspektrometrie verspricht für die Zukunft eine bessere Charakterisierung der einzelnen Antibiotica und vor allem ein brauchbares Mittel für die Identifizierung von Neuisolierungen, das bisher gefehlt hat.

Die geschilderten Mängel hatten zur Folge, daß zahlreiche Verbindungen in mehreren Laboratorien entdeckt und unter verschiedenen Namen beschrieben wurden, ohne daß ihre Identität sogleich erkannt werden konnte. Nach und nach sind zahlreiche Identitäten mit mehr oder weniger Sicherheit aufgedeckt worden. Es ist dabei u. a. die dem uneingeweihten Chemiker absurd erscheinende Lage eingetreten, daß in einem Fall zwei Antibiotica (Fungichromin und Lagosin, s. unten) erst als (wahrscheinlich!) gleich erkannt wurden, als auf mühsamem Wege für beide unabhängig die gleiche Konstitution abgeleitet worden war. Es muß mit Sicherheit angenommen werden, daß sich noch zahlreiche der gegen hundert Beschreibungen in der Literatur — bisher unerkannt — auf die gleiche Verbindung beziehen. Erst eine systematische Untersuchung mit den modernsten Methoden könnte die reichlich unübersichtliche Lage klären.

Aus diesen Gründen will ich im folgenden auf eine systematische Beschreibung aller in der Literatur beschriebenen Polyen-Makrolide verzichten. Näher eingegangen wird nur auf diejenigen, deren Individualität durch eine vollständige Konstitutionsformel oder wenigstens durch eine plausible Partialformel gesichert ist.

Eine auffällige Eigenschaft immerhin besitzen alle Antibiotica dieser Gruppe, die zwar nicht eine individuelle Charakterisierung, aber doch eine sichere Einteilung in eine von mehreren Untergruppen erlaubt: das UV.-Absorptionsspektrum. Dieses ist ausgezeichnet durch mehrere scharfe Banden, von denen die drei langwelligsten besonders intensiv sind. Ihre molaren Extinctionen kommen nahe an 100,000 heran und erlauben meist eine sichere Zuordnung zu einer Untergruppe schon in Rohextrakten auch dann, wenn der Antibioticagehalt recht gering ist. Vor knapp zehn Jahren ist in den „Fortschritten der Chemie organischer Naturstoffe" ein Referat von Oroshnik und Mebane (*235*) über die Polyenantibiotica erschienen, in dem die UV.-Absorptionsspektren und ihre Bedeutung für die Einteilung in Tetraene, Pentaene, Hexaene und Heptaene einläßlich dargestellt sind. Hier sei lediglich in der Tabelle 4 eine Auswahl von Spektren zusammengestellt, besonders darum, weil seither drei neue Gruppen hinzugekommen sind: die Triene, die Mycoticingruppe und das Dermostatin.

Die Werte in der Tabelle 4 sind meistens in alkoholischer oder methanolischer Lösung gewonnen worden. Sie mögen auch einen Anhaltspunkt geben über die beobachteten Streuungen. Bei Abweichung um 2—3 nm, wie sie z. B. in der Tetraenreihe auftauchen, dürfte es sich um experimentelle Ungenauigkeiten handeln. Hingegen lassen sich die Pentaene deutlich in zwei Untergruppen einteilen. Von der Gruppe B sind mehrere Strukturen aufgeklärt worden. Sie besitzen alle den Chromophor $-(CH=CH)_4-CH=C(CH_3)-$. Für die Gruppe A mit etwas kürzerwelligen Banden wird dagegen ein unverzweigter Pentaenchromophor $-(CH=CH)_5-$ angenommen. Zwischen den entsprechenden Untergruppen der Heptaene könnte ein analoger struktureller Unterschied bestehen.

Einige Anmerkungen sind zur Mycoticingruppe notwendig. Das unter normalen Bedingungen aufgenommene Spektrum mit einer breiten Bande bei 364 nm (mit angedeuteten Schultern bei etwa 350 und 374 nm) und einer etwa zehnmal weniger intensiven Bande bei 262 nm erinnert zunächst nicht an dasjenige der übrigen Polyene. Erst bei der Temperatur von

Tabelle 4. *UV.-Absorptionsmaxima von Polyen-Makroliden*

Gruppe		Antibioticum	Hauptmaxima in nm			Bemerk.	Literatur vgl. auch (*304*)
Triene		Trienin	278	267	257		(*13*)
		Mycotrienin	282	272	262		(*74*)
		M M 8	283	272	262		(*12*)
Tetraene		Lucensomycin	318	303	290		(*111*)
		Nystatin	318	304	292		(*93, 275*)
		Pimaricin	318	303	290		(*283*)
		P A 166	319	304	291		(*168*)
		Tetrine A u. B	318	303	290		(*265*)
		Ac$_2$ 435	320	305	288		(*237*)
Pentaene	A	Capacidin	350	332	318		(*43*)
		P A 153	349	332	317		(*168*)
		2814 P	351	333	317		(*293*)
	B	Filipin	354	337	321		(*21*)
		Chainin	357	338	324		(*120*)
		Fungichromin	356	338	322		(*296*)
Hexaene		Cryptocidin	380	358	341		(*270*)
		Hexamycin	376	356	336		(*98*)
		Mediocidin	378	357	339		(*299*)
Heptaene	A	Levorin A$_2$ (Candicidin)	400	378	358		(*34*)
		P A 150	397	377	358		(*168*)
		Hepcin	401	379	359		(*331*)
	B	Trichomycin	406	384	364		(*84*)
		2814 H	406	384	362		(*293*)
		Monicamycin	410	385	365		(*123*)
		Candidin	408	384	365		(*303*)
Mycoticin-Gruppe		Mycoticin	364 (breit)	262 (niedrig)		20°C	(*310*)
		Flavomycoin	363 (breit)	262 (niedrig)		20°C	(*274, 29*)
			390	369	351	−185°C	
Dermostatin			383 (breit)	282 (niedrig)		20°C	(*222*)
			422	398	381	−173°C	

flüssiger Luft erscheint das übliche Polyen-Spektrum mit den drei Hauptmaxima bei 390, 369 und 351 nm. Der zugehörige Chromophor wurde von SCHLEGEL und THRUM (*274*) auf einfache Weise demonstriert: Bei der Reduktion von Flavomycoin (Partialformel **106**) mit Lithiumaluminiumhydrid entsteht ein Produkt mit dem UV.-Absorptionsspektrum eines Pentaens (**107**) mit λ_{max} 355, 337 und 320 nm (vgl. auch

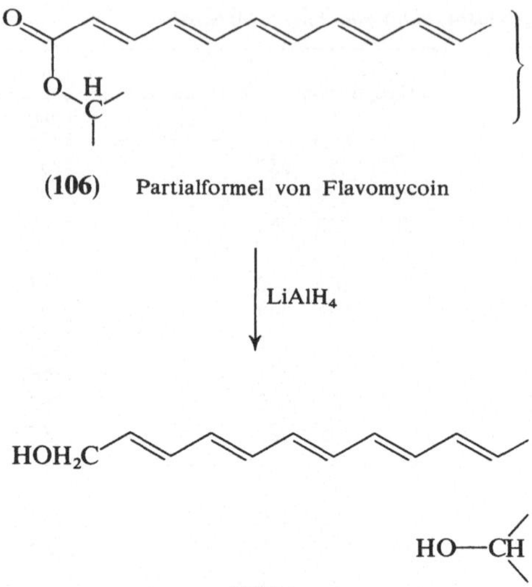

UV: 390, 369, 351 nm $(-185°)$

(106)   Partialformel von Flavomycoin

LiAlH$_4$

R

UV: 3.5, 337, 320 nm

(107)

Schema 22. Abwandlungen des Chromophors von Flavomycoin

Tabelle 4). Der Chromophor besteht demnach aus einer Pentaenkette und einer dazu konjugierten Carbonylgruppe. Daß dieses Carbonyl mit dem Lactoncarbonyl identisch ist, ergibt sich daraus, daß mit Natriumborhydrid der Chromophor nicht verändert wird (vgl. auch *29, 310*). Beim Dermostatin dürfte, analog zum Mycoticin (*222*) eine zum Lactoncarbonyl konjugierte Hexaengruppe vorliegen. Einige Polyenantibiotica besitzen außer der Polyengruppe noch zusätzliche Chromophore, die sich im kurzwelligen Teil des UV.-Absorptionsspektrums bemerkbar machen. So ist im Molekül des Pimaricins und des Lucensomycins eine α,β-ungesättigte Lactongruppe enthalten, im Trichomycin eine p-Aminophenylketon-Gruppe. Die entsprechenden Absorptionsbanden sind aber durchwegs von geringerer Intensität und stören das allgemeine Bild der Polyenabsorption nicht.

2. Konstitutionsaufklärung von Polyen-Makroliden

Die erwähnten ungünstigen Eigenschaften standen auch der Konstitutionsaufklärung lange Zeit hemmend im Weg. Erfolgreiche Arbeiten auf diesem Gebiet setzten daher wesentlich später ein als bei den früher besprochenen Makroliden. Als 1963 die Zusammenfassung von Oroshnik und Mebane in den „Fortschritten" erschien (*235*), war die Makrolid-Struktur der Polyen-Antibiotica gerade kurz vorher erkannt worden (*240, 78*), und von den drei damals vorliegenden Konstitutions-Vorschlägen hat gerade einer (Fungichromin/Lagosin) die Zeitdauer bis heute überstanden, während sich die beiden anderen (Filipin und Pimaricin) noch Korrekturen gefallen lassen mußten. Auch heute sind nur wenige der zahlreichen Verbindungen dieser Klasse vollständig oder weitgehend aufgeklärt. Doch scheint jetzt das Eis gebrochen zu sein. Es sind in der letzten Zeit mehrere Methoden entwickelt worden, die diese Verbindungen einer rationellen chemischen Bearbeitung zugänglich machen.

Der erste Durchbruch gelang wohl COPE und Mitarb. (*68*) beim Fungichromin. Das COPEsche Prinzip zur Aufklärung des Kohlenstoffgerüstes von Polyenmakroliden besteht in der totalen Reduktion der Polyhydroxy-Verbindungen bis zur Stufe eines gesättigten Kohlenwasserstoffs. Diese wurde in mehreren Stufen durch katalytische Hydrierung der Doppelbindungen, Reduktion der Lactongruppe und allfällig vorhandener Ketogruppen mittels Lithiumaluminiumhydrid, Reduktion mit Jodwasserstoffsäure und rotem Phosphor zu einem Polyjodkohlenwasserstoff und erneute Reduktion mit Lithiumaluminiumhydrid und katalytisch erregtem Wasserstoff durchgeführt. Die so erhaltenen gesättigten Kohlenwasserstoffe lassen sich durch Gaschromatographie gut reinigen und durch das Massenspektrum charakterisieren. Das Massenspektrum erlaubt meistens auch eine recht sichere Ableitung der Konstitution, da die C-C-Bindungen unmittelbar neben Verzweigungen besonders leicht fragmentieren und die entsprechenden Bruchstücke im Massenspektrum stark dominieren. Durch eine eindeutige Synthese des entsprechenden Kohlenwasserstoffs wird daraufhin die Struktur endgültig gesichert. Im Anschluß an die Untersuchungen von COPE hat sich vor allem CEDER (*51, 56*) mit der Verbesserung der Methode befaßt, die später auch von anderen Forschern übernommen worden ist (s. unten).

a) Die Konstitution des Lucensomycins

Über erste Untersuchungen zur Strukturaufklärung von Polyen-Makroliden ist bereits im Referat von OROSHNIK und MEBANE (*235*) berichtet worden. Als Beispiel soll hier eine neuere Arbeit, die Konstitutionsaufklärung des *Lucensomycins*, näher beschrieben werden. Sie macht eingehend von den Erfahrungen COPES und CEDERS Gebrauch und ist daher repräsentativ für die ganze Gruppe.

Das Lucensomycin (=Etruscomycin) wurde 1957 von ARCAMONE und Mitarb. zum erstenmal beschrieben (*11*). Es scheint von seinem Produzenten, *Streptomyces lucensis*, ohne verwandte Begleitsubstanzen gebildet zu werden und läßt sich daher nach einfacher Vorreinigung aus mit Butanol gesättigtem Wasser kristallisieren. Die farblosen Nadeln sind auf Grund der Eigenschaften und vor allem gemäß der Abbauergebnisse einheitlich. Durch das UV.-Absorptionsspektrum mit Maxima bei 290 nm (ε 56,000), 303 nm (82,000) und 318 nm (79,000) gibt sich das Lucensomycin als konjungiertes Tetraen mit *trans*-Anordnung an allen vier Doppelbindungen zu erkennen. Ein UV.-Absorptionsmaximum bei 218 nm (ε 21,000), das bei der Reduktion mit Natriumborhydrid bestehen

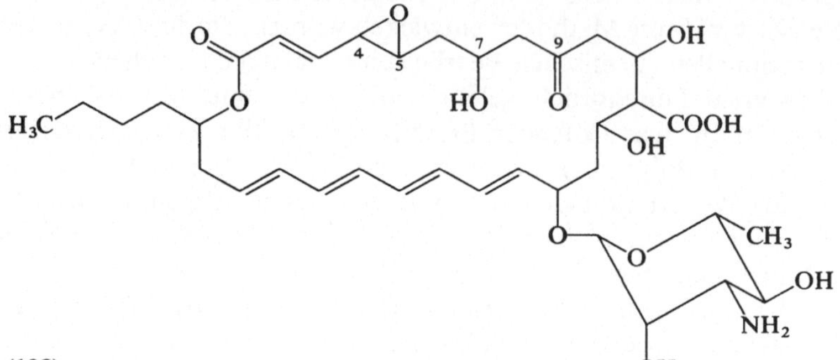

(108) Lucensomycin

RSH ($R{=}CH_3$ oder $n{-}C_{12}H_{25}$)

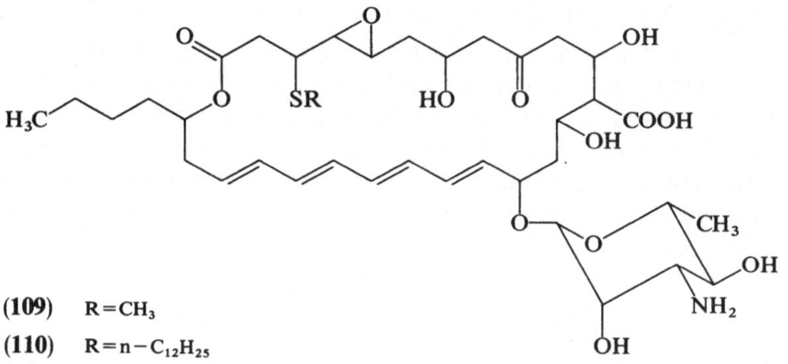

(109) $R{=}CH_3$

(110) $R{=}n{-}C_{12}H_{25}$

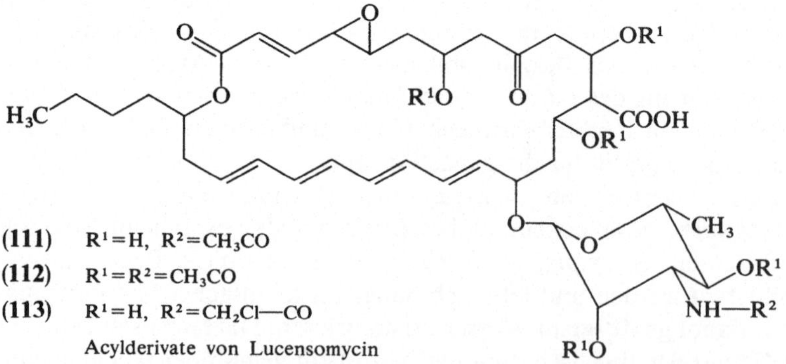

(111) $R^1{=}H$, $R^2{=}CH_3CO$

(112) $R^1{=}R^2{=}CH_3CO$

(113) $R^1{=}H$, $R^2{=}CH_2Cl{-}CO$

Acylderivate von Lucensomycin

Schema 23. Reaktionen des Lucensomycins

bleibt, zeigt die zusätzliche Anwesenheit einer α,β-ungesättigten Ester- oder Lactongruppe an (*111*). Die Anwesenheit einer solchen Gruppe wurde bestätigt durch die Bildung der MICHAEL-Additionsprodukte **109** und **110** mit Methyl- bzw. n-Dodecylmercaptan. Die beiden Produkte zeigten im UV. noch das Tetraenspektrum, aber nicht mehr die Bande bei 218 nm.

Das Lucensomycin ist amphother. Es ließ sich sowohl mit Perchlorsäure in Eisessig als auch mit Natriumhydroxid in Alkohol titrieren, wobei sich beidemale ein Äquivalentgewicht von etwa 700 ergab. Mit Säureanhydriden in Methanol ließen sich mehrere N-Monoacylderivate, z. B. das N-Monoacetylderivat **111** herstellen. Diese Verbindung löste sich leicht in Natriumbicarbonat und reagierte mit Diazomethan zu einem neutralen Methylierungsprodukt. Mit Essigsäureanhydrid in Pyridin entstand dagegen ein Hexaacetylderivat (**112**), das ebenfalls Reaktionen einer Carbonsäure zeigte. Ferner zeigte sich der amphothere Charakter durch das IR.-Absorptionsspektrum, das neben einer intensiven breiten Bande mit Schwerpunkt bei $3400\ cm^{-1}$ (mehrere Hydroxylgruppen) charakteristische Banden bei $1717\ cm^{-1}$ (breit, Lacton- und Ketogruppe) und $1572\ cm^{-1}$ besaß. Die letztere ist charakteristisch für Carboxylatgruppen (*111*).

Das NMR.-Spektrum des Lucensomycins ist sehr komplex und erlaubt wenig unmittelbare Schlußfolgerungen. Immerhin konnten einige negative Schlüsse gezogen werden, so z. B. das Fehlen von Aldehyd-, O-Methyl- und N-Methylgruppen.

Auch die Mikroanalysen waren anfangs wenig informativ. Einerseits war die Reproduzierbarkeit unbefriedigend, weil die Kristalle hartnäckig Lösungsmittel zurückhielten und die Proben oft Asche-Rückstände gaben. Zudem sind Analysen im Bereich von Molekulargewichten über etwa 500 bei Verbindungen mit zahlreichen Sauerstoffunktionen in der Regel mehrdeutig. Daher mußte die ursprünglich aus Analysen und Titrationen abgeleitete approximative Bruttoformel, $C_{36-37}H_{53-55}NO_{14}$, später nicht nur präzisiert, sondern auch in eine Formel mit 13 Sauerstoffatomen abgeändert werden. Dies gelang, als die Arbeiten zur Konstitutionsaufklärung bereits weit fortgeschritten waren, durch den Vergleich der Massenspektren von Trimethylsilyläthern verschiedener Derivate des Lucensomycins (*113*). Eine besonders wertvolle Rolle spielte dabei der Pentatrimethylsilyläther-trimethylsilylester des N-Chloracetyl-lucensomycins (**113**) sowie der Hexatrimethylsilyläther-trimethylsilylester **116** des N-Acetyl-dodecahydrolucensomycins (**115**), das durch katalytische Hydrierung des Mono-N-acetyllucensomycins **111** leicht zugänglich ist. Obwohl in den meisten Spektren keine Molekülionen erkennbar waren, ergab ein Vergleich der Zerfallsreihen eindeutig, daß sich die Derivate von einem Antibioticum $C_{36}H_{53}NO_{13}$ mit 5 alkoholischen Hydroxylgruppen ableiten. Die Anwesenheit von 5 Hydroxylgruppen neben einer Aminogruppe ist im Einklang mit der Bildung des Hexaacetylderivates **112**.

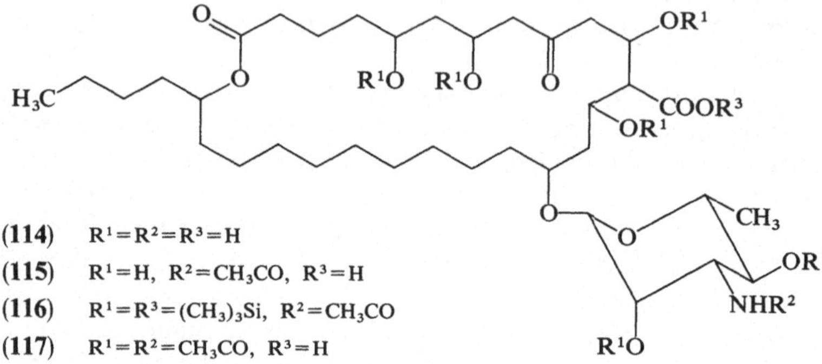

(114)	$R^1 = R^2 = R^3 = H$
(115)	$R^1 = H$, $R^2 = CH_3CO$, $R^3 = H$
(116)	$R^1 = R^3 = (CH_3)_3Si$, $R^2 = CH_3CO$
(117)	$R^1 = R^2 = CH_3CO$, $R^3 = H$

Hydrierungsprodukte von Lucensomycin

Bei der katalytischen Hydrierung verbraucht das Lucensomycin 6 Mole Wasserstoff unter Bildung eines Dodecahydroderivates (**114**). Außer dem Tetraensystem und der Doppelbindung in Konjugation zum Lacton-carbonyl ist demnach eine sechste hydrierbare Gruppe vorhanden. Es entsteht dabei eine neue Hydroxylgruppe, denn das Hydrierungsprodukt verbraucht bei der Acetylierung in Pyridin 7 Mole Essigsäureanhydrid und gibt ein Heptaacetylderivat (**117**). Daß diese hydrierbare Gruppe eine Epoxygruppe ist, ergab sich aus der Bildung von elementarem Jod bei der Reaktion von Lucensomycin mit Kaliumjodid in Essigsäure. Das Dodekahydroderivat (**114**) gab diese Reaktion nicht.

Ein tiefgreifender Abbau trat bei der Reaktion von Lucensomycin mit Essigsäureanhydrid in Eisessig und Schwefelsäure ein. Je nach den Versuchsbedingungen konnte dabei das Mycosamintriacetat (**118**) oder das α-Mycosamin-tetraacetat (**119**) (neben geringen Mengen des β-Anomeren) erhalten werden. Das Mycosamin ist ein Aminozucker, dessen Struktur bereits von Arbeiten mit anderen Polyen-Antibiotica gut bekannt war (s. Abschnitt VI/2). Ein repräsentatives Derivat des Aglykons ließ sich bei dieser Reaktion allerdings nicht fassen.

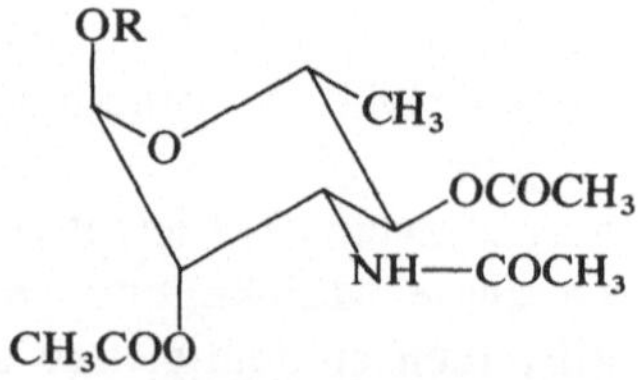

(118)	$R = H$
(119)	$R = CH_3CO$

Acetylderivate des Mycosamins

Literaturverzeichnis: SS. 445—460

Die bisher beschriebenen Eigenschaften und Reaktionen erlauben, für das Lucensomycin eine Partialformel A zu schreiben.

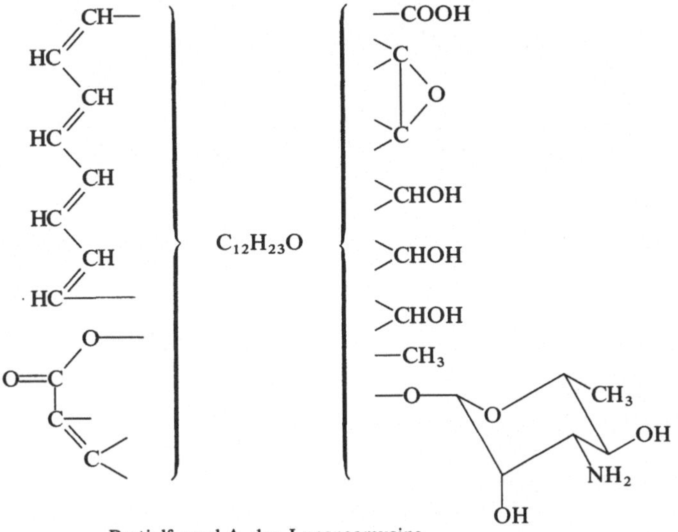

Partialformel A des Lucensomycins

Über den Bau des Kohlenstoffgerüstes im Bereiche der Tetraengruppe konnten aus den Ergebnissen eines Abbaus mit wasserfreiem Alkali wichtige Rückschlüsse gezogen werden. Nach kurzem Erhitzen mit Natriummethylat wurde ein Reaktionsprodukt $C_{17}H_{24}O_2$ als tiefgelbe Nadeln erhalten. Durch ein NMR.-Signal bei 9,5 ppm (Dublett) gab sich die Verbindung als Aldehyd zu erkennen. Die niedrige Carbonylfrequenz bei 1670 cm^{-1} im IR.-Absorptionsspektrum zeigte, daß die Aldehydgruppe zu einem Doppelbindungssystem konjugiert war. Das in Chloroform aufgenommene UV.-Absorptionsspektrum mit Maxima bei 267 nm (log ε 4,00) und 380 nm (4,76) ähnelt auffallend dem des Mycoticins (s. Tabelle 4) mit seiner konjugierten Pentaenongruppe. In Hexan ist die langwellige Bande in drei Spitzen (346, 362 und 382 nm) aufgelöst. In Methanol, dem eine Spur Mineralsäure zugefügt war, veränderte sich das Spektrum rasch zu dem eines unverzweigten Pentaens mit Maxima bei 314, 328 und 347 nm. Diese Veränderung ist auf die Bildung eines Dimethylacetals zurückzuführen. Eine IR.-Absorptionsbande bei 3450 cm^{-1} zeigt, daß das zweite Sauerstoffatom einer Hydroxylgruppe angehört.

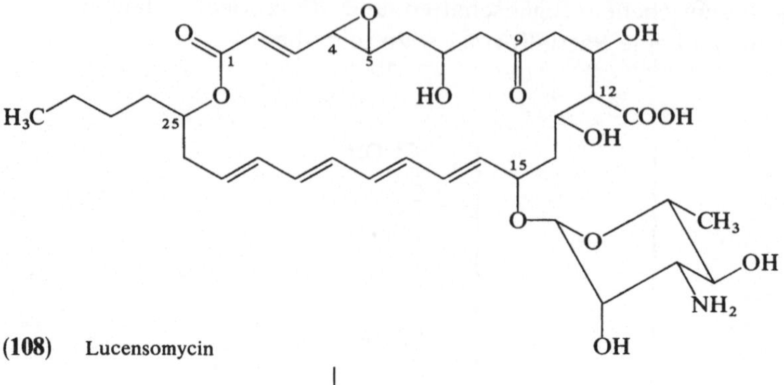

(108)   Lucensomycin

NaOCH$_3$

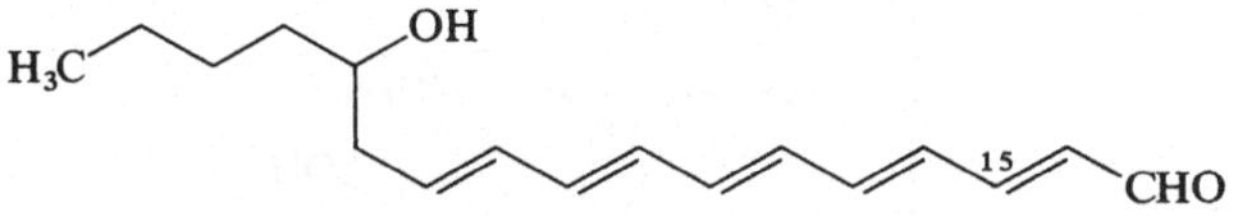

(120)   13-Hydroxy-heptadeca-2,4,6,8,10-pentenal

Kuhn-Roth

H$_2$, Pt (6 Mol H$_2$)

(121)   CH$_3$—COOH
          +
(122)   CH$_3$—CH$_2$—COOH
          +
(123)   CH$_3$—CH$_2$—CH$_2$—COOH
          +
(124)   CH$_3$—CH$_2$—CH$_2$—CH$_2$—COOH

CH$_3$—(CH$_3$)$_3$—CH—(CH$_2$)$_{11}$—CH$_2$OH
                OH

(125)

Heptadecan-1,13-diol

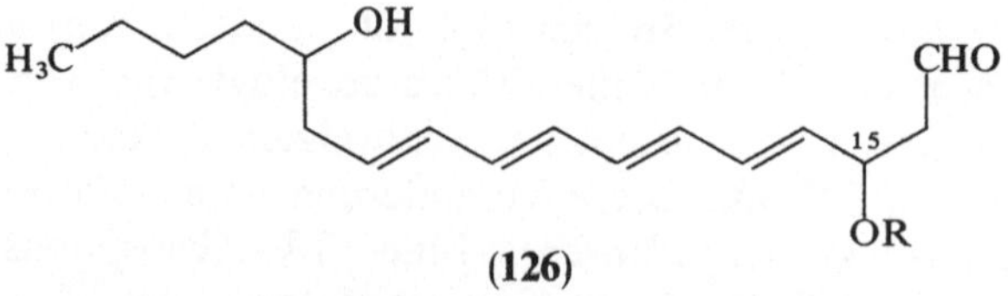

(126)

Schema 24. Retroaldolspaltung des Lucensomycins

Schema 24 (Fortsetzung)

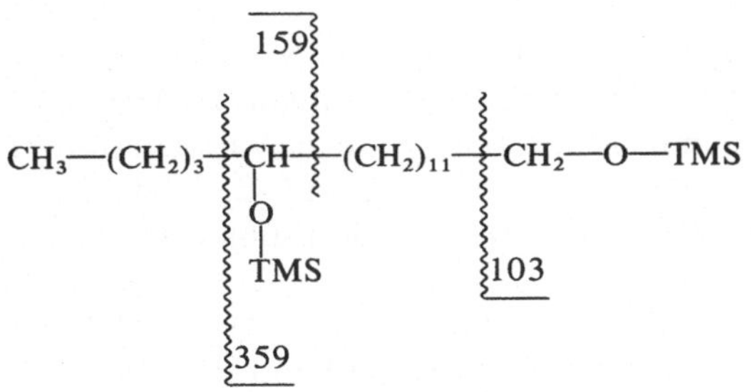

$$CH_3\text{---}(CH_2)_3\text{---}\underset{\underset{\underset{TMS}{|}}{|}{O}}{CH}\text{---}(CH_2)_{11}\text{---}CH_2\text{---}O\text{---}TMS$$

159 359 103

(**127**)   Massenspektrometrische Fragmentierung des Trimethylsilyläthers des Diols (**126**)
TMS $= (CH_3)_3\text{---}Si$

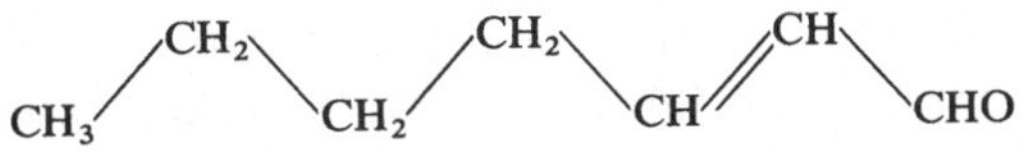

(**128**)   2-Heptenal

Bei der Hydrierung des Aldehyds mittels Platin wurden 6 Mol Wasserstoff verbraucht. Das Reaktionsprodukt ist ein kristallines gesättigtes, aliphatisches Diol $C_{17}H_{36}O_2$, dessen eine Hydroxylgruppe, aus einer Aldehydgruppe entstanden, primär sein muß. Auf die Lage des zweiten Hydroxyls ließ sich aus dem Massenspektrum des Bis-trimethylsilyläthers schließen, da bekannt ist, daß diese Äther bevorzugt an C-C-Bindungen neben den Sauerstoffatomen fragmentieren. Das Spektrum zeigt neben dem Molekülpeak ($M^+ = 416$) dominierende Fragmente bei m/e 359, 159 und 103, deren Interpretation sich aus der Formel **127** ergibt. Die Position 13 des zweiten Hydroxyls wird dadurch bestätigt, daß bei der Oxydation des Aldehyds **120** nach KUHN-ROTH ein Gemisch von wasserdampfflüchtigen Säuren entsteht, das neben Essigsäure (**121**) noch Propionsäure, Buttersäure und Valeriansäure (**122** bis **124**), aber keine höheren Fettsäuren enthält.

Die Entstehung des Aldehyds **120** ist als Retroaldolspaltung zu deuten. Wegen der Erweiterung des Tetraensystems zu einem Pentaen muß als Zwischenprodukt ein Aldehyd **126** angenommen werden, der durch Elimination von ROH ins Pentaenal **120** übergeht. In diesem Zwischenprodukt dürfte R nicht Wasserstoff sein, da sonst eher ein weiterer Abbau zu einem um zwei C-Atome ärmeren Tetraenal erwartet werden müßte, von dem aber im Reaktionsgemisch nicht eine Spur nachweisbar war.

Die relative Lage des Sauerstoffatoms an C-25 (Formel **108**) gegenüber dem Polyensystem ergibt sich auch aus einem anderen Abbau: Bei der Ozonisierung von Lucensomycin wird u. a. 2-Heptenal (**128**) gebildet. Die Aldehydgruppe muß durch Angriff an einer Doppelbindung, die Doppelbindung von **127** durch Elimination einer Sauerstoffunktion entstanden sein. Die energische Oxydation von Lucensomycin mit Salpetersäure oder Kaliumpermanganat gibt u. a. Valeriansäure (**124**), was einem erneuten Beweis der Butylseitenkette an C-25 (**108**) gleichkommt.

Das vollständige C-Gerüst von Lucensomycin konnte durch totale Reduktion erfaßt werden (*112*). Dodekahydrolucensomycin (**114**) wurde zunächst mit Lithiumaluminiumhydrid zu einem nicht näher charakterisierten Polyol reduziert und dieses mit Jod und rotem Phosphor in eine **Polyjodverbindung übergeführt. Nach der reduktiven Dejodierung mit** Lithiumaluminiumhydrid und anschließenden Hydrierung mit Platin wurde durch Chromatographie an Aluminiumoxid in etwa 20-prozentiger Ausbeute ein Kohlenwasserstoffgemisch isoliert, das durch präparative Gaschromatographie in zwei nahe miteinander verwandte Komponenten aufgetrennt wurde. Die etwas schwerer flüchtige zeigte im Massenspektrum ein Molekülion mit m/e 422 entsprechend einer Bruttoformel $C_{30}H_{62}$ und enthielt demnach sämtliche 30 C-Atome des Aglykons. Unter den Fragmentionen dominierten zwei sehr stark: m/e 267 für $C_{19}H_{39}$ und m/e 183 für $C_{13}H_{27}$. Dieser Befund legt die Lage der einzigen Verzweigung, einer Methylgruppe an C-12 und damit die Struktur des Reduktionsproduktes **129** fest; denn es ist bekannt, daß bei gesättigten Kohlenwasserstoffen die C-C-Bindungen neben Verzweigungen bevorzugt gespalten werden.

Der zweite, in etwas geringerer Menge erhaltene Kohlenwasserstoff ist gemäß dem Molekülion (m/e 408) eine Verbindung $C_{29}H_{60}$, und das Fragmentierungsmuster entspricht demjenigen eines unverzweigten Kohlenwasserstoffs (**130**) ohne dominierende Fragmente. Das Nonacosan (**130**) ist demnach durch Verlust der Seitenkette entstanden. Die Abspaltung hat vermutlich während der energischen Behandlung mit Jod und Phosphor stattgefunden. Ein ganz analoges Verhalten ist früher beim Pimaricin beobachtet worden (*51*).

Von den drei Methylgruppen des Kohlenwasserstoffs **129** muß auf Grund des Abbauproduktes **120** diejenige in Stellung 29 der ursprünglichen Methylgruppe entsprechen, die beiden anderen müssen durch Reduktion der Lactongruppe bzw. der Carboxylgruppe neu entstanden sein.

Eine weniger drastische Reduktion, Hochdruck-Hydrogenolyse bei etwa 250° und 250 Atm. mit Platin in Eisessig, gab relativ einfache sauerstoffhaltige Derivate, die sich aus den Kohlenwasserstoffen **129** und **130** ableiten und die Lage gewisser Sauerstoffunktionen anzeigen. Aus dem mit Diazomethan veresterten Hydrogenolyseprodukt wurde als Hauptkomponente in 10% Ausbeute der Nonacosansäuremethylester (**131**) in kristalliner Form erhalten und durch Vergleich mit einem synthe-

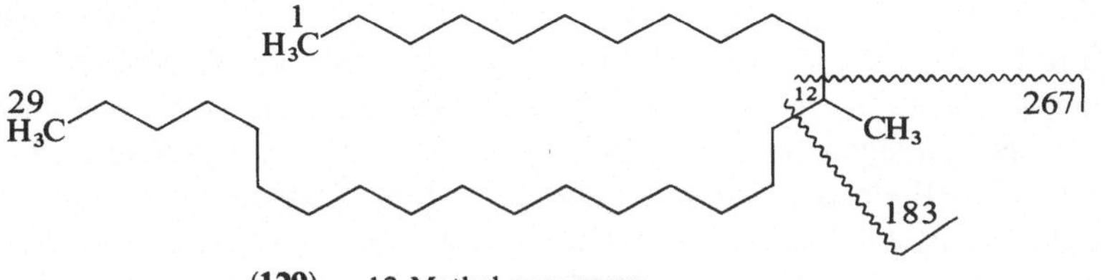

(129) 12-Methyl-nonacosan

$$CH_3 \text{—} (CH_2)_{27} \text{—} CH_3$$

(130) Nonacosan

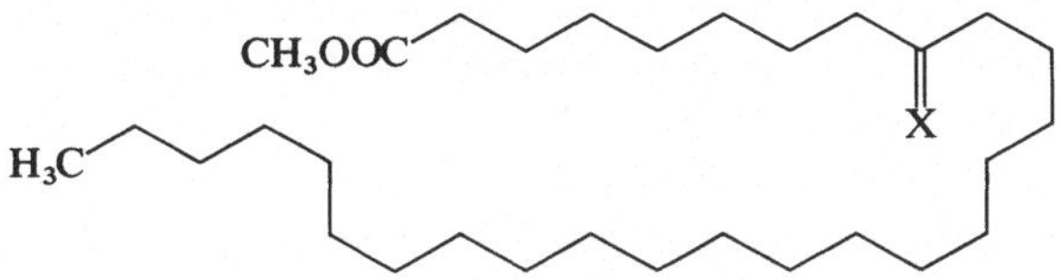

(131) X = 2 H : Nonacosansäure-methylester
(132) X = O : 9-Oxo-nonacosansäure-methylester

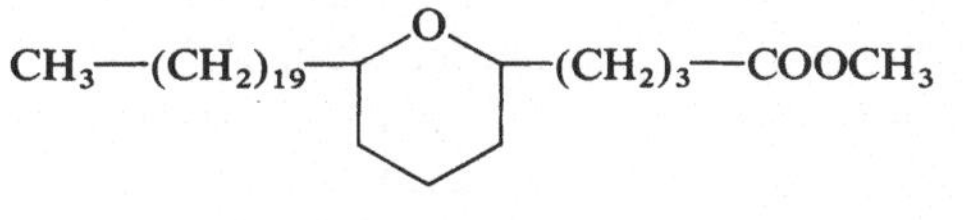

$$CH_3 \text{—} (CH_2)_{19} \text{—} \boxed{} \text{—} (CH_2)_3 \text{—} COOCH_3 \qquad CH_3 \text{—} (CH_2)_{16} \text{—} CH \text{—} (CH_2)_{10} \text{—} COOCH_3$$

(133) **(134)**

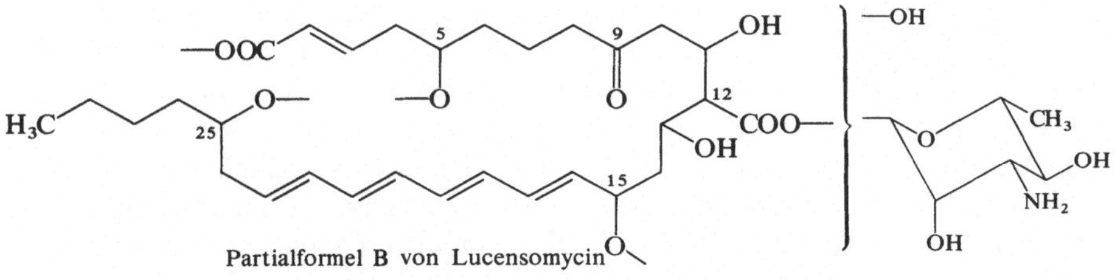

Partialformel B von Lucensomycin

Schema 25. Reduktionsprodukte von Lucensomycin

tischen Präparat identifiziert. Mit 6% Ausbeute wurde daneben das 9-Oxoderivat **132**, Smp. 77—79°, isoliert. IR.-Absorptionsmaxima bei 1745 und 1705^{-1} charakterisieren die drei Sauerstoffatome als Ester- bzw. Ketonsauerstoffe. Die Ketogruppe des Reaktionsproduktes muß einer ursprünglichen Ketogruppe des Lucensomycins entsprechen, für die damit die Stellung 9 bewiesen ist. Wenn das Antibioticum zuerst mit Natriumborhydrid reduziert und dann einer Hochdruckhydrierung unterzogen wird, kann die Ketosäure **132** im Produktegemisch nicht nachgewiesen werden, dagegen tritt dann als Hauptprodukt der verzweigte Dicarbonsäuredimethylester **134** auf. Dieses unterschiedliche Verhalten des Lucensomycins und seines Reduktionsproduktes ist aus der nachfolgenden Partialformel B leicht verständlich: Primäre Reaktion ist eine β-Elimination des Hydroxyls an C-11. Die resultierende vinyloge β-Ketosäure decarboxyliert spontan, worauf erst die Hydrogenolyse der verschiedenen Sauerstoffunktionen einsetzt. Wenn zuerst die Ketogruppe in Stellung 9 reduziert wird, ist die Elimination nicht mehr möglich; die Hydrogenolyse erfolgt unter Erhaltung der Carboxylgruppe an C-12.

Auch unter anderen Bedingungen erfolgt leicht eine Decarboxylierung von Lucensomycin. Sowohl Lucensomycin wie auch Dodekahydrolucensomycin, nicht aber das Reduktionsprodukt mit Natriumborhydrid, spalten beim Auflösen in Schwefelsäure ein Mol CO_2 ab. Diese Reaktion ist ebenfalls aus der Partialformel B über eine β-Elimination an C-11 leicht verständlich.

Als drittes kristallines Produkt (Smp. 47°) der Hochdruckhydrierung wurde in lediglich 1% Ausbeute ein Methylester $C_{30}H_{58}O_3$ erhalten. Der Ester zeigt im IR.-Absorptionsspektrum keine Hydroxylabsorption und lediglich eine scharfe Carbonylbande bei etwa 1745 cm^{-1}. Das dritte Sauerstoffatom ist demnach eine Ätherbrücke. Die beiden dem Äthersauerstoffatom benachbarten Wasserstoffatome (an C-5 und C-9) geben sich im NMR.-Spektrum durch ein breites Multiplett (Integral 2 H) bei 2,85 bis 3,35 ppm zu erkennen. Die Lage der Ätherbrücke an C-5 und C-9 gemäß Formel **133** wurde aus dem Fragmentierungsschema im Massenspektrum abgeleitet und fordert demnach eine weitere Sauerstoffunktion an C-5. Die Resultate der genannten Versuche sind in der Partialformel B zusammengefaßt, die nun auch die Bildung des Abbaualdehyds **120** über das Zwischenprodukt **126** durch eine zweifache Retroaldolspaltung der Bindungen C-10 − C-11 und C-12 − C-13 leicht verständlich macht.

Weitere Abbaureaktionen dienten der Fixierung der Lactongruppe. Bei der Oxydation nach Lemieux (Kaliumpermanganat und Kaliumperjodat) und der nachfolgenden Veresterung mit Diazomethan wurden zwei Methylester erhalten, die die ursprüngliche Lactonbindung jetzt als Esterbindungen enthielten. Der eine davon (**135**) war mit dem synthetischen Oxalsäurehalbester-dimethylester der 3-Hydroxy-heptansäure identisch (*110*). Das in geringerer Menge isolierbare Produkt war das entsprechende Formiat (**136**). Daß die Carboxylgruppe des Oxalsäurehalbesters (Vorstufe von **135**) durch Angriff an der zur Lactongruppe konjugierten Doppelbindung entstanden war, ergab sich aus einem ähnlichen Abbau des oben erwähnten Dodecylmercaptan-Additionsproduktes (**110**). Aus

(**108**)   Lucensomycin

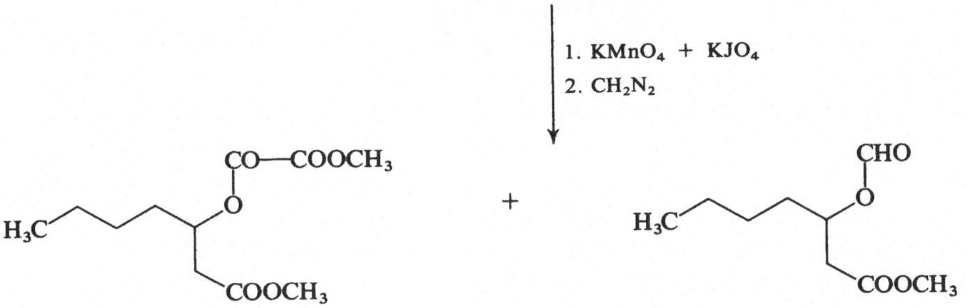

1. $KMnO_4$ + KJO_4
2. CH_2N_2

(**135**)

+

(**136**)

(**110**)   Dodecylmercaptan-Addukt

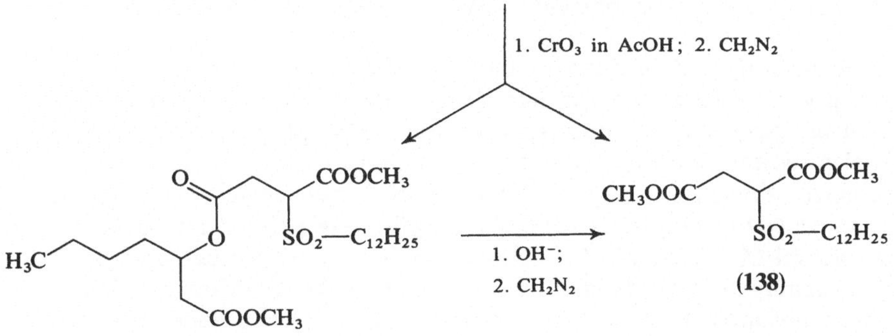

1. CrO_3 in AcOH; 2. CH_2N_2

(**137**)

1. OH^-;
2. CH_2N_2

(**138**)

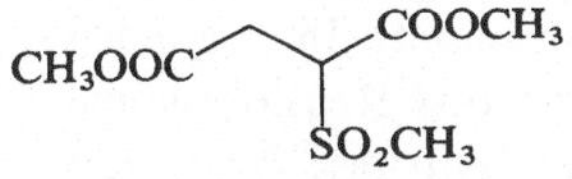

(**139**)

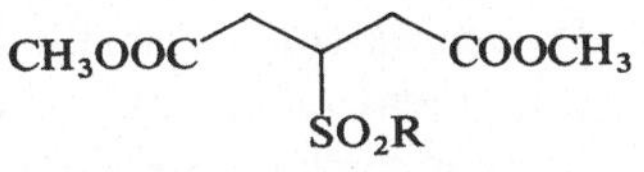

(**140**) $R = CH_3$ oder $C_{12}H_{25}$

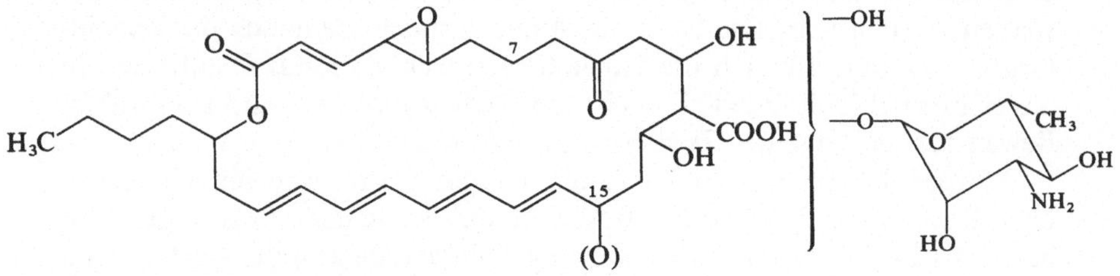

Partialformel C des Lucensomycins

Schema 26. Lage der Lactongruppe

den mit Chromsäure in Eisessig erhaltenen Oxydationsprodukten ließen sich nach der Veresterung mit Diazomethan zwei schwefelhaltige Methylester in reiner Form herauschromatographieren. Das eine Produkt (137) ging bei der Hydrolyse und erneuten Veresterung in das zweite (138) über. Die Schwefelfunktionen waren gleichzeitig zur Sulfonstufe oxydiert worden, denn beide Verbindungen zeigten die für Sulfone charakteristischen IR.-Absorptionsbanden bei 1325 und 1135 cm^{-1}. Ganz analog verlief die Oxydation des Methylmercaptan-Additionsproduktes (109). Als schwefelhaltiges Endprodukt wurde das Methylsulfon 139 isoliert. Beide Sulfone (138 und 139) erwiesen sich als identisch mit synthetischen Vergleichsproben (110). Vergeblich wurde in den Abbauprodukten nach dem entsprechenden Glutarsäurederivat 140 gesucht, das zum Vergleich ebenfalls synthetisiert worden war. Dies ist ein gutes Indiz dafür, daß die Stellung 4 im Lucensomycin durch eine Sauerstofffunktion besetzt ist.

Somit war für die beiden Stellungen 4 und 5 je eine Sauerstofffunktion nachgewiesen. Es stellte sich daher die Frage, ob das Aglykon ein 4,5-Diol sei, oder ob die früher nachgewiesene Epoxidgruppe diese beiden Stellen besetze. Titrationen mit Perjodat unter verschiedenen Bedingungen verliefen zunächst wenig aufschlußreich. Lucensomycin verbraucht, entsprechend der Gruppe $-CHOH-CHNH_2-CHOH-$ im Zuckerteil rasch zwei Mol Perjodat. Im Verlauf von 24 Stunden werden aber langsam weitere 2 Mol Oxydationsmittel verbraucht. N-Monopropionyllucensomycin (111, $R^1=H$, $R^2=CH_3-CH_2-CO$) verbraucht nur sehr langsam etwa 1 Mol Natriumperjodat. Daraus konnte aber nicht auf eine Diolgruppierung geschlossen werden, denn im Reaktionsprodukt konnte 2-Heptenal (127) nachgewiesen werden. Der Angriff erfolgte demnach am Polyensystem. Schließlich wurde Dodekahydrolucensomycin zunächst mit einem Überschuß Natriumperjodat oxydiert und das Reaktionsprodukt mit Platin in Eisessig bei 250° und 200 Atm. hydriert. Nach der Veresterung wurde mit etwa 20% Ausbeute der Ketoester 131 isoliert, d. h. daß das Kohlenstoffgerüst des Aglykons durch Perjodat nicht gespalten worden war. Die C-Atome 4 und 5 tragen demnach die Epoxid-Gruppe. Damit läßt sich die Partialformel weiter gemäß C auflösen. In dieser Formel fehlt lediglich noch eine Hydroxylgruppe, und ein direkter Beweis für die Lage des Zuckerrestes steht noch aus.

Sowohl Lucensomycin wie auch Dodekahydrolucensomycin geben beim Kochen mit 2 N Kalilauge Aceton, das, wenn auch in geringer Ausbeute, als 2,4-Dinitrophenylhydrazon gefaßt werden konnte. Aceton kann nur durch eine doppelte Retroaldolspaltung (C-7−C-8 und C-10−C-11) entstehen. Somit muß in Stellung 7 eine freie Hydroxylgruppe geschrieben werden, während C-8 und C-10 keine Sauerstoffatome tragen. Damit bleibt für den Zuckerrest allein noch das Sauerstoffatom an C-15 übrig, von dem schon früher festgestellt worden war, daß es wohl kaum als freie Hydroxylgruppe vorliegen könne.

Die Konstitutionsformel des Lucensomycins **108** wird damit vervollständigt. Nicht sicher bewiesen ist die β-Pyranosid-Konfiguration der Mycosaminid-Bindung. Für das nahe verwandte Pimaricin ist die Pyranosid-Form auf chemischem Wege bewiesen worden (*201*), die β-Konfiguration wurde in Analogie zum Amphothericin B geschrieben, von dem eine Röntgen-Strukturanalyse vorliegt (s. unten).

b) Weitere Polyen-Makrolide mit bekannter Struktur

Nahe verwandt mit dem Lucensomycin ist das Pimaricin (**141**) mit einer Methyl- statt einer n-Butylseitenkette. Die Konstitutionsaufklärung beruht auf denselben Prinzipien, wie sie oben für das Lucensomycin beschrieben wurden (*51, 52, 53, 54, 56, 57*). Die experimentellen Schwierigkeiten beim Arbeiten mit Substanzen dieser Reihe werden gerade an diesem Beispiel besonders deutlich: Als die Konstitutionsformel des Pimaricins bereits fertig abgeleitet zu sein schien (*53*), erwies sich eine Revision der Bruttoformel als notwendig. Die aus Elementaranalysen bestimmte Formel, $C_{33}H_{47}NO_{14}$, war mit massenspektrometrischen Befunden nicht vereinbar (*119*). Auf Grund der revidierten Formel, $C_{33}H_{47}NO_{13}$, mußte eine Korrektur der Konstitutionsformel vorgenommen werden.

Zwei weitere Glykoside des Mycosamins sind das Nystatin (=Fungicidin) und das Amphothericin B, die zwei verschiedenen spektralen Klassen angehören. Nystatin zeigt das UV.-Absorptionsspektrum eines konjugierten Tetraens (*93*), Amphothericin dasjenige eines Heptaens (*301*). Trotzdem erwiesen sich die beiden Antibiotica als strukturell nahe miteinander verwandt. Beim Nystatin ist die durchgehende Konjugation an einer Stelle unterbrochen, so daß zwei getrennte Doppelbindungssysteme, ein Tetraen und ein Dien, entstehen. Der im Amphothericin B (**143**) nachgewiesene Cyclo-hemiketal-Ring dürfte auch beim Nystatin — im kristallinen Zustand vollständig, in Lösung mindestens im Gleichgewicht mit der offenen Form (**142**) — vorhanden sein.

Die Konstitutionsformel des Nystatins ist das Produkt langjähriger chemischer und spektroskopischer Untersuchungen in verschiedenen Laboratorien (*26, 90, 154, 188*). Von der ersten Beschreibung des Antibioticums — Nystatin war das erste bekannte Polyenmakrolid (*93, 136, 137*) — bis zur Bestimmung der endgültigen Konstitution (*63*) vergingen volle zwanzig Jahre. Wiederum war es u. a. die Unsicherheit über die Bruttoformel, die einer raschen Lösung des Strukturproblems im Wege stand.

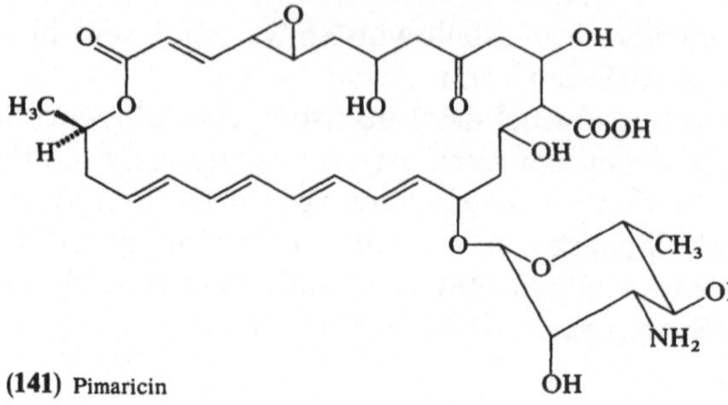

(141) Pimaricin

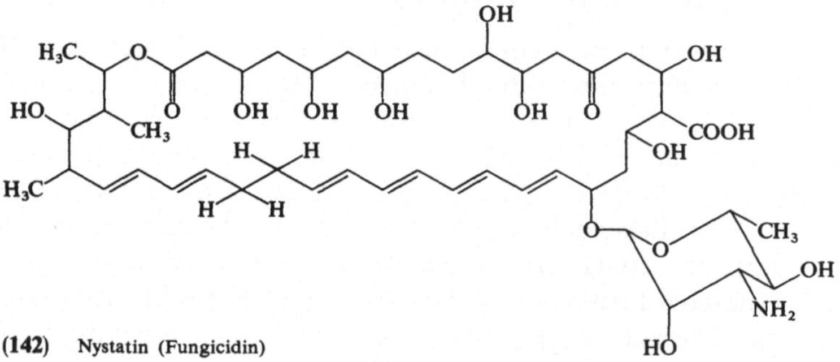

(142) Nystatin (Fungicidin)

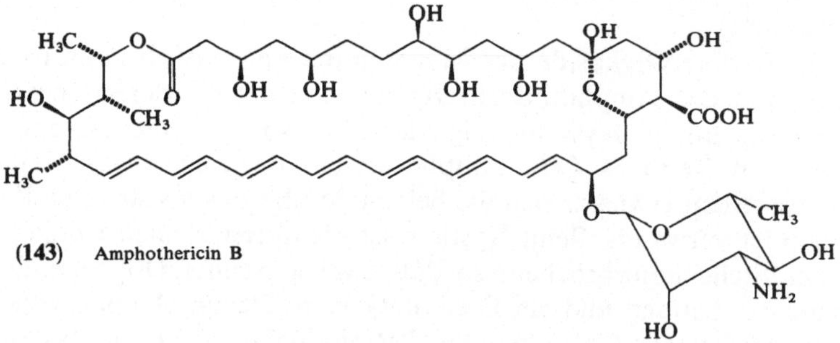

(143) Amphothericin B

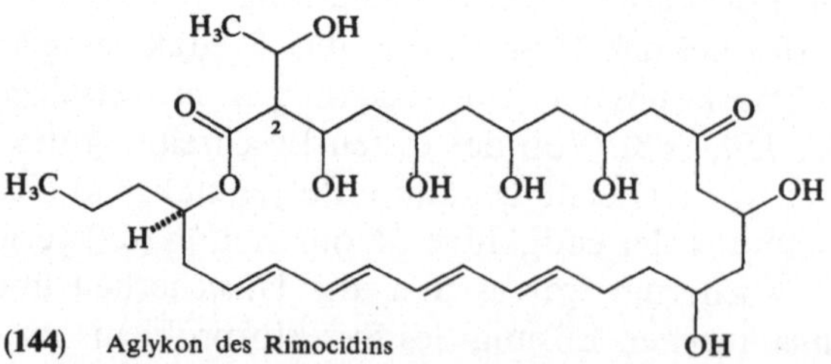

(144) Aglykon des Rimocidins

Schema 27. Weitere Polyen-Makrolide mit bekannten Strukturen

Literaturverzeichnis: SS. 445—460

Schema 27 (Fortsetzung)

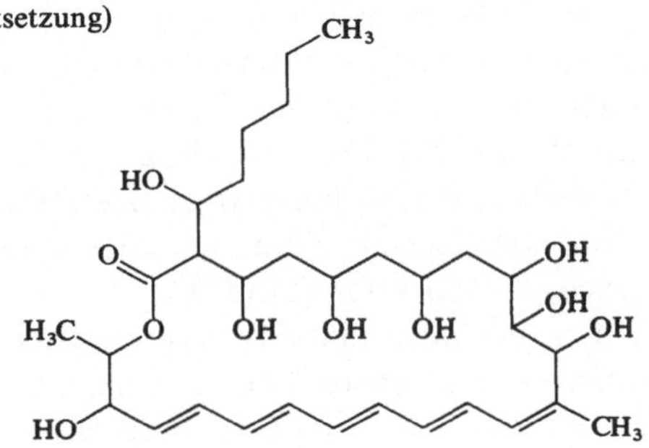

(145) Aurenin

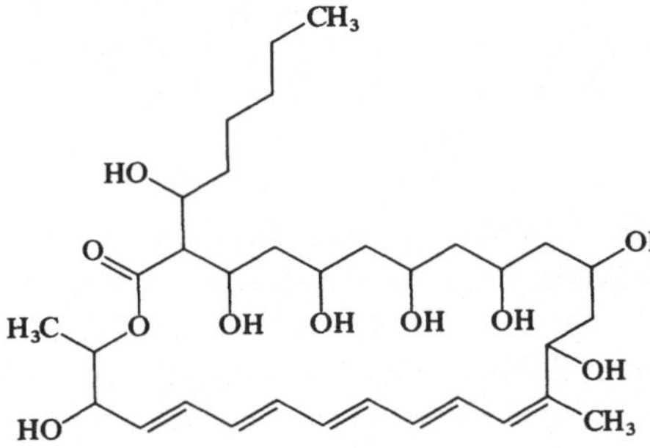

(146) Chainin

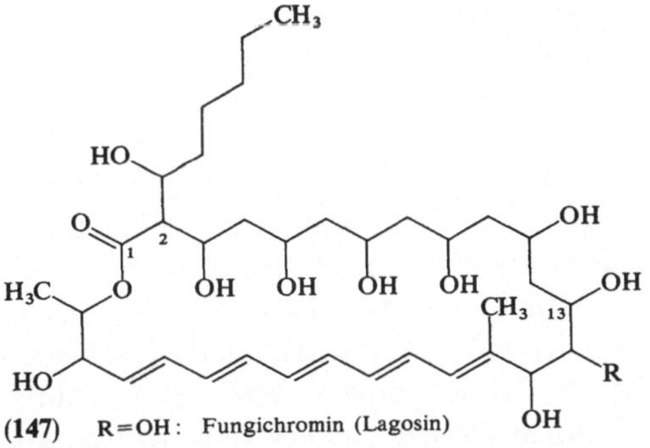

(147) R=OH: Fungichromin (Lagosin)
(148) R=H: Filipin III

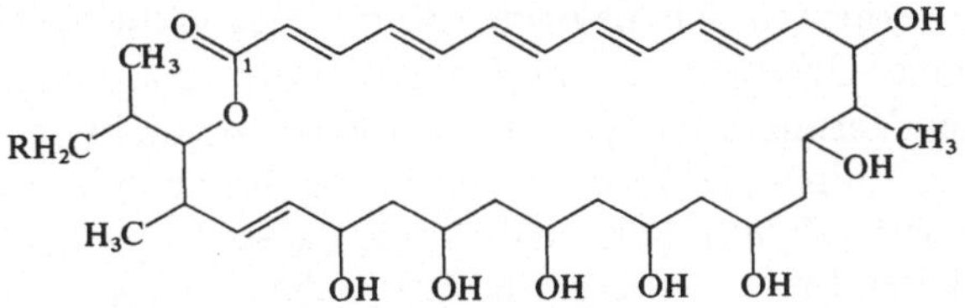

(149) R=H: Mycoticin A (Flavofungin A)
(150) R=CH₃: Mycoticin B (Flavofungin B)

Das Amphothericin B (**143**) ist bisher das einzige Polyenmakrolid, von dem ein geeignetes Derivat für eine Röntgen-Strukturanalyse zugänglich war. Es ist daher auch der erste Vertreter der Gruppe, dessen Stereochemie vollständig bekannt ist (*200*). Der kristallographischen Strukturaufklärung des Amphothericins B sind jahrelange Bemühungen um eine **Lösung des Problems auf chemischem Wege vorangegangen, die aber erst kürzlich abgeschlossen werden konnten** (*33*).

Nahe verwandt mit dem Nystatin und dem Amphothericin B ist das Candidin mit dem gleichen Kohlenstoffgerüst, einer durchgehenden Heptaen-Konjugation, aber zum Teil leicht veränderten Sauerstoffunktionen (*31*).

Beinahe vollständig aufgeklärt ist auch das Rimocidin (*67, 69*), ein Mycosaminid des Aglykons **144**. Es zeigt hinsichtlich der Aminoglykosid-Natur eine Verwandtschaft zu den oben besprochenen Lucensomycin und Pimaricin, andererseits leitet es hinüber zu den nichtglykosidischen Polyenen, mit denen es eine Hydroxyalkylseitenkette am C-Atom 2 und **das Fehlen einer Carboxylgruppe gemeinsam hat.**

Über die Konstitutionen der beiden einfachsten Vertreter der zuckerfreien Untergruppe, des Aurenins (**145**) mit einem 24gliedrigen und des Chainins (**146**) mit einem 26gliedrigen Lactonring, ist bisher nur in Vorträgen berichtet worden, so daß eine kritische Würdigung der Arbeiten noch nicht möglich ist (*298, 239*, vgl. *120, 288*).

Das Fungichromin (**147**) war das erste Polyenantibioticum, für das eine endgültige Konstitution abgeleitet wurde (*68, 70*). An ihm wurde die grundlegende Methodik der Chemie dieser Verbindungen zuerst durch Cope und Mitarb. entwickelt. Etwa gleichzeitig arbeiteten Whiting und Mitarb. unabhängig davon über das Antibioticum Lagosin und leiteten die gleiche Struktur (**147**) ab. Wegen geringfügiger Unterschiede in den spez. optischen Drehungen ($-174 \pm 4°$ für Fungichromin; $-160 \pm 4°$ für Lagosin, beide in Methanol) wurde die Möglichkeit stereochemischer Unterschiede — das Molekül besitzt 12 Chiralitätszentren — in Betracht gezogen. Doch ist es viel wahrscheinlicher, daß verschiedene Anteile von schwer nachweisbaren Beimengungen diese Abweichung hervorriefen. Denn im IR.-Absorptionsspektrum, im UV.-Absorptionsspektrum und bei der Dünnschichtchromatographie in vier verschiedenen Lösungsmittelsystemen verhielten sich die beiden Antibiotica völlig gleich (*68*), während geringfügige Unterschiede im Röntgen-Pulverdiagramm ebenfalls verschiedenen Verunreinigungen zugeschrieben werden können.

Für das Filipin (*317*) wurde, teils im Zusammenhang mit den Arbeiten über Lagosin (*79, 80*), teils unabhängig davon (*55, 89, 118*) die Struktur **148** abgeleitet. Erst später ergab es sich, daß das kristalline Filipin gar keine einheitliche Verbindung ist. Bergy und Eble (*21*) konnten durch Dünnschichtchromatographie und Verteilungschromatographie zahlreiche

Komponenten nachweisen. Durch Chromatographie an mit Phosphat gepuffertem Kieselgel ließen sich die drei Hauptkomponenten, die Filipine II, III und IV, die zusammen etwa 96% des Gemisches ausmachen, rein isolieren. Eine massenspektrometrische Untersuchung der Einzelkomponenten (238) ergab, daß die Konstitutionsformel **148**, die ursprünglich für „das Filipin" bestimmt worden war, für die Hauptkomponente, Filipin III, nach wie vor gültig ist. Die Komponenten Filipin I und II sind sauerstoffärmer, Filipin IV ist isomer zur Hauptkomponente.

Eine besondere Stellung innerhalb der Pentaen-Makrolide nehmen die Antibiotica Mycoticin und Flavofungin ein, indem der Polyen-Chromophor in die Konjugation zum Lactoncarbonyl verschoben ist. Auf die Art der biologischen Wirkung scheint dies keinen einschneidenden Effekt zu haben: Mycoticin und Flavofungin sind ebenfalls vorwiegend gegen Pilze und Hefen wirksam.

Die Konstitutionsaufklärung des Mycoticins (**149** + **150**) beruht auf denselben Methoden, wie sie für das Lucensomycin ausführlich geschildert wurden (310). Das kristalline Antibioticum ist ein bisher untrennbares Gemisch der beiden homologen Mycoticine A (**149**) und B (**150**). Eine Trennung von Homologen war erst auf der Stufe niedrigmolekularer Abbauprodukte sowie auf der Stufe des durch totale Reduktion erhaltenen Kohlenwasserstoffgemisches $C_{36}H_{74}$ bzw. $C_{37}H_{76}$ durch Gaschromatographie möglich.

Die gleichen Konstitutionsformeln (**149** und **150**) wurden auch für das Gemisch der beiden Flavofungine A und B abgeleitet (29). Trotz der ähnlichen Schmelzpunkte, nicht unterscheidbarer IR.-, UV.-, NMR.- und Massenspektren sowie identischem Verhalten mehrerer Derivate bei der Dünnschichtchromatographie sind die Flavofungine aber wahrscheinlich nicht identisch mit, sondern diastereomer zu den Mycoticinen, denn reines Flavofungin ist in Dioxan linksdrehend, während für das Mycoticin unter gleichen Bedingungen eine positive Drehung bestimmt wurde (29). Auch die Röntgen-Pulverdiagramme zeigen signifikante Unterschiede.

Obwohl bisher keine vollständige Konstitutionsformel angegeben werden kann, sind hier noch einige Worte über das Trichomycin (147) und einige verwandte (und wahrscheinlich zum Teil identische) Verbindungen, wie Ascosin (138), Candicidin (34), Levorin A (34) und Perimycin (32) anzufügen, weil hier ein neues strukturelles Element hinzukommt. Der oxydative Abbau von Trichomycin ergab je nach Reagens und Bedingungen — neben Abbauprodukten, die für ein Aglykon vom Typus der oben besprochenen zu erwarten sind — p-Aminobenzoesäure (**150**) oder p-Aminoacetophenon (**151**). Bei der energischen alkalischen Behandlung entsteht, wahrscheinlich über eine Retroaldolspaltung, ebenfalls p-Aminoacetophenon. Beim Perimycin ist das aromatische Produkt der alkalischen Spaltung N-Methyl-p-aminoacetophenon (153) (*180*). Für die Verknüpfung des aromatischen Bausteins mit dem Rest des Aglykons des Trichomycins postuliert HATTORI (*133*) auf Grund eingehender Abbauversuche die Partialformel 152. (Vgl. aber Nachtrag S. 460.)

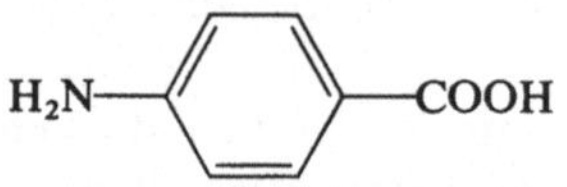

(150)

p-Aminobenzoesäure

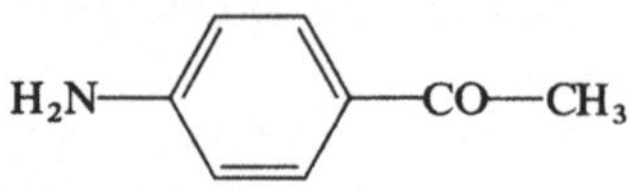

(151)

p-Aminoacetophenon

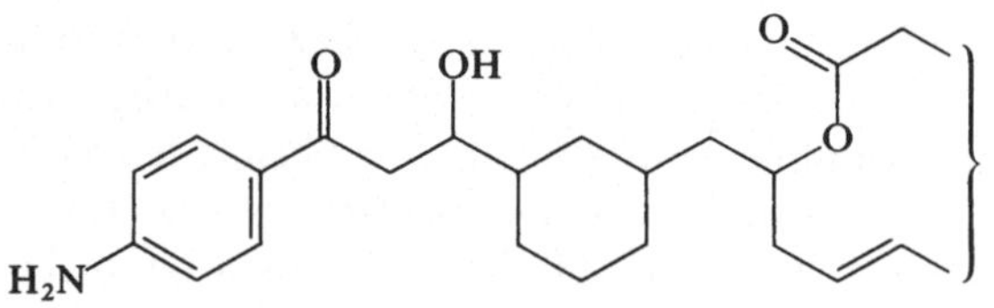

(152)   Partialformel des Trichomycins

(153)

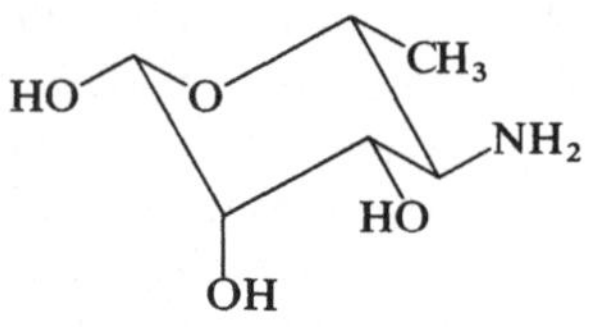

(154)   Perosamin

Schema 28. Ungewöhnliche Abbauprodukte des Trichomycins und Perimycins

Der Aminozuckerbaustein des Trichomycins ist, wie üblich in der Polyenreihe, das Mycosamin. Beim Perimycin tritt dagegen ein neuer Zuckerbaustein, das Perosamin (154) auf (*181, 280*).

Bei einem Vergleich der in der Struktur aufgeklärten und zahlreicher weiterer gut charakterisierter Polyenmakrolide fällt auf, daß mit der einzigen Ausnahme des Rimocidins (s. oben) alle Antibiotica entweder neutral sind und dann keinen Zucker enthalten, oder dann auf Grund der Anwesenheit eines Aminozuckers und einer Carboxylgruppe im Aglykonteil amphotheren Charakter besitzen. Ob dies hinsichtlich der

Biogenese oder im Hinblick auf die biologische Wirkung von Bedeutung ist, entzieht sich vorläufig unserer Kenntnis. Besonders interessant ist die weitgehende Uniformität in der biologischen Wirkung trotz beträchtlicher struktureller Unterschiede, die sich u. a. in der Ringgröße kundtut, die zwischen 24 (Aurenin) und 38 Gliedern (Amphothericin, Nystatin, Candidin) variieren kann.

IV. Primycin, Borrelidin, die Venturicidine und das Chlorothricin

In letzter Zeit sind einige Antibiotica aus Actinomyceten als Makrolide erkannt worden, die in keinem engeren Zusammenhang mit den drei großen Gruppen der voranstehenden Kapitel stehen. Obwohl sie auch unter sich keine nähere Verwandtschaft besitzen, sollen sie hier zusammen im gleichen Abschnitt behandelt werden.

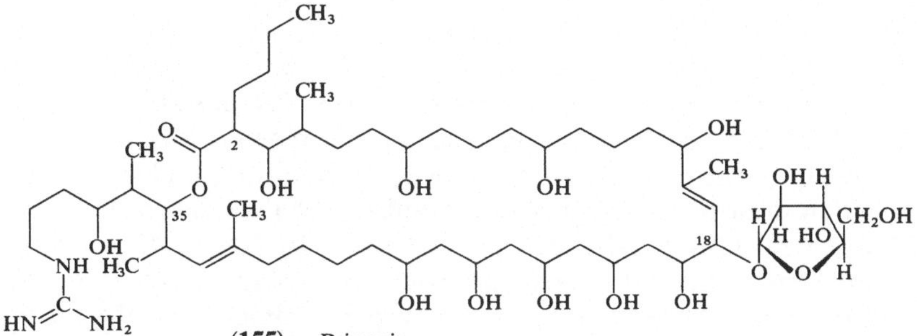

(155) Primycin

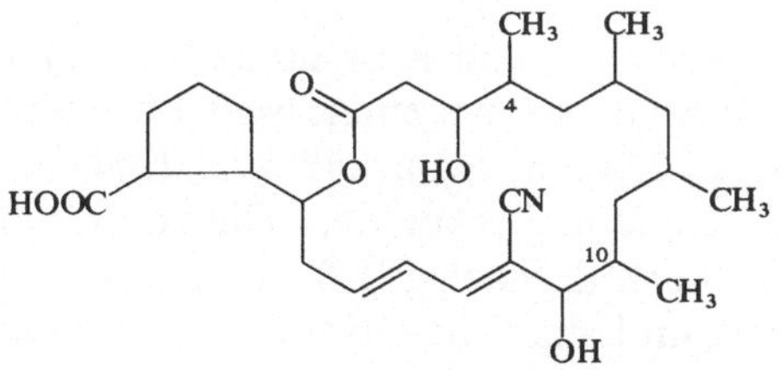

(156) Borrelidin

Schema 29. Primycin und Borrelidin

a) Primycin

Das *Primycin* (**155**) erinnert zwar in seiner Struktur noch stark an
die Polyenmakrolide, besonders was die Größe des Lactonringes (36
Ringglieder) betrifft. Aber auch andere strukturelle Eigenschaften sind solchen der Polyene ähnlich, so die Alkylseitenkette an C-2, der geringe Verzweigungsgrad der Kohlenstoffkette und die Verteilung der zahlreichen
freien Hydroxylgruppen. In seinen biologischen Eigenschaften ist es dagegen völlig verschieden von den Polyenen: es ist ausgesprochen hoch
wirksam gegen Gram-positive und säurefeste Bakterien, während Pilze
und Hefen nicht gehemmt werden.

Obwohl das Primycin schon 1954 erstmals beschrieben wurde (*300*),
ist über die Konstitutionsaufklärung erst kürzlich (*1*) in einer sehr knapp
gehaltenen Mitteilung berichtet worden. Sie beruht im wesentlichen auf
einem chemischen Abbau in Verbindung mit eingehenden instrumental-
analytischen Untersuchungen. Ungewöhnlich innerhalb der Reihe der
Makrolide ist die Guanidinogruppe, die dem Antibioticum einen stark
basischen Charakter verleiht. Der Zuckerbaustein ist die D-Arabinose, die
sonst im Bereiche der Antibiotica kaum je angetroffen worden ist.

b) Borrelidin

Ebenfalls vorwiegend antibakteriell wirkt das *Borrelidin* (*20, 184*).
Seine Wirkung beruht aber auf einem ganz anderen Mechanismus als
die der erythromycinartigen Antibiotica. Nach ZÄHNER und Mitarb. (*153*)
ist sie darauf zurückzuführen, daß der Einbau von Threonin in die Transfer-
RNA blockiert wird. BERGER und Mitarb. (*20*) fanden dazu eine ausgesprochene Aktivität gegen gewisse Spirochäten-Arten *(Borrelia)*,
DICKINSON und Mitarb. (*82*) eine solche gegen Viren. Die hohe Toxicität
gegenüber Warmblütern (*153, 184*) schon beim Hautkontakt mit Spuren
des Antibioticums hat eine eingehende chemische Untersuchung lange Zeit
verzögert.

Diese wurde eingeleitet von ANDERTON und RICKARDS (*7*), die durch
eine ausgedehnte spektroskopische Analyse sowie einige wenige chemische
Reaktionen den Chromophor (λ_{max} 254 nm, log ε 4,48; λ_{max} 2204 cm^{-1})
als eine konjugierte Diennitril-Gruppierung erkannten. Die vollständige
Konstitutionsaufklärung erfolgte kurz darnach (*159*) durch chemische
Abbaureaktionen, verbunden mit umfangreichen NMR.- und massenspektrometrischen Studien, und führte zur Konstitution **156**, an der neben
der ungewöhnlichen Nitrilgruppe der Cyclopentanring auffällt. Dieser ist
biogenetisch nicht ganz leicht erklärbar und tritt sonst im Bereich der
Makrolide nur noch beim Brefeldin A (s. unten) auf. Über die Stereo-

chemie des Borrelidins ist nur wenig bekannt. Der Cyclopentanring dürfte auf Grund von Abbauprodukten *trans*-substituiert sein, für die vier Verzweigungsstellen im Bereich der C-Atome 4 bis 10 wurde die relative *allo*-Konfiguration abgeleitet (*46, 160*).

c) Die Konstitutionsaufklärung der Venturicidine

Die Venturicidine A und B besitzen einen völlig anderen Charakter. Sie sind neutrale, lipophile Verbindungen, die — obwohl strukturell eher wieder an das Erythromycin erinnernd — eine ausgesprochen hohe Wirkung gegen Pilze aufweisen, darunter gegen zahlreiche Erreger von Pflanzenkrankheiten (*48, 257*). Das Besprühen von Pflanzen mit Lösungen von Venturicidin verleiht ihnen einen guten Schutz vor Ansteckung, doch verbieten phytotoxische Eigenschaften eine breite Anwendung im Pflanzenschutz.

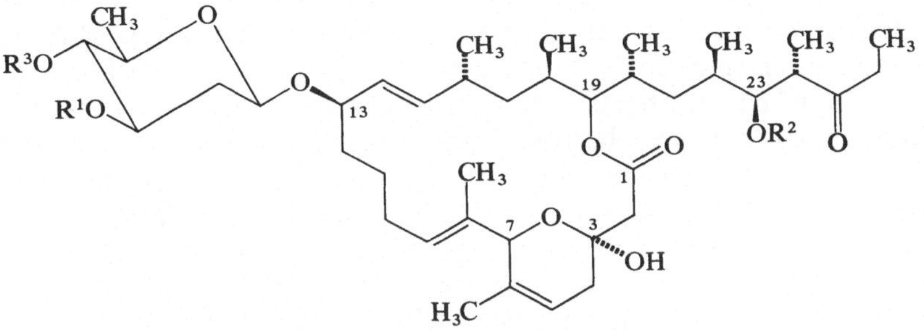

(**157**) $R^1 = -CONH_2$, $R^2 = R^3 = H$: Venturicidin A

(**158**) $R^1 = -CONH_2$, $R^2 = R^3 = -COCH_3$

(**159**) $R^1 = R^2 = R^3 = H$: Ventirucidin B

(**160**) $R^1 = R^2 = R^3 = -COCH_3$

(**161**) $R^1 = -CONH_2$, $R^2 = H$, $R^3 = J$—⬡—SO_2—

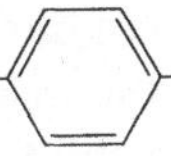

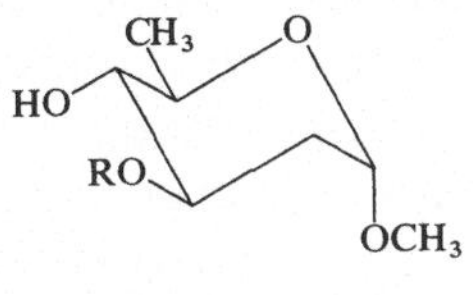

(**162**) $R = -CONH_2$

(**163**) $R = H$

Schema 30. Die Venturicidine A und B und einige Derivate

Das Venturicidin A wurde erstmals 1961 beschrieben (*257*). Später wurde aus Kulturen eines Stammes von *Streptomyces aureofaciens* ein Gemisch von Venturicidin A, Venturicidin B und dem mit den Oligomycinen verwandten Botrycidin erhalten (*48*). Die Auftrennung des Gemisches gelang mit Hilfe einer kombinierten Anwendung der CRAIG-Verteilung und der Säulenchromatographie an Kieselgel. Das Venturicidin A wurde in farblosen Kristallen erhalten, das Venturicidin B dagegen nur in Form eines amorphen Pulvers, das sich aber chromatographisch und spektroskopisch als einheitliche Substanz erwies. Das Venturicidin A enthielt Stickstoff, doch mußte die ursprünglich aus Analysen abgeleitete Bruttoformel ($C_{43}H_{71}NO_{12}$) später leicht korrigiert werden (*45*). Das Venturicidin B enthält keinen Stickstoff und unterscheidet sich vom Venturicidin A in der Elementarzusammensetzung durch den Mindergehalt von CHON. Die darauf gegründete Vermutung, daß das Venturicidin A ein Carbaminsäureester des Venturicidins B sei, wurde zunächst durch spektroskopische Befunde, später durch zahlreiche gemeinsame Abbauprodukte gestützt. Das Venturicidin A besitzt nämlich im CO-Gebiet IR.-Absorptionsbanden bei 1710 und 1605 cm^{-1}, die für Carbamate charakteristisch sind, das Venturicidin B dagegen nur ein scharfes Maximum bei 1705 cm^{-1}. Ebenfalls entsprechend dieser Annahme enthält das Venturicidin A eine acetylierbare Hydroxylgruppe weniger als das Venturicidin B, nämlich zwei (vgl. Formelschema 30). Die Lage des Carbaminsäurerestes am Sauerstoffatom in Stellung 3 eines 2-Desoxy-**D**-rhamnoserestes wurde durch die Isolierung des Methylglykosids **162** nach einer milden säurekatalysierten Methanolyse von Venturicidin A festgelegt (*48*). Das Venturicidin B gab bei der analogen Behandlung das stickstofffreie α-Methyl-2-desoxy-D-rhamnosid (**163**), das auch bei der alkalischen Hydrolyse des Carbaminsäureesters **162** entsteht. Bei der Konstitutionsaufklärung des Carbamats **162** spielte ferner die NMR.-Spektroskopie eine wesentliche Rolle.

Das Aglykon erlitt bei den erwähnten Methanolysen wie auch bei Hydrolyseversuchen sekundäre Umwandlungen. An seiner Stelle wurde ein schwer trennbares Gemisch mehrerer Zersetzungsprodukte isoliert, das keine wesentlichen Schlußfolgerungen erlaubte.

Die vollständige Strukturaufklärung der beiden Venturicidine ist besonders interessant, weil sie auf einer Kombination röntgenographischer und chemischer Methoden beruht. Diese Arbeit verdanken wir vor allem BRUFANI (*45*), der die chemischen, spektroskopischen und kristallographischen Methoden gleichermaßen beherrscht. Die Strukturen **157** und **159** sind denn auch durch keine der Methoden für sich allein bewiesen, hingegen erlauben alle zusammen, den Venturicidinen eindeutige Strukturen zuzuordnen.

Für die Röntgenstrukturanalyse wurden zunächst Versuche zur Herstellung eines kristallinen Schweratom-Derivates ausgeführt. Von mehreren, meist schlecht kristallisierenden Derivaten eignete sich das Mono-p-jodbenzolsulfonat (161) noch am besten, obwohl auch es nur recht kleine Kristallprismen gab. Sie erlaubten nur die Ausmessung von 1147 unabhängigen Reflexionen, also nur etwa halb so vieler, als für eine vollständige Strukturanalyse erwünscht wären. Dementsprechend war keine sehr gute Auflösung der Struktur zu erwarten. Tatsächlich war das Ergebnis der FOURIER-Synthesen zunächst eher enttäuschend. Es ist etwas vereinfacht (in ebener Projektion) in der Formel 164 wiedergegeben. Sie gibt zwar die Anordnung und die Verbundenheit aller mittelschweren Atome (C, N, O) bis auf eines (C-27) wieder, erlaubt aber nicht mit Sicherheit, zwischen Kohlenstoff-, Stickstoff- und Sauerstoffatomen zu unterscheiden. Ferner konnten wegen der Ungenauigkeit der Atomabstände Doppelbindungen nicht lokalisiert werden. Daß aber solche vorhanden sind, ergab sich aus dem NMR.-Spektrum des Venturicidins A mit einer etwa 4 Protonen entsprechenden Signalgruppe bei etwa 5,5 ppm.

Der Wert einer derart vagen Struktur mag auf den ersten Blick sehr problematisch erscheinen. Wie sich aber bald ergeben wird, hat sie die Konstitutionsaufklärung mit chemischen Methoden enorm erleichtert und ergänzt.

Die schweren Jod- und Schwefelatome ließen sich in der Formel 164 sofort lokalisieren, womit die Lage des p-Jodbenzolsulfonylrestes gegeben war. Im daran anschließenden Teil kann man ohne Schwierigkeit den 3-O-Carbaminyl-2-desoxy-**D**-rhamnoserest erkennen, der somit β-pyranosidisch an das Sauerstoffatom O-C-13 des Aglykons anschließt. Der große Ring ließ eine makrolidartige Struktur vermuten, konnte sie aber nicht sicher beweisen. Insbesondere blieb die Lage einer allfälligen Lactongruppe zunächst unbestimmt. In der langen Seitenkette wurde die Auflösung mit zunehmendem Abstand vom Ring weniger scharf, was wohl mindestens zum Teil auf eine gewisse konformative Variabilität in diesem Molekülteil zurückzuführen ist. Insbesondere konnte die später chemisch nachgewiesene Methylgruppe C-27 nicht erkannt werden. Die Raumfüllung im Kristallgitter ließ aber die Möglichkeit zu, daß das C-Atom 26 noch nicht das Ende der Kette war.

Der chemische Abbau wurde zunächst am Di-O-acetylventuricidin A (158) mittels Ozon durchgeführt und die rohen Ozonide mit Wasserstoffperoxid in Eisessig in Gegenwart von *Salzsäure* gespalten, was sich bald als bedeutungsvoll erwies. Das Rohprodukt wurde mit Diazomethan verestert und dann acetyliert, wodurch man ein gut chromatographierbares Produktegemisch erhielt.

Von den bisher vier in einheitlicher Form isolierten Produkten konnte eines sofort durch das NMR.-Spektrum als 1,4-Di-O-acetyl-3-O-carba-

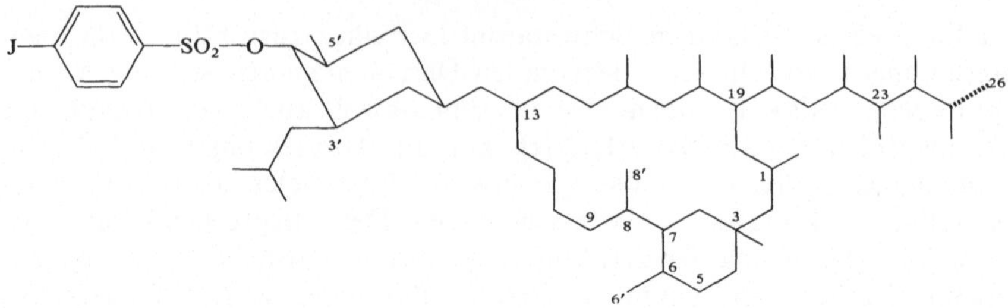

(164) Atomanordnung des Derivates **(161)** (gemäß Röntgenstrukturanalyse)

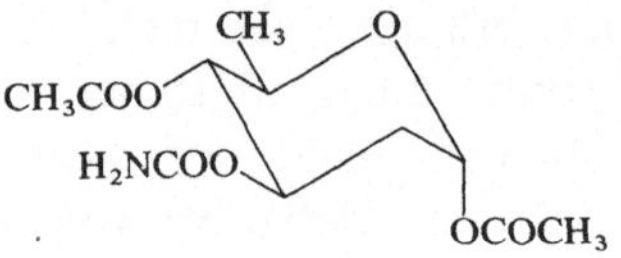

(165)

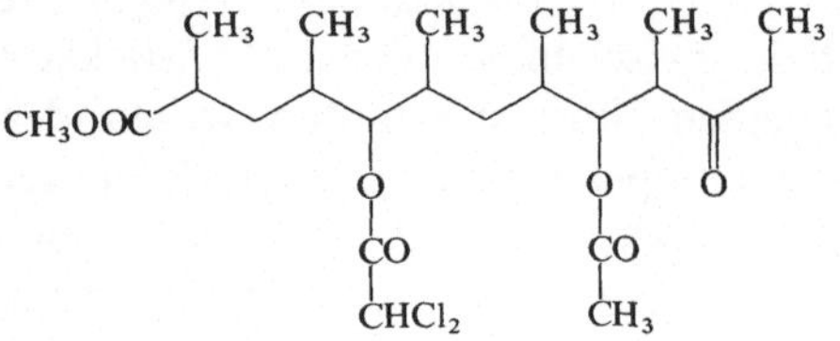

(166)

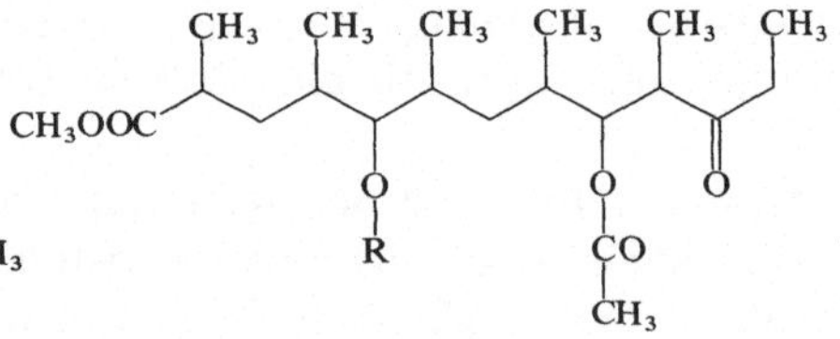

(167) R = —CO—CH$_2$—COOCH$_3$

(168) R = —CO—CH$_3$

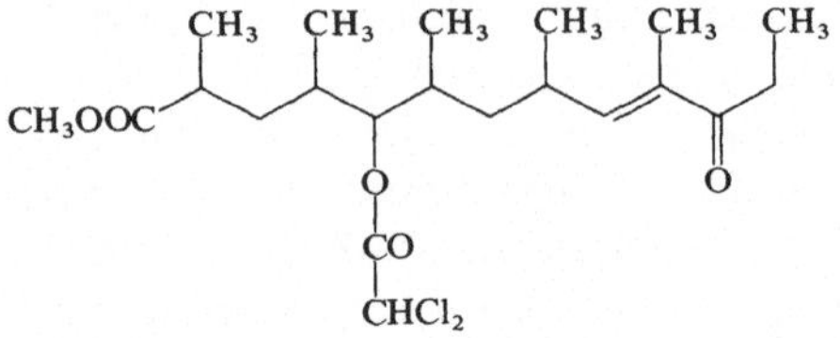

(169)

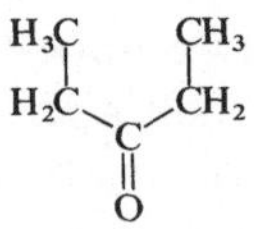

(170) Diäthylketon

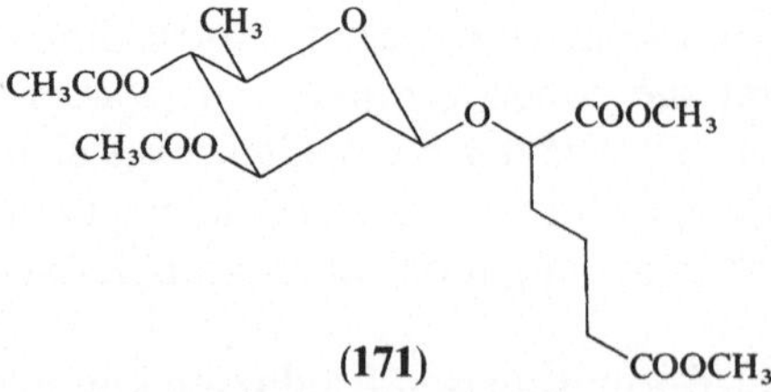

(171)

Schema 31. Röntgenstrukturanalyse und Abbau des Venturicidins A

Literaturverzeichnis: SS. 445—460

Schema 31 (Fortsetzung)

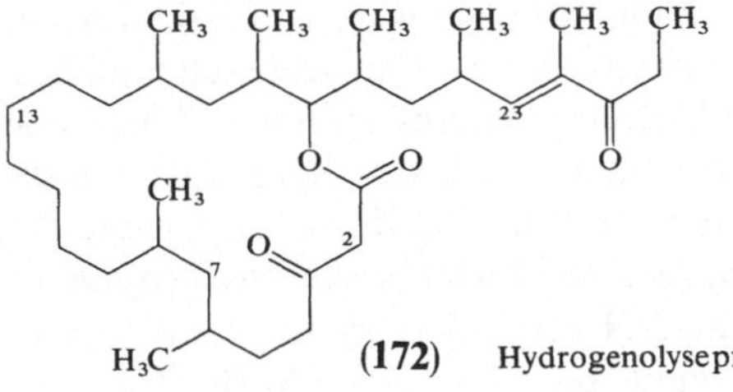

(**172**) Hydrogenolyseprodukt A

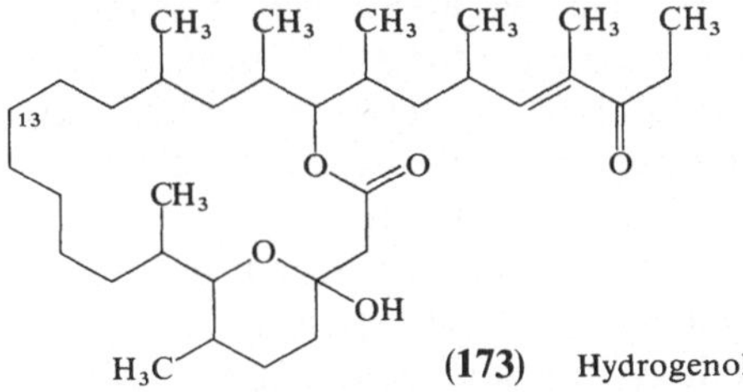

(**173**) Hydrogenolyseprodukt B

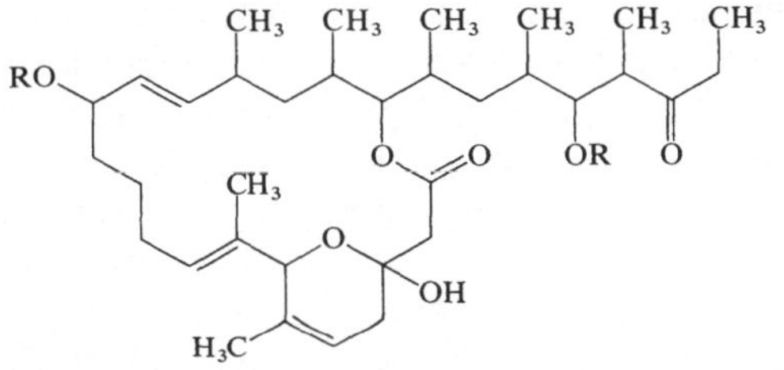

(**174**) R = H : Aglykon der Venturicidine
(**175**) R = —CO—CH$_3$: Di-O-acetyl-aglykon

minyl-2-desoxy-**D**-rhamnose (**165**) identifiziert werden. Ein zweites Produkt (**166**) erwies sich überraschenderweise als chlorhaltig, und die Isotopenverteilung im Bereich des schwachen Molekülions (m/e 496, $C_{23}H_{38}Cl_2O_7$) wies auf die Anwesenheit von zwei Chloratomen hin. Dem Massenspektrum konnte das Vorliegen einer O-Dichloracetylgruppe entnommen werden, die durch ein Singulett bei 5,93 ppm im NMR.-Spektrum (Cl_2CH-COO-) bestätigt wurde. Da das Ozonisierungsprodukt bei der Nachbehandlung und Aufarbeitung nie mit einem Derivat der Dichloressigsäure in Berührung gekommen war, muß die Dichloracetylgruppe aus einem Molekülteil des Venturicidins durch oxydativen

Abbau und Chlorierung entstanden sein, wobei sich die letztere bei der Ozonidspaltung in Gegenwart von Salzsäure in einem oxydativen Milieu abgespielt haben muß. Ein wahrscheinliches Zwischenprodukt dieses Abbaus bildet ein weiteres in reiner Form isoliertes Produkt (**167**). Ein spektroskopischer Vergleich zeigte, daß sich dieses vom eben beschriebenen Dichloracetat nur dadurch unterscheidet, daß die Dichloracetylgruppe durch eine O-Methylmalonylgruppe ersetzt ist (2 OCH_3-Signale, Singulett zu 2 H bei 3,41 ppm im NMR.-Spektrum; im Massenspektrum starkes Signal bei m/e 397 = M-89 für den Verlust von -CO-CH_2-$COOCH_3$). Übereinstimmend damit gaben die beiden Verbindungen (**166** und **167**) bei der alkalischen Verseifung, Veresterung mit Diazomethan und Reacetylierung das gleiche Diacetylderivat **168**. Es ist bekannt, daß Malonsäurederivate leicht halogenierbar, Halogenmalonsäuren leicht decarboxylierbar sind. Die Dichloracetylgruppe gibt demnach die Lage der Lactonbindung im Venturicidin A an.

Ein wichtiger Befund am Abbauprodukt **166**, $C_{23}H_{38}Cl_2O_7$, war die Anwesenheit von 6 C-Methylgruppen. Es war damit klar, daß sich dieses Produkt nur aus dem stark verzweigten Molekülteil etwa im Bereich der Atome 15 bis 26 (**164**) ableiten kann. Die Verbindung **166** besitzt eine einzige O-Methylgruppe, die offensichtlich einer Methylestergruppe angehört. Dazu sind im NMR.-Spektrum die Signale einer O-Acetylgruppe und von zwei zu Estersauerstoffen geminalen Wasserstoffatomen (δ 5,03 und 4,77 ppm) zu sehen. Sowohl die Acetyl- wie auch die Dichloracetylgruppe sind demnach an sekundäre Sauerstoffatome gebunden. Die Ketogruppe gibt sich nicht unmittelbar zu erkennen, da die vier Carbonylbanden im IR.-Absorptionsspektrum zu einer breiten Bandengruppe mit Schwerpunkt bei 1735 cm^{-1} zusammenfallen. Ihre Anwesenheit ergibt sich aber aus der leichten Elimination von Essigsäure bei der milden Behandlung des Esters **166** mit methanolischer Salzsäure. Das Produkt (**169**), das auch direkt aus dem veresterten Ozonisierungsprodukt isoliert wurde, besitzt im IR.-Absorptionsspektrum Banden bei 1730 und 1665 cm^{-1} und ein hohes UV.-Absorptionsmaximum bei 236 nm (log ε 4,00) und ist demnach ein α,β-ungesättigtes Keton. Gemäß dem Massenspektrum ($M^+ = 436$, starker Peak, $C_{21}H_{34}Cl_2O_5$) unterscheidet es sich vom Ausgangsmaterial (**166**) nur durch den Verlust von einem Molekül Essigsäure. Das letztere ist demnach ein β-Acetoxyketon. Ein weiterer intensiver Peak im Massenspektrum bei m/e 308 (M-128) ist durch den Verlust von Dichloressigsäure zu erklären, ein solcher bei m/e 407 (M-29) deutet auf eine Äthylketongruppe hin. Im NMR.-Spektrum des Eliminationsproduktes fehlt das Signal der O-Acetylgruppe; in der 1-ppm-Region ist ein Methylsignal weniger zu erkennen, dafür tritt neu ein Dreiprotonen-Singulett bei 1,78 ppm (Methylgruppe an der Doppelbindung) auf. Ein Dublett (J =

9,2 Hz., 1 H) bei 6,37 ppm zeigt ein einziges olefinisches Proton an. Doppelresonanzversuche ließen die Partialformel $-CH(CH_3)-CH=C(CH_3)-CO-$ ableiten. Wegen des Fehlens eines Singuletts in der 2-ppm-Gegend ist die Verbindung kein Methylketon.

Aus biogenetischen Überlegungen (Propionat-Regel) war eine gleichmäßige 1,3-Verteilung der C-Methylgruppen zu erwarten. Daraus, aus den erwähnten spektroskopischen Eigenschaften der Abbauprodukte **166** und **169** sowie unter Berücksichtigung der Atomanordnung **164** ergaben sich für diese Produkte, zunächst als Hypothese, die angegebenen Konstitutionsformeln, die nachfolgend durch weitere Versuche bestätigt wurden:

1. Das Auftreten eines starken Peaks in den Massenspektren beider Verbindungen bei m/e 88 (McLafferty-Umlagerung) und das Fehlen eines solchen bei m/e 74 beweist die Lage einer C-Methylgruppe in α-Stellung zur Carbomethoxygruppe und damit die Lage einer Doppelbindung zwischen den C-Atomen 14 und 15 im Venturicidin A (**157**).

2. Das Signal bei 5,03 ppm im NMR.-Spektrum des Esters **166** (Doppeldublett mit $J_1 = 3,4$ und $J_2 = 9,0$ Hz.) zeigte bei Doppelresonanzversuchen Spinentkopplung mit Signalen bei 2,8 ppm ($-CH-CO-$) und 1,9 ppm, welche beide wiederum mit Signalen in der C-Methylgegend gekoppelt sind. Es wird dadurch die Partialformel $-CH(CH_3)-CH(OAc)-CH(CH_3)-CO-$ eines β-Acetoxyketons bestätigt. Analoge Spinentkopplungen, ausgehend vom Signal bei 4,77 ppm, das in gleicher Form auch beim Eliminationsprodukt **169** auftritt und demnach zur Dichloracetoxygruppe geminal ist, bestätigten die Partialstruktur $-CH(CH_3)-CH(OCOCHCl_2)-CH(CH_3)-$.

Die Abbauprodukte **166**, **167** und **169** wurden auch bei einem analogen Abbau des Tri-O-acetylventuricidins B (**160**) erhalten, nicht aber das Zuckerderivat **165**. An dessen Stelle trat eine neue Verbindung auf, die auf Grund des Massenspektrums und des NMR.-Spektrums und unter Berücksichtigung der Atomanordnung **164** nur der Glycosidester **171** sein kann. Durch dieses Produkt wird zusätzlich die Lage einer Doppelbindung C-8 bis C-9 experimentell erfaßt.

Da die Äthylketon-Endgruppe, die durch diese Abbauprodukte für die Venturicidine gefordert wird, durch die Röntgenstrukturanalyse nicht erfaßbar war, wurde diese durch einen unabhängigen Abbau bestätigt. Beim Erwärmen der Venturicidine A und B mit verdünnter Natronlauge bildeten sich je etwa 60% eines Mols Diäthylketon, das als 2,4-Dinitrophenylhydrazon identifiziert wurde, als Folge einer Retroaldolspaltung zwischen den C-Atomen 23 und 24. Damit war die Äthylketon-Endgruppe der Seitenkette bewiesen.

Die Konstitution im übrigen Bereich des Moleküls (C-3 bis C-8) ergab sich aus folgenden Befunden: Das Tri-O-acetylventuricidin zeigt im

IR.-Absorptionsspektrum in Chloroform noch eine deutliche OH-Bande für eine nicht acetylierbare Hydroxylgruppe. Gemäß der Atomanordnung **164** kommt für eine tertiäre Hydroxylgruppe nur die Seitenkette am Atom 3 in Frage. Die Venturicidine A und B besitzen gemäß NMR.-Spektren je zwei Methylgruppen an Doppelbindungen (verbreiterte Singulette bei 1,45 und 1,52 ppm) für die nur noch die Verzweigungen an C-6 und C-8 (**164**) übrigbleiben. Da die beiden Doppelbindungen gemäß UV.-Absorptionsspektren nicht konjugiert sind, müssen sie zwischen den C-Atomen 5 und 6 bzw. 8 und 9 stehen. Eine Bestätigung dafür ergibt sich aus der Kristallstruktur, weniger aus den Atombeständen als daraus, daß einerseits die C-Atome 5, 6, 6' und 7, andererseits die Atome 7, 8, 8' und 9 je in einer Ebene liegen. Die Doppelbindung C-8 bis C-9 ergab sich schon früher aus dem Abbauprodukt **171**.

Zuletzt blieb noch übrig, ein zusätzliches Sauerstoffatom, das weder einer Hydroxyl- noch einer Carbonylgruppe angehört, zu lokalisieren. Aus biogenetischen Überlegungen wurde dafür die Brücke von C-3 nach C-7 in Betracht gezogen, die später röntgenographisch bestätigt wurde. Damit wurde die Strukturformel **157** vervollständigt. Diese weist zwei allylständige Sauerstoffatome in den Stellungen 7 und 13 auf. Es ist daher kein einheitlicher Verlauf von katalytischen Hydrierungen zu erwarten, da neben der Addition von Wasserstoff an die Doppelbindungen auch hydrogenolytische Spaltungen von C-O-Bindungen eintreten können. Tatsächlich nahmen die Venturicidine in Gegenwart von Palladiumkohle deutlich mehr als drei Mole Wasserstoff auf, und die Rohprodukte erwiesen sich als Gemische mehrerer Komponenten (*44*). Bisher konnten durch Chromatographie an Kieselgel zwei einheitliche Produkte abgetrennt werden. Das erste besaß die Bruttoformel $C_{34}H_{60}O_4$ (intensiver Molekülpeak bei m/e 532 im Massenspektrum) und besitzt keine freien Hydroxylgruppen. Das NMR.-Spektrum paßt ausgezeichnet auf die Konstitutionsformel **172**. Insbesondere ist das CH_2-Signal der β-Ketoestergruppe (H-C-2) als Singulett (2 H) bei 3,46 ppm gut zu erkennen. Das Produkt ist demnach durch zweifache Hydrogenolyse in den Stellungen 7 und 13 entstanden und bestätigt damit die Struktur im Bereiche des Cyclohemiketalringes. Das Produkt **172** zeigt ein UV.-Absorptionsspektrum eines α,β-ungesättigten Ketons, da offenbar bei der Aufarbeitung die Hydroxylgruppe an C-23 durch eine β-Elimination verlorengegangen ist. Damit im Einklang ist ein Signal bei 6,37 ppm für olefinisches Proton und bei 1,78 ppm (d, J = 1 Hz., 3 H) für die Methylgruppe an C-24.

Das zweite rein isolierte Produkt der Hydrierung (**173**) hat ebenfalls den Zuckerteil verloren. Das Massenspektrum ($M^+ = 548$) ist verträglich mit der Bruttoformel $C_{34}H_{60}O_5$. Es besitzt noch die tertiäre Hydroxylgruppe und zeigt das charakteristische NMR.-Signal für eine $-CO-$

CH_2—CO-Gruppe nicht. Der Cyclohemiketalring ist demnach erhalten; eine Hydrogenolyse ist nur an C-13 eingetreten. Auch diese Verbindung (**173**) hat die Hydroxylgruppe an C-23 durch eine β-Elimination im Anschluß an die Hydrierung verloren und ist daher ein α,β-ungesättigtes Keton (λ_{max} 234 nm, log ε 4,05; δ (CH_3) = 1,82 ppm).

Die Konfiguration entsprechend der Formel **157** ergab sich aus der Röntgenstrukturanalyse; der Chiralitätssinn folgt aus der bekannten Chiralität der 2-Desoxy-**D**-rhamnose.

Die Strukturen der Venturicidine A und B (**157** bzw. **159**) widersprechen den ursprünglich aus Analysen und osmometrischen Molekulargewichtsbestimmungen errechneten Bruttoformeln, $C_{43}H_{71}NO_{12}$ bzw. $C_{42}H_{70}O_{11}$ (*48, 257*), doch kann daraus kein stichhaltiger Einwand gegen die gut begründeten Strukturformeln abgeleitet werden. Wie bereits früher darauf hingewiesen wurde, sind die Elementaranalysen bei Verbindungen dieses Typus meist mehrdeutig. Tatsächlich weichen die theoretischen Elementarzusammensetzungen der revidierten Formeln, $C_{41}H_{67}NO_{11}$ bzw. $C_{40}H_{66}O_{10}$, von denen der ursprünglichen nur geringfügig ab. Trotzdem war eine exakte Bestätigung der neuen Bruttoformeln wünschenswert. Unglücklicherweise zeigten die Massenspektren der Venturicidine und ihrer Acetylderivate (**157** bis **160**) keine Molekülionen und keine aus den Molekülionen eindeutig ableitbaren großen Fragmente. Wie bereits erwähnt, war das Aglykon, das hätte aus dem Dilemma helfen können, durch Solvolysereaktionen nicht zugänglich. Auf einem Umweg konnte dieses schließlich doch gewonnen werden.

Das Venturicidin B (**159**) besitzt im Zuckerteil eine 1,2-Diolgruppierung und kann deshalb mit Perjodsäure oxidiert werden (*45*). Offensichtlich ist die Acetalgruppierung im Oxydationsprodukt viel empfindlicher als diejenige im intakten Glykosid, und wird schon bei der Aufarbeitung hydrolytisch gespalten. Die spektroskopischen Daten des einzigen isolierbaren und kristallisierbaren Produktes (Smp. 135—137°) waren in allen Einzelheiten mit der Struktur **174** vereinbar. Es muß sich demnach um das nicht weiter veränderte Aglykon handeln. Das Massenspektrum dieser Verbindung ist zwar immer noch schwer interpretierbar, doch das Diacetylderivat **175** gab endlich ein Massenspektrum mit einem deutlichen Signal bei m/e 642, das einem Ion $(C_{38}H_{58}O_8)^+$ = (M-18) entspricht. Diese Interpretation wird bestärkt durch intensive Peaks bei m/e 582 (M-18-60) und 522 (M-18-60-60) und bestätigt somit die revidierten Bruttoformeln der Venturicidine.

Zuletzt wurde die Situation auch von der kristallographischen Seite her wesentlich verbessert, als der Einsatz eines computergesteuerten automatischen Refraktometers möglich wurde (*45*). Dieser erlaubte eine viel genauere Ausmessung der Reflexe und damit eine bessere Auflösung der Struktur. Es war jetzt möglich, alle Sauerstoffatome gemäß Formel **157** durch die erhöhte Elektronendichte zu erkennen. Noch immer aber ließ sich die Methylgruppe C-27, die andererseits durch den chemischen Abbau sicher erwiesen ist, nicht nachweisen. Offensichtlich ist sie im Kristallgitter nicht streng fixiert, sondern kann verschiedene konformative Positionen einnehmen.

d) Chlorothricin

Ein ganz ungewöhnliches Makrolid stellt das *Chlorothricin* dar, in dem der makrocyclische Lactonring mit carbocyclischen Ringen und einem Tetronsäuresystem verbunden ist. Durch den Lactonringschluß direkt an die Doppelbindung der Tetronsäure erscheint die Carbonylfrequenz der Lactongruppe im IR.-Absorptionsspektrum bei etwa 1784 cm^{-1} und täuscht die Anwesenheit eines γ-Lactons vor.

Das Chlorothricin (**176**), eine farblose zweiwertige Säure, wird von einem Stamm von *Streptomyces antibioticus* im Gemisch mit wechselnden Anteilen seines Deschloroderivates (**177**) gebildet (*162*). Die beiden Antibiotica ließen sich trotz der Anwendung moderner Separierverfahren nicht voneinander trennen. Erst anhand von Abbauprodukten (**179, 181**)

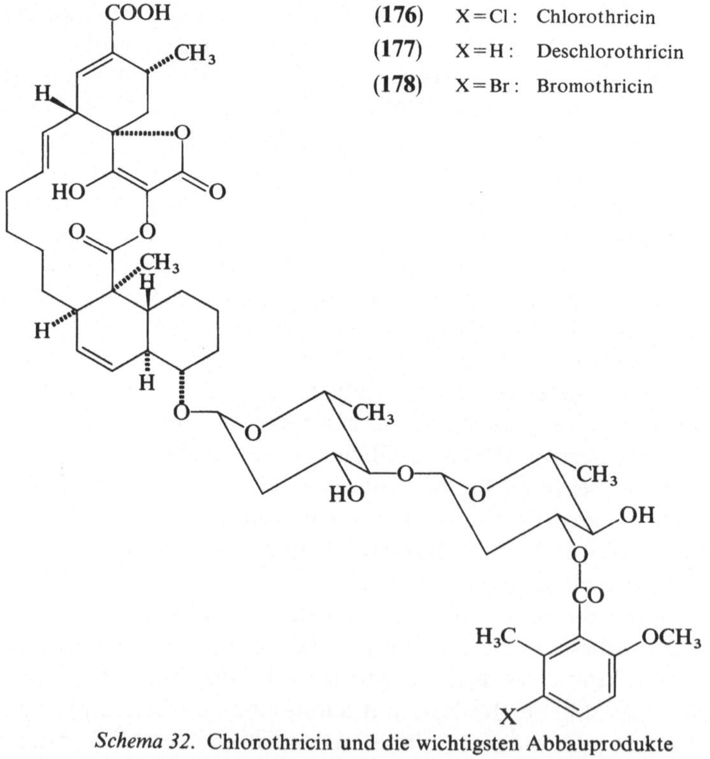

(**176**)   X = Cl :   Chlorothricin
(**177**)   X = H :   Deschlorothricin
(**178**)   X = Br :   Bromothricin

Schema 32. Chlorothricin und die wichtigsten Abbauprodukte

Schema 32 (Fortsetzung)

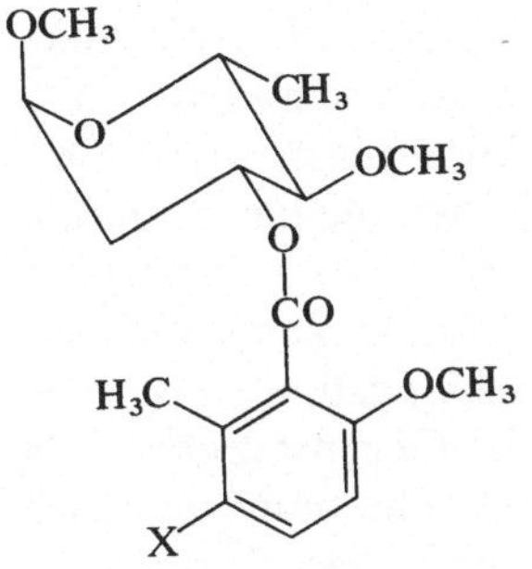

(179)

X = Cl oder H

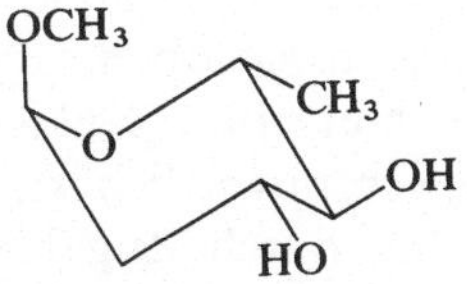

(180)

α-Methyl-2-desoxy-D-rhamnosid

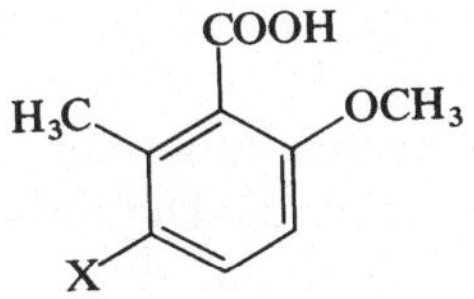

(181)

X = Cl oder H

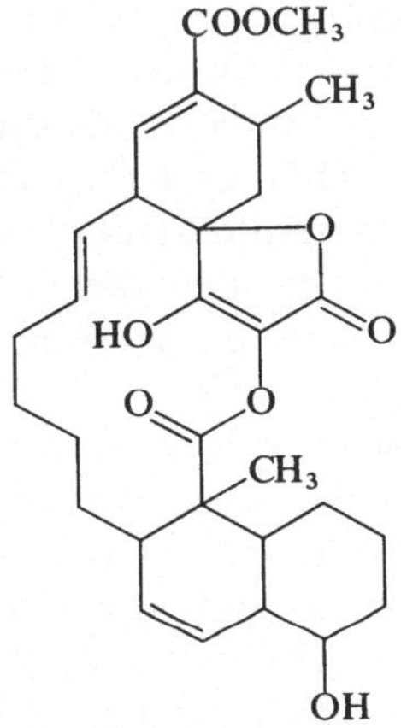

(182)

Aglykon-methylester des Chlorothricins

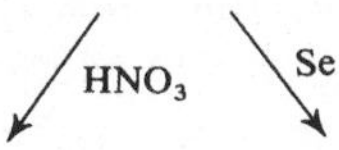

HNO₃   Se

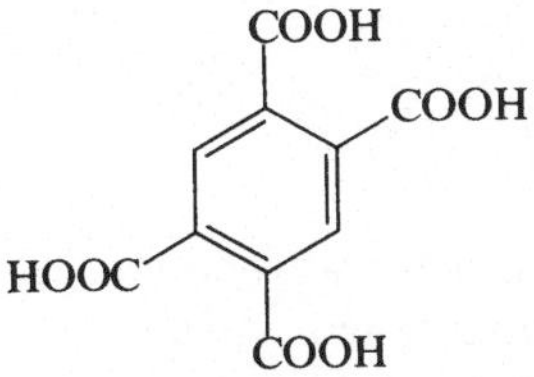

(183)

Hemimellitsäure

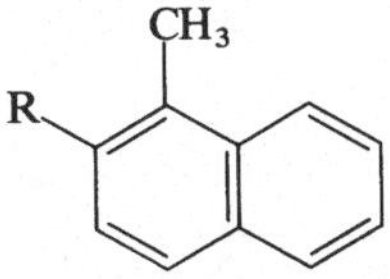

(184)

R = CH₃, C₂H₅, n-C₃H₇, n-C₄H₉, n-C₅H₁₁

wurde die Uneinheitlichkeit erkannt. Wenn der Stamm in Anwesenheit eines großen Überschusses an Bromidionen gezüchtet wird, erhält man anstelle des Chlorothricins das Bromanaloge, Bromothricin (**178**) (*161*).

Das Chlorothricin zeigt *in vitro* eine hohe Wirkung gegen Gram-positive Bakterien, wenn die Tests auf Nährböden aus „Minimalagar" durchgeführt werden. Durch Zugabe von komplexen Medien wie z. B. Fleischextrakt wird die Wirkung aufgehoben. Bisher konnte keine einzelne Verbindung (Aminosäure, Nukleotid, Nukleosid usw.) gefunden werden, die für die Desaktivierung verantwortlich ist, ganz im Gegensatz etwa zum Borrelidin, dessen Wirkung durch Zugabe von L-Threonin spezifisch aufgehoben wird (s. oben). Nach vorläufigen Befunden kommt die Desaktivierung hier dadurch zustande, daß das Chlorothricin mit Lipoproteinen eine Bindung eingeht (*162*).

Bei der Methanolyse des Chlorothricins entstehen zunächst ein Aglykonmethylester (**182**) und ein Mol eines Methylglykosids (**179**), das ein Ester des 5-Chlor-6-methylsalicylsäure-methyläthers (**181**) mit Methyl-2-desoxy-D-rhamnosid ist. Die Konstitution des Esters **179** wurde durch NMR.-Spektren und durch eine alkalische Hydrolyse zum α-Methyl-2-desoxy-D-rhamnosid (**180**) und der freien Säure **181** bewiesen. Die instrumentalanalytisch bestimmte Struktur der Säure wurde durch eine Synthese bestätigt (*213*). In den wasserlöslichen Anteilen der Methanolyseprodukte wurde später ein zweites Mol des Glykosids **180** aufgefunden.

Die Konstitution des Aglykon-methylesters wurde durch eine Röntgenstrukturanalyse seines Caesiumsalzes aufgeklärt (*212*). Es liegen ferner eine Menge spektroskopischer Befunde sowie Abbauergebnisse vor, die mit der Struktur **182** im Einklang sind. Die beiden wichtigsten Abbauprodukte, die Hemimellitsäure (**183**) und eine Reihe von homologen 1-Methyl-2-alkyl-naphthalinen (**184**), sind im Formelschema 32 aufgeführt.

Die Verknüpfung der Bausteine ergab sich aus einer vollständigen Methylierung des Antibioticums und anschließender Hydrolyse zu O-Methylzuckern (*212*). Die Lage der O-Methylgruppen in den letzteren wurde aus NMR.-Doppelresonanz-Versuchen mit den freien Methyläthern und ihren Acetylderivaten erschlossen.

V. Makrolide aus Pilzen

Die Fähigkeit, Makrolide zu synthetisieren, ist während längerer Zeit als eine Besonderheit der Actinomyceten angesehen worden, denn die Untersuchung von Makroliden aus Pilzen hat relativ spät und mit geringer Intensität eingesetzt, weil sie bisher nicht das Interesse der Mediziner gefunden haben. Bis heute ist daher auch die Liste pilzlicher

Makrolide recht klein geblieben. Sie sind durchwegs viel einfacher gebaut als ihre Verwandten aus Actinomyceten. Es fehlen ihnen die Zuckerbausteine, meistens alle Verzweigungen der Kohlenstoffkette, und in der Regel kommt nur eine geringe Zahl funktioneller Gruppen vor. Sie sind zwar durchwegs optisch aktiv, aber die Zahl der Chiralitätszentren beschränkt sich meist auf eines oder ganz wenige. Einzig das Brefeldin A mit 5 asymmetrischen Kohlenstoffatomen und die Cytochalasine sind etwas komplizierter gebaut.

Aus diesem Grund bietet die Konstitutionsaufklärung keine besonders schwierigen Probleme, und aus dem gleichen Grund finden wir in dieser Gruppe die einzigen Makrolide, von denen eine Totalsynthese vorliegt. In diesem Kapitel soll darum auf eine ausführliche Beschreibung von Konstitutionsaufklärungen verzichtet werden zugunsten einer etwas eingehenderen Diskussion synthetischer Arbeiten.

1. Das Zearalenon

Diese Verbindung wurde ursprünglich gar nicht als ein Antibioticum isoliert, sondern als ein Toxin, das vor allem an Schweinen, die mit verschimmeltem Mais gefüttert worden waren, schweren Schaden angerichtet hat. In mehreren Fällen konnte aus Überresten des verdorbenen Futters ein Pilz der Art *Gibberella zeae* isoliert werden, der auch in künstlicher Kultur ein Toxin produzierte, das an Versuchstieren die gleichen Vergiftungssymptome hervorrief: krankhafte Veränderung und Vergrößerung der Geschlechtsorgane bei weiblichen Tieren, Entwicklung sekundärer weiblicher Merkmale bei Männchen. In schwereren Fällen führten die Vergiftungen zum Tod der Tiere (*282*).

Die Konstitutionsaufklärung gemäß Formel **185** und die Bestimmung des Chiralitätssinnes wurde vier Jahre nach der Entdeckung bekanntgegeben (*178, 297*), und bald darauf wurde auch die erste Synthese beschrieben (*178, 291, 292*).

Der Synthese voran gingen präparative Studien an Umwandlungsprodukten des Naturstoffes (*292*), die sich in den Endphasen der Totalsynthese als sehr wertvoll erweisen sollten. Zearalenon-dimethyläther (**190**) wurde zur Hydroxysäure **186** hydrolysiert. Die alkalische Verseifung verlief aber nicht glatt, sondern führte zu einem Gemisch der beiden Cyclohemiketale **186a** und **186b**, die in alkalischer Lösung miteinander im Gleichgewicht stehen. Diese Gleichgewichtsreaktion führt auch zu einer vollständigen Racemisierung. Eine mechanistische Erklärung für die Entstehung des Isomeren **186b** durch Hydridverschiebung wird im Formelpaar **186c** wiedergegeben. Eine NMR.-Analyse des Hydrolyseproduktes zeigte ferner, daß die offenen Hydroxyketone im Gleichgewichtsgemisch nicht merklich in Erscheinung treten.

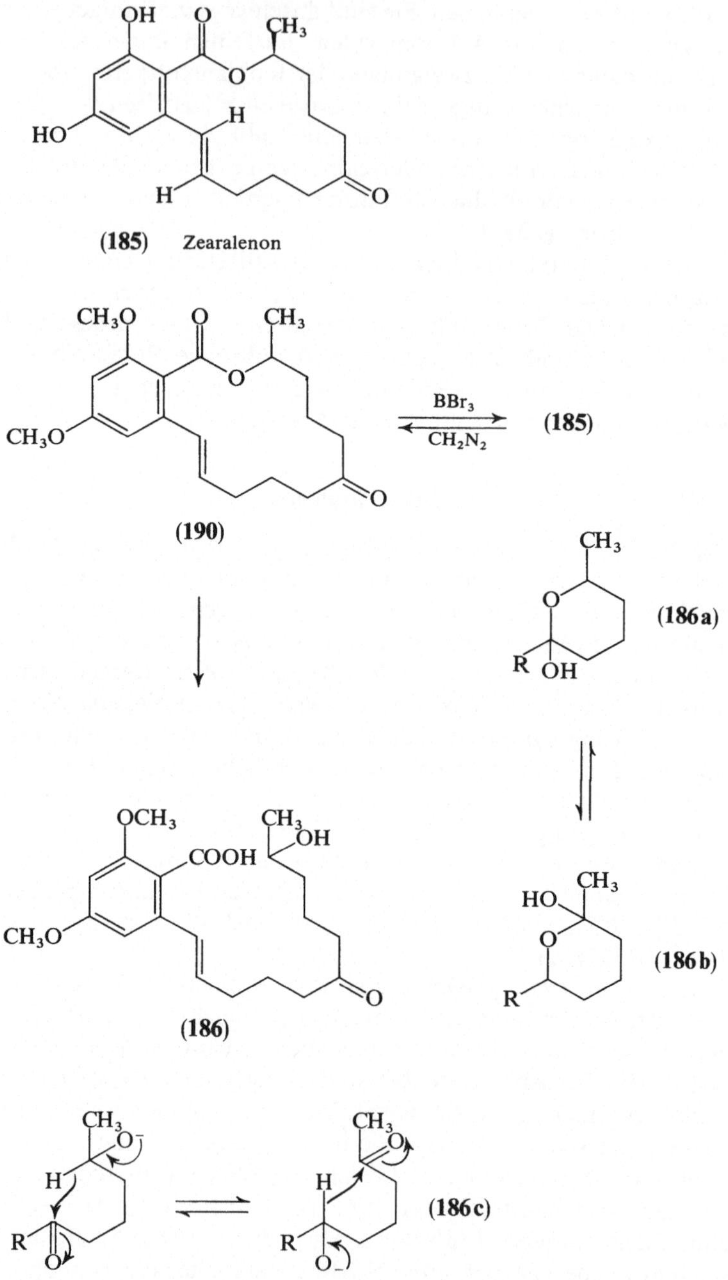
(185) Zearalenon
BBr3
CH2N2
(185)
(190)
(186a)
(186)
(186b)
(186c)

Schema 33. Zearalenon und einige Umwandlungsprodukte

Literaturverzeichnis: SS. 445—460

Schema 33 (Fortsetzung)

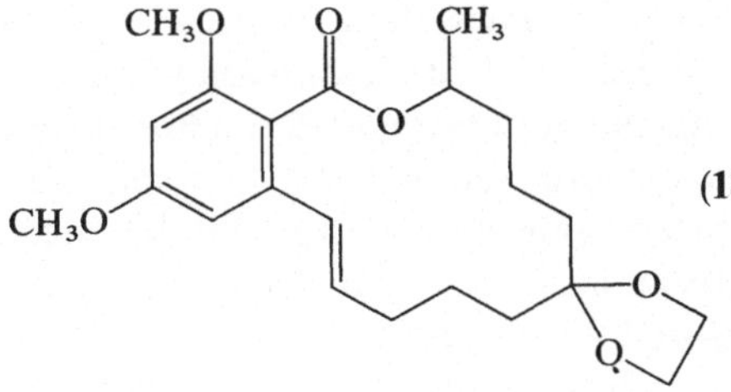

(187)

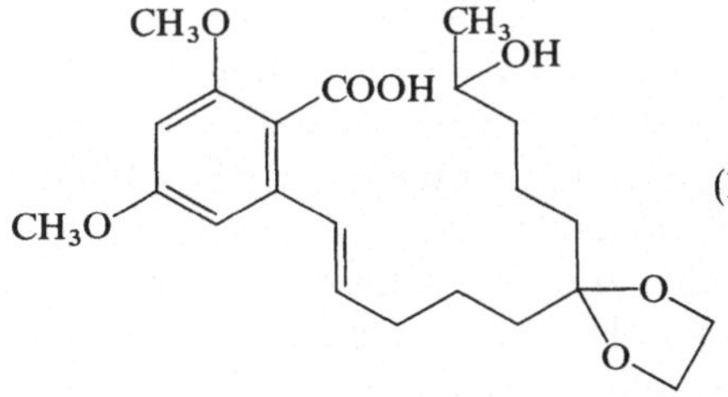

(188)

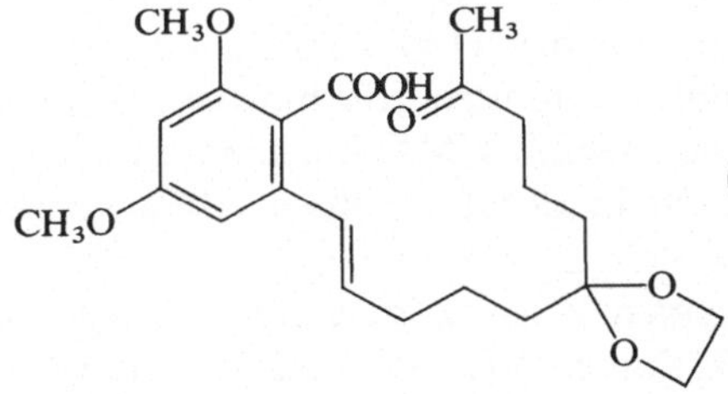

(189)

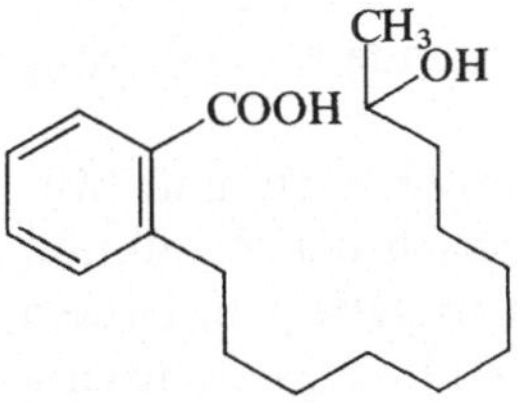

(191) 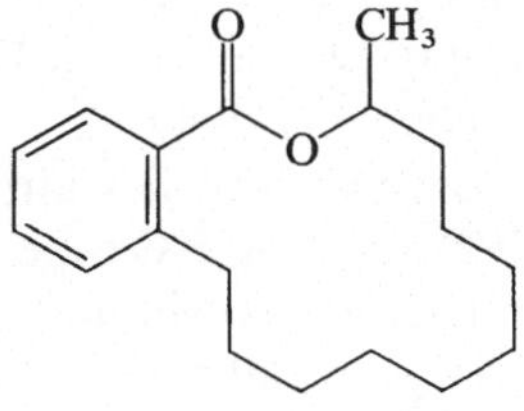 ⟶ (192)

Die Verseifung der Lactongruppe konnte jedoch ohne Racemisierung und Isomerisierung durchgeführt werden, wenn sie mit dem Äthylen-ketal **187** ausgeführt wurde. Aus dem optisch aktiven Ketal **188** konnte durch Oxydation mittels Chromsäureanhydrid und Reduktion des Methylketons **189** mit Natriumborhydrid die konstitutionell einheitliche racemische Hydroxysäure **188** bereitet werden, die durch saure Hydrolyse die reine racemische Ketosäure **186** lieferte. Die letztere wurde dagegen in optisch aktiver Form erhalten, wenn das optisch aktive Ketal **188** mit Säure hydrolysiert wurde.

Von allen geprüften Ringschlußreaktionen erwies sich die Umsetzung der Hydroxysäure **186** mit Trifluoressigsäureanhydrid in Benzol als vorteilhafteste. Sie erlaubte die Isolierung von Zearalenon-dimethyläther (**190**) in etwa 15% Ausbeute, je nach Ausgangsmaterial in optisch aktiver oder racemischer Form. Es ist interessant, daß Wehrmeister und Robertson (*314*) bei der Cyclisierung der Modellverbindung **191** mittels Trifluoressigsäureanhydrid wenig Erfolg hatten, dagegen mit Phosgen und Triäthylamin in Benzol bei hoher Verdünnung das Di-desoxy-zearalan (**192**) in etwa 8% Ausbeute erhalten konnten.

Der Zearalenon-dimethyläther (**190**) seinerseits konnte mittels Bortribromid in 34% Ausbeute zum kristallinen Zearalenon (**185**) gespalten werden. Daraus ergab sich, daß die einfache Methylschutzgruppe für die phenolischen Hydroxylgruppen verwendet werden konnte, was die anschließende Totalsynthese erheblich erleichterte (*292*).

Für die Herstellung des Ausgangsmaterials für den aromatischen Teil des Moleküls konnte auf Arbeiten von Brockmann und Mitarb. (*40*) zurückgegriffen werden, die für andersartige Naturstoffsynthesen ausgehend von 3,5-Dihydroxybenzoesäure (**193**) leicht größere Mengen 3,5-Dimethoxyphthalsäureanhydrid (**194**) herstellen konnten. Die Reduktion dieses Anhydrids mit Lithiumtritertiärbutoxyaluminiumhydrid lieferte den Aldehyd **195**, der gemäß NMR.-Spektrum im Gleichgewichtsgemisch fast ausschließlich in der Lactolform **195a** vorliegt, in guter Ausbeute.

Einen anderen bequemen Zugang zu einem Derivat des Aldehyds **195**, dem Äthylester **199**, fanden Vlattas und Mitarb. (*305*) ausgehend von der Orsellinsäure. Die Oxydation von Di-O-acetylorsellinsäureäthylester (**196**) mit Chromsäureanhydrid in Eisessig-Acetanhydrid gab das Tetraacetat **197**, das durch Umesterung und anschließende Methylierung den Aldehyd **199** lieferte.

Für die Synthese des aliphatischen Teilstücks gingen Taub und Mitarb. (*292*) von der 5-Oxohexansäure (**200**) aus, die nach der Reduktion mit Natriumborhydrid und Ansäuern direkt das Lacton **201** gab. Dessen Umsetzungsprodukt mit 4-Pentenylmagnesiumbromid (**202**) cyclisierte

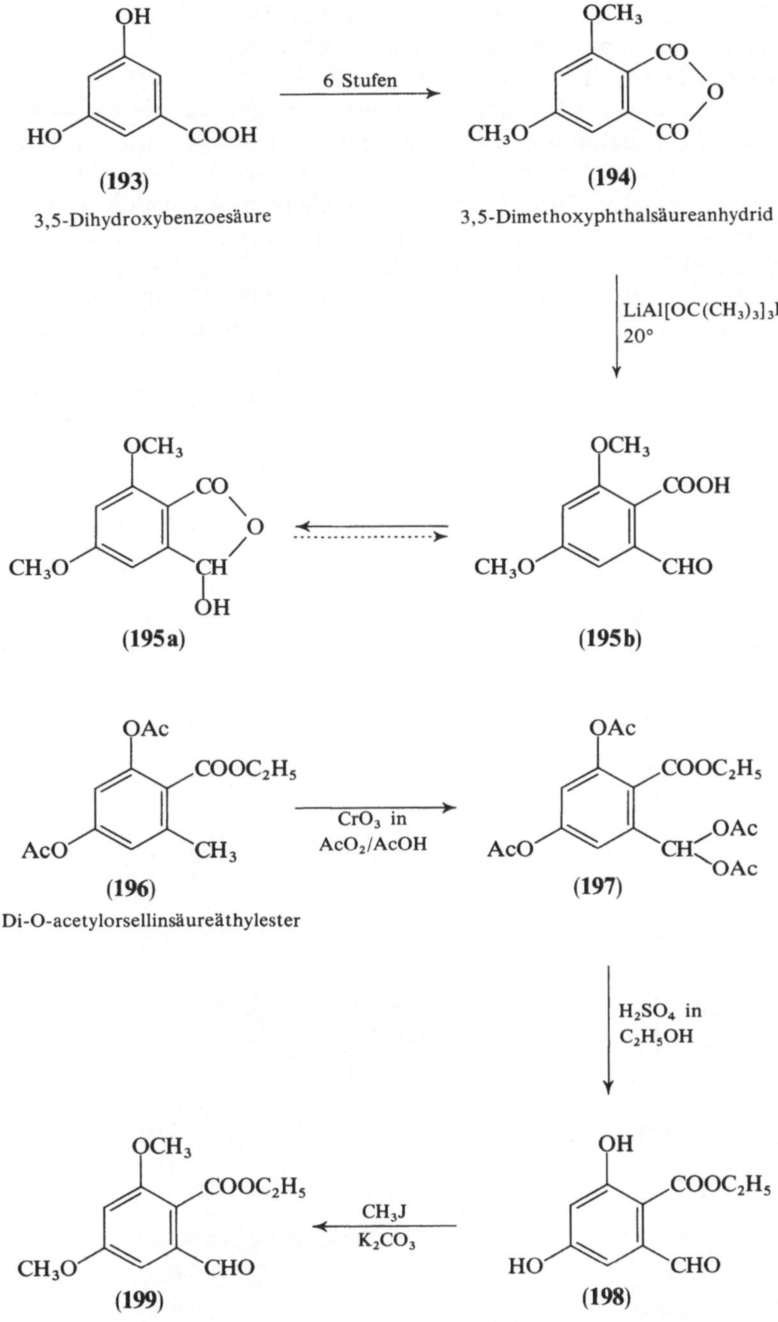

(193)

3,5-Dihydroxybenzoesäure

6 Stufen

(194)

3,5-Dimethoxyphthalsäureanhydrid

$LiAl[OC(CH_3)_3]_3H$
20°

(195a)

(195b)

(196)

Di-O-acetylorsellinsäureäthylester

CrO_3 in
$AcO_2/AcOH$

(197)

H_2SO_4 in
C_2H_5OH

(199)

CH_3J
K_2CO_3

(198)

Schema 34. Totalsynthese des Zearalenons (I). Synthese des aromatischen Bausteins

bei der Destillation im Vakuum zum Enoläther **203**, an den leicht Methanol angelagert werden konnte, wobei das Ketal **204** entstand. Das Ketal konnte auch direkt aus dem Hydroxyketon **202** mittels absoluter methanolischer Salzsäure gewonnen werden. Das Ketal **204** erwies sich als sehr unbeständig. Schon mit Spuren wässeriger Säure (z. B. beim Deuteriumaustausch einer NMR.-Probe mit D_2O in $CDCl_3$) bildete sich das Hydroxyketon **202** zurück. Bei peinlichem Ausschluß von Säure konnte dagegen die Methylen-Endgruppe mittels Ozon abgespalten werden. Durch reduktive Spaltung des Ozonids mit Natriumborhydrid gelangte man direkt zum primären Alkohol **205**, der über das Tosylat (**206**) und das Bromid (**207**) in das Phosphoniumbromid **208** übergeführt wurde. Dieses eignete sich gut zum Aufbewahren und wurde jeweils erst kurz vor Gebrauch mit Natriumhydrid in Dimethylsulfoxid ins Ylid **209** übergeführt.

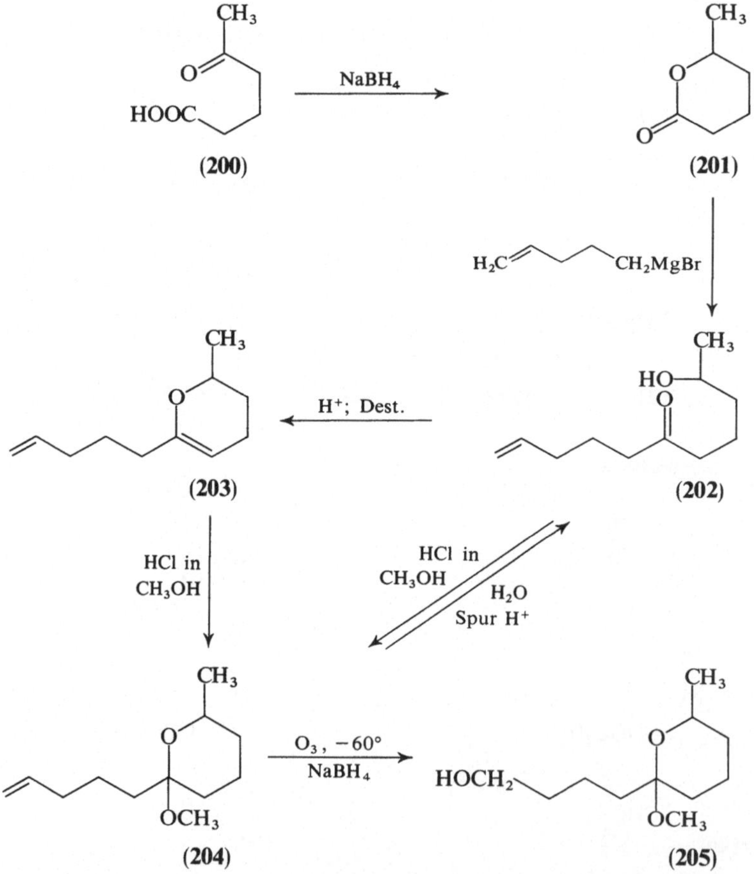

Schema 35. Totalsynthese des Zearalenons (II). Synthese des aliphatischen Teilstücks

Literaturverzeichnis: SS. 445—460

Schema 35 (Fortsetzung)

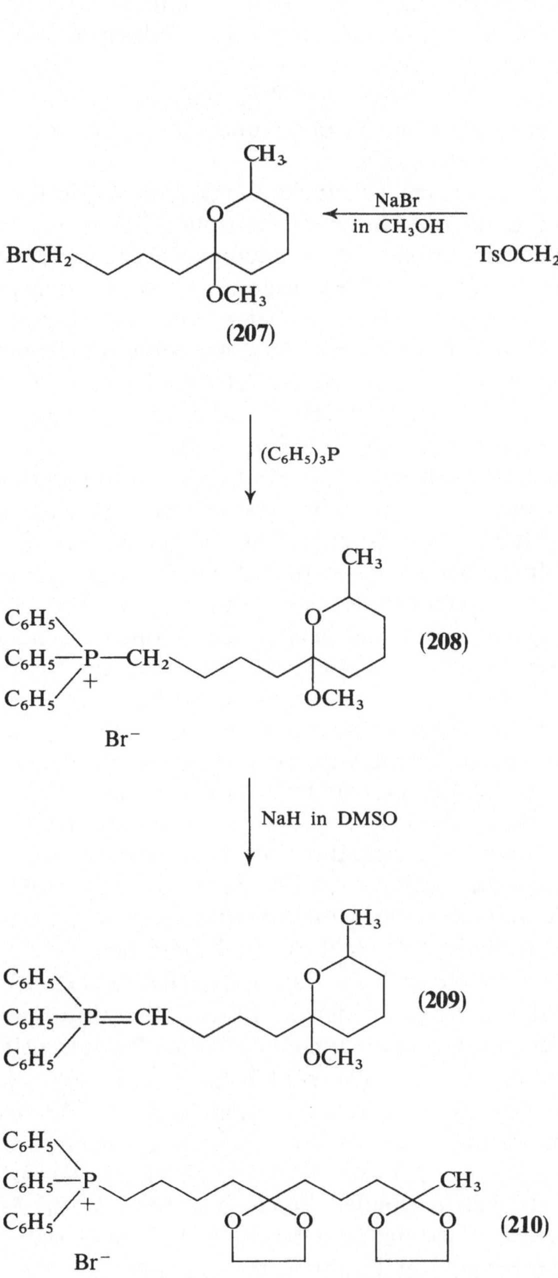

(207)

(206)

$(C_6H_5)_3P$

(208)

Br^-

NaH in DMSO

(209)

(210)

Br^-

26*

Wie die IR.- und NMR.-Spektren zeigten, trat jeweils bei der Aufarbeitung der Zwischenstufen **205** bis **207** eine teilweise Hydrolyse zu den entsprechenden Hydroxyketonen ein. Diese konnte aber in allen Fällen durch Nachbehandlung mit methanolischer Salzsäure wieder rückgängig gemacht werden.

Auf einen längeren und weniger ergiebigen Weg, der hier nicht im Einzelnen ausgeführt werden soll, stellten Vlattas und Mitarb. das ähnlich gebaute Phosphoniumsalz **210** her (*305*).

Für die Verknüpfung der beiden Teilstücke durch eine Wittig-Reaktion hatten Vorversuche (*292*) mit einfachen Phosphor-Yliden gezeigt, daß diese infolge von Nebenreaktionen mit schlechter Ausbeute verläuft, wenn von Estern der Säure **195** (z. B. **199**) ausgegangen wird. Hingegen konnte mit gutem Erfolg das Natriumsalz **211** der Säure **195** eingesetzt werden. Die Wittig-Reaktion lieferte mit etwa 60% Ausbeute ein Gemisch des *cis*- und des *trans*-Isomeren der Säure **212**. Das NMR.-Spektrum zeigte ein Isomerenverhältnis von 1:1 an, durch Gaschromatographie der Methylester (**213**) wurde ein solches von 48:52 gefunden.

Für den Ringschluß mit Trifluoressigsäureanhydrid wurde das Isomerengemisch **212** eingesetzt. Aus dem Reaktionsgemisch wurde der ($\pm$)-Zearalenon-dimethyläther (**214**) in kristalliner Form isoliert. Zum Zweck der Racematspaltung wurde dieser partiell an der zum Lactoncarbonyl o-ständigen Methyläthergruppe hydrolysiert, was mit Bortrichlorid bei tiefer Temperatur recht gut gelang. Der Monomethyläther wurde mit optisch aktivem Menthoxyessigsäurechlorid in Dioxan-Pyridin umgesetzt. Durch Kristallisation aus Aceton-Petroläther wurde das linksdrehende diastereomere Menthoxyacetat (Smp. 121—123°, $[\alpha]_D = -38,5°$ in Chloroform) abgetrennt, das in jeder Beziehung mit einem von natürlichen Zearalenon ausgehend bereiteten Präparat identisch war. Durch eine milde alkalische Hydrolyse (2,5 N Natronlauge bei 25°, 2 Stunden) wurde der Menthoxyessigsäureester zum optisch aktiven Zearalenon-4-monomethyläther (**215**) verseift. Spaltung der zweiten Methyläthergruppe mit Bortribromid gab schließlich optisch aktives Zearalenon, das sich vom Naturstoff nicht unterscheiden ließ.

Etwas umständlicher gestaltete sich die Variante nach Vlattas (*305*). Das Produkt der Wittig-Reaktion des Aldehyds **199** mit dem Bisäthylenketal **210** war ein stereochemisch wohl kaum einheitliches Produkt (**216**), das zwei gleichartige funktionelle Gruppen enthält, die aber in der Folge unterschiedlichen Reaktionsfolgen zugeführt werden mußten. Mit dem Äthylester **216** gelang eine partielle Ketalspaltung nicht. Interessanterweise konnte aber aus der freien Säure **217**, hergestellt durch milde alkalische Verseifung des Esters, in befriedigender Weise das Monoketon **218** erhalten werden; auf welche Weise die entfernt stehende freie Carboxylgruppe die Reaktivität der beiden Ketalgruppen dermaßen zu beeinflussen

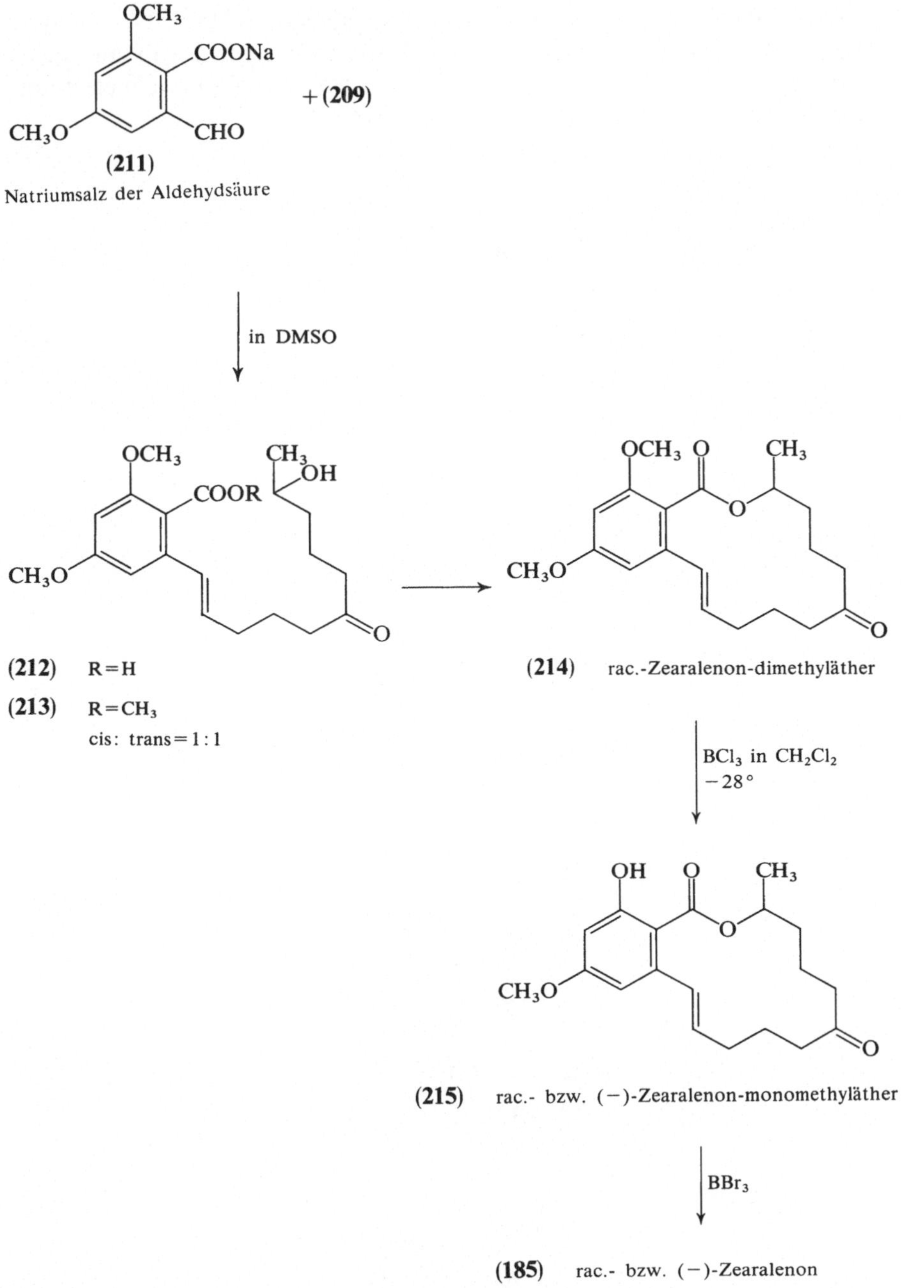
OCH3
COONa
CH3O
CHO
+ (209)
(211)
Natriumsalz der Aldehydsäure
in DMSO
OCH3
CH3
OH
COOR
CH3O
O
(212) R=H
(213) R=CH3
cis: trans=1:1
OCH3 O CH3
O
CH3O
O
(214) rac.-Zearalenon-dimethyläther
BCl3 in CH2Cl2
−28°
OH O CH3
O
CH3O
O
(215) rac.- bzw. (−)-Zearalenon-monomethyläther
BBr3
(185) rac.- bzw. (−)-Zearalenon

Schema 36. Totalsynthese des Zearalenons (III). Verknüpfung der Teilstücke und Racematspaltung vermag, läßt sich wohl zur Zeit nicht befriedigend erklären. Die Reduktion der Methylketongruppe mittels Natriumborhydrid gab aus dem Methylester **219** die Verbindung **212**, die direkt durch eine Umesterung mit Natrium-tertiäramylat in Toluol cyclisiert wurde.

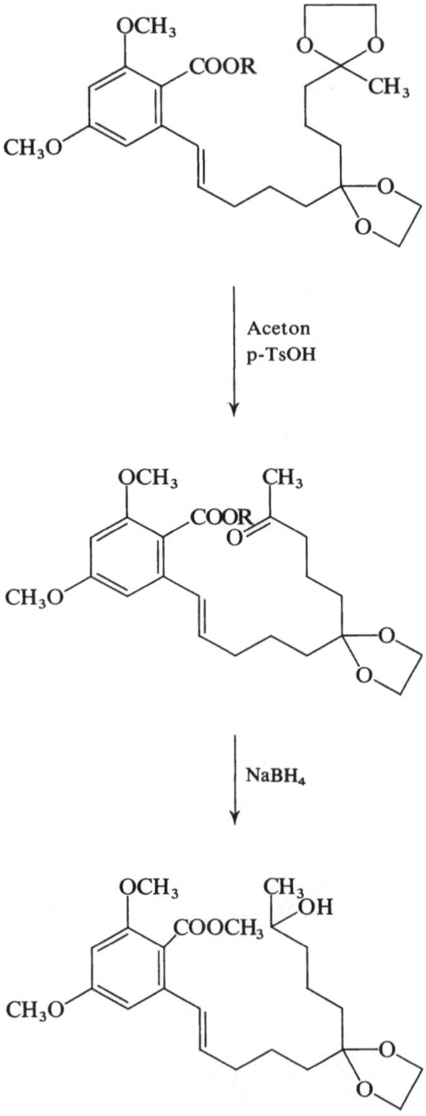

(216) $R = C_2H_5$
(217) $R = H$

Aceton
p-TsOH

(218) $R = H$
(219) $R = CH_3$

NaBH₄

(220)

Schema 37. Totalsynthese des Zearalenons (IV). Variante nach Vlattas u. Mitarb.

Literaturverzeichnis: SS. 445—460

Schema 37 (Fortsetzung)

Na—O—C(CH$_3$)(CH$_2$—CH$_3$)(CH$_3$)   in Toluol

↓

(±)–(**187**)

pTsOH
in Aceton

↓

(±)–(**190**)

↓

(±)–(**185**); Zearalenon

Wie wichtig ein genaues Studium der Reaktivitäten am Naturstoff und seinen Derivaten für den Erfolg einer komplizierten Totalsynthese sein kann, ergibt sich aus den experimentellen Angaben von VLATTAS und Mitarb. (*305*). Für die Cyclisierung standen noch etwa 10 mg des Hydroxyesters (**220**) zur Verfügung. Das mit 8% Ausbeute isolierbare Cylisierungsprodukt **187** wurde mit p-Toluolsulfosäure in Aceton zu 0,8 mg (±)-Zearalenondimethyläther (**190**) entketalisiert. Schließlich gab die Ätherspaltung 0,2 mg (±)-Zearalenon, das durch präparative Dünnschichtchromatographie rein isoliert und durch das Massenspektrum identifiziert wurde. Eine Racematspaltung war allerdings mit diesen minimalen Mengen nicht mehr durchführbar.

Eine vereinfachte Variante einer Zearalenonsynthese beschreiben GIROTRA und WENDLER (*117*). Der Aldehyd **222**, der durch Ozonolyse des Cycloketals **204** zugänglich ist (s. oben), kondensiert in Pyridin bei 25° glatt mit dem Homophthalsäureanhydrid **221**. Bei der Aufarbeitung des Rohproduktes mit verdünnter Säure wird das Cycloketal hydrolysiert zum Hydroxyketon **223**, das gemäß spektroskopischen Befunden ein Gleichgewichtsgemisch des offenen Hydroxyketons mit den diastereomeren Cyclohemiketalen darstellt. Beim Erhitzen in γ-Picolin auf 150° tritt eine Lactonöffnung durch Elimination und gleichzeitig eine Decarboxylierung ein. Die geometrische Einheitlichkeit des Produktes (**224**) ist allerdings fraglich. Die Cyclisierung der Hydroxysäure **224** zum Zearalenon-dimethyläther erfolgte nach dem bereits beschriebenen Verfahren mit Trifluoressigsäureanhydrid.

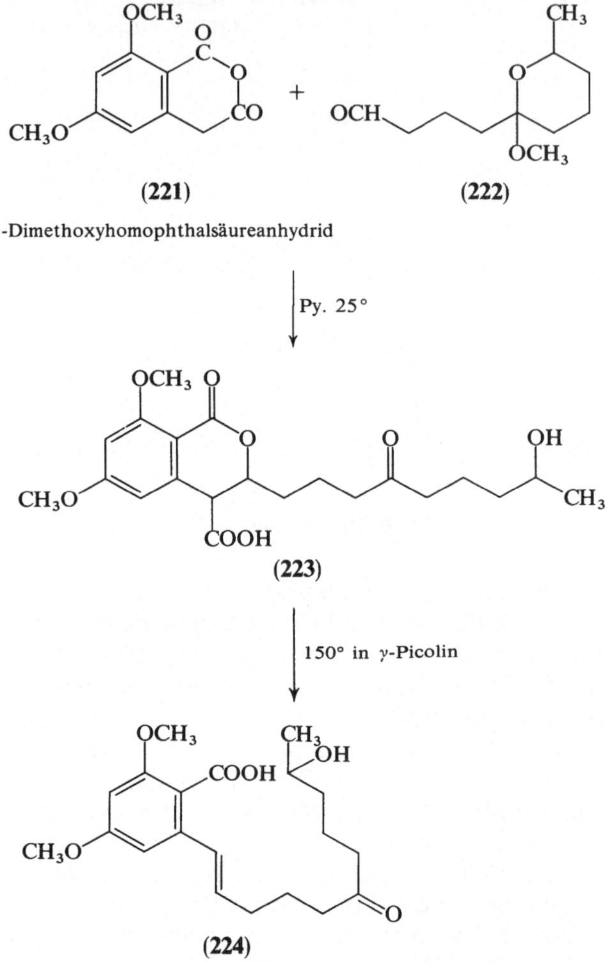
OCH_3
CH_3O
CH_3
OCH
OCH_3
(221)
(222)
3,5-Dimethoxyhomophthalsäureanhydrid
Py. 25°
COOH
OH
(223)
150° in γ-Picolin
COOH
(224)

Schema 38. Totalsynthese des Zearalenons (V). Variante nach Girotra und Wendler

Ein interessanter andersartiger Zugang zu makrocyclischen Lactonen vom Typus des Zearalenons wird von Immer und Bagli (*155*) beschrieben. Er hat zwar bisher nicht zur Synthese eines Naturstoffs geführt. Die bisher ausgeführten Modellreaktionen sind aber so vielversprechend, daß die Methode wenigstens im Formelschema 39 angedeutet werden soll. Ihr Vorteil liegt besonders darin, daß der allgemein schlecht verlaufende Schritt des Lactonringschlusses umgangen werden kann.

Literaturverzeichnis: SS. 445—460

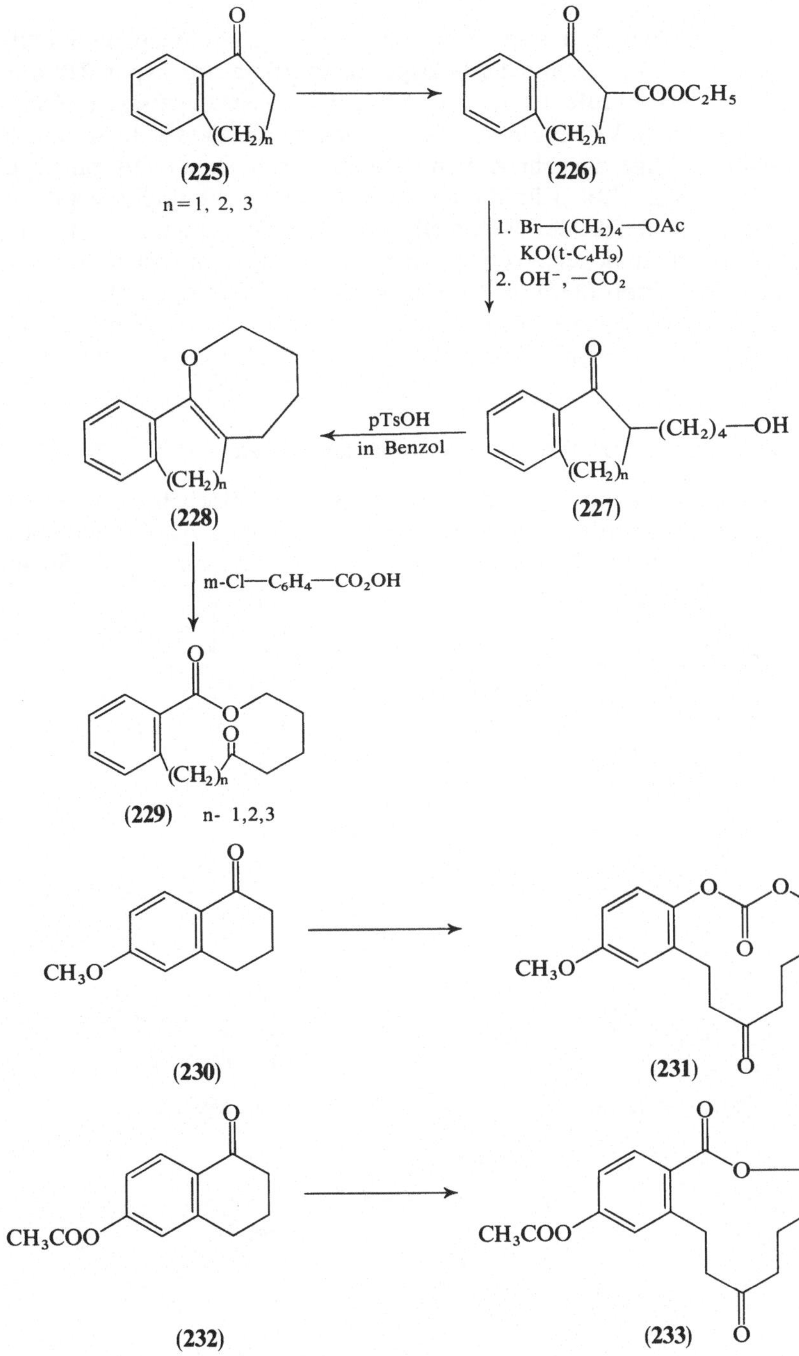
(225)
n=1, 2, 3
$COOC_2H_5$
(226)
1. $Br—(CH_2)_4—OAc$
$KO(t\text{-}C_4H_9)$
2. $OH^-, -CO_2$
pTsOH
in Benzol
$(CH_2)_4—OH$
(227)
(228)
$m\text{-}Cl—C_6H_4—CO_2OH$
(229) n- 1,2,3
CH_3O
(230)
(231)
CH_3COO
(232)
(233)

Schema 39. Makrolidsynthese nach IMMER und BAGLI

Ausgehend von Indanon, Tetralon und Benzocycloheptenon (225) wurden bisher 10-, 11- und 12gliedrige Lactonringe des Typus 229 aufgebaut. Die letzte Stufe, Oxydation mit m-Chlorbenzoepersäure, nimmt bei substituierten Verbindungen oft einen anderen Verlauf. So führte die Reaktionsfolge ausgehend von 6-Methoxytretralon (230) zum cyclischen Carbonat 231. Für diesen unerwünschten Verlauf scheint ein Nachbargruppeneffekt der Methoxylgruppe verantwortlich zu sein, denn ausgehend vom analogen Acetylderivat 232 gelangt man wieder zum entsprechenden Lacton (233), so daß nach dieser Methode durchaus auch Verbindungen mit Sauerstoffsubstituenten im aromatischen Ring zugänglich sind. Weitere Beispiele finden sich in der Originalarbeit von Immer und Bagli (*155*).

2. Weitere Makrolide aus Pilzen

Mit dem Zearalenon nahe verwandt ist das chlorhaltige Makrolid Radicicol (= Monorden), das zuerst als antifungisches Antibioticum aus Kulturen des Pilzes *Monosporium bonorden* isoliert wurde (*75*). Später

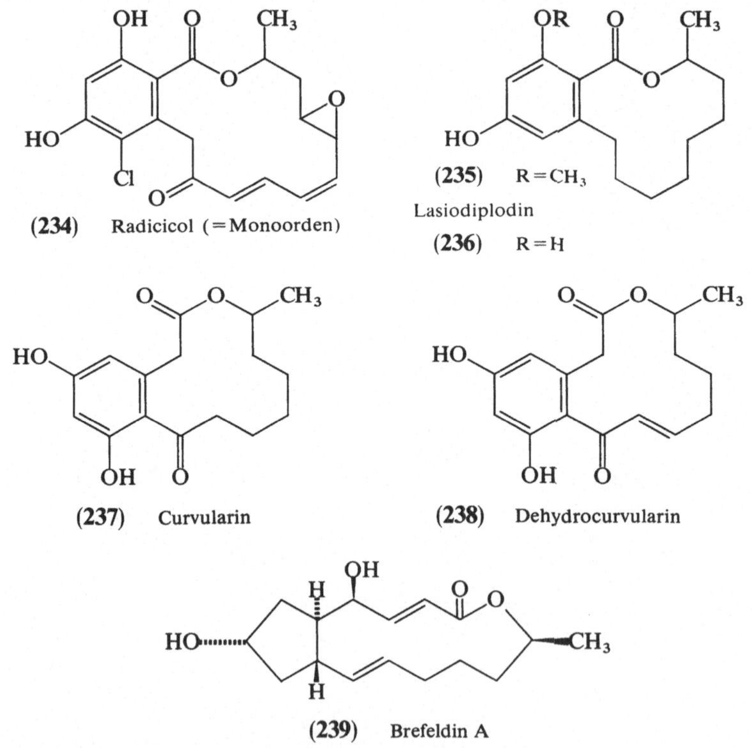

Schema 40. Weitere Makrolide aus Pilzen

wurde auch eine bemerkenswert hohe sedative Wirkung beobachtet (*196*). Die Konstitutionsaufklärung wurde unabhängig von zwei Arbeitsgruppen durchgeführt (*196, 204, 205*).

Ein verwandtes Makrolid mit einem zwölfgliedrigen Ring ist das Lasiodiplodin (**235**), das zusammen mit der Des-O-methylverbindung (**236**) vom phytopathogenen Pilz *Lasiodiplodia theobromae* neben anderen Stoffwechselprodukten gebildet wird (4).

Die gleiche Ringgröße wie Lasiodiplodin, aber ein anderes Cyclisierungsmuster besitzen das Curvularin (**237**) und das Dehydrocurvularin (**238**), die aus Pilzen der Gattung *Curvularia* isoliert wurden (*214*). Über eine biologische Wirkung der Curvularine und des Lasiodiplodins ist bisher nichts bekannt geworden. Die Konstitutionsaufklärung, die mit Hilfe spektroskopischer Methoden und durch einen chemischen Abbau durchgeführt wurde (*27, 28, 211, 215*), bedarf keiner weiteren Erläuterungen. Dagegen sei hier auf eine interessante Variante einer Makrolidsynthese hingewiesen (*15*). Da sich die Hydroxysäure **240** mit Methoden, die bei der Synthese des Zearalenons und verwandter Verbindungen erfolgreich waren, nicht zum Lacton schließen ließ, wurde die Reihenfolge der Verknüpfungsreaktionen umgekehrt und zunächst der Ester **241** aufgebaut. In einer Friedel-Crafts-ähnlichen Reaktion mit Trifluoressigsäure und Trifluoressigsäureanhydrid konnte mit immerhin 15% Ausbeute ein Ringschluß zum Keton **242** erzielt werden. Das racemische Produkt stimmte in seinen spektroskopischen und chromatographischen Eigenschaften mit dem Derivat des Naturstoffs überein.

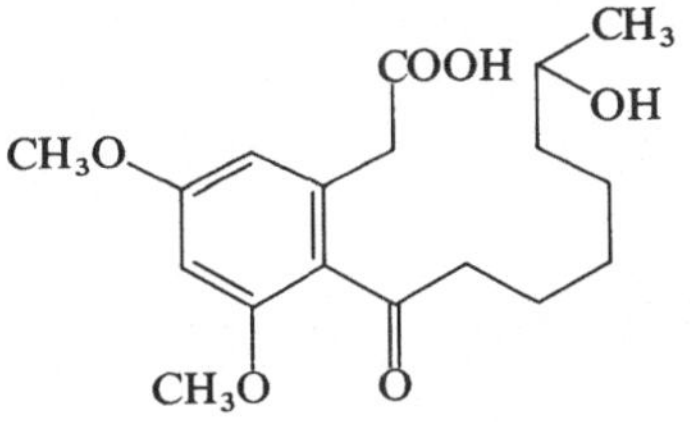

(**240**)

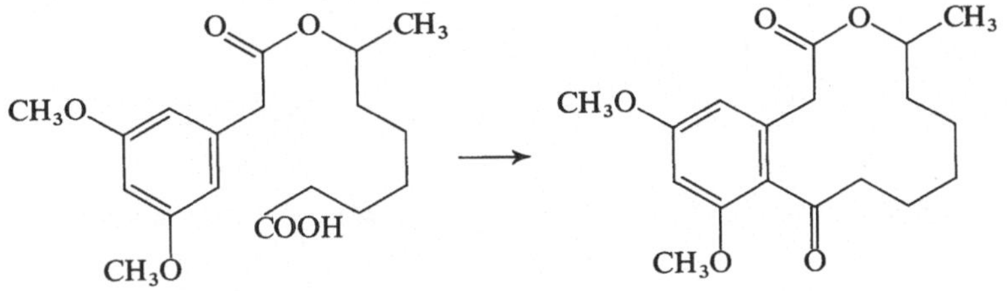

(**241**) → (**242**)

(±)-Curvularin-dimethyläther

Schema 41. Synthese des Curvularin-dimethyläthers

412 W. Keller-Schierlein:

Ein besonders interessantes Makrolid aus Pilzen ist das Brefeldin A,
das recht verbreitet zu sein scheint. Es wurde unter verschiedenen Namen*
(Brefeldin A, Decumbin, Cyanein, Ascotoxin) aus Pilzen der Gattungen
Penicillium (*18, 19, 127, 278, 279*), *Curvularia* (*66*) und *Ascochyta* (*287*)
beschrieben. Brefeldin A wirkt zwar nicht antibakteriell, doch sind neben
einer gewissen antifungischen Wirkung (*17*) bemerkenswerte cytotoxi-

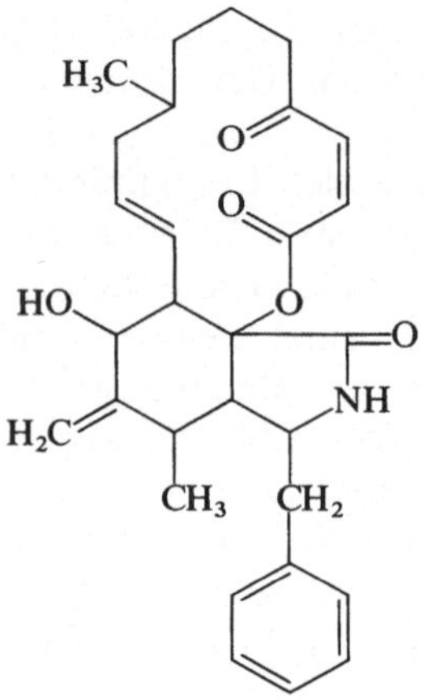

(243)   Cytochalasin A
(= 5-Dehydrophomin)

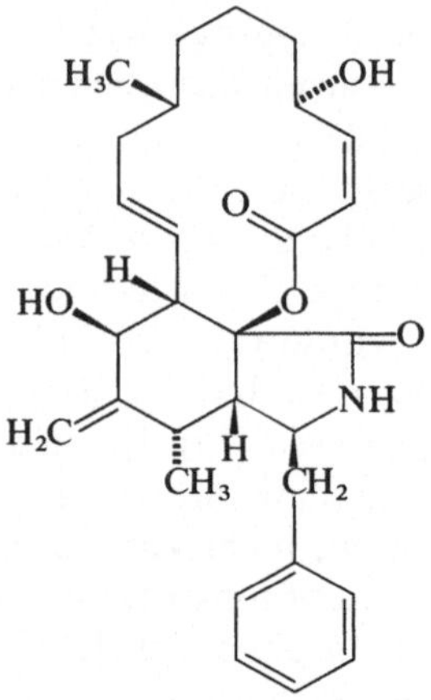

(244)   Cytochalasin B
(= Phomin)

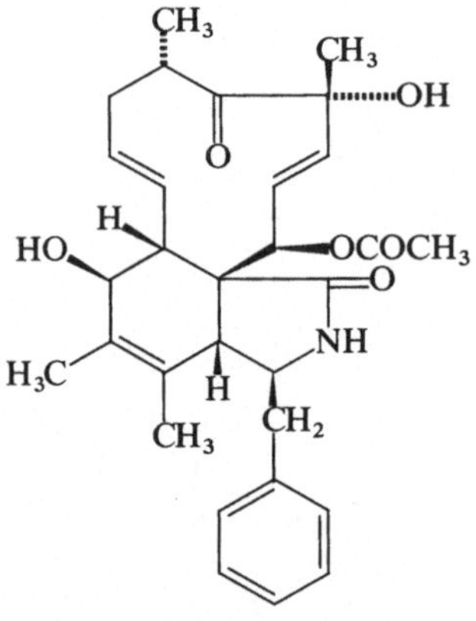

(245)   Cytochalasin C

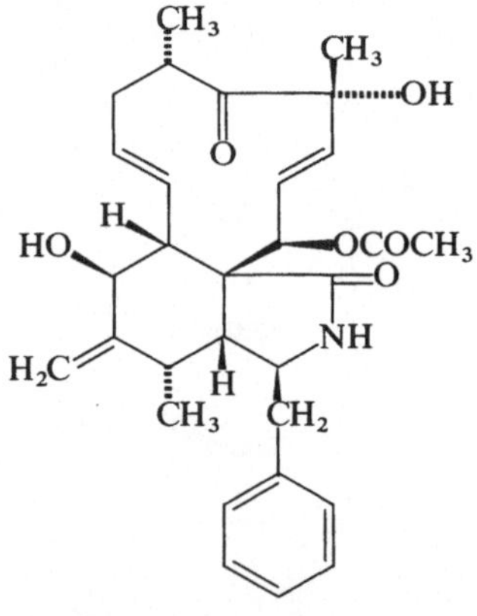

(246)   Cytochalasin D

Schema 42. Die Cytochalasine

* Aus Prioritätsgründen müßte der Verbindung wahrscheinlich der Name Decumbin
zukommen, doch hat sich „Brefeldin A" eingebürgert, da sie in den bedeutendsten Arbeiten
unter diesem Namen erscheint.

Literaturverzeichnis: SS. 445—460

sche Eigenschaften beobachtet worden (*19, 127*), die sich u. a. durch eine hohe Toxicität an Tieren (*19, 279*) und Pflanzen (*279*) sowie in einer Hemmung des Tumorwachstums (*127*) auswirkt. Eine antivirale Wirkung ist ebenfalls gefunden worden (*289*).

Die Strukturformel (**239**) ist recht ungewöhnlich wegen des Vorliegens eines Cyclopentanringes (*287, 277*), der sonst bei nicht mevalonoiden Naturstoffen nur selten gefunden wird (vgl. Borrelidin, Abschnitt IV) und vor allem aus biogenetischen Gründen von Interesse ist (*65, 125, 49*).

Gänzliche Außenseiter im Bereich der Makrolide sind die Cytochalasine aus verschiedenen Pilzarten. Dies gibt sich schon darin kund, daß von den acht bekannten Antibiotica dieser Gruppe nur zwei Makrolide sind, nämlich die Cytochalasine A (**243**) und B (**244**), während die Cytochalasine C (**245**) und D (**246**) sowie die Zygosporine D, E, F und G (*202*) an Stelle des Lactonrings einen elfgliedrigen carbocyclischen Ring besitzen. Sie sind weder antibakteriell noch ausgeprägt antifungisch, dagegen stark cytotoxisch, was vor allem an HeLa-Zellkulturen gut zu beobachten ist (*268*). CARTER (*50*) beobachtete, daß vor allem die Cytoplasma-Teilung gehemmt wird, während die Kernteilung weitergeht, so daß mehrkernige Zellen entstehen. Bei längerem Kontakt mit Cytochalasin B werden die Kerne aus den Zellen ausgestoßen.

Die Konstitutionsaufklärung wurde unabhängig in drei Laboratorien durchgeführt. TAMM und Mitarb. (*25, 268, 269*) beschäftigen sich vorwiegend mit den makrolidischen Cytochalasinen A und B (=Dehydrophomin und Phomin), MINATO und Mitarb. (*135, 202, 203*) mit dem carbocyclischen Cytochalasin D (=Zygosporin A) und mehreren ähnlich gebauten Nebenkomponenten, während ALDRIDGE und Mitarb. (*3, 5*) beide Gruppen bearbeiteten. Die Stereochemie der Cytochalasine B und D kennen wir aus Röntgenstrukturanalysen von Schweratom-Derivaten (*198, 295*).

VI. Die Zuckerbausteine der Makrolide

1. Die Aminozucker der Erythromycin- und Carbomycin-Gruppe

Nahezu alle Antibiotica dieser beiden Untergruppen sind basische Verbindungen auf Grund der Anwesenheit von glykosidisch gebundenen Aminozuckern, die sich konfigurativ von der D-Glucose ableiten. Als basischer Baustein fast aller Glieder der Carbomycin-Untergruppe tritt die Mycaminose auf. Sie wurde zuerst als Abbauprodukt des Carbomycins (*142*) und des Carbomycins B (*141*) isoliert und ist später als Baustein weiterer Antibiotica, wie der Spiramycine, des Tylosins, des Acumycins und der Leucomycine, erkannt worden (s. Abschnitt II, 2)

a) Die Struktur der Mycaminose

Die Konstitutionsaufklärung der Mycaminose, $C_8H_{17}NO_4$, erfolgte fast gleichzeitig und weitgehend unabhängig in zwei Laboratorien. Eine charakteristische Reaktion ist die leichte Abspaltung von Dimethylamin (**248**) in alkalischer Lösung, wodurch die Anwesenheit einer Dimethylaminogruppe leicht erkannt werden konnte. Diese ergibt sich auch aus der Anwesenheit von zwei nach ZEISEL bestimmbaren N-Methylgruppen bei nur einem Stickstoffatom (*142, 243*). Die Anwesenheit einer C-Methylgruppe, die sich durch eine KUHN-ROTH-Oxydation wie auch durch das NMR.-Spektrum (*144*) erfassen ließ, deutete auf eine Methylpentose hin. Die Mycaminose gibt sich durch ihre reduzierenden Eigenschaften, durch die Bildung von zwei diastereomeren Tri-O-acetylderivaten (**249** und **250**), sowie durch die Bildung von 1 Mol Ameisensäure bei der raschen Oxydation mit einem Mol Perjodsäure als Aldose zu erkennen, die wegen des Fehlens einer Carbonylbande im IR.-Absorptionsspektrum vollständig als Cyclohemiacetal (**247**) vorliegen muß.

Die vollständige Konstitutionsaufklärung ergab sich aus dem Abbau mit Perjodat. Mycaminose verbraucht rasch ein Mol des Oxydationsmittels, wobei in guter Ausbeute ein Mol Ameisensäure (**251**) und ein

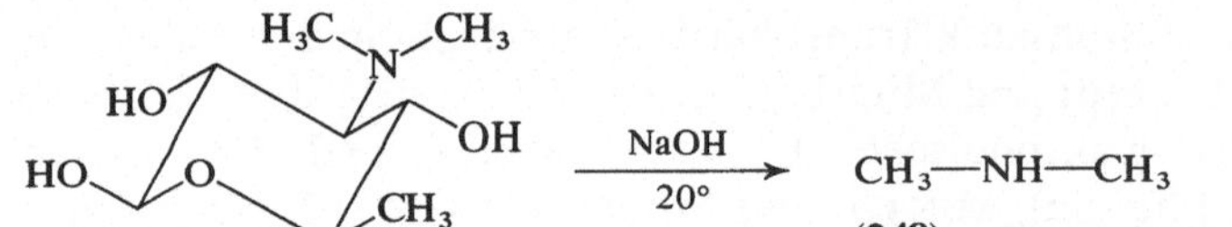

(**247**) Mycaminose (**248**) Dimethylamin

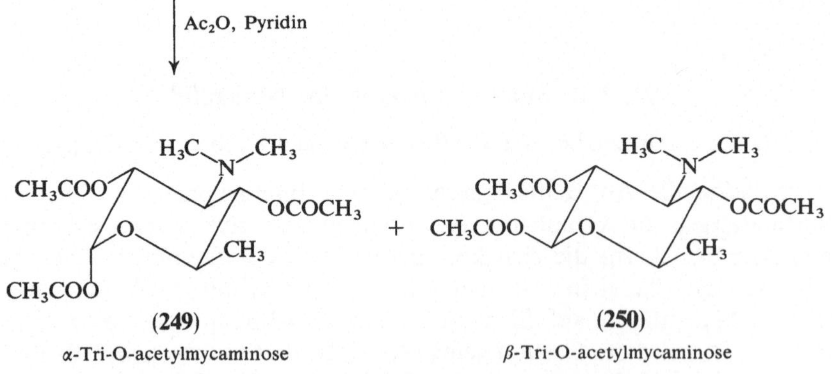

(**249**) α-Tri-O-acetylmycaminose

(**250**) β-Tri-O-acetylmycaminose

Schema 43. Umwandlungs- und Abbauprodukte der Mycaminose

Literaturverzeichnis: SS. 445—460

Schema 43 (Fortsetzung)

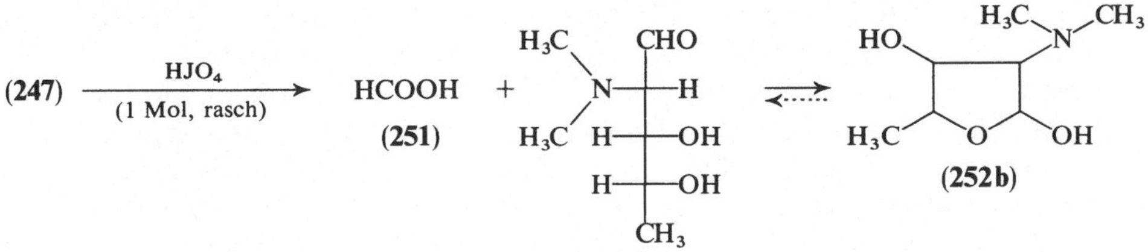

(252a)

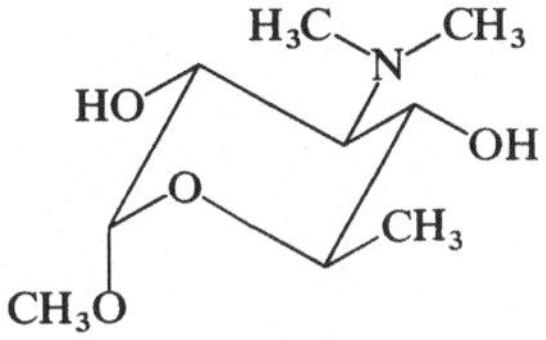

(253)

α-Methylmycaminosid

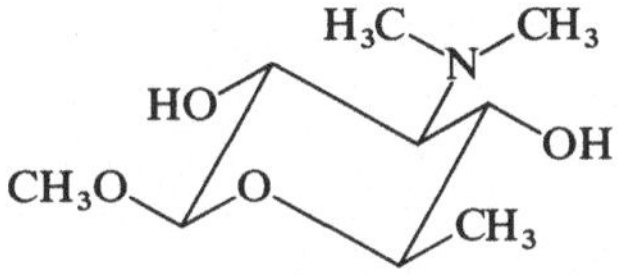

(254)

β-Methyl-mycaminosid

um ein C-Atom kleinerer Zucker (**252**) entsteht, der wiederum vollständig als Cyclohalbacetal (**252b**) vorliegt und in reiner Form isoliert und analysiert werden konnte. Der neue Zucker (**252**) verbraucht sehr viel langsamer drei weitere Mol Perjodsäure, wobei als Abbauprodukte Ameisensäure und Acetaldehyd nachgewiesen wurden, während Formaldehyd nicht auftrat. Diese Ergebnisse sprechen eindeutig für die Konstitutionsformel **247** für die Mycaminose.

Die Stellung der Dimethylaminogruppe am C-Atom 3 wird dadurch bestätigt, daß der Aminozucker **252** mit Alkali etwa 40mal langsamer Dimethylamin abspaltet als die Mycaminose selber. Es ist bekannt, daß β-Aminocarbonylverbindungen diese Reaktion besonders leicht eingehen. Interessant ist in diesem Zusammenhang, daß die beiden Methylmycaminoside (**253** und **254**), die beide in kristalliner Form bekannt sind (*243*), unter den gleichen Bedingungen kein Dimethylamin liefern, da ihnen die Möglichkeit, mit der Aldehydform ein Gleichgewicht zu bilden, fehlt.

Ungewöhnlich verlief der HOFMANNsche Abbau des quaternisierten α-Methylmycaminosids (**255**), das aus dem Methylglykosid **253** durch 15tägiges Stehen mit Methyljodid erhalten wurde. Das Produkt des Abbaus (**256**) zeigte im IR.-Absorptionsspektrum weder Banden von Carbonylgruppen noch solche von Doppelbindungen. Hingegen ließ sich mit Magnesiumbromid eine Epoxidgruppe nachweisen. Im Epoxid **256** ist die Methylglykosidbindung mit wässeriger Schwefelsäure erheblich leichter hydrolysierbar als die Epoxygruppe. Nach der Hydrolyse bei Zimmertemperatur wurde ein Produkt (**257**) erhalten, das lediglich ein Mol Perjodsäure verbrauchte und dabei etwa ein Mol Ameisensäure, nicht aber Acetaldehyd lieferte. Das Produkt der in der Wärme durchgeführten Hydrolyse verbrauchte dagegen 4 Mol Perjodsäure und im Oxydationsgemisch wurden 3,3 Mol Ameisensäure und gegen ein Mol Acetaldehyd nachgewiesen. Anderseits bildete sich bei der Hydrolyse des Epoxids in der Hitze mit 3 N Salzsäure eine mit Wasserdampf flüchtige Verbindung,

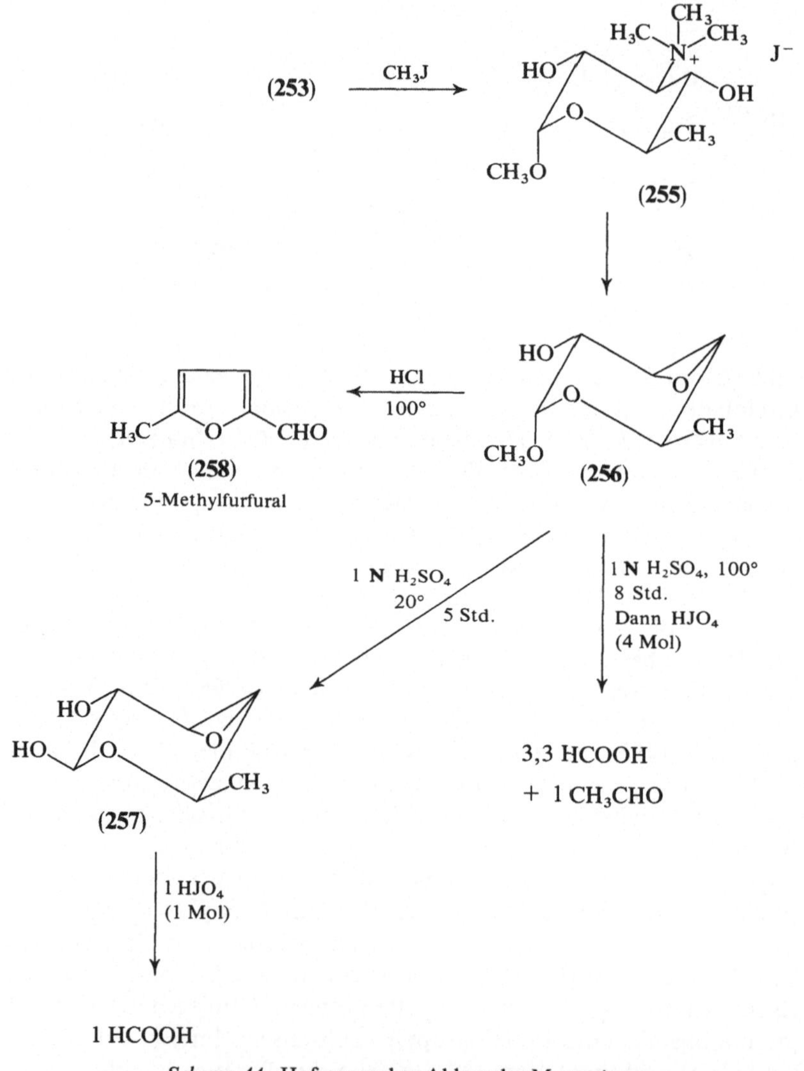

Schema 44. Hofmannscher Abbau der Mycaminose

Literaturverzeichnis: SS. 445—460

die durch ihr 2,4-Dinitrophenylhydrazon als 5-Methylfurfural (**258**) identifiziert wurde. Dadurch ist die Methylpentosestruktur der Mycaminose auf unabhängigem Wege bestätigt worden (*243*).

Die Stereochemie der Mycaminose wurde relativ spät richtig erkannt. Auf Grund von pK-Änderungen bei der partiellen Hydrolyse des Carbomycins wurde ursprünglich eine Altrose-Konfiguration in Betracht gezogen (*329*), doch konnte diese Annahme durch FOSTER und Mitarb. (*108*) auf Grund von Abbauprodukten sowie durch die Synthese der 3,6-Didesoxy-3-dimethylamino-L-altrose, die mit der Mycaminose weder identisch noch enantiomorph war, widerlegt werden.

Die eindeutige Zuordnung der *gluco*-Konfiguration gelang zuerst HOFHEINZ und GRISEBACH (*144*) auf Grund des NMR.-Spektrums der β-Tri-O-acetylmycaminose (**240**). Das in der Tabelle 5 angegebene Spektrum zeigt, daß alle Ringprotonen mit ihren Nachbar-Ringprotonen mit Kopplungskonstanten zwischen 8 und 10,2 Hz. in Wechselwirkung treten. Sie sind demnach alle axial angeordnet, was nur mit der *gluco*-Konfiguration im Einklang steht.

Tabelle 5. *NMR.-Spektrum der β-Tri-O-acetylmycaminose in CDCl₃ (60 MHz.)*

(ppm)	Multipl.	J (Hz.)		Integral	Zuordnung
1,16	d	6,4		3	CH$_3$-(CH) (Pos. 6)
1,99	s			3	
2,02	s			3	3 äquatoriale CH$_3$-COO
2,06	s			3	
2,37	s			6	$-$N(CH$_3$)$_2$
2,80	t	10,2		1	$-$CH-N; H-C-3
3,57	dq	6,3 (q)		1	H-C-5
		9,7 (d)			
4,74	dd	9,7		1	H-C-4
		10,2			
4,98	dd	8,0		1	H-C-2
		10,2			
5,52	d	8,0		1	H-C-1

Auf Grund ihrer chemischen Verschiebungen ließen sich die Signale der C-Methylgruppe, der drei Acetylgruppen, der N-Methylgruppen sowie der Ringprotonen H-C-1 (zwei Sauerstoffatomen anliegend) und H-C-5 leicht zuordnen. Hingegen besitzen die Signale der beiden Wasserstoffkerne an C-2 und C-4 ähnliche chemische Verschiebungen und überlappen teilweise. Ihre exakte Interpretation war erst nach Berücksichtigung der an andern Signalen erkannten Kopplungskonstanten möglich.

Fast gleichzeitig wurde das Problem auch durch zwei unabhängige Synthesen der Mycaminose geklärt, nachdem vergleichende Untersuchungen der chiroptischen Eigenschaften eine **D**-*gluco*-Konfiguration wahr-

418 W. Keller-Schierlein:

scheinlich gemacht hatten (*259*). Foster und Mitarb. (*105*) konnten für
die Herstellung des Ausgangsmaterials (**261**) ihrer Synthese auf ältere
Arbeiten von Richtmyer und Hudson (*262*) zurückgreifen. Das 4,6-O-
Benzyliden-α-methyl-D-glucosid (**259**) ist durch Umsetzung von α-Methylglucosid und Benzaldehyd in Gegenwart von wasserfreiem Zink-
chlorid leicht in größerer Menge zugänglich. Sein Ditosylat (**260**) geht
mit Natriummethanolat bei Einhaltung geeigneter Reaktionsbedingungen
in guter Ausbeute unter Umkehrung der Konfiguration an C-3 in das
2,3-Anhydroallose-Derivat **261** über. Unter milden sauren Bedingungen
läßt sich daraus die Benzylidengruppe selektiv abspalten unter Bildung
des α-Methyl-2,3-anhydro-allosids (**262**). Die Anlagerung von Dimethylamin an die Epoxidgruppe erfolgt erwartungsgemäß stereospezifisch,
aber nicht strukturspezifisch. Das Gemisch der Produkte (**263** und **264**)
wurde nicht physikalisch getrennt, sondern einer kurzen Oxydation mit
Perjodsäure unterworfen. Dabei wurde die Verbindung **264** mit einer 1,2-
Diolgruppe rasch abgebaut, während das gewünschte α-Methyl-3-des-
oxy-3-dimethylamino-D-glucosid (**263**) bei kurzer Einwirkungszeit nicht
verändert wurde und sich durch Chromatographie an einem Kationen-
austauscherharz rein gewinnen ließ. Die Reduktion der primären Hydroxylgruppe erfolgte in konventioneller Weise durch partielle Tosylierung
und Reduktion mit Lithiumaluminiumhydrid. Das Reduktionsprodukt
(**253**) war identisch mit α-Methylmycaminosid, sein Hydrolyseprodukt
mit der Mycaminose.

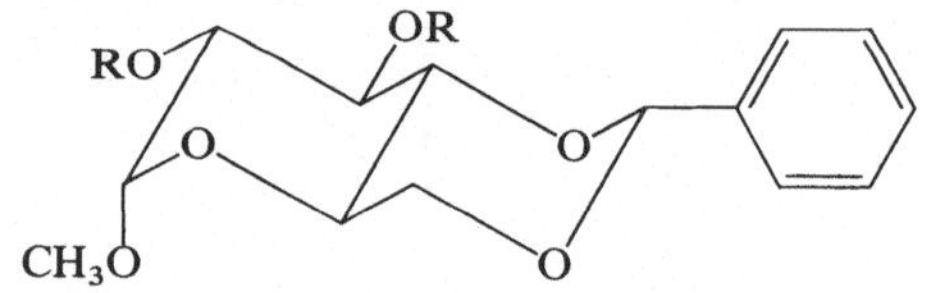

(**259**)   R = H, 4,6-O-Benzyliden-α-methyl-D-glucosid
(**260**)   R = pTs

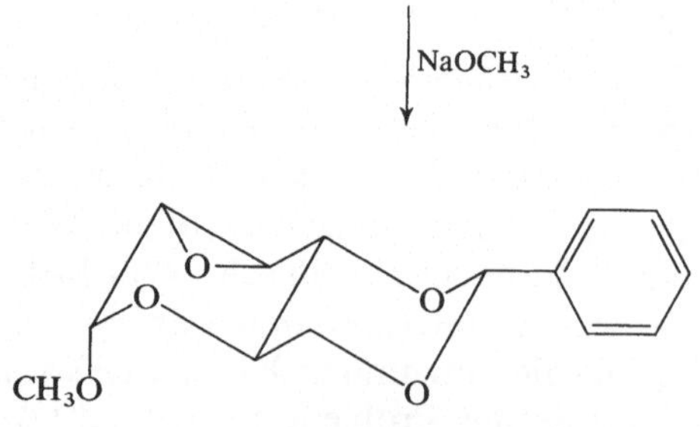

NaOCH₃

(**261**)

Schema 45. Synthese der Mycaminose nach Foster u. Mitarb.

Schema 45 (Fortsetzung)

0,05 N HCl, 60°, 2 Std.

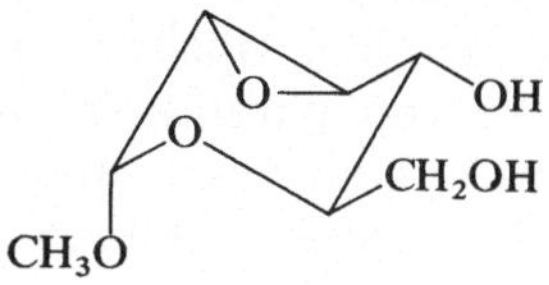

(262)

CH₃—NH—CH₃
in C₂H₅OH
170°, 18 Std.

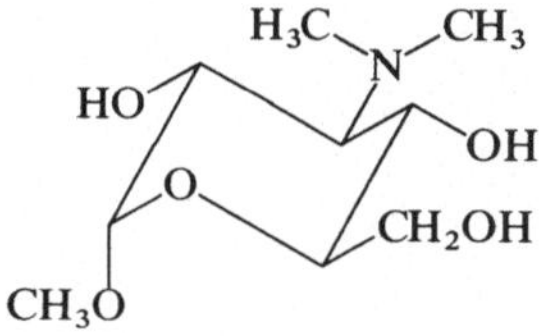

(263)

D-gluco-Konfiguration
resistent gegen HJO₄

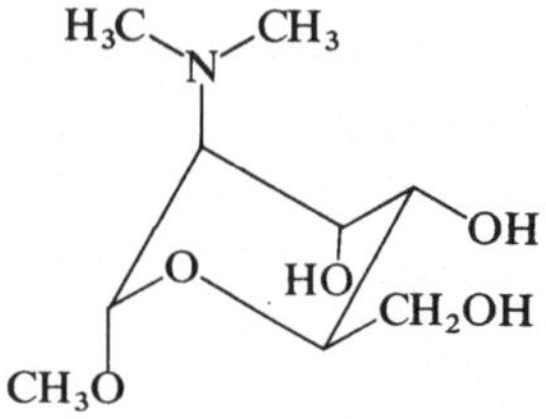

(264)

D-altro-Konfiguration
wird von HJO₄ rasch abgebaut

1. pTsCl (1 Mol)
2. LiAlH₄

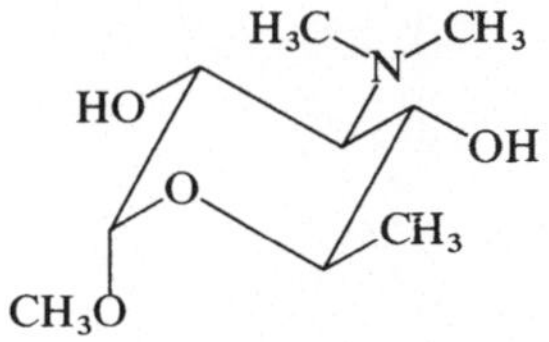

(253)

H₂O, H⁺

(247) Mycaminose (Hydrochlorid)

27*

Einen völlig anderen Weg zur Synthese der Mycaminose schlug Richardson ein (*258, 259*). Ausgehend von α-Methyl-D-rhamnosid (**265**) wurde zunächst mittels Perjodat der Dialdehyd **266** hergestellt, der durch basische Kondensation mit Nitromethan ein Diastereomerengemisch der Nitroverbindungen **267** lieferte. Die Abtrennung des reinen D-*gluco*-Isomeren in etwa 25 bis 30% Ausbeute gelang erst nach der Reduktion der Nitrogruppe. Die Konfigurationszuordnung ergab sich aus dem Vergleich der spezifischen Drehung des Aminozuckers mit denen der in Frage kommenden neutralen Zucker. Anhand zahlreicher Beispiele

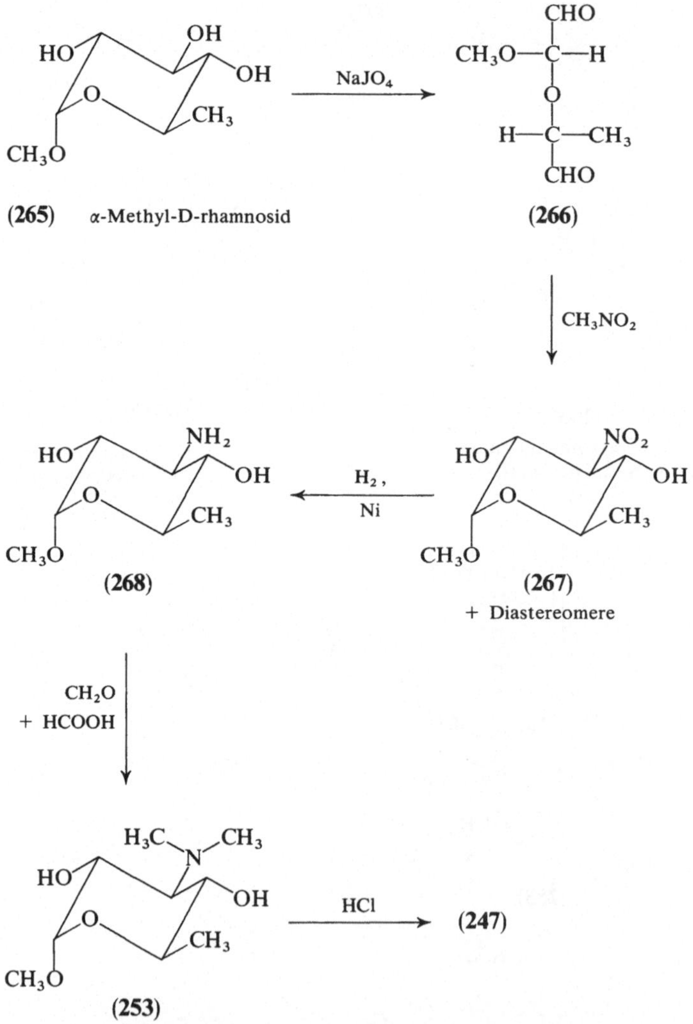

(**265**)   α-Methyl-D-rhamnosid

$NaJO_4$

(**266**)

CH_3NO_2

(**268**)

H_2, Ni

(**267**)

\+ Diastereomere

CH_2O + HCOOH

(**253**)

HCl

(**247**)

Schema 46. Synthese der Mycaminose nach Richardson

hatte man nämlich erkannt, daß der Ersatz einer Hydroxylgruppe durch eine Aminogruppe die chiroptischen Eigenschaften von Zuckern nur geringfügig verändert. Für die Einführung der N-Methylgruppen eignete sich ein Gemisch von Formaldehyd und Ameisensäure, das in 71% Ausbeute das α-Methylmycaminosid (**253**) gab, welches sich mit Salzsäure zur Mycaminose (**247**) verseifen ließ.

b) *Weitere Aminozucker aus antibakteriellen Makrolid-Glykosiden*

Das Angolosamin wurde bisher nur als Baustein des Angolamycins aufgefunden (*47*), eines Makrolids der Carbomycingruppe, dessen Aglykon noch nicht aufgeklärt ist, aber wahrscheinlich demjenigen des Tylosins nahesteht. Wegen der günstigen Position der verschiedenen Substituenten und deren all-*trans*-Anordnung konnten die Konstitution und Konfiguration (**269**) aus spektroskopischen Daten allein abgeleitet werden, der Chiralitätssinn ergab sich aus den chiroptischen Eigenschaften der beiden Di-O-acetylderivate gemäß den Regeln von HUDSON. Das Angolosamin ist demnach die 2-Desoxy-mycaminose.

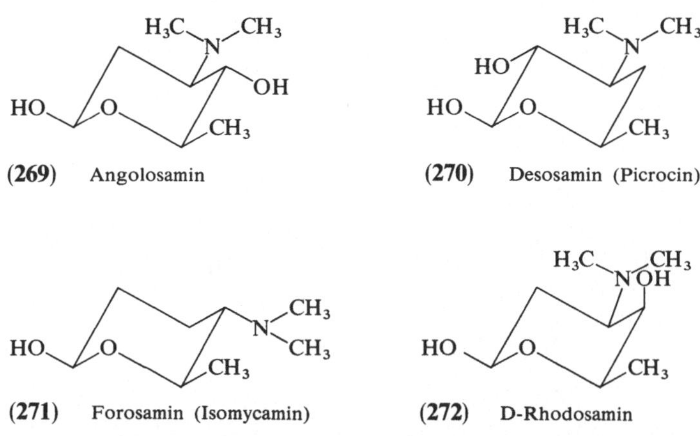

Schema 47. Weitere Aminozucker aus Makroliden

Der charakteristische Aminozucker der Erythromycine ist das Desosamin (=Picrocin), das außerdem noch im Pikromycin, Methymycin, Neomethymycin, Narbomycin und Oleandomycin vorkommt (vgl. Abschnitt II/1). Die Konstitutionsaufklärung dieses Aminozuckers (**270**) verdanken wir unabhängigen Untersuchungen von GERZON und Mitarb. (*204*), CLARK (*64*) sowie BROCKMANN und Mitarb. (*41*). Arbeiten zur Konfigurationsaufklärung durch eingehende NMR.-Analysen stammen von HOFHEINZ und GRISEBACH (*143*) sowie WOO und Mitarb. (*327*). Der

Chiralitätssinn wurde durch Abbau zum optisch aktiven (R)-Pentan-1,4-diol aufgeklärt (30) und steht mit den chiroptischen Eigenschaften des Desosamins und seiner Derivate im Einklang (143). Die β-pyranosidische Verknüpfung mit dem Aglykon wurde besonders sorgfältig am Beispiel des Oleandomycins mit Hilfe von NMR.-Spektren und mittels molekularen Drehungsverschiebungen abgeklärt (62). Beim Erythromycin geht sie aus der früher erwähnten Röntgenstrukturanalyse (128) hervor, und sie dürfte auch für alle anderen Antibiotica dieser Gruppe gelten.

Eine konfigurationsbeweisende Synthese wurde im Zusammenhang mit derjenigen der Mycaminose (s. oben) ausgeführt (260). Totalsynthesen, ausgehend von nicht zuckerartigen Ausgangsmaterialien, sind bisher zwei bekannt geworden (170, 223).

Die nahe chemische Beziehung des Desosamins zur Mycaminose findet ihren experimentellen Ausdruck in einer direkten Überführung des einen Zuckers in den anderen, die besonders darum bemerkenswert ist, weil sie nicht am isolierten Zucker, sondern am intakten Antibioticum durchgeführt wurde (157).

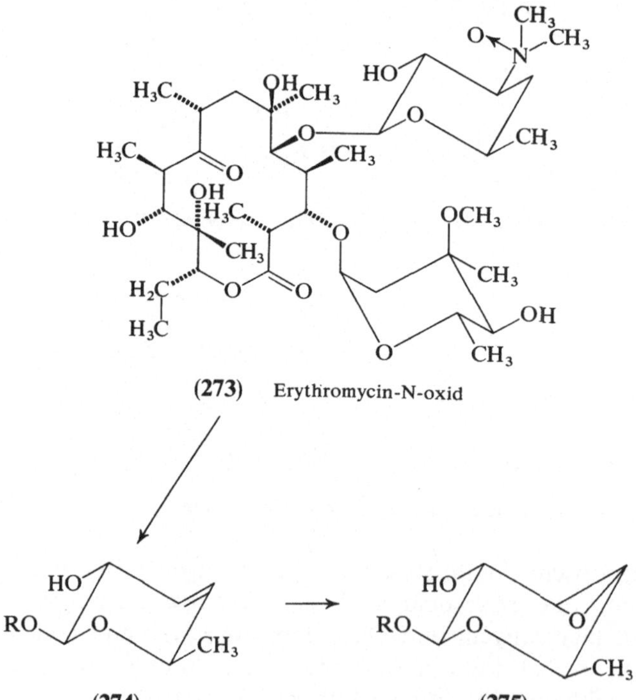

(273)   Erythromycin-N-oxid

(274)   (275)

R = Cladinosyl-erythronolid-Rest

Schema 48. Partialsynthese von 4'-Hydroxy-erythromycin A

Schema 48 (Fortsetzung)

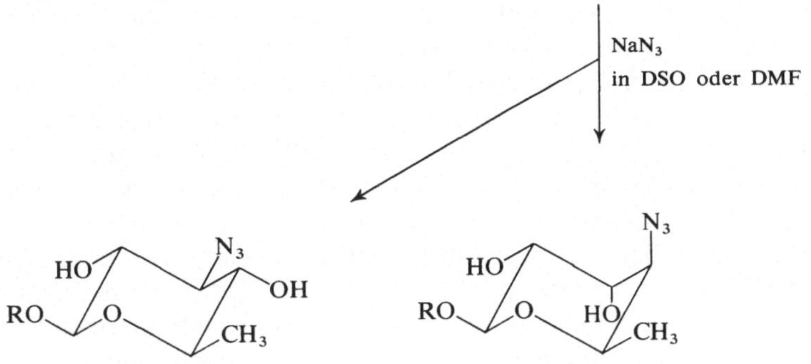

NaN$_3$
in DSO oder DMF

(276)
D-gluco-Konfiguration

(277)
D-gulo-Konfiguration

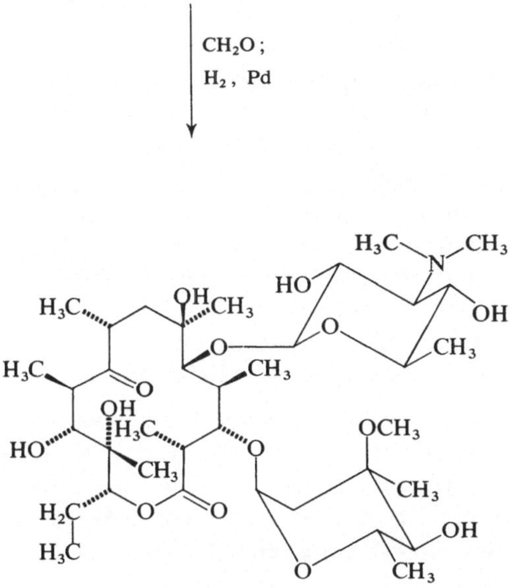

CH$_2$O;
H$_2$, Pd

(278)

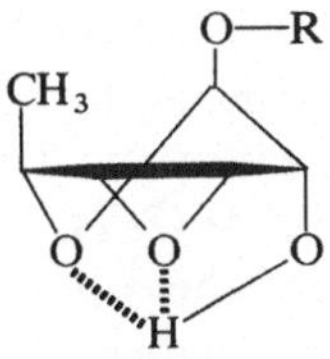

(279)

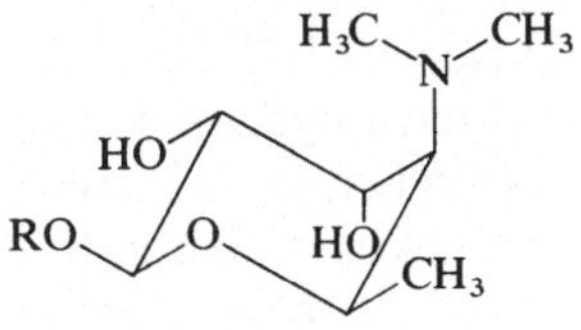

(280)

Das Erythromycin-N-oxid (**273**), das sich aus Erythromycin durch Oxydation mit Wasserstoffsuperoxid in guter Ausbeute herstellen ließ, und das als Dihydrat kristallisierte (*156*), erlitt beim Erhitzen auf 150° im Vakuum einen Abbau zum Allylalkohol **274**. Die Addition von Sauerstoff mittels m-Chlorbenzoepersäure erfolgte erwartungsgemäß in *cis*-Stellung zur Hydroxylgruppe und gab in hoher Ausbeute das Epoxid **275**. In Abwesenheit von Wasser in Dimethylformamid lagerte sich ein Azidion gemäß der Regel von Fürst und Plattner zu über 90% von der axialen Seite her an und führte vorherrschend zum unerwünschten Azid **277** mit der Azidgruppe in Stellung 4′ und der **D**-*gulo*-Konfiguration. Zugabe von etwa 5% Wasser zur Reaktionslösung veränderte das Verhältnis der Produkte entscheidend. Zwar herrschte noch immer die Verbindung **277** etwas vor, doch ließ sich jetzt auch das **D**-*gluco*-Derivat **276** in vergleichbarer Menge gewinnen. Es wurde vermutet (*157*), daß das Epoxid in wasserhaltiger Lösung teilweise in der Konformation **279** vorliegt, die durch intramolekulare Wasserstoffbrücken stabilisiert wird und gemäß Fürst und Plattner zur **D**-*gluco*-Verbindung **276** führen sollte. In wasserfreiem Lösungsmittel wird die Hydroxylgruppe dagegen vollständig für intermolekulare Wasserstoffbrücken zu Dimethylformamid-Molekülen beansprucht, so daß praktisch nur die Konformation des Epoxids mit drei äquatorialen Substituenten vorliegt.

Durch reduktive Methylierung (Formaldehyd, Wasserstoff und Palladiumkohle) konnte das Azid **276** mit 60% Ausbeute in das 4′Hydroxyerythromycin (**278**) übergeführt werden, das nun anstelle des Desosaminrestes den Mycosaminrest trägt. Dies ließ sich bestätigen durch Abbau des Produktes zu Dihydroerythronolid, Methylcladinosid und Mycaminose nach früher beschriebenen Methoden (Abschnitt II).

Das umgewandelte Antibioticum (**278**) erwies sich als etwa halb so wirksam wie Erythromycin A bei etwa gleicher Breite des Wirkungsspektrums. Mit den gleichen Methoden wurde auch das „falsche" Anlagerungsprodukt (**277**) in ein Erythromycinanalogon (**280**) übergeführt, das aber keine antibakterielle Wirkung zeigte (*157*).

Ein Dimethylaminozucker, der bisher nur als Baustein der Spiramycine (neben Mycaminose) aufgefunden wurde, erhielt zunächst die Bezeichnung *Forosamin* (*152*). Später wurde er auch unter dem Namen Isomycamin beschrieben (*245*). Seine Konstitution als 2-Hydroxy-5-dimethylamino-6-methyl-tetra-hydropyran (**271**) wurde schon bald nach der Entdeckung aufgeklärt (*242*), die Stereochemie jedoch erst im Zusammenhang mit einer stereoselektiven Synthese ausgehend von **D**-Glucose (*281*). Eine zweite, für sich allein nicht konfigurationsbeweisende Synthese aus **D**-Glucose wurde kurze Zeit später mitgeteilt (*2*). Andererseits haben später Umezawa und Mitarb. (*284*) ein Zwischenprodukt der Kasugaminsynthese in D,L-Forosamin übergeführt, um dessen relative Konfiguration zu beweisen.

Zusätzlich zum Desosamin tritt bei den Megalomycinen A, B, C und D (s. Abschnitt II/1) das D-Rhodosamin (**272**) auf, dessen Struktur durch Abbaureaktionen aufgeklärt wurde (*186*). Das enantiomorphe L-Rhodosamin ist ein häufiger Baustein von Anthracyclin-Antibiotica (*37, 164*).

2. Mycosamin und Perosamin

Als häufige Bausteine von Polyen-Makroliden treten ebenfalls Aminozucker auf. Von den beiden Zuckern dieser Gruppe spielt das Mycosamin weitaus die größere Rolle. Außer in den im Abschnitt III aufgeführten Antibiotica kommt es noch in einer Reihe von unvollständig charakterisierten Polyenen vor. Das Perosamin wurde dagegen bis heute nur als Hydrolyseprodukt des Perimycins nachgewiesen.

Das Mycosamin (**281**) konnte zunächst durch Hydrolyse von Amphothericin bzw. Nystatin nicht in reiner Form erhalten werden (*95, 96*), hingegen erhielt man durch Acetolyse ein Triacetyl- (**282**) und ein Tetraacetylderivat (**283**) des neuen Aminozuckers in kristalliner Form. Das Tetraacetat ging bei milder saurer Hydrolyse ins Triacetat (**282**) über, während beide bei der alkalischen Methanolyse das N-Monoacetat (**284**) lieferten, das im Carbonylgebiet nur noch IR.-Absorptionsbanden bei 1640 und 1550 cm^{-1} für die Amidgruppe besaß.

Durch Umsetzung des N-Monoacetylderivates mit salzsaurem Methanol konnte schließlich ein N-Monoacetyl-methylglykosid (**285**) erhalten werden, dem später die α-Konfiguration zugeordnet werden konnte. Das acetylfreie α-Methylmycosaminid (**286**) konnte später auch durch direkte Methanolyse des Nystatins gewonnen werden, unter Bedingungen allerdings, die zur völligen Zerstörung des Aglykons führten (*95*). Teerartige Zersetzungsprodukte des Aglykons scheinen auch schuld daran zu sein, daß die direkte Gewinnung des Mycosamins durch Hydrolyse ursprünglich nicht gelungen war. Durch Umsetzung eines Rohhydrolysates mit Chlorameisensäure-benzylester wurde aber das Carbobenzoxy-Derivat **287** in kristalliner Form erhalten, woraus sich durch Hydrogenolyse in salzsaurer Lösung das Mycosamin (**281**) als kristallines Hydrochlorid gewinnen ließ.

Aus Analysen der verschiedenen Derivate ergab sich für das freie Mycosamin die Bruttoformel $C_6H_{13}NO_4$. Durch Gruppenanalysen wurde das Vorliegen einer C-Methylgruppe sowie das Fehlen von N- und O-Methylgruppen festgestellt. Mycosamin gibt eine positive Jodoformreaktion und reduziert Tollens-Reagens sowie Fehlingsche Lösung. Aus diesen Befunden ergab sich die Konstitution einer Amino-desoxy-methylpentose. Die Lage der Aminogruppe am C-Atom 3 wurde durch Oxydation mit Perjodat abgeklärt. Während das N-Acetyl-methylgykosid (**285**) mit dem Oxydationsmittel nicht reagierte, verbrauchte das N-Acetylmycosamin (**284**) ein Mol Perjodat. Aus der Oxydationslösung wurde der Aldehyd **288** in amorpher Form erhalten. Bei der milden alkalischen Hydrolyse verbrauchte dieser ein Mol Lauge, als Reaktionsprodukte wurden Ameisensäure und eine kristalline Verbindung $C_7H_{13}NO_4$ isoliert. Die letztere zeigte die reduzierenden Eigenschaften einer Aldose und die IR.-Absorptionsbanden eines Amids. Es kann sich damit nur um das Pentosederivat **289** handeln.

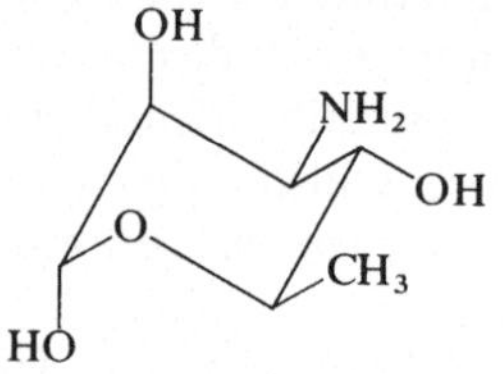

(**281**)

Mycosamin

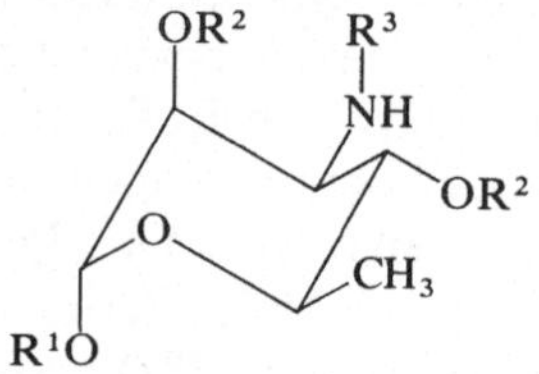

	R^1	R^2	R^3
(**282**)	H	Ac	Ac
(**283**)	Ac	Ac	Ac
(**284**)	H	H	Ac
(**285**)	CH_3	H	Ac
(**286**)	CH_3	H	H
(**287**)	H	H	$-COOCH_2-C_6H_5$

$$(284) \xrightarrow{\text{NaJO}_4} \overset{\overset{\displaystyle CH_3}{|}}{HCOO-CH}-CHOH-\overset{\overset{\displaystyle NHAc}{|}}{CH}-CHO$$

(**288**)

$$\Big\downarrow \text{NaOH}$$

$$CH_3-CHOH-CHOH-\overset{\overset{\displaystyle NHAc}{|}}{CH}-CHO$$

(**289**)

Schema 49. Mycosamin: Derivate und Abbau

Die Konfigurationsbestimmung an C-5 des Mycosamins folgte klassischen Arbeiten aus der Stereochemie der Zucker. Um die Position 3 des Mycosamins durch Perjodat angreifbar zu machen, wurde das α-Methyl-N-acetylmycosaminid (**285**) zunächst zum sekundären Amin **290** reduziert. Der Abbau mit Perjodat führte dann zu einer wohlbekannten Verbindung, D'-Methoxy-D-methyl-diglycolaldehyd (**291**), die im Gleichgewicht praktisch vollständig in der cyclischen Form (**291 b**) vorliegt (*185*). Die Konfigurationsaufklärung dieser Verbindung war seinerzeit durch Abbau zu Oxalsäure (**292**) und D-Milchsäure (**293**) bewiesen worden. Die Identifizierung des Abbauproduktes **291** bewies nicht nur die **D**-Konfiguration an C-5 des Mycosamins, sondern gleichzeitig die α-Konfiguration der Methylglykoside **285** und **286**.

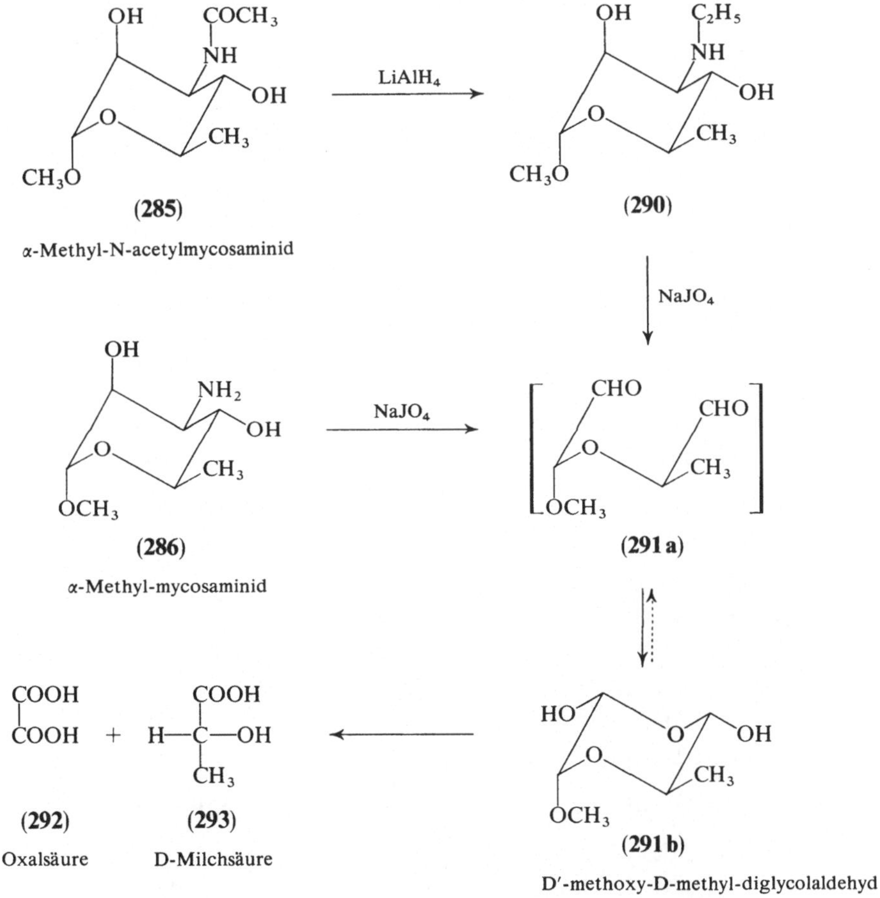

(**285**)
α-Methyl-N-acetylmycosaminid

LiAlH₄

(**290**)

NaJO₄

(**286**)
α-Methyl-mycosaminid

NaJO₄

(**291 a**)

(**291 b**)
D'-methoxy-D-methyl-diglycolaldehyd

(**292**)   (**293**)
Oxalsäure   D-Milchsäure

Schema 50. Konfiguration des Mycosamins (s. auch S. 428 und 429)

W. KELLER-SCHIERLEIN:

Schema 50 (1. Fortsetzung)

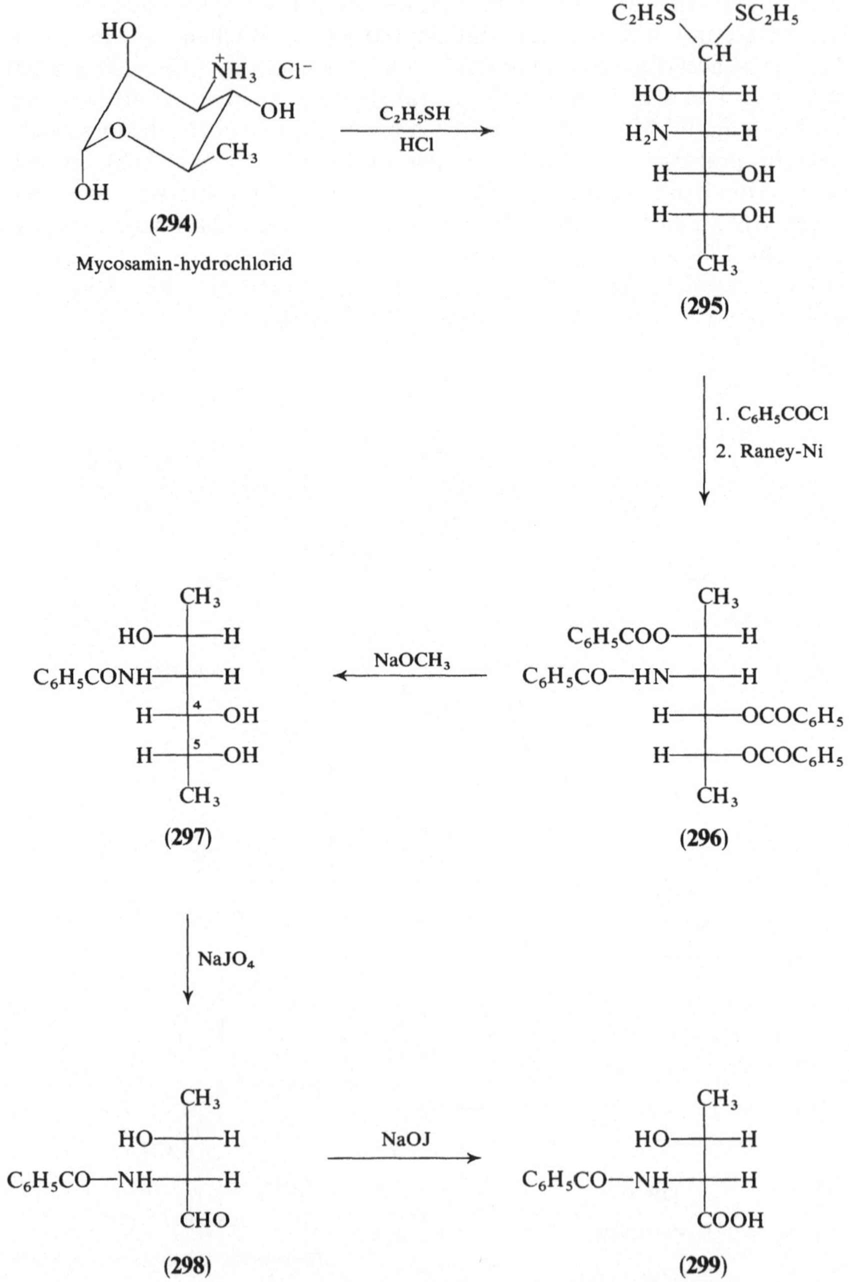

(294)

Mycosamin-hydrochlorid

$\xrightarrow{\begin{array}{c} C_2H_5SH \\ HCl \end{array}}$

(295)

1. C_6H_5COCl

2. Raney-Ni

(297)

$\xleftarrow{NaOCH_3}$

(296)

NaJO$_4$

(298)

$\xrightarrow{NaOJ}$

(299)

N-Benzoyl-D-allo-Threonin

Literaturverzeichnis: SS. 445—460

Schema 50 (2. Fortsetzung)

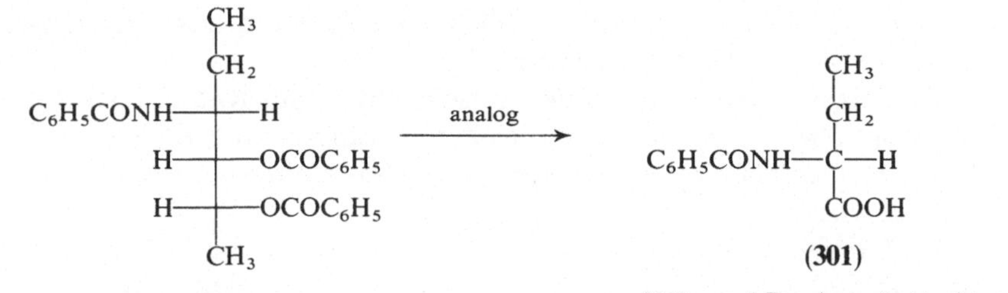

C_6H_5CONH — ... — H (300)

analog →

C_6H_5CONH — C — H (301)

N-Benzoyl-D-α-Aminobuttersäure

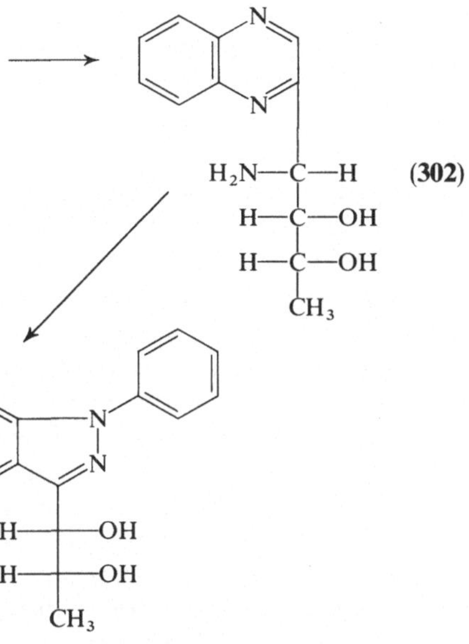

(294) ⟶ (302)

(303)   ein Phenylflavazol

Auch die Konfiguration an den C-Atomen 2 und 3 wurde durch einen klassischen Abbau ermittelt (*307*). Für den spezifischen Abbau zwischen den C-Atomen 4 und 5 mußte das Mycosamin zunächst umgeformt werden. Während das N-Acetylderivat **284** nur in schlechter Ausbeute ein Dithioacetal gab, entsprach das Hauptprodukt der Umsetzung von Mycosamin Hydrochlorid (**294**) mit Äthylmercaptan der erwarteten Formel (**295**). Nach dem Schutz der Hydroxyl- und Aminogruppen erfolgte die reduktive Entschweflung mit RANEY-Nickel zum Produkt **296**. Am Produkt der selektiven Abspaltung der O-Benzoylgruppen mit Natriummethanolat in absolutem Methanol (**297**) war nun ein spezifischer Abbau mit Natriumperjodat zwischen C-4 und C-5 möglich. Der amorphe Aldehyd (**298**) wurde nicht charakterisiert, sondern direkt zur N-Benzoyl-aminosäure **299** oxydiert, die in allen Eigenschaften dem bekannten Derivat des D-allo-Threonins entsprach.

Als Nebenprodukt war bei der Entschweflung mit Raney-Nickel das Tribenzoat **300** erhalten worden, das zur Bestätigung der Konfiguration an C-3 in analoger Weise zur N-Benzoyl-D-α-Amino-buttersäure (**301**) abgebaut wurde.

Die Konfiguration an C-4 wurde ebenfalls durch Verknüpfung mit einer Verbindung mit bekanntem Chiralitätssinn abgeleitet. Das Chinoxalinderivat **302**, das aus Mycosamin-hydrochlorid (**294**) und o-Phenylendiamin zugänglich ist, lieferte mit Phenylhydrazin das Phenylflavazol **303**, das enantiomer war zu einem aus L-Rhamnose hergestellten Präparat, hingegen diastereomer zum entsprechenden Derivat aus L-Fucose.

Diese Untersuchungen über die Stereochemie bildeten die Grundlage für synthetische Arbeiten (*308*) derselben Arbeitsgruppe, die sich ein kurz vorher von Baer und Fischer (*14*) entwickeltes Syntheseprinzip nutzbar machte. α-Methylglucosid (**304**) wurde mit Perjodat zum Dialdehyd **305** abgebaut. Dieser reagierte mit einem Mol Nitromethan in Gegenwart von Natriummethylat unter Ringschluß. Das als pulveriger Niederschlag anfallende Natriumsalz war wahrscheinlich ein Diastereomerengemisch, und wurde nicht getrennt. Auch aus dem Gemisch der Nitrozucker, das durch Ansäuern erhalten wurde, konnte keine einheitliche Verbindung gewonnen werden, obwohl es auf Grund späterer Beobachtungen das Diastereomere **306** in überwiegender Menge enthielt. Hingegen kristallisierte nach der katalytischen Reduktion in schwach salzsaurer Lösung das einheitliche Hydrochlorid **307** in etwa 35% Ausbeute aus. Durch Verseifung des Glykosids wurde darauf der Aminozucker **308** als Hydrochlorid erhalten. Da die Synthese nicht stereospezifisch ist, war eine Konfigurationsbestimmung notwendig. Diese erfolgte durch Abbau des N-Acetylderivates zur 2-Acetamido-2-desoxy-**D**-arabinose, die aus Arbeiten von R. Kuhn gut bekannt war und bereits als Ausgangsmaterial für eine nicht stereospezifische Synthese der 3-Amino-3-desoxy-D-mannose gedient hatte (*176*).

Für die Synthese des Mycosamins war im wesentlichen nur noch die selektive Reduktion an C-6 nötig, die auf dem im Formelschema 51 aufgezeigten gebräuchlichen Weg durchgeführt wurde (*308*). Das Methyl-glykosid-triacetat **285** und weitere daraus bereitete Derivate waren mit den entsprechenden Derivaten des Mycosamins identisch.

Eine spätere zweite Synthese (*261*) des Mycosamins macht sich das gleiche Syntheseprinzip dienstbar.

Der Aminozucker des antifungischen Heptaen-Makrolids Perimycin, dessen Aglykon noch nicht aufgeklärt worden ist, ist das Perosamin (**311**), ein Isomeres des Mycosamins mit der gleichen **D**-*manno*-Konfiguration aber vertauschten Substituenten an C-3 und C-4. Es gehört damit wie das Forosamin (s. oben) zu den relativ seltenen 4-Amino-4-desoxyhexosen. Die Strukturformel (311) wurde durch eine Totalsynthese des Methylglykosids bestätigt (*181, 280*).

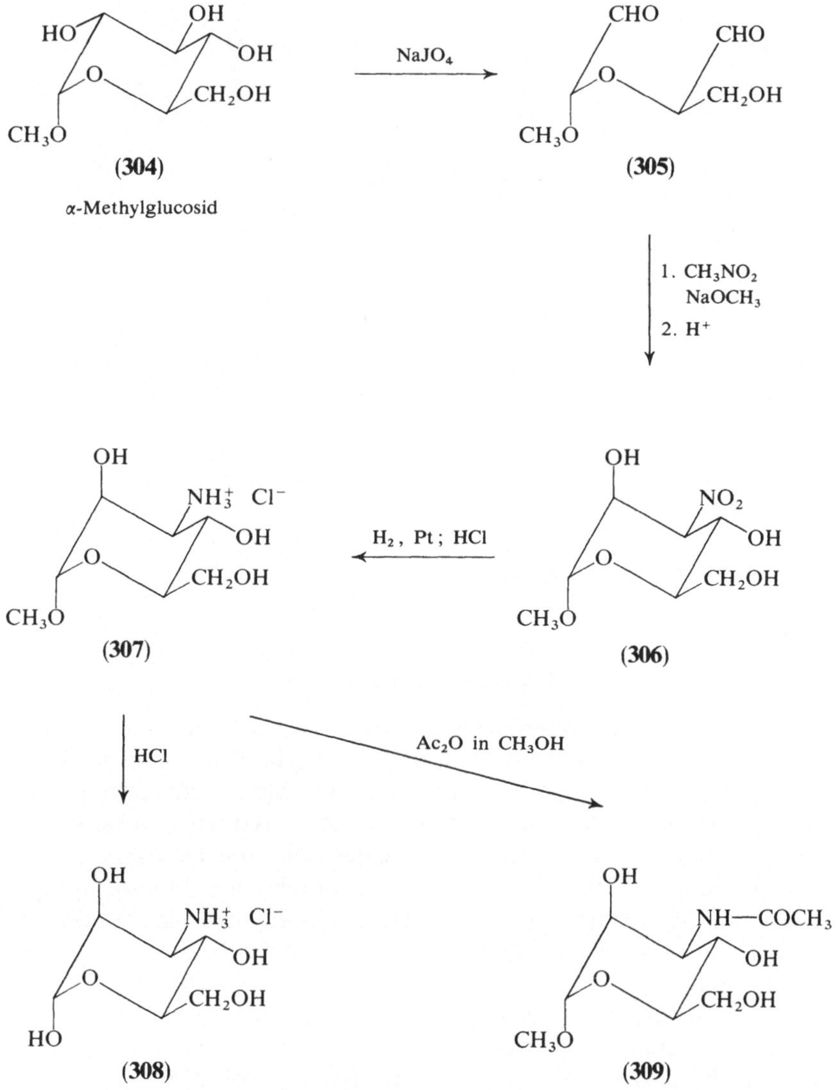

Schema 51. Synthese des Mycosamins (s. auch S. 432)

Schema 51 (Fortsetzung)

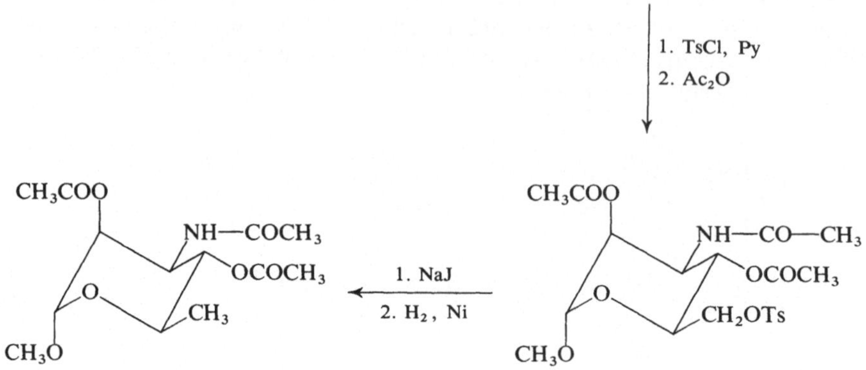

1. TsCl, Py
2. Ac$_2$O

CH$_3$COO

NH—COCH$_3$

OCOCH$_3$

O

CH$_3$

CH$_3$O

(285)

1. NaJ
2. H$_2$, Ni

CH$_3$COO

NH—CO—CH$_3$

OCOCH$_3$

O

CH$_2$OTs

CH$_3$O

(310)

OH

OH

NH$_2$

O

CH$_3$

HO

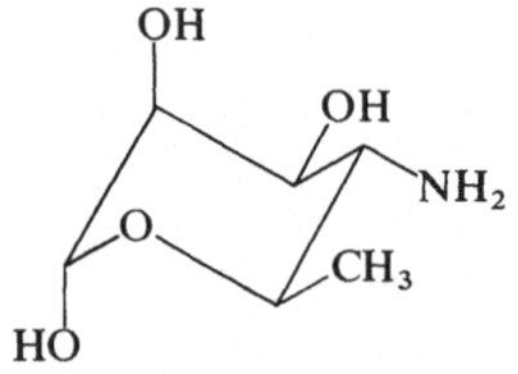

(311)

Perosamin

3. Lankavose (Chalcose)

Einigen Makroliden der Erythromycin- und Carbomycingruppe fehlt
der sonst für diese Verbindungen charakteristische Aminozucker (Mycaminose, Desosamin, Angolosamin). An dessen Stelle tritt dann ein neu-
traler Zucker, die Lankavose (= Chalcose), die strukturell zum Desosamin
in naher Beziehung steht. Sie unterscheidet sich vom Desosamin einzig
dadurch, daß die Dimethylaminogruppe durch eine Methoxylgruppe
ersetzt ist. Sie wurde bisher als Baustein der neutralen Makrolide
Lankamycin (*114, 165*), Chalcomycin (*325*) und Neutramycin (*177*)
nachgewiesen.

Die Konstitutionsaufklärung wurde unabhängig in zwei Laboratorien
durchgeführt und erfolgte durch chemische Abbaureaktionen, unter-
stützt durch spektroskopische Untersuchungen (*165, 325*). Die Stereochemie gemäß Formel **317** ergab sich aus einer eingehenden Analyse der
NMR.-Spektren der Lankavose und einiger Derivate sowie durch den

Abbau mit Salpetersäure, der die optisch aktive O-Methyl-**D**-Weinsäure (aus den C-Atomen 1 bis 4 des Zuckers) gab. Die **D**-Konfiguration des Zuckers folgt auch aus der Richtung der Mutarotation in Übereinstimmung mit der Isorotationsregel von HUDSON (*328*).

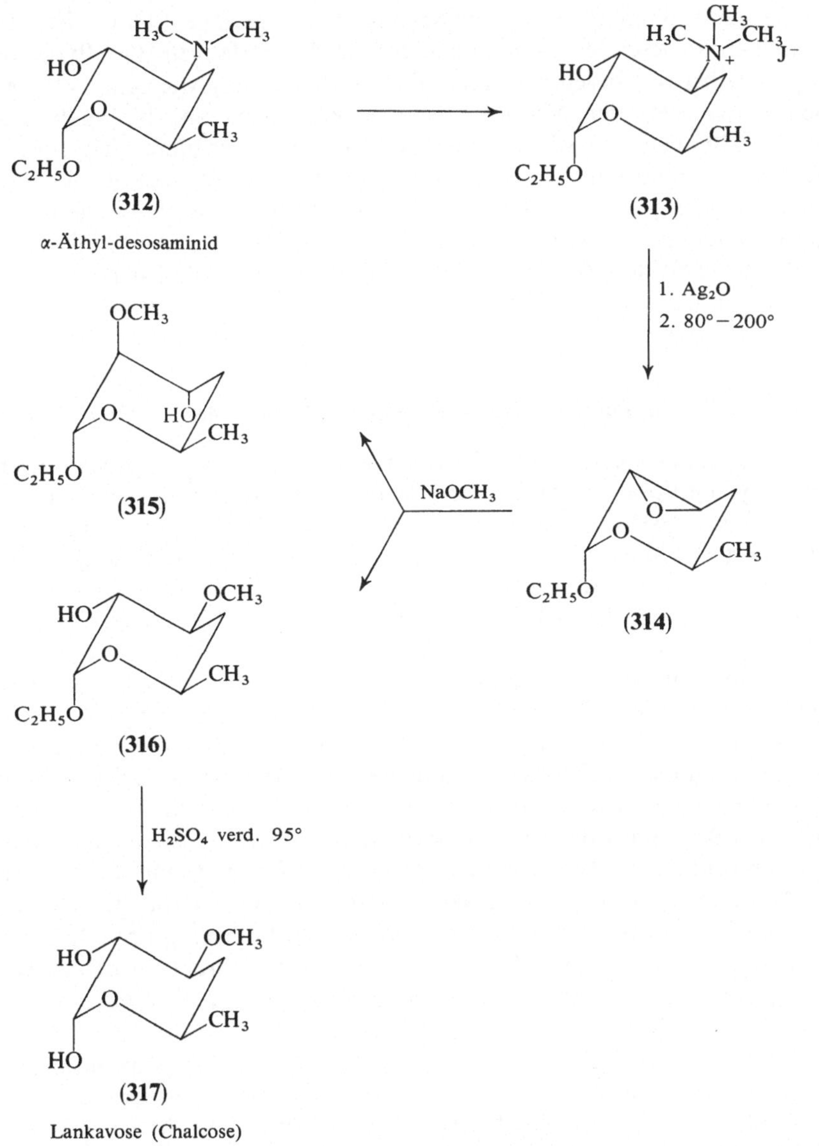

(312)

α-Äthyl-desosaminid

(313)

1. Ag$_2$O
2. 80° – 200°

(315)

NaOCH$_3$

(314)

(316)

H$_2$SO$_4$ verd. 95°

(317)

Lankavose (Chalcose)

Schema 52. Synthese der Lankavose aus Desosamin

Kurz nach der Strukturaufklärung folgte eine Bestätigung durch mehrere Synthesen (*109, 167, 199*). Im Formelschema 52 ist diejenige zusammengefaßt, die eine direkte Verknüpfung der Lankavose mit dem Desosamin darstellt. Ausgehend vom Äthyl-desosaminid (**312**) wurde das quaternäre Ammoniumjodid (**313**) in kristallisierter Form erhalten. Bei dessen Reinigung blieb wahrscheinlich die geringe Menge des ursprünglich anwesenden β-Glykosids in den Mutterlaugen. Die Pyrolyse der mit Silberoxid bereiteten freien Base verlief analog wie beim entsprechenden Derivat der Mycaminose (Formelschema 44) und gab das Epoxid **314**. Die Anlagerung des Methoxylanions erfolgte nicht spezifisch; es entstand ein Gemisch im Verhältnis von etwa 2:5 der beiden isomeren Methyläther **315** und **316**, das chromatographisch getrennt wurde. Die saure Hydrolyse des Glykosids **316** gab die Lankavose (**317**), die nach der Reinigung durch Sublimation kristallisierte und in allen Eigenschaften mit dem natürlichen Zucker übereinstimmte (*109*).

4. Die verzweigten Zucker aus Makroliden

Besonders charakteristisch für die Makrolide der Erythromycin- und Carbomycingruppe sind einige Zucker mit verzweigter Kohlenstoffkette: die Mycarose (**318**) in den meisten Antibiotica der Carbomycingruppe (Abschnitt II/2) sowie im Erythromycin C (*129*); die Cladinose (**319**) in den Erythromycinen A und B; die Arcanose (**320**) im Lankamycin. Während die Aminozucker (Abschnitt VI/1) sowie die Lankavose alle die **D**-Konfiguration besitzen und β-glykosidisch gebunden sind, gehören die verzweigten Zucker der **L**-Reihe an und sind α-glykosidisch mit dem Aglykon verbunden. Die Cladinose in den Erythromycinen trägt in Stellung 4 eine freie Hydroxylgruppe, die Mycarose in den Carbomycinen und verwandten Antibiotica ist dagegen in Position 4 meistens mit niederen Fettsäuren verestert, und auch die Arcanose kommt im Lankamycin als 4-O-Acetylderivat vor. Bemerkenswert ist das Auftreten der enantiomorphen D-Mycarose in einer anderen Gruppe von Antibiotica aus Actinomyceten, den Aureolsäuren (*16, 22*). Ein auffallender Unterschied zwischen der Erythromycin- und der Carbomycingruppe besteht darin, daß der verzweigte Zucker bei der ersteren direkt an das Aglykon (Stellung 3) anschließt, während er bei typischen Vertretern der letzteren die Stellung 4 des Aminozuckers besetzt.

Zu dieser Gruppe von Zuckern müssen noch zwei Verbindungen gezählt werden, die zwar eine unverzweigte C_6-Kette besitzen, aber in den Makrolidmolekülen die Stelle von Cladinose bzw. Mycarose einnehmen können und zu diesen Zuckern auch in einer biogenetischen Beziehung stehen. Es sind dies die Oleandrose (**321**), der neutrale

Zuckerbaustein des Oleandomycins (*62, 100*) sowie einiger Herzgiftglyko-
side, sowie die *Cinerulose A* (**322**), ein ungewöhnlicher Ketozucker, der
im Acumycin vorkommt (*286*). Ihren Namen hat die Cinerulose A von
ihrem erstmaligen Auffinden in den Hydrolysaten des Anthracyclin-
Antibioticums Cinerubin A (*164*).

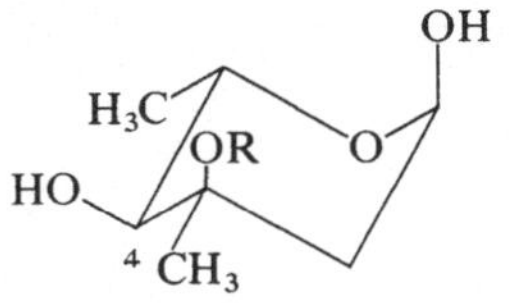

(**318**)   R = H: Mycarose

(**319**)   R = CH$_3$: Cladinose

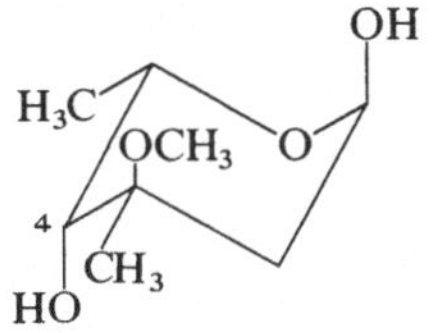

(**320**)   Arcanose

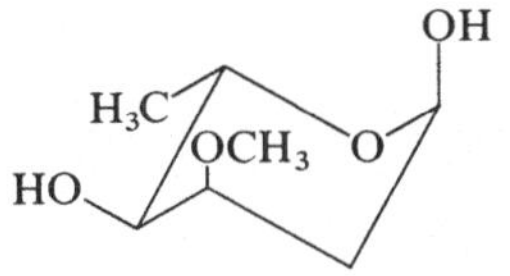

(**321**)   Oleandrose

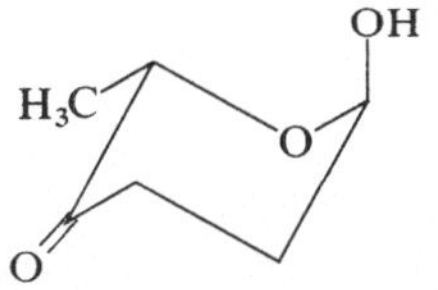

(**322**)   Cinerulose A

Schema 53. Die verzweigten Zucker aus Makroliden

Die Strukturaufklärung der Cladinose, Mycarose und Arcanose
wurde in mehreren Laboratorien bearbeitet und ist in zahlreichen Ab-
handlungen niedergelegt, deren Lektüre heute zum Teil verwirrend ist. Es
scheint mir daher angezeigt, die Lösung des Problems aus heutiger Sicht
vereinfacht darzustellen und auf die Wiedergabe von Doppelspurigkeiten
und Fehldeutungen zu verzichten.

Wir können heute davon ausgehen, daß die drei Zucker durch über-
sichtliche Reaktionen miteinander verknüpft sind (was zur Zeit der in-
tensivsten Bearbeitung des Problems keineswegs der Fall war).

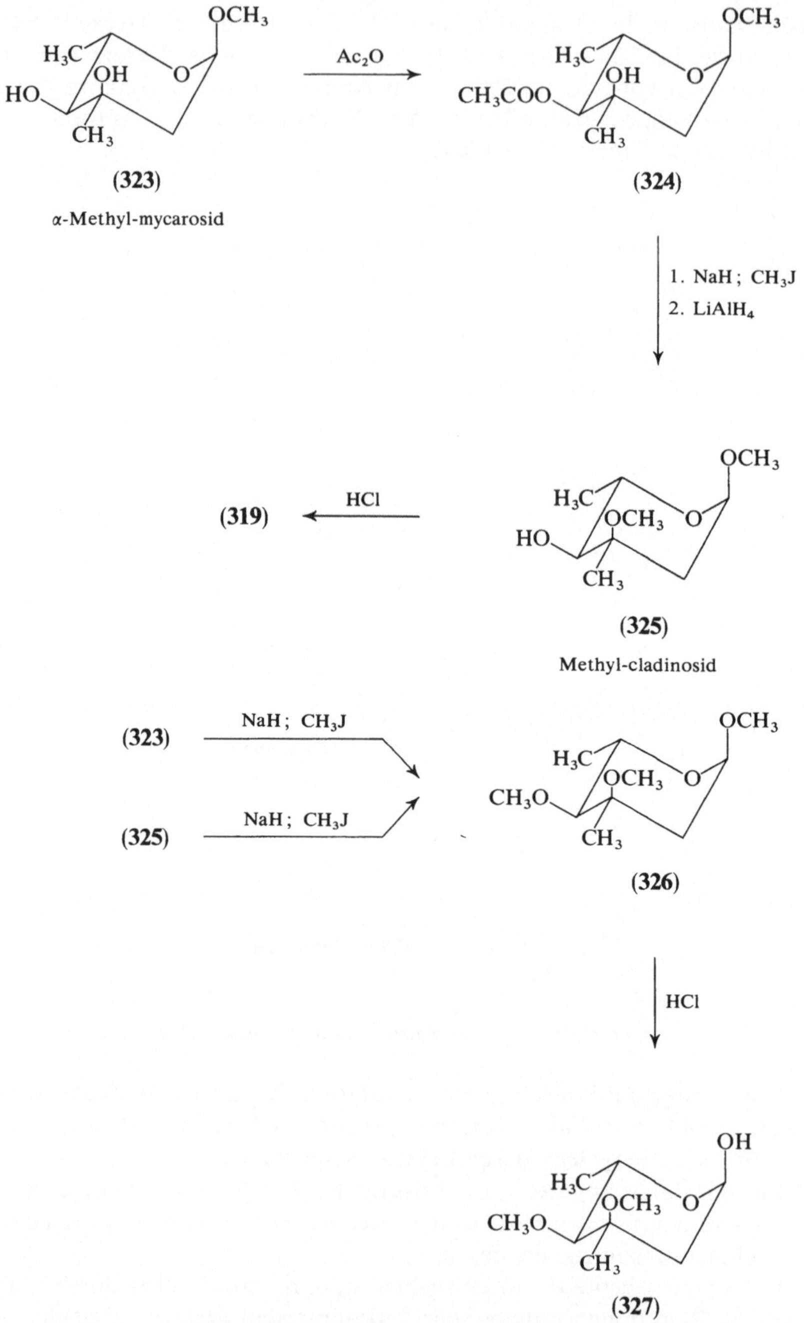
(323)
α-Methyl-mycarosid
Ac₂O
(324)
1. NaH; CH₃J
2. LiAlH₄
HCl
(319)
(325)
Methyl-cladinosid
(323)
NaH; CH₃J
(325)
NaH; CH₃J
(326)
HCl
(327)

Schema 54. Chemische Verknüpfung von Mycarose, Cladinose und Arcanose

Literaturverzeichnis: SS. 445—460

Schema 54 (Fortsetzung)

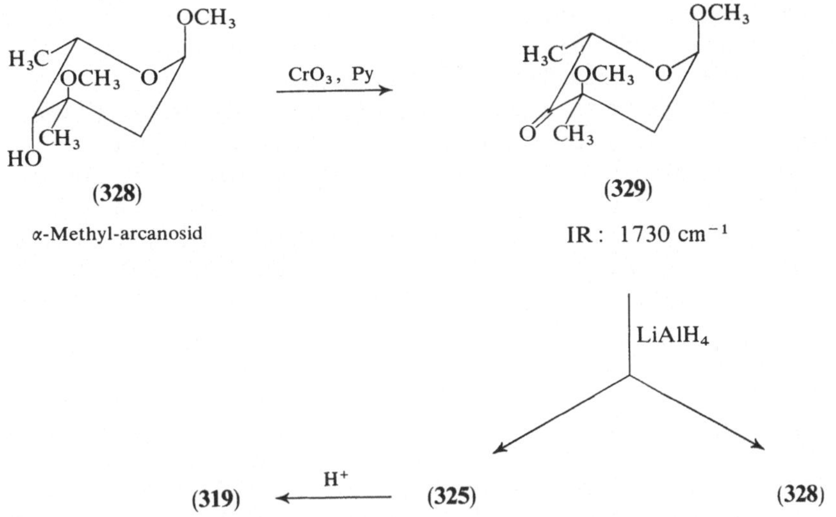

(328)

α-Methyl-arcanosid

CrO₃, Py →

(329)

IR : 1730 cm⁻¹

LiAlH₄

(319) ← H⁺ — (325) (328)

Durch eine Ätherspaltung mit Bortrichlorid wurde aus der Cladinose (319) ein flüssiger Zucker erhalten, der durch Papierchromatographie von der Mycarose (318) nicht unterschieden werden konnte (146). Umgekehrt gelang die Methylierung der tertiären Hydroxylgruppe im 4-O-Acetyl-methylmycarosid (324) mit Methyljodid und Natriumhydrid. Nach der reduktiven Abspaltung der Acetylgruppe zum Methylcladinosid (325) wurde das Glykosid mit verdünnter Säure hydrolysiert. Die partialsynthetische Cladinose (319) wurde allerdings nicht völlig rein erhalten, so daß keine ganz eindeutige Identifizierung des Produktes möglich war (183, vgl. auch 107). Es wurde daher die Herstellung eines gemeinsamen kristallinen Umwandlungsproduktes aus beiden Zuckern angestrebt, was auf folgendem Wege gelang (183):

Methylmycarosid (323; Gemisch mit geringeren Mengen des entsprechenden β-Glykosids) wurde mit Natriumhydrid und Methyljodid methyliert, wobei zwei neue O-Methylgruppen eintraten. Die gleiche Reaktion wurde mit Methylcladinosid (325) durchgeführt. Die beiden flüssigen Produkte (326) entsprachen einander offensichtlich nicht exakt im Anomerenverhältnis, so daß ihre Eigenschaften geringfügige Unterschiede zeigten. Dagegen wurde bei der sauren Hydrolyse aus beiden Glykosiden der gleiche kristalline Dimethyläther (327) (Smp. 83–86°; [α]_D = −20°) erhalten. Damit war bewiesen, daß die Cladinose ein Mycarosemonomethyläther ist. Die Mycarose besitzt eine nicht acetylierbare (tertiäre) Hydroxylgruppe und gibt ein Di-O-acetylderivat mit einer starken OH-Absorptionsbande im IR.-Spektrum. Die Cladinose gibt ebenfalls ein

Diacetat; das aber hydroxylfrei ist. Daraus geht hervor, daß die O-Methylgruppe der Cladinose am tertiären Sauerstoffatom sitzt. Mit diesen Befunden sind auch zahlreiche Übereinstimmungen in den NMR.-Spektren der beiden Zucker im Einklang (*146*).

Aus Methylarcanosid (**328**) konnte durch Oxydation mit Chromtrioxid in Pyridin das Keton **329** in 20% Ausbeute rein erhalten werden. Seine Konstitution wurde durch das IR.-und das NMR.-Spektrum gesichert. Die anschließende Reduktion mit Lithiumaluminiumhydrid verlief nicht stereospezifisch, doch überwog das Methylcladinosid (**325**) gegenüber dem Methylarcanosid (**328**) im Reaktionsprodukt deutlich. Das aus Arcanose bereitete flüssige Methylcladinosid besaß eine andere Anomerenzusammensetzung als das aus Cladinose direkt hergestellte und eignete sich daher schlecht für die Identifizierung. Dagegen wurde durch saure Hydrolyse ein Zucker erhalten, der in seinen spektroskopischen, chromatographischen und chiroptischen Eigenschaften völlig mit natürlicher Cladinose übereinstimmte. Demnach unterscheiden sich die Arcanose und die Cladinose nur in einer verschiedenen sterischen Anordnung der sekundären Hydroxylgruppe, womit auch die NMR.-Spektren gut vereinbar sind (*267*).

Von den zahlreichen, unabhängig voneinander an den drei Zuckern durchgeführten Abbaureaktionen (*104, 129, 145, 241, 255, 323*) sei hier auf den Abbau der Arcanose näher eingegangen (*165*).

Die Arcanose wie auch die 4-O-Acetylarcanose, die durch milde Hydrolyse aus Lankamycin erhalten wird, geben eine positive Farbreaktion nach Keller-Kiliani, woraus auf einen 2-Desoxyzucker geschlossen werden kann (*165*). Bei der energischen Einwirkung von Lithiumaluminiumhydrid auf die 4-O-Acetylarcanose wird die Halbacetalgruppe reduziert und gleichzeitig die O-Acetylgruppe reduktiv abgespalten. Das Triol (**330**) war jetzt einem Abbau mit Perjodat zugänglich. Als leichtflüchtiges Abbauprodukt wurde Acetaldehyd erhalten und als 2,4-Dinitrophenylhydrazon identifiziert. Das zweite Oxydationsprodukt war eine flüssige Verbindung $C_6H_{12}O_3$ (**331**), die wegen des Fehlens einer Carbonylbande im IR.-Absorptionsspektrum ein Cyclohalbacetal (**331 b**) sein muß. Mit Lithiumaluminiumhydrid ließ sie sich zu einem Methoxy-diol $C_6H_{14}O_3$ (**332**) reduzieren, dessen Konstitution sich eindeutig aus dem NMR.-Spektrum ergab: ein Singulett zu drei Protonen bei 1,15 ppm zeigte eine tertiäre C-Methylgruppe, ein gleiches Signal bei 3,24 ppm eine O-Methylgruppe an. Ein Methylensignal bei 1,80 ppm erschien als Triplett (J = 6 Hz.); die Methylengruppe ist demnach einer zweiten Methylengruppe benachbart, die eine Hydroxylgruppe trägt; denn die Signale der übrigen 6 Protonen erschienen als breiter Signalhaufe bei 3,5 bis 4,2 ppm. Das gleiche Abbauprodukt (**332**) wurde analog auch aus der Cladinose erhalten.

Literaturverzeichnis: SS. 445—460

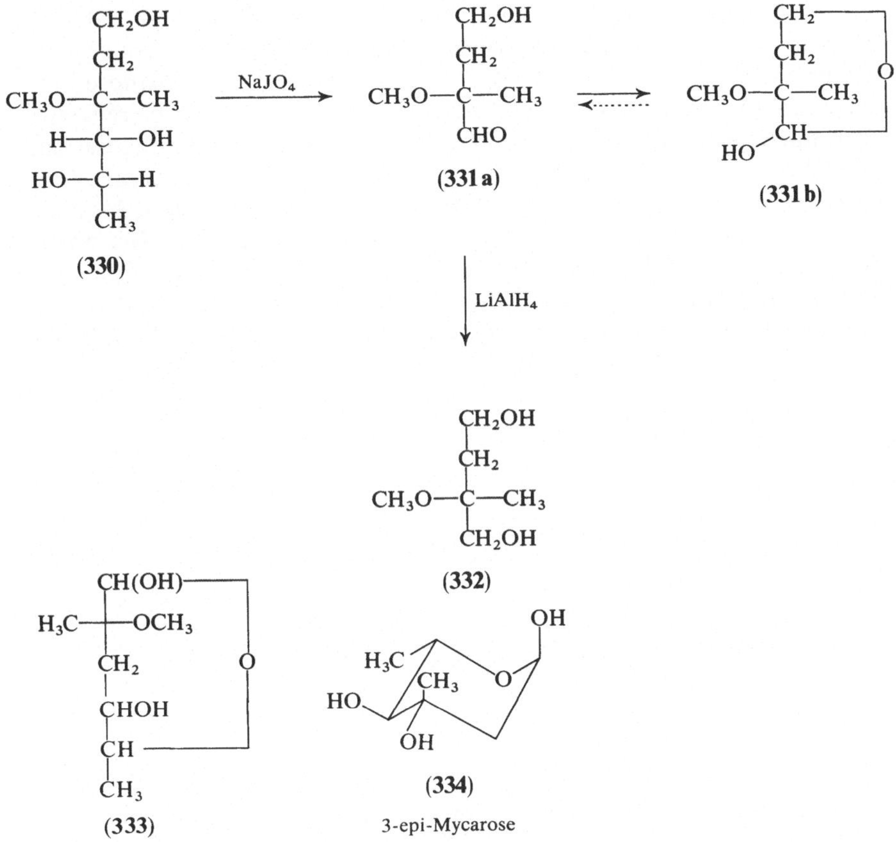

(330) → NaJO$_4$ → **(331a)** ⇌ **(331b)**

$$CH_2OH - CH_2 - C(OCH_3)(CH_3) - CH(OH) - CH(OH) - CH_3 \quad (330)$$

$$CH_2OH - CH_2 - C(OCH_3)(CH_3) - CHO \quad (331a)$$

(331b)

↓ LiAlH$_4$

$$CH_2OH - CH_2 - C(OCH_3)(CH_3) - CH_2OH \quad (332)$$

(333)

(334)

3-epi-Mycarose

Schema 55. Abbau der Arcanose

Auf Grund dieses Abbaus ließe sich für die Arcanose und die Cladinose auch noch die Konstitutionsformel **333** schreiben, doch ließ sich diese Möglichkeit ausschließen; denn sie steht zu den spektralen Eigenschaften im Widerspruch (*146, 267*). Die Multiplizität des Signals des Wasserstoffatoms an C-1 zeigt nämlich, daß diese in beiden Zuckern einer CH$_2$-Gruppe benachbart sein muß. Somit bleibt nur die Konstitutionsformel **320** übrig, die auch durch zahlreiche Abbaureaktionen der Cladinose und der Mycarose gesichert ist.

Die Aufklärung der Konfiguration gestaltete sich insofern schwierig, als durch die Verzweigung am C-Atom 3 die Kette benachbarter Protonen unterbrochen wird. Die NMR.-Spektroskopie allein ist daher zum Vorneherein nicht in der Lage, ein vollständiges Bild über die konformativen und konfigurativen Verhältnisse zu liefern (*146, 267*). Die einzige sichere Schlußfolgerung aus den NMR.-Spektren von Derivaten der Cladinose

und der Mycarose bestand darin, daß die Ringprotonen an C-4 und C-5
trans-diaxial sein müssen, denn sie stehen zueinander in einer Spin-Spin-
Wechselwirkung mit 9 bis 10 Hz. (*146*), während bei der Arcanose mit
axial-äquatorialer *cis*-Anordnung von H-C-5 und H-C-4 der betreffende
Wert nur etwa 1,5 Hz. beträgt (*267*). Für die Ableitung der Konfiguration
an C-3 wurden chemische Befunde herangezogen (*146*), wobei ein Vergleich der Mycarose mit synthetisch erhaltener 3-epi-Mycarose (**334**) und
ihren Derivaten (s. unten) sich als besonders nützlich erwies. Die Mycarose
bildet nämlich viel leichter cyclische Boratkomplexe als das 3-Epimere,
was sich mit Hilfe der Papierelektrophorese in Boratpuffer sowie durch
die Beeinflussung papierchromatographischer Rf-Werte durch Borat leicht

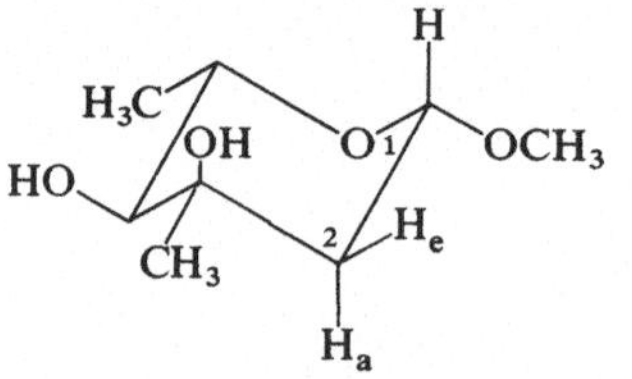

(**335**)

β-Methylmycarosid
$J_{1,2a}=9{,}0$; $J_{1,2e}=3{,}1$ Hz
$[\alpha]_D = +54°$

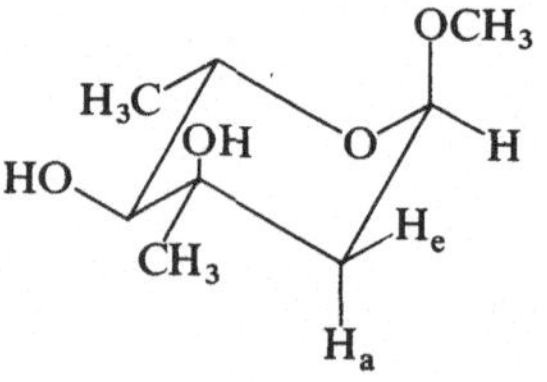

(**336**)

α-Methylmycarosid
$J_{1,2a}=J_{1,2e}=2{,}4$ Hz
$[\alpha]_D = -141°$

$$(\mathbf{336}) \xrightarrow{\text{NaJO}_4} (\mathbf{337}) \longrightarrow (\mathbf{338})$$

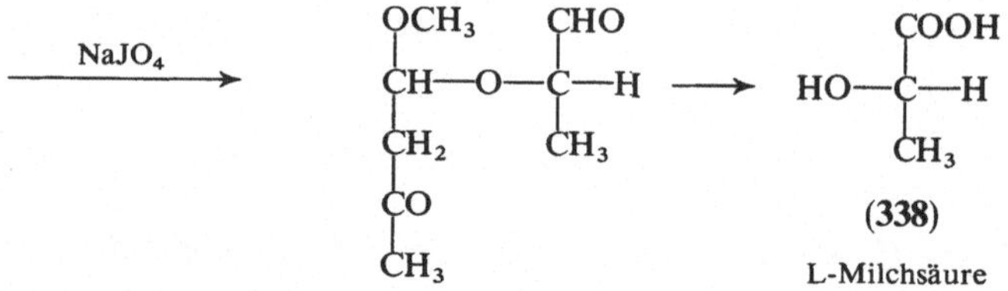

(**337**)

(**338**)

L-Milchsäure

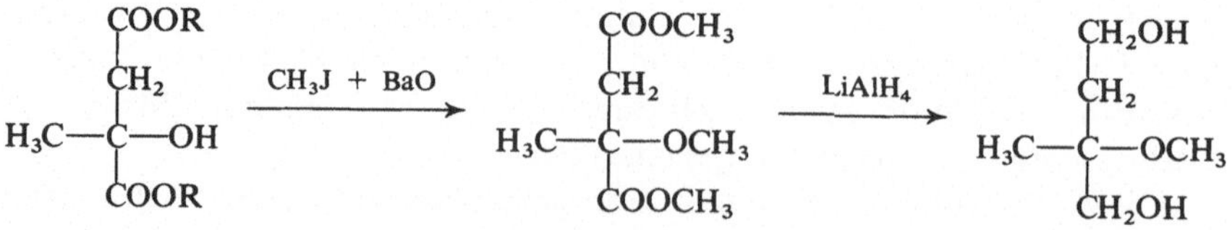

(**339**) R=H
(S)-Citramalsäure

(**340**) R=CH$_3$

$$(\mathbf{339/340}) \xrightarrow{\text{CH}_3\text{J} + \text{BaO}} (\mathbf{341}) \xrightarrow{\text{LiAlH}_4} (\mathbf{342})$$

(**341**)

(**342**)

Schema 56. Stereochemie der verzweigten Zucker

feststellen ließ. Daraus konnte geschlossen werden, daß die Anordnung der Hydroxylgruppen an C-3 und C-4 bei der Mycarose *cis*, bei der 3-epi-Mycarose aber *trans* ist. Damit im Einklang ist die relative Oxydierbarkeit der beiden Epimeren durch Perjodat. Die Mycarose wird etwa 17mal schneller oxydiert als die 3-epi-Mycarose (*146*). Damit war die relative Konfiguration dieser Zucker bewiesen.

Nachdem die Trennung der beiden anomeren Methylmycaroside schon vorher gelungen war, ergab sich aus den NMR.-Spektren eine Möglichkeit, die beiden Anomeren eindeutig zuzuordnen. Das eine mit $[\alpha]_D = +54°$ zeigte für das Wasserstoffatom an C-1 ein Doppeldublett mit Kopplungskonstanten von 9,0 und 3,1 Hz., woraus die axiale Anordnung von H-C-1 und damit die β-Konfiguration folgte (**335**). Das Anomere mit $[\alpha]_D = -141°$ zeigte dagegen für H-C-1 ein Triplett mit J=2,4 Hz., was nur mit der α-Konfiguration (**336**) verträglich ist. Aus den optischen Drehwerten folgt gemäß der Isorotationsregel von Hudson die L-Konfiguration.

Diese wurde bestätigt durch Abbau des Methylmycarosids (**336**) über die Zwischenstufe **337** zur L-Milchsäure (**338**), die allerdings nicht völlig optisch rein war (*183*, vgl. auch *106*). Andererseits wurde das Enantiomere (**342**) des Abbaudiols **332** aus Arcanose und Cladinose, ausgehend von der (S)-Citramalsäure (**339**) über die Zwischenprodukte **340** und **341** synthetisiert, wodurch unabhängig der Chiralitätssinn am C-Atom 3 bestimmt wurde (*267*).

Die Strukturen der Mycarose, Cladinose und Arcanose sind mehrfach durch Synthesen bestätigt worden. Als vollständig konfigurationsbeweisend sind nur diejenigen anzusehen, die ausgehend von Zuckern bekannter Stereochemie durch stereospezifische Reaktionen erzielt worden sind, wie etwa die Synthese der Mycarose und der Cladinose ausgehend von 2-Desoxy-L-glucose (*148*) und des Enantiomeren der Arcanose ausgehend von der D-Galactose (*149*). Auf ähnlichem Weg sind auch die Enantiomeren der Cladinose und Mycarose hergestellt worden (*103*). Mehrere Synthesen, ausgehend von nicht zuckerartigen Materialien, verliefen nur teilweise stereospezifisch (in bezug auf die C-Atome 4 und 5). Sie hatten aber den Vorteil, daß aus Zwischenprodukten mit sterisch anders verlaufenden Hydroxylierungsmethoden (*183, 172, 171*) oder als Nebenprodukt (*122*) die 3-epi-Mycarose erhalten wurde, die als Vergleichsmaterial bei der Konfigurationsbestimmung der Mycarose von Bedeutung war (s. oben). Im Formelschema 57 ist die am übersichtlichsten verlaufende Synthese von Grisebach und Mitarb. (*122*) aufgezeichnet.

Aus dem α,β-ungesättigten Keton **343** wurde nach Reformatzki der Ester **344** erhalten, der gemäß IR.-Absorptionsspektrum an der Doppelbindung ausschließlich *trans*-substituiert war. Demgemäß bestand das mit Phthalmonopersäure hergestellte Oxidationsprodukt (**345**) aus den

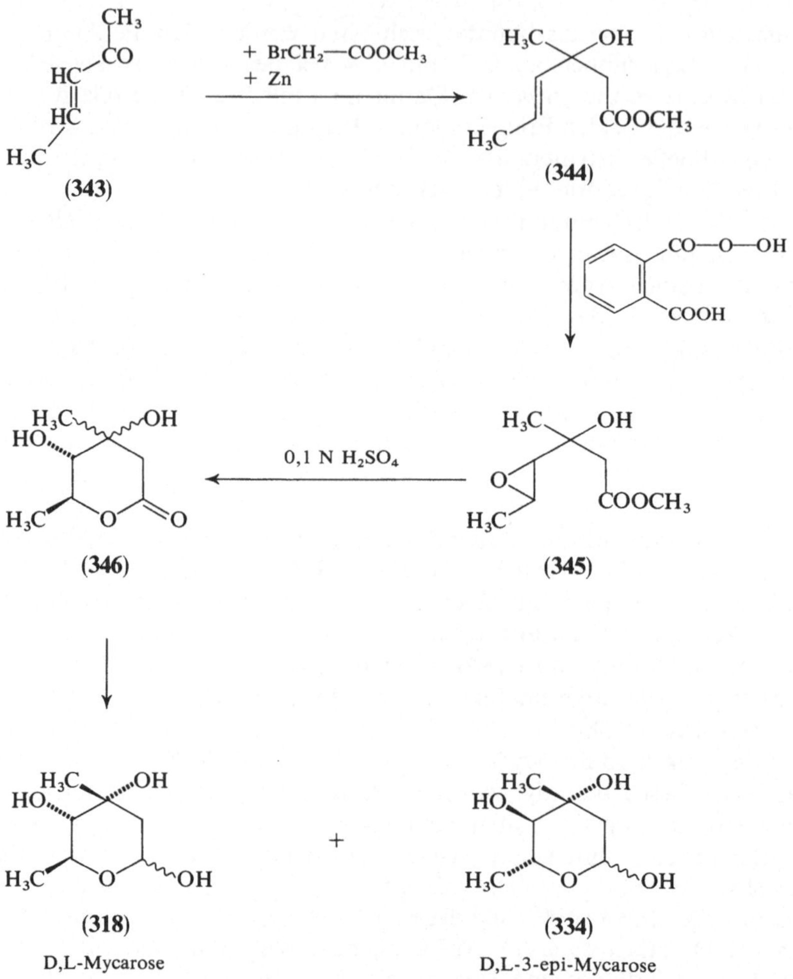

Schema 57. Synthese der D,L-Mycarose und der D,L-3-epi-Mycarose

beiden an C-3 epimeren *trans*-Epoxiden. Durch Ringspaltung mit verdünnter Schwefelsäure wurde ein Lactongemisch (346) erhalten, das auch noch einen gewissen Anteil von γ-Lactonen enthielt. Die Reduktion der Lactongruppe ließ sich am besten durch Hydroborierung nach H. C. Brown durchführen und führte zu einem Gemisch, das im wesentlichen aus den beiden racemischen Zuckern 318 und 334 bestand. Die Trennung erfolgte an Cellulose und gab dei beiden reinen racemischen Zucker in einem Verhältnis 334:318 = 2:1.

Korte (*171, 172*) erhielt die beiden Zucker 318 und 334 in zwei verschiedenen Synthesen und konnte nach seinem Aufbauprinzip auch die 4-epi-Mycarose herstellen. LEMAL und Mitarb. (*183*) gelang aus einem synthetischen Epimerengemisch die Abtrennung von reiner L-Mycarose mit Hilfe von Camphersulfonsäure.

5. Weitere Zucker aus Makroliden

Einige Makrolide der Carbomycingruppe enthalten zusätzlich zu typischen Makrolidzuckern einen weiteren neutralen Zucker, die Mycinose (**347**). Er wurde bisher bei den neutralen Makroliden Chalcomycin, Neutramycin und Aldgamycin E sowie bei den basischen Antibiotica Tylosin und Angolamycin nachgewiesen. Seine Konstitution und Konfiguration als 6-Desoxy-2,3-di-O-methyl-**D**-allose wurde durch chemischen

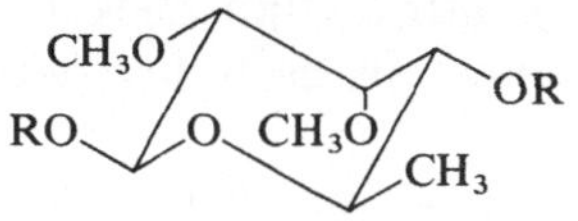

(**347**)   R = H; D-Mycinose
(**348**)   R = CH$_3$—CO—

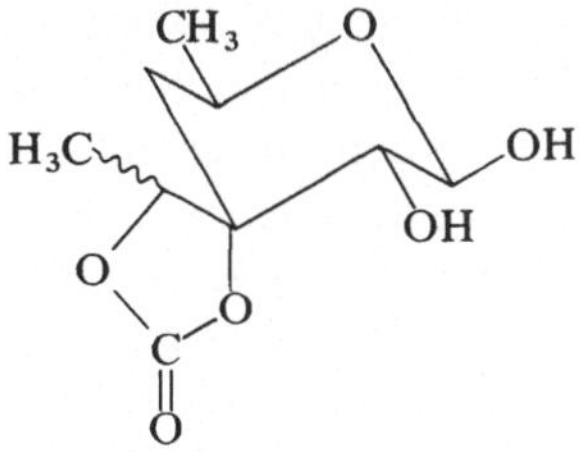

(**349**)   Aldgarose A

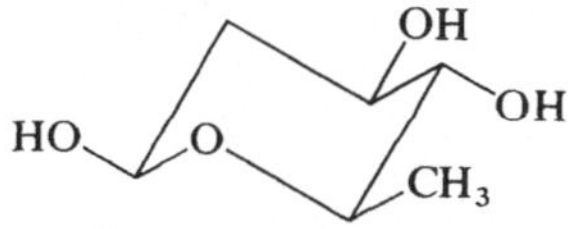

(**350**)   2-Desoxy-D-rhamnose

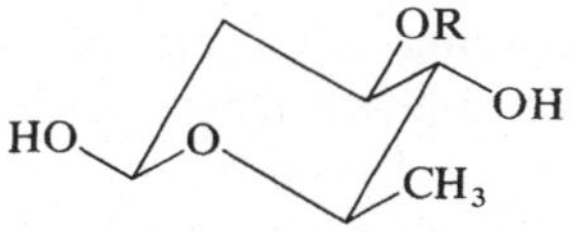

(**351**)   R = —CO—NH$_2$

(**352**)   R =

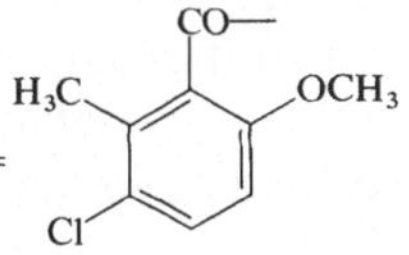

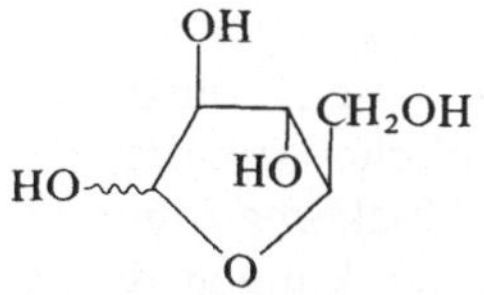

(**353**)   D-Arabinose

Schema 58. Weitere Zucker aus Makroliden

Abbau (*83*) und durch eine eingehende Analyse von NMR.-Spektren aufgeklärt. Da im β-Di-O-acetylderivat (**348**) alle H-Atome außer dem an C-3 axial angeordnet sind und ihre Signale durch die verschiedene chemische Umgebung genügend voneinander getrennt erschienen, gaben die Kopplungskonstanten erschöpfend Auskunft über die Stereochemie (*47*). Eine strukturbeweisende Synthese, ausgehend von L-Rhamnose, beschrieben Brimacombe und Mitarb. (*36*).

Aus dem neutralen Makrolid Aldgamycin E mit noch nicht aufgeklärtem Aglykon wurde ein ungewöhnlicher verzweigter Zucker mit einer cyclischen Kohlensäureester-Gruppe, die Aldgarose A (**349**) erhalten (*99*). Da die für cyclische Carbonate charakteristische IR.-Absorptionsbande bei etwa 1800 cm^{-1} auch bei anderen neutralen Makroliden, wie z. B. Bandamycin (*169*) und Megacidin (*101*) auftritt, sind Zucker vom Typus der Aldgarose möglicherweise verbreiteter, als es bisher den Anschein hatte.

Die keiner der großen Untergruppen angehörenden Antibiotica Venturicidin A, Venturicidin B und Chlorothricin enthalten einen Zucker, der bei Makroliden sonst nicht vorkommt, hingegen als Baustein anderer Antibioticagruppen sowie von Herzgiftglykosiden nachgewiesen worden ist: die 2-Desoxy-D-rhamnose (**350**), die auch unter den Namen D-Chromose C, D-Olivose und Canarose beschrieben wurde (Lit. s. bei *162*). Beim Venturicidin B ist der Zucker direkt β-glykosidisch an das Aglykon gebunden. Im Venturicidin A liegt er in Form des 3-O-Carbaminoylderivates (**351**) vor, und im Chlorothricin ist er in Stellung 3 mit einer chlorhaltigen aromatischen Säure verestert (**352**). (Vgl. auch Abschnitt IV.) Schließlich ist das Primycin (Abschnitt IV) ein Furanosid der D-Arabinose (**353**) (*1*).

VII. Schlußbemerkungen

Wenn man bedenkt, daß erst vor wenig mehr als 20 Jahren die ersten Isolierungen von Makrolidantibiotica beschrieben wurden und die ersten Konstitutionsaufklärungen erst etwa 15 Jahre zurückliegen, ist diese Stoffklasse in unglaublich kurzer Zeit zu einem nur noch schwer überblickbaren Feld angewachsen. Die vorliegende Zusammenfassung dürfte, trotz gewisser Lücken in den Einzelheiten, einen im wesentlichen vollständigen Überblick über unsere Kenntnisse der Chemie dieser Stoffe gewähren. Bereits kündigt sich aber an, daß diese „Vollständigkeit" nur vorübergehend sein wird. Es bestehen deutliche Anhaltspunkte dafür, daß die Scopamycine (*151*) neben einem Zuckerbaustein (*195*) einen makrocyclischen Lactonring enthalten. Ferner können die Antibiotica der Lankacidin-Gruppe als Abkömmlinge der Makrolide aufgefaßt werden, obwohl in ihnen ein großer carbocyclischer Ring mit einem lediglich sechsgliedrigen Lactonring verbunden ist (vgl. z. B. *126*). Im Botrycidin (*48*), das mit den Oligomycinen verwandt ist, kann die Anwesenheit eines großen Lactonringes vorläufig vermutet werden. Es sei ferner auf das Boromycin hingewiesen, das kürzlich als Borsäurekomplex eines makrocyclischen Dilactons erkannt wurde (*92, 150*). Über die Makrotetrolide, Antibiotica mit einem großen Tetralactonring, ist vor wenigen Jahren bereits ein Bericht in den „Fortschritten" erschienen (*161*). Und endlich läßt sich die Bedeutung großer Lactonringe nicht vollständig erfassen, ohne daß man sich vergegenwärtigt, daß auch eine große Zahl von Polypeptid-Antibiotica Makrolactone sind.

Literaturverzeichnis

1. ABERHART, J., T. FEHR, R. C. JAIN, P. DE MAYO, O. MOTL, L. BACZYNSKYI, D. E. F. GRACEY, D. B. MACLEAN, and I. SZILÁGYI: Primycin. J. Amer. chem. Soc. **92**, 5816 (1970).

2. ALBANO, E. L., and D. HORTON: A Synthesis of 2,3,4,6-Tetrahydroxy-4-(dimethylamino)-D-erythro-hexose (Forosamine) and its D-threo Epimer. Carbohydrate Research **11**, 485 (1969).

3. ALDRIDGE, D. D., J. J. ARMSTRONG, R. N. SPEAKE, and W. B. TURNER: The Structures of Cytochalasins A and B. J. Chem. Soc. (London) **1967**, 1667.

4. ALDRIDGE, D. C., S. GALT, D. GILES, and W. B. TURNER: Metabolites of *Lasiodiplodia theobromae*. J. Chem. Soc. (London) [C] **1971**, 1623.

5. ALDRIDGE, D. C., and W. B. TURNER: Structures of Cytochalasins C and D. J. Chem. Soc. (London) **1969**, 923.

6. — — The Identity of Zygosporin A and Cytochalasin D. J. Antibiotics (Tokyo) **22**, 170 (1969).

7. ANDERTON, K., and R. W. RICKARDS: Some Structural Features of Borrelidin, an Anti-viral Antibiotic. Nature **206**, 269 (1965).

8. ANLIKER, R., D. DVORNIK, K. GUBLER, H. HEUSSER und V. PRELOG: Stoffwechselprodukte von Actinomyceten, 5. Mitteil.; Über das Lacton der β-Hydroxy-α,α′,γ-trimethylpimelinsäure, ein Abbauprodukt von Narbomycin und Methymicin. Helv. chim. Acta **39**, 1785 (1956).

9. ANLIKER, R., und K. GUBLER: Stoffwechselprodukte von Actinomyceten, 6. Mitteil.; Über die Konstitution des Pikromycins. I. Helv. chim. Acta **40**, 119 (1957).

10. — —Stoffwechselprodukte von Actinomyceten, 10. Mitteil.; Die Konstitution des Kromycins, eines Abbauproduktes des Pikromycins. Helv. chim. Acta **40**, 1768 (1957).

11. ARCAMONE, F., C. BERTAZZOLI, G. CANEVAZZI, A. DI MARCO, M. GHIONE e A. GREIN: La Etruscomicina, nuovo antibiotico antifungino prodotto dallo *Streptomyces lucensis* n. sp. Giorn. Microbiol. **4**, 119 (1957).

12. ARMSTRONG, J. J., J. F. GROVE, W. B. TURNER, and G. WARD: An Antifungal Triene from a *Stroptomyces* sp. Nature **206**, 399 (1965).

13. ASZALOS, A., R. S. ROBINSON, P. LEMANSKI, and B. BERK: Trienine, an Antitumor Antibiotic. J. Antibiotics (Tokyo) **21**, 611 (1968).

14. BAER, H. H., and H. O. L. FISCHER: Cyclization of Dialdehydes with Nitromethane.

III. Preparation of 3-Amino-3-deoxy-D-mannose. J. Amer. chem. Soc. **82**, 3709 (1960).

15. Baker, P. M., B. W. Bycroft, and J. C. Roberts: Synthesis of ($\pm$)-Di-O-methyl-curvularin. J. Chem. Soc. (London) [C] **1967**, 1913.

16. Bakhaeva, G. P., Yu. A. Berlin, E. F. Boldyreva, O. A. Chuprunova, M. N. Kolosov, V. S. Soifer, T. E. Vasiljeva, and I. V. Yartseva: The Structure of Aureolic Acid (Mithramycin). Tetrahedron Lett. **1968**, 3595.

17. Betina, V., L. Drobnika, P. Nemec, and M. Zemanova: Study of the Antifungal Activity of the Antibiotic, Cyanein. J. Antibiotics (Tokyo), Ser. A **17**, 93 (1964).

18. Betina, V., J. Fusca, A. Kjaer, M. Kutkora, P. Nemec and R. H. Shapiro: Production of Cyanein by Penicillium simplicissimum. J. Antibiotics (Tokyo), Ser. A **19**, 115 (1966).

19. Betina, V., P. Nemec, S. Kováć, A. Kjaer, and R. H. Shapiro: The Identity of Cyanein and Brefeldin A. Acta chem. Scand. **19**, 519 (1965).

20. Berger, J., L. M. Jampolsky, and M. W. Goldberg: Borrelidin, a New Antibiotic with Antiborrelia Activity and Penicillin Enhancement Properties. Arch. Biochem. Biophys. **22**, 476 (1949).

21. Bergy, M. E., and T. E. Eble: The Filipin Complex. Biochemistry **7**, 653 (1968).

22. Berlin, Yu. A., M. N. Kolosov, I. V. Vasina, and I. V. Yartseva: The Structure of Chromocyclomycin. Chemical Communications **1968**, 762.

23. Berry, M. P., and M. C. Whiting: The Relative Configurations of Two Asymmetric Centres in Lagosin. J. Chem. Soc. (London) **1964**, 862.

24. Bickel, H., E. Gäumann, R. Hütter, W. Sackmann, E. Vischer, W. Voser, A. Wettstein, and H. Zähner: Stoffwechselprodukte von Actinomyceten, 37. Mitteil.; Acumycin. Helv. chim. Acta **45**, 1396 (1962).

25. Binder, M., J.-R. Kiechel, and Ch. Tamm: Zur Biogenese des Antibioticums Phomin. 1. Teil: Die Grundbausteine. Helv. chim. Acta **53**, 1797 (1970).

26. Birch, A. J., C. W. Holzapfel, R. W. Rickards, C. Djerassi, P. C. Seidel, M. Suzuki, J. W. Westley, and J. D. Dutcher: Nystatin. Part VI. Chemistry and Partial Structure of the Antibiotic. Tetrahedron Lett. **1964**, 1491.

27. Birch, A. J., B. Moore, and R. W. Rickards: Curvularin, Part IV. Synthesis of a Degradative Product. J. Chem. Soc. (London) **1962**, 220.

28. Birch, A. J., O. C. Musgrave, R. W. Rickards, and H. Smith: Studies in Relation to Biosynthesis. Part XX. Structure and Biosynthesis of Curvularin. J. Chem. Soc. (London) **1959**, 3146.

29. Bognár, R., B. O. Brown, W. J. S. Lockley, S. Makleit, T. P. Toube, B. C. L. Weedon, and K. Zsupán: The Structure of Flavofungin. Tetrahedron Lett. **1970**, 471.

30. Bolton, C. H., A. B. Foster, M. Stacey, and J. M. Webber: Carbohydrate Components of Antibiotics. Part. I. Degradation of Desosamine by Alkali; its Absolute Configuration at Position 5. J. Chem. Soc. (London) **1961**, 4831.

31. Borowski, E., L. Falkowski, J. Golik, J. Zielinski, W. Mechlinski, E. Jereczek, P. Koloodziejczyk, H. Adlercreutz, P. C. Schaffner, and S. Neelakantan: The Structure of Candidin A, a Polyene Macrolide Antifungal Antibiotic. Tetrahedron Lett. **1971**, 1987.

32. Borowski, E., C. P. Schaffner, H. Lechevalier, and B. S. Schwartz: Perimycin, a Novel Type of Heptaene Antifungal Antibiotic. Antimicrobial Agents Annual **1960**, 532.

33. Borowski, E., J. Zielinski, T. Ziminski, L. Falkowski, P. Koloodziejczyk, J. Golik, E. Jereczek, and H. Adlercreutz: Chemical Studies with Amphothericin B. III. Complete Structure of the Antibiotic. Tetrahedron Lett. **1970**, 3909.

34. Bosshardt, R., und H. Bickel: Zur Kenntnis von Levorin A und Candicidin. Experientia **24**, 422 (1968).

35. BREWSTER, J. H.: Some Applications of the Conformational Dissymmetry Rule. Tetrahedron **13**, 106 (1961).

36. BRIMACOMBE, J. S., M. STACEY, and L. C. N. TUCKER: Synthesis of Mycinose (6-Deoxy-2.3-di-O-methyl-D-allose. J. Chem. Soc. (London) **1964**, 5391.

37. BROCKMANN, H.: Anthracyclinone und Anthracycline (Rhodomycinone, Pyrromycinone und ihre Glykoside). Fortschr. Chem. org. Naturstoffe (Ed. L. Zechmeister) **21**, 121 (1963).

38. BROCKMANN, H., H. GENTH und R. STRUFE: Pikromycin, II. Mitteil. (Antibiotica aus Actinomyceten, IX. Mitteil.) Chem. Ber. **85**, 426 (1952).

39. BROCKMANN, H., und W. HENKEL: Pikromycin, ein bitter schmeckendes Antibioticum aus Actinomyceten. (Antibiotica aus Actinomyceten, VI. Mitteil.) Chem. Ber. **84**, 284 (1951).

40. BROCKMANN, H., F. KLUGE und H. MUXFELDT: Totalsynthese des Hypericins. Chem. Ber. **90**, 2302 (1957).

41. BROCKMANN, H., H.-B. KÖNIG und R. OSTER: Die Konstitution des Picrocins, eines stickstoffhaltigen Abbauproduktes des Pikromycins. (Pikromycin, IV. Mitteil.: Antibiotica aus Actinomyceten, XX. Mitteil.) Chem. Ber. **87**, 856 (1954).

42. BROCKMANN, H., und R. OSTER: Antibiotica aus Actinomyceten, XXXVIII. Zur Konstitution des Pikromycins und Kromycins (Pikromycin VI). Chem. Ber. **90**, 605 (1957).

43. BROWN, R., and E. L. HAZEN: Capacidin, a New Member of the Polyene Antibiotic Group. Antibiotics and Chemotherapy **10**, 702 (1960).

44. BRUFANI, M., L. CELLAI, S. CERRINI, W. FEDELI, W. KELLER-SCHIERLEIN e C. MUSU: Struttura delle Venturicidine. V° Convegno di Chimica organica; Riassunti delle relazioni e delle Communicazioni (Salice Terme, Pavia) **1971**, 8.

45. BRUFANI, M., S. CERRINI, W. FEDELI, C. MUSU, L. CELLAI, and W. KELLER-SCHIERLEIN: Structures of the Venturicidins A and B. Experientia **27**, 604 (1971).

46. BRUFANI, M., and W. FEDELI: Die Kristallstruktur der 2,4,6-Trimethylpimelinsäuren. 1. Mitteil.; Helv. chim. Acta **54**, 51 (1971).

47. BRUFANI, M., und W. KELLER-SCHIERLEIN: Stoffwechselprodukte von Mikroorganismen, 54. Mitteil.; Über die Zuckerbausteine des Angolamycins. L-Mycarose, D-Mycinose und D-Angolasamin. Helv. chim. Acta **49**, 1962 (1966).

48. BRUFANI, M., W. KELLER-SCHIERLEIN, W. LÖFFLER, I. MANSPERGER und H. ZÄHNER: Stoffwechselprodukte von Mikroorganismen. 69. Mitteil.; Über das Venturicidin B, das Botrycidin und die Zuckerbausteine der Venturicidine A und B. Helv. chim. Acta **51**, 1293 (1968).

49. BU'LOCK, J. D., and P. T. CLAY: Fatty Acid Cyclization in the Biosynthesis of Brefeldin A; a New Route to Some Fungal Metabolites. Chemical Communications **1969**, 237.

50. CARTER, S. B.: Effects of Cytochalasins on Mammalian Cells. Nature **213**, 261 (1967).

51. CEDER, O.: Pimaricin. I. Determination of the Carbon Skeleton. Acta chem. Scand. **18**, 77 (1964).

52. — Pimaricin. IV. Investigations on the Pattern of Unsaturation and Oxygenation, and the Empirical Formula. Acta chem. Scand. **18**, 103 (1964).

53. — Pimaricin. VI. Complete Structure of the Antibiotic. Acta chem. Scand. **18**, 126 (1964).

54. CEDER, O., and B. HANSSON: Pimaricin. VII. The Absolute Configuration at C-25. Tetrahedron **23**, 3753 (1967).

55. CEDER, O., and R. RYHAGE: The Structure of Filipin. Acta chem. Scand. **18**, 558 (1964).

56. CEDER, O., J. M. WAISVISZ, M. G. VAN DER HOEVEN, and R. RYHAGE: Pimaricin. II. High Pressure — High Temperature Hydrogenation Studies. Acta chem. Scand. **18**, 83 (1964).

57. Ceder, O., J. M. Waisvisz, M. G. van der Hoeven, and R. Ryhage: Pimaricin. V. Determination of the Size of the Macrocyclic Lactone Ring and the Position of Mycosamine. Acta chem. Scand. **18**, 111 (1964).

58. Celmer, W. D.: Biogenetic, Constitutional, and Stereo-chemical Unitarity Principles in Macrolide Antibiotics. Antimicrobial Agents and Chemotherapy **1965**, 144.

59. — Macrolide Stereochemistry. I. The Total Absolute Configuration of Oleandomycin. J. Amer. chem. Soc. **87**, 1797 (1965).

60. — Macrolide Stereochemistry. II. Configurational Assignments at Certain Centers in Various Macrolide Antibiotics. J. Amer. chem. Soc. **87**, 1799 (1965).

61. — Macrolide Stereochemistry. III. A Configurational Model for Macrolide Antibiotics. J. Amer. chem. Soc. **87**, 1801 (1965).

62. Celmer, W. D., and D. C. Hobbs: The α-L- and β-D-Pyranoside Linkages in Oleandomycin, Carbohydrate Research **1**, 137 (1965).

63. Chong, C. N., and R. W. Rickards: Macrolide Antibiotic Studies. XVI. The Structure of Nystatin. Tetrahedron Lett. **1970**, 5145.

64. Clark, R. K.: The Chemistry of Erythromycin, I. Acid Degradation Products. Antibiotics and Chemotherapy **3**, 663 (1953).

65. Coombe, R. G., P. S. Foss, and T. R. Watson: The Biosynthesis of Brefeldin A. Austral. J. Chem. **22**, 1943 (1909).

66. Coombe, R. G., J. J. Jacobs, and T. R. Watson: Constituents of some *Curvularia* Species. Austral. J. Chem. **21**, 783 (1968).

67. Cope, A. C., U. Axen, and E. P. Burrows: Rimocidin. II. Oxygenation Pattern of the Aglycone. J. Amer. chem. Soc. **88**, 4221 (1966).

68. Cope, A. C., R. K. Bly, E. P. Burrows, O. J. Ceder, E. Ciganek, B. T. Gillis, R. F. Porter, and H. E. Johnson: Fungichromine. Complete Structure and Absolute Configuration at C_{26} and C_{27}. J. Amer. chem. Soc. **84**, 2170 (1962).

69. Cope, A. C., E. P. Burrows, M. E. Derieg, S. Moon, and W. D. Wirth: Rimocidin I. Carbon Skeleton, Partial Structure and Absolute Configuration at C-27. J. Amer. chem. Soc. **87**, 5452 (1965).

70. Cope, A. C., and H. E. Johnson: Fungichromin. Determination of the Structure of the Pentaene Chromophore. J. Amer. chem. Soc. **80**, 1504 (1958).

71. Corbaz, R., L. Ettlinger, E. Gäumann, W. Keller-Schierlein, F. Kradolfer, E. Kyburz, L., Neipp, V. Prelog, P. Reusser und H. Zähner: Stoffwechselprodukte von Actinomyceten. 1. Mitteil.: Narbomycin. Helv. chim. Acta **38**, 935 (1955).

72. Corbaz, R., L. Ettlinger, E. Gäumann, W. Keller-Schierlein, F. Kradolfer, E. Kyburz, L. Neipp, V. Prelog, A. Wettstein und H. Zähner: Stoffwechselprodukte von Actinomyceten, 4. Mitteil.; die Foromacidine A, B, C und D. Helv. chim. Acta **39**, 304 (1956).

73. Corbaz, R., L. Ettlinger, E. Gäumann, W. Keller-Schierlein, L. Neipp, V. Prelog, P. Reusser und H. Zähner: Stoffwechselprodukte von Actinomyceten. 2. Mitteil. Angolamycin. Helv. chim. Acta **38**, 1202 (1955).

74. Coronelli, C., R. C. Pasquallucci, J. E. Thiemann, and G. Tamoni: Mycotrienin, a New Polyene Antibiotic Isolated from *Streptomyces*. J. Antibiotics (Tokyo), Ser. A. **20**, 329 (1967).

75. Delmotte, P., and J. Delmotte-Plaquee: A New Antifungal Substance of Fungal Origin. Nature **171**, 344 (1953).

76. Demarco, P. V.: NMR Study of some Erythromycin Aglycones. A Conformational and Configurational Analysis. J. Antibiotics (Tokyo) **22**, 327 (1969).

77. Demarco, P. V.: Pyridine Solvent Shifts in the NMR Analysis of Erythromycin Aglycones. Tetrahedron Lett. **1969**, 383.

78. Dhar, M. L., V. Thaller, and M. C. Whiting: A New Type of Macrolide Antibiotics. Proc. Chem. Soc. **1958**, 148.

79. — — — The Structure of Lagosin and Filipin. Proc. chem. Soc. **1960**, 310.

80. DHAR, M. L., V. THALLER, and M. C. WHITING: Researches on Polyenes. Part VIII. The Structures of Lagosin and Filipin. J. chem. Soc. (London) **1964**, 842.

81. DHAR, M. L., V. THALLER, M. C. WHITING, R. RYHAGE, S. STÄLLBERG-STENHAGEN, and E. STENHAGEN: The Carbon Skeleton of Lagosin (Antibiotic A 246). Proc. chem. Soc. **1959**, 154.

82. DICKINSON, L., A. J. GRIFFITHS, C. G. MARON, and R. F. N. MILLS: Anti-viral Activity of Two Antibiotics Isolated from a Species of *Streptomyces*. Nature **206**, 265 (1965).

83. DION, H. W., P. W. K. WOO, and Q. R. BARTZ: Chemistry of Mycinose. 6-Deoxy-2,3-di-O-methyl-D-allose. J. Amer. chem. Soc. **84**, 880 (1962).

84. DIVEKAR, P. V., V. C. VORA, and A. W. KHAN: Comparison of Hamycin with Trichomycin. J. Antibiotics (Tokyo), Ser. A **19**, 63 (1966).

85. DJERASSI, C., A. BOWERS, R. HODGES, and B. RINIKER: The Partial Structure of Methymycin. J. Amer. chem. Soc. **78**, 1733 (1956).

86. DJERASSI, C., A. BOWERS, and H. N. KHASTGIR: Methymycin. Reduction and Oxidation Studies. J. Amer. chem. Soc. **78**, 1729 (1956).

87. DJERASSI, C., and O. HALPERN: Macrolide Antibiotics. VII. The Structure of Neomethymycin. Tetrahedron **3**, 255 (1958).

88. DJERASSI, C., O. HALPERN, D. I. WILKINSON, and E. J. EISENBRAUN: Macrolide Antibiotics. VIII. The Absolute Configuration of Certain Centers in Neomethymycin, Erythromycin and Related Antibiotics. Tetrahedron **4**, 369 (1958).

89. DJERASSI, C., M. ISHIKAWA, H. BUDZIKIEWICZ, J. N. SCHOOLERY, and L. F. JOHNSON: The Structures of the Macrolide Antibiotic Filipin. Tetrahedron Lett. **1961**, 383.

90. DJERASSI, C., M. SUZUKI, J. WESTLEY, J. D. DUTCHER, and R. THOMAS: Nystatin. Part. V. Biosynthetic Definition of Some Structural Features. Tetrahedron Lett. **1964**, 1485.

91. DJERASSI, C., and J. A. ZDERIC: The Structure of the Antibiotic Methymycin. J. Amer. chem. Soc. **78**, 6390 (1956).

92. DUNITZ, J. D., D. M. HAWLEY, D. MIKLOS, D. N. J. WHITE, YU. BERLIN, R. MARUSIC, and V. PRELOG: Structure of Boromycin. Helv. chim. Acta. (1971) (im Druck).

93. DUTCHER, J. D., G. BOYACK, and S. FOX: The Preparation and Properties of Crystalline Fungicidin (Nystatin). Antibiotics Annual **1953—1954**, 191.

94. DUTCHER, J. D., J. VANDEPUTTE, S. FOX, and L. J. HEUSER: An Antiricketsial Antibiotic from a Streptomycete, M-4209, II. Chemical Characterization. Antibiotics and Chemotherapy **3**, 910 (1953).

95. DUTCHER, J. D., D. R. WALKERS, and O. WINTERSTEINER: Nystatin. III. Mycosamine. Preparation and Determination of Structure. J. Org. chem. **28**, 995 (1963).

96. DUTCHER, J. D., M. B. YOUNG, J. H. HERMAN, W. HIBBITS, and D. R. WALKERS: Chemical Studies on Amphothericin B. I. Preparation of the Hydrogenation Product and Isolation of Mycosamine, an Acetolysis Product. Antibiotics Annual **1956—1957**, 866.

97. EGAN, R. S., and J. R. MARTIN: Structure of Lankamycin. J. Amer. chem. Soc. **92**, 4129 (1970).

98. EISENBRANDT, K.: Untersuchungen über ein antifungales Hexaenantibiotikum, Hexamycin. II. Eigenschaften des Hexamycins. J. prakt. Chem. [4] **38**, 20 (1968).

99. ELLESTAD, G. A., M. P. KUNSTMANN, J. E. LANCASTER, L. A. MITSCHER, and G. MORTON: Structures of Methyl Aldgarosides A and B obtained from the Neutral Macrolide Antibiotic Aldgamycin E. Tetrahedron **23**, 3893 (1967).

100. ELS, H., W. D. CELMER, and K. MURAI: Oleandomycin (PA-105). II. Chemical Characterization. J. Amer. chem. Soc. **80**, 3777 (1958).

101. ETTLINGER, L., E. GÄUMANN, R. HÜTTER, W. KELLER-SCHIERLEIN, F. KRADOLFER, L. NEIPP, V. PRELOG, P. REUSSER und H. ZÄHNER: Stoffwechselprodukte von Actinomyceten, 11. Mitteil. Megacidin. Monatsh. Chem. **88**, 989 (1957).

102. FENNELL, D., K. B. RAPER, and F. H. STODOLA: Curvularin Formation by *Penicillium steckii.* Chemistry and Industry **1959,** 1382.
103. FLAHERTY, B., W. G. OVEREND, and N. R. WILLIAMS: Branched-chain sugars, Part VII. The Synthesis of D-Mycarose and D-Cladinose. J. Chem. Soc. (London) [C] **1966,** 398.
104. FLYNN, E. H., M. V. SIGAL, P. F. WILEY, and K. GERZON: Erythromycin, I. Properties and Degradation Studies. J. Amer. chem. Soc. **76,** 3121 (1954).
105. FOSTER, A. B., T. D. INCH, J. LEHMANN, M. STACEY, and J. M. WEBBER: Identification of Mycaminose as 3,6-Dideoxy-3-dimethyl-amino-D-glucose. Chemistry and Industry **1962,** 142.
106. FOSTER, A. B., T. D. INCH, L. LEHMANN, L. F. THOMAS, J. M. WEBBER, and J. A. WYER: The Absolute Configuration of Mycarose. Proc. Chem. Soc. **1962,** 254.
107. FOSTER, A. B., T. D. INCH, J. LEHMANN, and J. M. WEBBER: Relationship of Mycarose and Cladinose. Chemistry and Industry **1962,** 1619.
108. FOSTER, A. B., J. LEHMANN, and M. STACEY: Carbohydrate Components of Antibiotics. Part. II. Alkaline Degradation of Mycaminose and Synthesis of 3,6-Dideoxy-3-dimethyl-amino-L-altrose and Some Derivatives therefrom. J. Chem. Soc. (London) **1962,** 1396.
109. FOSTER, A. B., M. STACEY, J. M. WEBBER, and J. H. WESTWOOD: Carbohydrate Components of Antibiotics. Part IV. Configurational Correlation of Desosamine and Chalcose. J. Chem. Soc. (London) **1965,** 2318.
110. GAUDIANO, G., M. ACAMPORA e P. BRAVO: Struttura della lucensomicina. Nota IV. Sintesi de alcuni composti correlati ai prodotti di demolizione dell' antibiotico. Gazz. chim. Ital. **96,** 1492 (1966).
111. GAUDIANO, G., P. BRAVO e A. QUILICO: Struttura della Lucensomicina. Nota 1. Gazz. chim. Ital. **96,** 1322 (1966).
112. — — — Struttura della Lucensomicina. Nota 2. Gazz. chim. Ital. **96,** 1351 (1966).
113. GAUDIANO, G., P. BRAVO, A. QUILICO, B. T. GOLDING e R. W. RICKARDS: Struttura della Lucensomicina. Nota 3. Gazz. chim. Ital. **96,** 1470 (1966).
114. GÄUMANN, E., R. HÜTTER, W. KELLER-SCHIERLEIN, L. NEIPP, V. PRELOG und H. ZÄHNER: Stoffwechselprodukte von Actinomyceten. 21. Mitteil.; Lankamycin und Lankacidin. Helv. chim. Acta **43,** 601 (1960).
115. GERZON, K., E. H. FLYNN, M. V. SIGAL, P. F. WILEY, R. MONAHAN, and U. C. QUARCK: Erythromycin, VIII. Structure of Dihydroerythronolid. J. Amer. chem. Soc. **78,** 6396 (1956).
116. GERZON, K., R. MONAHAN, O. WEAVER, M. V. SIGAL, and P. F. WILEY: Erythromycin. IX. Degradation Studies of Erythromycin B. J. Amer. chem. Soc. **78,** 6412 (1956).
117. GIROTRA, N. N., and N. L. WENDLER: A New Total Synthesis of the Macrolide Zearalenone. Chemistry and Industry **1967,** 1493.
118. GOLDING, B. T., R. W. RICKARDS, and M. BARBER: The Determination of Molecular Formulae of Polyols by Mass Spectrometry of their Trimethylsilyl Ethers. The Structure of the Macrolide Antibiotic Filipin. Tetrahedron Lett. **1964,** 2615.
119. GOLDING, B. T., R. W. RICKARDS, W. E. MEYER, J. B. PATRICK, and M. BARBER: The Structure of the Macrolide Antibiotic Pimaricin. Tetrahedron Lett. **1966,** 3551.
120. GOPALKRISHNAN, K. S., N. NARASIMHACHARI, V. B. JOSHI, and M. J. THIRUMALACHAR: Chainin, a New Methylpentaene Antibiotic from a Species of *Chainia.* Nature **218,** 597 (1968).
121. GRISEBACH, H.: Biogenetic Patterns in Microorganisms and Higher Plants (E. R. Squibb Lectures on Chemistry of Microbial Products). New York: J. Wiley & Sons, Inc. 1967.

122. GRISEBACH, H., W. HOFHEINZ und N. DOERR: Eine Synthese der D,L-Mycarose und D,L-Epimycarose. Chem. Ber. **96**, 1823 (1963).

123. GUPTA, K. C.: Monicamycin, a New Polyene Antifungal Antibiotic. Antimicrobial Agents and Chemotherapy **1964**, 65.

124. HAMILL, R. L., M. E. HANEY, JR., M. STAMPER, and P. F. WILEY: Tylosin, a New Antibiotic II. Isolation, Properties, and Preparation of Desmycosin, a Microbiologically Active Degradation Product. Antibiotics and Chemotherapy **11**, 328 (1961).

125. HANDSCHIN, U., H. P. SIGG und CH. TAMM: Zur Biosynthese von Brefeldin A. Helv. chim. Acta **51**, 1943 (1918).

126. HARADA, S., T. KISHI, and K. MIZUNO: Studies on T-2636 Antibiotics. II. Isolation and Chemical Properties of T-2636 Antibiotics. J. Antibiotics (Tokyo) **24**, 13 (1971).

127. HÄRRI, E., W. LOEFFLER, H. P. SIGG, H. STÄHRLIN und CH. TAMM: Über die Isolierung neuer Stoffwechselprodukte aus *Penicillium brefeldianum* DODGE. Helv. chim. Acta **46**, 1235 (1963).

128. HARRIS, D. R., S. G. MCGEACHIN, and H. H. MILLS: The Structure and Stereochemistry of Erythromycin. Tetrahedron Lett. **1965**, 679.

129. HASBROUK, R. B., and F. C. GARVEN: The Chemistry of Erythromycin. II. Acid Degradation Products. Antibiotica and Chemotherapy **3**, 1040 (1953).

130. HATA, T., S. OMURA, M. KATAGIRI, H. OGURA, K. NAYA, J. ABE, and T. WATANABE: Structure of Leucomycin A_1. Chem. Pharm. Bull. (Tokyo) **15**, 358 (1967).

131. HATA, I., S. OMURA, A. MATSUMAE, M. KATAGIRI, and Y. SANO: Leucomycin A_3, a New Antibiotic from *Streptomyces kitasatoensis*. Antimicrobial Agents and Chemotherapy **1966**, 631.

132. HATA, T., Y. SANO, N. OHKI, Y. YOKOYAMA, A. MATSUMAE, and S. ITO: Leucomycin, a New Antibiotic. J. Antibiotics (Tokyo), Ser. A **6**, 87 (1953).

133. HATTORI, K.: Studies on Trichomycin. VIII. Chemical Structure of Trichomycin A. J. Antibiotics (Tokyo), Ser. B **15**, 39 (1962) vgl. (151).

134. HATTORI, K., H. NAKANO, M. SEKI, and Y. HIRATA: Studies on Trichomycin. IV. J. Antibiotics (Tokyo), Ser. A. **9**, 176 (1956).

135. HAYAKAWA, T., T. KIMURA, H. MINATO, and K. KATAGIRI: Zygosporin A, a New Antibiotic from Zygosporium mansonii. J. Antibiotics (Tokyo) **21**, 523 (1968).

136. HAZEN, E. L., and R. BROWN: Fungicidin, an Antibiotic Produced by a Soil Actinomycete. Proc. Soc. Exper. Biol. Med. **76**, 93 (1950).

137. — — Two Antifungal Agents Produced by a Soil Actinomycete. Science **112**, 423 (1950).

138. HICKEY, R. J., C. J. CORUM, P. H. HIDY, I. R. COHEN, U. F. B. NAGER, and E. KROPP: Ascosin, an Antifungal Antibiotic Produced by a Streptomycete. Antibiotics and Chemotherapy **2**, 472 (1952).

139. HIRAMATSU, M., A. FURUSAKI, T. NODA, K. NAYA, Y. TOMIE, I. NITTA, T. WATANABE, T. TAKE, and J. ABE: The Crystal and Molecular Structure of Demycarosyl Leucomycin A_3 Hydrobromide. Bull. Chem. Soc. Japan **40**, 2982 (1967).

140. HOCHSTEIN, F. A., H. ELS, W. D. CELMER, B. L. SHAPIRO, and R. B. WOODWARD: The Structure of Oleandomycin. J. Amer. chem. Soc. **82**, 3225 (1960).

141. HOCHSTEIN, F. A., and K. MURAI: Magnamycin B, a Second Antibiotic from *Streptomyces halstedii*. J. Amer. chem. Soc. **76**, 5080 (1954).

142. HOCHSTEIN, F. A., and P. P. REGNA: Magnamycins IV. Mycaminose, an Aminosugar from Magnamycin. J. Amer. chem. Soc. **77**, 3353 (1955).

143. HOFHEINZ, W., und H. GRISEBACH: Die Konfiguration des Desosamins. Tetrahedron Lett. **1962**, 377.

144. — — Die Stereochemie der Mycaminose. Z. Naturforschung **17b**, 355 (1962).

145. — — Zur Biogenese der Makrolide; X. Mitteil.: Über das Vorkommen von L-Mycarose in Erythromycin C. Z. Naturforschung **17b**, 852 (1962).

452   W. KELLER-SCHIERLEIN:

146. HOFHEINZ, W., H. GRISEBACH und H. FRIEBOLIN: Zur Biogenese der Makrolide. VIII. Die Stereochemie der Mycarose und Cladinose. Tetrahedron **18**, 1265 (1962).

147. HOSOYA, S., N. KOMATSU, M. SOEDA, T. YUWAGUCHI, and Y. SONODA: Trichomycin, a New Antibiotic with Trichomonadicidal and Antifungal Activities. J. Antibiotics (Tokyo) **5**, 564 (1952).

148. HOWARTH, G. B., and J. K. N. JONES: The Synthesis of L-Mycarose and L-Cladinose. Canad. J. Chem. **45**, 2253 (1967).

149. HOWARTH, G. B., W. A. SZAREK, and J. K. N. JONES: The Synthesis of D-Arcanose. Chem. Communications **1968**, 62.

150. HÜTTER, R., W. KELLER-SCHIERLEIN, F. KNÜSEL, V. PRELOG, G. C. RODGERS, P. SUTER, G. VOGEL, W. VOSER und H. ZÄHNER: Stoffwechselprodukte von Mirkoorganismen. 57. Mitteil. Boromycin. Helv. chim. Acta **50**, 1533 (1967).

151. HÜTTER, R., W. KELLER-SCHIERLEIN, J. NÜESCH und H. ZÄHNER: Stoffwechselprodukte von Mikroorganismen, 48. Mitteil. Scopamycine. Arch. Mikrobiol. **51**, 1 (1965).

152. HÜTTER, R., W. KELLER-SCHIERLEIN und H. ZÄHNER: Zur Systematik der Actinomyceten, 6. Die Produzenten von Makrolid-Antibiotica. Arch. Mikrobiol. **39**, 158 (1961).

153. HÜTTER, R., K. PORALLA, H. G. ZACHAU und H. ZÄHNER: Über die Wirkungsweise von Borrelidin — Hemmung des Threorin-Einbaues in sRNA. Biochem. Z. **344**, 190 (1966).

154. IKEDA, M., M. SUZUKI, and C. DJERASSI: Macrolide Antibiotics. XV. Nystatin. The Structure of the Aglycone. Tetrahedron Lett. **1967**, 3745.

155. IMMER, H., and J. F. BAGLI: Syntheses of Medium Ring Benzoic Acid Lactones. J. Org. Chem. **33**, 2457 (1968).

156. JONES, P. H., and E. K. ROWLEY: Chemical Modifications of Erythromycin. Antibiotics. I. 3'-De(dimethylamino)-erythromycin A and B. J. Org. Chem. **33**, 665 (1968).

157. JONES, P. H., K. S. IYER, and W. E. GRUNDY: Chemical Modifications of Erythromycin Antibiotics. II. Synthesis of 4'-Hydroxyerythromycin A. Antimicrobial Agents and Chemotherapy **1969**, 123.

158. KARPLUS, M.: Vicinal Proton Coupling in Nuclear Magnetic Resonance. J. Amer. chem. Soc. **85**, 2870 (1963).

159. KELLER-SCHIERLEIN, W.: Stoffwechselprodukte von Mikroorganismen, 55. Mitteil. Über die Konstitution des Borrelidins. Helv. chim. Acta **50**, 731 (1967).

160. KELLER-SCHIERLEIN, W., M. BRUFANI, R. MUNTWYLER und W. RICHLE: Stoffwechselprodukte von Mikroorganismen, 89. Mitteil. Synthese von zwei diastereomeren 2,4,6-Trimethylpimelinsäuren. Ein Beitrag zur Stereochemie des Borrelidins. Helv. chim. Acta **54**, 44 (1971).

161. KELLER-SCHIERLEIN, W., und H. GERLACH: Makrotetrolide. Fortschr. Chem. org. Naturstoffe (Ed. L. Zechmeister) **26**, 161 (1968).

162. KELLER-SCHIERLEIN, W., R. MUNTWYLER, W. PACHE und H. ZÄHNER: Stoffwechselprodukte von Mikroorganismen, 73. Mitteil. Chlorothricin und Des-chlorothricin. Helv. chim. Acta **52**, 127 (1969).

163. KELLER-SCHIERLEIN, W., R. MUNTWYLER und H. ZÄHNER: Stoffwechselprodukte von Mikroorganismen, 79. Mitteil. Bromothricin. Experientia **25**, 786 (1969).

164. KELLER-SCHIERLEIN, W., und W. RICHLE: Metabolic Products of Microorganisms. LXXXV. The Structure of Cinerubine A. Antimicrobial Agents and Chemotherapy **1970**, 68. Vgl. auch Chimia **24**, 35 (1970).

165. KELLER-SCHIERLEIN, W., und G. RONCARI: Stoffwechselprodukte von Actinomyceten, 33. Mitteil. Hydrolyseprodukte von Lankamycin. Lankavose und 4-O-Acetylarcanose. Helv. chim. Acta **45**, 138 (1962).

166. KELLER-SCHIERLEIN, W., und G. RONCARI: Stoffwechselprodukte von Mikroorganismen, 46. Mitteil. Die Konstitution des Lankamycins. Helv. chim. Acta **47**, 78 (1964).

167. KOCHETKOV, N. K., and A. I. USOV: The Synthesis of D-chalkose. Tetrahedron Lett. **1963**, 519.

168. KOE, B. K., F. W. TANNER, K. V. RAO, B. A. SOBIN, and W. D. CELMER: PA 150, PA 153, and PA 166, New Polyene Antifungal Antibiotics. Antibiotics Annual **1957—1958**, 897.

169. KONDO, S., J. M. J. SAKAMOTO, and H. YUMOTO: Bandamycin, a New Antibiotic. J. Antibiotics (Tokyo), Ser. A **14**, 365 (1961).

170. KORTE, F., A. BILOW und R. HEINZ: Zur Synthese des D,L-Picrocins. Tetrahedron **18**, 657 (1962).

171. KORTE, F., U. CLAUSSEN und K. GÖHRING: Synthese der D,L-Epimycarose und der D,L-Mycarose. Tetrahedron **18**, 1257 (1962).

172. KORTE, F., U. CLAUSSEN und G. SNATZKE: Zur Stereochemie der Mycarose und ihrer 3- und 4-Epimeren. Tetrahedron **20**, 1477 (1964).

173. KOSHIYAMA, H., M. OKANISHI, T. OHMORI, T. MIYAKI, H. TSUKIURA, M. MATZUZAKI, and H. KAWAGUCHI: Cirramycin, a New Antibiotic. J. Antibiotics (Japan), Ser. A **16**, 59 (1963).

174. KOSHIYAMA, H., H. TSUKIURA, K. FUJISAWA, M. KONISHI, M. HATORI, K. TOMITA, and H. KAWAGUCHI: Studies on Cirramycin A_1. I. Isolation and Characterization of Cirramycin A_1. J. Antibiotics (Japan) **22**, 61 (1969).

175. KUEHNE, M. E., and B. W. BENSON: The Structures of Spiramycin and Magnamycin. J. Amer. chem. Soc. **87**, 4660 (1965).

176. KUHN, R., und G. BASCHANG: Aminozucker-Synthesen XVIII. 3-Acetamino-3-desoxy-D-mannose. Liebigs Ann. Chem. **628**, 206 (1959).

177. KUNSTMANN, M. P., and L. A. MITSCHER: Some Additional Observations on the Chemical Nature of Neutramycin. Experientia **21**, 372 (1965).

178. KUO, C. H., D. TAUB, R. D. HOFFSOMMER, N. L. WENDLER, W. H. URRY, and G. MULLENBACH: The Resolution of ($\pm$) Zearalenone, Determinations of the Absolute Configuration of the Natural Enantiomorph. Chemical Communications **1967**, 761.

179. KURATH, P., and R. S. EGAN: Oxidation and Reduction of 8,9-Anhydroerythronolide B 6,9-Hemiacetal. Helv. chim. Acta **54**, 523 (1971).

180. LEE, CH.-H., and C. P. SCHAFFNER: Perimycin. The Structure of Some Degradation Products. Tetrahedron **25**, 2229 (1969).

181. LEE, CH.-H., and C. P. SCHAFFNER: Perimycin. Chemistry of Perosamin. Tetrahedron Lett. **1966**, 5837.

182. LEFEMINE, D. V., F. BARBATSCHI, M. DAUN, S. O. THOMAS, M. P. KUNSTMANN, L. A. MITSCHER, and N. BOHONOS: Neutramycin, a New Neutral Macrolide Antibiotic. Antimicrobial Agents and Chemotherapy **1963**, 41.

183. LEMAL, D. M., P. D. PACHT, and R. B. WOODWARD: The Synthesis of L-(-)-Mycarose and -(-)-Cladniose. Tetrahedron **18**, 1275 (1962).

184. LUMB, M., P. E. MACEY, J. SPYREE, J. M. WHITMARSH, and R. D. WRIGHT: Isolation of Vivomycin and Borrelidin, two Antibiotics with Antiviral Activity, from a Species of *Streptomyces*. Nature **206**, 263 (1965).

185. MACLAY, W. D., R. M. HANN, and C. S. HUDSON: The Cleavage of the Carbon Chains of Some Methyl Aldohexomethylopyranosides by Oxidation with Periodic Acid. J. Amer. chem. Soc. **61**, 1660 (1939).

186. MALLAMS, A. K.: The Megalomycins. I. D-Rhodosamine, a New Dimethylamino Sugar. J. Amer. chem. Soc. **91**, 7505 (1969).

187. MALLAMS, A. K., R. S. JARET, and H. REIMANN: The Megalomycins. II. The Structure of Megalomycin A. J. Amer. chem. Soc. **91**, 7506 (1969).

188. MANWARING, D. G., R. W. RICKARDS, and B. T. GOLDING: The Structure of the Aglycone of the Macrolide Antibiotic Nystatin. Tetrahedron Lett. **1969**, 5319.

454   W. KELLER-SCHIERLEIN:

189. MANWARING, D. G., R. W. RICKARDS, and R. M. SMITH: Macrolide Antibiotic Studies. XIV. The Total Absolute Configuration of Methymycin. Tetrahedron Lett. **1970,** 1029.   *a*

190. MARQUEZ, J., A. MURAWSKI, and G. H. WAGMAN: Isolation, Purification and Preliminary Characterization of Megalomycin. J. Antibiotics (Tokyo) **22,** 259 (1969).

191. MARTIN, J. R., and R. S. EGAN: 5,6-Dideoxy-5-oxoerythronolide B, a Shunt Metabolite of Erythromycin Biosynthesis. Biochemistry **9,** 3439 (1970).

192. MARTIN, J. R., and T. J. PERUN: Studies on the Biosynthesis of the Erythromycins. III. Isolation and Structure of 5-Deoxy-5-oxo-erythronolide B, a Shunt Metabolite of Erythromycin Biosynthesis. Biochemistry **7,** 1728 (1968).

193. MARTIN, J. J., T. J. PERUN, and R. L. GIROLAMI: Studies on the Biosynthesis of the Erythromycins. I. Isolation and Structure of an Intermediate Glycoside, 3-α-L-Mycaroyl-erythronolide B. Biochemistry **5,** 2852 (1966).

194. MARTIN, J. R., and W. ROSENBROOK: Studies on the Biosynthesis of the Erythromycins. II. Isolation and Structure of a Biosynthetic Intermediate, 6-Deoxyerythronolide B. Biochemistry **6,** 435 (1967).

195. MCALPINE, J. B., J. W. CORCORAN, and R. S. EGAN: Scopamycin. II. Identification of the Sugar Moiety of Scopamycin A as 2-O-Methyl-L-Rhamnose. J. Antibiotics (Tokyo) **24,** 51 (1971).

196. MCCAPRA, F., A. I. SCOTT, P. DELMOTTE, J. DELMOTTE-PLAQUÉE, and N. S. BHACCA: The Constitution of Monorden, an Antibiotic with Tranquillizing Action. Tetrahedron Lett. **1964,** 869.

197. MCGUIRE, J. M., R. L. BUNCH, A. C. ANDERSON, H. E. BOAZ, E. H. FLYNN, H. M. POWELL, and J. W. SMITH: "Ilotycin", a New Antibiotic. Antibiotics and Chemotherapy **2,** 281 (1952).

198. MCLAUGHLIN, G. M., G. A. SIM, J. R. KIECHEL, and CH. TAMM: The Absolute Stereochemistry of Phomins. X-Ray Analysis of the Phomin-Silver Fluoroborate Complex. Chemical Communications **1970,** 1398.

199. MCNALLY, S., and W. G. OVEREND: A Synthesis of Chalcose. Chemistry and Industry **1964,** 2021.

200. MECHLINSKI, W., C. P. SCHAFFNER, P. GANIS, and G. AVITABLE: Structure and Absolute Configuration of the Polyen Macrolide Antibiotic Amphothericin B. Tetrahedron Lett. **1970,** 3873.

201. MEYER, W. E.: PIMARICIN. The Glycosidic Form of Mycosamine. Chemical Communications **1968,** 471.

202. MINATO, H., and T. KATAYAMA: Studies on the Metabolites of *Zygosporium mansonii*. Part II. Structures of Zygosporins D, E, F and G. J. Chem. Soc. (London) [C] **1970,** 45.

203. MINATO, H., and M. MATSUMOTO: Studies on the Metabolites of *Zygosporicum mansonii*. Part I. Structure of Zygosporin A. J. Chem. Soc. (London) [C] **1970,** 38.

204. MIRRINGTON, R. N., E. RITCHIE, C. W. SHOPPEE, S. STERNHALL, and W. C. TAYLOR: Some Metabolites of *Nectria radicicola* Gerlach & Nilsson (Syn. Cylindro-carpon radiciola Wr.): The Structure of Radicicol (Monorden). Austral. J. Chem. **19,** 1265 (1966).

205. MIRRINGTON, R. N., E. RITCHIE, C. W. SHOPPEE, W. C. TAYLOR, and S. STERNHALL: The Constitution of Radicicol. Tetrahedron Lett. **1964,** 365.

206. MITSCHER, L. A., and M. P. KUNSTMANN: The Structure of Neutramycin. Experientia **25,** 12 (1969).

207. MITSCHER, L. A., B. J. SLATER, T. J. PERUN, P. H. JONES, and J. R. MARTIN: The Configuration of Macrolide Antibiotics. III. Circular Dichroism and the Conformations of Erythromycins. Tetrahedron Lett. **1969,** 4505.

208. MORIN, R., and M. GORMAN: Macrolide Antibiotics. Kirk-Othmer Encyclopedia of Chemical Technology, 2nd Ed. **12,** 632 (1967).

209. MORIN, R., and M. GORMAN: The Partial Structure of Tylosin, a Macrolide Antibiotic. Tetrahedron Lett. **1964,** 2339.

210. MORIN, R. B., M. GORMAN, R. L. HAMILL, and P. V. DEMARCO: The Structure of Tylosin. Tetrahedron Lett. **1970,** 4737.

211. MUNRO, H. D., O. C. MUSGRAVE, and R. TEMPLETON: Curvularin. Part. V. The Compound $C_{16}H_{18}O_5,\alpha,\beta$-Dehydro-curvularin. J. Chem. Soc. (London) [C] **1967,** 947.

212. MUNTWYLER, R., M. BRUFANI, W. FEDELI und W. KELLER-SCHIERLEIN: Die Struktur des Chlorothricins. Helv. chim. Acta (in Vorbereitung).

213. MUNTWYLER, R., J. WIDMER und W. KELLER-SCHIERLEIN: Stoffwechselprodukte von Mikroorganismen, 84. Mitteil. Synthese des 5-Chlor-6-methyl-salicylsäure-methyläthers, eines Abbauproduktes des Chlorithricins. Helv. chim. Acta **53,** 1544 (1970).

214. MUSGRAVE, O. C.: Curvularin. Part I. Isolation and Partial Characterization of a Metabolic Product from a New Species of *Curvularia*. J. Chem. Soc. (London) 1956, 4301.

215. — Curvularin. Part II. The Constitution of an Aromatic Degradation Product. J. Chem. Soc. (London) **1957,** 1104.

216. MUXFELDT, H., S. SHRADER, P. HAUSER, and H. BROCKMANN: The Structure of Picromycin. J. Amer. chem. Soc. **90,** 4748 (1968).

217. NAGAHAMA, N., M. SUZUKI, S. AWATAGUCHI, and T. OKUDA: Studies on a New Antibiotic, Albocycline. I. Isolation, Purification and Properties. J. Antibiotics (Tokyo) **20,** 261 (1967).

218. NAGAHAMA, N., I. TAKAMORI, K. KOTERA, and M. SUZUKI: Studies on an Antibiotic, Albocycline. III. Partial Structure of Albocycline. Chem. Pharm. Bull. (Japan) **19,** 649 (1971).

219. NAGAHAMA, N., I. TAKAMORI, and M. SUZUKI: Studies on an Antibiotic, Albocycline. IV. Catalytic Hydrogenation and Structure Elucidation of Albocycline. Chem. Pharm. Bull. (Japan) **19,** 655 (1971).

220. — — — Studies on an Antibiotic, Albocycline. V. High Pressure — High Temperature Hydrogenation Products, and Structure Proof of Albocycline. Chem. Pharm. Bull. (Japan) **19,** 660 (1971).

221. NAKANO, H.: Studies on Trichomycin V. J. Antibiotics (Tokyo) Ser. A **14,** 68 (1961).

222. NARASIMHACHARI, N., and M. B. SWAMI: Dermostatin. A Revised Hexaene Structure. J. Antibiotics (Tokyo) **23,** 566 (1970).

223. NEWMAN, H.: Degradation and Synthesis of Desosamine. J. Org. Chem. **29,** 1461 (1964).

224. OGURA, H., A. OTAGOSHI, Y. SANO, and T. HATA: Structure of Amaromycin. Chem. Pharm. Bull. (Japan) **15,** 682 (1967).

225. OMURA, S., Y. HIRONAKA, and T. HATA: Chemistry of Leucomycin. IX. Identification of Leucomycin A_3 with Josamycin. J. Antibiotics (Tokyo) **23,** 511 (1970).

226. OMURA, S., M. KATAGIRI, and T. HATA: The Chemistry of Leucomycins. VI. Structures of Leucomycins A_4, A_5, A_6, A_7, A_8 and A_9. J. Antibiotics (Tokyo) **21,** 272 (1968).

227. OMURA, S., M. KATAGIRI, T. HATA, M. HIRAMATSU, T. KIMURA, and K. NAYA: The Allylic Rearrangement of the Hydroxyl Group from C-9 to C-13 and the Absolute Configuration at C-9 of Leucomycin A_3. Chem. Pharm. Bull. **16,** 1402 (1968).

228. OMURA, S., M. KATAGIRI, H. OGURA, and T. HATA: The Chemistry of Leucomycins. Partial Structure of Leucomycin A_3. Chem. Pharm. Bull. (Japan) **15,** 1529 (1967).

229. — — — — The Chemistry of Leucomycins. II. Glycosidic Linkages of Mycaminose and Mycarose on Leucomycin A_3. Chem. Pharm. Bull. (Japan) **16,** 1167 (1968).

230. — — — — The Chemistry of Leucomycin. III. Structure and Stereochemistry of Leucomycin A_3. Chem. Pharm. Bull. (Japan) **16,** 1181 (1968).

231. OMURA, S., T. MURO, S. NAMIKI, M. SHIBATA, and J. SAWADA: Studies on the Antibiotics from *Streptomyces spinichromogenes* var. *Kujimyceticus*. III. The Structure of Kujimycin A and Kujimycin B. J. Antibiotics (Tokyo) **22**, 629 (1969).

232. Omura, S., A. Nakagawa, M. Katagiri, T. Hata, M. Hiramatsu, T. Kimura, and K. Naya: Chemistry of Leucomycins. VIII. Absolute Configuration of Leucomycin and Isoleucomycin. Chem. Pharm. Bull. **18**, 1501 (1970).

233. Omura, S., A. Nakagawa, M. Otani, T. Hata, H. Ogura, and K. Furuhata: Structure of the Spiramycins (Foromacidines) and their Relationship with the Leucomycins and Carbomycins (Magnamycins). J. Amer. chem. Soc. **91**, 3401 (1969).

234. Omura, S., S. Namiki, M. Shibata, T. Muro, H. Nakayoshi, and J. Sawada: Studies on the Antibiotics from *Streptomyces spinichromogenes* va. *Kujimyceticus*. II. Isolation and Characterization of Kujimycins A and B. J. Antibiotics (Tokyo) **22**, 500 (1969).

235. Oroschnik, W., and A. D. Mebane: Polyene Antifungal Antibiotics. Fortschr. Chem. org. Naturstoffe (Ed. L. Zechmeister) **21**, 17 (1963).

236. Osono, T., Y. Oka, S. Watanabe, Y. Numazaki, K. Moriyama, H. Ishida, and K. Suzuki: A New Antibiotic, Josamycin. I. Isolation and Physico-Chemical Characteristics. J. Antibiotics (Tokyo), Ser. A **20**, 174 (1967).

237. Pal, A., and P. Nandi: A New Antifungal Antibiotic Produced by *Streptomyces* sp.: Ac_2 435. Experientia **20**, 321 (1964).

238. Pandey, R. C., and K. L. Rinehart: Polyene Antibiotics. V. Characterization of Components of the Filipin Complex by Mass Spectrometry. J. Antibiotics (Tokyo) **23**, 414 (1970).

239. Pandey, R. C., K. L. Rinehart, and N. Narasimhachari: Structural Studies of Pentaene Antibiotics: Chainin and Filipin. Abstracts of Papers, 7th Symposium on the Chemistry of Natural Products (Riga 1970) E 157.

240. Patrick, J. B., R. P. Williams, and J. S. Webb: Pimaricin, II. The Structure of Pimaricin. J. Amer. chem. Soc. **80**, 6689 (1958).

241. Paul, R., et S. Tchelitcheff: Structure de la spiramycine. I. Etude des produits de dégradation, caractérisation du mycarose. Bull. Soc. Chim. France **1957**, 443.

242. — — Structure de la spiramycine. II. Etude des produits de dégradation: caractérisation du diméthyl-amino-5-méthyl-6-hydroxy-2-tétrahydro-pyranne. Bull. Soc. Chim. France **1957**, 734.

243. — — Structure de la spiramycine. III. Etude des produits de dégradation. Caractérisation du mycaminose. Bull. Soc. Chim. France **1957**, 1059.

244. — — Structure de la spiramycine. IV. Présence d'une lactone macrocyclique: Etude de ses produits d'oxydation. Bull. Soc. Chim. France **1960**, 150.

245. — — Structure de la spiramycine. V. Mode d'enchainement des divers produits d'hydrolyse. Bull. Soc.Chim. France **1965**, 189.

246. — — Structure de la spiramycine. VI. Etablissement de la formule développée. Bull. Soc. Chim. France **1965**, 650.

247. Perun, T. J.: Chemistry of Erythronolide B. Acid-Catalyzed Transformations of the Aglycone of Erythromycin B. J. Org. Chem. **32**, 2324 (1967).

248. Perun, T. J., and R. S. Egan: The Conformation of Erythromycin Aglycones. Tetrahedron Lett. **1969**, 387.

249. Perun, T. J., R. S. Egan, P. H. Jones, J. R. Martin, L. A. Mitscher, and B. Slater: Conformation of Macrolide Antibiotics-IV. Nuclear Magnetic Resonance, Circular Dichroism, and Chemical Studies of Erythromycin Derivatives. Antimicrobial Agents and Chemotherapy **1969**, 116.

250. Perun, T. J., R. S. Egan, and J. R. Martin: The Conformation of Macrolide Antibiotics. II. Configurational and Conformational Studies of Dihydroerythronolides. Tetrahedron Lett. **1969**, 4501.

251. PETTINGA, C. W., W. M. STARK, and F. R. VAN ABEELE: The Isolation of a Second Crystalline Antibiotic from *Streptomyces erythreus.* J. Amer. chem. Soc. **76,** 569 (1954).

252. PINNERT-SINDICO, S., L. NINET, J. PREND'HOMME, and C. COSAR: A New Antibiotic — Spiromycin. Antibiotics Annual **1954—1955,** 724 (1955).

253. POWELL, H. M., W. S. BONIECE, R. C. PITKUGER, R. C. STONE, and C. G. CULBERTSON: Laboratory Studies on "Ilotycin". Antibiotics and Chemotherapy **3,** 165 (1953).

254. PRELOG, V., A. M. GOLD, G. TALBOT und A. ZAMOJSKI: Stoffwechselprodukte von Actinomyceten, 31. Mitteil.: Über die Konstitution des Narbomycins. Helv. chim. Acta **45,** 4 (1962).

255. REGNA, P. P., F. A. HOCHSTEIN, R. L. WAGNER, and R. B. WOODWARD: Magnamycin. II. Mycarose, an Unusual Branched-chain Desoxysugar from Magnamycin. J. Amer. chem. Soc. **75,** 4625 (1953).

256. REUSSER, F.: Mode of Action of Albocycline, an Inhibitor of Nicotinate Biosynthesis. J. Bacteriol. **100,** 11 (1969).

257. RHODES, A., K. H. FANTES, B. BOOTHROYD, M. P. MCGONAGLE, and R. CROSSE: Venturicidin, a New Antifungal Antibiotic of Potential Use in Agriculture. Nature **192,** 952 (1961).

258. RICHARDSON, A. C.: Derivatives of 3-Amino-3,6-dideoxy-L-glucose and -L-talose. Proc. chem. Soc. (London) **1961,** 255.

259. — The Synthesis and Stereochemistry of Mycaminose. Proc. chem. Soc. (London) **1961,** 430.

260. — A Stereospecific Synthesis of Desosamine Hydrochloride. Proc. chem. Soc. (London) **1963,** 131.

261. — Selective Acylation of Methyl-3-Acetamido-3,6-dideoxy-α-L-glucupyranoside: its conversion into the L-galacto- and L-manno-Analogues. Carbohydrate Research **4,** 415 (1967).

262. RICHTMYER, N. K., and C. S. HUDSON: Crystalline α-Methyl-D-altroside and Some New Derivatives of D-Altrose. J. Amer. chem. Soc. **63,** 1727 (1941).

263. RICKARDS, R. W., and R. M. SMITH: Macrolide Antibiotic Studies. XIII. Partial Absolute Configurations of Methymycin, Neomethymycin, Narbomycin, and Picromycin. Tetrahedron Lett. **1970,** 1025.

264. RICKARDS, R. W., R. M. SMITH, and Z. MAJER: The Structure of the Macrolide Antibiotic Picromycin. Chem. Communications **1968,** 1049.

265. RINEHART, K. L., V. F. GERMAN, W. P. TUCKER und D. GOTTLIEB: Isolierung und Eigenschaften der Tetrine A und B. Liebigs Ann. Chem. **668,** 77 (1963).

266. RINEHART, K. L., JR., R. F. SCHIMBOR, and T. H. KINSTLE: Mass Spectrometric Behaviour of Sugars from Antibiotics. Antimicrobial Agents and Chemotherapy **1965,** 119.

267. RONCARI, G., und W. KELLER-SCHIERLEIN: Stoffwechselprodukte von Mikroorganismen, 50. Mitteil. Die Konfiguration der Arcanose. Helv. chim. Acta **49,** 705 (1966).

268. ROTHWEILER, W., and CH. TAMM: Isolation and Structure of Phomin. Experientia **22,** 750 (1966).

269. — — Isolierung und Struktur der Antibiotica Phomin und 5-Dehydrophomin. Helv. chim. Acta **53,** 696 (1970).

270. SAKAMOTO, J. M. J.: Etudes sur antibiotiques antifongiques II. La cryptocidine, un nouvel antibiotique produit par les streptomycètes. J. Antibiotics (Tokyo) Ser. A **12,** 21 (1959).

271. SANO, Y.: Studies on Leucomycin. V. Purification and Chemical Properties. J. Antibiotics (Tokyo) Ser. A **7,** 93 (1954).

272. — The Isolation of a Second Crystalline Antibiotic from *Streptomyces kitasatoensis.* J. Antibiotics (Tokyo) Ser. A **9,** 202 (1956).

273. SANO, Y., T. HOSHI, and T. HATA: Studies on Leucomycin, IV. Comparison of

Leucomycin with Erythromycin and Carbomycoin. J. Antibiotics (Tokyo) Ser. A **7**, 88 (1954).

274. Schlegel, R., und H. Thrum: Flavomycoin, ein neues antifungales Polyenantibioticum. Experientia **24**, 11 (1968).

275. Shenin, D., T. V. Kotenko, and O. N. Exzempliaroo: Composition of Nystatin. Antibiotiki (Moskau) **13**, 387 (1968).

276. Sigal, M. V., P. F. Wiley, K. Gerzon, E. H. Flynn, U. C. Quarck, and O. Weaver: Erythromycin, VI. Degradation Studies. J. Amer. chem. Soc. **78**, 388 (1956).

277. Sigg, H. P.: Die Konstitution von Brefeldin A. Helv. chim. Acta **47**, 1401 (1964).

278. Singleton, V. L., and N. Bohonos: Chemical Characterization of the Mold Product Decumbin. Agr. Biol. Chem. (Japan) **28**, 77 (1964).

279. Singleton, V. L., N. Bohonos, and A. J. Ullstrup: Decumbin, a New Compound from a Species of *Penicillium*. Nature **181**, 1072 (1958).

280. Stevens, C. L., S. K. Gupta, R. P. Glinski, K. G. Taylor, P. Blumbergs, C. P. Schaffner, and Ch.-H. Lee: Proof of Structure, Stereochemistry, and Synthesis of Perosamine (4-Amino-4,6-dideoxy-D-mannose) Derivatives. Carbohydrate Research **7**, 502 (1968).

281. Stevens, C. L., G. E. Gutowski, K. G. Taylor, and C. P. Bryant: The Stereochemistry and Synthesis of Forosamine, a Dioxyamino-sugar Moiety of the Spiramycin Antibiotics. Tetrahedron Lett. **1966**, 5717.

282. Stob, M., R. S. Baldwin, J. Tuite, F. N. Andrews, and K. G. Gilette: Isolation of an Anabolic, Uterotrophic Compound from Corn infected with *Gibberella zeae*. Nature **196**, 1318 (1962).

283. Struyk, A. P., I. Hoette, G. Drost, J. M. Waisvisz, T. van Eck, and J. C. Hoogerheide: Pimaricin, a New Antifungal Antibiotic. Antibiotics Annual **1957—1958**, 878.

284. Suhara, Y., F. Sasaki, K. Maeda, H. Umezawa, and M. Ohno: The Total Synthesis of Kasugamycin. J. Amer. chem. Soc. **90**, 6559 (1968).

285. Suzuki, M., I. Takamori, A. Kinumaki, Y. Sugawara, and T. Okuda: The Structures of Antibiotics YL-704 A and B. Tetrahedron Lett. **1971**, 435.

286. Suzuki, T.: The Structure of an Antibiotic, B-58941. Bull. Chem. Soc. (Japan) **43**, 292 (1970).

287. Suzuki, Y., H. Tanáka, H. Aoki, and T. Tamura: Ascotoxin (Decumbin), a Metabolite of *Ascochyta imperfecta* PECK. Agr. Biol. Chem. (Japan) **34**, 395 (1970).

288. Taig, M. M., N. K. Soloviera, and P. S. Braginskaya: Characteristics of Aureninproducing Organism. Antibiotiki (Moskau) **14**, 873 (1969).

289. Tamura, G., K. Audo, S. Suzuki, A. Takasaki, and K. Arima: Antiviral Activity of Brefeldin A and Verrucarin A. J. Antibiotics (Tokyo) **21**, 160 (1968).

290. Tanner, F. W., Jr., A. R. English, T. M. Lees, and J. B. Routien: Some Properties of Magnamycin, a New Antibiotic. Antibiotics and Chemotherapy **2**, 441 (1952).

291. Taub, D., N. N. Girotra, R. D. Hoffsommer, C. H. Kuo, H. L. Slates, S. Weber, and N. L. Wendler: Total Synthesis of the Macrolide, Zearalenone. Chemical Communications **1967**, 225.

292. — — — — — — — Total Synthesis of the Macrolide, Zearalenone. Tetrahedron **24**, 2443 (1968).

293. Thrum, H., und I.-D. Dcho: Die Antibiotica 2814 P und 2814 H, zwei neue antifungale Polyen-Antibiotica. Naturwiss. **47**, 474 (1960).

294. Tsukiura, H., M. Konishi, M. Saka, T. Naito, and H. Kawaguchi: Studies on Cirramycin A_1. III. Structure of Cirramycin A_1. J. Antibiotics (Japan) **22**, 100 (1969).

295. Tsukuda, Y., M. Matsumoto, H. Minato, and H. Koyama: Structure of Zygosporin A. X-Ray Analysis of Isozygosporin A p-Bromobenzoate. Chemical Communications **1969**, 41.

296. TYTELL, A. A., F. J. McCARTHY, W. P. FISHER, W. A. BOHLHOFER, and J. CHARNEY: Fungichromin and Fungichromatin. New Polyene Antifungal Agents. Antibiotics Annual **1954—1955**, 716.

297. URRY, W. H., H. L. WEHRMEISTER, E. B. HODGE, and P. H. HIDY: The Structure of Zearalenone. Tetrahedron Lett. **1966**, 3109.

298. USHAKOVA, T. A., V. I. FROLOVA, and A. D. KUZOVKOV: The Structure of Aurenin. Abstracts of Papers, 7th Symposium on the Chemistry of Natural Products (Riga 1970) E 153.

299. UTAHARA, R., Y. OKAMI, S. NAKAMURA, and H. UMEZAWA: On a New Antifungal Substance, Mediocidin, and Other Antifungal Substances of Streptomyces with three Characteristic Absorption Maxima. J. Antibiotics (Tokyo), Ser. A **7**, 120 (1954).

300. VALYI-NAGY, T., J. URI, and L. SZILAGYI: Primycin, a New Antibiotic. Nature **174**, 1105 (1954).

301. VANDEPUTTE, J., J. L. WACHTEL, and E. T. STILLER: Amphothericin A and B. Antifungal Antibiotics Produced by a Streptomycete. II. The Isolation and Properties of the Crystalline Amphothericins. Antibiotics Annual **1955—1956**, 587.

302. VAZQUEZ, D., T. STAEHELIN, M. L. CELMA, E. BATTANER, R. FERNANDEZ-MUNOZ, and R. E. MONRO: Inhibitors as Tools in Elucidating Ribosomal Function. Inhibitors, Tools in Cell Research (Ed. Th. Bücher and H. Sies). Berlin-Heidelberg-New York: Springer. 1969.

303. VINING, L. C., and W. A. TABER: Preparation and Properties of Crystalline Candidin. Canad. J. Chem. **34**, 1163 (1956).

304. VINING, L. C., W. A. TABER, and F. J. GREGORY: The Candidin-Candicidin Group of Antifungal Antibiotics. Antibiotics Annual **1954—1955**, 980.

305. VLATTAS, I., I. T. HARRISON, L. TÖKES, J. H. FRIED, and A. D. CROSS: The Synthesis of D,L-Zearalenone. J. Org. Chem. **33**, 4176 (1968).

306. VON EUW, J., und T. REICHSTEIN: Die Glykoside der Samen von Strophanthus Nicholsonii Holm-Glykoside und Aglykone. 34. Mitteil. Helv. chim. Acta **31**, 883 (1948).

307. VON SALTZA, M. H., J. D. DUTCHER, J. REID, and O. WINTERSTEINER: Nystatin. IV. The Stereochemistry of Mycosamine. J. Org. Chem. **28**, 999 (1963).

308. VON SALTZA, M. H., J. REID, J. D. DUTCHER, and O. WINTERSTEINER: Nystatin II. The Stereochemistry of Mycosamine. J. Amer. chem. Soc. **83**, 2785 (1961).

309. WAGNER, R. L., F. A. HOCHSTEIN, K. MURAI, N. MESSINA, and P. P. REGNA: Magnamycin, a New Antibiotic. J. Amer. chem. Soc. **75**, 4684 (1953).

310. WASSERMANN, H. H., J. E. VAN VERTH, D. J. McCAUSTLAND, I. J. BOROWITZ, and B. KAMBER: The Mycoticins, Polyene Macrolides from *Streptomyces ruber*. J. Amer. chem. Soc. **89**, 1535 (1967).

311. WATANABE, T.: Leucomycin. IV. Isolation of Mycominose from the Acid Hydrolysates. Bull. chem. Soc. Japan **34**, 15 (1961).

312. WATANABE, T., T. FUJI, and K. SATAKE: 4-O-Acetylmycarose, a New O-acetyl Sugar from Leucomycin Minor Components. J. Biochem. (Tokyo) **50**, 197 (1961).

313. WATANABE, T., H. NISHIDA, and K. SATAKE: Leucomycin. V. Isolation of Mycarose-4-isovalerate from Leucomycin A$_1$. Bull. chem. Soc. Japan **34**, 1285 (1961).

314. WEHRMEISTER, H. L., and D. E. ROBERTSON: Total Synthesis of the Macrocyclic Lactone, Dideoxyzearalane. J. Org. Chem. **33**, 4173 (1968).

315. WEINSTEIN, M. J., G. H. WAGMAN, J. A. MARQUEZ, R. T. TESTA, E. ODEN, and J. A. WAITZ: Megalomicin, a New Macrolide Antibiotic Complex Produced by *Micromonospora*. J. Antibiotics (Tokyo) **22**, 253 (1969).

316. WHALEY, H. A., E. L. PATTERSON, A. C. DORNBUSH, E. J. BACKUS, and N. BOHONOS: Isolation and Characterization of Relomycin, a New Antibiotic. Antimicrobial Agents and Chemotherapy **1963**, 45.

317. WHITFIELD, G. B., T. D. BROCK, A. AMMANN, D. GOTTLIEB, and H. E. CARTER:

Filipin, an Antifungal Antibiotic. Isolation and Properties. J. Amer. chem. Soc. **77**, 4799 (1955).

318. WILEY, P. F., R. GALE, C. W. PETTINGA, and K. GERZON: Erythromycin. XII. The Isolation, Properties, and Partial Structure of Erythromycin C. J. Amer. chem. Soc. **79**, 6074 (1957).

319. WILEY, P. F., K. GERZON, E. H. FLYNN, M. V. SIGAL, and U. C. QUARCK: Erythromycin, IV. Degradation Studies. J. Amer. chem. Soc. **77**, 3676 (1955).

320. — — — — — Erythromycin, V. Isolation and Structure of Degradation Products J. Amer. chem. Soc. **77**, 3677 (1955).

321. WILEY, P. F., K. GERZON, E. H. FLYNN, M. V. SIGAL, O. WEAVER, U. C. QUARCK, R. R. CHAUVETTE, and R. MONAHAN: Erythromycin, X. Structure of Erythromycin. J. Amer. chem. Soc. **79**, 6062 (1957).

322. WILEY, P. F., M. V. SIGAL, O. WEAVER, R. MONAHAN, and K. GERZON: Erythromycin. XI. Structure of Erythromycin B. J. Amer. chem. Soc. **79**, 6070 (1957).

323. WILEY, P. F., and O. WEAVER: Erythromycin. VII. The Structure of Cladinose. J. Amer. chem. Soc. **78**, 808 (1956).

324. WILHELM, J. M., N. L. OLEINICK, and J. W. CORCORAN: Interaction of Antibiotics with Ribosomes: Structure function relationships and a possible common mechanism for the antibacterial action of the macrolides and lincomycin. Antimicrobial Agents and Chemotherapy **1967**, 236.

325. WOO, P. W. K., H. W. DION, and Q. R. BARTZ: Chemistry of Chalcose, a 3-Methoxy-4,6-dideoxyhexose. J. Amer. chem. Soc. **83**, 3352 (1961).

326. — — — The Structure of Chalcomycin. J. Amer. chem. Soc. **86**, 2726 (1964).

327. WOO, P. W. K., H. W. DION, L. DURHAM, and H. S. MOSHER: The Stereochemistry of Desosamine: an NMR Analysis. Tetrahedron Lett. **1962**, 735.

328. WOO, P. W. K., H. W. DION, and L. F. JOHNSON: The Stereochemistry of Chalcose, a Degradation Product of Chalkomycin. J. Amer. chem. Soc. **84**, 1066 (1962).

329. WOODWARD, R. B.: Struktur und Biogenese der Makrolide. Eine neue Klasse von Naturstoffen. Angew. Chem. **69**, 50 (1957).

330. WOODWARD, R. B., L. S. WEILER, and P. C. DUTTA: The Structure of Magnamycin. J. Amer. chem. Soc. **87**, 4662 (1965).

331. ZYGANOV, W. A., L. O. BOLSHAKOWA, S. N. SOLOVEV, and M. A. MALYSHKINA: Hepcin, a New Pentaene from *Actinosporangium griseoroseum* n. sp. Antibiotiki (Moskau) **15**, 963 (1970).

Anmerkung bei der Korrektur: Seit der Fertigstellung des Manuskripts hat der Autor Kenntnis erhalten von zwei hervorragenden Arbeiten aus dem Laboratorium von F. BOHLMANN, die unsere Kenntnisse der Trichomycin-Untergruppe der Polyen-Makrolide wesentlich erweitern. Die Aufstellung von vollständigen Konstitutionsformeln zweier Aglykone dieser Reihe scheint teilweise die auf S. 382 beschriebenen Befunde von HATTORI zu widerlegen. Für Einzelheiten muß auf die Original-Literatur verwiesen werden:

BOHLMANN, F., E. V. DEHMLOW, H.-J. NEUHAHN, R. BRANDT und H. BETHKE: Neue Heptaen-Makrolide-II. Grundskelett, Stellung der funktionellen Gruppen und Struktur der Aglykone. Tetrahedron **26**, 2199 (1970).

BOHLMANN, F., E. V. DEHMLOW, H.-J. NEUHAHN, R. BRANDT und B. REINICKE: Neue Heptaen-Makrolide-I. Charakterisierung und Abbau. Tetrahedron **26**, 2191 (1970).

(Eingelaufen am 14. Oktober 1971)

Chemie und Biologie der Saponine

Von R. Tschesche und G. Wulff, Bonn

Inhaltsübersicht

I. Einleitung

Saponine stellen Naturstoffe dar, die durch eine Reihe gemeinsamer Eigenschaften gekennzeichnet sind. Als erstes Charakteristikum wurde das starke Schaumvermögen in wäßriger Lösung bemerkt, das dieser Gruppe den Namen gegeben hat. Weitere Eigenschaften sind die hämolytische Aktivität, die Toxizität für Fische, die Cholesterin-komplexbildung und, neuerdings beobachtet, eine antibiotische Aktivität, vornehmlich gegen niedere Pilze.

Im Verlauf der Strukturermittlung erkannte man, daß die oben genannten Merkmale keineswegs einen gleichartigen chemischen Aufbau bedingen. Zwar sind alle Saponine Glykoside, doch können die Aglykone sowohl der Steroid- als auch der Triterpen- oder der Steroidalkaloidreihe angehören. Dementsprechend unterscheidet man Steroid-, Triterpen- und Steroidalkaloidsaponine. Letztere Gruppe, auch als basische Steroidsaponine bezeichnet, wird hier nicht abgehandelt (siehe *389, 433*).

Während die Struktur vieler Aglykone schon seit längerer Zeit gut bekannt ist, konnte die Strukturermittlung der Saponine selbst wegen der schwierigen Reindarstellung und des komplexen Aufbaus erst in den letzten 10 Jahren vorankommen. Die genauere Untersuchung der Struktur und Eigenschaften vieler Vertreter ließ mehr und mehr die bisherige Definition des Saponinbegriffs problematisch erscheinen (*590, 595*). Während Saponine mit nur einer Zuckerkette vorwiegend am C-3 des Aglykons (Monodesmoside) die typischen Saponineigenschaften meist ausgeprägt zeigen, fehlen diese vielfach bei Glykosiden mit zwei unabhängigen Zuckerketten (Bisdesmoside) mit Ausnahme der Oberflächenaktivität (*516, 590*). Die Bisdesmoside können aber durch eine leicht vorzunehmende Zuckerabspaltung in die Monodesmoside mit den typischen Saponin-Eigenschaften übergehen. Es wäre daher wenig sinnvoll, sie nicht zu den Saponinen zu rechnen. Wir schlagen daher vor (*590, 595*), solange noch eine Definition allein aufgrund von chemischen Strukturmerkmalen nicht möglich ist, solche Glykoside mit ubiquitären Zuckern und terpenoiden penta- oder tetracyclischen Aglykonen als Saponine zu bezeichnen, die entweder selbst hämolytisch wirksam sind oder durch Hydrolyse in hämolytisch aktive Substanzen übergehen können. Damit wird die Hämolyse als eine leicht zu bestimmende Eigenschaft zu dem wesentlichen Saponincharakteristikum erhoben (siehe Abschnitt VI. 2.).

Saponine sind im Pflanzenreich außerordentlich verbreitet. So fanden Gubanov, Libizow und Gladkikh (*172*) bei der systematischen Untersuchung zentralasiatischer Pflanzen, daß von 1730 Spezies aus 104 Familien allein 627 Triterpen- und 127 Steroidsaponine aufwiesen. Insgesamt enthielten 79 (75,9%) der untersuchten Familien Saponine. In Anbetracht dieser weiten Verbreitung und des vergleichsweise hohen Gehalts von 0,1 bis 30% in den Pflanzen bzw. in einzelnen Pflanzenteilen dürften Saponine wohl zu der am häufigsten auftretenden Gruppe von sekundären Pflanzeninhaltsstoffen zählen.

In diesem Übersichtsartikel soll vornehmlich auf die Strukturermittlung der Saponine und auf den Zusammenhang von Struktur und Eigenschaften eingegangen werden. Ältere Untersuchungen über die Natur der Saponine und ihre pharmazeutische Verwendbarkeit finden sich vor allem in Koflers 1927 erschienenem Buch „Die Saponine" (*267*) sowie in den 1955 bzw. 1964 bekannt gewordenen Zusammenfassungen über Triterpene von Steiner und Holtzem (*468*) sowie von Boiteau, Pasich und Ratsimamanga (*40*) weitgehend vollständig, so daß hierauf in diesem Aufsatz nicht näher eingegangen zu werden braucht. Auch die Chemie der meist schon länger bekannten Aglykone ist in einigen ausgezeichneten Übersichten behandelt worden, so daß hier nur die bisher in Saponinen gefundenen Aglykone in tabellarischer Form mit ihren physikalischen Konstanten wiedergegeben zu werden brauchen.

II. Die Reindarstellung und Strukturermittlung der Saponine

1. Isolierung und Reindarstellung

a) Allgemeines

In saponinführenden Pflanzen kommen im allgemeinen recht erhebliche Mengen an Saponin vor, so daß die Rohsubstanz für eine Isolierung meist reichlich vorhanden ist. Die erheblichen Schwierigkeiten bei der Gewinnung reiner Verbindungen beruhen auf der starken Tendenz zur Komplexbildung der Saponine (z. B. mit phenolischen Substanzen), die eine Abtrennung ähnlicher Begleitsubstanzen oft erschwert. Zum anderen kommen die Saponine meist als komplizierte Mischungen sehr schwer trennbarer Einzelverbindungen in der Pflanze vor. Diese können sich sowohl im Aglykon als auch im Zuckerteil (Art, Anzahl und Verknüpfungsort der Zucker) unterscheiden. Gelegentlich sind die Komponenten des Gemisches einander so ähnlich, daß bis heute keine Auftrennung gelungen ist.

Eine für alle Saponinpflanzen gleichartige Isolierungs- und Reinigungsmethode läßt sich nicht angeben. Eine dem Einzelfall angepaßte Kombination der nachfolgend aufgeführten Methoden hat sich jedoch in vielen Fällen gut bewährt. Zweckmäßig unterteilt man sie in zwei Hauptgruppen:

A. Die Gewinnung eines Rohsaponins, das möglichst keine nichtsaponinischen Bestandteile mehr enthält.

B. Die Auftrennung des Saponingemisches in die Einzelkomponenten, die in der Regel die Anwendung chromatographischer Methoden erfordert.

b) Die Gewinnung des Rohsaponins

Die übliche Gewinnung eines Rohsaponins verläuft über die in geeigneter Weise durchgeführte Extraktion des Pflanzenmaterials, eine Verteilung zwischen verschiedenen Lösungsmitteln zur Anreicherung und eine sich anschließende Fällung oder eine ähnliche Reinigungsoperation (siehe Schema 1).

Bei der Extraktion muß von Anfang an streng darauf geachtet werden, daß auch wirklich die genuin in der Pflanze vorhandenen Inhaltsstoffe erfaßt werden. Von den möglichen Veränderungen der Saponine bei der Extraktion spielen in wäßrigem Milieu vor allem enzymatische Zuckerabspaltungen eine Rolle, wie dies u. a. an den Saponinen von *Hedera helix* (*590*), *Gypsophila paniculata* (*259*) und *Digitalis lanata* (*218*) nachgewiesen

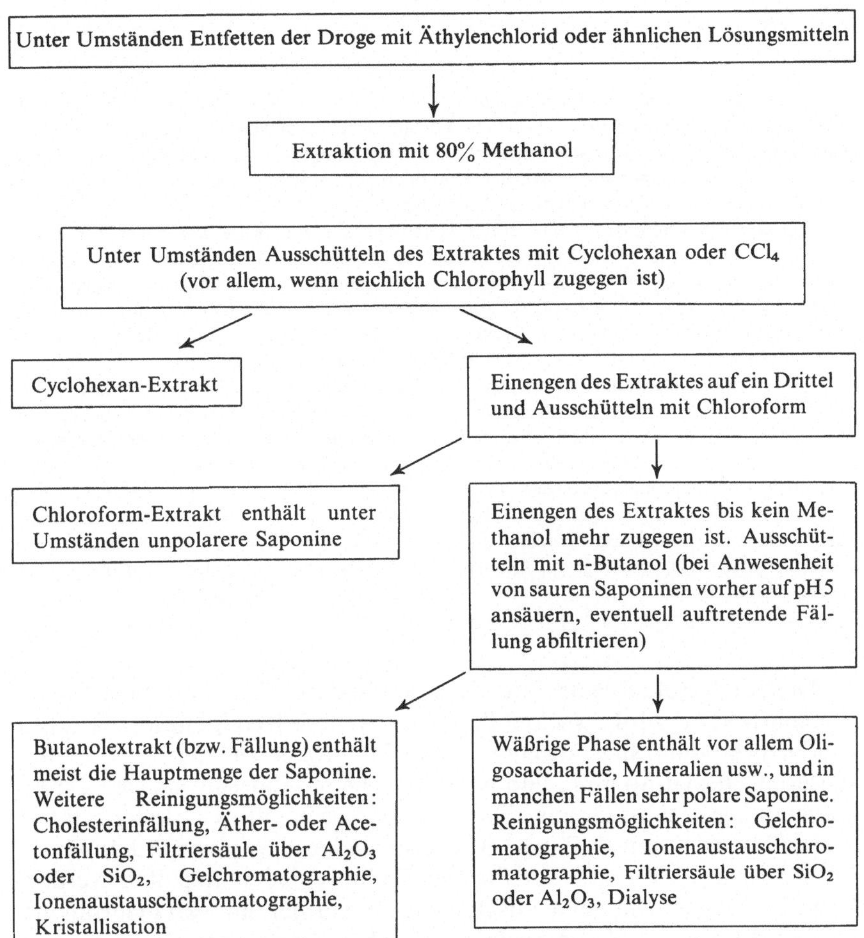

Schema 1. Typischer Gang der Isolierung einer Saponinfraktion (Rohsaponin)

wurde. Zur Extraktion benutzte Alkohole können saure Saponine in die Alkylester überführen (*579*), vorhandene Estergruppierungen können verseift werden (*594*) oder auch eine Acylwanderung erleiden (*564, 594*). Obgleich die Saponine als relativ stabile Pflanzeninhaltsstoffe gelten, ist daher bei der Aufarbeitung Vorsicht am Platze. Diese ist schon bei der Sammlung und Trocknung der Droge geboten. Es ist dabei empfehlenswert, mögliche Umwandlungen während der Trocknung und Lagerung der Pflanzen durch Vergleich mit Frischmaterial anhand papier- und dünnschichtchromatographischer Untersuchungen zu kontrollieren.

Die Extraktion der Saponine wird meist mit 70—90% Methanol oder Äthanol vorgenommen, in dem alle Saponine, zumindest im Gemisch vorliegend, gut löslich sind. Enzyme kommen unter diesen Bedingungen nicht zur Wirkung.

Bei stark ölhaltigen Drogen empfiehlt sich ein vorheriges Auspressen oder ein Entfetten mit Äthylenchlorid oder ähnlichen Lösungsmitteln. Stark chlorophyllhaltige Blattdrogen werden am besten frisch extrahiert und anschließend der Extrakt sofort mit Cyclohexan oder CCl_4 ausgeschüttelt. Bei thermostabilen Saponinen ist auch eine direkte Wasserextraktion bei $100°$ günstig, wobei das Chlorophyll ungelöst bleibt.

Der konzentrierte Extrakt wird durch Verteilung mit Chloroform und anschließend mit n-Butanol weiter aufgetrennt. Dabei erhält man die Hauptmenge der Saponine in die Butanolphase, unpolarere Anteile finden sich teilweise im Chloroform und sehr stark polare in der wäßrigen Phase. Aus der letzteren können sie mit Vorteil durch Gelchromatographie an Sephadex G-25 gewonnen werden; dabei werden sie mit der Front eluiert, während begleitende Oligosaccharide und Salze stärker zurückgehalten werden.

Saure Saponine können beim Ansäuern der konzentrierten wäßrigen Lösung oft in Form einer Fällung erhalten werden. Weiterhin kann man sie aus der wäßrigen Lösung mittels Ionenaustauschern reinigen. Anionenaustauscher adsorbieren die sauren Glykoside, durch Essigsäure oder NaCl können sie wieder eluiert werden (*142, 415*). Man kann sie aber auch über Kationenaustauscher filtrieren, wobei sie in die freie Säure übergehen und viele Verunreinigungen festgehalten werden (*221, 581*). Trennungen ähnlicher Saponine analog wie bei Peptiden und Proteinen sind an Ionenaustauschern bisher nicht beschrieben worden. Bei der Verwendung von Ionenaustauschern können gelegentlich Veränderungen am Saponin auftreten (*194, 360*), so wird z. B. eine Acylbindung zum Zucker in bisdesmosidischen Triterpensaponinen durch Dow-1 gespalten (*360*).

Die durch Butanolextraktion oder auf andere Weise gewonnene Rohsaponinfraktion wird zweckmäßig durch weitere Reinigungsoperationen von Nichtsaponin-Substanzen getrennt. Ein großer Teil der monodesmosidischen Steroidsaponine und auch manche monodesmosidische Triterpensaponine lassen sich aus alkoholischer Lösung mit Cholesterin fällen (*533, 578*). Die zum Teil gut kristallisierenden Cholesteride können in Pyridin/Äther (*432*) sowie in heißem Toluol wieder gespalten werden. Diese Fällung erbringt eine ausgezeichnete Abtrennung von nichtsaponinischem Material und von bisdesmosidischen Saponinen.

Auch die Fällung der Saponine mit Äther oder Aceton aus methanolischer Lösung erbringt einen Reinigungseffekt. Speziell unpolarere Substanzen werden abgetrennt, während polarere Verunreinigungen (z. B. Oligosaccharide) mit ausgefällt werden.

Die in manchen Fällen schwierige Befreiung von phenolischen Verunreinigungen (Flavonen, Flavonolen bzw. ihren Glykosiden usw.) läßt sich durch Filtration über durch Wasser desaktiviertes Al_2O_3 gut erreichen (Verhältnis Substanz/$Al_2O_3 = 5 - 10 : 1$).
In manchen Fällen ist eine sehr gute Reinigungsmöglichkeit in der Kristallisation gegeben. Die Kristallisationstendenz der Saponine ist sehr unterschiedlich. Während viele vor allem bisdesmosidische Saponine selbst in chromatographisch einheitlichem Zustand sich nicht zur Kristallisation bringen lassen, können andere aus angereicherten Fraktionen direkt

kristallin abgeschieden werden (z. B. Aescin, Theasaponin, Primulasaponin, Cyclamin, Digitonin, Lanatasaponin, α-Hederin, Parillin). Vielfach kristallisiert dabei aber eine Mischung von sehr ähnlichen Verbindungen aus, die durch Kristallisation allein nicht weiter aufgetrennt werden kann. Gelegentlich ist jedoch durch mehrfache Umkristallisation ein chromatographisch einheitliches und reines Saponin gewonnen worden [Cyclamin (**340**), Parillin (**103**), Primulasaponin (**352**)].

c) Auftrennung von Saponingemischen

Die Reindarstellung der einzelnen Saponine muß mit Ausnahme der wenigen Fälle, in denen diese durch Kristallisation möglich ist, durch chromatographische Methoden erfolgen. Dabei ist es wesentlich, ein möglichst weit vorgereinigtes Rohsaponin zu verwenden, da der Trenneffekt häufig durch Begleitsubstanzen ungünstig beeinflußt wird. Die Möglichkeiten der präparativen Trennung werden zweckmäßigerweise durch vorhergehende analytische Untersuchungen, z. B. mit Hilfe von Dünnschicht- (D.C.) und Papierchromatographie (P.C.), geprüft. Hierbei wird heute ganz überwiegend die Dünnschichtchromatographie an Kieselgel bevorzugt. Durch die Wahl geeigneter Lösungsmittelgemische (siehe Tabelle 1) gelingt es in den meisten Fällen, befriedigende Trennungen zu erhalten, die denen der Papierchromatographie in der Regel überlegen sind. Bei Verwendung der D.C. im Durchlaufverfahren mit einem vergleichsweise unpolaren Lösungsmittelgemisch oder mit Hilfe der Keilstreifentechnik (*466*) können die Auftrennungen in schwierigen Fällen weiter verbessert werden.

Für den Erfolg der Dünnschichtchromatographie ist die Vorbehandlung der Platten wesentlich. Da es bei der Trennung der vergleichsweise polaren Saponine sich überwiegend um eine Verteilungschromatographie handelt, ist ein gewisser Wassergehalt des Kieselgels erforderlich. Dieser läßt sich durch Trocknen der frischgestrichenen Platten an der Luft (6—8 Stunden) erreichen oder kann besser durch Aufbewahren der bei 130° aktivierten Platten an der Luft erzielt werden. Zur Sichtbarmachung der Saponine auf der Platte werden vor allem aggressivere Sprühreagentien (50% H_2SO_4, Trichloressigsäure, Chlorsulfonsäure/ Eisessig 1 : 2, Chromschwefelsäure usw.) verwendet. Auch mit Erythrocytensuspension oder Blutgelatine gelingt der Nachweis auf D.C.-Platten gut (*466*), jedoch werden auf diese Weise nur hämolytisch aktive Saponine erfaßbar (siehe Kapitel VI 2).

Die erreichten Trennungen auf der D.C.-Platte lassen sich im allgemeinen direkt auf die Säulenchromatographie an Kieselgel (Korngröße 63 bis 120 μ) übertragen und sind denen auf der Platte durchaus vergleichbar. Ein Verhältnis 1 : 100 (Substanz zu Kieselgel) ist empfehlenswert.

Tabelle 1. *Bewährte chromatographische Methoden zur Reindarstellung und Reinheitsprüfung*

Adsorbens	Eluens	besonders geeignet für	Literatur
Dünnschicht- und Säulenchromatographie			
Kieselgel	Chloroform/Methanol/Wasser 65 : 20(—30) : 10	unpolare Saponine	(*518, 530*)
Kieselgel	Chloroform/Methanol/Wasser 65 : 35 : 10	neutrale Saponine unpolare saure Saponine	(*224*)
Kieselgel	Essigester/Methanol/Wasser 70 : 15 : 15	saure und neutrale Saponine	(*561*)
Kieselgel	n-Butanol/Äthanol/Wasser	saure Saponine	(*236*)
Kieselgel	n-Butanol/Äthanol/25% NH$_3$-Lösung	polare saure Saponine	(*236*)
speziell hergestellte Kieselsäure	Chloroform/Methanol/Wasser 65 : 35 : 10	saure Saponine	(*453*)
Al$_2$O$_3$ wassergesättigt	Toluol/n-Butanol 2 : 1 bis 1 : 4	neutrale Saponine	(*549*)
Papier- und Säulenchromatographie (an Cellulose)			
Papier (Cellulose) Formamidimprägniert	Chloroform/Tetrahydrofuran/Pyridin 10 : 10 : 2/Formamid gesättigt	neutrale Saponine	(*531, 532*)
Papier	Isobutanol/Äthanol/Diäthylamin/Wasser 5 : 5 : 1 : 4	saure Saponine	(*236*)
Gelchromatographie			
Sephadex G-25 oder G-50	Wasser	Saponine verschiedener Molekülgröße	(*246, 548*)

Für Feintrennungen ist Kieselgel im allgemeinen dem Al$_2$O$_3$ als Trägermaterial überlegen. Trennungen an Cellulose sind mitunter mit sehr gutem Erfolg durchgeführt worden, doch ist die Chromatographie an Cellulose-Säulen häufig schwierig reproduzierbar.

Chromatographie an Molekularsieben nach der Molekulargröße oder an Ionenaustauschern aufgrund von Ladungsunterschieden führt im allgemeinen zur Auftrennung in einzelne Gruppen. Bei schwierigen Trennproblemen hat sich auch der Einsatz der arbeitsaufwendigen Craig-Verteilung bewährt (z. B. mit dem Lösungsmittelgemisch Chloroform/Methanol/Wasser 41,5 : 37,5 : 21,0) (*561, 562*).

Eine Zersetzung der Saponine während der Chromatographie an Cellulose oder Sephadex ist nicht zu befürchten, aber auch an Kieselgel sind bislang Veränderungen nicht beobachtet worden. An Al_2O_3 oder Ionen-austauschern muß die Stabilität der in Frage kommenden Verbindungen zunächst geprüft werden.

Die Charakterisierung der chromatographisch einheitlichen Saponine ist problematisch. In vielen Fällen gelingt eine Kristallisation nicht, es ist dann sehr schwierig, Reste von Farbstoffen, Mineralsalzen und anderen Begleitstoffen zu entfernen. Elementaranalysen und Molgewichts-bestimmungen geben in solchen Fällen keine zutreffenden Werte. Falls eine Kristallisation möglich ist, lassen sich Verunreinigungen vielfach leicht abtrennen und, abgesehen von ein oder mehreren Molekülen schwer entfernbaren Kristallwassers oder -alkohols, lassen sich zutreffende Elementaranalysen und Molgewichtsbestimmungen erzielen. Für die Mol-gewichtsbestimmungen hat sich in dem Bereich von 500 bis 2500 die Dampfdruckosmometrie sehr bewährt, die auf etwa 5% genaue Werte liefert. Von den nicht kristallisierbaren Saponinen lassen sich in manchen Fällen kristalline Peracetate erhalten (*258, 518*) und diese analy-sieren.

Schmelzpunkte, auch von den kristallinen Substanzen, sind meist wenig charakteristisch, da es sich häufig um Zersetzungsbereiche handelt. Die optischen Drehwerte jedoch sind unabhängig vom Aggregatszustand und werden durch geringfügige Verunreinigungen meist nicht stark be-einflußt. Doch sind die Unterschiede bei ähnlichen Glykosiden vielfach gering, so daß sie für Identifizierungen nur bedingten Wert besitzen. In die Tabellen der Saponine sind daher bei den isolierten Glykosiden keine physikalischen Konstanten aufgenommen worden. Zum Identitätsver-gleich bei zuckerreicheren Saponinen können im allgemeinen nur der direkte chromatographische Vergleich in der D.C. und P.C. mit verschie-denen Laufmitteln, sowie einige Abbaureaktionen (Bestimmung des Aglykons und der Zucker) dienen. Ob ein isoliertes Saponin wirklich ein-heitlich ist, kann genau genommen erst durch eine Strukturbestimmung festgestellt werden.

2. Strukturermittlung

a) Allgemeines

Die vollständige Strukturermittlung eines Saponins setzt die Klärung folgender Fragen voraus:

1. Struktur des genuinen Aglykons.
2. Art und Anzahl der vorhandenen Zucker.
3. Art und Anzahl eventuell vorhandener, esterartig gebundener Säuren.

4. Anknüpfungsort (bzw. -orte) der esterartig gebundenen Säuren.
5. Verknüpfungsart der Zucker untereinander.
6. Sequenz der Zucker in der Kette (bzw. der Ketten).
7. Ringgröße der Zucker.
8. Konfiguration der glykosidischen Bindungen.
9. Bindungsort (bzw. -orte) der Zucker am Aglykon.

Zur Klärung dieser Fragen wird im allgemeinen mit einer sauren oder enzymatischen Hydrolyse begonnen, danach können das Aglykon identifiziert (1.) und die Zucker qualitativ und quantitativ bestimmt werden (2.). Nach einer alkalischen Hydrolyse lassen sich die abgespaltenen Säuren ermitteln (3.). Aus dem Ergebnis der Methylierung und sauren Spaltung können Rückschlüsse auf die Verknüpfungsart der Zucker untereinander (5.), die Ringgröße der Zucker (7.) sowie den Anknüpfungsort der Zucker am Aglykon (9.) gezogen werden. Partialhydrolysen geben Aufschlüsse auf die Sequenz der Zucker (6.) und die Bestimmung der Molrotation, die enzymatische Spaltbarkeit und NMR-Untersuchungen zeigen, ob α- oder β-glykosidische Bindungen vorliegen (8.).

b) Vollständige Hydrolyse
(Bestimmung des Aglykons, der Zucker und eventuell der Carbonsäuren)

α) *Aglykone*. Die Saponine werden durch Kochen mit verdünnter Mineralsäure in die Monosaccharidbausteine und das Aglykon gespalten. Eventuell vorhandene esterartig gebundene Carbonsäuren werden unter diesen Bedingungen vielfach nur teilweise vom Aglykon abgetrennt und werden daher zweckmäßig durch eine nachfolgende alkalische Hydrolyse in Freiheit gesetzt.

Zur Identifizierung der Aglykone werden sie am besten in die Acetate bzw. bei sauren Sapogeninen in die Methylesteracetate übergeführt. Diese besitzen im allgemeinen niedrigere und schärfere Schmelzpunkte und eignen sich daher zur Identifizierung wesentlich besser. In den Tabellen 4, 5, 10 und 11 sind die bisher aus Saponinen abgespaltenen Aglykone, ihre physikalischen Konstanten und die ihrer Acetate bzw. Methylesteracetate verzeichnet. Sie gehören verschiedenen Steroidtypen sowie dem β-Amyrin-, α-Amyrin-, Lupeol-, Dammaran-, Lanostan- und Hopantyp an.

Handelt es sich um bislang unbekannte Aglykone, so müssen diese nach den üblichen Methoden der Steroid- (*137, 149*) und Triterpenchemie (*212, 342, 344, 456*) in der Struktur geklärt werden. Von besonderem Wert sind hierbei spektroskopische Untersuchungen, gegebenenfalls an geeigneten Derivaten (siehe Tabelle 2).

Chemische Untersuchungen galten vor 1960 fast immer den Aglykonen, die nach üblicher saurer Hydrolyse aus den Glykosiden gewonnen

Tabelle 2. *Spektroskopische Methoden zur Untersuchung von*
Steroidsapogeninen und Triterpenen

Methode	Substanzklasse	Literatur
IR	Steroidsapogenine	*(127, 219, 404, 461, 463, 479)*
	Triterpene	*(132, 463, 464)*
UV	Steroidsapogenine	*(434)*
	Triterpene	*(434)*
NMR	Steroidsapogenine	*(35, 42, 298, 407, 575)*
	Triterpene	*(95, 95a, 132, 203, 206, 206a, 302, 322, 421, 440, 542, 544a)*
Massenspektroskopie	Steroidsapogenine	*(50, 51, 52)*
	Triterpene	*(50, 51, 53, 136)*
ORD, CD	Steroidsapogenine	*(114, 123, 462)*
	Triterpene	*(114, 459a, 462)*
Röntgenstruktur-analyse	Steroidsapogenine	—
	Triterpene	*(4, 19, 196, 297, 367, 472, 486)*

wurden. Vielfach wurden auf diese Weise nicht die heute als genuin anzusehenden Aglykone erfaßt. In der Tabelle 3 sind einige typische Umwandlungen bei der sauren Hydrolyse aufgeführt, außerdem werden auch durch Nebenreaktionen entstandene Derivate erwähnt.

Die Isolierung der genuinen Aglykone gelang in solchen Fällen erst dann, als schonendere Methoden der Hydrolyse verwendet wurden, so z. B. eine mildere saure Hydrolyse *(186, 287, 509)* oder eine Kombination von Perjodatspaltung und Hydrolyse *(130, 446, 451)*.

Eine ausgezeichnete Möglichkeit zur schonenden Spaltung bietet auch die enzymatische Hydrolyse *(89, 215, 273, 458, 562, 594)*. Die Schwierigkeit besteht allerdings darin, daß nicht für alle Zucker geeignete Hydrolasen bekannt sind und daß verzweigte Zuckerketten enzymatisch oft schwerer angreifbar sind. Doch werden mit den komplexen Enzympräparaten aus *Helix pomatia* (Weinbergschnecke) *(89, 562, 594)* bzw. aus niederen Pilzen *(215, 458)* oft vollständige Hydrolysen erreicht.

In manchen Fällen können Zucker auch mit speziellen Bakterien-Kulturen abgespalten werden *(15, 526, 602, 604, 605)*, doch besteht hierbei die erhebliche Gefahr von sekundärer Umwandlung der Aglykone durch die Mikroorganismen.

Gelegentlich läßt sich das genuine Aglykon aus prinzipiellen Gründen nicht isolieren, da es nach Abspaltung eines Zuckers spontan cyclisiert, so z. B. geht das (25 S), 5β-Furostan-3β, 22α,26-triol-derivat (**102**) bei der Hydrolyse in Sarsasapogenin (**1**) über *(516)*.

In solchen Fällen muß durch chemische Reaktionen am Glykosid die Natur des Aglykons ermittelt werden *(516)*. Wesentliche Hilfen bieten hier spektroskopische (IR und NMR) Untersuchungen direkt am Glykosid, am Acetat oder am Persilylderivat *(263, 516, 594)*.

Tabelle 3. *Beispiele für Umwandlungen der genuinen Aglykone bei saurer Hydrolyse*

Genuines Aglykon	Isoliertes Artefakt	Typ der Reaktion	%	Literatur
Diosgenin (6)	$\Delta^{3,5}$-Spirostadien	Wasserabspaltung zur Doppelbindung	5—10	(596)
Digitogenin (62)	Neo-digitogenin (61)	Epimerisierung an C-25	~5	(537)
(25S)-5β-Furostantriol (3β,22α,26)	Sarsasapogenin (1)	Cyclisierung zum Spiroketal	~80	(516)
Polypodogenin (in 145)	kein Aglykon isolierbar	Zersetzung		(215)
Protoaescigenin (246)	Aescigenin (215)	Wasserabspaltung unter Ätherbildung	~80	(287)
Protoaescigenin (246)	Isoaescigenin	Wasserabspaltung zur Δ^{15}-Doppelbindung	~2	(491)
Protobassiasäure (249a)	Bassiasäure (230)	Wasserabspaltung zur Doppelbindung	~80	(251a)
Cyclamiretin A (273)	Cyclamiretin D (181)	Ätherringöffnung unter Doppelbindungsbildung	~90	(509)
Saikogenin G (271)	Saikogenin D (275)	Ätherringöffnung unter Doppelbindungsbildung	~70	(271)
Cyclamigenin B (278)	Cyclamigenin A$_2$ (282)	Decarbonylierung	~5	(128)
Gypsogenin (206)	Gypsogeninlacton (280)	Lactonbildung	~80	(263)
Presenegenin (259)	Senegeninsäure (346)	siehe S. 531		(130, 375)
Presenegenin (259)	Hydroxysenegenin (349)	siehe S. 531		(130, 451)
Presenegenin (259)	Senegenin (347)	siehe S. 531		(130, 131)
(20 S) Protopanaxadiol (308)	Panaxadiol (310)	Ätherbildung	~80	(361)
(20 S) Protopanaxadiol (308)	(20R) Protopanaxadiol (309)	Epimerisierung	~90	(361)
12α-Hydroxy-22,25-oxido-7,8-dihydro-holothurinogenin (325)	22,25-Oxido-holothurinogenin (324)	Wasserabspaltung zur Doppelbindung	~90	(89)
12α-Hydroxy-22,25-oxido-7,8-dihydro-holothurinogenin (325)	12β-Methoxy-22,25-oxido-7,8-dihydro-holothurinogenin (328)	Ätherbildung und Epimerisierung	~50	(80)
Echinocystsäure (187)	$\Delta^{13(18)}$-Oleanen-3β,16α-diol-28-säure	Doppelbindungswanderung	~5	(279)
Primulagenin A (157)	Aegiceradiol	Eliminierung von Wasser und Formaldehyd		(186, 188)
Protoaescigenin-21,22-diester	Protoaescigenin-21,16-diester	Acylwanderung		(594)
Oleanolsäure (167)	Oleanolsäuremethylester	Veresterung		

β) *Zucker*. Als Zucker sind bisher in Saponinen vornehmlich ubiquitäre Zucker wie D-Glucose, D-Galaktose, D-Arabinose, D-Xylose, D-Fucose, L-Rhamnose, D-Chinovose, D-Glucuronsäure und neuerdings auch Ribose (*100*) sicher nachgewiesen worden. Sie werden zweckmäßig mittels Papierchromatographie identifiziert (z. B. *112*). Bei Anwesenheit von Uronsäuren hat sich die Aufarbeitung nach FISCHER und DÖRFEL (*151*) bewährt.

Wesentlich schwieriger ist die quantitative Bestimmung der Zucker. Die früher übliche quantitative Auswertung von Papierchromatogrammen (z. B. *150*) liefert bestenfalls Werte mit etwa 5—10% Fehler für die zu bestimmenden Monosaccharide. Wesentlich günstiger ist die quantitative Bestimmung mittels Gaschromatographie in Form der persilylierten Zucker (*241, 423, 473, 488, 589*), der persilylierten Methylglykoside (*474, 589*) oder der peracetylierten Zuckeralkohole (*5, 424*). Günstig ist auch die automatisierbare Chromatographie am Ionenaustauscher in Äthanol/Wasser (*341*) oder in Boratpuffer (*197*). In allen Fällen ist die eigentliche Analyse auf 1,5−2,5% genau durchführbar, die Hauptfehlermöglichkeiten bestehen in unvollständiger Hydrolyse, teilweiser Zersetzung der Zucker durch die angewandte Säure oder entstehen bei der Neutralisation mit Ionenaustauschern (*197, 589*). Diese Einflüsse können durch Ermittlung von Korrekturkonstanten für jeden Zucker berücksichtigt werden (*197, 589*). Die Hydrolyse erfolgt am besten mit 2−3 n H_2SO_4 in Wasser oder Dioxan/Wasser unter Stickstoff (*197*).

Vorhandene Uronsäuren können ebenfalls gaschromatographisch untersucht werden; bewährt hat sich die Bestimmung einmal nach vorherigem Entfernen der Uronsäuren aus dem Zuckergemisch und zum anderen nach Reduktion der Uronsäure zur korrespondierenden Hexose (*594*).

γ) *Säuren*. Eventuell im Saponin esterartig gebundene Säuren bestimmt man am besten nach alkalischer Hydrolyse qualitativ und quantitativ durch Gaschromatographie der freien Säuren (*191, 289, 471, 594*) oder ihrer Methylester (*594*). Schwerer flüchtige Säuren wie z. B. die 4-Methoxyzimtsäure und die 3,4-Dimethoxyzimtsäure aus Senegin können auch dünnschicht- (*522*) und säulenchromatographisch (*451, 522*) getrennt werden.

Weitere Einzelheiten siehe Abschnitt IV 4.

c) Methylierung

Die wohl wichtigste Methode zur Ermittlung der Zuckerverknüpfung in einem Glykosid ist die Permethylierung, die anschließende saure Spaltung und die Identifizierung der gebildeten, ganz oder teilweise methylierten Monosaccharide. Die Hauptschwierigkeiten liegen dabei im

Erreichen einer wirklich vollständigen Permethylierung und in der einwandfreien Identifizierung der Methylzucker.

Es sind zahlreiche Methoden zur Permethylierung vorgeschlagen worden (Zusammenfassung siehe *571*). Besonders bewährt hat sich die Permethylierung nach Hakomori (*176*) mit Methyljodid in Dimethylsulfoxid in Gegenwart von Natriumhydrid (*8, 530*). Zu beachten ist hierbei jedoch, daß reduzierbare Gruppen im Aglykon (Aldehyd- oder Ketofunktion) teilweise oder vollständig reduziert werden.

Gute Permethylierungen werden auch nach Kuhn und Mitarb. in Dimethylformamid in Gegenwart von $BaO/Ba(OH)_2$ oder Ag_2O mit Methyljodid (*285, 532*) oder mit Dimethylsulfat (*292*) erzielt. Gelegentlich hat sich auch eine Nachmethylierung nach Purdie (*390*) mit CH_3J/Ag_2O als günstig erwiesen (*259*). Esterartig gebundene Carbonsäuren werden bei Methylierungen meist abgespalten und durch Methyl ersetzt. In diesem Fall ist die Verwendung von Diazomethan/BF_3 zur Methylierung günstig (*341a*); wobei Acylreste nicht abgespalten werden.

Zur vollständigen Methylierung muß häufig mehrfach methyliert werden. Es ist außerordentlich wichtig, die Vollständigkeit des Umsatzes sorgfältig zu kontrollieren. Hierfür eignen sich die IR-Spektroskopie zur Prüfung auf OH-Restschwingungen, die Dünnschichtchromatographie und die Methoxylgruppenbestimmung (*571*). Meist empfiehlt es sich, das Methylierungsprodukt säulenchromatographisch aufzutrennen und die unpolarste Hauptfraktion genauer zu untersuchen. Da in einigen Aglykonen bestimmte OH-Gruppen (z. B. 16α-OH in β-Amyrin-Derivaten) sehr schwer methyliert werden, gelingt in den entsprechenden Saponinen eine vollständige Methylierung sehr schlecht und in geringen Ausbeuten. Da eine vollständige Verschließung der freien OH-Gruppen des Aglykons jedoch oft nicht nötig ist, muß man in solchen Fällen lediglich prüfen, ob in der Zuckerkette alle Hydroxyle vollständig mit Methyl besetzt sind (siehe z. B. *512, 517*).

Die Methodik der Hydrolyse des Methylierungsproduktes und der Aufarbeitung der Methylzucker geht im wesentlichen auf die grundlegenden Arbeiten von Kuhn und Mitarb. an Steroidalkaloidglykosiden zurück (*290, 291*). Man unterwirft das permethylierte Glykosid einer Methanolyse in 5% methanolischer HCl. Anschließend wird nach Zusatz von Wasser das Methanol entfernt, das ausgefallene Aglykon abfiltriert und die Methylglykoside in der wäßrigen Lösung sauer hydrolysiert. Die Methylzucker werden nach üblichen Methoden säulen-, papier- oder gaschromatographisch aufgetrennt. Eine vorläufige Identifizierung kann mit Vergleichssubstanzen papier-, dünnschicht- und gaschromatographisch (*16, 184, 284, 373, 571*) sowie massenspektroskopisch (*38, 184, 189, 257, 266, 373*) erfolgen. Besondere Vorteile bringt die Kombination von Gaschromatographie und Massenspektroskopie (*306a*).

Literaturverzeichnis: SS. 576—606

Eine sichere Bestimmung erfordert meist die Isolierung des Methylzuckers in kristalliner Form bzw. als ein kristallisiertes Derivat und den Vergleich mit authentischer Substanz. Eine Zusammenstellung der bisher bekannten Methylzucker und einiger zur Identifizierung geeigneter Derivate findet man: D-Glucose (*43*), D-Galaktose (*32, 318*), D-Xylose (*300, 317*), L-Arabinose (*300, 317, 576*), D-Fucose (*300, 317, 465*), L-Rhamnose (*300, 317*), D-Chinovose (*402*) und D-Ribose (*300, 317*).

Die Methylzucker der Uronsäuren werden meist nach vorheriger Reduktion zur Hexose identifiziert, da die entsprechenden D-Glucose- und D-Galaktose-Derivate besser bekannt und charakterisiert sind.

Beim Auftreten unbekannter Methylzucker müssen diese mit den Methoden der Zuckerchemie in der Struktur geklärt werden. Hierfür sind insbesondere die NMR-Spektroskopie und die Massenspektroskopie (siehe z. B. *398a, 402, 549a*) von großem Nutzen.

Die Identifizierung der Methylzucker gibt Aufschluß über die Verknüpfungsweise der Zucker untereinander, die Natur der endständigen Zucker und über eventuell auftretende Verzweigungen in der Kette. Auch läßt sich beim Auftreten von zwei endständigen Zuckern und keiner nachweisbaren Verzweigung auf das Vorliegen von zwei unabhängigen Zuckerketten schließen. In vielen Fällen geben die isolierten Methylzucker auch Aufschluß über die Ringgröße der Zucker. Bemerkenswerterweise liegt nicht nur die Arabinose in manchen Saponinen furanoid vor (*259, 264*), sondern auch Glucose findet sich im Patrinosid C_1 in der ungewöhnlichen α-furanosidischen Verknüpfung (*60*).

Die Untersuchung des bei der Hydrolyse des Permethylierungsproduktes erhaltenen Aglykons läßt wesentliche Schlüsse auf den Anknüpfungsort (bzw. -orte) der Zucker am Aglykon zu. Zur Lokalisierung der freien OH-Gruppe(n) kann die Verbindung in das Acetat übergeführt und dann mittels NMR- oder Massenspektroskopie untersucht werden. Günstig ist auch die Oxidation der OH-Gruppe zum Keton und die Untersuchung des Oxidationsproduktes durch IR- (*259*), NMR- (*518*) oder Massenspektroskopie (*259, 518*) sowie mittels Circulardichroismus oder Optischer Rotationsdispersion (*246*).

Der Anknüpfungsort von esterartig gebundenen Carbonsäuren kann im allgemeinen mit Hilfe der Methylierung nicht geklärt werden (vgl. aber *341a, 453*). Auf die Problematik der Lokalisation von Estergruppen und die sie erschwerenden Acylwanderungen wird im Abschnitt über Estersaponine (IV 4) näher eingegangen.

d) Partialhydrolysen

Das Ergebnis der Methylierung läßt keine Schlüsse über die Sequenz der Monosaccharide in einer längeren Zuckerkette zu, sie wird am besten durch Partialhydrolyse ermittelt. So lassen sich Saponine mit verdünnten Mineralsäuren oder mit 10% Oxalsäure (*263*) so hydrolysieren, daß die Zuckerkette in mehr oder weniger große Bruchstücke zerfällt. Man kann dann sowohl die gebildeten Oligosaccharide (*532*) als auch die entstandenen zuckerärmeren Glykoside (*230, 513, 530*) oder auch beide (*263*) isolieren.

Die Trennung der Oligosaccharide erfolgt durch Papierchromatographie (z. B. *497*), Cellulose-Säulenchromatographie (z. B. *359*) oder an Kohle-Celite-Säulen (*574*). Sind die Oligosaccharide (z. B. Disaccharide) bekannt, liefert diese Kenntnis nicht nur eine Teilsequenz, sondern auch gleichzeitig partiell die Ringgröße, die Konfiguration der glykosidischen Bindung und eine Bestätigung der Verknüpfungsstelle am Zucker. Sind sie jedoch unbekannt, muß ihre Struktur mit den üblichen Methoden der Kohlenhydratchemie geklärt werden (z. B. *283, 532*). Die zuckerärmeren Glykoside werden nach Trennung mit den für Saponine beschriebenen Methoden untersucht.

Zur Gewinnung zuckerärmerer Glykoside eignet sich in vielen Fällen auch eine Partialhydrolyse mit Enzymen. Enzyme mit spezifischer β-Glucosidase- oder β-Galaktosidaseaktivität können zur gezielten Abspaltung von einzelnen Monosaccharideinheiten benutzt werden. Häufig ist es jedoch günstiger, komplexe Enzymgemische z. B. aus Mikroorganismen oder *Helix pomatia* einzusetzen und die nach bestimmten Zeiten erhaltenen Partialhydrolysate aufzutrennen (*215, 516, 517, 518, 526*).

Eine weitere Möglichkeit, zuckerärmere Glykoside zu gewinnen, besteht im Smith-Abbau (*1*), bei dem die Zuckerkette mit Perjodat gespalten, das Reaktionsprodukt anschließend mit $NaBH_4$ reduziert und mit sehr verdünnter Säure gespalten wird. Mit dieser Methode werden unverzweigte Zuckerketten abgebaut, lediglich Zuckerreste, in denen durch Verzweigung keine Glykolstrukturen existieren, bleiben erhalten. Diese Methode leistet vor allem bei zuckerreicheren Saponinen gute Dienste (*261, 262*). Längere Oligosaccharid-Bruchstücke können durch den Abbau des peracetylierten Saponins in HBr-Eisessig erhalten werden (*219, 497, 517, 532*). In manchen Fällen gelingt es hierbei, fast alle möglichen Bruchstücke der Zuckerkette zu erhalten, so daß die Sequenz leicht durch gedankliche Kombination der Bruchstücke ermittelt werden kann (*532*). Ein Beispiel hierfür ist bei der Strukturermittlung des Digitonins (siehe S. 495) beschrieben. Speziell zur Spaltung der in den bisdesmosidischen Triterpensaponinen vorkommenden Acyl-glykose-Bindung eignet sich die Reduktion mit $LiAlH_4$. Hierfür setzt man am besten das bereits permethylierte Glykosid ein und erhält ein reduziertes methyliertes Oligosaccharid, das vorher sich als Ester an der Carboxylgruppe befand und ein Restglykosid mit der methylierten Zuckerkette an C-3 (Beispiele siehe Strukturaufklärung von Kalopanaxsaponin B und Gypsosid A S. 540 und 553).

Bei den Glykosiden selbst läßt sich die Acylglykose-Bindung auch selektiv mit Alkali (*60, 109, 239, 258, 259*), mit Anionenaustauscher DOW-1 (*360*), mit Enzym (*590*) oder mit LiJ in Kollidin (*259*) spalten, jedoch erhält man auf diese Weise allein das monodesmosidische Saponin mit den Zuckern an C-3, während das Oligosaccharid von der Carboxylgruppe in der Regel nicht isolierbar ist.

e) Konfiguration der glykosidischen Bindung

Nach Feststellung der Verknüpfungsart, der Ringgröße und der Sequenz der Monosaccharide in der Zuckerkette bzw. den Zuckerketten bleibt die Konfiguration der jeweiligen glykosidischen Bindungen festzustellen. Ihre sichere Bestimmung ist insbesondere für zuckerreichere Glykoside methodisch bis heute nicht einwandfrei gelöst.

Bei Glykosiden mit ein bis zwei in der Konfiguration der Verknüpfung unbekannten Monosaccharid-Bausteinen läßt sich diese durch Vergleich der Molrotationen der Glykoside, der Aglykone und der α- und β-Methylglykoside der betreffenden Zucker nach KLYNE (*256*) ermitteln. Doch ist bei dieser Methode Vorsicht geboten, da bei verzweigten Zuckerketten die Drehwerte stärkere Abweichungen zeigen (siehe z. B. *517*). Für mehr als zwei in der Konfiguration unbekannte Monosaccharid-Bausteine in einer Zuckerkette ist das Verfahren im allgemeinen nicht mehr anwendbar und kann zu fehlerhaften Schlüssen führen (*306*).

Eine andere Möglichkeit zur Feststellung der Konfiguration der glykosidischen Bindung besteht in der Prüfung auf Spaltbarkeit durch spezifische Enzyme z. B. durch β-Glucosidase, β-Galaktosidase oder β-Glucuronidase (*215, 357, 516, 517, 526, 532, 594*). Nachteilig ist, daß nicht für alle Monosaccharide die erforderlichen Enzyme greifbar sind und daß verzweigte Zuckerketten enzymatisch oft schwerer angreifbar sind.

Sehr günstig ist eine Untersuchung von isolierten Di- oder Oligosacchariden. Soweit sie bekannt sind, erhält man direkt die Konfiguration an der glykosidischen Bindung (*259, 263, 497, 517, 530, 532*), sofern nicht, läßt sie sich meist leicht mit Hilfe enzymatischer Methoden, der Molrotation oder von NMR-Messungen bestimmen (*290, 291, 517, 530, 532*). Die NMR-Methode beruht auf dem Auftreten von verschiedenen Koppelkonstanten für das Signal des Protons am C-1 bei manchen Zuckern [z. B. bei α-glucosidischer Bindung (~ 3 Hz) und β-glucosidischer Bindung (6 − 7 Hz)]. Die Koppelkonstante läßt sich wegen Überlagerung der Signale nur bei bis zu zwei Zuckern im Molekül, also bei Disacchariden oder Diglykosiden, anwenden. Bis vor kurzem war auch bei Saponinen die Regel gültig, daß in pflanzlichen Glykosiden D-Zucker fast immer β- und L-Zucker α-glykosidisch gebunden sind. Neuerdings wurden allerdings α-D-glucosidische (*58, 60, 103, 106, 205, 553*), α-D-galaktosidische (*99, 110, 304, 305*) und β-L-arabinosidische Bindungen (*63*) festgestellt, die mit dieser Regel nicht im Einklang stehen. Leider ist jedoch in allen Fällen die Konfiguration nach KLYNE mit Hilfe der Molrotation ermittelt worden, eine Absicherung mit enzymatischen Methoden oder durch Isolierung von Oligosacchariden erscheint dringend erforderlich.

Nach der vollständigen Ermittlung des Aufbaus der Saponine bleibt noch die Frage nach der Konformation dieser Verbindungen offen. Während eine größere Zahl von Aglykonen röntgenspektroskopisch untersucht sind, liegen solche Arbeiten für die komplizierter gebauten Glykoside noch nicht vor. Kenntnisse der Konformation der Saponine könnten unter Umständen Hinweise zur Erklärung spezifischer Eigenschaften dieser Stoffklasse liefern.

III. Die Struktur der neutralen Steroidsaponine

1. Allgemeine Bemerkungen

Neutrale Steroidglykoside mit Saponincharakter werden nach Struktur und Eigenschaften zweckmäßig in zwei Hauptgruppen unterteilt. Die erste, schon lange bekannte und wohl am weitesten verbreitete, enthält als Aglykon Steroide vom Spirostantyp und eine meist komplexer gebaute Zuckerkette am OH vom C-3 des Aglykons. Sie zeigt die typischen Eigenschaften der Saponine, wie Schaumvermögen, hämolytische Aktivität usw. (siehe Abschnitt VI), in ausgeprägtem Maße. Die zweite Hauptgruppe enthält Steroide als Aglykon, die sich formal durch Öffnen des Ringes F der Spirostanole bilden können und die dem sogenannten Furostanoltyp angehören. Sie besitzen neben einer Zuckerkette analog derjenigen der Saponine des Spirostanoltyps am OH an C-3 eine zweite an einem OH an C-26, die in den bisher untersuchten Fällen nur aus einer D-Glucose-Einheit besteht. Sie wurden von uns als bis-desmosidische Glykoside bezeichnet. Sie gehen bei der Abspaltung der Glucose unter gleichzeitigem Ringschluß in die monodesmosidischen Spirostanolglykoside über. Der Furostanoltyp zeigt außer dem Schaumbildungsvermögen die anderen typischen Eigenschaften der Saponine kaum ausgeprägt. Neben diesen beiden Hauptgruppen finden sich im Pflanzenreich noch zwei Nebengruppen, bei denen im Aglykon weitere Variationen in bezug auf Ringbildungen aus den 8 C-Atomen an C-17 im Ring D der Steroide vorliegen (siehe S. 492 und 502).

Das Vorkommen offenkettiger Glykoside in der Natur hatten schon 1947 R. E. MARKER und J. LOPEZ (*331*) ohne nähere Angaben postuliert, ihre Existenz blieb jedoch fraglich. Das Problem konnte erst geklärt werden, nachdem es gelang, derartige Verbindungen zu isolieren und ihre Eigenschaften zu studieren. Wir haben gegenüber der üblichen Bezeichnung der Saponine mit der Endung -in, wie Digitonin, Parillin usw. zur Charakterisierung des neuen Furostanoltyps die Endung -osid gewählt wie Sarsaparillosid, Convallamarosid usw. (*516*), während die japanischen Autoren S. KIYOSAWA *et al.* (*254*) sich für die Vorsilbe Proto entschlossen haben, die dem Namen des zugehörigen Spirostanolsaponins vorangestellt wird, z. B. Protogracillin.

Die nahe verwandten Steroidalkaloidglykoside vom Typ des Solanins und Tomatins sollen hier außer Betracht bleiben; sie bilden trotz naher chemischer Verwandtschaft durch den basischen Charakter des Stickstoffs eine Sondergruppe.

2. Tabellen der neutralen Steroidsapogenine

In den Tabellen 4 und 5 sind die in der Natur aufgefundenen neutralen Steroidsapogenine nach steigendem Sauerstoffgehalt verzeichnet. Die Tabelle 4 enthält Sapogenine vom Spirostantyp, die Tabelle 5 abgewandelte Verbindungen dieses Typs. Früher erschienene Zusammenstellungen dieser Art (137, 149) wurden hierfür mit herangezogen. Bei den Sapogeninen mit einer Hydroxylgruppe an C-25 ist nicht sicher, ob sie genuiner Natur sind, da bei der Säurehydrolyse mit ihrer Bildung aus Verbindungen des Nuatigenin-Typs gerechnet werden muß.

3. Furostanolglykoside

Der erste bekannte Vertreter dieser Stoffklasse ist das Sarsaparillosid (**102**, *515, 516*), das zusammen mit Parillin (**103**) und Desglucoparillin (**129**) in den Wurzeln von *Radix sarsaparillae* aufgefunden wurde. Während Parillin und sein Desglucoderivat (siehe S. 499f.) zum Spirostanoltyp gehören, ist in den Furostanolderivaten der endständige Sauerstoffring geöffnet und die nunmehr freie Hydroxylgruppe glykosidisch durch D-Glucose verschlossen. Bei der Abspaltung des Zuckers, sei es durch Säure oder enzymatisch durch β-Glucosidasen, schließt sich der Ring sofort und ein wahres Spirostanolderivat entsteht. Da in Pflanzen Enzyme derartiger Wirkung nachgewiesen worden sind, ist zur Isolierung der Furostanolglykoside ein Arbeiten unter möglichst enzymhemmenden Bedingungen notwendig, um die Umwandlung zu vermeiden. Die Furo-

Tabelle 4. *Natürlich vorkommende Spirostanole*

Name	Konfiguration		Substituenten			Sapogenin		Acetat[b]		Literatur
	$C_{(25)}$	$C_{(5)}$	OH	Keto-funktion	Doppel-bindung	Schmp.	$(\alpha)_D^a$	Schmp.	$(\alpha)_D^a$	
$C_{27}H_{44}O_3$										
1 Sarsasapogenin	S	β	3β	—	—	198—199°	−75°	143—144°	−70°	(73, 338)
2 Smilagenin	R	β	3β	—	—	187—188°	−66°	150°	−60°	(338)
3 Neotigogenin	S	α	3β	—	—	202—203°	−75°	179°	−79°	(73, 168)
4 Tigogenin	R	α	3β	—	—	205—208°	−67°	206—208°	−74°	(211, 338, 494)
$C_{27}H_{42}O_3$										
5 Yamogenin	S	—	3β	—	5	201°	−129°	182°	−119°	(338, 339)
6 Diosgenin	R	—	3β	—	5	208°	−129°	199—202°	−121°	(334)
7 Macranthogenin	—	β	3β	—	25 (27)	162,6°	−69,5°	138—144°	−63,5°	(380)
$C_{27}H_{40}O_3$										
8 Sceptrumgenin	—	—	3β	—	5, 25 (27)	182—184°	−122°	191—193°	−119°	(152)
$C_{27}H_{44}O_4$										
9 Rhodeasapogenin	S	β	1β 3β	—	—	293—295°	−72°	185—187°	−70,5°	(363, 364)
10 Isorhodeasapogenin	R	β	1β, 3β	—	—	245—248°	−71°	205°	−73°	(265, 363)
11 Markogenin	S	β	2β, 3β	—	—	257°	−70°	185—186°	−84°	(126, 569)
12 Samogenin	R	β	2β, 3β	—	—	205—207°	−86°	198°	−75°	(126, 339)
13 Yonogenin	R	β	2β, 3α	—	—	240—243°	−53°	212°	−29°	(347, 484)
14 Neoyonogenin	S	β	2β, 3α	—	—	198—199°	−63,7°	205—208°	−19,6°	(3)
15 Neogitogenin	S	α	2α 3β	—	—	248°	—	218—220°	—	(330)
16 Gitogenin	R	α	2α, 3β	—	—	264—267°	−78°	242—243°	−97°	(503)
17 Neochlorogenin	S	α	3β, 6α	—	—	261—265°	−65°	200—202°	−51°	(403)
18 Chlorogenin	R	α	3β, 6α	—	—	276°	−45°	155°	−38°	(327, 328)
19 Neonogiragenin	S	β	3β, 11α	—	—	128—131°	−75°	142—144°	−78°	(346)

#	Name										
20	Nogiragenin	R	β	3β, 11α	—	—	200—201°	−71	208—209	−74°	(475)
21	Rockogenin	R	α	3β, 12β	—	—	216—220°	−63°	206—209°	−65°	(338)
22	Neodigalogenin	S	α	3β, 15β	—	—	223—226°	—	211—212°	−88°	(537)
23	Digalogenin	R	α	3β, 15β	—	—	218—219°	−75,5	238—239	−70	(531)
24	Isoplexigenin A	R	α	3β, 23S	—	—	227°	−61°	194—195°	−57°	(152)
25	Isocholegenin	—	β	3α, 25	—	—	250—257°	−65°	187°	−42°	(10, 343)
	C$_{27}$H$_{42}$O$_4$										
26	Ruscogenin	R	—	1β, 3β	—	5	205—210°	−127°	192—194°	−85°	(33, 69, 301)
26a	(25S)-Ruscogenin	S	—	1β, 3β	—	5	212—214°	−112°	182—185°	−88°	(166)
27	Lilagenin	S	—	2α, 3β	—	5	246°	—	155°	—	(137)
28	Yuccagenin	R	—	2α, 3β	—	5	248°	−122°	178°	−139°	(333)
29	Heloniogenin	R	—	3β, 12α	—	5	212—213°	−91°	184—185°	−58°	(371)
30	Chiapagenin	S	—	3β, 12β	—	5	249—251°	−126°	191—193°	−127°	(183, 439)
31	Isochiapagenin	R	—	3β, 12β	—	5	236—237°	−121°	206—207°	−120°	(183)
32	Isoplexigenin B	R	—	3β, 23S	—	5	205—207°	−96°	187—190°	−97°	(152)
33	Isonuatigenin	—	—	3β, 25S	—	5	215—218°	−140°	211°	−109°	(519, 520)
34	Narthogenin	R	—	3β, 27	—	5	214—216°	−112°	144—146°	−98°	(346)
35	Isonarthogenin	S	—	3β, 27	—	5	240—242°	−110°	160—162°	−94°	(346)
36	Convallamarogenin	—	β	1β, 3β	—	25 (27)	259—261°	−79°	214—215°	−88°	(483, 523)
37	Dehydrogitogenin	—	α	2α, 3β	—	25 (27)	266—267°	−80,1°	226—228°	−97,1°	(483)
38	Laxogenin	R	α	3β	6	—	210—212°	−86°	219—222°	−89°	(2, 372)
39	Willagenin	S	β	3β	12	—	166—168°	+5°	183—185°	−1°	(235)
40	Sisalagenin	S	α	3β	12	—	244—246°	−4,5°	228—232°	−12°	(73)
41	Hecogenin	R	α	3β	12	—	268°	+10°	245°	−1°	(566)
42	Hispidogenin	R	α	12β	3	—	208°	−46°	212°	−51°	(319)
	C$_{27}$H$_{40}$O$_4$										
43	Neoruscogenin	—	—	1β, 3β	—	5,25(27)	198—201°	−118°	132—134°	−64°	(321, 418)

Tabelle 4 (Fortsetzung)

Name	Konfiguration		Substituenten			Sapogenin		Acetat[b]		Literatur
	$C_{(25)}$	$C_{(5)}$	OH	Keto-funktion	Doppel-bindung	Schmp.	$(\alpha)_D^a$	Schmp.	$(\alpha)_D^c$	
44 7-Oxodiosgenin	R	—	3β	7	5	215—220°	−170°	197—198,5°	−170°	(167)
45 Tamusgenin	R	—	3β	11	5	180—182°	−75°	209—213°	−81°	(153)
46 Neobotogenin (= Correllogenin)	S	—	3β	12	5	209—211°	−69°	213—215°	−67°	(567, 568)
47 Botogenin (= Gentrogenin)	R	—	3β	12	5	215—216°	−57°	227°	−56°	(567, 568)
48 9-Dehydrohecogenin	R	α	3β	12	9	235°	−10°	227°	−9°	(565)

$$C_{27}H_{44}O_5$$

Name	Konfiguration		Substituenten			Sapogenin		Acetat[b]		Literatur
	$C_{(25)}$	$C_{(5)}$	OH	Keto-funktion	Doppel-bindung	Schmp.	$(\alpha)_D^a$	Schmp.	$(\alpha)_D^c$	
49 Tokorogenin	R	β	1β, 2β, 3α	—	—	266—268°	−50°	255°	−8°c	(350, 368)
50 Neotokorogenin	S	β	1β, 2β, 3α	—	—	250°	−55°	185—190°	−20,9°	(3)
51 Convallagenin A	S	β	1β, 3β, 5β	—	—	268—269°	−28°	208—210°°	−78°	(244)
52 Diotigenin	S	β	2β, 3α, 4β	—	—	280—281°	−59°	—	—	(369, 480)
53 Agapanthagenin	R	α	2α, 3β, 5α	—	—	285°	—	298—299°	−101°	(469)
54 Magogenin	R	α	2, 3, 6	—	—	284°	—	214°	—	(324)
55 Metagenin	R	β	2β, 3β, 11α	—	—	273—274°	−82°h	250—252°	−77,5°	(178, 179)
56 Agavogenin	R	α	2α, 3β, 12	—	—	242°	−62°f	230°	−98°i	(338)
57 Neodigitogenin	S	α	2α, 3β, 15β	—	—	277—279°	−82°k	229—232°g	−114°	(255, 537)
58 Digitogenin	R	α	2α, 3β, 15β	—	—	296°	−80°	235—236°g	−103°	(125, 335, 503)
59 Isoplexigenin C	R	α	2α, 3β, 23S	—	—	272—273°	−61°	245—247°	−89°	(152)
60 Isoplexigenin D	R	α	2α, 3β, 23R	—	—	280—281°	−74°	198—200°	−91°	(152)
61 Paniculogenin	S	α	3β, 6α, 23S	—	—	214—216°	−45,4°	200—202°	−51°	(403)
62 Asperagenin	S	β	3β, 6α, 25	—	—	265—268°	−135,9°	185—186°	−89°	(18, 391)
63 Reineckiagenin	R	β	1β, 3β, 25	—	—	278—280°	−71°i	198—200°	−82°	(482)

64	Isoreineckiagenin	S	β	1β, 3β, 25	—	—	240—242°	−66°	202—204°[n]	−87°	(*482*)
65	Isocarneagenin	S	β	1β, 3β, 27	—	—	242—244°	−63°	215—218°[n]	−75°	(*482*)
66	Carneagenin	R	β	1β, 3β, 27	—	—	262—264°	−71,6°	—	—	(*482*)
67	Igagenin	R	β	2β, 3α, 27	—	—	248—249°	−43,6°	191—192°	−15°	(*597*)
	$C_{27}H_{42}O_5$										
68	Neomanogenin	S	α	2α, 3β	12	—	242°	—	222°	—	(*330*)
69	Manogenin	R	α	2α, 3β	12	—	242°	−5°	259°	−45°	(*573*)
70	Neomexogenin	S	β	2β, 3β	12	—	221—222°	—	162—164°	—	(*331*)
71	Mexogenin	R	β	2β, 3β	12	—	237—238°	−6°	208°	—	(*126*)
72	Crestagenin	R	α	2α, 3β, 27	—	—	255—260°	−44°[b]	196—201°	−94°	(*121*)
	$C_{27}H_{40}O_5$										
73	Neokammogenin	S	—	2α, 3β	12	5	230°	—	203—205°	—	(*330*)
74	Kammogenin	R	—	2α, 3β	12	5	245°	−53°	259°	−83°	(*333*)
75	9-Dehydromanogenin	R	α	2α, 3β	12	9	240°	−16°[l]	263°	−62°[i]	(*570*)
76	25-Hydroxy-tamusgenin	S	—	3β, 25	11	5	245—247°	−76°	210—213°	−87°	(*167*)
77	Lowegenin	R	—	3β, 16α	11	5	223—226°	−55°	200—203°[e]	−62°	(*153*)
78	25(27)-Dehydromano-genin	—	α	2α, 3β	12	25(27)	238—240°	—	—	—	(*483, 565*)
79	Sapogenin aus *Helleborus odorus*, bzw. *niger* L.	—	—	1β, 3β, 11α	—	5,25(27)	236—240°	−86,9°	190—193°	−131,6°	(*307*)
80	Sansevierigenin	—	—	1β, 3β, 23 (S)	—	5,25(27)	234—237°	−125°	—	—	(*166*)
	$C_{27}H_{38}O_5$										
81	Δ⁹ ²⁵⁽²⁷⁾-Bisdehydro-manogenin	R	α	2α, 3β	12	9,25(27)	230—232°	−36,1°	246—248°	—	(*483*)
82	7-Keto-tamusgenin	R	—	3β	7, 11	5	250—253°	−131°	191—194°	−133°	(*167*)
	$C_{27}H_{44}O_6$										
83	Kogagenin	R	β	1β, 2β, 3α, 5β	—	—	318—322°	−27°	249—252°[l]	−26°	(*478*)

Tabelle 4 (Fortsetzung)

Name	Konfiguration		Substituenten			Sapogenin		Acetat[b]		Literatur
	$C_{(25)}$	$C_{(5)}$	OH	Keto-funktion	Doppel-bindung	Schmp.	$(\alpha)_D^a$	Schmp.	$(\alpha)_D^c$	
84 Convallagenin B	S	β	1β, 3β, 4β, 5β —	—	—	277—278°	−43°	228—230°[m]	−46,5°	(245)
85 Kitigenin	R	β	1β, 3β, 4β, 5β —	—	—	298°	−35°	219—220°	−54°	(419, 476)
$C_{27}H_{42}O_6$										
86 Cacogenin	R	α	2, 3, 6	12	—	278°	—	248°	—	(324)
$C_{27}H_{44}O_6$										
87 Dracogenin	—	—	1β, 3β, 23, 24 —		5,25(27)	215—217°	−144°	Tetraacetat amorph		(165)
$C_{27}H_{44}O_7$										
88 Pentelogenin	R	—	1, 2, 3, 4, 5	—	—	320°	−54,5°[k]	165—168°	—	(476)

Weitere in Herkunft und Eigenschaften wenig definierte Spirostanolderivate sind z. B. *Gloriogenin* (**89**) (*116*), *Tenuipegenin* (**90**) (*3*), *Jimogenin* (**91**) und *Cologenin* (**92**) (*477*).

a   in Chloroform, wenn keine anderen Angaben
b   Peracetat, wenn keine anderen Angaben
c   Lösungsmittel nicht angegeben
d   in Aceton
e   3-Monoacetat
f   in Äthanol
g   2,3-Diacetat
h   in Chloroform-Methanol
i   in Dioxan
k   in Pyridin
l   1,2,3-Triacetat
m   1,3,4-Triacetat
n   1,2,3,4-Tetraacetat
o   1,3-Diacetat

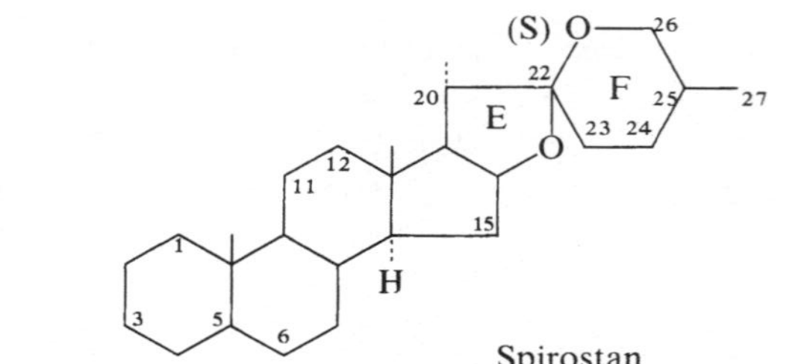

Tabelle 5. *Natürlich vorkommende modifizierte Spirostanole*

Ring E und F geöffnet:

Nr.	Name	Formel										
93	Kryptogenin	$C_{27}H_{42}O_4$	R	—	3β, 27	16, 22	5	192—193°	−200°	152—153°	−182°	*(164)*
94	Ricogenin	$C_{27}H_{40}O_5$	R	—	3β, 27	12,16,22	5	225—227°	—	195—197°	—	*(325)*
95	Eduligenin	$C_{27}H_{36}O_5$	R	—	3β, 27	11,16,22	—	233—237°	−139°	187—189°	−125°	*(153)*

Furostan-Derivate:

Nr.	Name	Formel										
96	Nologenin	$C_{27}H_{42}O_5$	R	—	3β, 16α, 20, 27	—	5	265—267°	—	179—180°	—	*(332, 337)*
97	Cholegenin	$C_{27}H_{44}O_4$	S	β	3α, 27	—	—	193°	−26°	174—175°	−20°	*(10, 343)*
98	Nuatigenin	$C_{27}H_{42}O_4$	S	—	3β, 27	—	5 22,25-oxid	210—214°	−93°	156—159°	−95°	*(519, 520)*
99	Afurigenin	$C_{27}H_{40}O_5$	R	—	3β, 27	11	5	240—243°	−50°	147—151°	—	*(167)*

Ring A aromatisch, Ring E und F Spirostanol-Struktur:

Nr.	Name	Formel										
100	Luvigenin	$C_{27}H_{38}O_2$	R, CH_3 an C-4	—	—	—	—	183—184°	−35°	—	—	*(481)*
101	Meteogenin	$C_{27}H_{38}O_3$	R, CH_3 an C-1	—	11α	—	—	157—158°	−174°	147—148°	−163°	*(199, 200)*

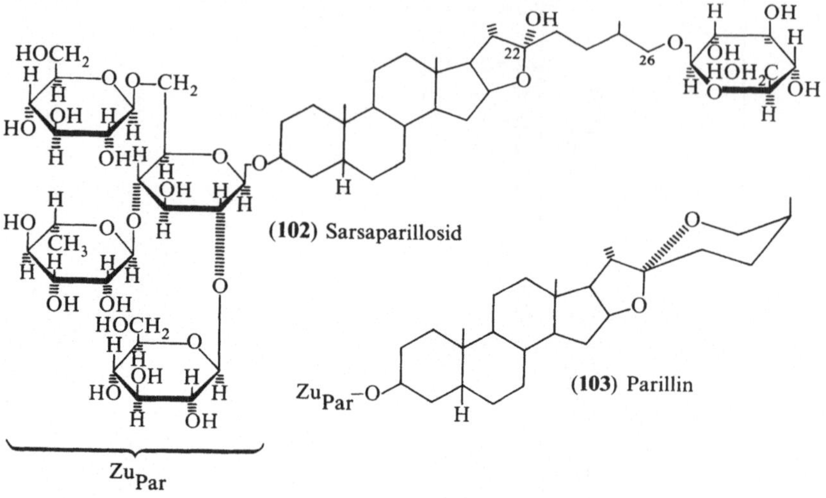

(102) Sarsaparillosid

(103) Parillin

Zu_{Par}

stanolglykoside finden sich vornehmlich in den assimilierenden Teilen der Pflanze, während in Samen, Wurzeln oder Knollen die Spirostanolglykoside meist überwiegen.

Zur Unterscheidung von Furostanol- und Spirostanolglykosiden lassen sich folgende Beobachtungen heranziehen:

22-Hydroxy-furostanolglykoside vom Typ des Sarsaparillosids reagieren mit Alkoholen leicht zu den entsprechenden 22-Alkoxy-derivaten **104**,

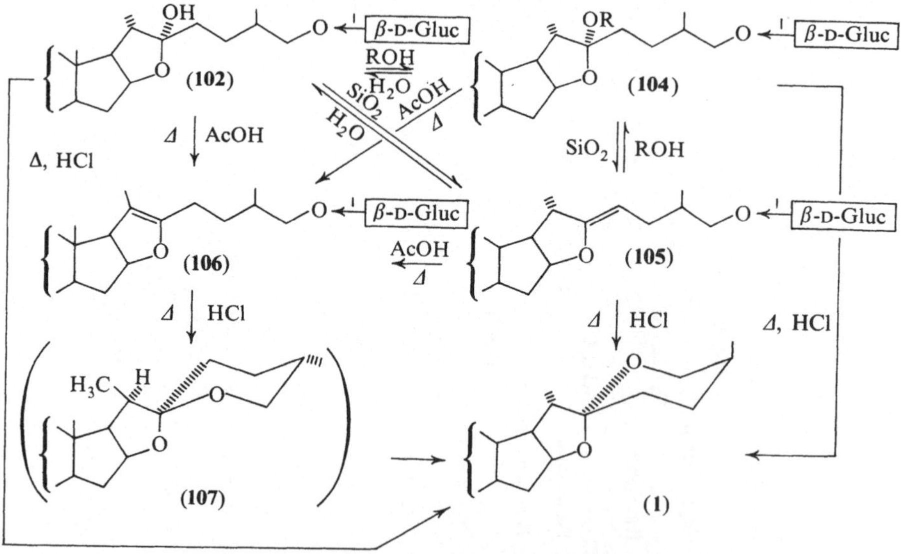

Schema 2. Umwandlungen des Sarsaparillosids

Literaturverzeichnis: SS. 576—606

die ihrerseits leicht wieder zu **102** hydrolisierbar sind (Schema 2). Ein Gleichgewicht zwischen beiden Formen läßt sich durch Dünnschichtchromatographie leicht nachweisen (*516*), es erschwert die Reindarstellung der Verbindungen außerordentlich.

An Kieselgel tritt leicht eine Eliminierung von Wasser zu Δ^{22}-Furostenderivaten **105** ein, die reversibel ist. Durch Kochen mit Eisessig erhält man eine Wasserabspaltung zu den $\Delta^{20(22)}$-Furostenverbindungen **106**, die recht stabile Verbindungen darstellen.

Beim Abspalten des Zuckerrestes von C-26 cyclisieren **102**, **104** und **105** zu den zugehörigen Spirostanolen (z. B. **1**); aus **106** entsteht zunächst die Cyclopseudoform **107**, die erst bei längerem Erhitzen in Säure in das Spirostanol (z. B. **1**) übergeht (siehe Schema 2). Diese Reaktionen entsprechen denen, die H. HIRSCHMAN und F. B. HIRSCHMAN (*195*) schon früher an entsprechenden Furostanolacetaten beobachtet haben.

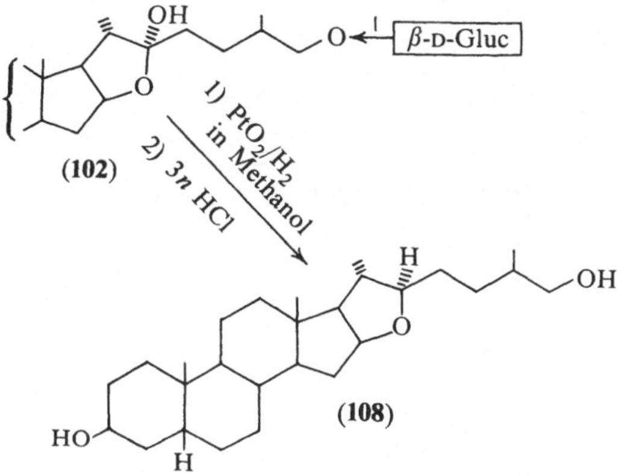

Die 22-Hemiketal-gruppierung läßt sich in Methanol mit Platinoxid als Katalysator leicht hydrieren, wobei die OH- bzw. OCH_3-Gruppe an C-22 durch H ersetzt wird. Die saure Hydrolyse liefert danach ein wesentlich polareres Aglykon, das sogenannte Dihydrosapogenin (**108**), als es unter den gleichen Bedingungen ohne vorherige Hydrierung entsteht. In letzterem Fall wäre ein wahres Spirostanolderivat (**1**) unter Ringschluß entstanden. Liegen nur Spirostanolverbindungen vor, so tritt in neutralem Medium keine Hydrierung ein, die Hydrolyse führt dann zur Abspaltung der Zucker und Freilegung des Spirostanols. Die Entstehung polarerer Aglykone nach Hydrierung und Hydrolyse zeigt daher Furostanolverbindungen an (*516*).

Einfacher ist noch die Anwendung von Farbreaktionen, so geben Furostanolderivate mit Ehrlichs-Reagenz eine Rotfärbung und mit Anisaldehyd eine gelbe Anfärbung, während Spirostanolderivate nicht reagieren (*254*). Auch die Komplexbildung speziell mit Cholesterin kann zur Unterscheidung herangezogen werden, da, soweit bekannt, damit nur Spirostanolglykoside Molekülverbindungen eingehen.

Im IR zeigen Furostanolderivate nicht die Banden bei 845, 892 und 912/cm, wie sie für Spirostanole typisch sind (*149*), dagegen tritt eine schwache verbreiterte Bande bei $\sim$900/cm auf (*516*).

Der Beweis für den Aufbau der Seitenkette in den Furostanolglykosiden wurde auf folgendem Wege geführt (*515, 516*) (Schema 3): Sarsaparillosid als Peracetat lieferte bei der Behandlung mit Eisessig die $\Delta^{20\,(22)}$-ungesättigte Verbindung (**109**), die mit Chromsäure in Eisessig zu **110** oxydiert werden konnte. Bei der Spaltung mit Kalium-tert. butylat entstand neben dem glykosidierten Pregnenolon-Derivat **111** die Verbindung **112,** eine glucosidierte (γS) δ-Hydroxy-γ-methyl-valeriansäure. Aus ihr wurde enzymatisch die freie Säure erhalten und in Form des Lactons **113** charakterisiert.

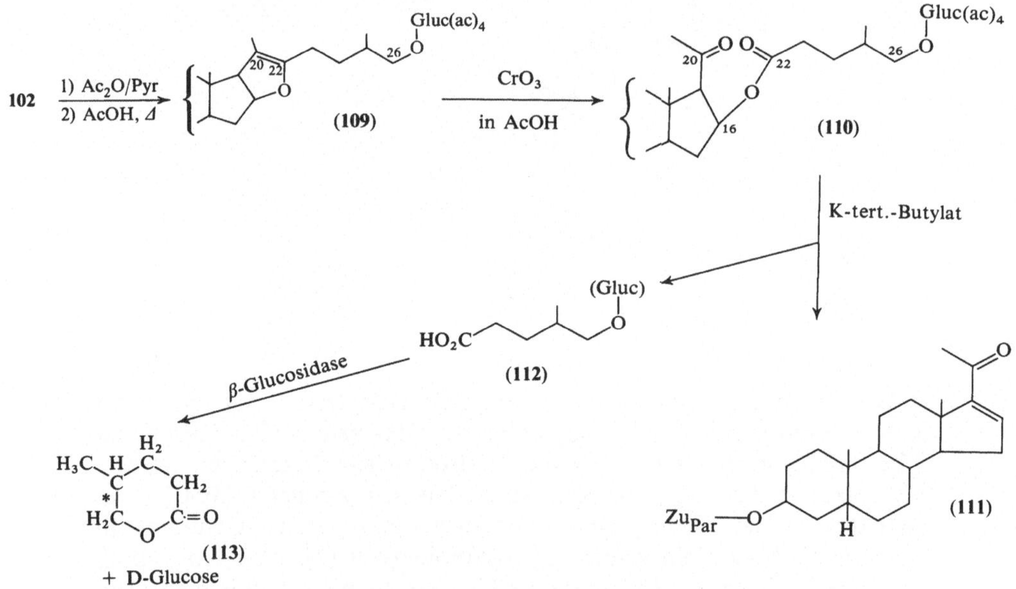

Schema 3. Chromsäureabbau des Sarsaparillosids

Während die Furostanolglykoside von Blättern der *Smilax*-Arten Mittelamerikas, die *Radix sarsaparillae* liefern, noch nicht untersucht worden sind, liegen an Blättern und Samen von *Digitalis lanata* Ehrh. bereits Ergebnisse vor. Sie zeigen, daß in der Pflanze nicht allein eine Umwandlung der offenkettigen Furostanol-Form in Spirostanolderivate eintritt, sondern daß dabei auch ein teilweiser Umbau in der Zucker-kette am Hydroxyl an C-3 erfolgt. Diese Arbeiten zeigen, daß in den Samensaponinen (siehe S. 499) die Glucose III mit der Galaktose IV vorwiegend (1→3) verknüpft ist, während im Lanatigosid (116) und Lanagitosid (117) der Blätter hier eine (1→4)-Bindung vorliegt (524).

Entsprechendes dürfte für die Blattsaponine aus *Digitalis purpurea* gelten, da das durch enzymatischen Abbau der Blattsaponine gewonnene F-Gitonin (125) (228) eine andere Zuckerverknüpfung als das aus den Samen gewonnene Gitonin (124) (532) aufweist.

Protogracillin (114), früher Kikubasaponin genannt, wurde von der Arbeitsgruppe um T. KAWASAKI (230, 254) aus *Dioscorea gracillima* isoliert und in der Struktur geklärt. Die gleiche Verbindung wurde von R. A. JOLY und Mitarb. (218) auch aus *D. floribunda* erhalten, die in dieser Pflanze auch eine β-Glucosidase nachwiesen, die Protogracillin schnell in Gracillin (127), ein Spirostanol-Saponin, umwandelt.

Auch das schon länger bekannte Dioscin (128) findet sich in den Blättern von *Dioscorea*-Arten als Proto-dioscin (115) (524).

Zum gleichen Typ von Furostanolderivaten gehört auch das Convalla-marosid (118) (524, 529) aus den Wurzeln und Rhizomen von *Convallaria majalis* (Maiglöckchen). Das Glykosid wurde erstmals von W. VOSS und G. VOGT (560) untersucht und Convallamarin benannt. Die Säure-hydrolyse lieferte das Spirostanol-aglykon Convallamarogenin (36) (523). Neuere Untersuchungen (505) zeigten, daß als Zucker 2 Mol. L-Rhamnose, 2 Mol. D-Glucose und 1 Mol. D-Chinovose im Glykosid vorhanden sind. Es konnte nachgewiesen werden, daß mit Ausnahme eines Mol. D-Glucose die Zucker über die OH-Gruppen am C-1 und C-3 gebunden sind. Ein Abbau des Convallamarosids, analog dem beim Sarsaparillosid erwähnten, führte zu einem Pregnenolonderivat, in dem noch beide Hydroxyle am C-1 und C-3 durch Zucker verschlossen sind, und zu dem Glucosid einer ungesättigten Valeriansäure. Convallamarosid muß daher ein tris-desmosidisches Glykosid sein. Als Spaltprodukt der Seitenkette wurde nach Hydrolyse der Glucose γ-Methylen-valerolacton erhalten (529). Die genaue Anordnung der Zucker an den OH-Gruppen an C-1 und C-3 ist noch nicht bekannt.

Ein weiteres Glykosid dieses Typs ist Asperosid (119) (379), das aus den Blättern von *Smilax aspera* L. gewonnen wurde. Es kristallisiert ebenfalls nicht und zeigt die erwähnten Eigenschaften der Furostanol-glykoside. β-Glucosidase spaltet 1 Mol. D-Glucose aus der Seitenkette des Aglykons ab, danach steigt der hämolytische Index auf das Dreifache an (etwa 1 : 70000). Die Säurehydrolyse liefert Diosgenin (6) und 2 Mol. D-Glucose und 3 Mol. L-Rhamnose. Die Verknüpfung der Zucker in der Kette am OH an C-3 ist noch nicht ganz vollständig geklärt. Zucker am C-1 und C-26 des Furostanols enthält Ruscosid (118a) (42b).

R. Tschesche und G. Wulff:

Tabelle 6. *Bisdesmosidische Furostanolsaponine und Nuatigenin-Glykoside*

102 Sarsaparillosid

β-D-Gl 1→2
β-D-Gl 1→6 β-D-Gl 1→3 [(25S)-5β-Furostan-3β,22α,26-triol] 26←1 β-D-Gl   (*515, 516*)
α-L-Rh 1→4

Smilax aristolochiaefolia
(Wurzeln, Rhizome)

114 Proto-gracillin

β-D-Gl 1→3
α-L-Rh 1→2 β-D-Gl 1→3 [(25R)-Δ⁵-Furosten-3β,22α,26-triol] 26←1 β-D-Gl   (*230, 254*)

Dioscorea gracillima (Wurzeln)

115 Proto-dioscin

α-L-Rh 1→2
α-L-Rh 1→4 β-D-Gl 1→3 [(25R)-Δ⁵-Furosten-3β,22α,26-triol] 26←1 β-D-Gl   (*254*)

Dioscorea septemloba (Wurzeln)

116 Lanatigosid

β-D-Gl 1→3 β-D-Ga 1→2
β-D-Xy 1→3 β-D-Gl 1→4 β-D-Ga 1→3 [(25R)-5α-Furostan-3β,22α,26-triol] 26←1 β-D-Gl   (*524*)

Digitalis lanata (Blätter)

117 Lanagitosid

β-D-Gl 1→3 β-D-Ga 1→2
β-D-Xy 1→3 β-D-Gl 1→4 β-D-Ga 1→3 [(25R)-5α-Furostan-2α,3β,22α,26-tetraol] 26←1 β-D-Gl   (*524*)

Digitalis lanata (Blätter)

118 Convallamarosid

L-Rh 1→3
β-D-Gl 1→x L-Rh 1→2 D-Chi 1→1 [5β-Δ²⁵-Furosten-1β,3β,22α,26-tetraol]   (*505, 529*)
26←1 β-D-Gl

Convallaria majalis (Wurzeln)

118a *Ruscosid*

$\boxed{\beta\text{-D-Gl}}$ $1\to3$ $\boxed{\alpha\text{-L-Rh}}$ $1\to2$ $\boxed{\alpha\text{-L-Ar}}$ $1\to1$ $[\Delta^{5,25\,(27)}$-Furostadien-1β,3β,22α,26-tetraol] $26\leftarrow1$ $\boxed{\beta\text{-D-Gl}}$ *(42b)*

Ruscus aculeatus (Rhizome)

119 Asperosid $\boxed{\text{D-Gl}}$ $1\to$ 2 (4) / 4 (2) $\boxed{\text{D-Gl}}$ $1\to3$ [(25 R)-Δ^5-Furosten-3β, 22α, 26-triol] $26\leftarrow1$ $\boxed{\beta\text{-D-Gl}}$ *(379, 503a)*

$\boxed{\text{L-Rh}}$ $1\to4$ $\boxed{\text{L-Rh}}$ $1\nearrow$

Smilax aspera (Blätter)

120 Avenacosid A $\boxed{\text{L-Rh}}$ $1\to4$ / $\boxed{\beta\text{-D-Gl}}$ $1\to2$ $\boxed{\beta\text{-D-Gl}}$ $1\to3$ [Nuatigenin (**98**)] $26\leftarrow1$ $\boxed{\beta\text{-D-Gl}}$ *(526)*

Avena sativa (oberirdische Teile)

121 Avenacosid B $\boxed{\text{L-Rh}}$ $1\to4$ / $\boxed{\beta\text{-D-Gl}}$ $1\to2$ $\boxed{\beta\text{-D-Gl}}$ $1\to3$ [Nuatigenin (**98**)] $26\leftarrow1$ $\boxed{\beta\text{-D-Gl}}$ *(514)*

$\boxed{\beta\text{-D-Gl}}$ $1\to3$ $\boxed{\beta\text{-D-Gl}}$ $1\nearrow$

Avena sativa (oberirdische Teile)

Die Struktur der in dieser und den folgenden Tabellen verzeichneten Glykoside ist in einer Kurzschreibweise dargestellt. Dabei bedeuten: Rest der D-Glucose D-Gl, D-Galaktose D-Ga, D-Xylose D-Xy, L-Arabinose L-Ar, L-Rhamnose L-Rh, D-Fucose D-Fu, D-Chinovose D-Chi, D-Glucuronsäure D-Glr.

Die Konfiguration der glykosidischen Bindung wird, soweit bekannt, innerhalb des Kästchens mit α oder β angegeben. Sofern nicht anders angegeben, handelt es sich um pyranoide Strukturen. Eine furanoide Form wird durch ein nachgestelltes f bezeichnet, z. B. α-L-Ar f. Die Verknüpfung der Zucker untereinander ist aus den Zahlen zwischen den Kästchen zu ersehen.

Für den Aglykonanteil ist jeweils der Name des Aglykons in Klammern gesetzt. Zwischen geninständigem Zucker und Aglykon ist die Position der funktionellen Gruppe im Aglykon, welche die Bindung vermittelt, angegeben.

Außerdem ist der Name der Pflanze und der Pflanzenteil verzeichnet, aus dem das Glykosid für die Strukturermittlung isoliert wurde.

492   R. Tschesche und G. Wulff:

Zahlreiche weitere Pflanzen enthalten bisdesmosidische Furostanolsaponine, wie eine Untersuchung mit Hilfe von Farbreaktionen aufzeigte
(*254*), doch sind diese Glykoside bisher nicht rein isoliert worden.

4. Nuatigeninglykoside

Nuatigenin (**98**), ein Aglykon, in dem der Ring F gegenüber den Spirostanolen zum Fünfring verengt ist und das dann am C-26 eine OH-Gruppe
enthält, wurde zuerst aus den Wurzeln von *Solanum sisymbrifolium* nach
vorsichtiger Säurehydrolyse erhalten (*519*). Es wandelt sich sehr leicht in

$$H_3O^{\oplus}$$

(**98**) Nuatigenin   (**33**) Isonuatigenin

Isonuatigenin **33** um, das ein wahres Spirostanolderivat ist und bei dem
an C-25 nun eine tertiäre Hydroxylgruppe erscheint. Während die Glykoside aus *Solanum sisymbrifolium* noch nicht rein erhalten wurden, konnten
solche des Nuatigenins aus den oberirdischen Teilen des Hafers *(Avena
sativa)* rein isoliert werden. Es handelt sich dabei um die Avenacoside A,
B und C (*514, 520, 526*), von denen die Struktur der beiden ersteren bereits bestimmt wurde.

Avenacosid A (**120**) ist das Desglucoderivat von Avenacosid B (**121**),
die Verknüpfung der Zucker wurde durch stufenweise enzymatische
Hydrolyse mit β-Glucosidasen verschiedener Herkunft, durch Hydrolyse
der permethylierten Verbindungen und unter Anwendung der Massenspektrometrie von geeigneten Derivaten ermittelt. Avenacosid C dürfte
ein um eine Glucose reicheres Avenacosid B sein.

Nuatigeninglykoside scheinen die Ursache des Auftretens von Cholegenin (**122**) in der Rindergalle zu sein, die Umkehr der sterischen Verhältnisse der OH-Gruppe an C-3 in α und die Hydrierung der Doppelbindung
Δ^5 zu einem Ring A/B *cis*-Derivat dürften auf die Aktivität der Pansen-
bakterien zurückgehen. Hierfür spricht, daß der Gehalt der Rindergalle

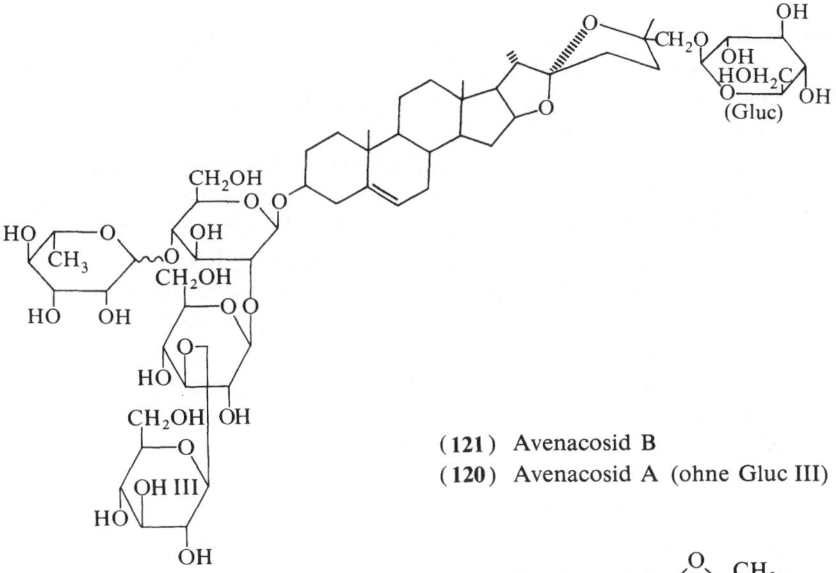

(Gluc)

(**121**) Avenacosid B
(**120**) Avenacosid A (ohne Gluc III)

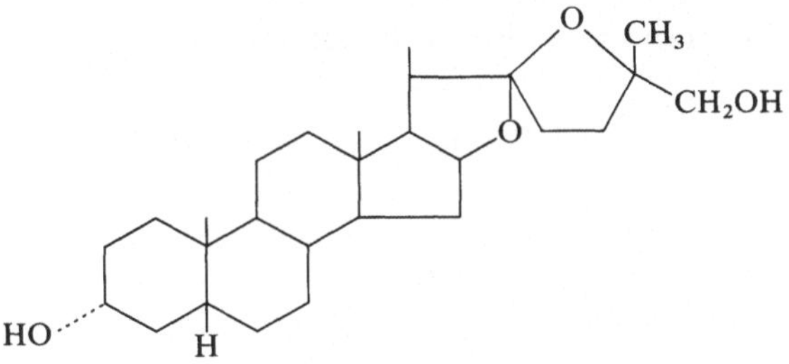

(**122**) Cholegenin

an Cholegenin von der Nahrung der Tiere abhängig ist (*10*). Es scheint daher möglich, daß Nuatigeninglykoside außer im Hafer auch noch anderweitig in Futtermitteln verbreitet sind. Cholegenin erleidet die gleiche leichte Umwandlung zu Isocholegenin, die vom Nuatigenin bekannt ist.

5. Spirostanolglykoside

a) Allgemeines

Obwohl mehr als 90 verschiedene Aglykone vom Spirostanoltyp bekannt sind (siehe Tabelle 4), ist die Zahl der entsprechenden in der Struktur aufgeklärten Glykoside vergleichsweise gering. Die Ursache dürfte vor allem auf die bislang schwierige Reindarstellung zurückzuführen sein. Manche der hier beschriebenen Verbindungen können Artefakte sein, die bei der Aufarbeitung aus zuckerreicheren, z. B. bisdesmosidischen Furostanolsaponinen, entstanden sind. Genuin finden sich Spirostanol- saponine vor allem in Samen, wo sie als Hemmstoffe gegen niedere Pilze (siehe S. 569f.) von Bedeutung sind.

In den meisten Glykosiden vermittelt das Hydroxyl am C-3 des Agly- kons die Verknüpfung mit dem Zuckeranteil. Es liegen meist mehrere und verschiedene Hexosen und Pentosen vor, die oft in verzweigten Ketten angeordnet sind. In den bisher aufgeklärten Glykosiden enthalten diese maximal 5 Monosaccharideinheiten. In einigen Glykosiden kommen auch andere Verknüpfungsorte als das OH am C-3 des Aglykons vor. So sind Verknüpfungen der Zucker über OH-Gruppen an C-1, C-2, C-5, C-6 und C-11 bekannt geworden. Auch Anknüpfungen von Zuckern über zwei verschiedene OH-Gruppen des Spirostanolgerüstes wurden in *Convallaria*-Arten aufgefunden. Diese Verbindungen sind formal Bis- desmoside, doch greifen beide Ketten an OH-Gruppen in den Ringen A und B an; das dürfte der Grund sein, warum sie in ihren Eigenschaften sich wie monodesmosidische Spirostanol- und nicht wie bisdesmosidische Furostanolsaponine verhalten.

b) Glykoside mit Zuckern am OH an C-3 des Aglykons

Die Samen von *Digitalis purpurea* enthalten ein Saponingemisch, das als „Digitonin" bezeichnet und wegen der Schwerlöslichkeit seiner Komplexverbindung mit Cholesterin und ähnlichen Sterolen zu deren Abtrennung und zur quantitativen Bestimmung benutzt wird. Das zu etwa 2% aus den Samen gewinnbare, gut kristallisierende Saponin ist ein Gemisch verschiedener Glykoside, von denen das eigentliche Digitonin zu 40% den Hauptanteil stellt. Durch Verteilungschromatogra- phie an Cellulose konnten die Komponenten rein isoliert werden (siehe Tabelle 7) (*531, 532*).

Tabelle 7. *Im „Digitonin" enthaltene Glykoside*

Glykosid	Aglykon	Zucker	%-Gehalt
Digitonin **123**	Digitogenin	2 D-Gl, 2 D-Ga, 1 D-Xy	40
Digalonin	Digalogenin	2 D-Gl, 2 D-Ga, 1 D-Xy	15
Desglucodigitonin	Digitogenin	1 D-Gl, 2 D-Ga, 1 D-Xy	25
Gitonin **124**	Gitogenin	1 D-Gl, 2 D-Ga, 1 D-Xy	15

Durch Papier- und durchlaufende Dünnschichtchromatographie ließ sich ferner ein Gitogeninglykosid nachweisen, das die gleichen Zucker wie Digitonin enthielt (*593*). Auch ein Tigogeninglykosid kommt in geringer Menge vor. Die Strukturermittlung erfolgte am Digitonin (*532*). Neben

Tabelle 8. *Auftretende Oligosaccharidspaltstücke des HBr-Eisessig-Abbaus von Digitonin*

Substanz	Bausteinanordnung							
Solabiose	β-D-Gl	1→3	β-D-Ga	→				
Lycobiose					β-D-Gl	1→4	β-D-Ga	→
2 β-D-Galaktosido-D-glucose			β-D-Ga	1→2	β-D-Gl	→		
Digitotriose B			β-D-Ga	1→2	β-D-Gl	1→4	β-D-Ga	→
Digitotriose C	β-D-Gl	1→3	β-D-Ga	1→2	β-D-Gl	→		
Digitotetraose	β-D-Gl	1→3	β-D-Ga	1→2	β-D-Gl	1→4	β-D-Ga	→

dem Ergebnis der Methylierung erbrachte vor allem die Untersuchung von Teilstücken der Zuckerkette in Form von Oligosacchariden entscheidende Aufschlüsse über den Aufbau. Sie konnten durch Abbau des peracetylierten Digitonins mit HBr/Eisessig in Chloroform erhalten werden (über die isolierten Oligosaccharide vgl. Tabelle 8). Xylosehaltige Oligosaccharide entstanden nur in sehr geringer Menge. Die Strukturaufklärung der Oligosaccharide offenbarte nicht nur Verknüpfungsart und Sequenz der vorliegenden Zucker, sondern bei Anwendung spezifischer Enzyme auch die Konfiguration der glykosidischen Bindungen. Digitonin hat danach die Struktur **123**. Gleichen Bau der Zuckerkette dürfte Digalonin besitzen, während dem Gitonin und Desglucodigitonin die endständige Glucose fehlen.

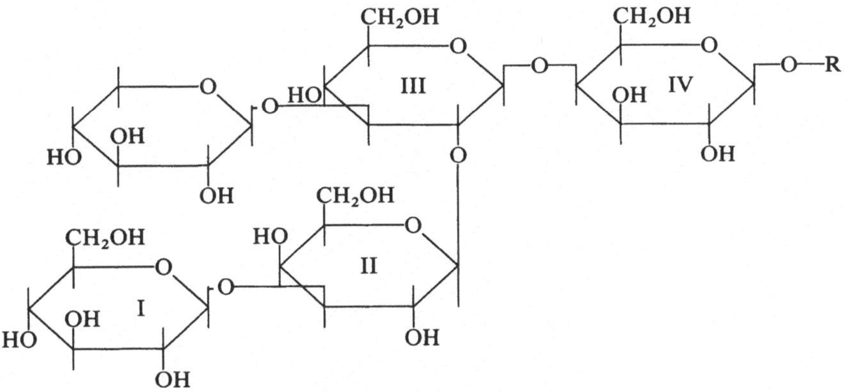

R = Rest des Digitogenin (**58**)

(**123**) Digitonin

Tabelle 9. *Spirostanolglykoside*

123 Digitonin β-D-Gl 1→3 β-D-Ga 1→2 / β-D-Xy 1→3 → β-D-Gl 1→4 β-D-Ga 1→3 [Digitogenin (**58**)] *(531, 532)*
Digitalis purpurea (Samen)

124 Gitonin β-D-Ga 1→2 / β-D-Xy 1→3 → β-D-Gl 1→4 β-D-Ga 1→3 [Gitogenin (**16**)] *(524)*
Digitalis purpurea (Samen)

125 F-Gitonin β-D-Gl 1→2 / β-D-Xy 1→3 → β-D-Gl 1→4 β-D-Ga 1→3 [Gitogenin (**16**)] *(228)*
Digitalis purpurea (Blätter)

126 Lanatigonin I β-D-Gl 1→3 β-D-Ga 1→2 / β-D-Xy 1→3 → β-D-Gl 1→3 β-D-Ga 1→3 [Tigogenin (**4**)] *(497)*
Digitalis lanata (Samen)

126a Desgluco-Lanatigonin II β-D-Ga 1→2 / β-D-Xy 1→3 → β-D-Gl 1→4 β-D-Ga 1→3 [Tigogenin (**4**)] *(524)*
Digitalis lanata (Blätter)

127 Gracillin α-L-Rh 1→2 / β-D-Gl 1→3 → β-D-Gl 1→3 [Diosgenin (**6**)] *(230)*
Dioscorea gracillima und andere (Wurzeln)

128 Dioscin α-L-Rh 1→2 / α-L-Rh 1→4 → β-D-Gl 1→3 [Diosgenin (**6**)] *(230)*
Dioscorea spec. (Wurzeln)

128a Timosaponin A III [β-D-Gl] 1→2 [β-D-Ga] 1→3 [Sarsasapogenin (**1**)] *Anemarrhena asphodeloides* (Rhizome) (*232, 233*)

103 Parillin [β-D-Gl] 1→2, [α-L-Rh] 1→4 [β-D-Gl] 1→3 [Sarsasapogenin (**1**)], [β-D-Gl] 1→6 *Smilax aristolochiaefolia* (Wurzeln) (*513*)

129 Desglucoparillin [α-L-Rh] 1→4, [β-D-Gl] 1→6 [β-D-Gl] 1→3 [Sarsasapogenin (**1**)] *Smilax aristolochiaefolia* (Wurzeln) (*513*)

130 Desgluco-desrhamnoparillin [β-D-Gl] 1→6 [β-D-Gl] 1→3 [Sarsasapogenin (**1**)] *Smilax aristolochiaefolia* (Wurzeln) (*513*)

131 Convallasaponin E [α-L-Ar] 1→2 [α-L-Ar] 1→2 [α-L-Ar] 1→3 [Diosgenin (**6**)] *Convallaria keisukei* (Blüten) (*247*)

132 Convallasaponin A [α-L-Ar] 1→3 [Convallagenin A (**51**)] *Convallaria keisukei* (Blüten) (*242, 246*)

133 Gluco-convalla-saponin A [β-D-Gl] 1→2 [α-L-Ar] 1→3 [Convallagenin A (**51**)] *Convallaria keisukei* (Blüten) (*246*)

134 Convallasaponin C [α-L-Rh] 1→3 [α-L-Rh] 1→2 [α-L-Ar] 1→3 [Isorhodeasapogenin (**10**)] *Convallaria keisukei* (Blüten) (*242, 243*)

135 Tokoronin [α-L-Ar] 1→1 [Tokorogenin (**49**)] *Dioscorea Tokoro* Makino (*231, 348*)

136 Tokorogenin-glucosid [β-D-Gl] 1→1 [Tokorogenin (**49**)] *Dioscorea Tokoro* Makino (*347*)

Tabelle 9 (Fortsetzung)

137   Yononin   [α-L-Ar]   1→2 [Yonogenin (13)]   (225, 231)
Dioscorea Tokoro Makino

138   Convallasaponin B   [α-L-Ar]   1→5 [Convallagenin B (84)]   (242, 246)
Convallaria keisukei (Blüten)

139   Paniculonin A   [β-D-Xy]   1→3 [β-D-Chi]   1→6 [Paniculogenin (61)]   (402)
Solanum paniculatum (Blätter)

140   Paniculonin B   [α-L-Rh]   1→3 [β-D-Chi]   1→6 [Paniculogenin (61)]   (402)
Solanum paniculatum (Blätter)

140a   Ophiopoganin B   [α-L-Rh]   1→2 [β-D-Fu]   1→1 [Ruscogenin (26)]   (474a)
Ophiopogon japonicus (Holz)

140b   Ruscin   [β-D-Gl]   1→3 [α-L-Rh]   1→2 [α-L-Ar]   1→1 [Ruscogenin (26)]   (42a)
Ruscus aculeatus (Rhizome)

141   Metanartheticum-Prosapogenin   [2,3,4-Tri-O-acetyl-α-L-Ar]   1→11 [2-O-Acetyl-metagenin (55)]   (605)
Metanarthetikum luteo-viride

142   Convallasaponin D   [α-L-Rh]   1→2 [β-D-Xy]   1→3 [α-L-Rh]   1→3 / [β-D-Gl]   1→1 [Rhodeasapogenin (9)]   (246)
Convallaria keisukei (Blüten)

143   Glucoconvalla-saponin B   [β-D-Gl]   1→3 / [α-L-Ar]   1→5 [Convallagenin B (84)]   (246)
Convallaria majalis (Blüten)

144   Convallamarogenin-triglykosid   [α-L-Rh]   1→3 / [α-L-Rh]   1→2 [β-D-Chi]   1→1 [Convallamarogenin (36)]   (505)
Convallaria majalis (Blüten)

Durch eine etwas andersartige Auftrennungsmethode konnten KAWA-SAKI und NISHIOKA (*226*) später die gleichen Saponine ebenfalls aus den Samen von *Digitalis purpurea* gewinnen. Aus den Blättern der Pflanze isolierten sie jedoch (*227, 229*) nach enzymatischer Spaltung zwei Spirostanolglykoside, Desgalakto-tigonin und F-Gitonin, die sich in der Zuckerkette von den Samensaponinen unterscheiden. Die Strukturaufklärung erfolgte mit den üblichen Methoden. Um die Verknüpfung der Zuckerkette über das OH am C-3 des Gitogenins zu sichern, wurde das nach der Methylierung und Hydrolyse erhaltene Aglykon mit CrO_3 zu einem Monoketon oxidiert und diese Verbindung mit synthetisch dargestellten 3-Dehydro-2-O-methyl- und 2-Dehydro-3-O-methyl-gitogenin verglichen. Danach besitzt F-Gitonin die Struktur **125**. Genuin dürften diese Glykoside als bisdesmosidische Furostanolsaponine vorliegen.

Sehr ähnliche Saponine wie die aus *Digitalis purpurea* enthalten die Samen von *Digitalis lanata* Ehrh., jedoch findet sich in keinem der 5 Hauptglykoside Digitogenin als Aglykon. In der Struktur konnte das Lanatigonin I (**126**) bestimmt werden. Es enthält Tigogenin als Aglykon, und die Zuckerkette unterscheidet sich nur durch eine $1 \rightarrow 3$-Bindung der Glucose zur geninständigen Galaktose von der des Digitonins, die an dieser Stelle eine $1 \rightarrow 4$-Bindung aufweist. Lanatigonin II dürfte dagegen die Zuckerkette des Digitonins tragen. Die bisher nicht aufgetrennten Lanadigalonin I und II könnten zueinander in gleichem Verhältnis stehen (*497*).

In zahlreichen *Dioscorea*-Arten finden sich die beiden Diosgeninglykoside Gracillin (**127**) und Dioscin (**128**), die KAWASAKI und Mitarb. (*230*) in der Struktur bestimmt haben. Gracillin enthält 2 Moll. D-Glucose und 1 Mol. L-Rhamnose, während Dioscin 1 Mol. D-Glucose und 2 Moll. L-Rhamnose enthält. Diosgenin (**6**) spielt als Ausgangsmaterial für die Steroidhormonproduktion eine bedeutende Rolle. Es wird meist durch saure Hydrolyse der Pflanzenextrakte gewonnen und kann aus manchen Wurzeln (*Dioscorea* spec. „barbasco") in bis zu 7% Ausbeute erhalten werden. Wieweit in den zur Diosgenin-Gewinnung herangezogenen Pflanzen Dioscin bzw. Gracillin oder Protodioscin und Protogracillin (Kikubasaponin) oder noch unbekannte Glykoside vorkommen, ist nicht bekannt.

Aus den Wurzeln und Rhizomen bestimmter mittelamerikanischer *Smilax*-Arten (vor allem *Smilax artistolochiaefolia*, Handelsbezeichnung *Radix sarsaparillae*), wurde neben dem schon erwähnten Sarsaparillosid (**102**) ein zweites Hauptsaponin, das Parillin (**103**) isoliert (*513*). Es entsteht auch nach Abspaltung der D-Glucose aus der Seitenkette und Cyclisierung aus Sarsaparillosid. Demgemäß enthalten beide Verbindungen die gleiche Zuckerkette am OH an C-3 des Aglykons. Die

Strukturaufklärung dieser Kette erfolgte am Parillin und ergab, daß an einer geninständigen D-Glucose 2 Moll. D-Glucose und 1 Mol. L-Rhamnose gebunden sind. Die Zuckerkette besitzt also einen Verzweigungsgrad, wie er bisher nur hier gefunden worden ist. Die Strukturermittlung wurde durch das gleichzeitige Auftreten zuckerärmerer Nebenglykoside sehr erleichtert. So fand man Desglucoparillin (**129**), Desgluco-desrhamnoparillin (**130**) und Sarsasapogenin-β-D-glucosid. Ob diese Verbindungen und auch Parillin selbst genuin in den Wurzeln vorliegen oder erst bei der Trocknung oder Lagerung entstanden sind, ist nicht geklärt.

Aus *Convallaria keisukei*, dem japanischen Maiglöckchen, isolierten Yoshizawa und Mitarb. (*242, 243, 247*) 4 Glykoside mit Zuckerketten am OH an C-3 des Aglykons. Als Aglykone treten in **131** Diosgenin, in **132** und **133** Convallagenin A und in **134** Isorhodeasapogenin auf. Über daneben vorliegende Glykoside mit zwei getrennten Zuckerketten siehe S. 501.

c) Glykoside mit einem Zuckeranteil, der nicht an das OH am C-3 des Aglykons gebunden ist

Eine Verknüpfung von Kohlenhydrat mit dem *OH an C-1* findet sich im Tokoronin (**135**), im Tokorogenin-glucosid (**136**) (*231, 347, 348*), im Ophiopoganin B (**140a**) (*474a*) und im Ruscin (**140a**) (*42a*). Als Aglykone liegen Tokorogenin (**49**) und Ruscogenin (**26**) vor. In Covallaria-Arten treten Verbindungen mit zwei Zuckerketten auf, die über OH-Gruppen am C-1 und C-3 gebunden sind (siehe S. 501).

Im Yononin (**137**) ist eine L-Arabinose über das *OH am C-2* des Yonogenins gebunden. Yononin konnte aus *Dioscorea tokoro* isoliert werden (*225*). Eine Verknüpfung über ein *OH am C-5* findet sich im Convallasaponin B (**138**) (*246*). Eine Zuckerkette am *OH an C-6* tragen die von Ripperger und Schreiber aus Solanum paniculatum isolierten Paniculonine A und B (*402*). Beide Glykoside enthalten den Zucker D-Chinovose, der geninständig ist. Besondere Schwierigkeiten bereitete die Ermittlung der Verknüpfung der D-Xylose bzw. L-Rhamnose mit der D-Chinovose, da die Methylzucker der Chinovose nicht ausreichend bekannt sind. Der isolierte Methylzucker konnte durch massen- und NMR-spektroskopische Untersuchungen als 2,4-Di-O-methyl-D-chinovose bestimmt werden. Der Anknüpfungsort am Aglykon ließ sich aus dem nach Spaltung des Permethylierungsproduktes erhaltenen Aglykon-Anteil ermitteln. Demgemäß besitzt A die Struktur **139** und B **140**. Ein sehr ungewöhnliches Glykosid (**141**) mit einem 2,3,4-Tetra-O-acetyl-α-L-arabinopyranosyl-Rest an einem *OH an C-11* des Metagenins erhielten I. Yosioka und Mitarbb. (*605*) bei der Spaltung der Saponine aus

Metanartheticum luteo-viride mittels einer Kultur von Bodenbakterien. Ob dieses Glykosid **141** wirklich als Teilstück eines Saponins dieser Pflanze anzusehen ist oder erst unter dem Einfluß der Bodenbakterien aus Metagenin entstanden ist, ist unentschieden.

Es dürfte sicherlich noch weitere Glykoside geben, die Zucker an anderen OH-Gruppen als am C-3 des Aglykons geknüpft enthalten. So kann im Hispidin aus *Solanum hispidum* die Zuckerkette nicht über ein OH am C-3 gebunden sein, da das Aglykon Hispidogenin **42** eine Ketofunktion an dieser Stelle enthält (*319*).

d) Glykoside mit mehr als einer Zuckerkette am Spirostanol-Aglykon

Die bisher einzige Quelle für Spirostanol-saponine mit zwei Zuckerketten am Ringsystem des Aglykons sind *Convallaria*-Arten. So isolierten YOSZIZAWA und Mitarb. (*246*) aus *Convallaria keisukei* das Convallasaponin D (**142**), welches neben einer Zuckerkette am C-3 des Rhodeasapogenins (**9**) einen β-D-Glucopyranosyl-Rest am OH an C-1 trägt.

Zur Strukturermittlung wurde aus **142** die Glucose mit β-Glucosidase abgespalten und das Restglykosid methyliert und hydrolysiert. Der Aglykonanteil enthielt eine O-Methylgruppe, seine freie OH-Gruppe wurde mit CrO_3 zum Keton oxidiert. Das ORD-Spektrum wies auf ein 3-Keton hin, womit die Oligosaccharidkette aus 2 Moll. L-Rhamnose und 1 Mol. L-Arabinose über die OH-Gruppe am C-3 gebunden sein muß.

Eine saure Partialhydrolyse von **142** ergab ein Monoglucosid, das nach Methylierung und Spaltung ein gegenüber dem vorhergehenden Versuch verschiedenes Mono-O-methylderivat ergab. Das nach Oxidation erhaltene Keton zeigte eine für 1-Ketone typische ORD-Kurve. Damit ist für das Convallasaponin D die in der Struktur **142** gegebene Anknüpfung der Zuckerketten bewiesen.

Im Glucoconvallasaponin B (**143**) befindet sich neben einem β-D-Glucopyranosyl-Rest am OH an C-3 ein α-L-Arabino-pyranosyl-Rest am C-5 des Convallagenins B. Die Strukturaufklärung gestaltete sich verhältnismäßig einfach, da Glucoconvallasaponin B sich leicht in Convallasaponin B überführen ließ. Die Untersuchung der nach der Permethylierung erhaltenen teilmethylierten Aglykone erlaubte wiederum die Anknüpfungsorte für beide Zucker eindeutig festzulegen (*246*).

Auch *Convallaria majalis,* das europäische Maiglöckchen, enthält neben dem Convallamarosid ein Convallamarogenin-triglykosid (**144**), das 2 Moll. L-Rhamnose und 1 Mol. D-Chinovose enthält. Diese Zucker sind über die OH-Gruppen am C-1 und C-3 gebunden, dabei ist die Struktur 3-O-[α-L-Rhamnopyranosyl] 1-O-[α-L-rhamnopyranosyl [1 → 2] β-D-chinovopyranosyl]-convallamarogenin (*524*).

6. Polypodosaponin-Typ

Bei der Untersuchung (*216*) der Rhizome von *Polypodium vulgare* L. (Tüpfelfarn oder Engelsüß) fanden J. Jizba und Mitarbb. (*215*) den ersten Vertreter eines neuen Typs von Saponinen, dessen Aglykon nicht mehr die übliche Spirostanol-, bzw. deren geöffnete Form in Gestalt der Furostanol-Derivate aufwies. In ihm sind die 8 C-Atome der Seitenkette des Cholesterols zu einem Cyclo-halbacetal-Sechsring umgebildet, ohne daß im *Polypodosaponin* (**145**) die typischen Saponin-Eigenschaften

(**145**) R = H Polypodosaponin

(**146**) R = CH₃

verloren gegangen wären. Das Methyl-vollacetal (**146**) zeigt starkes Schaumbildungsvermögen in wäßriger Lösung und eine hämolytische Aktivität im Verhältnis von etwa 1:80000 bis 100000. Die Zucker-kette, mit der Hydroxylgruppe am C-3 verbunden, besteht aus L-Rhamnose und D-Glucose im Verhältnis 1:1. Die Untersuchungen wurden im wesentlichen an dem 26-O-Methyl-polypodosaponin **146** vorgenommen, das sich sehr leicht in methanolischer Lösung unter Vollacetalbildung aus dem genuinen **145** bildet. Durch saure Hydrolyse konnte weder aus **145** noch aus **146** das Aglykon gewonnen werden, da Zersetzung eintrat. Erst die Anwendung von Enzymen ergab aus **146** in guter Ausbeute das 26-O-Methyl-polypodogenin **147**, dessen Struktur durch chemische und spektroskopische Methoden aufgeklärt wurde. Vor allem die massenspektroskopische Fragmentierung erbrachte wesentliche Aufschlüsse zur Struktur (*215*).

Literaturverzeichnis: SS. 576—606

(146) $\xrightarrow{\text{Enzym}}$

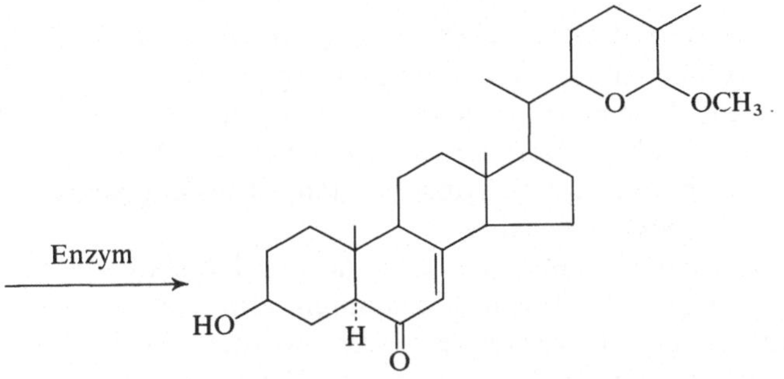

(147) 26-O-Methyl-polypodogenin

Bemerkenswerterweise enthält die Pflanze auch eine bisdesmosidische Form, Osladin (**148**) (*214*), in der das acetalische OH am C-26 glucosyliert ist. Im Aglykon des Osladin (**149**) fehlt jedoch gegenüber dem Polypodogenin (**146**) die Δ^7-Doppelbindung. Osladin ist die Ursache des ausgesprochen süßen Geschmacks der Rhizome dieser Pflanze.

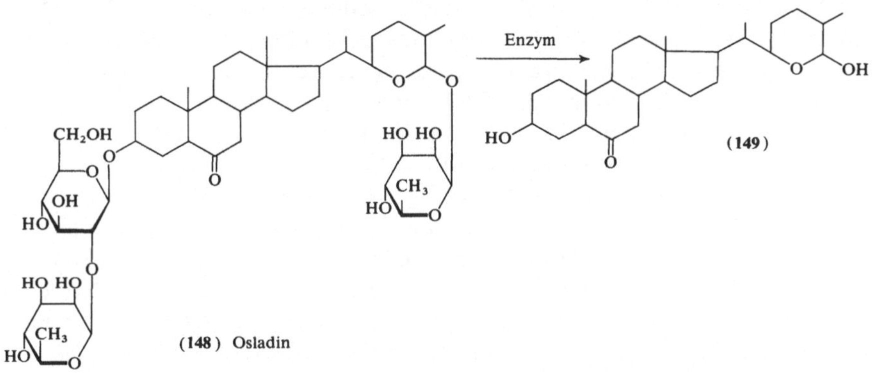

(148) Osladin

(149)

IV. Die Struktur der Triterpensaponine

1. Allgemeines

Die weitaus größte Zahl der in der Natur vorkommenden Saponine enthält Aglykone vom Triterpentyp. Das drückt sich auch in der Zahl von mehr als 110 in der Struktur bestimmter Triterpensaponine und -prosapogenine aus. Vor allem russische Arbeitsgruppen haben sich an den Strukturaufklärungen sehr wesentlich beteiligt.

Überwiegend Oleanan-Derivate (105 Vertreter) stellen in den bisher strukturell bekannten Saponinen die Aglykone, während Ursan- (8 Vertreter) und Dammaran-Derivate (3 Vertreter) eine wesentlich geringere Bedeutung haben. Von den Saponinen, die Hopan-, Lanostan- oder Lupeol-Derivate als Aglykon enthalten, ist bislang keines in der Struktur vollständig geklärt worden.

Die bisher aufgefundenen Δ^{12}-Oleanen-Derivate sind in der Tabelle 10 mit ihren physikalischen Konstanten und denen ihrer Acetate bzw. Methylesteracetate zusammengestellt worden. In der Tabelle 11 finden sich die entsprechenden Daten der Vertreter anderer Triterpentypen, jedoch nur, insoweit als sie auch als Aglykon in Saponinen auftreten. Weitere Einzelheiten über Chemie und Vorkommen finden sich in der Literatur (*30, 40, 121a, 138, 139, 177, 192, 212, 251, 342, 456, 468, 583, 615*). Auch Zusammenfassungen über Triterpensaponine liegen bereits zahlreich vor (*30, 40, 192, 258, 259, 468, 533, 583, 590, 615*).

In Analogie zu den Steroidsaponinen findet man auch bei den Triterpenen Glykoside mit einer (Monodesmoside) und mit zwei Zuckerketten (Bisdesmoside). Im Unterschied zu den Steroiden ist bei den Triterpenbisdesmosiden die zweite Zuckerkette jedoch fast immer über die Carboxylgruppe am C-17 des Aglykons gebunden. Für Verbindungen dieses Typs ist auch der Name Acyloside vorgeschlagen worden (*360*). Lediglich bei den Dammaran-Glykosiden aus *Panax ginseng* sind in Analogie zu den Steroid-Saponinen jeweils beide Zuckerketten über OH-Gruppen glykosidisch verknüpft.

Wie sich bei der Behandlung der Eigenschaften der Saponine erweisen wird (Abschnitt VI), zeigen Monodesmoside typische Saponineigenschaften, während Bisdesmoside solche entweder gar nicht oder in abgeschwächter Form aufweisen. Dabei werden die Eigenschaften offensichtlich auch von dem Vorliegen saurer Gruppierungen beeinflußt, die entweder im Aglykon, im Zuckerteil als Uronsäure oder als beide Möglichkeiten in einem Glykosid vorhanden sein können. Saponine mit Estergruppierungen zeigen einige Besonderheiten, die eine getrennte Abhandlung angezeigt erscheinen lassen. Insgesamt ergibt sich hieraus die folgende Einteilung für die Triterpensaponine (*591*):

A. Monodesmoside

    1. Neutrale Glykoside (9 Vertreter)
    2. Estersaponine (4 Vertreter)
    3. Durch eine Uronsäure saure Glykoside (4 Vertreter)
    4. Durch das Aglykon saure Glykoside (40 Vertreter)
    5. Durch Aglykon und Uronsäure saure Glykoside (13 Vertreter)
    6. Acylglykosen (6 Vertreter)

B. Bisdesmoside
   1. Neutrale Glykoside (30 Vertreter)
   2. Saure Glykoside (18 Vertreter)
C. Tierische Saponine

Von diesen Glykosiden gehen die neutralen Bisdesmoside (B 1) mit Ausnahme der Ginseng-Glykoside leicht in die im Aglykon sauren Monodesmoside (A 4) und die sauren Bisdesmoside (B 2) leicht in die durch Aglykon und Uronsäure sauren Monodesmoside (A 5) über. Bei den sauren Monodesmosiden (A 4) scheinen lediglich die Musennine und Mubenine in größerer Menge als Monodesmoside genuin vorzukommen, während die übrigen gegenüber den entsprechenden Bisdesmosiden nur in untergeordneter Menge auftreten oder auch aus ihnen erst bei der Aufarbeitung gebildet werden. Entsprechendes gilt für die Gruppe A 5, von der nur vom Glycyrrhizin nicht auch eine bisdesmosidische Form bekannt ist.

Die Gruppen A 1 und A 3 können naturgemäß keine Acylglykose-Bindung im Aglykon eingehen. Interessant sind die Vertreter der Gruppe A 6 mit lediglich einer Zuckerkette an der Carboxylgruppe am C-17, die in ihren Eigenschaften den Bisdemosid-Glykosiden ähneln (siehe S. 565).

Eine besondere Gruppe stellen die aus niederen Meerestieren (Echinodermata) isolierten Verbindungen dar. Sie zeigen typische Saponineigenschaften, scheinen aber im Strukturtyp etwas von den pflanzlichen Saponinen abzuweichen. Eine in allen Einzelheiten vollständige Strukturaufklärung einer dieser Verbindungen steht noch aus.

Die bisher in der Struktur bestimmten Triterpensaponine enthalten bis zu 11 Monosaccharideinheiten. Bei den zuckerreicheren Glykosiden findet sich, worauf KOCHETKOV und KHORLIN (*258, 259*) hinwiesen, ein allmählicher Übergang zu den Eigenschaften der Polysaccharide, deshalb führten sie für Glykoside mit mehr als 4 Zuckerresten den Begriff „Oligoside" ein.

Die Monosaccharide können dabei sowohl in Form langer unverzweigter als auch verzweigter oder mehrfach verzweigter (**340, 341**) Zuckerketten miteinander verbunden sein.

Im folgenden sollen (nach den Tabellen der Aglykone) charakteristische Vertreter der einzelnen Typen der Triterpen-Saponine abgehandelt werden, eine vollzählige Besprechung ist wegen der großen Zahl nicht möglich, jedoch sind alle bisher im Aufbau geklärten Vertreter in den Tabellen nach Struktur und Herkunft verzeichnet. Im Rahmen dieser Abhandlung sollen auch typische Methoden der Strukturermittlung derartiger Glykoside erwähnt werden. Die Eigenschaften der Saponine werden im Abschnitt VI zusammenfassend dargestellt.

2. Tabellen der Triterpenaglykone

Tabelle 10. Δ^{12}-*Oleanen-Typ*

Nach steigendem Sauerstoffgehalt geordnete Tabelle der bisher bekannten Triterpene vom Δ^{12}-Oleanen-Typ. Angegeben sind Schmelzpunkt und Drehung der Aglykone und der Acetate bzw. Methylesteracetate. Das für die Bestimmung der Drehung verwendete Lösungsmittel ist in Klammer angegeben (C = Chloroform, B = Benzol, M = Methanol, Ä = Äthanol, P = Pyridin, D = Dioxan, Ac = Aceton)

Δ^{12}-Oleanen

Nr.	Name	OH	C=O	COOH	Sonstiges	Schmp.	$[\alpha]_D$	Literatur
		$C_{30}H_{50}O$						
150	β-Amyrin	3β	—	—	—	199—200°	+87° (C)	(40, 139, 177,
		mono-acetat	—	—	—	235°	+78° (B)	344)
		$C_{30}H_{50}O_2$						
151	Maniladiol	3β, 16β	—	—	—	220—22°	+66° (C)	(40, 139, 177,
		diacetat	—	—	—	203—4°	+84,5° (C)	344)
152	Sophoradiol	3β, 22β (oder 7β) —	—	—	219—20°	+113° (C)	(40, 139, 177,
		diacetat	—	—	—	219—21°	+71,8° (C)	34)
153	Erythrodiol	3β, 28	—	—	—	235—7°	+76° (C)	(40, 139, 177,
		diacetat	—	—	—	186—8°	+63° (C)	344)
		$C_{30}H_{48}O_2$						
154	Aegiceradiol	3β, 28	—	—	Δ^{15}	236—8°	+40,3° (C)	(396)
		diacetat	—	—	Δ^{15}	214—5°	+52,7° (C)	

155	Sojasapogenol C	3β, 24	—	—	Δ^{21}	259—60°	+65° (C)	(40, 177, 344)
		diacetat	—	—	Δ^{21}	202—3°	+59,5° (C)	
156	Oleanolaldehyd	3β	28	—	—	112—86°	—	(40, 177, 344)
		acetat	28	—	—	225—7°	—	
	$C_{30}H_{50}O_3$							
157	Primulagenin A	3β, 16α, 28	—	—	—	248—50°	+55° (C)	(40, 139, 177,
		triacetat	—	—	—	152—60°	−11° (C)	344)
158	Longispinogenin	3β, 16β, 28	—	—	—	247—9°	+53° (C)	(40, 177, 344)
		triacetat	—	—	—	219—21°	+73° (C)	
159	Sojasapogenol B	3β, 21α, 24	—	—	—	260—1°	+90° (C)	(40, 177, 344)
		triacetat	—	—	—	179—80°	+77,5° (C)	
160	22α-Hydroxyerythrodiol	3β, 22α, 28	—	—	—	279—82°	+37° (P)	(609)
	$C_{30}H_{48}O_3$							
161	Saikogenin B	3β, 16β, 28	—	—	$\Delta^{9\,(11)}$	267—9°	+285° (C)	(281)
		triacetat	—	—	$\Delta^{9\,(11)}$	209—10°	+213° (C)	
162	Glycyrrhetol	3β, 30	11	—	—	252—4°	+87° (C)	(76)
		diacetat	11	—	—	253—5°	+110° (C)	
163	Sojasapogenol E	3β, 24	21	—	—	—	—	(577)
		diacetat	21	—	—	234—6°	+31° (C)	
164	Primulagenin D	3β, 16α	28	—	—	217—8°	+25° C)	(539)
		diacetat	28	—	—	162—3°	+24,2° (C)	
165	Gummosogenin	3β, 16β	28	—	—	251—2°	+28° (C)	(40, 177, 344)
		diacetat	28	—	—	219—21°	+66° (C)	
166	α-Boswellinsäure	3α	—	24	—	289°	+114° (C)	(40, 177, 344)
		acetat	—	methylester	—	229—230°	+68° (C)	
166a	β-Peltoboykinolsäure (β-peltoboykinolic acid)	3β	—	27	—	220—2°	+114° (C)	(361b)
		acetat	—	methylester	—	205—6°	+131° (C)	
167	Oleanolsäure (oleanolic acid)	3β	—	28	—	305—10°	+80° (M)	(40, 139, 177,
		acetat	—	methylester	—	217—9°	+70,5° (C)	344)
168	3-Epi-oleanolsäure (3-epi-oleanolic acid)	3α	—	28	—	297—9°	+68° (C)	(40, 177, 344)
		acetat	—	methylester	—	158—9°	+36° (C)	
169	11-Desoxoglycyrrhetinsäure (11-desoxo-glycyrrhetinic acid)	3β	—	30	—	330°	+148° (C)	(80)

Tabelle 10 (Fortsetzung)

Nr.	Name	OH	C=O	COOH	Sonstiges	Schmp.	$[\alpha]_D$	Literatur
170	Katonsäure	3α	—	30	—	285—7°	+47° (C)	(40, 177, 344)
	(katonic acid)	acetat	—	methylester	—	201—2°	+14° (C)	
	$C_{30}H_{46}O_3$							
171	Gymnosporal	6β	3, 11	—	—	313—5°	+65° (C)	(171)
172	Stryphnodendronsapogenin B	3β	—	—	28,21β-lacton	240—3°	−16°	(546)
173	11-Desoxo-glabrolid	3β	—	—	30,22β-lacton	274—8°	+59,6° (Ä)	(81)
	$C_{30}H_{50}O_4$							
174	Barringtogenol A	2α, 3β, 23, 28	—	—	—	290—1°	+18° (P)	(7)
	[Barringtogenol]	tetraacetat	—	—	—	269—270°	+15° (P)	
175	Castanogenol	2β, 3β, 23, 28	—	—	—	263—5°	+90° (Ä)	(397)
		tetraacetat	—	—	—	127—9°	—	
176	Dihydropriverogenin A	3β, 16α, 22α, 28	—	—	—	290—3°	+20° (Ä)	(207, 209, 527,
	[Camelliagenin A]	tetraacetat	—	—	—	133—4°	−19° (C)	528)
177	Chichepegenin	3β, 16β, 22α, 28	—	—	—	321—3°	+43° (C)	(40, 177, 344)
		tetraacetat	—	—	—	280—2°	+26° (C)	
178	Sojasapogenol A	3β, 21α, 22α, 24	—	—	—	310—3°	+103° (C)	(40, 177, 344)
		tetraacetat	—	—	—	228—9°	+85,5° (C)	
179	16-Desoxybarringtogenol C	3β, 21β, 22α, 28	—	—	—	288—90,5°	+50° (Ä)	(604)
		tetraacetat	—	—	—	225—6°	+50° (C)	
179a	Cyclamiretin E	3β, 16α, 28, 30	—	—	—	247—9°	—	(206a)
		tetraacetat	—	—	—	210—1°	—	
	$C_{30}H_{48}O_4$							
180	Priverogenin A	3β, 16α, 22α	28	—	—	198—202°	−4,2° (C)	(527, 528)
		triacetat	28	—	—	227—31°	−35,7° (C)	
181	Cyclamiretin D	3β, 16α, 28	30	—	—	242—4°	+45° (Ä)	(21, 509, 525;
181a	Armillargenin	3β, 16α, 28	21	—	—	290—3°	+20° (Ä)	(316)
		triacetat	21	—	—	206—8°	−35° (C)	
		diacetat	30	—	—	249—252°	+41° (Ä)	
182	Commisäure C	2β, 3β	—	23	—	343—5°	—	(490)
	(commic acid C)	diacetat	—	methylester	—	189—90°	+46° (C)	

183	Erythrodiol-27-säure	3β, 28	—	27	—	184—7°	+124° (P)	(499)
		diacetat	—	27	—	224—6°	+83° (P)	
184	Crataegolsäure [Maslinsäure]	2α, 3β	—	28	—	267—9°	+43° (P)	(70)
	(crataegolic acid [maslinic acid])	diacetat	—	methylester	—	184—6°	+34° (C)	
185	Bredemolsäure	2β, 3α	—	28	—	288—92°	+100,5° (P)	(504)
	(bredemolic acid)	diacetat	—	methylester	—	220—3°	+74° (P)	
185a	2α,3α-Dihydroxy-	2α, 3α	—	28	—	278—82°	+64,6° (C)	(95a)
	olean-12-en-28-säure	2α, 3α	—	methylester	—	296—9°	+58° (C)	
186	Sumaresinolsäure	3β, 6β	—	28	—	298—9°	+54° (C)	(40, 177, 344)
	(sumaresinolic acid)	diacetat	—	methylester	—	258°	+25,3° (C)	
186a	Rubusasäure	3β, 7α	—	28	—	—	—	(37a)
	(rubusic acid)							
187	Echinocystsäure	3β, 16α	—	28	—	305—12°	+33° (C)	(40, 177, 344)
	(echinocystic acid)	diacetat	—	methylester	—	205—8°	−15° (C)	
188	Cochalsäure	3β, 16β	—	28	—	303—6°	+58° (D)	(40, 177, 344)
	(cochalic acid)	diacetat	—	methylester	—	194—6°	+58° (C)	
189	Siaresinolsäure	3β, 19α	—	28	—	279—80°	+39,2° (Ä)	(40, 177, 344)
	(siaresinolic acid)	3-mono-acetat	—	methylester	—	155—7°	+47,5° (C)	
190	Spinosasäure A	3β, 19β	—	28	—	270—2°	+12,5° (C)	(11)
	(spinosic acid A)	diacetat	—	methylester	—	247—50°	+78,1° (C)	
191	Machaerinsäure	3β, 21β	—	28	—	256—8°	+82,4°	(550)
	(machaerinic acid)	diacetat	—	methylester	—	278—80°	+90°	
192	Procersäure	3β, 21β	—	28	18α-H (?)	268—70°	—	(550)
	(proceric acid)	diacetat	—	methylester	—	280—1°	+80,2° (C)	
193	22β-Hydroxyoleanolsäure	3β, 22β	—	28	—	—	—	
	Dihydro-rehmannsäure	22-Angeloyl	—	28	—	296—8°	+70° (C)	(9)
194	Hederagenin	3β, 23	—	28	—	332—4°	+81° (P)	(40, 177, 344)
		diacetat	—	methylester	—	190—3°	+75,3° (C)	
195	Mesembryanthemoidigensäure	3β, 29	—	28	—	305—9°	+70,7°	(544)
	(mesembryanthemoidigenic acid)	diacetat	—	methylester	—	242—3°	+52°	
196	Queretarosäure	3β, 30	—	28	—	318—23°	+77,9° (P)	(40, 177, 344)
	(queretaroic acid)	diacetat	—	methylester	—	211—2°	+68,5° (C)	

Tabelle 10 (Fortsetzung)

Nr.	Name	OH	C=O	COOH	Sonstiges	Schmp.	$[\alpha]_D$	Literatur
197	24-Hydroxy-11-desoxy-glycyrrhetinsäure	3β, 24 3β, 24	— —	30 methylester	— —	— 263—4°	— +134° (C)	(78)
	C₃₀H₄₆O₄							
198	Isomacedonsäure (isomacedonic acid)	3β, 21α diacetat	— —	29 methylester	$\Delta^{9(11)}$ $\Delta^{9(11)}$	— 229—30°	— —	(250, 616a)
199	Lantanolsäure (lantanolic acid)	3α 3α	— —	28 methylester	3β, 25-oxid 3β, 25-oxid	306—9° 197—8°	+151° (C) +156° (C)	(24)
200	Proceragenin A	3β, 16α diacetat	— —	— —	28, 15β-lacton	294—6° 246—8°	−13° (C) −38° (C)	(412)
201	Dumortierigenin	3β, 22α diacetat	— —	— —	28:15β lacton	292—5° 318—21°	−18,6° (C) −10° (C)	(40, 177, 344)
202	Stryphnodendron-Sapogenin F	2β, 3β diacetat	— —	— —	28:21β-lacton	265—7° 222—6°	+5° −48°	(546)
203	Momordicasäure (momordic acid)	3β 3β	1 1	28 methylester	— —	274—6° 217—20°	— —	(353)
204	Desacyl-rehmannsäure Rehmannsäure (rehmannic acid)	22β 22-Angeloyl	3 3	28 28	— —	— 295—300°	— +85° (C)	(40, 177, 344)
205	Machaersäure (machaeric acid)	3β acetat	21 21	28 methylester	— —	309—12° 256—60°	+20° (C) +17° (C)	(40, 177, 344)
206	Gypsogenin	3β acetat	23 23	28 methylester	— —	274—6° 190—2°	+91° (Ä) +82,5° (C)	(40, 139, 177, 344)
207	Liquiritiasäure (liquiritic acid)	3β acetat	11 11	29 methylester	— —	298—303° 245—6°	+86° (Ä) +58° (C)	(80)
208	Glycyrrhetinsäure (glycyrrhetic acid)	3β acetat	11 —	30 methylester	— —	297—8° 299—303°	+161° (C) +140° (C)	(40, 177, 344)
	C₃₀H₄₄O₄							
209	Glabrolid	3β acetat	11 11	— —	30, 22β-lacton	360—5° 332—5°	+76,5° (Ä) —	(81)

Nr.	Name							
210	Isoglabrolid	3β	11	—	29, 18α-	318—25°	+46° (Ä)	*(75)*
	(18α-Oleanon-Derivat)	acetat	11	—	lacton	326—30°	—	
210a	Isomeristotropsäure	3β	6	29	$\Delta^{9\,(11)}$	263—5°	+85° (D)	*(249a)*
	(isomeristotropic acid)	acetat	6	methylester	$\Delta^{9\,(11)}$	271—3°	+109° (C)	
C$_{30}$H$_{50}$O$_5$								
211	A$_1$-Barrigenol	3β, 15α, 16α, 22α, 28	—	—	—	300—2°	+4° (C)	*(147, 208)*
		pentaacetat	—	—	—	279—80°	—	
212	Barringtogenol C	3β, 16α, 21β 22α, 28	—	—	—	286—9°	+11° (P)	*(23, 594, 610)*
	[Theasapogenol B] [Aescinidin] [Jegosapogenol]	3, 21, 22, 28-tetraacetat	—	—	—	222—4°	+20° (C)	
213	Gymnestrogenin	3β, 16β, 21β, 23, 28	—	—		288—9	+53,5 (M)	*(470)*
		pentaacetat	—	—	—	nicht kristallin	+78,2° (C)	
214	Camelliagenin C	3β, 16α, 22α, 23, 28	—	—	—	280—3°	+25,4° (Ä)	*(207, 209, 210)*
C$_{30}$H$_{48}$O$_5$								
215	Aescigenin	3β, 22α, 24, 28	—	—	16α, 21α-oxid	306—12°	+56,5° (Ä)	*(72, 594)*
		tetraacetat	—	—		204—7°	+57,3° (C)	
217	Camelliagenin B	3β, 16α, 22α, 28	23	—	—	200—5°	+48° (Ä)	*(207, 209, 210)*
218	Arjunasäure	2α, 3β, 19α	—	28	—	335	+20° (Ä)	*(409)*
	(arjunic acid)	2,3-diacetat	—	methylester	—	235°	+35° (C)	
219	Stryphnodendron	2α, 3β, 21β	—	28	—	236—39°	—	*(489, 543)*
	Sapogenin K	2α, 3β, 21β	—	methylester	—	228—30°	—	
220	Arjunolsäure	2α, 3β, 23	—	28	—	337—40°	+63,5° (Ä)	*(40, 177, 344)*
	(arjunolic acid)	triacetat	—	28	—	182—3°	+50°	
221	Bayogenin	2β, 3β, 23	—	28	—	324—6°	+85° (P)	*(133)*
		triacetat	—	methylester	—	196—7°	+79°	
222	Entagensäure	3β, 15α, 16α	—	28	—	310—5°	+35° (Ä)	*(22, 25)*
	(entagenic acid)	triacetat	—	28	—	192°	+6° (Ä)	

Tabelle 10 (Fortsetzung)

Nr.	Name	OH	C=O	COOH	Sonstiges	Schmp.	$[\alpha]_D$	Literatur
223	Akaciasäure	3β, 16β, 21β	—	28	28:21-	268—72°	—	(552)
	(acacic acid)	3, 16-diacetat	—	28	lacton	235—6°	—	
224	Acerogensäure	3β, 21β, 22α	—	28	—	308—10°	+66° (Ä)	(296, 297)
	(acerogenic acid)	triacetat	—	methylester	—	212—13°	+54° (C)	
225	Treleasegensäure	3β, 21β, 30	—	28	—	290°	—	(40, 177, 344)
	(treleasegenic acid)	triacetat	—	methylester	—	244—5°	+88° (C)	
226	22β,24-Dihydroxy-	3β, 22β, 24	—	28	—	—	—	(9)
	oleanolsäure	22-Angeloyl	—	28	—	294—8°	+67° (C)	
227	Myrtillogensäure	3β, 16β, 28	—	29	—	288—93°	+84° (M)	(40, 177, 344)
	(myrtillogenic acid)	triacetat	—	methylester	—	146—9°	+77° (C)	
228	Liquiridiolsäure	3β, 21α, 24	—	29	—	285—7°	+45° (C+M)	(77)
	(liquiridiolic acid)	triacetat	—	methylester	—	165—9°	+22° (C)	
229	Jaquiniasäure	3β, 16α, 28	—	30	—	310—2°	+49,3° (C)	(175)
	(jaquinic acid)	3β, 16α, 28	—	methylester	—	245—7°	+32,7° (C)	
	$C_{30}H_{46}O_5$							
230	Bassiasäure	2α, 3β, 23	—	28	Δ^5	316°	+82,4° (P)	(40, 177, 344)
	(bassic acid)	triacetat	—	methylester	Δ^5	148—50°	—	
231	Desacyl-icterogenin	22β, 24	3	28		—	—	(40, 177, 344)
	Icterogenin	22-Angeloyl	3	28	—	239—41°	+64° (C)	
232	Quillajasäure	3β, 16α	23	28	—	292—4°	+56,1° (P)	(40, 177, 344)
	(quillaic acid)	diacetat	23	28	—	180—7°	+4° (C)	
233	24-Hydroxy-liquiritiasäure	3β, 24	11	29	—	282—6°	+79° (C+M)	(77)
	(24-hydroxy-liquiritic acid)	diacetat	11	methylester	—	218—20°	+53° (C)	
234	18α-Hydroxy-	3β, 16α	11	30	—	252—5°	+60° (Ac)	(74)
	glycyrrhetinsäure							
	(18α-Oleanan-Derivat!)	3-acetat	11	methylester	—	237—9°	+61° (C)	
235	24-Hydroxy-	3β, 24	11	30	—	—	—	(543)
	glycyrrhetinsäure	3β, 24	11	methylester	—	247—8°	+142,5°	
236	Glabrinsäure	3β, 26	11	30	—	329—33°	—26°	(31, 134)

Nr.	Name					Smp.	$[\alpha]_D$	Lit.
237	Cincholsäure (cincholic acid)	3β acetat	— —	27, 28 dimethylester	— —	265—8° 248—51°	+118° (P) +98° (P)	(499)
238	Serratagensäure (serratagenic acid)	3β 3β	— —	28, 29 dimethylester	— —	>310° 202—4°	+37,1° (P) +35,7 (C)	(392)
239	Spergulagensäure (spergulagenic acid)	3β acetat	— —	28, 30 28, 30	— —	320—3° 301—3°	+116° (P) +86° (C)	(83, 422)
239a	Gypsogensäure (gypsogenic acid)	3β acetat	— —	23, 28 dimethylester	— —	>380° 179—80°	— —	(65, 68, 138)
240	Liquoriasäure (liquoric acid)	3β acetat	11 11	30 methylester	16α, 21α- oxid	239—41° 256—8°	+7,9° (C) +166,6° (C)	(135, 134)
241	21α-Hydroxy-isoglabroid (18α-Oleanan-Derivat!)	3β, 21α diacetat	11 11	— —	29, 18α- lacton	304—5° 235—6°	−11° (P) +34° (C)	(76)
	$C_{30}H_{50}O_6$							
242	Tanginol	3β, 6β, 7β, 16β, 23, 28 hexaacetat	— —	— —	— —	283—4° 151—3°	+9° (Ä) −63° (Ä)	(420)
243	R₁-Barrigenol [Phyllosapogenol]	3β, 15α, 16α, 21β, 22α, 28 hexaacetat	— —	— —	— —	308—10° 186—7,5°	+37° (C) +28° (C)	(147, 208)
244	Theasapogenol A	3β, 16α, 21β, 22α, 23, 28 3, 21, 22, 23, 28-pentaacetat	— —	— —	— —	301—3° 174—8°	+14° (P) +29° (C)	(610)
245	Gymnemagenin	3β, 16β, 21β, 22α, 23, 28 hexaacetat	— —	— —	— —	329,5—31° 290—1°	— —	(394)
246	Protoaescigenin	3β, 16α, 21β, 22α, 24, 28 hexaacetat	— —	— —	— —	303—11° 140—1°	+31,5° (D) −3,5° (C)	(287, 594)
	$C_{30}H_{48}O_6$							
247	Theasapogenol E [Camelliagenin E]	3β, 16α, 21β, 22α, 28 3, 21, 22, 28-tetraacetat	23 23	— —	— —	237—9° (270—3°) 184—7°	+34° (P) (+64° P) —	(207, 608, 611)

Tabelle 10 (Fortsetzung)

Nr.	Name	OH	C=O	COOH	Sonstiges	Schmp.	$[\alpha]_D$	Literatur
248	Camelliagenin D	$3\beta, 16\alpha, 21\beta$ $22\alpha, 28$	24	—	—	250—8°	+39° (Ä)	(207)
		pentaacetat	diacetat	—	—	221,5—22°	+2,6° (C)	
249	Terminolsäure (terminolic acid)	$2\alpha, 3\beta, 6\beta, 23$ tetraacetat	— —	28 methylester	— —	347° 187—9°	+42° (C) −13° (C)	(40, 177, 344)
249a	Protobassiasäure (protobassic acid)	$2\beta, 3\beta, 6\beta, 23$ $2\beta, 3\beta, 6\beta, 23$	— —	28 methylester	— —	310—2° 198—201°	+22,7° (P) +42,4° (M)	(251a)
250	Platycogensäure C (platycogenic acid C)	$2\beta, 3\beta, 16\beta, 21\beta$ — $2\beta, 3\beta, 16\beta, 21\beta$ —		28 methylester	— —	282—8° 175—8°	+77,8° (Ä) +74,9° (C)	(280)
251	Polygalasäure (polygalacic acid)	$2\beta, 3\beta, 16\alpha, 23$ tetraacetat	— —	28 methylester	— —	300—5° 169—74°	+74,1° (P) +12,9° (Ä)	(277, 406)
252	Tomentosasäure (tomentosic acid)	$2\alpha, 3\beta, 19\beta, 23$ $2\alpha, 3\beta, 19\beta, 23$	— —	28 methylester	— —	328—30° 221—2°	+64° +72° (C)	(410)
253	Caccigenin	$2\alpha, 3\beta, 21\beta, 23$ $2\alpha, 3\beta, 21\beta, 23$	— —	28 methylester	— —	230—57° 273—5°	— +77° (M)	(17, 489)
254	24-Hydroxy-acerogensäure (24-hydroxy-acerogenic acid)	$3\beta, 21\beta, 22\alpha, 24$ — tetraacetat	—	28 methylester	— —	341,5—42° 230—2°	+70° (Ä) +48° (C)	(296, 297)
	$C_{30}H_{46}O_6$							
255	Barringtogensäure (barringtogenic acid)	$2\alpha, 3\beta$ diacetat	— —	23, 28 dimethylester	— —	332—4° 239—40°	+72° (M) +32° (M)	(40, 177, 344)
256	Medicagensäure (medicagenic acid)	$2\beta, 3\beta$ diacetat	— —	23, 28 dimethylester	— —	352—3° 235—8°	+106° (Ä) +87° (C)	(40, 177, 344)
	$C_{30}H_{48}O_7$							
257	Platycodigenin	$2\beta, 3\beta, 16\alpha, 23, 24$ — $2\beta, 3\beta, 16\alpha, 23, 24$ —		28 methylester	— —	250—2° 246—50°	+46,5° (Ä) +44,7° (Ä)	(4, 278)
	$C_{30}H_{46}O_7$							
258	Barringtoniasäure A (barringtonic acid A)	$2\alpha, 3\beta, 19$ triacetat	— —	23, 28 23, 28	— —	285° 146—9°	+19,6° (M) +5° (C)	(28, 411)

258a	Barriniasäure (barrinic acid)	$2\alpha, 3\beta, 19\alpha$	—	23, 28	—	—	—	(*28a*)
		—	—	—	—	—	—	
259	Presenegenin	$2\beta, 3\beta, 27$	—	23, 28	—	$310-11°$	$+91°$ (M)	(*130, 451*)
		$2\beta, 3\beta, 27$	—	dimethylester	—	$223—5°$	$+80°$ (C)	
260	Phytolaccageninsäure	$2\beta, 3\beta, 23$	—	28, 30	—	—	—	(*472*)
	Phytolaccagenin	$2\beta, 3\beta, 23$	—	30-mono-methylester	—	$317—8°$	—	
	$C_{30}H_{46}O_8$							
261	Platycogensäure B (platycogenic acid B)	$2\beta, 3\beta, 16\beta, 21\beta$	—	24, 28	—	$274—7°$	$+101,1°$ (Ä)	(*280*)
262	Platycogensäure A (platycogenic acid A)	$2\beta, 3\beta, 16\alpha, 23$	—	24, 28	—	$243—9°$	$+73,9°$ (Ä)	(*280*)

Tabelle 11. *Triterpenaglykone, die nicht dem Δ^{12}-Oleanentyp angehören*

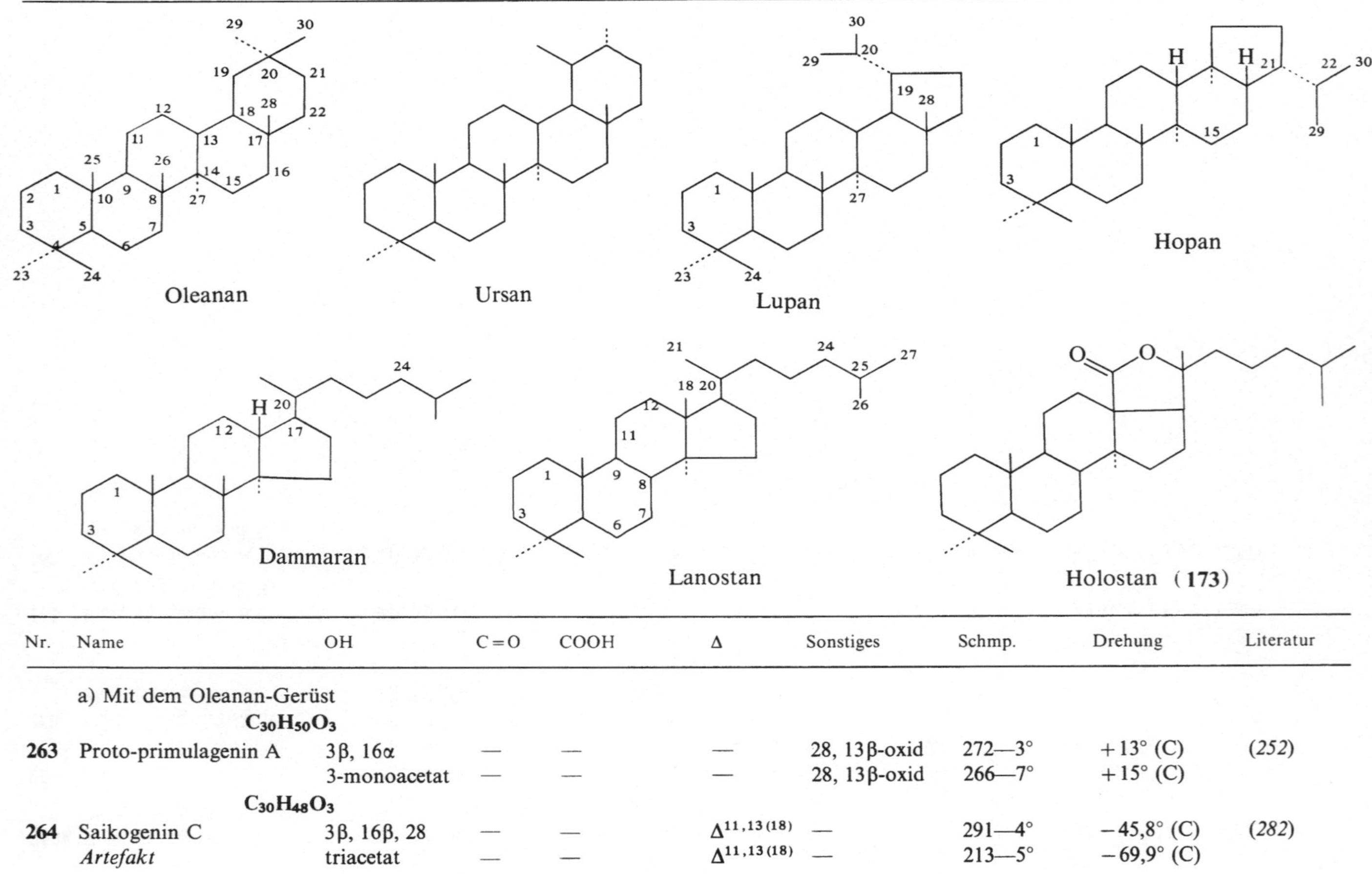

Nr.	Name	OH	C=O	COOH	Δ	Sonstiges	Schmp.	Drehung	Literatur
	a) Mit dem Oleanan-Gerüst								
	$C_{30}H_{50}O_3$								
263	Proto-primulagenin A	3β, 16α	—	—	—	28, 13β-oxid	272—3°	+13° (C)	(252)
		3-monoacetat	—	—	—	28, 13β-oxid	266—7°	+15° (C)	
	$C_{30}H_{48}O_3$								
264	Saikogenin C	3β, 16β, 28	—	—	$\Delta^{11,13\,(18)}$	—	291—4°	−45,8° (C)	(282)
	Artefakt	triacetat	—	—	$\Delta^{11,13\,(18)}$	—	213—5°	−69,9° (C)	

Nr.	Name								
265	Smithiandienol	3β, 23, 28	—	—	$\Delta^{11,13(18)}$	—	295—9°	+65,4° (P)	(46)
		triacetat	—	—	$\Delta^{11,13(18)}$	—	180—1°	−85° (C)	
266	Morolsäure	3β	—	28	Δ^{18}	—	309°	+32° (C)	(40, 177,
	(morolic acid)	monoacetat	—	methylester	Δ^{18}	—	263—4°	+37° (C)	344)
267	Aegicerin	3β	16	—	—	3β, 28-oxid	254—6°	−23,6° (C)	(395)
		monoacetat	16	—	—	3β, 28-oxid	273—5°	−17,7° (C)	
268	Saikogenin E	3β, 16β	—	—	Δ^{11}	13β, 28-oxid	283°	+108° (C)	(276)
		diacetat	—	—	Δ^{11}	13β, 28-oxid	207—11°	+111° (C)	
	C₃₀H₅₀O₄								
269	Priverogenin B	3β, 16α, 22α	—	—	—	13β, 28-oxid	280—3°	−7,9° (M)	(527, 528)
		3, 22-diacetat	—	—	—	13β, 28-oxid	234—6°	−17,3° (C)	
270	Dihydrocycla-miretin A	3β, 16α, 30	—	—	—	13β, 28-oxid	256—9,5°	+15° (Ä)	(509, 525)
		3,30-diacetat	—	—	—	13β, 28-oxid	219—20°	+3° (Ä)	
	C₃₀H₄₈O₄								
271	Saikogenin G	3β, 16α, 23	—	—	Δ^{11}	13β, 28-oxid	238—45°	+83° (C+M)	(276)
272	Saikogenin F	3β, 16β, 23	—	—	Δ^{11}	13β, 28-oxid	265—73°	+96° (C)	(276)
		triacetat	—	—	Δ^{11}	13β, 28-oxid	205—8,5°	+110° (C)	
273	Cyclamiretin A	3β, 16α	30	—	—	13β, 28-oxid	202—4°	+8° (C)	(509, 525)
		3-monoacetat	30	—	—	13β, 28-oxid	142—5°	+5° (P)	
274	Anagalligenon B	3β, 23	16	—	—	13β, 28-oxid	300—10°	−25° (P)	(185)
		diacetat	16	—	—	13β, 28-oxid	235—42°	−23° (C)	
275	Saikogenin D *Artefakt*	3β, 16α, 23, 28	—	—	$\Delta^{11,13(18)}$	—	256—61°	−47,9 (Ä)	(282)
276	Saikogenin A *Artefakt*	3β, 16β, 23, 28	—	—	$\Delta^{11,13(18)}$	—	287—90°	−43,3° (Ä)	(282, 445)
277	Albigensäure (albigenic acid)	3β, 16α	—	28	$\Delta^{13(18)}$	—	246—8°	−13° (Ä)	(29)
		diacetat	—	methylester	$\Delta^{13(18)}$	—	189—90°	−54° (C)	
	C₃₀H₄₆O₄								
278	Cyclamigenin B	3β	16, 30	—	—	13β, 28-oxid	327—8°	+1,5° (C)	(129)
		acetat	16, 30	—	—	13β, 28-oxid	273—5°	−17,7°	
279	Macedonsäure	3β, 21α	—	29	$\Delta^{11,13(18)}$	—	—	—	(248, 616a)
		diketon	—	methylester	$\Delta^{11,13(18)}$	—	242—4°	−50° (C)	

Tabelle 11 (Fortsetzung)

Nr.	Name	OH	C=O	COOH	Δ	Sonstiges	Schmp.	Drehung	Literatur
280	Gypsogeninlacton	3β	23	—	—	28:13β-lacton	330—2°	+30° (Ä)	*(40, 177,*
		monoacetat	23	—	—	28:13β-lacton	262—3°	+65° (C)	*263, 344)*
	C₃₀H₄₄O₄								
281	Meristotropsäure	3β	6	29	$\Delta^{11,13(18)}$	—	—	—	*(249)*
		3β	6	methylester	$\Delta^{11,13(18)}$	—	277—9°	—	
	C₂₉H₄₆O₃								
282	Cyclamigenin A²	3β	16	—	—	13β, 28-oxid, 30-Nor.	208—9°	−17° (C)	*(128)*
	Artefakt	monoacetat	16	—	—	13β, 28-oxid, 30-Nor.	231—3°	−2° (C)	
	C₂₉H₄₂O₄								
283	Eupteleogenin	3β	—	—	Δ^{29}	11α, 12α-oxid 28, 13β-lacton 30-Nor	268—72°	+83° (C)	*(355, 356, 367)*
		monoacetat	—	—	Δ^{29}	11α, 12α-oxid 28, 13β-lacton 30-Nor	321—2°	+87° (C)	
	C₃₁H₅₂O₃								
284	Sojasapogenol D	3β, 24	—	—	$\Delta^{13(18)}$	21α-OCH₃	297—9°	−56,2°	*(40,71,177,*
	Artefakt	diacetat	—	—	$\Delta^{13(18)}$	21α-OCH₃	191—2°	−44°	*344)*

b) Mit dem Ursangerüst

Nr.	Name	OH	C=O	COOH	Δ	Sonstiges	Schmp.	Drehung	Literatur
	C₃₀H₅₀O								
285	α-Amyrin	3β	—	—	Δ^{12}	—	183—7°	+82° (C)	*(40, 177,*
		monoacetat	—	—	Δ^{12}	—	220—7°	+83° (C)	*344)*
	C₃₀H₄₈O₃								
286	Ursolsäure (ursolic acid)	3β	—	28	Δ^{12}	—	285—92°	+72° (C)	*(40, 177,*
		monoacetat	—	methylester	Δ^{12}	—	244—7°	+58° (C)	*344)*
287	Commisäure B (commic acid B)	3β	—	23	Δ^{12}	—	—	—	*(490, 177)*
		acetat	—	methylester	Δ^{12}	—	236—7°	+39° (C)	

Chemie und Biologie der Saponine 519

C₃₀H₄₆O₃

288	Tomentosolsäure (tomentosolic acid) [sanguisorbigenin] *Artefakt*	3β acetat	— —	28 methylester	$\Delta^{12,19}$ $\Delta^{12,19}$	— —	285° 246—8°	+18° (C) +9° (C)	(20)

C₃₀H₄₈O₄

289	Commisäure D (commic acid D)	2β, 3β diacetat	— —	23 methylester	Δ^{12} Δ^{12}	— —	— 170—1°	— +49° (C)	(490, 177)
290	Pomolsäure (pomolic acid)	3β, 19α 3-monoacetat	— —	28 methylester	Δ^{12} Δ^{12}	— —	301—3° 248—9°	+37° (T) +40° (C)	(48)
291	20β-Hydroxy-ursolsäure (20β-Hydroxy-ursolic acid)	3β, 20β 3-acetat	— —	28 methylester	Δ^{12} Δ^{12}	— —	— 252—3°	— +43,7°	(303)

C₃₀H₄₆O₄

292	Pomonsäure	19α 19α	3 3	28 methylester	Δ^{12} Δ^{12}	— —	— 204°	— +50° (C)	(48)

C₃₀H₄₈O₅

293	Commisäure E (commic acid E)	1β, 2β, 3β triacetat	— —	23 methylester	Δ^{12} Δ^{12}	— —	328—30° 225—8°	+104° (P) +56° (C)	(490, 170, 344)
294	Madasiatsäure (madasiatic acid)	2α, 3β, 6β triacetat	— —	28 methylester	Δ^{12} Δ^{12}	— —	248—50° 105—7°	— —	(381)
295	Asiatsäure (asiatic acid)	2α, 3β, 23 3, 23-diacetat	— —	28 methylester	Δ^{12} Δ^{12}	— —	325—7° 124—7° [α]₅₇₈	+56,1° (M) +20° (C)	(40, 384)
296	Centsäure (centic acid)	3β, 5α, 6β	—	28	Δ^{12}	—	250—5°	+38,5° (Ä)	(36)
297	Tormentsäure (tormentic acid)	2α, 3β, 19α 2, 3-diacetat	— —	28 methylester	Δ^{12} Δ^{12}	— —	273° 162°	−20° (P) +5° (C)	(386)

C₃₀H₄₆O₅

298	Chinovasäure (chinovic acid)	3β monoacetat	— —	27, 28 dimethylester	Δ^{12} Δ^{12}	— —	298—303° 219—22°	+99° (P) +98,1° (C)	(40, 499)

C₃₀H₄₈O₆

299	Madecassiasäure (madecassic acid)	2α, 3β, 6β, 23 2, 3, 23-triacetat	— —	28 methylester	Δ^{12} Δ^{12}	— —	265—8° 142—5°	+31° (M) −5,9° (C)	(382)

Tabelle 11 (Fortsetzung)

Nr.	Name	OH	C=O	COOH	Δ	Sonstiges	Schmp.	Drehung	Literatur
300	Indocentosäure (indo-centoic acid)	3β, 5β, 6α, 23 (24)	—	28	Δ^{12}	—	272°	+38,6° (P)	(37)
		3β, 5β, 6α, 23 (24)	—	methylester	Δ^{12}	—	173—5°	—	
301	Centosäure (centoic acid)	3β, 5α, 6β, 23 (24)	—	28	Δ^{12}	—	234—6°	+35,7° (Ä)	(36)
		3, 6, 23-triacetat	—	methylester	Δ^{12}	—	160°	+7,0° (Ä)	
	c) Mit dem Lupan-Gerüst								
	$C_{30}H_{50}O$								
302	Lupeol	3β	—	—	$\Delta^{20\,(29)}$	—	212—6°	+28° (C)	(40, 177,
		monoacetat	—	—	$\Delta^{20\,(29)}$	—	215—20°	+45° (C)	344, 551)
	$C_{30}H_{50}O_2$								
303	Betulin	3β, 28	—	—	$\Delta^{20\,(29)}$	—	260—1°	+15° (C)	(40, 177,
		diacetat	—	—	$\Delta^{20\,(29)}$	—	220—4°	+21°	222, 344)
	$C_{30}H_{48}O_3$								
304	Betulinsäure (betulinic acid)	3β	—	28	$\Delta^{20\,(29)}$	—	323—5°	+12°	(40, 177,
		acetat	—	methylester	$\Delta^{20\,(29)}$	—	200—3°	+18° (C)	198, 344)
	$C_{30}H_{46}O_3$								
305	Thurberogenin	3β	—	—	$\Delta^{20\,(29)}$	28:19β-lacton	283—5°	+11° (P)	(124, 344)
		acetat	—	—	$\Delta^{20\,(29)}$	28:19β-lacton	249—52°	+22° (C)	
	$C_{30}H_{48}O_4$								
306	Stellatogenin	3β, 20	—	—	—	28:19β-lacton	317—9°	+36°	(122, 344)
		3-monoacetat	—	—	—	28:19β-lacton	323—5°	+49°	
	d) Mit dem Dammaran-Gerüst								
	$C_{30}H_{50}O_2$								
307	Genuines Sapogenin F	3β, 12β	—	—	$\Delta^{13\,(17),24}$	—	—	—	(144)
	$C_{30}H_{52}O_3$								
308	(20S)-Protopanaxadiol	3β, 12β, 20 (S)	—	—	Δ^{24}	—	197—200°	+26,7° (C)	(361, 361a)

Nr.									
309	(20R)-Protopanaxadiol	$3\beta, 12\beta, 20\,(R)$	—	—	Δ^{24}	—	236—8	+20,5 (C)	(447, 344)
	Artefakt	diacetat	—	—	Δ^{24}	—	125—7°	−5,6° (C)	
310	Panaxadiol	$3\beta, 12\beta$	—	—	—	20, 25-oxid	250°	+1,0°	(444)
	Artefakt	3-monoacetat	—	—	—	20, 25-oxid	215°	+12°	
	$C_{30}H_{50}O_3$								
311	Genuines Sapogenin A	$3\beta, 6\alpha, 12\beta$	—	—	$\Delta^{13(17),24}$	—	—	—	(144)
	$C_{30}H_{54}O_4$								
312	Panaxgenin F_1	$3\beta, 12\beta, 20\,, 25$	—	—	—	—	257—61°	+15° (M)	(141, 144)
	Artefakt	3, 12-diacetat	—	—	—	—	191—8°	+35,5° (C)	
	$C_{30}H_{52}O_4$								
313	(20S)-Protopanaxatriol	$3\beta, 6\alpha, 12\beta,$	—	—	Δ^{24}	—	—	—	(361)
		20 (S)	—	—	—	—			
314	(20R)-Protopanaxatriol	$3\beta, 6\alpha, 12\beta,$	—	—	Δ^{24}	—	—	—	(448)
	Artefakt	20 (R)	—	—	—	—			
315	Panaxatriol	$3\beta, 6\alpha, 12\beta$	—	—	—	20, 25-oxid	238—9°	+14,2° (C)	(140, 448)
	Artefakt	diacetat	—	—	—	—	268—9°	+24,9° (C)	
	$C_{30}H_{50}O_4$								
316	Bacogenin A_1	$3\beta, 19, 20$	16	—	Δ^{24}	—	242°	−70° (C)	(90)
		3,19-diacetat	16	—	Δ^{24}	—	220°	—	
	$C_{30}H_{54}O_5$								
317	Panaxgenin A_1	$3\beta, 6\alpha, 12\beta,$	—	—	—	—	252—4°	+30° (M+C)	(145)
	Artefakt	$20\zeta, 25$							

e) Mit dem Lanostan-Gerüst

Nr.									
	$C_{30}H_{46}O_3$								
318	Seychellogenin	3β	—	—	$\Delta^{7,9(11)}$	18, 20-lacton	234—8°	−7°	(405)
		monoacetat	—	—	$\Delta^{7,9(11)}$	18, 20-lacton	211—3°	—	
	$C_{30}H_{46}O_4$								
319	Koellikerigenin	$3\beta, 25$	—	—	$\Delta^{7,9(11)}$	18, 20-lacton	213—4°	−8°	(405)
		3-monoacetat	—	—	$\Delta^{7,9(11)}$	18, 20-lacton	213—6°	—	
	$C_{30}H_{44}O_4$								
320	Stichopogenin A_2	$3\beta, 17\alpha$	—	—	$\Delta^{5,8,24}$	18, 20-lacton	238—40°	−48° (C)	(143)
		3-monoacetat	—	—	$\Delta^{5,8,24}$	18, 20-lacton	216—9°	−36,3° (C)	

Tabelle 11 (Fortsetzung)

Nr.	Name	OH	C=O	COOH	Δ	Sonstiges	Schmp.	Drehung	Literatur
321	17-Desoxy-22,25-oxido-holothurinogenin (22,25-oxido-holosta-7,9(11)-dien-3β-ol) *Artefakt*	3β	—	—	$\Delta^{7,9(11)}$	22, 25-oxid 18, 20-lacton	285—6,5°	−9,3° (C)	(86)
		acetat	—	—	$\Delta^{7,9(11)}$	22, 25-oxid 18, 20-lacton	266—6,5°	+21,3° (C)	
	$C_{30}H_{46}O_5$								
322	Stichopogenin A_4	3β, 17α, 25	—	—	$\Delta^{5,8}$	18, 20-lacton	238—40°	—	(143)
		3, 25-diacetat	—	—	$\Delta^{5,8}$	18, 20-lacton	212—6°	—	
	$C_{30}H_{46}O_5$								
323	Griseogenin	3β, 17α, 22	—	—	$\Delta^{7,9(11)}$	18, 20-lacton	285—7°	−22° (C)	(545)
		3, 22-diacetat	—	—	$\Delta^{7,9(11)}$	18, 20-lacton	259—61°	—	
	$C_{30}H_{44}O_5$								
324	22,25-Oxido-holothurinogenin (22,25-Oxido-holosta-7,9(11)-dien-3β,17α-diol) *Artefakt*	3β, 17α	—	—	$\Delta^{7,9(11)}$	22, 25-oxid 18, 20-lacton	315—6°	−21,2° (C)	(86)
		3-monoacetat	—	—	$\Delta^{7,9(11)}$	22, 25-oxid 18, 20-lacton	289—90°	+5,5° (C)	
	$C_{30}H_{46}O_6$								
325	22,25-Oxido-holost-9 (11)-en-3β,12α,17α-triol	3β, 12α, 17α	—	—	$\Delta^{9(11)}$	22, 25-oxid 18, 20-lacton	—	—	(89)
		3, 12-diacetat	—	—	$\Delta^{9(11)}$	22, 25-oxid 18, 20-lacton	240—3	—	
	$C_{31}H_{48}O_4$								
326	Ternaygenin *Artefakt*	3β	—	—	$\Delta^{7,9(11)}$	18, 20-lacton 25-methoxyl	239—42°	+2°	(405)
	$C_{31}H_{48}O_5$								
327	Praslinogenin *Artefakt*	3β, 17α	—	—	$\Delta^{7,9(11)}$	18, 20-lacton 25-methoxyl	290—1,5°	—	(541)
		3-monoacetat	—	—	$\Delta^{7,9(11)}$	18, 20-lacton 25-methoxyl	271—4°	—	

C₃₁H₄₆O₆

328	22,25-Oxido-holost-9(11)-en-3β,17α-diol-12β-methoxyl	3β, 17α	—	—	$\Delta^{7,9\,(11)}$	22, 25-oxid 18, 20-lacton 12β-methoxyl	—	—	(89)
	Artefakt	3-monoacetat	—	—	$\Delta^{7,9\,(11)}$	22, 25-oxid 18, 20-lacton 12β-methoxyl	273—4°	−60° (C)	

f) Mit dem Hopan-Gerüst

C₃₀H₄₈O₂

329	Mollugogenol B	3β, 6α	—	—	$\Delta^{15,17\,(21)}$	—	222—5°	+91° (C)	(85)
		diacetat	—	—	$\Delta^{15,17\,(21)}$	—	194—7°	—	

C₃₀H₄₆O₂

330	Mollugogenol C	3β	6	—	$\Delta^{15,17\,(21)}$	—	215—20°	+39,8° (C)	(84a)

g) Mit dem Isohopan-Gerüst

C₃₀H₅₂O₄

331	Mollugogenol A	3β, 6α, 16β, 22	—	—	—	—	250—2°	+58,5° (P)	(82)

C₃₀H₅₀O₄

332	Mollugogenol E	3β, 16β, 22	6	—	—	—	317—9°	+51° (P)	(84)
		3, 16-diacetat	6	—	—	—	258—60°	+59,3° (C)	

3. Neutrale, monodesmosidische Glykoside (A 1) (siehe Tabelle 12)

Glykoside der insgesamt gesehen seltener vorkommenden neutralen Triterpenaglykone können entweder zu den neutralen Monodesmosiden (A 1), den im Zuckerteil sauren (A 3) oder den sich von beiden Gruppen ableitenden Estersaponinen (A 2) gehören. In allen bisher bekannten Fällen findet sich die Zuckerkette am OH an C-3 des Aglykons. Vertreter der neutralen Monodesmoside sind in der Natur bisher seltener beobachtet worden. In ihrem chemischen Aufbau und in ihren Eigenschaften erscheint diese Gruppe den monodesmosidischen Steroidsaponinen am meisten ähnlich. So bilden sie ebenfalls recht schwerlösliche Cholesterinkomplexe und zeigen starke hämolytische Wirksamkeit. Mit Ausnahme der Glykoside aus *Panax ginseng* bzw. *P. japonica* sind von dieser Gruppe keine zugehörigen bisdesmosidischen Vertreter bekannt geworden. Es ist ungewiß, ob es hier auch eine andersartige, weniger cytotoxische Transportform gibt.

Besonders interessant in dieser Gruppe erscheint das schon lange in kristalliner Form bekannte Cyclamin aus den Wurzelknollen von *Cyclamen europaeum* L. (europäischen Alpenveilchen). Es hat im Vergleich zu anderen Saponinen einen bemerkenswert hohen hämolytischen Index.

Schwierigkeiten bereitet bei der Strukturermittlung in diesem Fall die Isolierung des genuinen Aglykons. Bei der üblichen sauren Hydrolyse erhält man Cyclamiretin D (**181**) (*21*), erst durch eine sehr schonende saure Hydrolyse läßt sich das genuine Aglykon Cyclamiretin A (**273**) (*509, 525*) neben einer Reihe von Sekundärprodukten gewinnen. Cyclamiretin A enthält eine 13β,28-Oxid-Gruppierung, die mit Säure leicht zum Δ^{12}-28-Alkohol geöffnet wird. Ähnliche Verhältnisse werden auch bei den Glykosiden von Aegicerin (**267**) (*395*), Protoprimulagenin A (**263**) (*252*), Saikogenin F (**272**), E (**268**) und G (**271**) (*275, 276*), Priverogenin B (**269**) (*527, 528*), Cyclamigenin B (**278**) (*129*) und Anagalligenon B (**274**) (*185*) beobachtet. Es konnte jedoch nachgewiesen werden, daß nicht alle isolierten Δ^{12}-28-Hydroxy-aglykone Artefakte sind, die bei der Hydrolyse erst aus 13β,28-Oxiden entstehen. So kommen Barringtogenol C (**212**) und Protoaescigenin (**245**) in *Aesculus hippocastanum* und *A. turbinata* genuin in den Glykosiden vor (*594, 607*).

Daß im Cyclamin tatsächlich Cyclamiretin A das Aglykon bildet, konnte durch RuO_4-Oxidation des per-trimethylsilylierten Saponins zum 28,13β-Lacton bewiesen werden (*509*).

Die Strukturaufklärung der Zuckerkette erfolgte weitgehend analog der des Digitonin. Durch HBr/Eisessig-Abbau erhielt man vor allem ein Tetrasaccharid (Cyclamotetraose), ein Trisaccharid (Cyclamotriose A) und ein bereits bekanntes Disaccharid, die Laminaribiose. Enzymatische

Tabelle 12. *Neutrale Monodesmoside*

| 333 | Ginseng-Prosapogenin | β-D-Gl | 1→2 | β-D-Gl | 1→3 [Protopanaxadiol (308)] *Panax ginseng* (Rhizome, Wurzeln) | (201, 361a) |

| 334 | Chikusetsusaponin III | β-D-Gl | 1→2 | β-D-Xy | 1→6 | β-D-Gl | 1→3 [20(S)-Protopanaxadiol (308)] *Panax japonica* (Wurzeln, Rhizome) | (268) |

335 Saikosaponin c

β-D-Gl 1→6
α-L-Rh 1→4   β-D-Gl 1→3 [Saikogenin E (268)]
Bupleurum falcatum (Wurzeln) (275)

336 Saikosaponin a   β-D-Gl 1→3 β-D-Fu 1→3 [Saikogenin F (272)]
Bupleurum falcatum (Wurzeln) (275)

337 Saikosaponin d   β-D-Gl 1→3 β-D-Fu 1→3 [Saikogenin G (271)]
Bupleurum falcatum (Wurzeln) (275)

338 Desgluco-cyclamin I   β-D-Xy 1→2 β-D-Gl 1→4
β-D-Gl 1→2   α-L-Ar 1→3 [Cyclamiretin A (273)]
Cyclamen europaeum (Knollen) (517)

339 Desgluco-cyclamin II   β-D-Gl 1→3
β-D-Xy 1→2   β-D-Gl 1→4 α-L-Ar 1→3 [Cyclamiretin A (273)]
Cyclamen europaeum (Knollen) (517)

Tabelle 12 (Fortsetzung)

340 Cyclamin

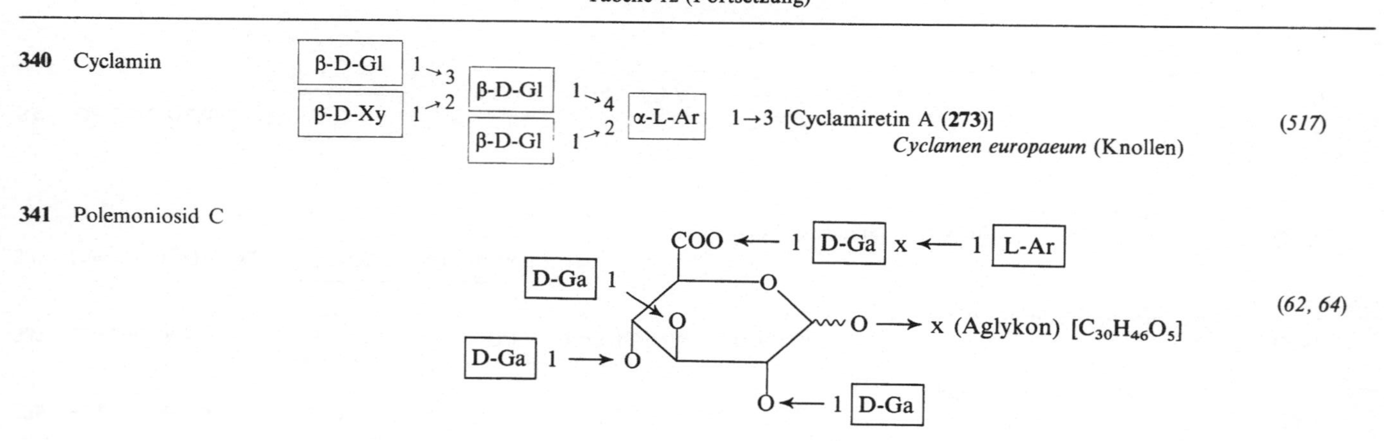

(517)

341 Polemoniosid C

(62, 64)

Spaltung des Cyclamins ergab ein Desglucocyclamin und ein Bisdesglucocyclamin. Die Ergebnisse der Methylierung und Spaltung der Desglucoprodukte und der Oligosaccharide führten zur Aufstellung der Struktur der Zuckerkette des Cyclamins (**340**) (*517*). Bemerkenswert ist das Vorliegen einer zweimaligen Verzweigung, die hier zum ersten Mal beobachtet und die inzwischen auch im Saponarosid D (**446**) (*305*) und im Dianthosid C (**445**) (*67*) nachgewiesen wurde. Neben dem Cyclamin finden sich in der Pflanze in geringer Menge auch zuckerärmere Derivate (**338, 339**) sowie Glykoside mit anderen Aglykonen wie z. B. Cyclamigenin B (**278**) (*129*). Aglykone mit 13β,28-Oxidogruppierung enthalten auch die neutralen Saikosaponine a, c und d (**336, 335, 337**) aus *Bupleurum falcatum,* die jedoch zuckerärmer sind (*275*). Die aus *Panax ginseng* isolierten Glykoside dieser Gruppe (**333, 334**) stellen unpolarere Nebenglykoside dar. Die genuinen Hauptglykoside sind zu den neutralen Bisdesmosiden zu rechnen.

4. Estersaponine (A 2) (siehe Tabelle 13)

Eine recht kleine Gruppe stellen die Estersaponine dar. Jedoch gibt die geringe Zahl von strukturell geklärten Vertretern ein falsches Bild von der Verbreitung dieses Saponintyps. Da diese Gruppe wohl zu den kompliziertest aufgebauten Saponinen gehört, ist die in manchen Fällen schon längere Zeit betriebene Strukturermittlung vielfach noch nicht abgeschlossen. Estergruppierungen sind in Saponinen zahlreich nachgewiesen worden (z. B. *9, 26, 27, 191, 193, 296, 297, 320, 370, 435, 453, 457, 471, 527, 606, 612a*).

An esterartig gebundenen Säuren wurden bisher Ameisensäure, Essigsäure, n-Buttersäure, Isobuttersäure, Isovaleriansäure, α-Methylbuttersäure, Angelicasäure, Tiglinsäure, 3,3-Dimethylacrylsäure (*191*), *cis*-Hexen-(2)-säure (*612a*), *trans, trans-* und *cis, trans*-Nonadien-(2,4)-säure (*296, 297*), Benzoesäure (*27*), Zimtsäure (*370*), p-Methoxyzimtsäure (*451*), 3,4-Dimethoxyzimtsäure (*453*), Ferulasäure (*457*) und N-Methylanthranilsäure (*320*) in Saponinen nachgewiesen.

Vertreter dieses Typs zeichnen sich oft durch bemerkenswerte physiologische Effekte aus und stellen zum Teil wertvolle Arzneimittel dar. So haben die in diese Gruppe gehörenden Aescin (**342**) (*557*) und Theasaponin (**343**) (*558*) eine ausgeprägte antiexsudative Wirkung, Acer-Saponin (*296*) eine spezifisch hemmende Wirksamkeit gegen Tumore, Senegin (**344**) (*467*) eine erhebliche expectorierende Wirksamkeit und die Gymnemasäure (*471*) stellt den die Empfindung „süß" blockierenden Wirkstoff aus *Gymnema silvestre* dar.

Tabelle 13. *Estersaponine*

342 Aescin
(Hauptglykosid)

β-D-Gl 1→4
β-D-Gl 1→2   β-D-Glr 1→3 [Protoaescigenin (**246**)]-21-angelat(tiglat)-22-acetat   (*594*)
Aesculus hippocastanum (Früchte)

343 Theasaponin

β-D-Ga 1→3
β-D-Xy 1→2 α-L-Ar 1→2   β-D-Glr 1→3 [Theasapogenol À (**244**)]-21-angelat(tiglat)-22-acetat (*530*)
Thea sinensis (Früchte)

344 Senegin II

β-D-Gl 1→3 [Presenegin (**259**)] 28←1 D-Fu 2←1 L-Rh 4←1 D-Xy 4←1 β-D-Ga   (*453*)

Polygala senega (Wurzeln)

CH_3O—CH=CH—CO_2 (4)
CH_3O—

345 Sanguisorba-Saponin

β-L-Ar 1→3 [Ursolsäure (**286**)] 28←1 β-D-Gl   (*63*)
3
↑
O_2C-CH_3

Sanguisorba officinalis (Wurzeln)

Das erste in dieser Gruppe in der Struktur bestimmte Saponin war das „Aescin" aus *Aesculus hippocastanum*, an dessen Strukturaufklärung verschiedene Arbeitsgruppen (R. KUHN und I. LÖW; J. WAGNER; R. TSCHESCHE und G. WULFF) beteiligt waren. Aufgrund aller vorhandenen Daten gelang es WULFF und TSCHESCHE (*594*) 1969, eine Strukturformel aufzustellen, die auch von den anderen Gruppen akzeptiert werden konnte (*564*).

Da „Aescin" trotz guter Kristallisationstendenz und scheinbarer Homogenität eine sehr komplexe Mischung kaum trennbarer Verbindungen darstellt, mußte die Strukturermittlung mit dem Gemisch durchgeführt werden. Durch Spaltungsreaktionen konnten einheitliche Teilstücke erhalten werden, aus deren Strukturaufklärung auf den Bau der Einzelkomponenten und die Zusammensetzung des Glykosidgemisches geschlossen werden konnte. Diese Art der statistischen Strukturermittlung (*594*) war möglich, da allen im „Aescin" vorliegenden Komponenten ein gemeinsames Bauprinzip zugrunde lag.

An die im „Aescin" vorhandenen Aglykone [Protoaescigenin (**246**)] und Barringtogenol C (**212**) sind Essigsäure, α-Methylbuttersäure, Isobuttersäure, Tiglinsäure und Angelicasäure laut gaschromatographischer Analyse im Verhältnis 8 : 1 : 1 : 6 : 4 bis 10 : 1,5 : 1 : 4 : 3,5 (*287, 564, 594*) esterartig gebunden. Die Lokalisierung dieser Säuren im Glykosid bereitete außerordentliche Schwierigkeiten, da bereits unter recht milden Bedingungen Acylwanderungen an den OH-Gruppen im Ring D und E eintraten (*309, 562, 564, 594*). Auch bei ähnlich gebauten Aglykonen wurden mehrfach entsprechende Wanderungen beobachtet (*187, 527, 528, 612*). Insgesamt kennt man bereits folgende Wanderungstendenzen:

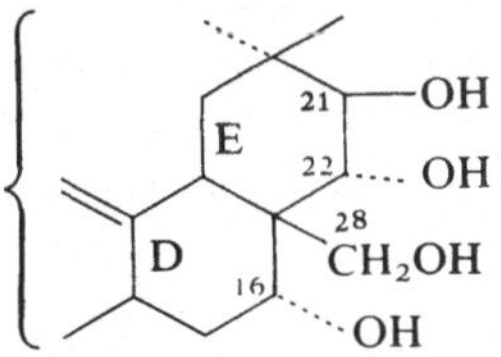

$$21\,\beta \rightleftharpoons 28, \quad 22\,a \rightleftharpoons 28, \quad 16\,a \rightleftharpoons 28$$
$$22\,a \rightleftharpoons 16\,a, \quad 21\,a \rightleftharpoons 22\,\beta.$$

Eine Hydrolyse des „Aescins" ohne die Gefahr der Acylwanderung erreichte man mit einem Enzymgemisch aus *Helix pomatia*, eine einwandfreie Lokalisierung der Säuren gelang durch Röntgenstrukturanalyse (*196*) oder NMR-Spektroskopie (*594*) des zur Hauptsache vorliegenden Acylaglykons. Demnach befinden sich Angelica- und Tiglinsäure am OH an C-22 und Essigsäure an C-21. Die als Nebenbestandteile vorhandenen Säuren konnten nicht einwandfrei lokalisiert werden. Unabhängig und auf andere Weise kamen WAGNER, HOFFMANN und LÖW (*562*) zum gleichen Ergebnis.

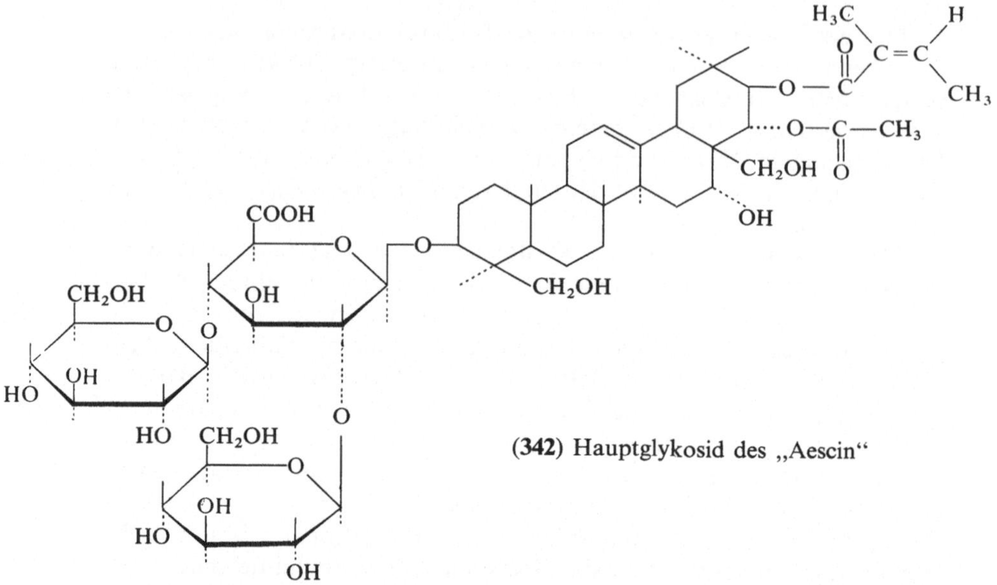

(**342**) Hauptglykosid des „Aescin"

Als Zuckerbestandteile findet man nach saurer Hydrolyse im „Aescin"
D-Glucuronsäure, D-Glucose, D-Xylose und D-Galaktose im Verhältnis
1,00 : 1,58 : 0,23 : 0,19 (*287, 594*). Nach dem Ergebnis der Methylierung
und Spaltung besteht die Zuckerkette in den Hauptglykosiden aus einem
Mol. D-Glucuronsäure, an die in 2- und 4-Stellung 2 Moll. D-Glucose
geknüpft sind, in den Nebenglykosiden kann eine D-Glucose zum Teil
durch D-Galaktose oder durch D-Xylose ersetzt sein. Damit ergibt sich
für den Hauptbestandteil im Aescin die Struktur **342** (*594*).

Insgesamt hat man nach Untersuchungen an partiell aufgetrennten
Glykosiden (*287, 561, 594*) bei zwei Aglykonen, drei verschiedenen Zuckerketten und 5 verschiedenen Säuren weit mehr als 30 verschiedene Einzel-
individuen im „Aescin" zu erwarten, von denen die häufigsten in Tabelle 14
angegeben sind.

Tabelle 14. *Die Zusammensetzung der im „Aescin" vorhandenen Hauptbestandteile*

Triterpenanteil	Zucker	Säuren	%-Gehalt im Aescin nach 594	nach 564
Protoaescigenin	D-Glr, D-Gl, D-Gl	Angelica- + Essigsäure	23%	19%
Protoaescigenin	D-Glr, D-Gl, D- Gl	Tiglin- + Essigsäure	15%	22%
Protoaescigenin	D-Glr, D-Gl, D-Xy	Angelica- + Essigsäure	9%	5%
Protoaescigenin	D-Glr, D-Gl, D-Xy	Tiglin- + Essigsäure	6%	6%
Protoaescigenin	D-Glr, D-Gl, D-Ga	Angelica- + Essigsäure	7%	1%
Protoaescigenin	D-Glr, D-Gl, D-Ga	Tiglin- + Essigsäure	5%	1%
Barringtogenol C	D-Glr, D-Gl, D-Gl	Angelica- + Essigsäure	5%	< 1%
Barringtogenol C	D-Glr, D-Gl, D-Gl	Tiglin- + Essigsäure	3%	< 1%

Literaturverzeichnis: SS. 576—606

Das früher beschriebene wasserlösliche α-Aescin besteht aus einer Mischung von normalem Aescin (β-Aescin) und Kryptoaescin; letzteres stellt ein Artefakt dar, das durch Acylwanderung aus dem β-Aescin gebildet wird (*564*).

Offenbar ist dieser im Aescin neu aufgefundene Saponintyp in der Natur weiter verbreitet. Ganz allgemein besteht er aus einem hochhydroxylierten Aglykon mit zwei Moll. esterartig gebundenen Säuren am C-21 und C-22 und meistens einer Glucuronsäure am C-3. Hierzu gehört auch das dem Aescin sehr ähnlich gebaute Theasaponin (**343**) (*530*), ebenfalls eine komplexe Mischung. Weitere in der Struktur nicht vollständig geklärte Vertreter dieses Typs dürften in *Gymnema silvestre* (*471*), *Aesculus turbinata* (*603*), *Styrax japonica* (*606*), *Styrax officinalis* (*435*), *Barringtonia acutangula* (*26, 27*), *Acer negundo* (*296, 297*) und *Sanicula officinalis* (*191*) vorkommen.

Daneben finden sich auch strukturell meist noch nicht bekannte Estersaponine, die sicherlich einen andersartigen Aufbau besitzen. So enthält das in der Struktur aufgeklärte Senegin II aus *Polygala senega (Radix senegae)*, eine 3,4-Dimethoxyzimtsäure im Zuckerteil an eine Fucose gebunden, die an die Carboxylgruppe an C-17 geknüpft ist (**344**) (*453*).

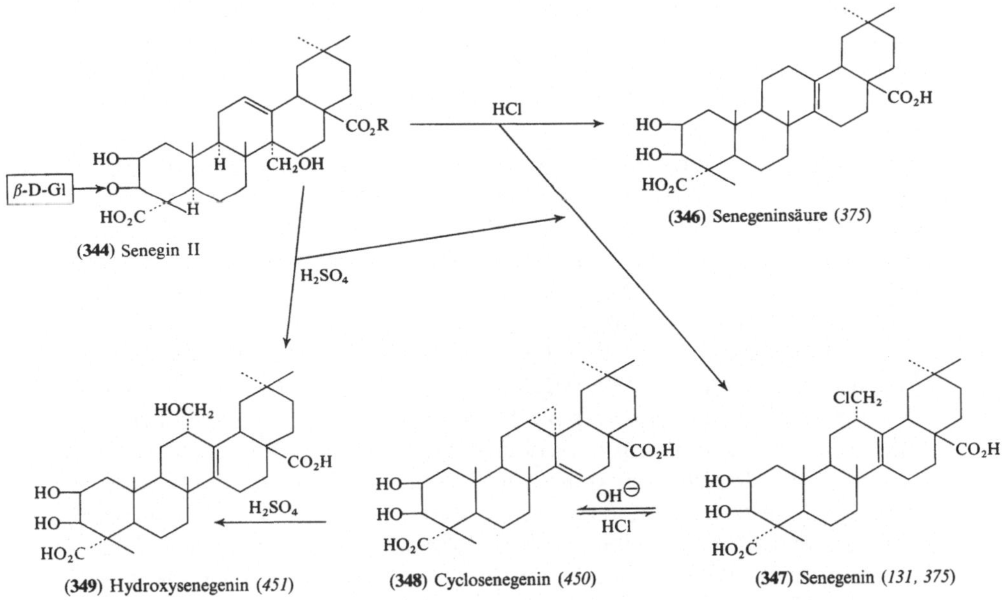

Schema 4. Bei saurer Hydrolyse von Senegin erhaltene Umwandlungsprodukte des Presenegenins

Im Falle der Saponine aus *Radix senega* hat besonders die Ermittlung des genuinen Aglykons erhebliche Schwierigkeiten bereitet, da es sich sehr leicht mit Säure strukturell verändert und bei Verwendung von HCl sogar Chlor in die Molekel eingeführt wird. Erst die schonende Hydrolyse (*130, 451, 602*) erbrachte das genuine Presenegin (**259**). Das Schema 4 zeigt die beobachteten Umwandlungen.

Im Falle des Senegins ist der saure Charakter im Gegensatz zum Aescin durch eine Carboxylgruppe im Aglykon bedingt.

Avenacin ist ein neutrales Estersaponin (*320*) aus den Wurzelspitzen von *Avena sativa*, das aus dem strukturell noch ungeklärten Avenagenin, N-Methylanthranilsäure, 2 Moll. D-Glucose und 1 Mol. L-Arabinose aufgebaut ist (*508*).

5. Durch Uronsäure saure Monodesmoside (A 3) (siehe Tabelle 15)

Zur Gruppe A 3 der Saponine gehören Glykoside mit neutralen Aglykonen, die durch eine Uronsäure im Zuckerteil sauren Charakter aufweisen. Bei den bisher in der Struktur geklärten Saponinen dieser Gruppe handelt es sich immer um Derivate der D-Glucuronsäure, die direkt an das OH am C-3 des Aglykons gebunden ist. Es erscheint zweifelhaft, ob D-Galakturonsäure überhaupt in Saponinen vorkommt, nachdem die als solche angesehene Uronsäure aus Primula-Arten tatsächlich D-Glucuronsäure ist (*327*).

Zahlreiche Saponine dieses Typs tragen zusätzlich Estergruppierungen und sind daher im vorigen Abschnitt abgehandelt worden. Eine größere Verbreitung dürfte der hier besprochene Typ in der Familie der Leguminosen haben. Die vor allem in den Samen vorhandenen Saponine verdienen als Bestandteile der menschlichen Nahrung und als mögliche Abwehrstoffe der Pflanzen gegen Insekten eine gewisse Beachtung (*12, 13*). Seit langem schon werden die Saponine aus Soja-Bohnen untersucht, eine vollständige Strukturaufklärung ist indes noch nicht gelungen (*160, 161*).

Vor kurzem wurde von V. J. Chirva und Mitarb. über die vollständige Strukturaufklärung von Phaseolosid D (**350**) (*110*) und E (**351**) (*99*) aus den Samen von *Phaseolus vulgaris* (Bohne) berichtet. Es handelt sich um relativ zuckerreiche Glykoside, wobei im Phaseolosid E (**351**) mit 8 Zuckerresten die längste bisher bekannte Einzelzuckerkette vorliegt. Die Strukturaufklärung gestaltete sich daher recht schwierig. Durch Partialhydrolyse konnten verschiedene zuckerärmere Glykoside gewonnen werden, deren Spaltung nach Methylierung dann die zur Aufstellung einer Strukturformel wichtigen Ergebnisse lieferte. Aus dem Vergleich der Molrotationen nach Klyne (*256*) wurden dann auch die Konfigurationen an den glykosidischen Bindungen bestimmt. Danach soll in beiden Glykosiden

Tabelle 15. *Durch eine Uronsäure saure Monodesmoside*

350 Phaseolosid D

α-L-Rh 1→6 / α-D-Ga 1→2 β-D-Gl 1→4 α-D-Ga 1→2 α-L-Ar 1→3 β-D-Glr 1→ → 3 [Soyasapogenol C (**155**)] *Phaseolus vulgaris* (Samen) (*110, 304*)

351 Phaseolosid E

α-L-Rh 1→6 β-D-Gl 1→4 α-D-Ga 1→2 α-L-Ar 1→ β-D-Gl 1→4 β-D-Gl 1→4 α-D-Ga 1→2 → 3 β-D-Glr 1→3 [Soyasapogenol C (**155**)] *Phaseolus vulgaris* (Samen) (*99, 304*)

352 Primulasaponin

α-L-Rh 1→3 (2) / β-D-Gl 1→3 β-D-Ga 1→2 (3) β-D-Glr 1→3 [Primulagenin A (**157**)] *Primula elatior* (Wurzeln, Rhizome) (*498*)

353 Polemoniosid B

D-Ga 1→2 / D-Ga 1→3 / D-Ga 1→4 D-Glr 1→ [$C_{30}H_{46}O_5$] *Polemonium coeruleum* (*64*)

die Galaktose α-glykosidisch vorliegen. Es muß jedoch betont werden, daß bei einer so langen Zuckerkette die Berechtigung der Verwendung der Additivität der Molrotationen zur Klärung derartiger Fragen zweifelhaft ist, daher sollten nach Möglichkeit andere Methoden zur Kontrolle derartiger Aussagen herangezogen werden.

Ebenfalls in diese Gruppe gehören die Saponine aus den Wurzeln und Rhizomen von Primula-Arten. Das Hauptsaponin aus *Primula elatior* konnte durch Kristallisation gereinigt werden (*498*). Eine früher vorgeschlagene Struktur (*539*) erwies sich als korrekturbedürftig (*527*) und mußte durch eine neue ersetzt (**352**) (*498*) werden.

6. Durch das Aglykon saure Monodesmoside (A 4) (siehe Tabelle 16)

Die mit Abstand stärkste Gruppe von Triterpensaponinen sind die sauren Monodesmoside, deren acider Charakter durch eine Carboxylgruppe im Aglykon hervorgerufen wird. Mit wenigen Ausnahmen handelt es sich immer um in recht geringer Menge vorkommende, bei der Aufarbeitung erst entstandene oder durch chemische Reaktion aus den entsprechenden neutralen Bisdesmosiden (B 1) gebildete Verbindungen.

Wie im Fall der Saponine des Efeus *(Hedera helix)* nachgewiesen wurde, finden sich als genuine Bestandteile in den Blättern die neutralen Bisdesmoside, Hederacosid B und C. Daneben liegen in nur geringer Menge die zugehörigen sauren Monodesmoside α- und β-Hederin (*521*) vor. Arbeitet man jedoch die Blätter wäßrig bei Zimmertemperatur auf und läßt die zerkleinerten Blätter über Nacht in Wasser stehen, so läßt sich kein Hederacosid A und B mehr nachweisen. An ihrer Stelle isoliert man entsprechende Mengen α- und β-Hederin (*590*). Offenbar finden sich in den Blättern Enzyme, welche bei ihrer Freisetzung die Zucker an der Carboxylgruppe abspalten. Diese Tatsache ist vor allem deswegen interessant, weil die Monodesmoside im Gegensatz zu den Bisdesmosiden hämolytisch und antibiotisch aktiv sind.

Eine große Rolle für die Strukturermittlung der zugehörigen Bisdesmoside spielen die durch alkalische Hydrolyse hergestellten Verbindungen dieses Typs [z. B. Patrinosid D_1 (**361**), Clematosid A' (**362**), Helianthosid A (**380**)]. An ihnen ist es möglich, die Zuckerkette am C-3 getrennt von der am Carboxyl an C-17 in der Struktur zu ermitteln. Hierzu sind die üblichen Methoden geeignet, die jedoch beim Vorliegen längerer, vor allem unverzweigter Ketten nur mit erheblichem Aufwand zu einwandfreien Ergebnissen führen.

Von den in der Tabelle 16 verzeichneten Vertretern scheinen lediglich die Musennine und Mubenine in der überwiegenden Menge als Monodesmosid vorzuliegen, wenngleich in diesen Fällen auch die entsprechenden Bisdesmoside vorhanden sein dürften.

Literaturverzeichnis: SS. 576—606

Tabelle 16. *Durch das Aglykon saure Monodesmoside*

354 Oleanolsäure-glucosid	β-D-Gl 1→3 [Oleanolsäure (**167**)]	*Beta vulgaris* (Wurzeln)	*(372)*
355 β-Hederin (Eleutherosid K)	α-L-Rh 1→2 α-L-Ar 1→3 [Oleanolsäure (**167**)]	*Hedera helix* (Blätter) *(Eleutherococcus senticosus)* (Blätter)	*(521, 590)* *(156a)*
356 Mubenin B	L-Rh 1→4 L-Ar 1→3 [Oleanolsäure (**167**)]	*Stauntonia hexaphylla* (Samen)	*(485)*
356a Eleutherosid J	α-L-Rh 1→4 α-L-Ar 1→3 [Oleanolsäure (**167**)]	*Eleutherococcus senticosus* (Blätter)	*(156a)*
357 Scabiosid B	β-D-Gl 1→4 α-L-Ar 1→3 [Oleanolsäure (**167**)]	*Patrinia scabiosofolia* (Wurzeln)	*(59)*
358 Calendulosid A	β-D-Ga 1→4 β-D-Gl 1→3 [Oleanolsäure (**167**)]	*Calendula officinalis* (Wurzeln)	*(554, 584)*
359 Patrinosid C$_1$	β-D-Xy 1→2 α-D-Gl f 1→2 α-L-Rh 1→3 [Oleanolsäure (**167**)]	*Patrinia intermedia* (Wurzeln)	*(59a, 60)*
360 Mubenin A	D-Gl 1→6 D-Gl 1→2 L-Rh 1→4 L-Ar 1→3 [Oleanolsäure (**167**)]	*Stauntonia hexaphylla* (Samen)	*(485)*
361 Patrinosid D$_1$	β-D-Gl 1→3 / β-D-Xy 1→2 \ α-L-Rh 1→4 β-D-Xy 1→3 [Oleanolsäure (**167**)]	*Patrinia intermedia* (Wurzeln)	*(60)*
362 Clematosid A′	D-Gl 1→4 D-Xy 1→2 L-Ar 1→2 L-Ar 1→4 L-Rh 1→3 [Oleanolsäure (**167**)]	*Clematis mandschurica* (Wurzeln)	*(109)*
363 Ursolsäure-arabinosid	β-L-Ar 1→3 [Ursolsäure (**286**)]	*Sanguisorba officinalis* (Wurzeln)	*(63)*

Tabelle 16 (Fortsetzung)

364	Empetrosid C	L-Ar 1→4 D-Gl 1→3 [Ursolsäure (286)]	*Empetrum sibiricum*	*(61)*
365	Koelreuteria-Saponin A (Leontosid A) (Scabiosid A) (Caulophyllum Saponin A) (Akebia-Saponin A)	α-L-Ar 1→3 [Hederagenin (194)]	*Koelreuteria paniculata* (Früchte) *(Leontice eversmannii)* (Knollen) *(Patrinia scabiosaefolia)* (Wurzeln) *(Caulophyllum robustum)* *(Akebia quinata)* (Samen)	*(101)* *(357)* *(59)* *(354)* *(189b)*
366	Hederagenin-rhamnosid	α-L-Rh 1→3 [Hederagenin (194)]	*Hedera pastuchovii* (Blätter)	*(204)*
366a	Akebia-Saponin B	β-D-Xy 1→2 α-L-Ar 1→3 [Hederagenin (194)]	*Akebia quinata* (Samen)	*(189b)*
367	α-Hederin (Kalopanaxsaponin A) (Sapindosid A)	α-L-Rh 1→2 α-L-Ar 1→3 [Hederagenin (194)]	*Hedera helix* (Blätter) *(Kalopanax septemlobum)* (Wurzeln) *(Sapindus mukorossi)* (Früchte)	*(521)* *(259)* *(239)*
368	Caulophyllum-Saponin B (Calcosid D) (Akebia-Saponin C)	β-D-Gl 1→2 α-L-Ar 1→3 [Hederagenin (194)]	*Caulophyllum robustum* (Wurzeln, Rhizome) *(Caltha silvestris)* *(Akebia quinata)* (Samen)	*(354)* *(472a)* *(189b)*
369	Scabiosid C (Leontosid B)	β-D-Gl 1→4 α-L-Ar 1→3 [Hederagenin (194)]	*Patrinia scabiosaefolia* *(Leontice eversmannii)* (Knollen)	*(59)* *(357)*
370	Hederacosid A	β-D-Gl 1→5 α-L-Ar f 1→3 [Hederagenin (194)]	*Hedera helix* (Holz)	*(427)*
371	Sapindosid B	β-D-Xy 1→3 α-L-Rh 1→2 α-L-Ar 1→3 [Hederagenin (194)] *Sapindus mukorossi* (Früchte)		*(105)*

372 Leontosid C

β-D-Gl $1 \to 4$ / β-D-Gl $1 \to 3$ α-L-Ar $1 \to 3$ [Hederagenin (**194**)]

Leontice eversmannii (Knollen) (*358*)

373 Pastuchosid B

α-D-Gl $1 \to 4$ α-D-Gl $1 \to 3$ α-L-Rh $1 \to 3$ [Hederagenin (**194**)

Hedera pastuchovii (Blätter) (*205*)

374 Sapindosid C

β-D-Gl $1 \to 4$ β-D-Xy $1 \to 3$ α-L-Rh $1 \to 2$ α-L-Ar $1 \to 3$ [Hederagenin (**194**)]

Sapindus mukorossi (Früchte) (*102*)

375 Mubenin C

D-Ga $1 \to 6$ D-Gl $1 \to 2$ L-Rh $1 \to 4$ L-Ar $1 \to 3$ [Hederagenin (**194**)]

Stauntonia hexaphylla (Samen) (*485*)

376 Sapindosid D

α-L-Rh $1 \to 6$ / α-D-Gl $1 \to 2$ β-D-Gl $1 \to 4$ β-D-Xy $1 \to 3$ α-L-Rh $1 \to 2$ α-L-Ar $1 \to 3$ (Hederagenin (**194**)]

Sapindus mukorossi (Früchte) (*106*)

377 Musennin A

L-Ar $1 \to 2$ L-Ar $1 \to 2$ L-Ar $1 \to 3$ [Echinocystsäure (**187**)]

Albizzia anthelmintica (Wurzelrinde) (*511, 512*)

378 Mar-Saponin A

D-Xy $1 \to 6$ D-Gl $1 \to 2$ L-Ar $1 \to 3$ [Echinocystsäure (**187**)]

Chenopodium anthelminticum (Wurzeln) (*96*)

379 Musennin B

β-D-Gl $1 \to 3$ α-L-Ar $1 \to 2$ α-L-Ar $1 \to 2$ α-L-Ar $1 \to 3$ [Echinocystsäure (**187**)]

Albizzia anthelmintica (Wurzelrinde) (*511, 512*)

380 Helianthosid A

L-Rh $1 \to 4$ L-Rh $1 \to 3$ / D-Xy $1 \to 6$ D-Gl $1 \to 3$ [Echinocystsäure (**187**)]

Helianthus annuus (Blütenblätter) (*93*)

381 Gypsogeninglucosid

β-D-Gl $1 \to 3$ [Gypsogenin (**206**)]

Dianthus superbus var. *longicalicinus* (Wurzeln) (*452*)

Tabelle 16 (Fortsetzung)

382	Saponarosid	β-D-Xy 1→3 [Gypsogensäure (239a)]	*Saponaria officinalis* (Wurzeln)	*(65)*
383	Ziyu-Glykosid II	α-L-Ar 1→3 [Pomolsäure (290)]	*Sanguisorba officinalis* (Wurzeln)	*(613)*
384	Dianthus-Saponin A	β-D-Gl 1→3 [Gypsogensäure (239a)]	*Dianthus deltoides* (Wurzeln)	*(68)*
385	Dianthus-Saponin B	β-D-Gl 1→6 β-D-Gl 1→3 [Gypsogensäure (239a)]	*Dianthus deltoides* (Wurzeln)	*(68)*
386	Cinchona-Glykosid B	β-D-Chi 1→3 [Cincholsäure (237)]	*Cinchona calisaya* (Rinde)	*(499)*
387	Cinchona-Glykosid A	β-D-Chi 1→3 [Chinovasäure (298)]	*Cinchona calisaya* (Rinde)	*(499)*
388	Cinchona-Glykosid C	β-D-Gl 1→3 [Chinovasäure (298)]	*Cinchona calisaya* (Rinde)	*(499)*
388a	Bassiasaponin A	β-D-Gl 1→2 / β-D-Gl 1→3 [Bassiasäure (230)]	*Bassia latifolia* (Samen)	*(182b)*
389	Presenegenin-glucosid	β-D-Gl 1→3 [Presenegenin (259)]	*Polygala senega, Polygala tenuifolia* (Wurzeln)	*(377)*
390	Luzerne-Saponin	β-D-Gl 1→6 β-D-Gl 1→3 β-D-Gl 1→3 [Medicagensäure (256)]	*Medicago sativa* (Stengel und Blätter)	*(159)*

Extrakte aus der Wurzelrinde von *Albizzia anthelmintica* werden in der Volksmedizin Afrikas vielfach als Anthelmintikum gegen Bandwürmer verwendet. Die anthelmintisch aktiven Bestandteile sind die Saponine (*500*), beide Hauptbestandteile Musennin (**379**) und Desglucomusennin (**377**) sind rein erhalten worden (*512*). Das Ergebnis der Methylierung und Spaltung ergab direkt Sequenz und Verknüpfung der Zuckerkette (*511*). Der Anknüpfungsort der Zuckerkette könnte am Aglykon Echinocystsäure das 3β-OH, 16α-OH oder das 17β-COOH sein. Da beide Saponine einen Methylester bilden, scheidet die 17β-Carboxylgruppe als Verknüpfungsstelle aus. Zur Entscheidung zwischen 3β- und 16α-OH wurden die Methylester der Saponine bei Raumtemperatur acetyliert. Unter diesen Bedingungen wird die 16α-OH-Gruppe nicht acetyliert. Anschließend wurde mit CrO_3 in Aceton oxidiert, das Oxidationsprodukt mit Methanol/ HCl gespalten und 16-Dehydro-echinocystsäure isoliert, die durch ihren charakteristischen Circulardichroismus und den Vergleich mit authentischer Substanz identifiziert wurde. Die Zuckerkette im Musennin bzw. Desglucomusennin ist demnach (*511, 512*) über die OH-Gruppe am C-3 und nicht, wie früher angenommen (*500*), am C-16 mit dem Aglykon verbunden.

7. Neutrale Bisdesmoside (B 1) (siehe Tabelle 17)

Wie im vorhergehenden Abschnitt 6 ausgeführt wurde, stehen die sauren Monodesmoside (A 4) in einem engen Zusammenhang mit den neutralen Bisdesmosiden (B 1), der wohl bedeutendsten Gruppe der Triterpensaponine. Von den 30 bisher in der Struktur bekannten Vertretern enthalten allein 13 Oleanolsäure und 11 Hederagenin als Aglykon. Glykoside dieser beiden Aglykone kommen auch oft nebeneinander in der gleichen Pflanze vor, vor allem in solchen der Familie der Araliaceen.

Der erste in der Struktur bestimmte Vertreter dieses Typs war das von KOCHETKOV, KHORLIN und Mitarb. geklärte Kalopanaxsaponin B (**405**) (*240, 259*) aus den Wurzeln von *Kalopanax septemlobum*. Es enthält Hederagenin als Aglykon und 2 Moll. D-Glucose, 2 Moll. L-Rhamnose und 1 Mol. L-Arabinose. Die Art der Strukturermittlung findet sich im Schema 5 dargestellt.

Die alkalische Hydrolyse lieferte das zugehörige Monodesmosid Kalopanaxsaponin A (**367**), das sich später mit dem schon länger in reiner Form bekannten α-Hederin als identisch erwies. Die Methylierung und Spaltung dieses Derivats ergab Art, Sequenz und Verknüpfung der Zuckerkette am OH an C-3.

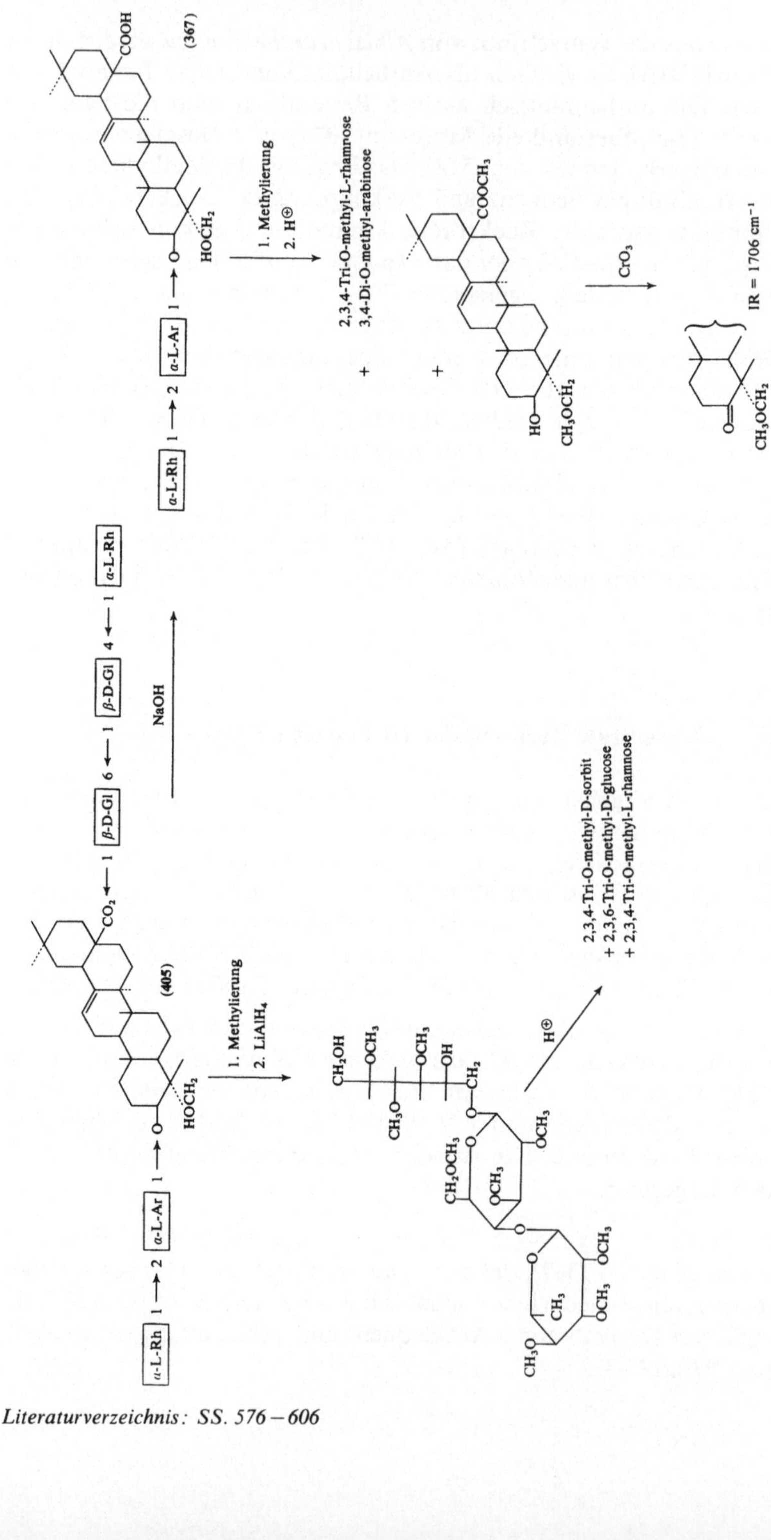

COOH

(367)

HOCH$_2$

α-L-Ar 1 $\longrightarrow$ O

α-L-Rh 1 $\longrightarrow$ 2

$\xrightarrow{\text{1. Methylierung} \\ \text{2. H}^{\oplus}}$

2,3,4-Tri-O-methyl-L-rhamnose
3,4-Di-O-methyl-arabinose

+

+

COOCH$_3$

HO   CH$_3$OCH$_2$

$\xrightarrow{\text{CrO}_3}$

CH$_3$OCH$_2$

IR = 1706 cm^{-1}

CO$_2$

(405)

HOCH$_2$

α-L-Ar 1 $\longrightarrow$ O

α-L-Rh 1 $\longrightarrow$ 2

$\boxed{\beta\text{-D-Gl}}$ 6 $\longleftarrow$ 1 $\boxed{\beta\text{-D-Gl}}$ 4 $\longleftarrow$ 1 $\boxed{\alpha\text{-L-Rh}}$

$\xrightarrow{\text{NaOH}}$

$\xrightarrow{\text{1. Methylierung} \\ \text{2. LiAlH}_4}$

CH$_2$OH
OCH$_3$
OCH$_3$
OH
CH$_2$
CH$_3$O

CH$_2$OCH$_3$
OCH$_3$
OCH$_3$
OCH$_3$
CH$_3$O
CH$_3$
OCH$_3$
OCH$_3$
CH$_3$O

$\xrightarrow{\text{H}^{\oplus}}$

2,3,4-Tri-O-methyl-D-sorbit
+ 2,3,6-Tri-O-methyl-D-glucose
+ 2,3,4-Tri-O-methyl-L-rhamnose

Schema 5. Strukturermittlung des Kalopanaxsaponin B

Tabelle 17. *Neutrale Bisdesmoside*

391 Scabiosid D $\boxed{\beta\text{-D-Gl}}$ $1\rightarrow4$ $\boxed{\alpha\text{-L-Ar}}$ $1\rightarrow3$ [Oleanolsäure (**167**)] $28\leftarrow1$ $\boxed{\beta\text{-D-Xy}}$ *(56)*
Patrinia scabiosofolia (Wurzeln)

392 Calendulosid B $\boxed{\beta\text{-D-Ga}}$ $1\rightarrow4$ $\boxed{\beta\text{-D-Gl}}$ $1\rightarrow3$ [Oleanolsäure (**167**)] $28\leftarrow1$ $\boxed{\alpha\text{-D-Gl}}$ *(553)*
Calendula officinalis (Wurzeln)

393 Scabiosid E $\boxed{\beta\text{-D-Gl}}$ $1\rightarrow4$ $\boxed{\alpha\text{-L-Ar}}$ $1\rightarrow3$ [Oleanolsäure (**167**)] $28\leftarrow1$ $\boxed{\beta\text{-D-Xy}}$ $4\leftarrow1$ $\boxed{\alpha\text{-L-Rh}}$ *(56)*
Patrinia scabiosofolia (Wurzeln)

394 Scabiosid F $\boxed{\beta\text{-D-Gl}}$ $1\rightarrow4$ $\boxed{\alpha\text{-L-Ar}}$ $1\rightarrow3$ [Oleanolsäure (**167**)] $28\leftarrow1$ $\boxed{\beta\text{-D-Xy}}$ $4\leftarrow1$ $\boxed{\alpha\text{-L-Rh}}$ $3\leftarrow1$ $\boxed{\beta\text{-D-Xy}}$ *(57)*
Patrinia scabiosofolia (Wurzeln)

395 Hederacosid B $\boxed{\alpha\text{-L-Rh}}$ $1\rightarrow2$ $\boxed{\alpha\text{-L-Ara}}$ $1\rightarrow3$ [Oleanolsäure (**167**)] $28\leftarrow1$ $\boxed{\beta\text{-D-Gl}}$ $6\leftarrow1$ $\boxed{\beta\text{-D-Gl}}$ $4\leftarrow1$ $\boxed{\alpha\text{-L-Rh}}$ *(521, 590)*
Hedera helix (Blätter)
Eleutherococcus senticosus (Blätter)   *(156a)*

(Eleutherosid M)

395a Eleutherosid L $\boxed{\alpha\text{-L-Rh}}$ $1\rightarrow4$ $\boxed{\alpha\text{-L-Ar}}$ $1\rightarrow3$ [Oleanolsäure (**167**)] $28\leftarrow1$ $\boxed{\beta\text{-D-Gl}}$ $6\leftarrow1$ $\boxed{\beta\text{-D-Gl}}$ $4\leftarrow1$ $\boxed{\alpha\text{-L-Rh}}$ *(156a)*
Eleutherococcus senticosus (Blätter)

396 Scabiosid G $\boxed{\beta\text{-D-Gl}}$ $1\rightarrow4$ $\boxed{\alpha\text{-L-Ar}}$ $1\rightarrow3$ [Oleanolsäure (**167**)] $28\leftarrow1$ $\boxed{\beta\text{-D-Xy}}$ $4\leftarrow1$ $\boxed{\alpha\text{-L-Rh}}$ $3\leftarrow1$ $\boxed{\beta\text{-D-Xy}}$ *(58)*
Patrinia scabiosofolia (Wurzeln) $4\uparrow1$ $\boxed{\alpha\text{-D-Gl}}$

Tabelle 17 (Fortsetzung)

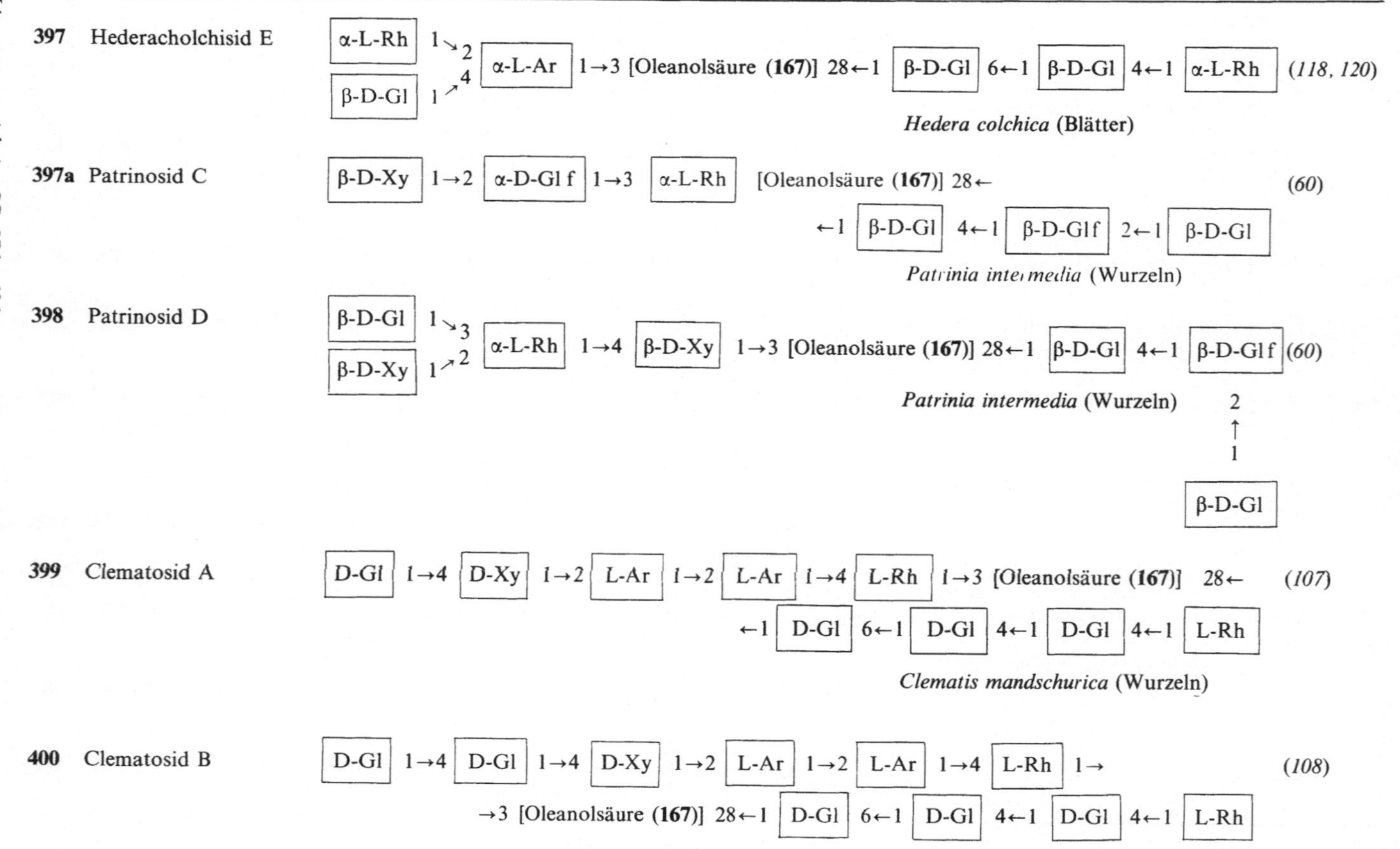

401 Clematosid C
[L-Rh] 1→6 [D-Gl] 1→4 [D-Gl] 1→4 [D-Xy] 1→2 [L-Ar] 1→2 [L-Ar] 1→4 [L-Rh] 1→
→3 [Oleanolsäure (**167**)] 28←1 [D-Gl] 6←1 [D-Gl] 4←1 [D-Gl] 4←1 [L-Rh] (*111, 237, 109a*)
Clematis mandschurica (Wurzeln)

402 Sibirosid A
[β-D-Gl] 1→3 [Hederagenin (**194**)] 28←1 [L-Ar] *Patrinia sibirica* (Wurzeln) (*54*)

402a Akebia-Saponin D
[α-L-Ar] 1→3 [Hederagenin (**194**)] 28←1 [β-D-Gl] 6←1 [β-D-Gl] (*189a*)
Akebia quinata (Samen)

403 Dipsacosid B
[α-L-Rh] 1→2 [α-L-Ar] 1→3 [Hederagenin (**194**)] 28←1 [β-D-Gl] 6←1 [β-D-Gl] (*352*)
Dipsacus azureus

403a Akebia-Saponin E
[β-D-Xy] 1→2 [α-L-Ar] 1→3 [Hederagenin (**194**)] 28←1 [β-D-Gl] 6←1 [β-D-Gl] (*189a*)
Akebia quinata (Samen)

403b Akebia-Saponin F
[β-D-Gl] 1→2 [α-L-Ar] 1→3 [Hederagenin (**194**)] 28←1 [β-D-Gl] 6←1 [β-D-Gl] (*189a*)
Akebia quinata (Samen)

404 Leontosid D
[β-D-Gl] 1→4 [α-L-Ar] 1→3 [Hederagenin (**194**)] 28←1 [β-D-Gl] 6←1 [β-D-Gl] 4←1 [α-L-Rh] (*359*)
Leontice eversmannii (Knollen)

405 Kalopanaxsaponin B
[α-L-Rh] 1→2 [α-L-Ara] 1→3 [Hederagenin (**194**)] 28←1 [β-D-Gl] 6←1 [β-D-Gl] 4←1 [α-L-Rh] (*240, 259*)
Kalopanax septemlobum (Wurzeln)
(*Hedera helix*) (Blätter) (*521*)

Tabelle 17 (Fortsetzung)

406 Sibirosid C

β-D-Gl 1→4 α-L-Ar 1→4 β-D-Gl 1→3 [Hederagenin (**194**)] 28←1 α-L-Ara 4←
←1 D-Gl 4←1 D-Gl

Patrinia sibirica (Wurzeln)

406a Akebia Saponin G

β-D-Gl 1→2, α-L-Rh 1→4 α-L-Ar 1→3 [Hederagenin (**194**)] 28←1 β-D-Gl 6←1 β-D-Gl 4←1 α-L-Rh (*189a*)

Akebia quinata (Samen)

407 Leontosid E

β-D-Gl 1→4, β-D-Gl 1→3 α-L-Ar 1→3 [Hederagenin (**194**)] 28←1 β-D-Gl 6←1 β-D-Gl 4←1 α-L-Rh (*360*)

Leontice eversmannii (Knollen)

408 Sapindosid E

β-D-Xy 1→3 α-L-Rh 1→2 α-L-Ar 1→3 [Hederagenin (**194**)] 28←
←1 α-L-Ar 2←1 α-L-Rh 3←1 β-D-Xy 4←1 β-D-Gl 6←1 α-L-Rh, 2←1 α-D-Gl (*103*)

Sapindus mukorossi (Früchte)

409 Mar-Saponin B

D-Xy 1→6 D-Gl 1→2 L-Ar 1→3 [Echinocystsäure (**187**)] 28←1 L-Rh 4←1 D-Gl (*96*)

Chenopodium anthelminticum (Wurzeln)

410 Helianthosid B

D-Xy 1→4 L-Rh 1→3 [Echinocystsäure (**187**)] 28←1 L-Ar 2←1 L-Rh 4←1 D-Gl (*92*)

Helianthus annuus (Blütenblätter)

411 Helianthosid C   [L-Rh] 1→4 [L-Rh] $\begin{smallmatrix} 1\searrow 3 \\ \ \end{smallmatrix}$ [D-Gl] 1→3 [Echinocystsäure(**187**)] 28←1 [L-Ar] 2←1 [L-Rh] 4←1 [D-Gl] (*94*)
[D-Xy] $1\nearrow 6$

Helianthus annuus (Blütenblätter)

411a Bassiasaponin B   [β-D-Gl] $\begin{smallmatrix} 1\searrow 2 \\ \ \end{smallmatrix}$ [Bassiasäure (**230**)] 28←1 [α-L-Ar] 2←1 [β-D-Xy] 4←1 [α-L-Rh]   (*182b*)
[β-D-Gl] $1\nearrow 3$

Bassia latifolia (Samen)

412 Ziyu-Glykosid I   [α-L-Ar] 1→3 [Pomolsäure (**290**)] 28←1 [β-D-Gl]   *Sanguisorba officinalis* (Wurzeln)   (*613*)

413 Ginsengosid Rg₁   [β-D-Gl] 1→6 [Protopanaxatriol (**313**)] 20←1 [β-D-Gl]   *Panax ginseng* (Wurzeln)   (*362*)

Die gleiche Zuckerkette mußte auch Kalopanaxsaponin B am C-3 tragen, den Bau der zweiten Zuckerkette aus 2 Moll. D-Glucose und 1 Mol. L-Rhamnose ermittelte man duch Methylierung von **405** und LiAlH$_4$-Reduktion des methylierten Produktes. Dabei wurde diese Zuckerkette als reduziertes, permethyliertes Trisaccharid abgespalten. Aus den bei der Hydrolyse auftretenden Methylzuckern ergibt sich der Bau der Kette und damit die Gesamtstruktur des Kalopanaxsaponins B (siehe Schema 5). Die Verknüpfung der letztgenannten Zuckerkette mit der Carboxylgruppe am C-17 ergibt sich aus der Tatsache, daß Kalopanaxsaponin B neutral ist und erst bei alkalischer Spaltung unter Verlust von drei Molekülen Zucker in ein saures Saponin übergeht. Auch die LiAlH$_4$-Reduktion unter Abspaltung dieser Zuckerkette entspricht dieser Auffassung.

Die gleichen Verbindungen fanden sich auch in *Hedera helix* (*521*) und anderen Hedera-Arten (z. B. *117*). Das in *Dipsacus azureus* aufgefundene Dipsacosid B (**403**) stellt ein Desrhamno-kalopanaxsaponin B dar, während im Leontosid D (**404**) aus *Leontice eversmannii* die gleiche Zuckerkette an C-17, an C-3 jedoch ein $\boxed{\beta\text{-D-Gl}}$ 1→4 $\boxed{\alpha\text{-L-Ar}}$ Rest sich befindet.

Ebenfalls ähnlich aufgebaut wie **405,** jedoch wesentlich zuckerreicher, sind die Clematoside A, B und C, die aus *Clematis mandschurica* isoliert wurden. Die von Kochetkov, Chirva und Mitarb. (*237*) durchgeführte Strukturbestimmung gestaltete sich vor allem deswegen schwierig und langwierig, da sehr lange unverzweigte Zuckerketten in der Sequenz zu klären waren. Erst die Kombination verschiedener Methoden zur Partialhydrolyse und die Untersuchung der gewonnenen zuckerärmeren Glykoside führte zum erfolgreichen Abschluß (*109a*). Clematosid C (**401**) ist mit 11 Monosaccharidresten das zur Zeit zuckerreichste in der Struktur bekannte Saponin. Es ist allerdings bis heute noch nicht gelungen, die Konfiguration der glykosidischen Bindungen festzulegen.

Bemerkenswert ist in dieser Gruppe noch das Patrinosid C (**397a**), in dessen Zuckerkette am C-3 (siehe auch Patrinosid C$_1$ **359**) eine α-furanoid gebundene D-Glucose enthalten sein soll (*60*). Diese Tatsache, würde sie sich bestätigen lassen, zeigt, daß man in Saponinen mit recht ungewöhnlichen Strukturmerkmalen zu rechnen hat und daß Analogieschlüsse mit Vorsicht zu betrachten sind.

Während die meisten Saponine dieser Gruppe unverzweigte Zuckerketten enthalten, kommen doch in einigen Fällen Verzweigungen vor. Wieweit sich dadurch die physiologischen Eigenschaften (z. B. die hämolytische Wirksamkeit) ändern, ist noch unbekannt. Die untersuchten Vertreter mit unverzweigten Ketten zeigten bisher keine hämolytische Wirksamkeit.

In die Gruppe der neutralen Bisdesmoside gehören auch die genuinen Saponine aus den Wurzeln von *Panax ginseng* und *Panax japonica*. Extrakte aus diesen Drogen spielen in der chinesischen Medizin und zum Teil auch in der anderer Länder eine wichtige Rolle. Sie sollen gegen viele Gebrechen, insbesondere Altersbeschwerden und ähnliches wirksam sein (siehe z. B. *378, 417*). Als wirksame Bestandteile werden die darin enthaltenen Saponine angesehen.

Die seit mehr als 10 Jahren vor allem in den Arbeitskreisen von ELYAKOV und von SHIBATA betriebene Strukturermittlung hat mit außerordentlichen Schwierigkeiten zu kämpfen, so daß bis heute kein genuines Hauptsaponin in der Struktur ermittelt worden ist. Das Aglykon erwies sich als sehr säurelabil und erst sehr schonende Methoden (Perjodatspaltung und enzymatische Spaltung mit Bakterien *602*) ergaben die Voraussetzung für die Struktur der in den Hauptsaponinen vorliegenden Aglykone Protopanaxadiol (**308**) und -triol (**313**) (*361, 448*) als (20 S). Es handelt sich um Dammaran-Derivate. Die zunächst isolierten (20 R) Protopanaxadiol (**309**) bzw. Panaxadiol (**310**) wurden später als Artefakte durch Säureeinwirkung erkannt.

(20 S)-Protopanaxadiol (**308**)

(20 R)-Protopanaxadiol (**309**)

Panaxadiol (**310**)

Tabelle 18. *Monodesmosidische Acylglykosen*

414 Hederacaucasid B	[Oleanolsäure (**167**)] 28←1 β-D-Gl 6←1 β-D-Gl 4←1 α-L-Rh *Hedera caucasigena* (Blätter)	(*119*)
415 Asiaticosid	[Asiatsäure (**295**)] 28←1 β-D-Gl 6←1 β-D-Gl 4←1 α-L-Rh *Centella asiatica* (Madagaskar) (ganze Pflanze)	(*383*)
416 Arjunetin	[Arjunasäure (**218**)] 28←1 β-D-Gl *Terminalia arjuna* (Rinde)	(*409*)
417 Mollugo-Saponin	[Spergulagensäure (**239**)] 28←1 β-D-Xy 4←1 β-D-Xy *Mollugo spergula* (Wurzeln)	(*182*)
418 Hernaria-Saponin I	[Medicagensäure (**256**)] 28←1 β-D-Gl 6←1 β-D-Gl *Hernaria glabra*	(*66*)
419 Hernaria-Saponin II	[Medicagensäure (**256**)] 28←1 α-L-Rh 2←1 β-D-Fu / 4←1 β-D-Gl *Hernaria glabra*	(*66*)

Neben (20 S)-Protopanaxadiol und -triol sind noch mehrere andere Aglykone in der sehr komplexen Saponinmischung der Wurzeln aufgefunden worden. Die bisher mitgeteilten Ergebnisse der Strukturermittlung sind vielfach widersprüchlich. Einige zum Teil auch längere Zuckerketten konnten strukturell geklärt werden. Es handelt sich offenbar jeweils um zwei Zuckerketten, die über zwei OH-Gruppen, wahrscheinlich am C-3 und C-20 (*361a*), am Aglykon gebunden sind (z. B. *146, 441a, 443, 486, 547*). Die bisher in der Struktur erkannten Glykoside stellen offensichtlich nur Nebensaponine oder Kunstprodukte dar.

8. Monodesmosidische Acylglykosen (A 5) (siehe Tabelle 18)

In einigen Fällen wurden Glykoside nur mit einer Zuckerkette isoliert, die in Form einer Acylglykosebindung über die Carboxylgruppe am C-17 an das Aglykon gebunden ist. Verbindungen dieses Typs zeigen nicht die den monodesmosidischen Saponinen eigentümlichen Eigenschaften. Wie noch bei der Diskussion über hämolytische Aktivität und chemische Struktur dargelegt werden wird, verhindert eine stark polare Gruppierung an der Rückseite des Aglykons die Ausbildung typischer Saponineigenschaften, z. B. eine hämolytische Wirksamkeit. Diese Substanzgruppe würde ihren Eigenschaften entsprechend daher besser zu den Bisdesmosiden gezählt werden. Zum Teil besitzen die zugrundeliegenden Aglykone (Asiatsäure und Medicagensäure) stark polare Gruppierungen am Ring A, die in der Polarität einem Zucker wie in den Bisdesmosiden entsprechen dürfte.

Genetisch könnte diese Gruppe von Glykosiden sowohl mit den neutralen (B 1) wie den sauren Bisdesmosiden (B 2) in Zusammenhang stehen. Im Fall des Hederacaucosid B (**414**) ist dies sicherlich der Fall; es ginge durch Anknüpfung einer $\boxed{\alpha\text{-L-Rh}}$ 1→2 $\boxed{\alpha\text{-L-Ar}}$ 1→Gruppierung am C-3 in Hederacosid B (**395**) über.

Dagegen steht das in seiner Struktur schon länger untersuchte und medizinisch genutzte Asiaticosid (**415**) soweit bekannt nicht mit einem echten Bisdesmosid in Zusammenhang. Es wurde aus *Centella aciatica* isoliert und von POLONSKY und Mitarb. (*383*) in seiner Struktur geklärt. Es stellte das erste Triterpen mit einer Acylglykosebindung dar.

Centella asiatica tritt offenbar in verschiedenen Chemovarietäten auf, da Asiaticosid nur aus der in Madagaskar wachsenden Varietät isoliert werden kann. Pflanzenmaterial indischer oder ceylonesischer Herkunft liefert andere Glykoside, die zum Teil auch unterschiedliche Aglykone enthalten (*36, 37, 398*), doch gehören alle, wie die Asiatsäure, dem Ursan-Typ an.

Tabelle 19. *In Aglykon und Zuckerteil saure Monodesmoside*

420 Calendula-Glykosid F   β-D-Glr   1→3 [Oleanolsäure (**167**)]   *Calendula officinalis* (Blüten)   (*223*)
                                                          (*Beta vulgaris*)   (*272*)

421 Spinasaponin A   β-D-Gl   1→3   β-D-Glr   1→3 [Oleanolsäure (**167**)]   *Spinacia oleracea* (Wurzeln)   (*518*)

422 Calendula-Glykosid E   D-Gl   1→4   β-D-Glr   1→3 [Oleanolsäure (**167**)]   *Calendula officinalis* (Blüten)   (*223*)

423 Calendula-Glykosid D   D-Ga   1→3   β-D-Glr   1→3 [Oleanolsäure (**167**)]   *Calendula officinalis* (Blüten)   (*223*)

424 Putranjia-Saponin I   α-L-Rh   1→2   β-D-Gl   1→3   β-D-Glr   1→3 [Oleanolsäure (**167**)]   (*393*)
                                               Putranjia roxburghii (Blätter)

424a Achyranthes-Saponin A   α-L-Rh   1→4   β-D-Gl   1→4   β-D-Glr   1→3 [Oleanolsäure (**167**)]   (*182a*)
                                               Achyranthes aspera (Samen)

425 Calendula-Glykosid B   D-Ga   1→3
                                  β-D-Glr   1→3 [Oleanolsäure (**167**)]   *Calendula officinalis* (Blüten)   (*223*)
                           D-Gl   1→4

426 Putranjia-Saponin II   α-L-Rh   1→2   β-D-Gl   1→3
                                           β-D-Glr   1→3 [Oleanolsäure (**167**)]   (*393*)
                                    β-D-Xy   1→4   *Putranjia roxburghii* (Blätter)

427 Spinasaponin B   β-D-Gl   1→3   β-D-Glr   1→3 [Hederagenin (**194**)]   *Spinacia oleracea* (Wurzeln)   (*518*)

428 Koelreuteria-Saponin B

| α-L-Rh | 1→3 | α-L-Ar | 1→4 |
| β-D-Ga | 1→3 |

β-D-Glr 1→3 [Gypsogenin (**206**)] (*101*)

Koelreuteria paniculata (Früchte)

429 Xanthoceras-Saponin

| β-D-Ga | 1→6 | β-D-Gl | 1→3 |
| β-L-Ar | 1→4 |

β-D-Glr 1→3 [Gypsogenin (**206**)] (*104*)

Xanthoceras sorbifolium (Früchte)

430 Glycyrrhizin

β-D-Glr 1→2 β-D-Glr 1→3 [Glycyrrhetinsäure (**208**)] *Glycyrrhiza glabra* (Wurzeln) (*313, 340*)

431 Quillajasäureglucuronid

β-D-Glr 1→3 [Quillajasäure (**232**)] *Quillaja saponaria* (Rinde) (*299*)

9. Durch Aglykon und Zuckeranteil saure Monodesmoside (A 6)
(siehe Tabelle 19)

Die Glykoside dieser Gruppe enthalten im Gegensatz zur Gruppe
A 4 auch im Zuckerteil eine saure Gruppierung. In Analogie zur Gruppe
A 3 handelt es sich dabei stets um D-Glucuronsäure, die direkt an das
Aglykon am C-3 gebunden ist. Analog zu den im Aglykon sauren Monodesmosiden besteht bei den meisten Verbindungen dieses Typs ein enger
Zusammenhang mit den entsprechenden sauren Bisdesmosiden, auch
wenn diese wie im Fall der Spinasaponine (**421, 427**), des Koelreuteria-
Saponins (**428**) und des Xanthocerassaponins (**429**) in der Struktur noch
nicht geklärt sind. Es liegt nahe, daß zunächst die unpolareren, leichter
rein darzustellenden und in der Struktur aufzuklärenden Glykoside bearbeitet werden, auch wenn es sich um Nebenglykoside oder gar um Abbauprodukte handelt.

Eine Ausnahme bildet das Glycyrrhizin (**430**) aus *Glycyrrhiza glabra*,
das aus mehreren Gründen eine Sonderstellung einnimmt. Einmal weicht
das Aglykon, die Glycyrrhetinsäure (**208**), in seiner Struktur mit einer
11-Ketofunktion und einer Carboxylgruppe am C-20 von derjenigen der
übrigen Vertreter ab, zum anderen besteht die Zuckerkette aus zwei
Glucuronsäuremolekülen, die den sehr polaren Charakter dieser Verbindung hervorrufen. Glycyrrhizin besitzt daher kaum noch Saponineigenschaften. Das Glykosid kann verhältnismäßig leicht rein und kristallin erhalten werden und spielt eine gewisse Rolle in der Medizin (siehe
S. 574 f.). Auch das Aglykon **208** und daraus hergestellte Umwandlungsprodukte werden pharmazeutisch genutzt.

Die Wurzeln von *Glycyrrhiza glabra* enthalten ein außerordentlich
komplexes Gemisch weiterer Glykoside mit einer Anzahl anderer Aglykone **162, 169, 173, 197, 207, 208, 209, 210, 228, 233, 234, 235, 236** und **241**,
die vor allem von Canonica und Mitarb. (Zusammenfassung siehe *413*)
bearbeitet worden sind.

10. Saure Bisdesmoside (B 2) (siehe Tabelle 20)

Zu dieser wichtigen Saponingruppe gehören vor allem die stark
polaren, zuckerreichen Saponine, wie sie z. B. aus den Seifenwurzeln ver-
schiedener Provenienz in großer Menge gewonnen werden können.
Wegen ihrer leichten Zugänglichkeit stellen sie die technisch verwendeten
Saponine dar. Insbesondere in früheren Jahrhunderten wurden sie als
indifferente Waschmittel, als Detergentien zur Schaumerzeugung in
Feuerlöschern und zu ähnlichen Zwecken viel verwendet. Auch das in der
Filmindustrie bei der Herstellung der lichtempfindlichen Schicht viel ver-

wendete Quillajasaponin aus *Quillaja saponaria* dürfte diesem Strukturtyp zuzurechnen sein. Die Auftrennung der meist sehr komplexen Saponingemische macht in dieser Gruppe ganz besondere Schwierigkeiten. Die den sauren Charakter der Zuckerkette verursachende D-Glucuronsäure trägt fast stets in den genuinen Hauptsaponinen eine Verzweigung.

Das bisher am häufigsten beobachtete Saponin dieser Gruppe ist das zuerst aus *Gypsophila pacifica* isolierte Gypsosid A. Von 62 untersuchten Arten der Familie Caryophyllaceen wurde es in 44 nachgewiesen. Vor allem in den Gattungen *Gypsophila* (24 Arten) und *Acantophyllum* (8 Arten) ist es fast stets vertreten, dagegen fehlt es in den *Dianthus*-Arten ganz (*614*).

Die Strukturaufklärung wurde im Arbeitskreis von KOCHETKOV und KHORLIN (*259, 263*) ausgeführt und soll als Beispiel für die Untersuchung eines zuckerreichen Saponins näher erörtert werden (siehe Schema 6):

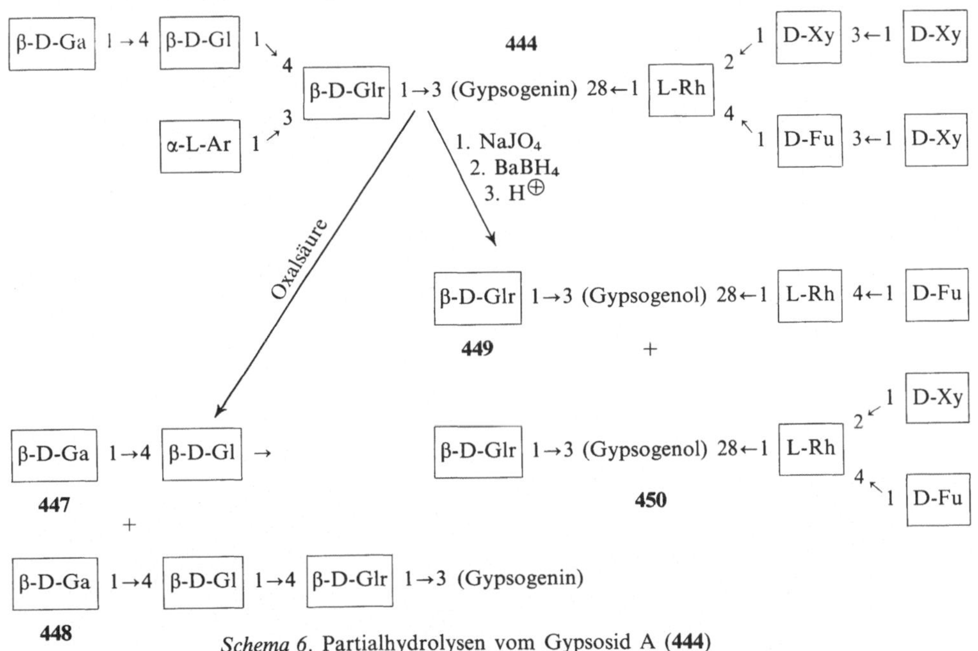

Schema 6. Partialhydrolysen vom Gypsosid A (**444**)

Die saure Hydrolyse ergab Gypsogeninlacton und Glucose, Galaktose, Glucuronsäure, Rhamnose, Fucose, Arabinose und Xylose, deren Verhältnis zunächst nicht festgelegt werden konnte. Nach der Methylierung erhielt man als Spaltprodukte 2,3,4,6-Tetra-O-methyl-D-galaktose, 2,3,4-Tri-O-methyl-L-arabinose, 2,3,6-Tri-O-methyl-D-glucose, 2,4-Di-O-methyl-D-fucose, 3-O-Methyl-L-rhamnose, 2,4-Di-O-methyl-D-xylose, 2 Moll. 2,3,4-Tri-O-methyl-D-xylose und den Methylester der 2-O-Methyl-D-glucuronsäure. Damit war sicher, daß der Kohlenhydratanteil zwei Verzweigungen (an Glucuronsäure und Rhamnose) und vier endständige Zucker (2 Moll. Xylose, je 1 Mol. Galaktose und Arabinose) haben mußte.

Tabelle 20. *Saure Bisdesmoside*

432 Aralosid A
(Chikusetsusaponin IV)

α-L-Ar f $1\rightarrow4$ β-D-Glr $1\rightarrow3$ [Oleanolsäure (**167**)] $28\leftarrow1$ β-D-Gl   *(259, 264)*

Aralia mandschurica (Wurzeln)
(Panax japonicum) (Wurzeln)   *(270)*

434 Calendula-Glykosid C

D-Ga $1\rightarrow3$ (?) β-D-Glr $1\rightarrow3$ [Oleanolsäure (**167**)] $28\leftarrow1$ D-Gl   *(223)*

Calendula officinalis (Blüten)

435 Chikusetsusaponin V

β-D-Gl $1\rightarrow2$ β-D-Glr $1\rightarrow3$ [Oleanolsäure (**167**)] $28\leftarrow1$ β-D-Gl   *(269)*

Panax japonicum (Rhizome)

435a Achyranthes-Saponin B

α-L-Rh $1\rightarrow4$ β-D-Gl $1\rightarrow4$ β-D-Glr $1\rightarrow3$ [Oleanolsäure (**167**)] $28\leftarrow1$ β-D-Ga   *(182a)*

Achyranthes aspera (Samen)

436 Calendula-Glykosid A

D-Ga $1\searrow3$ (?)
D-Gl $1\nearrow4$ β-D-Glr $1\rightarrow3$ Oleanolsäure (**167**)] $28\leftarrow1$ D-Gl   *(223)*

Calendula officinalis (Blüten)

437 Aralosid B

α-L-Ar f $1\searrow3$
α-L-Ar f $1\nearrow4$ β-D-Glr $1\rightarrow3$ [Oleanolsäure (**167**)] $28\leftarrow1$ β-D-Gl   *(259, 264)*

Aralia mandschurica (Wurzeln)

438 Aralosid C

β-D-Xy 1→4 β-D-Glr 1→3 [Oleanolsäure (**167**)] 28←1 β-D-Gl
β-D-Ga 1→3

(259, 264)
Aralia mandschurica (Wurzeln)

439 Putranjia-Saponin III

α-L-Rh 1→2 β-D-Gl 1→3 β-D-Glr 1→3 [Oleanolsäure (**167**)] 28←1 β-D-Gl

(393)
Putranjia roxburghii (Blätter)

439a Bersamasaponosid

β-L-Ar 1→4 β-D-Xy 1→4 β-D-Gl 1→3 [Oleanolsäure (**167**)] 28←1 β-D-Gl

(549a)
Bersama yangambiensis

440 Putranjia-Saponin IV

α-L-Rh 1→2 β-D-Gl 1→3 β-D-Glr 1→3 [Oleanolsäure (**167**)] 28←1 β-D-Gl
β-D-Xy 1→4

(393)
Putranjia roxburghii (Blätter)

441 Trichosid A

β-D-Gl 1→3 β-D-Glr 1→3 [Gypsogenin (**206**)] 28←1 β-D-Ga

(312)
Gypsophila trichotoma (Wurzeln)

442 Trichosid B

D-Gl 1→3 D-Glr 1→3 [Gypsogenin (**206**)] 28←1 D-Ga 4←1 D-Gl

(311)
Gypsophila trichotoma (Wurzeln)

443 Saponasid A

β-D-Glr 1→3 [Gypsogenin (**206**)] 28←1 D-Gl 6←1 D-Gl
3←1 β-D-Gl

(97)
Saponaria officinalis (Wurzeln)

Tabelle 20 (Fortsetzung)

443a Trichosid C   β-D-Gl 1→3 β-D-Glr 1→3 [Gypsogenin (**206**)] 28←1 β-D-Fu 4←1 α-L-Rh 4←1 β-D-Ga   *(311a)*

Gypsophila trichotoma

444 Gypsosid   β-D-Ga 1→4 β-D-Gl 1↘4 / α-L-Ar 1↗3 β-D-Glr 1→3 [Gypsogenin (**206**)] 28←   *(259, 263, 614)*

←1 L-Rh 2←1 D-Xy 3←1 D-Xy / 4←1 D-Fu 3←1 D-Xy

Gypsophila pacifica (Wurzeln)
Gypsophila und *Acantophyllum*-Spezies

444a Nutanosid   α-L-Ar 1↘6 / β-D-Xy 1↗4 β-D-Ga 1→3 β-D-Glr 1→3 [Gypsogenin (**206**)] 28←   *(63a)*

←1 α-L-Rh 4←1 β-D-Gl / 3←1 β-D-Ga 2←1 β-D-Fu 4←1 β-D-Gl

Silene nutans

445 Dianthosid C

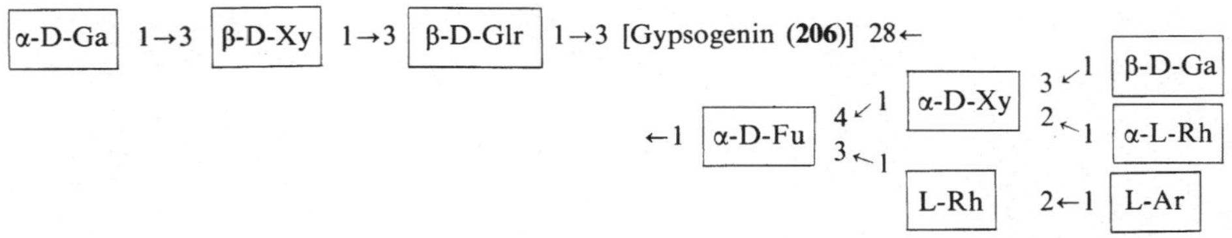

α-D-Ga 1→3 β-D-Xy 1→3 β-D-Glr 1→3 [Gypsogenin (**206**)] 28←

←1 α-D-Fu 4←1 α-D-Xy 3←1 β-D-Ga / 2←1 α-L-Rh ; 3←1 L-Rh 2←1 L-Ar

(67)

Dianthus deltoides (Wurzeln)

446 Saponasid D

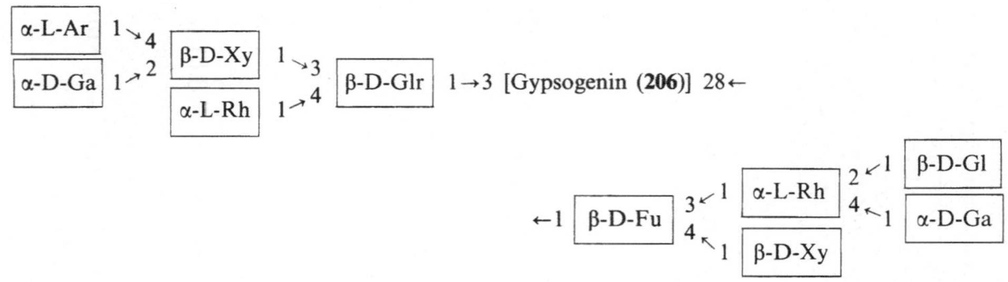

α-L-Ar 1→4 / α-D-Ga 1→2 β-D-Xy 1→3 / α-L-Rh 1→4 β-D-Glr 1→3 [Gypsogenin (**206**)] 28←

←1 β-D-Fu 3←1 α-L-Rh 2←1 β-D-Gl / 4←1 α-D-Ga ; 4←1 β-D-Xy

(98, 305)

Saponaria officinalis (Wurzeln)

Reduktion des permethylierten Gypsosid A mit LiAlH$_4$, wie beim Kalopanaxsaponin B erörtert, ergab einen permethylierten, tetraglykosidierten Hexit, der somit durch eine Acylglykose-Bindung über der Carboxylgruppe an C-17 mit dem Aglykon verbunden gewesen sein mußte. Dessen Hydrolyse offenbarte die Art und die Verknüpfung der Zucker in dieser Kette. Die entsprechende Information erhielt man nach Hydrolyse des permethylierten Tetraglykosids vom reduzierten Gypsogenin für die Zuckerkette an C-3.

Nähere Kenntnis der Sequenz der Zucker in den einzelnen Ketten wurden durch Partialhydrolyse erhalten (siehe Schema 6). Bei der Hydrolyse von Gypsosid A mit 10% Oxalsäure erhielt man Lactose (**447**) und das Triosid des Gypsogenins (**448**). Der Abbau mit Perjodat nach Smith lieferte ein Triosid (**449**) und ein Tetraosid (**450**).

Durch die Identifizierung der Lactose und die Ergebnisse der Methylierung und Spaltung der Glykoside **448, 449** und **450** ließ sich zweifelsfrei die Sequenz der Zucker in den beiden Ketten und damit die Struktur des Gypsosids A festlegen.

Gypsosid kann je nach pH-Wert in zwei verschiedenen Formen auftreten. Für dieses Verhalten wird von Kochetkov und Khorlin (*259, 263*) die Bildung eines Hemiacylals der Gypsogenin-aldehydgruppe mit der Carboxylgruppe der Uronsäure verantwortlich gemacht, die bei einer Bootform der Uronsäure möglich erscheint.

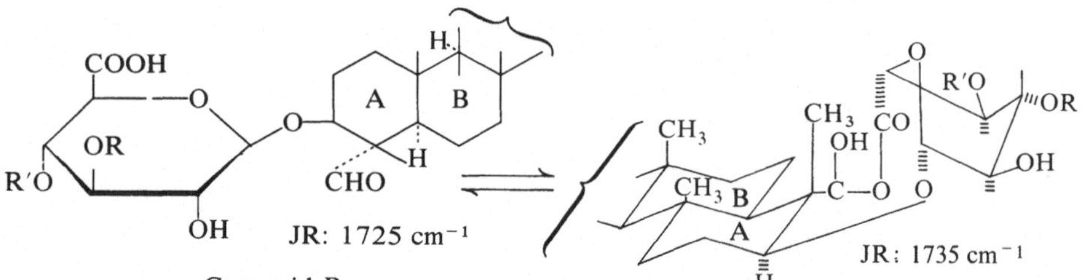

Sehr ähnlich ist auch das von Chirva und Mitarb. aufgeklärte Saponasid D (**446**) (*98, 305*) gebaut, das Hauptsaponin aus *Saponaria officinalis* (Seifenkraut). Es enthält 10 Monosaccharidreste mit zwei Verzweigungen in jeder Zuckerkette. Zur Strukturermittlung wurden ähnliche Methoden wie beim Gypsosid angewendet.

11. Tierische Saponine

Überraschend sind aus niederen Meerestieren in den letzten Jahren Glykoside isoliert worden, die alle Eigenschaften der Saponine aufweisen (Übersichten siehe *171a, 428, 429*). Neben dem Schaumvermögen zeigen sie ausgesprochene hämolytische Aktivität, bilden Cholesterinkomplexe und weisen zum Teil auch fungicide Wirksamkeit (*449*) auf. Damit müssen diese tierischen Inhaltsstoffe zu den Saponinen gerechnet werden. Zunächst blieb offen, ob es sich um Verbindungen tierischer Syntheseleistung handelt oder ob sie nur mit der pflanzlichen Nahrung aufgenommen und eventuell verändert werden. Da aber zum Teil nachgewiesen werden konnte (*460*), daß die Tiere die Aglykone selbst aus Mevalonsäure aufbauen, müssen diese Substanzen als tierische Saponine angesprochen werden. Ihr Vorkommen ist auf zwei Klassen der *Echinodermata* (Stachelhäuter) beschränkt, lediglich in den *Holothurioidea* (Seewalzen oder Seegurken) und den *Asteroidea* (Seesternen) konnten derartige Saponine gefunden werden (*601*).

Bislang ist die Struktur keines dieser Saponine vollkommen geklärt und auch die Struktur der genuinen Aglykone steht noch zur Diskussion. Allen bisher isolierten Glykosiden ist jedoch ein Aglykon vom Lanostan- oder Cholestantyp gemeinsam; außerdem enthalten alle eine Zuckerkette am Aglykon (wahrscheinlich am C-3) und einen Natriumhydrogensulfat-Rest im Zuckerteil.

Dem am häufigsten vorkommenden Glykosid aus den Seegurken, dem Holothurin A, könnte nach den bisherigen Ergebnissen die Struktur **451** zukommen (*87, 89*). Holothurin A zeigt eine starke physiologische Aktivität (*155, 156, 366*).

Mögliche Struktur von Holothurin A (**451**)

Bei der sauren Hydrolyse von Holothurin A erhält man aus dem Aglykon unter Ausbildung eines $\Delta^{7,9(11)}$-Diensystems 22,25-Oxido-holothurinogenin (**324**) und die Zucker D-Xylose, D-Glucose, 3-O-Methyl-D-glucose und D-Chinovose sowie Schwefelsäure. Bei der Spaltung in Methanol/HCl entfällt die Bildung des Diensystems, man erhält unter Verätherung und Inversion am C-12 12β-Methoxy-22,25-oxido-7,8-dihydro-holothurinogenin (**328**) (*88, 89*). Die Entstehung von Methyläthern in methanolischer Lösung verläuft auch in anderen Positionen des

Aglykons bemerkenswert leicht, so daß häufig methylierte Verbindungen erhalten werden. Genuine Aglykone konnten erst nach Spaltung mit Enzymen aus *Helix pomotia* isoliert werden (*89*).

Neben Holothurin A wurde aus Holothurioiden noch Holothurin B (*600*), Holotoxin (*449*) und Stichoposid A und B (*143*) erhalten. Allen Holothurien-saponinen ist wahrscheinlich ein 18,20-Lactonring im Lanostanringsystem gemeinsam. Dieser dürfte auch der bemerkenswerteste Unterschied zu den Asterosaponinen sein, soweit man dies aus zwei in der Struktur geklärten Aglykonen [Marthasteron (**452**) (*540*) und 24,25-Dihydromarthasteron (*540*)] vom Cholestantyp schließen kann. Doch zeigen auch die aus Seesternen isolierten Saponine Asterosaponin A (*598*) und B (*599*) sowie die Marthasterias-Saponine (*314, 540*) im IR keine Lacton-Absorption.

Die biologische Bedeutung dieser Toxine könnte sein, daß sie einen Schutz gegenüber anderen Tieren ausüben. So besitzen sie wie alle Saponine eine toxische Wirkung gegen Fische. Man nimmt ferner an, daß Seesterne mit Hilfe dieser Wirkstoffe Mollusken fern zu halten vermögen (*540*). Die physiologische Aktivität der Hothurien- und der Asterosaponine unterscheidet sich dabei etwas (*154*).

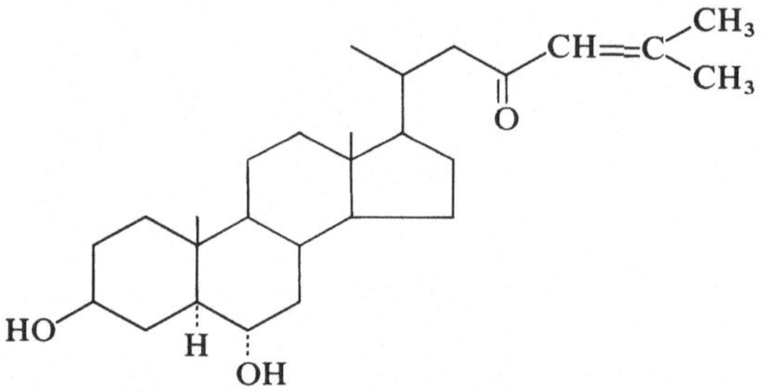

(**452**) Marthasteron

V. Biosynthese der Saponine

1. Die Biogenese der Steroidsapogenine (*169, 496*)

Alle bisher bekannten Steroidsapogenine enthalten 27 C-Atome im Molekül und ihre Ringe E und F können durch Cyclisierung der 8 C-Atome der Cholesterolseitenkette zusammen mit dem Ring D gebildet werden. Es lag daher nahe anzunehmen, daß dieses Sterol oder eine sehr nahe verwandte Verbindung der biogenetische Vorläufer sein könnte

(*506*). Dem stand entgegen, daß Cholesterol lange als ein typisches Zoosterol galt, und erst die letzten Jahre haben gezeigt, daß es im Pflanzenreich weit verbreitet ist, meist aber nur in geringer Menge. Durch die Versuche von HANAHAN und WAKIL (*181*) wurde gezeigt, daß für die Pflanzensterole mit 28 bzw. 29 C-Atomen ein Vorläufer mit 27 C-Atomen existiert und daß die zusätzlichen 1 bzw. 2 C-Atome in der Seitenkette an C-24 aus Methylgruppen des Methionins stammen. Diese letztere Feststellung machten ALEXANDER und SCHWINK (*6*) bei der Verwendung in der CH_3-Gruppe radioaktiv markierten Methionins. Daß speziell Cholesterol Precursor der Spirostanole ist, konnten TSCHESCHE und HULPKE (*506*) an überlebenden *Digitalis lanata*-Blättern mit ^{14}C-markiertem Cholesterol-β-D-glucosid zeigen. Es ließ sich nach 14 Tagen radioaktives Tigogenin und Gitogenin nach Hydrolyse in einer Ausbeute von 1 bis 3% des aufgenommenen Cholesterols nachweisen. Dieser Versuch zeigt sogleich, daß die Biosynthese in den Blättern abläuft. Bestätigt wird dieses Ergebnis mit ganzen Pflanzen von *Dioscorea spiculiflora* durch BENNETT und HEFTMANN (*217*). Sie boten freies ^{14}C markiertes Cholesterol über die Blattoberfläche an und konnten radioaktives Diosgenin nach Hydrolyse gewinnen. Die gleichen Autoren zeigten weiter (*218*), daß das bisdesmosidische Protodioscin ebenfalls aus Cholesterol entsteht und durch eine β-Glucosidase aus *Dioscorea floribunda* in Dioscin überführbar ist. Pflanzensterole mit 28 bzw. 29 C-Atomen scheinen nicht mehr in Spirostanole umgewandelt zu werden, eine Wiederabspaltung der zusätzlichen C-Atome findet nicht statt, wie Versuche mit markierten ^{14}C-Sitosterol an *Digitalis lanata* (*510*) ergaben.

Die Absättigung der Δ^5-Doppelbindung des Cholesterols zum Ring A/B *trans*-Derivat dürfte ein relativ früher Schritt der Biogenese sein, dabei wird die 3-Ketostufe durchlaufen. Bietet man geeigneten Pflanzen in $3\alpha\ ^3$H-substituiertes Desmosterol an, so läßt sich nach Absättigung der Doppelbindung in der Seitenkette radioaktives Cholesterol isolieren, aber keine Tritium-markierten Spirostanole. Cholestanol und ebenso Cholestanon sind verwertbare Vorläufer für die Spirostanolbiogenese, für beide Verbindungen wurde ein Gleichgewicht in *Digitalis lanata* Pflanzen festgestellt (*502, 507*). Die Einführung der Sauerstoffe in die Seitenkette beginnt mit einer Hydroxylierung an C-26, der nächste Schritt ist die Einführung einer OH-Gruppe an C-16, erst nachher wird an C-22 eine Hydroxylgruppe eingeführt. Diese Reihenfolge wurde mit geeignet markierten Verbindungen bei Pflanzenversuchen ermittelt und mit Gewebekulturen aus *Dioscorea tokoro* Makino durch TOMITA und Mitarb. (*492*) bestätigt. Sie zeigten, daß auch bei Vorliegen der Δ^5-Doppelbindung die Hydroxylierungsschritte analog ablaufen und daß 3,16,22,26-Tetrahydroxy-Δ^5-cholesten Zwischenstufe der Diosgenin-Biosynthese ist. Neuerdings jedoch konnte gezeigt werden, daß auch 16β-Hydroxycholesterol zur Spirostanolsynthese verwertet wird. Canonica und Mitarb. (*79*) stellten 2,2′,4,4′ $^3H_4$-Furostan-3β,26-diol her, das sich bei *Digitalis lanata* als Precursor für Tigogenin erwies.

So ergibt sich das nachfolgende Schema der Biosynthese:

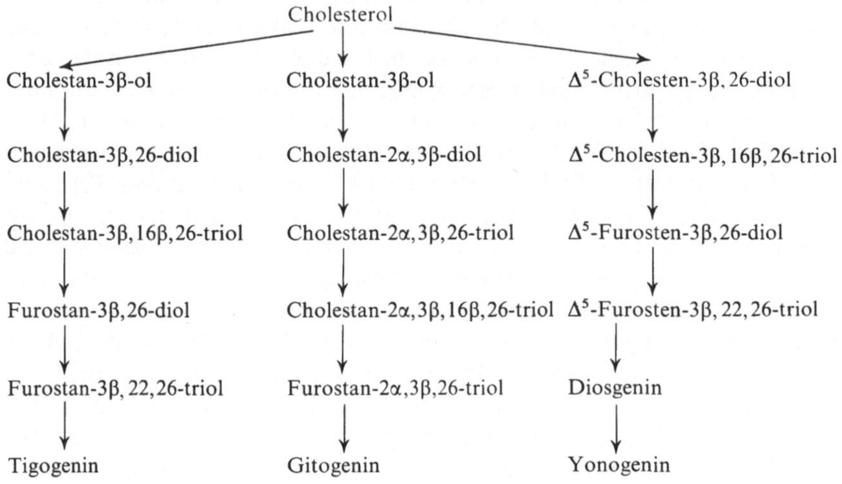

Schema 7. Biosynthese der Spirostanole

Wie weit Cholesterol selbst stets Zwischenstufe der Biogenese ist, erscheint zweifelhaft, nachdem gezeigt werden konnte, daß Convalla-marogenin mit zusätzlicher OH-Gruppe am C-1 zu der am C-3 und einer Doppelbindung $\Delta^{25\,(27)}$ nicht aus Cholesterol entsteht (*510*). Auch die Stufe der Hydroxylierung an C-2 (α) beim Gitogenin ist unklar, jedenfalls wird dabei nicht die Stufe des Tigogenins durchlaufen. Tomita und Mitarb. (*492*) nehmen an, daß Yonogenin eine 2(β),3(α)-Verbindung mit Ring A/B-*cis*-Konfiguration aus Diosgenin hervorgeht.

2. Die Biogenese der pentacyclischen Triterpensapogenine

Die Triterpensapogenine, vorwiegend der β-Amyringruppe zuge-hörend, werden ebenfalls aus Squalen gebildet. Während bei den Steroid-sapogeninen der Weg über das Cycloartenol führt (*400*), das schließlich über mehrere Zwischenstufen in Cholesterol übergeht, läuft die Cyclisie-rung des Squalens nach T. W. Goodwin und Mitarb. (*345, 399, 405*) nach untenstehendem Schema ab, das an Erbsen- und Maiskeimlingen mit (4 R) $^3H_1$-Mevalonsäure ermittelt wurde. Auch hier läuft die Biogenese

Literaturverzeichnis: SS. 576−606

über das Squalenepoxid, wie E. J. COREY und Mitarb. (*113*) an zellfreien
Erbsenkeimling-Homogenaten feststellten.

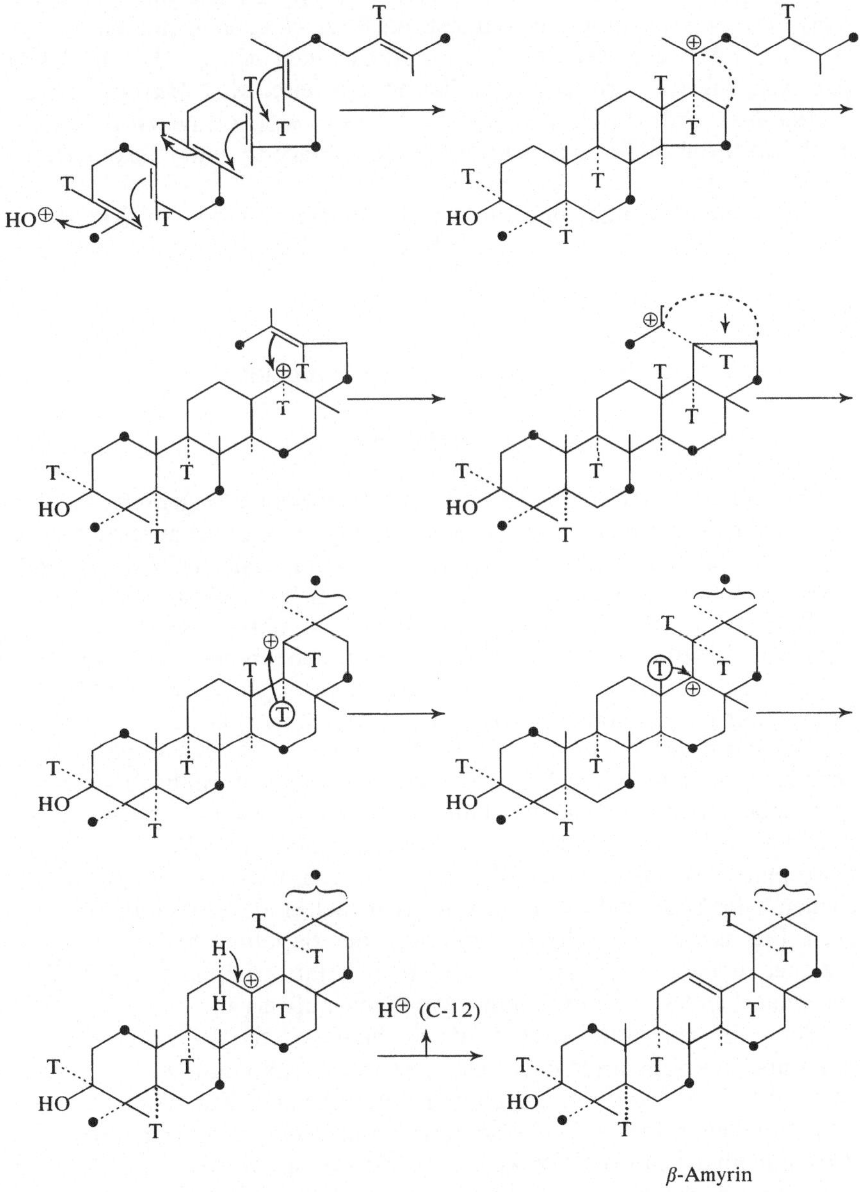

Schema 8. Cyclisierung des Squalens zu β-Amyrin (*169*)

Die Einführung der Methylgruppen an C-4 erfolgt bei der Biogenese nach Arigoni (*14*) stereospezifisch. Er zeigte, daß eine Markierung am C-2 der Mevalonsäure ausschließlich in der Methylgruppe C-23 des Sojasapogenols erscheint (aequatoriales Methyl), so daß die terminalen CH₃-Gruppen sich im Cyclisierungsprozeß verschieden formieren.

Die Einführung weiterer Sauerstoffunktionen neben OH am C-3 in das Amyringerüst erfolgt vermutlich stufenweise. Aus *Machaerium incorruptible* wurde durch Magalhaes Alves *et al.* (*315*) neben β-Amyrin auch Erythrodiol-3-acetat, Oleanolaldehyd-3-acetat und Oleanolsäure-3-acetat isoliert.

Über den Zeitpunkt und die Art der Einführung der Zuckerreste ist für Steroid- wie Triterpensaponine bislang noch nichts näheres bekannt.

VI. Eigenschaften der Saponine

1. Allgemeines

Auf die sehr umfangreiche frühere Literatur über Eigenschaften der Saponine soll hier nicht eingegangen werden, da es einige ausgezeichnete Zusammenfassungen gibt. Die ältere Literatur findet sich bei Kofler (*267*) zusammengestellt, neue Angaben findet man vor allem bei Steiner und Holtzem (*468*), Boiteau, Pasich und Ratsimamanga (*40*) sowie Hiller und Mitarb. (*192, 583*). Im nachfolgenden Abschnitt soll vor allem auf die Eigenschaften der Reinsaponine eingegangen werden, und zwar speziell auf den Zusammenhang mit der Molekülstruktur.

Die bekannteste Eigenschaft der Saponine ist das Schaumbildungsvermögen; systematische Untersuchungen zur Strukturabhängigkeit liegen jedoch nicht vor. Es ist daher auch nicht bekannt, ob dieses in Analogie zu den Seifen nur auf dem Vorliegen eines hydrophoben und eines hydrophilen Teiles im Molekül beruht oder ob auch spezielle Strukturelemente der Saponine dafür mit verantwortlich sind. Insgesamt erscheint diese Eigenschaft zur Charakterisierung der Saponine nicht besonders geeignet, insbesondere da auch andere Substanzgruppen ein zum Teil hervorstechendes Schaumbildungsvermögen aufweisen.

Sehr viel charakteristischer ist die hämolytische Wirksamkeit, die zwar auch einigen anderen Substanzgruppen zukommt, die sich aber strukturell stark von den Saponinen unterscheiden. Zur hämolytischen Aktivität liegen inzwischen eine Reihe von Untersuchungen über den Zusammenhang mit der Struktur vor. Für den im vorliegenden Aufsatz verwendeten Saponinbegriff wurde daher auch diese Eigenschaft zur Definition herangezogen (siehe Abschnitt I und VI 2).

Literaturverzeichnis: SS. 576—606

Als weitere charakteristische Fähigkeit wurde die Cholesterinkomplex-bildung in Abhängigkeit von der Struktur der Saponine untersucht (*533, 590*). Dabei wurde eine Technik angewandt, die auch die Existenz löslicher Komplexe zu erkennen gestattet (*533*). Starke und auch in 96%igem Äthanol schwer lösliche Komplexe werden vor allem bei Spirostanolglykosiden beobachtet, während bisdesmosidische Furostanolsaponine nur sehr schwache oder keine Komplexbildung aufweisen. Triterpensaponine zeigen im allgemeinen eine mittlere bis schwächere Komplexbildungs-tendenz, doch lassen auch sie sich häufiger als Sterolkomplexe fällen. Entsprechend sollen Saponine bei Verwendung in Arzneimitteln den Cholesterinblutspiegel senken (*172a*).

Als neue Saponineigenschaft wurde ihre antibiotische, vor allem fungistatische Wirksamkeit (Abschnitt VI 3) und ihre Wirksamkeit gegen Insekten (Abschnitt VI 4) erkannt. Beide Eigenschaften können als Schutzmechanismen von Pflanzen gegen mikrobiellen Angriff bzw. gegen Insektenfraß angesehen werden. Ob damit die langdauernde Diskussion über die Funktion der Saponine in der Pflanze abgeschlossen ist, sei dahingestellt. Die in einigen Fällen beobachtete Wachstumsförderung bzw. -hemmung durch Saponine (*323, 555*) dürfte wohl keine normale Funktion der Saponine in der Pflanze sein.

Die früher sehr verbreitete Verwendung von Saponinen in der Medizin, speziell der Volksmedizin, ist heute weitgehend verlassen. Nur noch wenige Saponine werden bei früher üblichen Indikationen eingesetzt. Dagegen haben sorgfältige pharmakologische Untersuchungen neue Anwendungsbereiche erschlossen, so für Aescin, Glycyrrhizin und Asiaticosid. Auch durch die Auffindung weiterer Aktivitäten bei einigen Saponinen, wie der cancerostatischen und fungiciden, scheinen weitere Anwendungsmöglichkeiten (Abschnitt VI 6) nicht ausgeschlossen.

Als wesentliche Voraussetzung für die pharmazeutische Anwendung der Saponine in Pflanzenextrakten ist eine eindeutige Charakterisierung und Wertbestimmung der als aktiv erkannten Inhaltsstoffe notwendig, dafür wären zum Teil neue Bestimmungsmethoden erforderlich (Abschnitt VI 7).

2. Hämolyse und Saponinbegriff

Als charakteristische Eigenschaft der Saponine muß die Fähigkeit angesehen werden, Hämolyse hervorzurufen. Bereits in sehr geringen Konzentrationen zerstören viele Saponine die Erythrozytenmembran, so daß Hämoglobin in die Lösung austreten kann. Diese Eigenschaft der Saponine ist zu ihrer quantitativen Bestimmung herangezogen worden (siehe Abschnitt VI) (ältere Literatur zur Hämolyse siehe *267, 385, 468*).

Nachdem erstmalig eine größere Zahl reiner Saponine mit bekannter Struktur zur Verfügung stand, konnte auch die Frage der Strukturspezifität dieser Fähigkeit näher untersucht werden (*431, 533, 590*). Überraschenderweise stellte sich heraus, daß die hämolytische Aktivität in Abhängigkeit von der Struktur stark variiert.

Erhebliche hämolytische Aktivität weisen die monodesmosidischen Steroidsaponine und die monodesmosidischen Triterpensaponine der Gruppen A 1, A 2, A 3 und A 4 auf. Keine oder nur sehr geringe Wirksamkeit findet sich bei den Acylglykosen (A 5), den bisdesmosidischen Furostanolsaponinen und den neutralen Triterpen-Bisdesmosiden (B 1). Mittlere bis schwache Hämolyse wird von den Gruppen A 6 und B 2 hervorgerufen.

Vermutlich verhindert eine stark polare Gruppierung an der Rückseite des Moleküls in den Ringen D und E die hämolytische Aktivität, daher sind die Bisdemoside und die Acylglykosen (A 5) hämolytisch inaktiv. Auch Verbindungen mit vielen freien OH-Gruppen in diesen Ringen, wie z. B. Didesacyl-aescin und -theasaponin, zeigen keinerlei hämolytische Aktivität (siehe Tabelle 21). Eigentlich sollten auch die sauren Bisdesmoside (B 2) nicht hämolytisch sein. Ob hier durch den sauren Charakter die Acylglykose-Bindung gespalten wird oder ob noch andere Strukturmerkmale wie Verzweigungen oder ähnliches eine Rolle spielen, ist ungeklärt.

Tabelle 21. *Hämolytische Aktivität von Saponinen (431, 592)*

Saponine mit polarer Gruppierung in den Ringen D und E		Zugehörige Saponine ohne polare Gruppierung in den Ringen D und E	
Sarsaparillosid (**102**)	>250	Parillin (**103**)	4,0
Convallamarosid (**118**)	>250	Convallamarogenin-triglykosid (**144**)	2,0
Lanatigosid (**116**)	>250	Lanatigonin	0,6
Methyl-protogracillin (**114**)	>200	Gracillin (**127**)	2,0
Hederacosid C (**405**)	>250	α-Hederin (**367**)	3,0
Hederacosid B (**395**)	>250	β-Hederin (**355**)	0,9
Clematosid C (**401**)	>250	—	—
Gypsosid A (**444**)	30	—	—
Asiaticosid (**415**)	>250	—	—
Aescinol (Didesacylaescin)	>250	Aescin (**342**)	2,5
Didesacyltheasaponin	>250	Theasaponin (**343**)	20

Vollständige Hämolyse gemessen an gewaschenen Erythrozyten (Konzentration in μg/ml).

Literaturverzeichnis: SS. 576–606

Die hämolytische Aktivität der Saponine ist von der Struktur des Aglykons abhängig (*413, 436, 592*); der Zuckeranteil hat nur einen verstärkenden oder auch abschwächenden Einfluß (*431, 590*). Auch die Aglykone selbst zeigen eine hämolytische Wirkung. Da diese leichter durch strukturchemische Änderungen variiert werden können, bieten sie eine günstigere Basis zur Untersuchung der Strukturspezifität der Hämolyse. Störend ist hierbei lediglich die geringe Löslichkeit in Wasser, die das Arbeiten mit Dimethylformamid bzw. Dimethylsulfoxid als Lösungsvermittler erforderlich macht und die vielleicht zum Teil für die widersprüchlichen Ergebnisse (*431, 436*) verantwortlich sein könnte.

Nach den bisherigen Ergebnissen (*431*) sollten hämolytisch stark wirksame Aglykone eine polare Gruppierung im Ring A enthalten und eine schwächer polare im Ring D und E, wobei eine 16α-OH-Gruppe eine erhebliche Wirkungsverstärkung hervorruft. Stark polare Gruppierungen im Ring D und E führen in Analogie zu bisdesmosidischen Glykosiden zum Verlust der Aktivität. Mit dem Einfluß von Estergruppierungen auf die Hämolyse befaßten sich eingehend R. SEGAL und Mitarb. (*436, 437, 438*).

Ähnliche Ergebnisse ließen sich auch für Steroidsaponine erhalten (*592*). Wegen der Schwerlöslichkeit der Genine wurde in diesem Falle von dem gut löslichen Parillin (**103**) ausgegangen, das systematisch in der Seitenkette des Aglykons verändert wurde (Tabelle 22). Daß die Hämolyse nicht allein auf dem Vorhandensein eines hydrophilen und hydrophoben Molekülteils zurückzuführen ist, zeigt die Inaktivität des Pregnenolonglykosids (**111**). Offenbar muß die Seitenkette bestimmte Strukturvoraussetzungen für eine hohe Aktivität erfüllen. Eine polare Gruppierung an C-26 (Glucose oder OH) verhindert die Wirksamkeit. Da **454** eine Aktivität gleich derjenigen der Spirostanole aufweist, scheint es, daß dem 16,22-Äthersauerstoff eine wesentliche Bedeutung zukommt, der auch etwa den gleichen Abstand vom C-3 wie die wirkungsverstärkende 16α-OH-Gruppe in Triterpenen aufweist.

Wenn in Zukunft eine eindeutige Festlegung der Strukturerfordernisse für die Saponinhämolyse möglich sein sollte, könnte vielleicht einmal darauf eine chemische Definition des Saponinbegriffs basiert werden. Solange dies nicht möglich ist, muß mit der in der Einleitung gegebenen Definition des Saponinbegriffs oder einer ähnlichen gearbeitet werden.

Über den Mechanismus der Saponinhämolyse besteht ebenso Unklarheit. Es ist wahrscheinlich, daß die Saponine Bestandteile der Erythrocytenmembran durch Komplexbildung herauslösen. Als derartige Bestandteile kommen vor allem Cholesterin, auch Albumin oder andere Eiweißstoffe in Betracht. Eine Parallelität zwischen hämolytischer Aktivität und Cholesterinkomplexbildung besteht jedoch nicht (*533*), doch läßt sich aussagen, daß bei guten Komplexbildnern (vor allem bei den monodesmosidischen Spirostanolsaponinen) die Ursache der Hämolyse zum wesentlichen Teil, wenn nicht ganz durch das Herauslösen des Cholesterins aus der Erythrocytenmembran zustandekommt.

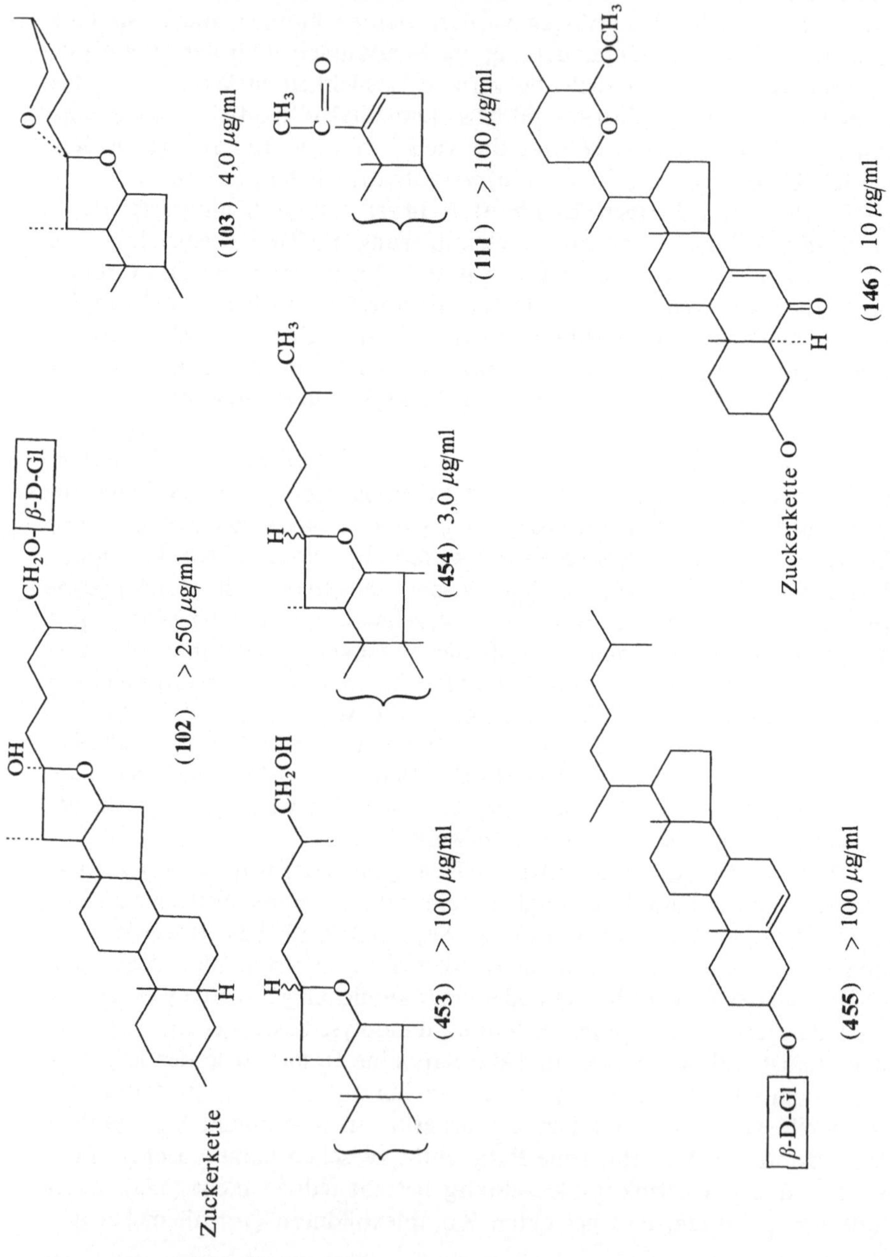

Tabelle 22. *Hämolytische Aktivität von in der Seitenkette des Aglykons variierten Glykosiden (592)*

3. Die antibiotische Wirksamkeit

Erst in den letzten Jahren wurde eine antibiotische Wirksamkeit als allgemeine Saponineigenschaft erkannt. Schon vorher ergaben sich, vor allem bei basischen Saponinen, vereinzelt Hinweise auf antibiotisches Verhalten (*40, 190, 459*). R. Tschesche und G. Wulff (*535*) untersuchten erstmalig systematisch die Wirksamkeit einer größeren Anzahl von Reinsaponinen und Gemischen gegenüber verschiedenen Mikroorganismen [Zusammenfassungen (*495, 590*)] und konnten zeigen, daß die meisten Saponine neben einer schwächeren antibakteriellen eine bemerkenswerte antimykotische Wirksamkeit aufweisen. Zahlreiche Untersuchungen von Wolters (*585, 586, 587, 588*) erweiterten die vorhandenen Kenntnisse und zeigten, daß zumindest die antimykotische Wirksamkeit als eine typische Saponineigenschaft anzusehen ist. Nach Wolters (*586*) ist die fungistatische Wirksamkeit auf eine Komplexbildung mit den Sterinen der Pilzmembranen zurückzuführen, doch dürften andere diese überlagernde Wirkungsmechanismen möglich sein. Beim Vergleich der Aktivität verschiedener Saponine (*202, 535, 588, 590*) lassen sich drei Gruppen unterschiedlichen Wirkungsspektrums unterscheiden (siehe Tabelle 23). Zur ersten Gruppe gehören die monodesmosidischen Spirostanolsaponine (analog verhalten sich auch Spirosolanolglykoside (z. B. α-Tomatin), die eine starke fungistatische Wirksamkeit zeigen, gegen Bakterien jedoch kaum Wirkung aufweisen. Die stärkste Aktivität zeigen Glykoside mit 4–5 Molekülen Zucker, solche mit nur 2–3 reagieren infolge ihrer geringeren Wasserlöslichkeit deutlich schwächer. Auffallend ist die geringe antibiotische Aktivität der bisdesmosidischen Saponine vom Steroid- wie vom Triterpentyp, die in bemerkenswerter Parallelität zur schwachen hämolytischen Wirksamkeit steht.

Monodesmosidische Triterpensaponine sind insgesamt weniger aktiv als die Spirostanolsaponine, besitzen aber von allen geprüften Verbindungen das breiteste Wirkungsspektrum. So ist insbesondere auch eine Wirkung gegen Bakterien signifikant. Zahlreiche Berichte über die antimikrobielle Wirksamkeit der Saponine sind inzwischen bekannt geworden (*148, 157, 170, 202, 320, 355, 425, 442, 493*). Auch eine antivirale Wirksamkeit ist beschrieben (*457*). Interessant sind besonders die Untersuchungen von Ferenczy und Mitarb. (*148*), die Extrakte von etwa 2500 verschiedenen Pflanzenarten nach dünnschichtchromatographischer Trennung auf antimykotische Substanzen untersuchten. Die in zahlreichen Pflanzen nachgewiesenen wirksamen Substanzen waren fast ausnahmslos auch hämolytisch aktiv, stellten daher offenbar Saponine dar. Die antibiotische Wirksamkeit der Saponine könnte bei der Frage nach einer möglichen Funktion der Saponine in der Pflanze von Bedeutung sein; es ließe sich eine Schutzfunktion gegenüber Mikroorganismen annehmen (*430a*).

In Anbetracht der Konzentration der Saponine in vielen Pflanzen oder
Pflanzenteilen von 0,1 bis über 10% dürfte eine Hemmwirkung gegen
Bakterien und vor allem pflanzenpathogene Pilze gegeben sein. In diese
Richtung weist auch die bevorzugte Lokalisierung der Saponine in ge-
fährdeten Teilen der Pflanze, wie in Wurzeln, Rinden und Samen, hin. An
diesen Stellen finden sich auch die aktiveren monodesmosidischen Formen, die als ein Depot fungieren mögen. Die besser wasserlöslichen bisdesmosidischen Saponine der Blätter, für die Pflanzenzelle selbst kaum
toxisch, sind nur wenig antibiotisch aktiv und stellen wohl vorwiegend die
Transportform dar. Bei Verletzung der Pflanze können sie schnell in
Gegenwart nachgewiesener pflanzeneigener Enzyme in die antibiotisch
wirksamen monodesmosidischen Saponine übergehen (*218, 590*).

Die antibiotische Aktivität der Saponine könnte eine mögliche
Erklärung für die medizinische Anwendung der früher in reichem Maße
verordneten Saponindrogen sein. Bei Infektionen der Haut und des
Verdauungstraktes wäre bei äußerlicher bzw. peroraler Applikation eine
Wirkung zu erwarten. Gegenüber den heute gebrauchten Antibiotika aus
Mikroorganismen und spezifischen Synthetika dürften die Saponine
unterlegen sein (*493, 590*), da sie in der Aktivität durch Cholesterin
inhibiert werden.

Tabelle 23. Antibiotische Wirksamkeit von Saponinen (535)

Substanz	*Staphylococcus aureus* 2438	*Escherichia coli*	*Candida albicans*	*Trichodermes mentagrophytes*
Parillin (**103**)	∅	∅	6 mm (16 γ)	15 mm (4 γ)
Digitonin (**122**)	∅	∅	7 mm (16 γ)	14 mm (4 γ)
Sarsaparillosid (**102**)	∅	∅	∅	∅
Hederacosid C (**395**)	8 mm (>1000 γ)	∅	∅	10 mm (250 γ)
Gypsosid A (**444**)	∅	∅	∅	3 mm (>1000 γ)
Primulasaponin (**352**)	4 mm (>1000 γ)	4 mm (1000 γ)	6 mm (1000 γ)	10 mm (250 γ)
Cyclamin (**340**)	7 mm (>1000 γ)	∅	3 mm (250 γ)	6 mm (62 γ)

Durchmesser des Hemmhofes im Agardiffusionstest bei 10 mg/ml. In Klammer ist die
Menge Saponin in γ/ml angegeben, die im Reihenverdünnungstest das Wachstum des Mikro-
organismus noch vollständig unterdrückt.

4. Wirksamkeit gegen Insekten

Von den basischen Saponinen wie α-Tomatin und Demissin ist bekannt, daß sie als Fraßhemmer (Repellents) z. B. gegen die Larven des
Kartoffelkäfers wirksam sind und daß eine gewisse Konzentration dieser

Substanzen auf Blätter aufgebracht Schutz gegen den Befall dieses Schädlings gewährt (*286, 433*). Noch wesentlich wirksamer sind die entsprechenden Acetylderivate der 23-Hydroxy-alkaloidglykoside, die Leptine (*286*). Erst seit kurzem ist auch für die übrigen Saponine eine analoge Wirksamkeit nachgewiesen worden. Besonders interessant ist in diesem Zusammenhang die Wirksamkeit von Saponinen gegen Termiten, da diese vor allem in tropischen Gebieten große Schäden durch Holzzerstörung anrichten. KONDO und Mitarb. konnten nachweisen, daß die Resistenz des Holzes von *Ternstroemia japonica* (*572*) und von *Kalopanax septemlobum* (*271*) gegen Termitenfraß auf dem Vorhandensein solcher Glykoside beruht. Auch Tempelhölzer aus der Maya-Zeit (600—900 n. Chr.) haben nach den Untersuchungen von SANDERMANN und FUNKE (*416*) aufgrund ihres Saponingehalts in feuchten, termitenverseuchten Urwäldern Jahrhunderte überdauern können, solange kein Auswaschen oder ein Abbau der Saponine im Holz erfolgte. Eine fraßhemmende Wirkung gegen Termiten kommt nach den Ergebnissen von TSCHESCHE, SCHMIDT und Mitarbb. (*538*) den meisten Saponinen in Konzentrationen von 0,5 bis 3% zu. Die Wirksamkeit geht vermutlich vorwiegend auf die Aglykone der Saponine zurück. Da die Termiten als Holzverwerter über eine starke Glykosidase-Aktivität im Verdauungstrakt verfügen, zeigen Mono- und Bisdesmoside sowie die entsprechenden Aglykone etwa gleiche Aktivität. In jüngster Zeit wurden vor allem Leguminosen untersucht, deren Saponine nicht nur eine Wirksamkeit gegen Termiten (*115*), sondern auch gegen viele Schadkäfer und ihre Larven zeigen (*12, 13, 159, 441*). Im Arbeitskreis von BIRK und GESTETNER (*12, 13*) wurden in für die menschliche Ernährung wichtigen Leguminosensamen (Soja-Bohnen, Erbsen, Kichererbsen, Bohnen, Linsen und Erdnüssen) Saponine nachgewiesen, die eine mehr oder weniger große Fraßhemmung bei Insekten bewirken. Dabei zeigten besonders solche Saponine eine relativ hohe Aktivität, die aus gegen die betreffenden Insekten weitgehend resistenten Varietäten isoliert worden waren.

Inwieweit der beschriebene Effekt einer direkten insektiziden Wirksamkeit entspricht, auf einer fungiziden Aktivität gegen die Symbionten der Insekten (*220*) oder auf einem Vergällungseffekt (Repellent-Wirkung) (*433*) beruht, ist bisher nicht geklärt worden.

5. Toxizität der Saponine

Die Saponine sind gegenüber verschiedenen Tierklassen je nach Applikation recht unterschiedlich toxisch. Über die Wirkung auf Mikroorganismen und Insekten ist in den vorangegangenen Abschnitten bereits berichtet worden. Stark giftig sind die Saponine gegen Kiemenatmer. So sind Giftwirkungen auf Mollusken (*387, 540*), auf im Wasser lebende Würmer (*351*), auf Kaulquappen (*267, 468*) und auf Fische (*267, 468*) beschrieben. Nach Vogel (*556*) beruht diese Giftwirkung auf einer pathologischen Permeabilitätserhöhung in den Kiemenepithelien, wodurch im Plasma gelöste lebenswichtige Elektrolyte ins Wasser austreten. Bereits bei Konzentrationen von 1 : 200000 im Wasser wirken viele Saponine auf Kiemenatmer tödlich. Daher wurden Saponine auch als Molluscizide in der Bekämpfung des Zwischenwirts der Bilharziaerreger empfohlen (*387*). Die Verwendung saponinhaltiger Pflanzen zum Fischfang ist seit langem bekannt (*267, 468*), sie beeinträchtigen die Genießbarkeit der Fische nicht.

Die Toxizität gegenüber Warmblütlern ist bei peroraler Aufnahme gering, sie liegt bei 50—100 mg/kg oder darüber. Als Ursache hierfür ist wohl die geringe Resorption anzusehen. Auch das Agrostemma-Saponin führt bei peroraler Anwendung, entgegen früheren Angaben, nicht zu toxischen Erscheinungen (*192*). Dagegen ist hierbei für das Saponin aus *Phytolacca americana* eine starke Toxizität beschrieben (*472*).

Diese Ergebnisse sind deswegen von Interesse, weil in zahlreichen für die menschliche Ernährung verwendeten Pflanzen bzw. Pflanzenteilen Saponine enthalten sind, so in vielen Leguminosen (Bohnen, Erbsen, Kichererbsen, Soja-Bohnen, Erdnüsse, Linsen) (*12, 13*). Auch Spinat (*Spinacia oleraceae*) (Wurzeln) und Haferflocken (*Avena sativa*) enthalten Saponine. Größere Mengen Saponine im Nahrungsmittel bewirken vielfach einen bitteren Geschmack. Kulturpflanzen sind daher häufig saponinärmer als Wildpflanzen. Interessanterweise nimmt damit aber die Anfälligkeit für Insekten und phytopathogene Pilze zu (*49*).

Im Gegensatz zur Harmlosigkeit bei peroraler Applikation haben viele Saponine eine erhebliche Toxizität bei parenteraler, speziell intravenöser Injektion. Die Toxizität allerdings schwankt erheblich; zwischen Quillajosid aus *Quillaja saponaria* mit 0,67 mg/kg und Hederacosid C mit 50 mg/kg DL_{50} sind alle möglichen Abstufungen beobachtet worden (siehe Tabelle 24). Die Giftwirkung zeigt keine Parallelität zur Hämolyse, diese ist nach Vogel (*556*) auch nur eine mittelbare Ursache für die Vergiftungserscheinungen, die vor allem von Sekundärprodukten der Hämolyse hervorgerufen werden. Für das Ausmaß der Toxizität spielt auch die unterschiedliche Entgiftung durch Komplexbildung mit Albuminen bzw. Cholesterin eine erhebliche Rolle.

Neben dieser unspezifischen Giftwirkung der Saponine sind auch spezifisch toxische Erscheinungen z. B. beim Agrostemma-Saponin bekannt (*192*). Spezielle Wirkungen im nichttoxischen Bereich werden in einer Reihe von Fällen zu Heilzwecken ausgenutzt, sie werden im nächsten Abschnitt behandelt.

Literaturverzeichnis: SS. 576—606

6. Pharmakologie der Saponine

Extrakte aus Saponindrogen spielen heute in der Medizin nur noch eine sehr begrenzte Rolle (*158, 426*). Die Entwicklung zielt vorwiegend auf die Verwendung von Reinsubstanzen ab, die exakt pharmakologisch untersucht und standardisierbar sind. Untersuchungen an Reinsaponinen oder auch an gereinigten Saponinen haben wertvolle Erkenntnisse über gewisse pharmakologische Wirkungen der Saponine erbracht.

Die pharmakologischen Eigenschaften der Saponine werden am besten in unspezifische, der ganzen Gruppe mehr oder minder zukommende, und in spezifische Eigenschaften unterteilt, die nur einem oder einigen Vertretern zukommen. Nach VOGEL (*556*) besteht eine der beachtenswerten unspezifischen Wirkungen in einer allgemeinen Gewebsreizung, insbesondere der Schleimhäute. Solche Reizungen können im Nasen-Rachen-Raum, den Bronchien, im Lungengewebe und möglicherweise in den Nierenepithelien auftreten. Hierauf scheint im wesentlichen die expectorierende Wirkung (z. B. von Saponinen aus *Radix senegae, Radix primulae, Radix liquiritiae* u. a.) zu beruhen, dabei dürfte die Erniedrigung der Oberflächenspannung zusätzlich zu einer Verflüssigung des Schleimes beitragen. Die den meisten Saponinen zukommende diuretische Wirkung könnte auf die erwähnte Reizung der Nierenepithelien zurückgehen.

Einige Saponindrogen finden in der Volksmedizin mancher Länder als Anthelmintikum Verwendung. Eine anthelmintische Wirksamkeit kommt den meisten Saponinen zu, doch steht vielfach die starke Schleimhautreizung einer praktischen Verwendung im Wege (*213*).

Für viele den Saponinen zugeschriebene Wirkungen fehlt bislang der exakte Nachweis. Gerade bei oral applizierten Saponinen kommt hinzu, daß die Saponine im allgemeinen schlecht resorbiert und meist unverändert wieder ausgeschieden werden, so daß eine Wirkung kaum vorstellbar ist. Allerdings ließ sich im Fall des Asiaticosids nachweisen, daß dieses zum Teil im Magen-Darm-Trakt in Zucker und Aglykon gespalten und das gleichfalls wirksame Aglykon resorbiert wird (*91*). Möglicherweise liegt bei anderen Saponinen ein ähnliches Verhalten vor, doch muß betont werden, daß gerade im Asiaticosid eine besonders leicht spaltbare Acylglykose-Bindung vorliegt. Daß Saponine ihrerseits die Resorption anderer Substanzen fördern, ist vielfach diskutiert, aber selten nachgewiesen worden.

Die schon im Abschnitt 3 erwähnte antibiotische Wirkung der Saponine ist ebenfalls eine unspezifische Gruppeneigenschaft. Das gleiche gilt für die cytostatische Aktivität, die 1965 von TSCHESCHE und WULFF für die meisten der untersuchten Saponine (auch der basischen) in vitro nachgewiesen wurde (*535*). In vivo Untersuchungen an Ratten mit Walker-carcinom bestätigten diesen Effekt (*536*), jedoch war die Toxizität in allen Fällen so hoch, daß eine praktische Verwendung nicht in Frage kommt (siehe Tabelle 24). Ähnliche Ergebnisse erbrachten Prüfungen des β-Solamarins (*293*), des Saponins aus *Saponaria officinalis* (*617*) und des Saponins aus *Myrsene africana*, das dem Primula-saponin sehr ähnlich gebaut ist (*295*). Auch die Saponine aus *Hedera helix* (*41*) und aus *Entada phaseoloides* (*308*) sollen eine anticancerogene Wirkung ausüben.

Tabelle 24. *Cancerostatische Wirksamkeit und Toxizität von Saponinen (536)*

Substanz	DE_{50} (mg/kg)	DL_{50} (mg/kg)
Aescin (**342**)	20	36
Cyclamin (**340**)	10	20
Senegin (Gemisch) (u. a. **344**)	1,5	3
Primulasaponin (**352**)	40	70
α-Hederin (**367**)	4	10
Parillin (**103**)	50	80
Quillajosid	—	0,68
Hederacosid C (**405**)	—	> 50

DE_{50} = Dosis, die nach parenteraler Zufuhr das Wachstum von Walker-Carcinom an Ratten um 50% gegenüber unbehandelten Vergleichstieren hemmt.

Die Messungen wurden freundlicherweise im Cancer Chemotherapy National Service Center, National Cancer Institute, Bethesda, USA, durchgeführt.

Ein spezieller Fall liegt offenbar bei dem von Kupchan und Mitarb. (*294, 297*) aus *Acer negundo* (Ahorn) isolierten Saponin vor, das bei sehr guter Wirksamkeit auch einen großen therapeutischen Index besitzt. Diese sehr vielversprechende Substanz befindet sich bereits in klinischer Prüfung.

Zu den spezifiischen Eigenschaften, die nur wenigen Saponinen in ausgeprägtem Maße zukommen, gehört der antiexsudative und oedemhemmende Effekt des Aescins (**342**), der pharmakologisch sorgfältig untersucht worden ist (*163, 180, 310, 388, 455, 557, 559, 616*).

Bei diesen Versuchen wurde das Aescin parenteral appliziert. Es entfaltete seine antiexsudativen Eigenschaften nur bei intakter Nebennierenrinde, ohne dabei aber Corticoide freizusetzen. Eine Wirkung trat erst auf, wenn Aescin mindestens 8 Stunden vor einer experimentell hervorgerufenen Entzündung gegeben wurde. Durch Untersuchung der Exsudation an verschiedenen Entzündungsmodellen konnten Vogel und Mitarb. (*557*) nachweisen, daß die am Aescin zu beobachtenden antiexsudativen und antiinflammatorischen Eigenschaften in erster Linie durch die Normalisierung der im Initialstadium der Entzündung gestörten Kapillarpermeabilität zu verstehen sind. Auch das dem Aescin sehr ähnlich gebaute Theasaponin (**343**) zeigt verwandte Eigenschaften (*530, 558*).

Literaturverzeichnis: SS. 576—606

Kristallisiertes Aescin wird in Europa in gewissem Ausmaß in der Humanmedizin als Antiexsudativum bei intravenöser Applikation eingesetzt. Die Toxizität ist im Vergleich zu anderen Saponinen verhältnismäßig gering, da Aescin relativ fest an das Serum-Albumin gebunden wird.

Breite Verwendung finden auch aescinhaltige Aesculus-Extrakte, die bei Durchblutungsstörungen, Thrombose und ähnlichen Indikationen wirksam sein sollen. Wegen der peroralen Anwendung und der nicht sicher nachgewiesenen Resorptionsquote besteht über diese Therapie keine einhellige Meinung, immerhin berichtete NEHRING (*365*), bei peroraler Anwendung eine Wirksamkeit auf den Venentonus beobachtet zu haben.

Glycyrrhizin (**430**) und die dieses Glykosid enthaltenden Extrakte aus *Glycyrrhiza glabra*-Wurzeln *(Radix liquiritiae)* finden breite Anwendung als Expectorans und werden wegen ihres süßen Geschmacks auch als Geschmackskorrigens benutzt. Süßer Geschmack ist eine Besonderheit unter den Saponinen, die außerdem noch das Osladin (**148**) aus *Polypodium vulgare* zeigt (*214*). Eine antisaccharine Wirkung weist eine der Gymnemasäuren aus *Gymnema silvestre* auf, die für gewisse Zeit die Fähigkeit zur süßen Geschmacksempfindung blockiert (*471*).

Daneben haben Glycyrrhizin, sein Aglykon Glycyrrhetinsäure und viele Derivate dieser Substanzen eine Reihe weiterer wertvoller Eigenschaften, die Gegenstand zahlreicher Patente sind. Es wurden Nebennierenrinden-Wirksamkeit in Form von Entzündungshemmung und erhöhter Wasserretention beschrieben. Auch wegen der ihnen zugeschriebenen günstigen Beeinflussung von Ulcus werden sie verwendet (Lit. siehe *30, 467, 583*).

Asiaticosid aus *Centella asiatica* (Madagaskar) wird ebenfalls in kristallisierter Form vor allem in Frankreich in größeren Mengen medizinisch angewandt (*40, 91*). Es soll bei peroraler Applikation leprawirksam sein und vor allem eine gute wundheilende und antiphlogistische Aktivität entfalten (*383*). Daß es bei peroraler Applikation zu 50% metabolisiert wird (*91*), wurde bereits erwähnt.

7. Quantitative Bestimmung der Saponine

Die Verwendung von saponinhaltigen Pflanzenextrakten in der Medizin macht eine quantitative Bestimmung der wirksamen Bestandteile erforderlich. Da bis vor einigen Jahren mit wenigen Ausnahmen Saponine nicht rein dargestellt werden konnten und auch in der Struktur weitgehend unbekannt waren, konnten exakte chemische Bestimmungsverfahren kaum angewandt werden. Man führte Wertbestimmungen aufgrund charakteristischer Saponineigenschaften, wie Änderung der

Oberflächenspannung, Schaumbildungsvermögen (Schaumindex), hämolytischer Aktivität (hämolytischer Index), Toxizität für Fische (Fischindex), durch (siehe z. B. *267, 467, 468*). Diese Verfahren können heute als überholt gelten und haben nur noch einen bedingten Wert. Sehr gut wären z. B. der Schaumindex bzw. der hämolytische Index zur quantitativen Bestimmung von Saponinen nur dann geeignet, wenn das Produkt z. B. als Schäummittel in Feuerlöschern bzw. als Hämolytikum benutzt werden sollte. Bei einer andersartigen, z. B. medizinischen Verwendung müßte die Voraussetzung gemacht werden können, daß hämolytische Wirksamkeit bzw. Schaumvermögen parallel der betrachteten pharmakologischen Wirksamkeit verlaufen.

Nachdem die chemische Untersuchung eine starke Abhängigkeit der Eigenschaften der Saponine von der Struktur gezeigt hat, sollte diese Art von Wertbestimmung nur mit äußerster Vorsicht angewendet werden. Es ist z. B. bekannt, daß die leicht erfolgende Abspaltung eines Zuckers (*590*) oder von Säureresten (*431*) oder eine Acylwanderung (*180*) außerordentliche Unterschiede in bestimmten physiologischen Eigenschaften bewirken. So kann ein niedriger hämolytischer Index einer sehr schonend aufgearbeiteten Droge mit hohem Bisdesmosidgehalt einen geringen Saponingehalt vortäuschen.

Als Maximalfall sollte die Bestimmung aller Einzelsaponine einer Droge angestrebt werden, und es sollte von jedem dieser Bestandteile auch eine pharmakologische Prüfung vorgenommen worden sein. Dieses Ziel ist in der Praxis zumeist nicht erreichbar. Die allgemeine Entwicklung zum Ziele der quantitativen Bestimmung der Saponine geht heute dahin, sie durch Ausschütteloperationen, durch Papier- oder Dünnschichtchromatographie aus dem Extrakt abzutrennen und anschließend nach Überführung in einen geeigneten Farbstoff photometrisch zu bestimmen. Auf diese Weise sind je nach Abtrennungsmethode das Gesamtsaponingemisch, Saponingruppen oder Einzelsaponine quantitativ bestimmbar. In vielen Fällen besteht allerdings die Schwierigkeit, daß über die quantitative Wirksamkeit der Einzelkomponenten keine Angaben vorliegen, so daß der eigentliche Wert einer Droge oder eines Extraktes schwer beurteilt werden kann.

Quantitative Bestimmungen nach obengenannten Schemata wurden unter anderem bereits für *Aesculus hippocstanum* (*430, 580*), *Glycyrrhiza glabra* (*47, 174*), *Primula elatior* (*374*) und *Soja hispidus* (*162*) ausgearbeitet (siehe auch *274*).

Literaturverzeichnis

1. Abdel-Akher, M., M. Hamilton, J. K. Montgomery, and F. Smith: A New Procedure for the Determination of the Fine Structure of Polysaccharides. J. Amer. Chem. Soc. **74**, 4970 (1952).

2. AKAHORI, A., and F. YASUDA: Laxogenin, a New Steroidal Sapogenin Isolated from *Smilax sieboldi*. Yakugaku Zasshi **83**, 557 (1963); Chem. Abstr. **59**, 7782 (1963).

3. AKAHORI, A., F. YASUDA, and T. OKANISHI: Studies on the Steroidal Components of Domestic Plants, LI. Steroidal Compounds Contained in the Rokkô Population of *Dioscorea tennuipes* Complex. Chem. Pharm. Bull. (Tokyo) **16**, 498 (1968).

4. AKIYAMA, T., Y. LITAKA, and O. TANAKA: Structure of Platicodigenin, a Sapogenin of *Platycodon grandiflorum* A. de Candolle. Tetrahedron Letters **1968**, 5577.

5. ALBERSHEIM, P., D. J. NEVINS, P. D. ENGLISH, and A. KARR: A Method for the Analysis of Sugars in Plant Cell-Wall Polysaccharides by Gas Liquid Chromatography. Carbohydr. Res. **5**, 340 (1967).

6. ALEXANDER, G., and E. SCHWENK: Biogenesis of Yeast Sterols IV. Transmethylation in Ergosterol Synthesis. J. Biol. Chem. **232**, 611 (1958).

7. ANANTARAMAN, R., and K. S. MADHAVAN PILLAI: Barringtogenol and Barringtogenic Acid, Two New Triterpenoid Sapogenins. J. Chem. Soc. (London) **1956**, 4369.

8. ANDERSON, D. M. W., and G. M. CREE: Studies on Uronic Acid Materials. XIV. Methylation with the Sodium Hydride-Methyl Iodide-Dimethyl Sulphoxide System. Carbohydrate Res. **2**, 162 (1966).

9. ANDERSON, L. A. P., W. T. DE KOCK, and P. R. ENSLIN: Constitution of Two Physiol.-Active Triterpenoids from *Lippia rehmanni*. J. South African Chem. Inst **14**, 58 (1961); Chem. Abstr. **57**, 15163 (1962).

10. ANTIA, N. J., Y. MAZUR, R. R. WILSON, and F. S. SPRING: Steroids XI, Isolation of Cholegenin and iso-Cholegenin from Oxbile. J. Chem. Soc. (London) **1954**, 1218.

11. APLIN, R. T., W. H. HUI, C. T. HO, and C. W. YEE: An Examination of the *Rubiaceae* of Hong Kong. Part VIII. The Structure of Spinosic Acid A, a New Triterpenoid Sapogenin from *Randia spinosa* (Thunb.) Poir. J. Chem. Soc. (C) **1971**, 1067.

12. APPLEBAUM, S. W., B. GESTETNER, and Y. BIRK: J. Insect Physiol. **11**, 611 (1965).

13. APPLEBAUM, S. W., S. MARCO, and Y. BIRK: Saponins as Possible Factor of Resistence of Legume Seeds to the Attack of Insects. Agr. Food Chem. **17**, 618 (1969).

14. ARIGONI, D.: Biosynthese von pentacyclischen Triterpenen in höheren Pflanzen. Experientia **14**, 153 (1958).

15. ARNAUDI, C. A., and A. SCHIESSER: The Microbial Degradation of Digitonin. Congr. intern. biochim. Résumés communs., 2e Congr., Paris **1952**, 118; Chem. Abstr. **49**, 7649 (1955).

16. ASPINALL, G. O.: Gas Liquid Partition Chromatography of Methylated and Partially Methylated Methyl Glycosides. J. Chem. Soc. (London) **1963**, 1676.

17. AYENGAR, K. N. N., and S. RANGASWAMI: Structure of Caccigenin, a New Triterpenoid Sapogenin from *Caccinia glauca* Savi. Tetrahedron Letters **1966**, 1947.

18. AYENGAR, K. N. N., and S. RANGASWAMI: Asperagenin, a Rare Type of Steroidal Sapogenin with 25 Hydroxyl Group. Current Science Bangalore **36**, 654 (1967).

19. BARCLAY, G. A., R. A. EADE, H. V. SIMES, J. J. H. SIMES, and J. C. TAYLOR: Ebelin Lactone, a Triterpene. Chem. and Ind. **1963**, 1206.

20. BARTON, D. H. R., H. T. CHEUNG, P. J. L. DANIELS, K. G. LEWIS, and J. F. McGHIE: Triterpenoids. Part XXVI. The Triterpenoids of *Vangueria tomentosa*. J. Chem. Soc. (London) **1962**, 5163.

21. BARTON, D. H. R., A. HAMEED, and J. F. McGHIE: The Constitution and Stereochemistry of Cyclamiretin. J. Chem. Soc. (London) **1962**, 5176.

22. BARUA, A. K.: Triterpenoids XXV. The Constitution of Entagenic Acid — A New Triterpenoid Sapogenin from *Entada phaseoloides* Merrill. Tetrahedron **23**, 1499 (1967).

23. BARUA, A. K., and P. CHAKRABARTI: The Constitution of Barringtogenol C — A New Triterpenoid Sapogenin from *Barringtonia acutangula* Gaertn. Tetrahedron **21**, 381 (1965).

24. BARUA, A. K., P. CHAKRABARTI, S. P. DUTTA, D. K. MUKHERJEE, and B. C. DAS:

Triterpenoids — XXXVII Stereochemistry and Structure of Lantanolic Acid, a New Triterpenoid from *Lantana camara*. Tetrahedron **27**, 1141 (1971).

25. Barua, A. K., P. Chakrabarti, S. K. Pal, and B. Das: Structure and Stereochemistry of Entagenic Acid. J. Indian Chem. Soc. **47**, 195 (1970); Chem. Abstr. **72**, 133021 (1970).

26. Barua, A. K., S. P. Dutta, and B. C. Das: Triterpenoids — XXIX. The Structure of Barringtogenol B — A New Triterpenoid Sapogenin from *Barringtonia acutangula* Gaertn. Tetrahedron **24**, 1113 (1968).

27. Barua, A. K., S. P. Dutta, and S. K. Pal: Triterpenoids XXX. The Structure of Barringtogenol E. A New Triterpenoid Sapogenol from *Barringtonia acutangula*. J. Indian Chem. Soc. **44**, 991 (1967); Chem. Abstr. **68**, 49885 (1968).

28. Barua, A. K., S. K. Pal, and S. P. Dutta: Triterpene Acid Isolated from *Barringtonia acutangula*. Sci. Cult. (Calcutta) **34**, 259 (1968); Chem. Abstr. **70**, 68560 (1969).

28a. Barua, A. K., S. K. Pal, and S. P. Dutta: Triterpenoids XXXIX. Constitution of Barrinic Acid, a New Triterpene Acid from *Barringtonia acutangula*. J. Indian Chem. Soc. **44**, 519 (1972); Chem. Abstr. **77**, 1268912 (1972).

29. Barua, A. K., and S. P. Raman: Triterpenoids — X. The Constitution of Albigenic Acid, a New Triterpenoid Sapogenin from *Albizzia lebbeck*. Benth. Tetrahedron **7**, 19 (1959).

30. Basu, N., and R. P. Rastogi: Triterpenoid Saponins and Sapogenins. Phytochem. **6**, 1249 (1967).

31. Beaton, J. M., and F. S. Spring: Triterpenoids Part LI. The Isolation and Characterisation of Glabric Acid, a New Triterpenoid Acid from Liquorice Root. J. Chem. Soc. (London) **1956**, 2417.

32. Bell, D. J.: The Methyl Ethers of D-Galactose. Advances of Carbohydr. Chem. **6**, 11 (1951).

33. Benn, W. R., F. Colton, and R. Pappo: The Structure of Ruscogenin. J. Amer. Chem. Soc. **79**, 3920 (1957).

34. Berger, F.: Handbuch der Drogenkunde, Bd. 1—8. Wien: W. Maudrich.

35. Bhacca, N. S., and D. H. Williams: Application of NMR-Spectroscopy in Organic Chemistry. Illustrations from Steroid Field. Holden-Day, San Francisco 1964.

36. Bhattacharyya, S. C.: Constituents of *Centella asiatica* I. Examination of the Ceylonese Variety II. Structures of the Terpene Acids. J. Indian Chem. Soc. **33**, 579, 630 (1956); Chem. Abstr. **51**, 4511 and 8700 (1957).

37. Bhattacharyya, S. C.: Constituents of *Centella asiatica* III. Examination of the Indian Variety. J. Indian Chem. Soc. **33**, 893 (1956); Chem. Abstr. **51**, 9808 (1957).

37a. Bhattacharya, A. K., and H. K. Dutta: Rubusic Acid, a New Triterpen from *Rubus moluccanus*. J. Indian. Chem. Soc. **46**, 381 (1969); Chem. Abstr. **71**, 50268 (1969).

38. Biemann, K., D. C. De Jongh, and H. K. Schnoes: Application of Mass Spectrometry to Structure Problems. XIV. Acetates of Partially Methylated Pentoses and Hexoses. J. Amer. Chem. Soc. **85**, 2289 (1963).

39. Biglino, G.: Constituents of the Root of *Bryonia dioica* IV. Constitution of the Bryonol Acid. Chem. Abstr. **61**, 8351 (1964).

40. Boiteau, P., B. Pasich et A. R. Ratsimamanga: Les Triterpenoides en physiologie végétale et animale. Paris: Gauthier-Villars. 1964.

41. Balansard, J. J. J., and P. J. L. Bernard: Extraction of Saponins from Plants. Neth. Pat. Appl. 6513079 12. Apr. 1966; Chem. Abstr. **66**, 14034 (1967).

42. Boll, P. M., and W. v. Philipsborn: NMR Studies and the Absolute Configuration of Solanum Alkaloids (Spiro-aminoketal Alkaloids). Acta Chem. Scand. **19**, 1365 (1965).

42a. Bombardelli, E., A. Bonati, B. Gabetta, and G. Mustich: Glykosides from Rhizomes of *Ruscus aculeatus* L. I. Fitoterapia **42**, 127 (1971).

42b. BOMBARDELLI, E., A. BONATI, B. GABETTA, and G. MUSTICH: Glycosides from Rhizomes of *Ruscus aculeatus* L. II. Filoterapia **43**, 3 (1972).

43. BOURNE, E. J., and S. PEAT: The Methyl Ethers of D-Glucose, Advances Carbohydrate Chem. **5**, 145 (1950).

44. BOWDEN, K., J. M. HEILBRON, E. R. H. JONES, and B. C. L. WEEDON: Researches on Acetylenic Compounds, Part I. Preparation of Acetylenic Ketones by Oxidation of Acetylenic Carbinols and Glykols. J. Chem. Soc. (London) **1946**, 39.

45. BOYER, J. P., R. A. EADE, H. LOCKSLEY, and J. J. H. SIMES: Extractives of Austr. Timbers II. Emmolic Acid, Triterpen Acid from *Emmenospermum alphitonioides*. Austral. J. Chem. **11**, 236 (1958); Chem. Abstr. **53**, 436 (1959).

46. BRETON, J. L., and A. G. GONZALEZ: Glucosides and Aglycones from Canary Scrophulariaceae. Part VI. Structure of Two New Triterpenes from *Scrophularia smithii* Wydler. J. Chem. Soc. (London) **1963**, 1401.

47. BRIESKORN, C. H., und W. WALLENSTÄTTER: Quantitative Bestimmung von Glycyrrhizinsäure in Wurzel und Extrakt des Süßholzes mittels Acetanhydrid-Schwefelsäure. Arch. Pharmaz. **300**, 717 (1967).

48. BRIESKORN, C. H., und H. WUNDERER: Pomol- und Pomonsäure. Über den chem. Aufbau der Apfelschale. Chem. Ber. **100**, 1252 (1967).

49. BRÜCHER, H.: Proteinreichere Indianer-Gewächse durch Bestrahlung. Umschau Wiss. Techn. **70**, 308 (1970).

50. BUDZIKIEWICZ, H., C. DJERASSI, and D. H. WILLIAMS: Structure Elucidation of Natural Products by Mass Spectrometry, Volume II. Steroids, Terpenoids, Sugars, and Miscellaneows Classes. San Francisco: Holden-Day. 1964.

51. — — — Mass Spectrometry of Organic Compounds. San Francisco: Holden-Day. 1967.

52. BUDZIKIEWICZ, H., J. M. WILSON und C. DJERASSI: MS und ihre Anwendung auf strukturelle und stereochemische Probleme. 15. Mitt. Steroidsapogenine. Monatshefte Chem. **93**, 1033 (1962).

53. — — — MS in Structural and Stereochemical Problems XXXII. Pentacyclic Triterpenes. J. Amer. Chem. Soc. **85**, 3688 (1963).

54. BUKHAROV, V. G., and V. V. KARLIN: *Patrinia sibirica* Glycosides I.; Khim. Prir. Soedin. **6**, 60 (1970); Chem. Abstr. **73**, 56361 (1970).

55. — — *Patrinia sibirica* Glycosides II.; Khim. Prir. Soedin. **6**, 64 (1970); Chem. Abstr. **73**, 56359 (1970).

56. — — Triterpenic Glycosides from *Patrinia scabiosofolia* II.; Khim. Prir. Soedin. **6**, 211 (1970); Chem. Abstr. **73**, 66857 (1970).

57. — — Structure of Scabioside F.; Khim. Prir. Soedin. **6**, 372 (1970); Chem. Abstr. **73**, 99167 (1970).

58. — — Structure of Scabioside G.; Khim. Prir. Soedin. **6**, 373 (1970); Chem. Abstr. **73**, 110059 (1970).

59. BUKHAROV, V. G., V. V. KARLIN, and T. N. SIDOROVICH: Triterpene Glycosides of *Patrinia scabiosofolia* I.; Khim. Prir. Soedin **6**, 69 (1970); Chem. Abstr. **73**, 99162 (1970).

59a. BUKHAROV, V. G., V. V. KARLIN, and V. A. TALAN: Triterpene Glycosides of *Patrinia intermedia* II. Structure of the Carbohydrate Chain of Patrinosid C_1. Khim. Prir. Soedin. **5**, 22 (1969); Chem. Abstr. **71**, 30664 (1969).

60. — — — Triterpene Glycosides of *Patrinia intermedia* III.; Structure of the Carbohydrate Chain of Patrinoside D. Triterpene Glycosides of *Patrinia intermedia* IV.; Structures of the Carbohydrate Chains of Patrinosides C and D; Khim. Prir. Soedin. **5**, 84 and 89 (1969); Chem. Abstr. **71**, 70877 and 70878 (1969).

61. BUKHAROV, V. G., and L. N. KARNEEVA: Glycosides of Ursolic Acid from *Empetrum sibiricum*. Izv. Akad. Nauk. SSSR, Ser. Khim. **1970**, 171; Chem. Abstr. **73**, 4146 (1970).

62. — — New Type of Glycoside Bond. Izv. Akad. Nauk. SSSR, Ser. Khim **1970**, 1916; Chem. Abstr. **74**, 76627 (1971).

63. Bukharov, V. G., and L. N. Karneeva: Triterpenoid Glycosides from *Sanguisorba officinalis*. Izv. Akad. Nauk. SSSR, Ser. Khim. **1970**, 2402; Chem. Abstr. **74**, 121340 (1971).

63a. — — Structure of Nutanoside. Khim. Prir. Soedin **7**, 412 (1971); Chem. Abstr. **75**, 152040 (1971).

64. Bukharov, V. G., L. N. Karneeva, and V. A. Talan: Structure of Carbohydrate Chains of Polemoniosides B and C. Khim. Prir. Soedin. **5**, 511 (1969); Chem. Abstr. **73**, 15158 (1970).

65. Bukharov, V. G., and S. P. Shcherbak: Triterpene Glycosides from *Saponaria officinalis*. Khim. Prir. Soedin **5**, 389 (1969); Chem. Abstr. **72**, 79404 (1970).

66. — — Triterpenoid Glycosides of *Herniaria glabra* I.; Khim. Prir. Soedin. **6**, 307 (1970); Chem. Abstr. **73**, 110070 (1970).

67. — — Triterpenoid Glycosides from *Dianthus deltoides* II. Structure of Dianthoside C. Khim. Prir. Soedin. **7**, 420 (1971); Chem. Abstr. **76**, 14854 (1972).

68. Bukharov, V. G., S. P. Shcherbak, and A. P. Beshchekova: Triterpenoid Glycosides of *Dianthus deltoides* I. Khim. Prir. Soedin. **7**, 33 (1971); Chem. Abstr. **74**, 112253 (1971).

69. Burn, D., B. Ellis, and V. Petrow: The Structure of Ruscogenin. J. Chem. Soc. (London) **1958**, 795.

70. Caglioti, L., C. Cainelli, and F. Mirutilli: Isolation of a Triterpenic Acid from the Cakes of *Olea europea*. Gazz. Chim. Ital. **91**, 1387 (1961); Chem. Abstr. **56**, 7369 (1962).

71. Cainelli, G., J. J. Britt, D. Arigoni und O. Jeger: Zur Kenntnis der Triterpene 196. Mitt. Zur Konstitution der Sojasapogenole A, B, C und D. Helv. Chim. Acta **41**, 2053 (1958).

72. Cainelli, G., A. Melera, D. Arigoni und O. Jeger: Zur Kenntnis der Triterpene. Konstitution des Äscigenins. Helv. Chim. Acta **40**, 2390 (1957).

73. Callow, R. K., and V. H. T. James: Epimerisation of C-25 of Steroid Sapogenins: Sarsasapogenin, neo-Tigogenin and Sisalagenin. J. Chem. Soc. (London) **1955**, 1671.

74. Canonica, L., B. Danieli, R. Giovanni, and A. Bonati: Triterpenes from *Glycyrrhiza glabra* IV. 18α-Hydroxyglycyrrhetic Acid. Gazz. Chim. Ital. **97**, 769 (1967); Chem. Abstr. **68**, 13206 (1968).

75. Canonica, L., B. Danieli, P. Manitto, and G. Russo: Triterpenes of *Glycyrrhiza glabra* III. Structure of Isoglabrolide. Gazz. Chim. Ital. **96**, 843 (1966); Chem. Abstr. **65**, 15436 (1966).

76. Canonica, L., B. Danieli, P. Manitto, G. Russo, and E. Bombardelli: *Glycyrrhiza glabra* Triterpenes. V. Glycyrrhetol and 21α-Hydroxy-isoglabrolide. Gazz. Chim. Ital. **97**, 1347 (1967); Chem. Abstr. **68**, 49810 (1968).

77. Canonica, L., B. Danieli, P. Manitto, G. Russo, E. Bombardelli, and A. Bonati: Triterpenes of *Glycyrrhiza glabra* VII. 24-Hydroxyliquiritic Acid and Liquiridiolic Acid. Gazz. Chim. Ital. **98**, 712 (1968); Chem. Abstr. **69**, 59439 (1968).

78. Canonica, L., B. Danieli, P. Manitto, G. Russo, and A. Bonati: *Glycyrrhiza glabra* Triterpenes. VI. 24-Hydroxy-glycyrrhetic Acid and 24-Hydroxy-11-deoxyglycyrrhetic Acid. Gazz. Chim. Ital. **97**, 1359 (1967); Chem. Abstr. **68**, 49811 (1968).

79. Canonica, L., F. Ronchetti, and G. Russo: Incorporation of 5α-Furostan-3β,26-diol into Tigogenin by *Digitalis lanata*. Phytochem. **11**, 243 (1972).

80. Canonica, L., G. Russo, and E. Bombardelli: Triterpenes of *Glycyrrhiza glabra* II. Liquiritic Acid. Gazz. Chim. Ital. **96**, 833 (1966); Chem. Abstr. **65**, 15435 (1966).

81. Canonica, L., G. Russo, and A. Bonati: Triterpenes of *Glycyrrhiza glabra* I. Two New Lactones with an Oleanane Structure. Gazz. Chim. Ital. **96**, 772 (1966); Chem. Abstr. **65**, 15435 (1966).

82. Chakrabarti, P.: Constitution of Mollugogenol A: A New Sapogenin from *Mollugo hirta*. J. Indian Chem. Soc. **46**, 98 (1969); Chem. Abstr. **70**, 93943 (1969).

83. CHAKRABARTI, P., D. K. MUKHERJEE, A. K. BARUA, and B. C. DAS: The Structure and Stereochemistry of Spergulagenic Acid. Tetrahedron **24**, 1107 (1968).

84. CHAKRABARTI, P., and A. K. SANYAL: Constitution of Mollugogenol E — New Sapogenin from *Mollugo hirta*. Indian J. Chem. **8**, 1042 (1970); Chem. Abstr. **75**, 20692 (1971).

84a. — — The Constitution of Mollugogenol C. A New Sapogenin from *Mollugo hirta*. J. Indian. Chem. Soc. **46**, 1061 (1969).

85. CHAKRABARTI, P., P. K. SANYAL, and A. K. BARUA: Constitution of Mollugogenol B: A New Sapogenin from *Mollugo hirta*. J. Indian Chem. Soc. **46**, 96 (1969); Chem. Abstr. **70**, 93943 (1969).

86. CHANLEY, J. D., T. MEZZETTI, and H. SOBOTKA: The Holothurinogenins. Tetrahedron **22**, 1857 (1966).

87. CHANLEY, J. D., J. PERLSTEIN, R. F. NIGRELLI, and H. SOBOTKA: Further Studies on the Structure of Holothurin. Ann. New York Acad. Sci. **90**, 902 (1960).

88. CHANLEY, J. D., and C. ROSSI: The Holothurinogenins — II. Methoxylated Neo-Holothurinogenins. Tetrahedron **25**, 1897 (1969).

89. — — The Neo-Holothurinogenins — III. Neo-Holothurinogenins by Enzymatic Hydrolysis of Desulfated Holothurin A. Tetrahedron **25**, 1911 (1969).

90. CHATTERJI, N., R. P. RASTOGI, and M. L. DHAR: Chemical Examination of *Bacopa monniera* II. The Constitution of Bacoside A. Indian J. Chem. **3**, (24), 5483 (1968).

91. CHAUSSEAUD, L. F., B. J. FRY, D. R. HAWKINS, J. D. LEWIS, I. P. SWORD, T. TAYLOR, and D. E. HATHWAY: The Metabolism of Asiatic Acid, Madecassic Acid and Asiaticosid in the Rat. Arzneimittelforsch. **21**, 1379 (1971).

92. CHEBAN, P. L., and V. J. CHIRVA: Structure of Helianthoside B. Khim. Prir. Soedin. **5**, 327 (1969); Chem. Abstr. **73**, 45777 (1970).

93. CHEBAN, P. L., V. J. CHIRVA, and G. V. LAZUREVSKII: Structure of Helianthoside A. Khim. Prir. Soedin. **5**, 59 (1969); Chem. Abstr. **71**, 3623 (1969).

94. — — — Structure of Helianthoside C, a Sunflower Saponin; Khim. Prir. Soedin. **5**, 129 (1969); Chem. Abstr. **71**, 70889 (1969).

95. CHEUNG, H. T., and D. G. WILLIAMSON: NMR-Signals of Methyl Groups of Triterpenes with Oxygen Functions at Positions 2, 3, and 23. Tetrahedron **25**, 119 (1969).

95a. CHEUNG, H. T., and T. C. YAN: Nuclear Magnetic Resonance Signals of Methyl Groups in Structural Determination of Triterpenes. $2\alpha,3\alpha$- and $2\beta,3\beta$-Dihydroxy-oleanen-12-en-28-oic Acids. Chem. Commun. **1970**, 369.

96. CHIRVA, V. J., P. L. CHEBAN, P. K. KINTYA, and V. A. BOBEIKO: Structure of Triterpenoid Glycosides from *Chenopodium anthelminticum* Roots. Khim. Prir. Soedin. **7**, 27 (1971); Chem. Abstr. **74**, 112384 (1971).

97. CHIRVA, V. J., and P. K. KINTYA: Structure of Saponoside A. Khim. Prir. Soedin. **5**, 188 (1969); Chem. Abstr. **72**, 13007 (1970).

98. — — Structure of Saponaside D. Khim. Prir. Soedin. **6**, 214 (1970); Chem. Abstr. **73**, 88123 (1970).

99. CHIRVA, V. J., P. K. KINTYA, and L. G. KRETSU: Triterpenoid Glycosides of Legumes III. Structure of the Minor Glycoside of Beans. Khim. Prir. Soedin. **6**, 563 (1970); Chem. Abstr. **74**, 31935 (1971).

100. CHIRVA, V. J., P. K. KINTYA, and V. N. MELVIKOV: Ribose in Triterpene Glycosides of *Clematis vitalba* I. Khim. Prir. Soedin. **7**, 297 (1971); Chem. Abstr. **75**, 115918 (1971).

101. CHIRVA, V. J., P. K. KINTYA, and V. A. SOSNOVSKII: Triterpenoid Glycosides of *Koelreuteria paniculata* IV. Structure of Koelreuteria Saponins A and B. Khim. Prir. Soedin. **6**, 328 (1970); Chem. Abstr. **73**, 110063 (1970).

102. — — — Triterpenoid Glycosides of *Sapindus mukorossi* III. Structure of Sapindoside C. Khim. Prir. Soedin. **6**, 374 (1970); Chem. Abstr. **73**, 110062 (1970).

582 R. Tschesche und G. Wulff:

103. Chirva, V. J., P. K. Kintya, and V. A. Sosnovskii: Triterpenoid Glycosides of *Sapindus mukorossi* V. Structure of Sapindoside E. Khim. Prir. Soedin. **6**, 431 (1970); Chem. Abstr. **74**, 13384 (1971).

104. — — — Structure of Xanthoceras-saponin. Khim. Prir. Soedin. **7**, 442 (1971); Chem. Abstr. **75**, 152047 (1971).

105. Chirva, V. J., P. K. Kintya, V. A. Sosnovskii, P. E. Krivenchuk, and N. V. Zykova: Triterpene Glycosides of *Sapindus mukorossi* II. Structure of Sapindosides A and B. Khim. Prir. Soedin. **6**, 218 (1970); Chem. Abstr. **73**, 77544 (1970).

106. Chirva, V. J., P. K. Kintya, V. A. Sosnovskii, and B. M. Zolotarev: Triterpenoid Glycosides of *Sapindus mukurossi* V. Structure of Sapindoside D. Khim. Prir. Soedin. **6**, 316 (1970); Chem. Abstr. **73**, 110071 (1970).

107. Chirva, V. J., and V. P. Konyukhov: Structure of Clematoside A. Khim. Prir. Soedin. **4**, 140 (1968); Chem. Abstr. **69**, 77681 (1968).

108. — — Structure of Clematoside B. Khim. Prir. Soedin. **4**, 141 (1968); Chem. Abstr. **69**, 77682 (1968).

109. — — Structure of Clematoside A′. Khim. Prir. Soedin. **5**, 60 (1969); Chem. Abstr. **71**, 13314 (1969).

109a. Chirva, V. J., V. P. Konyukhov, P. L. Cheban, and G. V. Lazurevskii: New Data on the Structure of Triterpene Saponins. Khim. Biokhim. Uglevodov, Mater. Vses. Konf., 4th (1967), 98 Chem. Abstr. **73**, 88099 (1970).

110. Chirva, V. J., L. G. Kretsu, and P. K. Kintya. Triterpenoid Glycosides of Legumes II. Structure of the Major Glycoside of Beans. Khim. Prir. Soedin. **6**, 559 (1970); Chem. Abstr. **74**, 31934 (1971).

111. Chirva, V. J., A. I. Usov, and V. P. Konjukhov: Methylation and Periodate Oxidation of Clematosides A′, A, and B. Izvest. Akad. Nauk. SSSR, Ser. Khim. **1968**, 2336; Chem. Abstr. **70**, 47797 (1969).

112. Colombo, P., D. Corbetta, A. Pirotta, G. Ruffini, and A. Sartori: A Solvent for Qualitative and Quantitative Determination of Sugars Using Paper Chromatography. J. Chromatography **3**, 343 (1960).

113. Corey, E. J., and P. R. Ortiz de Montellano: Enzymic Synthesis of β-Amyrin from 2,3-Oxidosqualene. J. Amer. chem. Soc. **89**, 3362 (1967).

114. Crabbé, P.: Optical Rotatory Dispersion and Circular Dichroism in Organic Chemistry. San Francisco: Holden-Day. 1965. Erweiterte französische Edition: Gauthier-Villars 1968.

115. Crombie, L., P. J. Ham, and D. A. Whiting: Sapogenin of the Termite-repellent Fruit of *Swartzia madagascariensis*. Chem. Ind. (London) **1971**, 176.

116. Dávila, C. A., and F. M. Panizo: National Sources of Steroids, VIII. Sapogenins of *Asparagus acutifolius, Asparagus stipularis, Agave americana, Yucca gloriosa* and *Ruscus aculeatus*. An. real. Soc. espan. Fisica Quim. Ser. B **54**, 697 (1958); Chem. Abstr. **54**, 6029 (1960).

117. Dekanosidze, G. E., T. T. Gorovits, T. A. Pkheidze, and E. P. Kemertelidze: Calopanax-saponin B from *Hedera caucasigena*. Khim. Prir. Soedin. **6**, 489 (1970); Chem. Abstr. **74**, 1055 (1971).

118. Dekanosidze, G. E., T. A. Pkheidze, T. T. Gorovits, and E. P. Kemertelidze: Triterpenoid Oligoside, Hederacolchiside E from *Hedera colchica*. Khim. Prir. Soedin. **6**, 484 (1970); Chem. Abstr. **74**, 10352 (1971).

119. Dekanosidze, G. E., T. A. Pkheidze, and E. P. Kemertelidze: Structure of Hederacaucaside B, Saponin from *Hedera caucasignea*. Khim. Prir. Soedin. **6**, 491 (1970); Chem. Abstr. **74**, 23091 (1971).

120. Dekanosidze, G. E., T. A. Pkheidze, E. P. Kemertelidze, L. J. Mikhailova, A. Z. Tolokneva, and N. K. Fruentov: Chemical and Pharmacological Study of the Triterpenoid Glycosides of Ivy, *Hedera colchica*. Soobshch. Akad. Nauk. Gruz. SSR. **61**, 609 (1971); Chem. Abstr. **75**, 45630 (1971).

121. DELGADO, J. M., VELAZQUEZ, J. L. BRETÓN, and A. GONZÁLES: Glucosides and Aglukons of *Scrophulariaceae,* XIII. Aglukons of *Digitalis canariensis.* Ann. Chim. **65,** 817 (1969); Chem. Abstr. **72,** 75610 (1970).

121a. DEVON, T. K., and A. I. SCOTT: Handbook of Naturally Occurring Compounds, Vol. II. Terpenes. New-York-London: Academic Press. 1972.

122. DJERASSI, C., A. BOWERS, S. BURSTEIN, H. ESTRADA, J. GROSSMANN, J. HERRAIN, A. J. LEMIN, A. MANJARREZ, and S. C. PAKRASHI: Terpenoids XXII Triterpenes from Some Mexican and South American Plants. J. Amer. Chem. Soc. **78,** 2312 (1956).

123. DJERASSI,C., and R. EHRLICH: Optical Rotatory Dispersion Studies IV Steroidal Sapogenins. J. Amer. Chem. Soc. **78,** 440 and 3163 (1956).

124. DJERASSI, C., L. E. GELLER, and A. J. LEMIN: Terpenoids I. The Triterpenes of the Cactus. *Lemaireocereus thurberi.* J. Amer. Chem. Soc. **75,** 2254 (1953).

125. DJERASSI, C., T. T. GROSSNICKLE, and L. B. HIGH: Constitution and Stereochemistry of Digitogenin. J. Amer. Chem. Soc. **78,** 3166 (1956).

126. DJERASSI, C., and J. FISHMAN: Constitution and Stereochemistry of Samogenin, Markogenin and Mexogenin. J. Amer. chem. Soc. **77,** 4291 (1955).

127. DOBRINER, K., E. R. KATZENELLENBOGEN, and R. N. JONES: Infrared Absorption Spectra of Steroids. An Atlas. New York: Interscience Publishers. 1953.

128. DORCHAÍ, R. Ó., H. E. RUBALCAVA, J. B. THOMSON, and B. ZEEH: Triterpenoids IV. The Cyclamigenins A_1, A_2, C, and D. Tetrahedron **24,** 5649 (1968).

129. DORCHAÍ, R. Ó., and J. B. THOMSON: Triterpenoids III. The Structure of Cyclamigenin B. Tetrahedron **24,** 1377 (1968).

130. DUGAN, J. J., and P. DE MAYO: Terpenoids X. Pre-Senegenin, a Quite Normal Triterpenoid. Canad. J. Chem. **43,** 2033 (1965).

131. DUGAN, J. J., P. DE MAYO, and A. N. STARRATT: Senegenin. Proc. Chem. Soc. **1964,** 264; Chem. Abstr. **61,** 13352 (1964).

132. DZIZENKO, A. K., M. J. NEFEDOVA, and G. B. ELYAKOV: IR- and NMR-Spectra of Triterpenoids Isolated from Ginseng. Dokladi Akad. Nauk. SSSR **162,** 569 (1965); Chem. Abstr. **63,** 5684 (1965).

133. EADE, R. A., J. J. H. SIMES, and B. STEVENSON: Extractives of Australian Timbers IV. Castanogenin (Medicagenic Acid) and Bayogenin $C_{30}H_{48}O_5$ from *Castanospermum australe.* Austral. J. Chem. **16,** 900 (1963); Chem. Abstr. **60,** 592 (1964).

134. ELGAMAL, M. H. A., and M. B. E. FAYEZ: Structure of Glabric Acid, a Further Triterpenoid Constituent of *Glycyrrhiza glabra.* Acta chim. Acad. Sci. hung. **58,** 75 (1968); Chem. Abstr. **70,** 29115 (1969).

135. ELGAMAL, M. H. A., M. B. F. FAYEZ, and G. SNATZKE: Constituents of Local Plants — VI. Liquoric Acid, a New Triterpenoid from the Roots of *Glycyrrhiza glabra* L. Tetrahedron **21,** 2109 (1965).

136. ELKIN, J. N., A. K. DZIZENKO, and G. B. ELYAKOV: Mass-spectrometric Study of Some Triterpenes of the Dammarane Series. Khim. Prir. Soedin. **7,** 286 (1971); Chem. Abstr. **75,** 110463 (1971).

137. ELKS, J.: Steroid Saponins and Sapogenins. In: COFFEY, S. (Edit.) Rodd's Chemistry of Carbon Compounds, Vol. II. Amsterdam: E. Elsevier. 1971.

138. Elsevier's Encyclopedia of Organic Chemistry, Bd. 14. Amsterdam: E. Elsevier. 1940.

139. Elsevier's Encyclopedia of Organic Chemistry, Vol. 14 Supplement. Amsterdam: E. Elsevier. 1952.

140. ELYAKOV, G. B., A. K. DZIZENKO, and J. N. ELKIN: On the Structure of Panaxatriol. Tetrahedron Letters **1966,** 141.

141. ELYAKOV, G. B., A. K. DZIZENKO, and E. V. SHAPKINA: Structure of the Acid Hydrolysis Products of Panaxosides D, E, and F. Khim. Prir. Soedin. **3,** 164 (1967); Chem. Abstr. **67,** 100268 (1967).

142. ELYAKOV, G. B., A. J. KHORLIN, L. I. STRIGINA, and N. K. KOCHETKOV: Triterpene

584   R. TSCHESCHE und G. WULFF:

Saponins III. Araloside A from *Aralia schmidtii.* Izv. Akad. Naukk. SSSR Ser. Khim. Nauk. **1962,** 1605; Chem. Abstr. **58,** 4604 (1963).

143. ELYAKOV, G. B., T. A. KUZNETSOVA, A. K. DZIZENKO, and J. N. ELKIN: A Chemical Investigation of the Trepang (*Stychopus japonicus* Selenka): The Structure of Triterpenoid Aglycones Obtained from Trepang Glycosides. Tetrahedron Letters **1969,** 1151.

144. ELYAKOV, G. B., L. I. STRIGINA, E. V. SHAPKINA, N. T. ALADYINA, S. A. KORNILOVA, and A. K. DZIZENKO: The Probable Structure of the True Aglycones of Ginseng Glycosides. Tetrahedron **24,** 5483 (1968).

145. ELYAKOV, G. B., L. I. STRIGINA, N. J. UVAROVA, V. E. VASKOVSKY, A. K. DZIZENKO, and N. K. KOCHETKOV: Glycosides from Ginseng Roots. Tetrahedron Letters **1964,** 3591.

146. ELYAKOV, G. B., N. J. UVAROVA, and R. R. GORSHKOVA: The Structure of Carbohydrate Chains of Panaxisides D, E, and F. Tetrahedron Letters **1965,** 4669.

147. ERRINGTON, S. G., D. E. WHITE, and M. W. FULLER: The Structures of A_1-Barrigenol and R_1-Barrigenol. Tetrahedron Letters **1967,** 1289.

148. FERENCZY, L., K. HORVATH, and J. ZSOLT: Thin-Layer Chromatography of Antifungal Compounds from Higher Plants. Herb. Hung. **5,** 88 (1966); Chem. Abstr. **68,** 46992 (1968).

149. FIESER, L. F., und M. FIESER: Steroide. Weinheim: Verlag Chemie. 1961; Steroids. New York: Reinhold Publishing Corp. 1959.

150. FISCHER, F. G., und H. DÖRFEL: Die quantitative Bestimmung reduzierender Zucker auf Papierchromatogrammen. Z. physiol. Chem. **297,** 164 (1954).

151. FISCHER, F. G., und H. DÖRFEL: Die papierchromatographische Trennung und Bestimmung der Uronsäuren. Z. physiol. Chem. **301,** 224 (1955).

152. FREIRE, R., A. G. GONZÁLEZ, and E. SUÁREZ: New Sources of Steroid Sapogenins VIII, Sceptrumgenin and Isoplexigenins A, B, C and D from *Isoplexis sceptrum.* Tetrahedron **26,** 3233 (1970).

153. FREIRE, R., A. G. GONZÁLEZ, J. A. SALAZAR, and E. SUÁREZ: Eduligenin and Lowegenin. Two New Steroidal Sapogenins from *Tamus edulis.* Phytochem. **9,** 1641 (1970).

154. FRIESS, S. L., R. C. DURANT, and J. D. CHANLEY: Further Studies on Biological Actions of Steroidal Saponins Produced by Poisonous Ehinoderms. Toxicon **6,** 81 (1968).

155. FRIESS, S. L., R. C. DURANT, J. D. CHANLEY, and T. MEZZETTI: Some Structural Requirements Underlying Holothurin A Interactions with Synoptic Chemoreceptors. Biochem. Pharmacol. **14,** 1237 (1965).

156. FRIESS, S. L., G. STANDAERT, E. R. WHITCOMB, R. F. NIGRELLI, J. D. CHANLEY, and H. SOBOTKA: Some Pharmacological Properties of Holothurin A, a Glycosidic Mixture from the Sea Cucumber. Ann. New York Acad. Sci. **90,** 893 (1960).

156a. FROLOVA, G. M., and J. S. OVODOV: Triterpenoid Glycosides of *Eleutherococcus senticosus* Leaves. Khim. Prir. Soedin. **1971,** 618; Chem. Abstr. **76,** 59965 (1972).

157. GÁL, J. E.: Capsicidin, eine neue Verbindung mit antibiotischer Wirksamkeit aus Gewürzpaprika. Z. Lebensmittelunters. Forsch. **124,** 333 (1964).

158. GERLACH, H.: Saponine und ihre Bedeutung für die Pharmazie. Pharm. Zentralhalle **105,** 553 (1966).

159. GESTETNER, B.: Lucerne Saponins V. Structure of a Saponin from *Medicago sativa* (Alfalfa). Phytochem. **10,** 2221 (1971).

160. GESTETNER, B., Y. BIRK, and A. BONDI: Soya Bean Saponins — VI. Composition of Carbohydrate and Aglycone Moieties of Soya Bean Saponin Extract and of its Fractions. Phytochem. **5,** 799 (1966).

161. GESTETNER, B., Y. BIRK, A. BONDI, and Y. TENCER: Soya Bean Saponins — VII. A Method for the Determination of Sapogenin and Saponin Contents in Soya Beans. Phytochem. **5,** 803 (1966).

162. GESTETNER, B., I. ISHAAYA, Y. BIRK, and A. BONDI: Soybean Saponins III. Fractionation and Characterization. Israel J. Chem. 1, 460 (1963).
163. GIRERD, R. J., G. D. PASQUALE, B. G. STEINETZ, V. L. BEACH, and W. PEARL: Antiedema Properties of Escin. Arch. Intern. Pharmacodynamic 133, 127 (1961); Chem. Abstr. 56, 5341 (1942).
164. GOLDSMITH, D. P. J., P. R. ULSHAFER, and C. H. RUOF: Sterols CLIII, Sapogenins LXV, Kryptogenin, a New Type of Sapogenins from Beth Root. J. Amer. Chem. Soc. 65, 739 (1943).
165. GONZÁLEZ, A. G., R. FREIRE, M. G. GARCIA-ESTRADA, J. A. SALAZAR, and E. SUÁREZ: New Natural Sources of Steroidal Sapogenins XI, Dracogenin, New *Dracaena draco* Spirostan Tetrahydroxy Sapogenin. Ann. Chim. 67, 557 (1971); Chem. Abstr. 75, 130017 (1971).
166. GONZÁLES, A. G., R. FREIRE, M. G. GARCIA-ESTRADA, J. A. SALAZAR, and E. SUÁREZ: New Sources of Steroid Sapogenins XIV. 25 S-Ruscogenin and Sansevierigenin, two New Spirostan Sapogenins from *Sansevieria trifasciata*. Tetrahedron 28, 1289 (1972).
167. GONZÁLEZ A. G., R. FREIRE, J. A. SALÁZAR, and E. SUÁREZ: 7-Ketotamusgenin, 7-Ketodiosgenin, 25 S-Hydroxytamusgenin and Afurigenin. Four new Steroidal Sapogenins from Tamus edulis. Phytochem. 10, 1339 (1971).
168. GOODSON, L. H., and C. R. NOLLER: Saponins and Sapogenins XI, Neotigogenin, a New Steroid Sapogenin. J. Amer. Chem. Soc. 61, 2420 (1939).
169. GOODWIN, T. W.: The Biogenesis of Terpenes and Steroids. In: COFFEY S. (Edit.): Rodd's Chemistry of Carbon Compounds, Vol. II E, p. 54. Amsterdam: E. Elsevier. 1971.
170. GOTO, M., S. IMAI, T. MURATA, T. NOGUCHI, and S. FUYIOKA: Antimicrobial Glycosides of *Euptelea polyandra* A. Isolation, Constitutions and Antimicrobial Activity of Eupteleoside A and B. Yakugaku Zasshi 90, 736 (1970); Chem. Abstr. 73, 84655 (1970).
171. GOVINDACHARI, T. R., P. A. MOHAMED, P. C. PARTHASARATHY: Gymnosporal, a New Pentacyclic Triterpene from *Gymnosporia rothiana*. Indian J. Chem. 8, 395 (1970); Chem. Abstr. 73, 63176 (1970).
171a. GROSSER, J. S.: Natural Products of Echinoderms. Chem. Soc. Rev. 1, 1 (1972).
172. GUBANOV, I. A., N. J. LIBIZOV, and A. S. GLADKIKH: Search for Saponincontaining Plants among the Flora of Central Asia and Southern Kazakhstan. Farmatsiya (Moscow) 19, 23 (1970); Chem. Abstr. 73, 95408 (1970).
172a. GUSEINOV, D. Y., and G. B. ISKENDEROV: Chemical Composition and Biological Value of Saponins of Some Plants of Azerbeidzhan. Biol. Nauki 15, 85 (1972); Chem. Abstr. 77, 16529 (1972).
173. HABERMEHL, G., and G. VOLKWEIN: Aglycones of the Toxines from the Cuverian Organs of *Holothuria forskali* and a New Nomenclature for the Aglycons from Holothuriodeae. Toxicon 9, 319 (1971); Chem. Abstr. 75, 141016 (1971).
174. HADA, H., and M. INAGAKI: Determination of Glycyrrhizinic Acid in Glycyrrhiza. Yakugaku Zasshi 78, 795 (1958); Chem. Abstr. 52, 17357 (1958).
175. HAHN, L. R., C. SANCHEZ, and J. ROMO: Isolation and Structure of Jacquinic Acid. Tetrahedron 21, 1735 (1965).
176. HAKOMORI, S.: Permethylation of Glycolipids and Polysaccharides Catalyzed by Methylsulfinyl Carbanion in Dimethylsulfoxide. J. Biochemistry (Tokyo) 55, 205 (1964); Chem. Abstr. 60, 15959 (1964).
177. HALSALL, T. G., and R. T. APLIN: A Pattern of Development in the Chemistry of Pentacyclic Triterpenes. Fortschr. Chem. organ. Naturstoffe 22, 153 (1964).
178. HAMAMOTO K.: Studies on the Steroidal Components of Domestic Plants, XXIII. Structure of Metagenin. Chem. Pharm. Bull. (Tokyo) 8, 1099 (1960); Chem. Abstr. 57, 8635 (1962).
179. — Studies on the Steroidal Components of Domestic Plants, XXIV. Structure of Metagenin (4). Chem. Pharm. Bull. (Tokyo) 9, 32 (1961).
180. HAMPEL, H., G. HOFRICHTER, H.-D. LIEHN und W. SCHLEMMER: Zur Pharmakologie

586 R. Tschesche und G. Wulff:

der Aescin-Isomere unter besonderer Berücksichtigung von α-Aescin Arzneimittel-forsch. **20,** 209 (1970).

181. Hanahan, D. J., and S. J. Wakil: The Origin of Some of the Carbon Atoms of the Side Chain of ^{14}C-Ergosterol. J. Amer. Chem. Soc. **75,** 273 (1952).

182. Hariharan, V., and S. Rangaswami: Steroids and Triterpenoids from the Roots of Mollugo spergula. Phytochem. **10,** 621 (1971).

182a. — — Structure of Saponine A and B from the Seeds of *Achyranthes aspera.* Phytochemistry **9,** 409 (1970).

182b. Hariharan, V., S. Rangaswami, and S. Sarangan: Saponins of the Seeds of *Bassia latifolia.* Phytochem. **11,** 1791 (1972).

183. Harrison, L. T., M. Velasco, and C. Djerassi: Chiapagenin and Isochiapagenin. Two New Steroidal Sapogenins from *Dioscorea chiapasensis.* J. Org. Chem. **26,** 155 (1961).

184. Haworth, S., J. G. Roberts, and B. F. Sagar: Quantitative Determination of Mixtures of Alkyl Ethers of D-Glucose. Carbohyd. Res. **9,** 491 (1969).

185. Heitz, S., D. Billet, et D. Raulais: Triterpènes d'*Anagallis arvensis* L. (Primulacée). I. Structure de l'anagalligénone B. Bull. Soc. Chim. France **1971,** 2320.

186. Hensens, O. D., and K. G. Lewis: Reactions of Primula Genin A — Part I. Tetrahedron Letters **1965,** 4639.

187. — — Acetyl Migration between the 28- and the 16-Hydroxyl Groups of Primulagenin A. Tetrahedron Letters **1968,** 3213.

188. — — Reactions of Primulagenin A II. Acid Catalyzed Transformations. Austr. J. Chem. **24,** 2117 (1971); Chem. Abstr. **75,** 118421 (1971).

189. Heyns, K., K. R. Sperling und H. F. Grützmacher: Massenspektrometrische Untersuchungen. XIX. Mitt.: Kombination von Gaschromatographie und Massen-spektroskopie zur Analyse partiell methylierter Zuckerderivate. Die Massenspektren von partiell methylierten Methylglucosiden. Carbohydr. Res. **9,** 79 (1969).

189a. Higuchi, R., and T. Kawasaki: Seed Saponins of *Akebia quinata* II. Herderagenin 3,28-Bisglykosides. Chem. Pharm. Bull. (Tokyo) **20,** 2143 (1972).

189b. Higuchi, R., K. Miyahara, and T. Kawasaki: Seed Saponins of *Akebia quinata* I. Hederagenin 3-O-Glycosides. Chem. Pharm. Bull. (Tokyo) **20,** 1935 (1972).

190. Hiller, K.: Antimikrobielle Stoffe in Blütenstoffen. Eine Übersicht. Pharmazie **19,** 167 (1964).

191. Hiller, K., und E. Hallstein: Saniculagenin F — ein neues Estersapogenin. Pharmazie **25,** 790 (1970).

192. Hiller, K., M. Keipert und B. Linzer: Triterpensaponine. Pharmazie **21,** 713 (1966).

193. Hiller, K., B. Linzer, S. Pfeifer, L. Lökes und J. Murphy: Über die Saponine von *Sanicula europaea* L. 9. Mitteil. Zur Kenntnis der Inhaltsstoffe einiger Saniculoideaé. Pharmazie **22,** 376 (1967).

194. Hirschmann, H., and F. B. Hirschmann: C-22-Ketals Related to the Sapogenins. Tetrahedron **3,** 243 (1958).

195. Hoppe, W., A. Gieren, N. Brodherr, R. Tschesche und G. Wulff: Die Struktur des Hauptaglykons aus dem Roßkastaniensaponin. Angew. Chem. **80,** 563 (1968); Angew. Chem. (Int. Ed.) **7,** 547 (1968).

196. Hough, L., J. V. S. Jones, and P. Wusteman: On the Automated Analysis of Neutral Monosaccharides in Glycoproteins and Polysaccharides. Carbohyd. Res. **21,** 9 (1972).

197. Hui, W. H., and C. W. Yee: Rubiaceae of Hong Kong IV. The Occurrence of Triter-penoids, Steroids and Triterpenoid Saponins in *Adina pilulifera.* Austr. J. Chem. **21,** 543 (1968); Chem. Abstr. **68,** 75706 (1968).

198. Huneck, S.: Triterpene — IV. Die Triterpensäuren des Balsams von *Liquidambar orientalis* Miller. Tetrahedron **19,** 479 (1963).

199. IGARASHI, K.: Studies on the Steroidal Components of Domestic Plants XXXV, Structure of Meteogenin (1). Chem. Pharm. Bull. (Tokyo) **9**, 722 (1961).

200. IGARASHI, K.: Studies on the Steroidal Components of Domestic Plants XXXVI, On Meteogenin (2). Chem. Pharm. Bull. (Tokyo) **9**, 729 (1961).

201. IIDA, Y., O. TANAKA, and S. SHIBATA: Studies on the Saponins of Ginseng: The Structure of Ginsenoside-Rg$_1$. Tetrahedron Letters **1968**, 5449.

202. IMAI, S., S. FUJIOKA, F. MURATA, M. GOTO, T. KAWASAKI, and T. YAMAUCHI: Bioassay of Crude Drugs and Oriental Crude Drug Prepns. XXII. Biological Active Plant Ingredients by Antimicrobial Tests, Antifungal Activity of Dioscin and Related Compounds. Ann. Rep. Takeda Res. Lab. **26**, 76 (1967); Chem. Abstr. **68**, 76007 (1968).

203. INUBUSHI, Y., and T. SANO: Applications of NMR to Org. Chemistry, esp. in Relation to the Structure Problems of Triterpenoids. Kagaku No Ryoiki **19**, 217 and 281 (1965); Chem. Abstr. **63**, 16392 (1965).

204. ISKENDEROV, G. B.: Triterpenic glycosides from *Hedera pastuchovii* II. Structure of Pastuchoside C. Khim. Prir. Soedin. **6**, 376 (1970); Chem. Abstr. **73**, 127733 (1970).

205. — Triterpenoid Glycosides from *Hedera pastuchovii* IV. Structure of Pastuchoside B. Khim. Prir. Soedin. **7**, 425 (1971); Chem. Abstr. **75**, 152041 (1971).

206. ITÔ, S., M. KODAMA, and M. SUNAGAWA: Correlation of Methyl Signals in 3β-Hydroxyoleanenes. Tetrahedron Letters **1967**, 3989.

206a. ITÔ, S., M. KODAMA, M. SUNAGAWA, T. ÔBA, and H. HIKINO: Substituent Effects on the Methyl Signals in the NMR-Spectra of Oleanan-12-en-3β-ols. Effect of the Hydroxyl and Acetoxyl Groups. Tetrahedron Letters **1969**, 2905.

207. ITÔ, S., and T. OGINO: Two New Triterpenes from *Camellia sinensis* and *Camellia sasanqua*. Tetrahedron Letters **1967**, 1127.

208. ITÔ, S., T. OGINO, H. SUGIYAMA, and M. KODAMA: Structures of A$_1$-Barrigenol and R$_1$-Barrigenol. Tetrahedron Letters **1967**, 2289.

209. ITOKAWA, H., N. SAWADA, and T. MURAKAMI: The Structures of Camelliagenin A, B, and C obtained from *Camellia japonica* L. Tetrahedron Letters **1967**, 597.

210. — — — The Structures of Camelliagenin A, B, and C obtained from *Camellia japonica* L. Chem. Pharm. Bull. (Tokyo) **17**, 474 (1969).

211. JACOBS, W. A., and E. E. FLECK: Tigogenin, a Digitalis Sapogenin. J. Biol. Chem. **88**, 525 (1930).

212. JEGER, O.: Über die Konstitution der Triterpene. Fortschritte Chem. organ. Naturstoffe **7**, 1 (1950).

213. JENTSCH, K.: Vergleichende Untersuchungen der anthelminthischen Wirksamkeit von Saponinen in vitro. Arzneimittelforsch. **11**, 413 (1961).

214. JIZBA, J., L. DOLEJŠ, V. HEROUT, and F. ŠORM: The Structure of Osladin — The Sweet Principle of the Rhizomes of *Polypodiums vulgare* L. Tetrahedron Letters **18**, 1329 (1971).

215. JIZBA, J., L. DOLEJŠ, V. HEROUT, F. ŠORM, H.-W. FEHLHABER, G. SNATZKE, R. TSCHESCHE und G. WULFF: Polypodosaponin, ein neuer Saponintyp aus *Polypodium vulgare* L. Chem. Ber. **104**, 837 (1971).

216. JIZBA, J., and V. HEROUT: Plant Substances XXVI. Isolation of Constituents of Common Polypody Rhizomes. Collect. Czech. Communications **32**, 2867 (1967).

217. JOLY, R. A., J. BONNER, R. D. BENNETT, and E. HEFTMANN: Conversion of Cholesterol to an Open-chain Saponin by *Dioscorea floribunda*. Phytochem. **8**, 857 (1969).

218. JOLY, R. A., J. BONNER, R. D. BENNETT, and E. HEFTMANN: Conversion of an Open-chain Saponin to Dioscin by a *Dioscorea floribunda* Homogenate. Phytochem. **8**, 1445 (1969).

219. JONES, N., E. KATZENELLENBOGEN, and K. DOBRINER: The Infrared Absorption Spectra of the Steroid Sapogenins. J. Amer. Chem. Soc. **75**, 158 (1953).

220. JURSITZA, G., und B. WOLTERS: Untersuchungen über die Wirkung sekundärer Pflan-

588   R. TSCHESCHE und G. WULFF:

zeninhaltsstoffe auf die Pilzsymbiose des Tabakkäfers *Lasioderma serricorne* F. I. Wachstum der Larven in Drogenpulvern. Z. angew. Entomol. **59**, 397 (1967).

221. KAMPHUIS, G., P. PATT und W. WINKLER: Verfahrenstechnische Gewinnung einiger Naturstoffe durch Ionenaustausch. Österr. Apotheker Z. **24**, 3 (1970).

222. KARIYONE, T., S. ISHIMASA, and T. SHIOMI: Triterpenoids VIII. Triterpenoids Contained in *Sophora japonica.* J. Pharmac. Soc. Japan **76**, 1210 (1956); Chem. Abstr. **51**, 3513 (1957).

223. KASPRZYK, Z., and Z. WOJCIECHOWSKI: The Structure of Triterpenic Glycosides from the Flowers of *Calendula officinalis* L. Phytochem. **6**, 69 (1967).

224. KAWASAKI, T., and K. MIYAHARA: Thin Layer Chromatography of Steroid Saponins and Their Derivatives. Chem. Pharm. Bull. (Tokyo) **11**, 1546 (1963).

225. — — Structure of Yononin. A Novel Type of Spirostanol Glycoside. Tetrahedron **21**, 3633 (1965).

226. KAWASAKI, T., and I. NISHIOKA: Digitalis Saponins I, Seed Saponins of *Digitalis purpurea* L. (Commercial Digitonin). Chem. Pharm. Bull. (Tokyo) **12**, 1250 (1964).

227. — — Digitalis Saponins (II) Leaf Saponins of *Digitalis purpurea* L. Chem. Pharm. Bull. (Tokyo) **12**, 1311 (1964).

228. KAWASAKI, T., I. NISHIOKA, T. KOMORI, T. YAMAUCHI, and K. MIYAHARA: Digitalis Saponins IV, Structure of F-Gitonin. Tetrahedron **21**, 299 (1965).

229. KAWASAKI, T., I. NISHIOKA, T. YAMAUCHI, K. MIYAHARA, and M. ENBUTSU: Digitalis Saponins (III). Enzymatic Hydrolysis of Leaf Saponins of *Digitalis purpurea* L. Chem. Pharm. Bull. (Tokyo) **13**, 435 (1965).

230. KAWASAKI, T., and T. YAMAUCHI: Structures of Dioscin, Gracillin and Kikuba Saponin (2). Saponins of Japanese *Dioscoreaceae* (XI). Chem. Pharm. Bull. (Tokyo) **10**, 703 (1962).

231. — — Yononin and Tokoronin, New Steroid Saponins in Rhizome of *Dioscorea tokoro.* Yakugaku Zasshi **83**, 757 (1963); Chem. Abstr. **59**, 15535 (1963).

232. — — Saponins of Timo. *Anemarrhena* Rhizoma. II. Structure of Timosaponin A-III. Chem. Pharm. Bull. (Tokyo) **11**, 1221 (1963).

233. KAWASAKI, T., T. YAMAUCHI, and N. HAKURA: Saponins of *Anemarrhena* Rhizomes. Yakugaku Zasshi **83**, 892 (1963); Chem. Abstr. **60**, 592 (1964).

234. KAWASAKI, T., T. YAMAUCHI, and R. YAMAUCHI: Kikubasaponin (1). Saponins of Japanese *Dioscoreaceae* (X). Chem. Pharm. Bull. (Tokyo) **10**, 698 (1962).

235. KENNEY, H. E., and M. E. WALL: Steroidal Sapogenins, XLI, Willagenin, a New 12-Ketosapogenin. J. Org. Chem. **22**, 468 (1957).

236. KHORLIN, A. J., L. V. BAKINOVSKII, V. E. VASKOVSKII, A. G. VENJAMINOVA, and J. S. OVODOV: Triterpensaponins VI. Distributive Chromatography of Triterpene Saponins. Izv. Akad. Nauk. SSSR Ser. Khim. Nauk. **1963**, 2008; Chem. Abstr. **60**, 8113 (1964).

237. KHORLIN, A. J., V. J. CHIRVA, and N. K. KOCHETKOV: Triterpenic Saponins XXI. Structure of Clematoside C. Izv. Akad. Nauk. SSSR, Ser. Khim. **1967**, 1306; Chem. Abstr. **67**, 117188 (1967).

238. KHORLIN, A. J., S. OVODOV, and R. G. OVODOVA: Identity of Gypsoside and Triterpensaponin from *Gypsophila paniculata.* Izv. Akad. Nauk. SSSR, Ser. Khim. **1963**, 1521; Chem. Abstr. **59**, 15535 (1963).

239. KHORLIN, A. J., A. G. VENJAMINOVA, and N. K. KOCHETKOV: Structure of *Kalopanax septemlobus* Saponin A. Dokladi Akad. Nauk. SSSR **115**, 619 (1964); Chem. Abstr. **60**, 15964 (1964).

240. KHORLIN, A. J., A. G. VENJAMINOVA, and N. K. KOCHETKOV: Triterpensaponins XVIII. Structure of Kalopanax Saponin B. Izv. Akad. Nauk. SSSR, Ser. Khim. **1966**, 1588; Chem. Abstr, **66**, 65803 (1967).

241. KIMURA, M., J. HATTORI, I. JOSHIZAWA, and M. TOHMA: Constituents of Convallaria

IX. Quantitative Analysis of Sugar Components in Steriodal Saponins by Gas-Chromatography. Chem. Pharm. Bull. (Tokyo) **16**, 613 (1968).

242. KIMURA, M., M. TOHMA, and I. YOSHIZAWA: Constituents of Convallaria IV, Isolation of Convallasaponin A, B, and C. Chem. Pharm. Bull. (Tokyo) **14**, 50 (1966).

243. — — — Constituents of Convallaria, V. On the Structure of Convallasaponin-C. Chem. Pharm. Bull (Tokyo) **14**, 55 (1966).

244. — — — Constituents of Convallaria VII. Structure of Convallagenin A. Chem. Pharm. Bull. (Tokyo) **15**, 1204 (1967).

245. — — — Constituents of Convallaria VIII. Structure of Convallagenin B. Chem. Pharm. Bull. (Tokyo) **15**, 1713 (1967).

246. KIMURA, M., M. TOHMA, I. JOSHIZAWA, and H. AKIYAMA: Constituents of Convallaria X. Structures of Convallasaponin A, B, and Their Glucosides. Chem. Pharm. Bull. (Tokyo) **16**, 25 (1968).

247. KIMURA, M., M. TOHMA, I. YOSHIZAWA, and A. FUJINO: Constituents of Convallaria XII, Convallasaponin E, Diosgenin Triarabinoside. Chem. Pharm. Bull. (Tokyo) **16**, 2191 (1968).

248. KIRYALOV, N. P.: Structure of Macedonic Acid. Khim. Prir. Soedin **5**, 448 (1969); Chem. Abstr. **72**, 67136 (1970).

249. KIRYALOV, N. P., and G. S. AMIROVA: Structure of Meristotropic Acid. Khim. Prir. Soedin. **4**, 87 (1968); Chem. Abstr. **69**, 87231 (1968).

249a. — — The Structure of Isomeristotropic Acid. Khim. Prir. Soedin. **4**, 150 (1968); Chem. Abstr. **69**, 8723 (1968).

250. KIRYALOV, N. P., and V. F. BOGATKINA: Isomacedonic Acid from *Glycyrrhiza echinata* Roots. Khim. Prir. Soedin. **7**, 123 (1971); Chem. Abstr. **75**, 36381 (1971).

251. KITAGAWA, I.: Chemistry of Triterpenoids — from the Recent Topics. Jap. J. Pharm. Chem. **38**, 343 (1968).

251a. KITAGAWA, I., A. INADA, I. YOSIOKA, R. SOMANATHAN, and M. U. S. SULTANHAWA: Protobassic Acid, a Genuine Sapogenol of Seed Kernels of *Madhuca longifolia*. Chem. Pharm. Bull. (Tokyo) **20**, 630 (1972).

252. KITAGAWA, I., A. MATSUDA, and I. YOSIOKA: Genuine Sapogenins of Three Primulaceous Plants. Tetrahedron Letters **1968**, 5377.

253. KITAGAWA, I., K. KITAZAWA, and I. YOSIOKA: Transformation of Spergulagenic Acid, a Dicarboxyclic Triterpenoid, to Eupteleogenin, a Unique Nortriterpenoid. Tetrahedron Letters **1970**, 1905.

254. KIYOSAWA, S., M. HUTOH, T. KOMORI, T. NOHARA, I. HOSOKAWA, and T. KAWASAKI: Detection of Proto-type Compounds of Diosgenin- and Other Spirostanol-Glycosides. Chem. Pharm. Bull. (Tokyo) **16**, 1162 (1968).

255. KLASS, D. L., M. FIESER, and L. F. FIESER: Digitogenin. J. Amer. Chem. Soc. **77**, 3829 (1955).

256. KLYNE, W.: The Configuration of the Anomeric Carbon Atoms in Some Cardiac Glycosides. Biochem. J. **47**, XLI (1950).

257. KOCHETKOV, N.K., and O. S. CHIZHOV: Mass Spectrometry of Methylated Methyl Glycosides. Principles and Analytical Application. Tetrahedron **21**, 2029 (1965).

258. KOCHETKOV, N. K., and A. J. KHORLIN: Oligosides, a Type of Plant Glycosides Dokladi. Akad. Nauk. SSSR **150**, 1289 (1963); Chem. Abstr. **59**, 11886 (1963).

259. — — Oligoside, ein neuer Typ von Pflanzenglykosiden. Arzneimittelforsch. **16**, 101 (1966).

260. KOCHETKOV, N. K., A. J. KHORLIN, and V. J. CHIRVA: Clematoside C — Triterpenic Oligoside from *Clematis manshurica*. Tetrahedron Letters **1965**, 2201.

261. KOCHETKOV, N. K., A. J. KHORLIN, and J. S. OVODOV: The Structure of Gypsoside-Triterpenic Saponin from *Gypsophila pacifica* Kom. Tetrahedron Letters **1963**, 477.

262. — — — Triterpen Saponins VII. Monosaccharide Decompositions of the Carbohydrate Portion of Gypsoside. VIII. Some Data on the Structure of the Carbohydrate

590 R. Tschesche und G. Wulff:

Part of Gypsoside. Izv. Akad. Nauk. SSSR Ser. Khim. **1964**, 83, 90, 143; Chem. Abstr. **60**, 10776 (1964).

263. Kochetkov, N. K., A. J. Khorlin, and J. S. Ovodov: Structure of Gypsoside from *Gypsophila pacifica*. Izv. Akad. Nauk. SSSR, Ser. Khim. **1964**, 1436.

264. Kochetkov, N. K., A. J. Khorlin, and V. E. Vaskovskii: The Structures of Aralosides A and B. Tetrahedron Letters **1962**, 713.

265. Kochetkov, N. K., A. J. Khorlin, V. E. Vaskovskii, and I. P. Gudkova: Triterpene Saponins XVI. Structure of Araloside C. Izv. Akad. Nauk. SSSR, Ser. Khim. **1965**, 1214; Chem. Abstr. **63**, 14965 (1965).

266. Kochetkov, N. K., N. S. Wulfson, O. S. Chizhov, and B. M. Zolotarev: Mass Spectrometry of Carbohydrate Derivatives. Tetrahedron **19**, 2209 (1963).

267. Kofler, L.: Die Saponine. Wien: Springer. 1927.

268. Kondo, N., K. Aoki, H. Ogawa, R. Kasai, and J. Shoji: Constituents of Panacis japonici Rhizoma III. Structure of Chikusetsusaponin III. Chem. Pharm. Bull. (Tokyo) **18**, 1558 (1970); Chem. Abstr. **73**, 120858 (1970).

269. Kondo, N., Y. Marumoto, and J. Shoji: Studies on the Constituents of Panacis Japonici Rhizoma IV. The Structure of Chikusetsusaponin V. Chem. Pharm. Bull. (Tokyo) **19**, 1103 (1971).

270. Kondo, N., J. Shoji, N. Nagumo, and N. Komatsu: Constituents of Panax japonicum Rhizomes II. Structure of Chikusetsusaponin IV and Some Observations on the Structural Relation with Araloside A. Yakugaku Zasshi **89**, 846 (1969); Chem. Abstr. **71**, 102180 (1969).

271. Kondo, T., S. Kurotori, M. Teshima, and M. Sumimoto: The Termiticidal Wood-Ectractive from *Kalopanax septemlobus* Koedz; Nippon Mokuzai Gakkaishi **9**, 125 (1963); Chem. Abstr. **60**, 2273 (1964).

272. Kretsu, L. G., and G. V. Lazurevskii: Triterpenic Glycosides of *Beta vulgaris*. Khim. Prir. Soedin. **6**, 272 (1970); Chem. Abstr. **73**, 63200 (1970).

273. Krider, M. M., and M. E. Wall: Steroidal Sapogenins V. Enzymatic Hydrolysis of Steroidal Saponins. J. Amer. Chem. Soc. **74**, 3201 (1952).

274. Kubeczka, K. H.: Beitrag zur Analytik von Saponindrogen. Apotheker-Ztg. **110**, 1707 (1970).

275. Kubota, T., and H. Hinoh: The Constitution of Saponins Isolated from *Bupleurum falcatum* L. Tetrahedron Letters **1968**, 303.

276. — — Triterpenoids from *Bupleurum falcatum* L. — III. Isolation of Genuine Sapogenins, Saikogenins E, F, and G. Tetrahedron **24**, 675 (1968).

277. Kubota, T., and H. Kitatani: Revised Structure of the Triterpene Polygalacic Acid. Configuration of the C-16 Hydroxy-Group. Chem. Commun. **1968**, 1005.

278. — — The Structure of Platycodigenin, a 4,4-Di(hydroxy-methyl)-triterpene. Chem. Commun. **1969**, 190.

279. Kubota, T., H. Kitatani, and H. Hinoh: Isomerisation of Quillaic Acid and Echinocystic Acid with Hydrochloric Acid. Tetrahedron Letters **1969**, 771.

280. — — — The Structure of Platycogenic Acids A, B, and C, Further Triterpenoid Constituents of *Platycodon grandiflorum* A. De Candolle. J. Chem. Soc. (D) **1969**, 1313.

281. Kubota, T., F. Tonami, and H. Hinoh: The Structure of Saikogenins A, B, C, and D, Triterpenoid Alcohols of *Bupleurum falcatum* L. Tetrahedron Letters **1966**, 701.

282. — — — Triterpenoids from *Bupleurum falcatum* L. — I. The Structure of Saikogenins A, C, and D. — II. Isolation of Saikogenin B and Longispinogenin. Tetrahedron **23**, 3333 and 3353 (1967).

283. Kuhn, R., H. H. Baer und A. Gauhe: Kristallisation und Konstitutionsermittlung der Lacto-N-fucopentaose I. Chem. Ber. **89**, 2514 (1956).

284. Kuhn, R., und H. Egge: Über Ergebnisse der Permethylierung der Ganglioside G_1 und G_2. Chem. Ber. **96**, 3338 (1963).

285. KUHN, R., H. EGGE, R. BROSSMER, A. GAUHE, H. TRISCHMANN, P. KLESSE, W. LO-CHINGER, E. RÖHM und D. TSCHAMPEL: Über die Ganglioside des Gehirns. Angew. Chem. **72**, 805 (1960).

286. KUHN, R., und I. LÖW: Zur Konstitution der Leptine. Chem. Ber. **94**, 1088 (1961).

287. — — Über Protoäscigenin, die Stammsubstanz der Äscine. Liebigs Ann. Chem. **669**, 183 (1963).

288. — — Äscinidin, ein Pentahydoxy-Triterpen aus Äscinpräparaten. Tetrahedron Letters **1964**, 891.

289. — — Über die Protoäscigeninester aus Äscin. Tetrahedron **22**, 1899 (1966).

290. KUHN, R., I. LÖW und H. TRISCHMANN: Die Konstitution des Solanins. Chem. Ber. **88**, 1492 (1955).

291. — — —Die Konstitution der Lycotetraose; Chem. Ber. **90**, 203 (1957).

292. KUHN, R., und H. TRISCHMANN: Permethylierung von Inosit und anderen Kohlehydraten mit Dimethylsulfat. Chem. Ber. **96**, 284 (1963).

293. KUPCHAN, S. M., S. J. BARBOUTIS, J. R. KNOX, and C. A. LAN CAM: Beta-Solamarine: Tumor Inhibitor Isolated from *Solanum dulcamara*. Science **150**, 1827 (1965).

294. KUPCHAN, S. M., R. J. HEMINGWAY, J. R. KNOX, S. J. BARBOUTIS, D. WERNER, and M. A. BARBOUTIS: Tumor Inhibitors. XXI. Active Principles of *Acer negundo* and *Cyclamen persicum*. J. Pharm. Sci. **56**, 603 (1967); Chem. Abstr. **67**, 31349 (1967).

295. KUPCHAN, S. M., P. S. STEYN, M. D. GROVE, S. M. HORSFIELD, and S. W. MEITNER: Tumor Inhibitors XXXV. Myrsene Saponin, the Active Principle of *Myrsene africana* L. J. Med. Chem. **12**, 167 (1969).

296. KUPCHAN, S. M., M. TAKASUGI, R. M. SMITH, and P. S. STEYN: Acerotin and Acerocin, Novel Triterpene Ester Aglycones from the Tumourinhibitory Saponins of *Acer negundo*. J. Chem. Soc. (D) **1970**, 969.

297. — — — — Tumor Inhibitors LXII. The Structures of Acerotin and Acerocin, Novel Triterpene Ester Aglycones from the Tumor Inhibitory Saponins of *Acer negundo*. J. Org. Chem. **36**, 1972 (1971).

298. KUTNEY, J. P.: A NMR Study in the Steroidal Saponin Series, The Stereochemistry of the Spiroketal System. Steroids **2**, 225 (1963).

299. LABRIOLA, R. A., and V. DEULOFEU: The Structure of Prosapogenin from Quillaja Saponin. Experientia **25**, 124 (1969).

300. LAIDLAW, R. A., and E. G. V. PERCIVAL: The Methyl Ethers of the Aldopentoses and of Rhamnose and Fucose. Advances Carbohydr. Chem. **7**, 1 (1952).

301. LAPIN, H., and C. SANNIÉ: Studies on Sterolic Sapogenins, V. Ruscogenin, 22a-Spirostane-3β,?-diol, a New Sterolic Sapogenin of *Ruscus aculeatus* L. Bull. soc. chim. France **1955**, 1552; Chem. Abstr. **50**, 10117 (1956).

302. LAVIE, D., Y. SHVO, and E. GLOTTER: Triterpenoids II. The NMR-Spectra of Tetracyclic Triterpenes. Tetrahedron **19**, 2255 (1963).

303. LAWRIE, W., J. MCLEAN, and M. EL-GARBY-YOUNES: Triterpenoids in Apple Peel. J. Chem. Soc. (C) **1967**, 851.

304. LAZUREVSKII, G. V., V. J. CHIRVA, P. K. KINTYA, and L. G. KRETSU: Triterpenoid Glycosides of Beans. Dokl. Akad. Nauk. SSSR **199**, 226 (1971); Chem. Abstr. **75**, 115903 (1971).

305. LAZUREVSKII, G. V., P. K. KINTYA, and V. J. CHIRVA: Structure of Saponoside D, a Triterpenglycoside from Soapwort *(Saponaria officinalis)* Roots. Dokladi. Akad. Nauk. SSSR **188**, 1060 (1969); Chem. Abstr. **72**, 67224 (1970).

306. LEWBART, M. L., W. WEHRLI, H. KAUFMANN und T. REICHSTEIN: Die Cardenolide von *Gongronema gazense* (S. MOORE) Bullock. Helv. Chim. Acta **46**, 517 (1963).

306a. LINDBERG, B., H. BJÖRNDAL, C. G. HELLERQUIST und S. SVENSSON: Gas-Flüssigkeits-Chromatographie und Massenspektroskopie bei der Methylierungsanalyse von Polysacchariden. Angew. Chem. **82**, 643 (1970).

307. LINDE, H. F. G., D. ISAAC, H. H. A. LINDE und D. ZIVANOV: Über Sapogenine (1).

592 R. Tschesche und G. Wulff:

Über die Konstitution eines neuen Sapogenins aus *Helleborus odorus* Waldst. et Kit. und *Helleborus niger* L. Helv. chim. Acta **54**, 1703 (1971).

308. Liu, W. C., M. Kugelman, R. A. Wilson, and K. V. Rao: A Crystalline Saponin with Anti-tumor Activity from *Entados phaseoloides*. Phytochem. **11**, 171 (1972).

309. Löw, I.: Über Acylwanderungen bei Protoäscigeninestern. Z. physiol. Chem. **348**, 839 (1967).

310. Lorenz D., und M. L. Marek: Das therapeutisch wirksame Prinzip der Roßkastanie *(Aesculus hippocastanum)* I. Aufklärung des Wirkstoffes. Arzneimittelforsch. **10**, 263 (1960).

311. Luchanskaya, V. N., E. S. Kondratenko, and N. K. Abubakirov: Triterpenoid Glycosides from *Gypsophila trichotoma* III. Structure of Trichoside B. Khim. Prir. Soedin. **7**, 431 (1971); Chem. Abstr. **75**, 152039 (1971).

311a. — — — Triterpenoid Glycosides of *Gypsophila trichotoma*. IV. Structure of Trichoside C. Khim. Prir. Soedin. **1971**, 608; Chem. Abstr. **76**, 59967 (1972).

312. Luchanskaya, V. N., E. S. Kondratenko, T. T. Gorovits, and N. K. Abubakirov: Triterpenoid Glycosides from *Gypsophila trichotoma* II. Structure of Trichoside A. Khim. Prir. Soedin. **7**, 151 (1971); Chem. Abstr. **75**, 49495 (1971).

313. Lythgoe, B., and S. Trippet: The Constitution of the Disaccharide of Glycyrrhinic Acid. J. Chem. Soc. (London) **1950**, 1983.

314. Mackie, A. M., and A. B. Turner: Partial Characterization of a Biologically Active Steroid Glycose Isolated from the Starfish *Marthasterias glacialis*. Biochem. J. **117**, 543 (1970).

315. Magalhães- Alves, H., and V. H. Arndt: Triterpenoids isolated from *Machaerium incorruptibile*. Phytochem. **5**, 1327 (1966).

316. de Maheas, M. R., D. Billet, D. Raulais et M. Chaigneau: Structure de l'armillarigénine, nouveau triterpène de *Jacquinia armillaris* Jacq. Bull. Soc. Chim. France **1969**, 226.

317. Maher, G. G.: The Methyl Ethers of the Aldopentoses and of Rhamnose and Fucose. Advances Carbohydr. Chem. **10**, 257 (1955).

318. — The Methyl Ethers of D-Galactose. Advances Carbohydr. Chem. **10**, 273 (1955).

319. Maiti, P. C., and S. Mookherjea: Hispidogenin. Chem. and Ind. **1965**, 1653.

320. Maizel, J. V., H. J. Burkhardt, and H. K. Mitchell: Avenacin, an Antimicrobial Substance Isolated from *Avena sativa* I. Isolation and Antimicrobial Activity. Biochemistry **3**, 424 (1964).

321. Mandell, L., A. L. Nussbaum, and E. P. Oliveto: The Structure of Neoruscogenin. Tetrahedron Letters **1960**, 25.

322. Mani, J. C.: Structure of Triterpenes by Nuclear Magnetic Resonance. Ann. Chim. (France) **10**, 533 (1965); Chem. Abstr. **65**, 2312 (1966).

323. Marchaim, U., Y. Birk, A. Dovrat, and T. Bergman: Alfalfa Saponins as Inhibitors of Cotton Seed Germination. Plant Cell. Physiol. **11**, 511 (1970); Chem. Abstr. **73**, 32444 (1970).

324. Marker, R. E.: Steroidal Sapogenins, Nr. 169. Magogenin and Cacogenin and their Biogenesis to Chlorogenin and Tigogenin. J. Amer. Chem. Soc. **69**, 2399 (1947).

325. Marker, R. E.: Steroidal Sapogenins, 173: 16-Dehydro-pregnen-5-ol-3-dione-12,20 from Ricogenin. A New Steroidal Sapogenin. J. Amer. Chem. Soc. **71**, 3856 (1949).

326. Marker, R. E., and K. Applezweig: Steroidal Sapogenins as a Source for Cortical Steroids. Chem. Eng. News **27**, 3348 (1949); Chem. Abstr. **44**, 2000 (1950).

327. Marker, R. E., E. M. Jones, and D. L. Turner: Sterols CII, Chlorogenin. J. Amer. Chem. Soc. **62**, 2537 (1940).

328. — — — Sterols CX. The Position of the Hydroxyl Groups in Chlorogenin. J. Amer. Chem. Soc. **62**, 3006 (1940).

329. Marker, R. E., and J. Krueger: Sterols, CXII, Sapogenins XLI, The Preparation of Trillin and its Conversion in Progesterone. J. Amer. Chem. Soc. **62**, 3349 (1940).

330. MARKER, R. E., and J. LOPEZ: Steroidal Sapogenins. Nr. 161. The Seasonal Variation of Sapogenins in Plants. J. Amer. Chem. Soc. **69**, 2375 (1947).

331. — — Steroidal Sapogenins. Nr. 163. The Biogenesis of Steroidal Sapogenins in Plants. J. Amer. Chem. Soc. **69**, 2383 (1947).

332. — — Steroidal Sapogenins. Nr. 164. Nologenin and its Degradation Products. J. Amer. Chem. Soc. **69**, 2386 (1947).

333. — — Steroidal Sapogenins. Nr. 170. The Position of the Double Bond in Yuccagenin and Kammogenin. J. Amer. Chem. Soc. **69**, 2401 (1947).

334. MARKER, R. E., T. TSUKAMOTO, and D. L. TURNER: Sterols, C, Diosgenin. J. Amer. Chem. Soc. **62**, 2525 (1940).

335. MARKER, R. E., D. L. TURNER, and P. R. ULSHAFER: Sterols, CL Sapogenins LXIII. The Position of the Hydroxyl Groups in Digitogenin. J. Amer. Chem. Soc. **64**, 1843 (1942).

336. MARKER, R. E., D. L. TURNER, and E. L. WITTBECKER: Sterols CXXXVI. Sapogenins LVII. The Structure of the Side-chain of Chlorogenin. J. Amer. Chem. Soc. **64**, 809 (1942).

337. MARKER, R. E., R. B. WAGNER, D. P. J. GOLDSMITH, P. R. ULSHAFER, and C. H. RUOF: Sterols CLV, Sapogenins LXVII, Pennogenin, Nologenin and Fesogenin. Three New Sapogenins from Beth Root. J. Amer. Chem. Soc. **65**, 1248 (1943).

338. MARKER, R. E., R. B. WAGNER, P. R. ULSHAFER, E. L. WITTBECKER, D. P. J. GOLDSMITH, and C. R. RUOF: Sterols CLVII. Sapogenins LXIX. Isolation and Structures of Thirteen New Steroidal Sapogenins. New Sources for Known Sapogenins. J. Amer. Chem. Soc. **65**, 1199 (1943).

339. MARKER, R. E., R. B. WAGNER, P. R. ULSHAFER, E. L. WITTBECKER, D. P. J. GOLDSMITH, and C. H. RUOF: Steroidal Sapogenins. J. Amer. Chem. Soc. **69**, 2167 (1947).

340. MARSH, C. A., and G. A. LEVVY: Glucuronide Metabolism in Plants III. Triterpene Glucuronides. Biochem. J. **63**, 9 (1956).

341. MARTINSSON, E., and O. SAMULSON: Partition Chromatography of Sugars on Ion-Exchange Resins. J. Chromatogr. **50**, 429 (1970).

341a. MASTRONARDI, J. O., S. M. FLEMATTI, J. O. DEFFERRARI, and E. G. GROS: Methylation of Carbohydrates Bearing Base-labile Substituents, with Diazomethan-Boron Trifluoride Etherate. Carbohydr. Res. **3**, 177 (1966).

342. MAYO, DE P.: The Chemistry of Natural Products, Vol. III, New York 1959.

343. MAZUR, G., and F. S. SPRING: Steroids XII, The Structures of Cholegenin and iso-Cholegenin. J. Chem. Soc. (London) **1954**, 1223.

344. McCRINDLE, R., and K. H. OVERTON: The Diterpenoids, Sesquiterpenoids and Triterpenoids. In: COFFEY, S. (Edit.) Rodd's Chemistry of Carbon Compounds, Vol. II. Amsterdam: Elsevier. 1969.

345. MERCER, E. I., and T. W. GOODWIN: Incorporation of (2-^{14}C) Mevalonic Acid into Steroids of Maize Seedlings. Biochem. J. **88**, 46 P (1963).

346. MINATO, T., and A. SHIMAOKA: Studies on Steroidal Components of Domestic Plants XLII, Narthogenin, Isonarthogenin and Neonogiragenin of *Metanarthecium luteoviride* Maxim., Pharm. Bull. Tokyo **11**, 876 (1963).

347. MIYAHARA, K., F. ISOZAKI, and T. KAWASAKI: Isolation and Structure of a New Tokorogenin Glycoside. Chem. Pharm. Bull. (Tokyo) **17**, 1735 (1969).

348. MIYAHARA, K., and T. KAWASAKI: Structure of Tokoronin. Chem. Pharm. Bull. (Tokyo) **17**, 1369 (1969).

349. MOCKLE, J. A.: Trillioside. Canad. Pharm. J. **90**, 162 (1957); Chem. Abstr. **54**, 16477 (1960).

350. MORITA, K.: Steroids IX, Sapogenins of *Dioscorea tokoro*, II. The Structure of Tokorogenin. Bull. Chem. Soc. Japan **32**, 791 (1959); Chem. Abstr. **54**, 16476 (1960).

351. MÜHLEMANN, H., und W. SCHEIDEGGER: Biologische Bestimmung von Saponinen mit Tubifex-Würmern. Pharmac. Acta Helvetiae **22**, 147 (1947).

352. Mukhamedziev, M. M., P. K. Alimbaeva, T. T. Gorovits, and N. K. Abubakirov: Structure of the Triterpenoid Glycoside from *Dipsacus azureus* Dipsacoside B. Khim. Prir. Soedin. **7**, 153 (1971); Chem. Abstr. **75**, 49494 (1971).

353. Murakami, T., M. Nagasawa, H. Itokawa, and Y. Tachi: The Structure of a New Triterpene, Momordic Acid, obtained from *Momordica cochinchinensis* Sprenger. Tetrahedron Letters **1966**, 5137.

354. Murakami, T., M. Nagasawa, S. Urayama, and T. Satake: New Triterpenoid Saponins in the Rhizome and Roots of *Caulophyllum robustum.* Yakugaku Zasshi **88**, 321 (1968); Chem. Abstr. **69**, 67661 (1968).

355. Murata, T., S. Imai, M. Imanishi, M. Goto, and K. Morita: The Structure of Eupteleogenin, the Aglycone of Anti-Fungal Glycosides, Eupteleoside A and Eupteleoside B. Tetrahedron Letters **1965**, 3215.

356. Murata, T., S. Imai, M. Imanishi, and M. Goto: Antimicrobial Glycosides of *Euptelea polyandra.* 2. Structure of Eupteleogenin. Yakugaku Zasshi **90**, 744 (1970); Chem. Abstr. **73**, 66754 (1970).

357. Mzhelskaja, L. G., and N. K. Abubakirov: Triterpenic Glycosides of *Leontice eversmanni* II. The Structures of Leontoside A and Leontoside B. Khim. Prir. Soedin. **3**, 101 (1967); Chem. Abstr. **67**, 100383 (1967).

358. —— The Structure of Leontoside C. Khim. Prir. Soedin. **3**, 218 (1967); Chem. Abstr. **67**, 117181 (1967).

359. — — Triterpenic Glycosides of *Leontice eversmanni* IV. Structure of Leontoside D. Khim. Prir. Soedin. **4**, 153 (1968); Chem. Abstr. **69**, 77678 (1968).

360. — — Triterpenic Glycosides of *Leontice eversmanni* V. Structure of Leontoside E. Khim. Prir, Soedin. **4**, 216 (1968); Chem. Abstr. **70**, 88199 (1969).

361. Nagai, M., T. Ando, O. Tanaka, and S. Shibata: 20-Epiprotopanaxadiol, a Genuine Sapogenin of Ginsenosides R_{b-1}, R_{b-2}, and R_c. Tetrahedron Letters **1967**, 3579.

361a. Nagai, M., T. Ando, N. Tanaka, O. Tanaka, and S. Shibata: Chemical Studies on the Oriental Drugs XXVIII. Saponins and Sapogenins of *Ginseng.* Stereochemistry of the Sapogenins of Ginsenosides R_{b-1}, R_{b-2} and R_c. Chem. Pharm. Bull. (Tokyo) **20**, 1212 (1972).

361b. Nagai, M., K. Izwa, and T. Inoue: Studies on the Constituents of Saxifragaceous Plants I. Two Triterpene Acids of *Peltoboykinia tellimoides.* Chem. Pharm. Bull. (Tokyo) **17**, 1438 (1969).

362. Nagai, J., O. Tanaka, and S. Shibata: Chemical Studies on the Oriental Plant Drugs— XXIV. Structure of Ginsenoside-Rg_1, a Neutral Saponin of Ginseng Root. Tetrahedron **27**, 881 (1971).

363. Nawa, H.: Rhodeasapogenin, a New Steroidal Sapogenin of *Rhodea japonica.* Proc. Japan Acad. **29**, 214 (1953); Chem. Abstr. **49**, 2459 (1955).

364. — Components of *Rhodea japonica,* XI. Structure of Rhodeasapogenin. Chem. Pharm. Bull. (Tokyo) **6**, 255 (1958); Chem. Abstr. **53**, 2282 (1959).

365. Nehring, U.: Zum Nachweis der Wirksamkeit von Roßkastanienextrakt auf den Venentonus nach oraler Applikation. Med. Welt (Stuttgart) **17**, 1662 (1966).

366. Nigrelli, R. F., and S. Yakowska: Effects of Holothurin, a Steroid Saponin from the Bahamian Sea Cucumber *(Actinopyga agassizi)* on Various Biological Systems. Ann. New York Acad. Sci. **90**, 884 (1960).

367. Nishikawa, M., K. Kamiya, T. Murata, Y. Tomiie, and I. Nitta: The X-Ray Investigation of Eupteleogenin Iodoacetate. Tetrahedron Letters **1965**, 3223.

368. Nishikawa, M., K. Morita, H. Hagiwara, and M. Inoue: Steroids, II, Tokorogenin, a New Sapogenin of *Dioscorea tokoro.* Yakugaku Zasshi **74**, 1165; Chem. Abstr. **49**, 14785 (1955).

369. Ogata, M., F. Yasuda, and K. Takeda: Studies on the Steroidal Components of Domestic Plants, Part LIII. Structure of Diotigenin. J. Chem. Soc. (London) C **1967**, 2397.

370. Ogino, T., T. Hayasaka, S. Ito, and T. Takahashi: Triterpenic Constituents of *Stewartia monadelpha*. Sieb. et Zucc., Camelliagenin A 16-Cinnamate as a Sapogenin. Chem. Pharm. Bull. (Tokyo) **16**, 1846 (1968).

371. Okanishi, T., A. Akahori, and F. Yasuda: Studies on the Steroidal Components of Domestic Plants, XL. Constituents of *Heloniopsis orientalis* (Thunb.). Chem. Pharm. Bull. (Tokyo) **10**, 1195 (1962).

372. Okanishi, T., A. Akahori, and F. Yasuda: Studies on the Steroidal Compounds of Domestic Plants XLVII, Constituents of the Stem of *Smilax sieboldi* Miq., The Structure of Laxogenin. Chem. Pharm. Bull. (Tokyo) **13**, 545 (1965).

373. Ovodov, Y. S., and A. F. Pavlenko: Gas-liquid Chromatography of Methylated D-Galactose Derivatives. J. Chromatogr. **36**, 531 (1968).

374. Pasich, B.: Absorptiometric Determination of Saponins in Plant Materials. Planta Med. (Stuttgart) **11**, 16 (1963).

375. Pelletier, S. W., N. Adityachaudhury, M. Tomasz, J. J. Reynolds, and R. Mechoulam: The Structure of Senegenic Acid, a Nortriterpene Artifact from *Polygala senega*. J. Org. Chem. **30**, 4234 (1965).

376. — — — — — Senegenic Acid, a Pentacyclic Nor-Triterpene Acid. Tetrahedron Letters **1964**, 3065.

377. Pelletier, S. W., and S. Nakamura: A Prosapogenin from *Polygala senega* and *Polygala tenuifolia*. Tetrahedron Letters **1967**, 5303.

378. Petkov, V.: Panax Ginseng, ein Reaktivitätsregler. Pharm. Ztg. **113**, 1281 (1968).

379. Petričić, J., und A. Radošević: Asperosid, ein neues bisdesmosidisches 22-Hydroxyfurostanolsaponin aus *Smilax aspera* L., Farmaceutski Glasnik **XXV**, 91 (1969).

380. Petričić, J., D. Tarle, and M. Kupinic: Structure of Macranthogenin, the Main Aglycon of Saponin in Subsurface Parts of *Helleborus macranthus*. Acta Pharm. Jugoslav. **19**, 149 (1969); Chem. Abstr. **73**, 131210 (1970).

381. Pinhas, H.: Structure de l'acide madasiatique, nouvel acide triterpénique, isolé de *Centella asiatica* L. Bull. Soc. Chim. France **1969**, 3592.

382. Pinhas, H., D. Billet, S. Heitz et M. Chaigneau: Structure de l'acide madécassique nouveau triterpène de *Centella asiatica* de Madagascar. Bull. Soc. Chim. France **1967**, 1890.

383. Polonsky, J., E. Sach et E. Lederer: Sur la constitution chimique de la parie glucidique de l'asiaticoside. Bull. Soc. chim. France **1959**, 880.

384. Polonsky, J., et I. Zylber: Détermination de la configuration de l'hydroxyle primaire de l'acide asiatique. Bull. Soc. Chim. France **1961**, 1586.

385. Ponder, E.: Hemolysis and Related Phenomena. London: J. & A. Churchill. 1948.

386. Potier, P., B. C. Das, A. M. Bui, M. M. Janot, A. Pourrat et H. Pourrat: Structure de l'acide tormentique, acide triterpénique pentacyclique isolé des racines de *Potentilla tormentilla* Neck. (Rosacées). Bull. Soc. Chim. France **1966**, 3458.

387. Powell, J. W., and W. B. Whalley: Triterpenoid Saponins from *Phytolacca dodecandra*. Phytochem. **8**, 2105 (1969).

388. Preciosi, P., und P. Manca: Die antioedematische und die antiinflammatorische Wirkung von Aescin und ihre Beziehung zu Hypophysen-Nebennieren-Achse. Arzneimittelforsch. **15**, 404 (1965).

389. Prelog, V., and O. Jeger: in R. H. F. Manske: The Alkaloids, Bd. VII, S. 346. New York: Academic Press. 1960.

390. Purdie, T., and I. C. Irvine: The Alkylation of Sugars. J. Chem. Soc. (London) **83**, 1021 (1903).

391. Rangaswami, S., K. N. N. Ayengar: Chemical Components of *Smilax aspera* (Sarsaparilla). J. Res. Indian Med. **3**, (1) 1 (1968).

392. Rangaswami, S., and S. Sarangan: Sapogenins of *Clerodendron serratum*. Constitution of a New Pentacyclic Triterpene Acid, Serratagenic Acid. Tetrahedron **25**, 3701 (1969).

393. Rangaswami, S., and T. Seshadri: Constitution of Putranjiva Saponins A, B, C, and D. Indian J. Chem. **9,** 189 (1971); Chem. Abstr. **75,** 1259 (1971).

394. Rao, G. S., and J. E. Sinsheimer: Structure of Gymnemagenin. Chem. Commun. **1968,** 1681.

395. Rao, K. V.: Chemistry of *Aegiceras majus* Gaertn. — V. Structure of the Triterpene Aegicerin. Tetrahedron **20,** 973 (1964).

396. Rao, K. V., and P. K. Bose: Chemistry of *Aegiceras majus* Gaertn III. Structure of Aegiceradiol. Tetrahedron **18,** 461 (1962).

397. Rao, M. G., L. R. Row, and C. Rukmini: New Triterpene Sapogenin, Castano-genol from the Bark of *Castanospermum australe.* Indian J. Chem. **7,** 1203 (1969); Chem. Abstr. **72,** 67135 (1970).

398. Rastogi, R. P., and M. L. Dhar: Examination of *Centella asiatica* II. Brahmoside and Brahminoside. Indian J. Chem. **1,** 267 (1963); Chem. Abstr. **59,** 11641 (1963).

398a. Rathbone, E. B., A. M. Stephen, and K. G. R. Pachler: PMR Spectroscopy of O-Methyl Derivatives of D-Galactopyranose and related Compounds in Solution in Deuterium Oxide. Carbohyd. Res. **23,** 275 (1972).

399. Rees, H. H., G. Britton, and T. W. Goodwin: The Biosynthesis of β-Amyrin. Biochem. J. **106,** 659 (1968).

400. Rees, H. H., L. J. Goad, and C. T. Goodwin: Cyclization of 2,3-Oxidosqualene to Cycloartenol in a Cell-free System from Higher Plants. Tetrahedron Letters **1968,** 723.

401. Rees, H. H., E. I. Mercer, and T. W. Goodwin: The Stereospecific Biosynthesis of Plant Sterols and α- and β-Amyrin. Biochem. J. **99,** 726 (1966).

402. Ripperger, H., und K. Schreiber: Struktur von Paniculonin A und B, zwei neuen Spirostanglykosiden aus *Solanum paniculatum* L. Chem. Ber. **101,** 2450 (1968).

403. Ripperger, H., K. Schreiber und H. Budzikiewicz: Isolierung von Neochlorogenin und Paniculogenin aus *Solanum paniculatum* L. Chem. Ber. **100,** 1741 (1967).

404. Roberts, G., B. S. Gallagher, and R. N. Jones: Infrared Absorption Spectra of Steroids. An Atlas, Volume II. New York: Interscience Publishers. 1958.

405. Roller, P., C. Djerassi, R. Cloetens, and B. Tursch: Terpenoids LXIV. Chemical Studies of Marine Invertebrates V. The Isolation of Three New Holothurinogenins and Their Chemical Correlation with Lanosterol. J. Amer. Chem. Soc. **91,** 4918 (1969).

406. Rondest, J. et J. Polonsky: Structure de l'acide polygalacique, nouvel acid triterpé-nique, isolé de *Polygala paenea* L. Bull. Soc. chim. France **1963,** 1253.

407. Rosen, W. E., J. B. Ziegler, A. C. Shabica, and J. N. Shoolery: The Stereochemi-stry of Steroidal Sapogenins, III. NMR Spectra. J. Amer. Chem. Soc. **81,** 1687 (1959).

408. Rothman, E. S., M. E. Wall, and H. A. Walens: Steroidal Sapogenins IV. Hydroly-sis of Steroidal Saponins. J. Amer. Chem. Soc. **74,** 5791 (1952).

409. Row, R. L., P. S. Murty, G. S. R. S. Rao, C. S. P. Sastry, and K. V. J. Rao: Terminalia Species XII. Isolation and Structure Determination of Arjunic Acid, a New Trihydroxytriterpene Carbocyclic Acid from *Terminalia arjuna* Bark. Indian J. Chem. **8,** 716 (1970); Chem. Abstr. **74,** 1020 (1971).

410. Row, R. L., and G. S. R. S. Rao: Chemistry of *Terminalia* Species — VI. The Consti-tution of Tomentosic Acid, a New Triterpene Carboxyclic Acid from *Terminalia tomen-tosa* Wight et ARN. Tetrahedron **18,** 827 (1962).

411. Row, R. L., and C. S. P. Sastry: Isolation of Barringtonic Acid, a New Triterpene Dicarboxylic Acid from *Barringtonia acutangula.* Indian J. Chem. **2,** 463 (1964); Chem. Abstr. **62,** 7811 (1965).

412. Roy, S., and A. Roy: Constitution of Proceragenin A — A Triterpenoid Sapogenin from *Albizzia procera* Benth. Tetrahedron Letters **1966,** 5743.

413. Russo, G.: I triterpeni della *Glycyrrhiza glabra* L. Fitoterapia **38,** 98 (1967).

414. Ruzicka, L., W. Baumgartner und V. Prelog: Zur Kenntnis über Triterpene. Zur Konstitution des Äscigenins. Helv. Chim. Acta **32,** 2057 (1944).

415. SANDER, H., M. ALKEMEYER und R. HÄNSEL: Methode zur Trennung der Glyko-alkaloide von *Solarum dulcamara.* Pharm. Acta Helv. **35**, 30 (1960); Chem. Abstr. **54,** 15830 (1960).

416. SANDERMANN, W., und H. FUNKE: Termitenresistenz alter Tempelhölzer aus dem Mayagebiet durch Saponine. Naturwiss. **57,** 407 (1970).

417. SANDLBERG, F.: Die Ginsengwurzel. Neue Untersuchungen an einer alten Arznei-pflanze. Österr. Apotheker-Ztg. **25,** 205 (1971).

418. SANNIE, C., and H. LAPIN: Sterolic Sapogenins, VII, Neoruscogenin (3 β, 1-Dihydroxy-22β, 25 L-5-spirostene), a New Sapogenin from *Ruscus aculeatus.* Bull. soc. chim. France **1957,** 1237; Chem. Abstr. **52,** 2965 (1958).

419. SASAKI, K., and K. TAKEDA: Studies on the Steroidal Components of Domestic Plants, XXXII. Constituents of *Reineckia carnea* Kunth. (4). Structure of Kitigenin (1). Chem. Pharm. Bull. (Tokyo) **9,** 684 (1961), XXXIII (5) (2) ibid. **9,** 693 (1961).

420. SASTRY, C. S. P., and L. R. ROW: New Triterpenes from *Barringtonia acutangula* Gaertn — III. Tetrahedron **23,** 3837 (1967).

421. SAVOIR, R., R. OTTINGER, B. TURSCH, and G. CHIURDOGLU: Triterpenes. XII. NMR-Spectroscopy of Triterpenes. Methyl Groups in the Ursene Series. Bull. Soc. Chim. Belges **76,** 371 (1967); Chem. Abstr. **68,** 73884 (1968).

422. SAVOIR, R., B. TURSCH, and M. KAISIN: Triterpenes XIII. Serjanic Acid. A new Tri-terpene from *Sapindaceae.* Tetrahedron Letters **1967,** 2129.

423. SAWARDEKER, J. S., and J. H. SLONEKER: Quantitative Determination of Mono-saccharides by Gas Liquid Chromatography. Analyt. Chem. **37,** 945 (1965).

424. SAWARDEKER, J. S., J. H. SLONEKER, and A. JEANES: Quantitative Determination of Monosaccharides as Their Alditol Acetates by Gas Liquid Chromatography. Analyt. Chem. **37,** 1602 (1965).

425. SCARDAVI, A., and F. C. ELLIOT: A Review of Saponins in Alfalfa and their Bioassay Utilizing Trichoderma Species. Mich. State Univ. Agr. Exp. Sta. Quart. Bull. **50,** 163 (1967); Chem. Abstr. **68,** 57350 (1968).

426. v. SCHANTZ, M.: Die Saponindrogen. Dansk Tidsskr. Farm. **40,** 140 (1966).

427. SCHEIDEGGER, J. J., et E. CHERBULIEZ: L'hédéracoside A, un nouvel hétéroside extrait du lierre (*Hedera helix* L.). Helv. Chim. Acta **38,** 547 (1955).

428. SCHEUER, P. J.: The Chemistry of some Toxins Isolated from Marine Organism. Fortschr. Chem. org. Naturstoffe **27,** 322 (1969).

429. — Toxins from Marine Invertebraten. Naturwissenschaften **58,** 549 (1971).

430. SCHLEMMER, W.: Bestimmung von Aescin. Apotheker-Ztg. **106,** 1315 (1966).

430a. SCHLÖSSER, E.: Cyclamin, an Antifungal Resistence Factor in *Cyclamen* species. Acta Phytopathol. **6,** 85 (1971).

431. SCHLÖSSER, E., und G. WULFF: Über die Strukturspezifität der Saponinhämolyse. I. Triterpensaponine und -aglykone. Z. Naturforsch. **24 b,** 1284 (1969).

432. SCHÖNHEIMER und H. DAM: Über die Spaltbarkeit und Löslichkeit von Sterindigito-niden. Z. physiol. Chem. **215,** 59 (1933).

433. SCHREIBER, K.: Steroid Alkaloids: The Solanum Group. In: MANSKE, R. H. F. (Edit.): The Alkaloids. Chemistry and Physiology, Vol. X, p. 1. New York and London: Academic Press. 1968.

434. SCOTT, A. J.: Interpretation of the Ultraviolet Spectra of Natural Products. Oxford: Pergamon Press. 1964.

435. SEGAL, R., H. GOVRIN, and D. V. ZAITSCHEK: A New Type of Saponin from Styrax officinalis L. Tetrahedron Letters **1964,** 527.

436. SEGAL, R., M. MANSOUR, and D. V. ZAITCHEK: Effects of Ester Groups on the Hemolytic Action of Some Saponins and Sapogenins Biochem. Pharmacol. **15,** 1411 (1966).

437. SEGAL, R., and I. MILO-GOLDZWEIG: On the Similarity between the Hemolytic Proper-ties of Bile Acids and of Neutral Steroids. Life Sciences **10,** 685 (1971).

438. Segal, R., I. Milo-Goldzweig, H. Schupper, and D. V. Zaitchek: Effect of Ester Groups on the Haemolytic Action of Sapogenins II. Esterfication with Bifunctional Acids. Biochem. Pharmacol. **19**, 2501 (1970).

439. Serota, S., R. P. Koob, and M. E. Wall: Steroidal Sapogenins. LXIII, Chiapagenin, a New Normal (25 L) Δ^5-12 β-Hydroxysapogenin. J. Org. Chem. **25**, 1768 (1960).

440. Severini, R. G., and G. Russo: Relation between NMR. Methyl Signals and Molecular Structure of Olean-12-ene Triterpenes. Gazz. Chim. Ital. **98**, 602 (1968); Chem. Abstr. **69**, 59438 (1968).

441. Shany, S., B. Gestetner, J. Birk, and A. Bondi: Lucerne Saponins. III. Effect of Lucerne Saponins on Larval Growth and their Detoxification by Various Sterols. J. Sci. Food Agr. **21**, 508 (1970); Chem. Abstr. **74**, 29398 (1971).

441a. Shaposhnikova, G. J., N. A. Ferens, N. I. Uvarova, and G. B. Elyakov: Structure of the Carbohydrate Chains of Panaxosides B1 and B from *Panax ginseng.* Carbohydrate Res. **15**, 319 (1970).

442. Shcherbanovskii, L. R.: Antimicrobial Properties of Saponins and Steroid Glyco-alkaloids. Rast. Resur **7**, 133 (1971); Chem. Abstr. **74**, 108568 (1971).

443. Shibata, S., T. Ando, and O. Tanaka: Oriental Plant Drugs XVIII. Prosapogenin of the Ginseng Saponins. Chem. Pharm. Bull. (Tokyo) **14**, 1157 (1966).

444. Shibata, S., M. Fujita, H. Itokawa, O. Tanaka, and T. Ishii: The Structure of Panaxadiol, a Sapogenin of Ginseng. Tetrahedron Letters **1962**, 419.

445. Shibata, S., I. Kitagawa, and H. Tujimato: The Chemical Studies on Oriental Plant Drugs XV. On the Constituents of *Bupleurum* Spp. The Structure of Saikogenin A. Chem. Pharm. Bull. (Tokyo) **14**, 1023 (1966).

446. Shibata, S., O. Tanaka, T. Ando, M. Sado, S. Tsushima, and T. Ohsawa: Chemical Studies on Oriental Plant Drugs XIV. Protopanaxadiol, a Genuine Sapogenin of Ginseng Saponins. Chem. Pharm. Bull. (Tokyo) **14**, 595 (1966).

447. Shibata, S., O. Tanaka, M. Sado, and S. Tsushima: On Genuine Sapogenin of Ginseng. Tetrahedron Letters **1963**, 795.

448. Shibata, S., O. Tanaka, K. Soma, Y. Iida, T. Ando, and H. Nakamura: Studies on Saponins and Sapogenins of Ginseng. The Structure of Panaxatriol. Tetrahedron Letters **1965**, 207.

449. Shimada, S.: Antifungal Steroid Glycoside from Sea Cucumber. Science **163**, 1462 (1969).

450. Shimizu, Y., and S. W. Pelletier: Cyclosenegenin, a Derivative of Senegenin in *Polygala senega.* J. Amer. Chem. Soc. **87**, 2065 (1965).

451. — — Elucidation of the Structures of the Sapogenins of *Polygala senega* by Correlation with Medicagenic Acid. J. Amer. Chem. Soc. **88**, 1544 (1966).

452. Shimizu, M., and T. Takemoto: Constituents of Caryophyllaceous Plants. I. Saponin of *Dianthus superbus* var. *longicalycinus.* Yakugaku Zasshi **87**, 250 (1967); Chem. Abstr. **67**, 8716 (1967).

453. Shoji, S., S. Kawanishi, and Y. Tsukitani: On the Structure of Senegin II of Senegae Radix. Chem. Pharm. Bull. (Tokyo) **19**, 1740 (1971).

454. — — — Constituents of Senegae Radix I. Isolation and Qualitative Analysis of the Glycosides. Yakugaku Zasshi **91**, 198 (1971); Chem. Abstr. **74**, 121322 (1971).

455. Siering, H.: Die Permeabilität von Zellmembranen für Ionen unter dem Einfluß von Aescin. Arzneimittelforsch. **12**, 376 (1962).

456. Simonsen, J., and W. C. J. Ross: The Terpenes, Vol. IV and V. Cambridge: University Press. 1957.

457. Sinsheimer, J. E.: Isolation and Antiviral Activity of the Gymnemic Acids. Experientia (Basel) **24**, 302 (1968).

458. Sinsheimer, J. E., and G. S. Rao: Constituents of *Gymnema sylvestre* Leaves VI. Acylated Genins of the Gymnemic Acids. Isolation and Preliminary Characterization. J. Pharm. Sci. **59**, 629 (1970); Chem. Abstr. **73**, 42374 (1970).

459. SKINNER, F. A.: Antibiotics in: K. PEACH, und M. V. TRACEY, Moderne Methoden der Pflanzenanalyse, Bd. III, S. 626. Berlin-Göttingen-Heidelberg: Springer. 1955.

459a. SLIWOWSKI, J., and Z. KASPRZYK: ORD and CD Studies on the Structure of Carbonyl Derivatives of Pentacyclic Triterpenes. Tetrahedron 28, 991 (1972).

460. SMITH, A. G., and L. J. GOAD: The Metabolism of Cholesterol by Echinoderms *Aterias rubens* and *Solaster papposus*. FEBS Letters 12, 233 (1970).

461. SMITH, A. M., and C. R. EDDY: High Resolution Infrared Spectra of Steroids in the Carbon-Hydrogen Stretching Region. Analyt. Chem. 31, 1539 (1959).

462. SNATZKE, G. (Ed.): Optical Rotatory Dispersion and Circular Dichroism in Organic Chemistry. London: Heyden. 1967.

463. SNATZKE, G., und H.-W. FEHLHABER: Eine einfache Methode zur Lokalisierung isolierter Doppelbindungen in Steroiden und Triterpenen. Liebigs Ann. Chem. 663, 123 (1963).

464. SNATZKE, G., F. LAMPERT und R. TSCHESCHE: Über Triterpene VIII. Zuordnung von Triterpenen zu den Grundtypen durch IR-Spektroskopie. Tetrahedron 18, 1417 (1962).

465. SPRINGER, G. F., P. R. DESAI, and B. KOLECKI: Synthesis and Immunochemistry of Fucose Methyl Ethers and Their Methylglycosides. Biochemistry 3, 1076 (1964).

466. STAHL, E.: Dünnschichtchromatographie, 2. Aufl. Berlin-Heidelberg-New York: Springer. 1967.

467. STEINEGGER, E., und R. HÄNSEL: Lehrbuch der Allgemeinen Pharmakognosie. Berlin-Göttingen-Heidelberg: Springer. 1963.

468. STEINER, M., und R. HOLTZEM: Triterpene und Triterpen-Saponine in: Paech-Tracey, Moderne Methoden der Pflanzenanalyse, Bd. III. Berlin-Göttingen-Heidelberg: Springer. 1955.

469. STEPHEN, T.: Saponins and Sapogenins, Part IV, Agapanthagenin (22α-Spirostan-2α,3β,5α-triol, a New Sapogenin from *Agapanthus* Species. J. Chem. Soc. (London) 1956, 1167.

470. STÖCKLIN, W.: Gymnestrogenin, ein neues Pentahydroxytriterpen aus den Blättern von *Gymnema sylvestre* R. Br. Helv. Chim. Acta 51, 1235 (1968).

471. STÖCKLIN, W., E. WEISS und T. REICHSTEIN: Gymnemasäure, das antisaccharine Prinzip von *Gymnema sylvestre* R. Br. Isolierungen und Identifizierungen. Helv. Chim. Acta 50, 474 (1967).

472. STOUT, G. H., B. M. MALOFSKY, and V. F. STOUT: Phytolaccagenin: A Light Atom X-Ray Structure Proof Using Chemical Information. J. Amer. Chem. Soc. 86, 957 (1964).

472a. STRIGINA, L. J., T. M. REMENNIKOVA, A. P. SHCHERDRIN, and G. B. ELYAKOV: Triterpenic Glycosides of *Caltha silvestris*. Khim. Prir. Soedin 1972, 303; Chem. Abstr. 77, 161938 (1972).

473. SWEELEY, C. C., R. BENTLEY, M. MAKITA, and W. W. WELLS: Gas Liquid Chromatography of Trimethylsilyl Derivatives of Sugars and Related Substances. J. Amer. Chem. Soc. 85, 2497 (1963).

474. SWEELEY, C. C., and B. WALKER: Determination of Carbohydrates in Glycolipides and Gangliosides by Gas Chromatography. Analyt. Chem. 36, 1461 (1964).

474a. TADA, A. J., and J. SKOJI: Constituents of *Ophiopogenis* Tuber II. Structure of Ophiopoganin B. Chem. Pharm. Bull. (Tokyo) 20, 1729 (1972).

475. TAKEDA, K.: Studies on the Steroidal Components of Domestic Plants, XXX. Structure of Nogiragenin, a New Sapogenin Isolated from *Metanarthecium luteo-viride* Maxim. Chem. Pharm. Bull. (Tokyo) 9, 388 (1961).

476. — Studies on the Steroidal Components of Domestic Plants, XXXI, Constituents of *Reineckia carnea* Kunth. (3), Pentelogenin and Kitigenin. Chem. Pharm. Bull. (Tokyo) 9, 631 (1961).

477. TAKEDA, K., and K. HAMAMOTO: Studies on the Steroidal Components of Domestic Plants. Structure of Metagenin. Chem. Pharm. Bull. 8, 1004 (1960).

478. Takeda, K., T. Kubota, and A. Shimaoka: Studies on the Steroidal Components of Domestic Plants XIX. The Structure of Kogagenin, a Sapogenin from *Dioscorea tokoro* Makino. Tetrahedron **7**, 62 (1959).

479. Takeda, A. K., H. Minato, A. Shimaoka, and Y. Matsui: IR-Spectra of Steroidal Sapogenins having Oxygen Substituents in the F-Ring. J. Chem. Soc. (London) **1963**, 4815.

480. Takeda, K., T. Okanishi, A. Akahori, and F. Yasuda: Structure of Diotigenin (1). Chem. Pharm. Bull. (Tokyo) **16**, 421 (1968).

481. Takeda, K., T. Okanishi, K. Igarashi, and A. Shimaoka: Studies on the Steroidal Components of Domestic Plants — XXXIV. The Structure of Luvigenin, a New Sapogenin from *Metanarthecium luteo-viride*. Maxim. Tetrahedron **15**, 183 (1961).

482. Takeda, K., T. Okanishi, M. Minato, and A. Shimaoka: Studies on the Steroidal Components of Domestic Plants — XLI. Constituents of *Reineckia carnea* Kunth. (6). Structure of Reineckiagenin, Isoreineckiagenin and Isocarneagenin. Tetrahedron **19**, 759 (1963).

483. — — — — Studies on the Steroidal Components of Domestic Plants, XLVI. Constituents of *Hosta* Species (3). $\Delta^{25\,(27)}$-Sapogenins. Tetrahedron **21**, 2089 (1965).

484. Takeda, K., T. Okanishi, and A. Shimaoka: Steroidal Components of Domestic Plants, XVII. Structure of Yonogenin, a New Steroidal Sapogenin from *Dioscorea tokoro*. Chem. Pharm. Bull. (Tokyo) **6**, 532 (1958); Chem. Abstr. **53**, 16206 (1959).

485. Takemoto, T., und K. Kometani: Triterpenglykoside (Mubenine) aus Samen von *Stauntonia hexaphylla*. Liebigs Ann. Chem. **685**, 237 (1965).

486. Tanaka, O., N. Tanaka, T. Ohsawa, Y. Iitaka, and S. Shibata: Stereochemistry of the Side Chain of Dammarane Type Triterpenes. Tetrahedron Letters **1968**, 4235.

487. Tandon, J. S., K. P. Agarwal, and M. M. Dhar: Triterpenoids from *Randia dumetorum*. Indian J. Chem. **4**, 483 (1966); Chem. Abstr. **66**, 85884 (1967).

488. Tesarik, K.: The Separation of Some Monosaccharides by Capillary Gas-Chromatography. J. Chromatogr. **65**, 295 (1972).

489. Tewari, N. C., K. N. N. Ayengar, and S. Rangaswami: Structure of Sapogenins of *Caccinia glauca*. Caccigenin, Caccigenin Lactone, and 23-Deoxycaccigenin. Indian J. Chem. **8**, 593 (1970); Chem. Abstr. **73**, 127719 (1970).

490. Thomas, A. F., and B. Willhalm: The Triterpenes of *Commiphora* IV (1). Mass Spectra and Organic Analysis V (2). Mass Spectroscopic Studies and the Structure of Commic Acids A and B. Tetrahedron Letters **1964**, 3177.

491. Thomson, J. B.: The Structure of Isoaescigenin. Some Derivatives of Aescigenin. Tetrahedron **22**, 351 (1966).

492. Tomita, Y., and A. Uomori: Biosynthesis of Sapogenins in Tissue Cultures of *Dioscorea tokoro* Makino. Chem. Commun. **1971**, 284.

493. Trzenschik, U., R. Przyborowski, K. Hiller und B. Linzer: Antimikrobielle Eigenschaften der Sanicula Saponine. Pharmazie **22**, 715 (1967).

494. Tschesche, R.: Über neutrale Saponine, Überführung von Digitogenin, Gitogenin und Tigogenin in identische Derivate. Ber. dtsch. chem. Ges. **68**, 1090 (1935).

495. — Advances in the Chemistry of Antibiotic Substances from Higher Plants. In: Wagner, H., and L. Hörammer (Edit.): Pharmacognosy and Phytochemistry, p. 274. Berlin-Heidelberg-New York: Springer. 1971.

496. — Biosynthesis of Cardenolides, Bufadienolides and Steroid Sapogenins. Proc. Roy. Soc. London B **180**, 187 (1972).

497. Tschesche, R., und G. Balle: Über Saponine der Spirostanolreihe X. Zur Konstitution der Samensaponine von *Digitalis lanata* Ehrh. Tetrahedron **19**, 2323 (1963).

498. Tschesche, R., und L. Ballhorn: 1972, unveröffentlicht.

499. Tschesche, R., J. Duphorn und G. Snatzke: Über Triterpene, IX. Über die Konstitution der Cincholsäure und die sauren Triterpenglykoside der Chinarinde. Liebigs Ann. Chem. **667**, 151 (1963).

500. Tschesche, R., und D. Forstmann: Über Triterpene, IV. Musennin, ein wurmwirksames Saponin aus der Rinde von *Albizzia anthelmintica*. Chem. Ber. **90**, 2383 (1957).

501. Tschesche, R., und R. Fritz: Zur Biosynthese von Steroid-Derivaten im Pflanzenreich, 16. Mitt.: Beiträge zur Spirostanol-Biogenese in *Digitalis lanata* und Spirosolanol-Biogenese in *Solanum lycopersicum*. Z. Naturforsch. **25b**, 590 (1970).

502. Tschesche, R., R. Fritz und G. Josst: Vergleich der Spirostanolbiogenese aus Cholestanol und Cholestanon. Koprostanol ist keine Vorstufe für Cardenolide oder Sapogenine. Phytochem. **9**, 371 (1970).

503. Tschesche, R., und A. Hagedorn: Über Saponine der Cyclopentanohydrophenanthren-Gruppe IV. Zur Konstitution des Gitogenins und Digitogenins. Ber. dtsch. Chem. Ges. **69**, 797 (1936).

503a. Tschesche, R., A. Harz und J. Petričić: 1973 unveröffentlicht.

504. Tschesche, R., E. Henckel und G. Snatzke: Über Triterpene XIV. Die Konstitution von Bredemol- und Crataegolsäure. Liebigs Ann. Chem. **676**, 175 (1964).

505. Tschesche, R., K. Hermann und R. Langlais: Unveröffentlicht.

506. Tschesche, R., und H. Hulpke: Zur Biosynthese von Steroidderivaten im Pflanzenreich, 3. Mitt., Spirostanol-Biogenese aus Cholesterin-glucosid. Z. Naturforsch. **21b**, 494 (1966).

507. Tschesche, R., H. Hulpke und R. Fritz: Zur Biosynthese von Steroidderivaten im Pflanzenreich — X. Zur Spirostanol-Biogenese aus Δ^4-Cholestenon und aus anderen möglichen Vorstufen. Phytochem. **7**, 2021 (1968).

508. Tschesche, R., H. C. Iha und G. Wulff: Über Triterpene, XXIX. Zur Struktur des Avenacins. Tetrahedron **29**, 629 (1973).

509. Tschesche, R., F. Inchaurrondo und G. Wulff: Über Triterpene XVII. Das Aglykon des Cyclamins. Liebigs Ann. Chem. **680**, 107 (1964).

510. Tschesche, R., und G. Josst: 1971, unveröffentlicht.

511. Tschesche, R., und F.-J. Kämmerer: Über Triterpene XXVII. Die Struktur von Musennin und Desglucomusennin. Liebigs Ann. Chem. **724**, 183 (1969).

512. Tschesche, R., F.-J. Kämmerer und G. Wulff: Über Triterpene XXII. Über die Saponine aus *Albizzia anthelmintica* (Brongn). Z. Naturforsch. **21b**, 596 (1966).

513. Tschesche, R., R. Kottler und G. Wulff: Über Saponine der Spirostanolreihe, XII. Über Parillin, ein Saponin mit stark verzweigter Zuckerkette. Liebigs Ann. Chem. **699**, 212 (1966).

514. Tschesche, R., und P. Lauven: Avenacosid B, ein zweites bisdesmosidisches Steroidsaponin aus *Avena sativa*. Chem. Ber. **104**, 3549 (1971).

515. Tschesche, R., G. Lüdke und G. Wulff: Über Sarsaparillosid, ein Saponinderivat mit bisglykosidischer Furostanolstruktur. Tetrahedron Letters **1967**, 2785.

516. — — — Steroidsaponine mit mehr als einer Zuckerkette, II. Sarsaparillosid, ein bisdesmosidisches 22-Hydroxy-furostanol-saponin. Chem. Ber. **102**, 1253 (1969).

517. Tschesche, R., H.-J. Mercker und G. Wulff: Über Triterpene XXIV. Die Konstitution der Zuckerkette des Cyclamins. Liebigs Ann. Chem. **721**, 194 (1969).

518. Tschesche, R., H. Rehkämper und G. Wulff: Über Triterpene XXVIII. Über die Saponine des Spinats (*Spinacia oleracea* L.). Liebigs Ann. Chem. **726**, 125 (1969).

519. Tschesche, R., und K. H. Richert: Über Saponine der Spirostanolreihe, XI. Nuatigenin, ein Cholegenin-analogon des Pflanzenreiches. Tetrahedron **20**, 387 (1964).

520. Tschesche, R., und W. Schmidt: Über Saponine der Spirostanolreihe — XIII. Zwei neue Saponine der oberirdischen Teile des Hafers (*Avena sativa*) mit Nuatigenin als Aglykon. Z. Naturforsch. **21b**, 896 (1966).

521. Tschesche, R., W. Schmidt und G. Wulff: Über Triterpene, XIX. Reindarstellung und Strukturermittlung der Saponine des Efeus (*Hedera helix* L.). Z. Naturforsch. **20b**, 708 (1965).

522. Tschesche, R., H. Schulze und I. Goebels: 1972, unveröffentlicht.

523. Tschesche, R., H. Schwarz und G. Snatzke: Über Saponine der Spirostanolreihe, VI. Die Konstitution des Convallamarogenins. Chem. Ber. **94,** 1699 (1961).

524. Tschesche, R., W. Seidel, S. C. Sharma und G. Wulff: Chem. Ber. **105,** 3397 (1972).

525. Tschesche, R., H. Striegler und H. W. Fehlhaber: Über Triterpene, XIX. Die Struktur des Cyclamiretins A. Liebigs Ann. Chem. **691,** 165 (1966).

526. Tschesche, R., M. Tauscher, H.-W. Fehlhaber und G. Wulff: Avenacosid A, ein bisdesmosidisches Steroidsaponin aus *Avena sativa.* Chem. Ber. **102,** 2072 (1969).

527. Tschesche, R., B. T. Tjoa und G. Wulff: Über Triterpene, XXI. Über die Sapogenine aus den Wurzeln von *Primula veris* L. Liebigs Ann. Chem. **696,** 160 (1966).

528. — — — Über Triterpene, XXIII. Über Struktur und Chemie der Priverogenine. Tetrahedron Letters **1968,** 183.

529. Tschesche, R., B. T. Tjoa, G. Wulff und R. Noronha: Steroidsaponine mit mehr als einer Zuckerkette, III. Convallamarosid, ein weiteres 22-Hydroxy-furostanolsaponin. Tetrahedron Letters **49,** 5141 (1968).

530. Tschesche, R., A. Weber und G. Wulff: Über Triterpene, XXV. Über die Struktur des „Theasaponins", eines Gemisches von Saponinen aus *Thea sinensis* L. mit stark antiexsudativer Wirksamkeit. Liebigs Ann. Chem. **721,** 209 (1969).

531. Tschesche, R., und G. Wulff: Über Saponine der Spirostanol-Reihe, VII. Über Digalogenin, ein neues Saponin aus den Samen von *Digitalis purpurea* L. Chem. Ber. **94,** 2019 (1961).

532. — — Über Saponine der Spirostanolreihe, IX. Die Konstitution des Digitonins. Tetrahedron **19,** 621 (1963).

533. — — Konstitution und Eigenschaften der Saponine. Planta Med. **12,** 272 (1964).

534. — — Über Triterpene, XVIII. Über die Struktur des Aescinidins und seine Identität mit Barringtogenol C. Tetrahedron Letters **1965,** 1569.

535. — — Über die antimikrobielle Wirksamkeit von Saponinen. Z. Naturforsch. **20b,** 543 (1965).

536. — — Unveröffentlicht.

537. Tschesche, R., G. Wulff und G. Balle: Über Saponine der Spirostanolreihe, VIII. Über das gemeinsame Vorkommen von 25α- und 25β-Sapogeninen in den Saponinen von *Digitalis purpurea* L. und *Digitalis lanata* Ehrh. Tetrahedron **18,** 959 (1963).

538. Tschesche, R., G. Wulff, A. Weber und H. Schmidt: Fraßhemmende Wirkung von Saponinen auf Termiten (Isoptera, Reticulitermes). Z. Naturforsch. **25b,** 999 (1970).

539. Tschesche, R., und F. Ziegler: Über Triterpene, XII. Über die Saponine der Wurzeln von *Primula elatior* L. Schreber. Liebigs Ann. Chem. **674,** 185 (1964).

540. Turner, A. B., D. S. H. Smith, and A. M. Mackie: Characterization of the Principal Steroidal Saponins of the Starfish *Marthasterias glacialis:* Structure of the Aglycones. Nature **233,** 209 (1971).

541. Tursch, B., R. Cloetens, and C. Djerassi: Chemical Studies of Marine Invertebrates, VI. Terpenoids, LXV. Praslinogenin, a New Holothurinogenin from the Indian Ocean Sea Cucumber *Bohadschia koellikeri.* Tetrahedron Letters **1970,** 467.

542. Tursch, B., D. Daloze, R. Ottinger, J. Reisse, and G. Chiurdoglu: Triterpenes, IV, Conformational Effects on the NMR-Frequency of the 12C Olefinic Proton in the β-Amyrin Group of Triterpenes. Bull. Soc. Chim. Belges **75,** 191 (1966); Chem. Abstr. **65,** 5495 (1966).

543. Tursch, B., D. Daloze, E. Tursch, and G. Chiurdoglu: Triterpenes, III. Sapogenin K from *Stryphnodendron coriaceum.* Bull. Soc. Chim. Belges **75,** 127 (1966); Chem. Abstr. **64,** 19692 (1966).

544. Tursch, B., J. Leclercq et G. Chiurdoglu: Structure de l'acide Mesembryanthemoidigenique, Triterpene nouveau des cactacées. Tetrahedron Letters **1965,** 4161.

544a. Tursch, B., R. Savoir, R. Ottinger, and G. Chiurdoglu: Triterpenes VIII. NMR-Spectra of Triterpenes. Effect of Substitution on the Chemical Shifts of Methyl Groups in the Δ^{12}-Oleanene Series. Tetrahedron Letters **1967,** 539.

545. TURSCH, B., I. S. DE SOUZA GUIMARÃES, B. GILBERT, R. T. APLIN, A. M. DUFFIELD, and C. DJERASSI: Chemical Studies of Marine Invertebrates, II. Terpenoids, LVIII. Griseogenin, a New Triterpenoid Sapogenin of the Sea Cucumber *Halodeima grisea* L. Tetrahedron **23**, 761 (1967).

546. TURSCH, B., E. TURSCH, J. T. HARRISON, BRAZÃO DA SILVA, H. J. MONTEIRO, B. GILBERT, W. B. MORS, and C. DJERASSI: Terpenoids, LIII. Demonstration of Ring Conformational Changes in Triterpenes of the β-Amyrin Class Isolated from *Stryphnodendron coriaceum*. J. Org. Chem. **28**, 2390 (1963).

547. UVAROVA, N. J., N. A. FERENS, G. I. SHAPOSHNIKOVA, and G. B. ELYAKQV: Carbohydrate Chains of Panaxosides B' and C, Glycosides from *Panax ginseng* Roots. Khim. Prir. Soedin **6**, 312 (1970); Chem. Abstr. **73**, 99157 (1970).

548. UVAROVA, N. J., R. P. GORSHKOVA, and G. B. ELYAKOV: Separation of Glycosides of *Panax ginseng* C. Mey on Sephadex. Izvest. Akad. Nauk SSSR, Ser. Khim. Nauk. **1963**, 1850.

549. UVAROVA, N. J., R. P. GORSHKOVA, L. U. STRIGINA, G. B. ELYAKOV, and N. K. KOCHETKOV: Khim. Prir. Soedin. **1**, 82 (1965); Chem. Abstr. **63**, 7249 (1965).

549a. VANHAELEN, M.: New Saponoside from *Bersama yangambiensis*. Phytochemistry **11**, 1111 (1972).

550. VARSHNEY, I. P., and S. Y. KHAN: Saponins and Sapogenins, XVIII. Isolation of Proceric Acid, a Triterpenic Acid, from Maharashtrian *Albizzia procera* Seeds. J. Pharm. Soc. **53**, 1532 (1964); Chem. Abstr. **62**, 6720 (1965).

551. VARSHNEY, I. P., and M. LOGANI: Saponins and Sapogenins from *Blepheris edulis*. Indian J. Appl. Chem. **32**, 72 (1969); Chem. Abstr. **75**, 16036 (1971).

552. VARSHNEY, I. P., and K. M. SHAMSUDDIN: Saponins and Sapogenins, XXV. The Sapogenin of *Acacia concinna* d.c. Pods and the Constitution of Acacic Acid. Tetrahedron Letters **1964**, 2055.

553. VECHERKO, L. P., V. S. KABANOV, and E. P. ZINKERICH: Structure of Calenduloside B from *Calendula officinalis* Roots. Khim. Prir. Soedin **7**, 533 (1971); Chem. Abstr. **75**, 152042 (1971).

554. VECHERKO, L. P., E. P. ZINKEVICH, N. J. LIBIZOV, and A. J. BANKOVSKII: Structure of Calenduloside A. Khim. Prir. Soedin **7**, 22 (1971); Chem. Abstr. **74**, 112385 (1971).

555. VENDRIG, J. C.: Growth-regulating Activity of some Saponins. Nature **203**, 1301 (1964).

556. VOGEL, G.: Die Pharmakologie der Saponine. Planta Med. **11**, 362 (1963).

557. VOGEL, G., M. L. MAREK und R. OERTNER: Zur Frage der Wertbestimmung antiexsudativ wirkender Pharmaka. Arzneimittelforsch. **18**, 426 (1968).

558. — — — Zur Pharmakologie des Teesamen-Saponins, eines Saponingemisches aus *Thea sinensis* L. Arzneimittelforsch. **18**, 1466 (1968).

559. VOGEL, G., und H. UEBEL: Das therapeutisch wirksame Prinzip der Roßkastanie, III. Zur Frage von Angriffspunkt und Wirkungsmechanismus. Arzneimittelforsch. **10**, 275 (1960).

560. VOSS, W., und W. VOGT: Über Convallamarin. Chem. Ber. **69**, 2333 (1936).

561. WAGNER, J., und H. HOFFMANN: Über Inhaltsstoffe des Roßkastaniensamens, V. Untersuchungen an Triterpenglykosiden. Z. physiol. Chem. **348**, 1697 (1967).

562. WAGNER, J., H. HOFFMANN und I. LÖW: Über Inhaltsstoffe des Roßkastaniensamens, VI. Die Acylaglyka der Äscine. Tetrahedron Letters **1968**, 4387.

563. — — —Über Inhaltsstoffe des Roßkastaniensamens, VIII. Die Acylaglyka des Kryptoäscins und α-Äscins. Z. physiol. Chem. **351**, 1133 (1970).

564. WAGNER, J., W. SCHLEMMER und H. HOFFMANN: Über Inhaltsstoffe des Roßkastaniensamens, IX. Struktur und Eigenschaften der Triterpenglykoside. Arzneimittelforsch. **20**, 205 (1970).

565. WAGNER, R. B., R. F. FORKER, and P. F. SPITZER, JR.: The Δ^9-12-Keto Steroidal Sapogenins. J. Amer. Chem. Soc. **73**, 2494 (1951).

604 R. Tschesche und G. Wulff:

566. Wagner, R. B., J. A. Moore, and P. R. Forker: The C-12 Position of the Carbonyl Group in Hecogenin. J. Amer. Chem. Soc. **72**, 1856 (1950).

567. Walens, H. A., S. Serota, and M. E. Wall: Steroidal Sapogenins, XXXI. Gentrogenin and Corellogenin, New Sapogenins from *Dioscorea spiculiflora*. J. Amer. Chem. Soc. **77**, 5196 (1955).

568. Walens, H. A., S. Serota, and M. E. Wall: Steroidal Sapogenins, XXXV. Gentrogenin (Botogenin) and Corellogenin, New Sapogenins from *Dioscorea spiculiflora*. J. Org. Chem. **22**, 182 (1957).

569. Wall, M. E., C. R. Eddy, S. Serota, and R. F. Mininger: Steroidal Sapogenins, VIII. Markogenin (22b-Spirostane-2α,3β-diol). A New Sapogenin Isolated from Yucca. J. Amer. Chem. Soc. **75**, 4437 (1953).

570. Wall, M. E., C. R. Eddy, J. J. Willaman, D. S. Correll, B. G. Schubert, and H. S. Gentry: Steroidsapogenins, XII. Survey of Plants for Steroidal Sapogenins and other Constituents. J. Amer. Pharm. Ass. **43**, 503 (1954).

571. Wallenfels, K., G. Bechtler, R. Kuhn, H. Trischmann und H. Egge: Permethylierung von oligomeren und polymeren Kohlenhydraten und quantitative Analyse der Spaltungsprodukte. Angew. Chem. **75**, 1014 (1963).

572. Watanabe, N., J. Saeki, M. Sumimoto, T. Kondo, and S. Kurotori: On the Antitermic Substance of *Ternstroemia japonica* Wood I. Nippon Makuzai Gakkaishi **12**, 236 (1966); Chem. Abstr. **66**, 102477 (1967).

573. Wendler, N. L., H. L. Slates, and M. Tishler: The Transformation of Manogenin in Hecogenin. J. Amer. Chem. Soc **74**, 4849 (1952).

574. Whistler, R. J., and D. F. Durso: Chromatographic Separation of Sugars on Charcoal. J. Amer. Chem. Soc. **72**, 677 (1950).

575. Williams, D. H., and N. S. Bhacca: Solvent Effects in NMR Spectroscopy, II. Solvent Shifts in some Steroidal Sapogenins. Tetrahedron **21**, 1641 (1965).

576. Williams, S. C., and J. K. N. Jones: The Synthesis, Separation and Identification of the Methyl Ethers of Arabinose and their Derivatives. Canad. J. Chem. **45**, 275 (1967).

577. Willner, D., B. Gestetner, D. Lavie, Y. Birk, and A. Bondi: Soya Bean Saponins. Part V. Soyasapogenol E. J. Chem. Soc. (London) **1964**, 5885.

578. Windaus, A.: Über die quantitative Bestimmung des Cholesterins und der Cholesterinester in einigen normalen und pathologischen Nieren. Z. physiol. Chem. **65**, 110 (1910).

579. Winkler, W.: Stabilität von Aescin in Methanol-Lösung-Esterbildung. Kolloid-Z. **177**, 63 (1961).

580. Winkler, W.: Die Aescinbestimmung in Roßkastanienextrakten und in Extrakt bzw. Aescin enthaltenden Zubereitungen. Arzneimittelforsch. **18**, 1031 (1968).

581. Winkler, W., und P. Patt: Zur Gewinnung saurer Saponine durch Behandlung mit Ionenaustauschern. Naturwiss. **47**, 83 (1960).

582. Winterstein, A., und G. Stein: Untersuchungen in der Saponinreihe, XI. Über die Oxy-triterpensäure-glykoside der Araliaceen. Z. Physiol. Chem. **211**, 5 (1932).

583. Woidke, H.-D., J. -P. Kayser und K. Hiller: Fortschritte in der Erforschung der Triterpensaponine I und II. Pharmazie **25**, 133 und 213 (1970).

584. Wojciechowski, Z., A. Jelonkiewicz-Konador, M. Tomaszewski, J. Jankowski, and Z. Kasprzyk: The Structure of Glycosides of Oleanolic Acid Isolated from the Roots of *Calendula officinalis*. Phytochem. **10**, 1121 (1970).

585. Wolters, B.: Der Anteil der Steroidsaponine an der antibiotischen Wirkung von *Solanum dulcamara*. Planta Med. **13**, 189 (1965).

586. — Zur antimikrobiellen Wirksamkeit pflanzlicher Steroide und Triterpene. Planta Med. **14**, 392 (1966).

587. — Die Wirkung einiger Triterpensaponine auf Pilze. Naturwiss. **53**, 253 (1966).

588. — Saponine als pflanzliche Pilzabwehrstoffe. Planta (Berlin) **79**, 77 (1968).

589. Wulff, G.: Über die quantitative Zuckerbestimmung in Glykosiden und Oligosacchariden mit Hilfe der Gaschromatographie. J. Chromatography **18**, 285 (1965).

590. WULFF, G.: Neuere Ergebnisse auf dem Saponingebiet. Apotheker-Ztg. **108**, 797 (1968).
591. — Konstitution und Eigenschaften der Saponine (Vortragsreferat). Pharm. Ztg. **116**, 228 (1971).
592. WULFF, G., und E. SCHLÖSSER: Unveröffentlicht.
593. WULFF, G., und G. STROH: Unveröffentlicht.
594. WULFF, G., und R. TSCHESCHE: Über Triterpene, XXVI. Über die Struktur der Roßkastaniensaponine und die Aglykone verwandter Glykoside. Tetrahedron **25**, 415 (1969).
595. WULFF, G., R. TSCHESCHE, G. LÜDKE und B. T. TJOA: Steroidsaponine mit geöffneter Spiroketalseitenkette. Apotheker-Ztg. **107**, 718 (1967).
596. YAMAUCHI, T.: Saponins of Japanese Dioscoreaceae, IX. Hydrolysis of Diosgenine Glycosides. Chem. Pharm. Bull. (Tokyo) **7**, 343 (1959).
597. YASUDA, F., Y. NAKAGAWA, A. AKAHORI, and T. OKANISHI: Studies on the Steroidal Components of Domestic Plants — LVII. The Structure of Igagenin. Tetrahedron **24**, 6535 (1968).
598. YASUMOTO, T., and Y. HASHIMOTO: Properties and Sugar Components of Asterosaponin A Isolated from Starfish. Agr. Biol. Chem. (Tokyo) **29**, 804 (1965); Chem. Abstr. **63**, 18540 (1965).
599. — — Properties of Asterosaponin B Isolated from a Starfish *Asterias amurensis*. Agr. Biol. Chem. (Tokyo) **31**, 368 (1967); Chem. Abstr. **67**, 22100 (1967).
600. YASUMOTO, T., K. NAKAMURA, and Y. HASHIMOTO: A New Saponin, Holothurin B, Isolated from Sea Cucumber, *Holothuria vagabunda* and *Holothuria lubrica*. Agr. Biol. Chem. (Tokyo) **31**, 7 (1967); Chem. Abstr. **67**, 9023 (1967).
601. YASUMOTO, T., M. TANAKA, and Y. HASHIMOTO: Distribution of Saponin in Echinoderms. Bull. Jap. Soc. Sci. Fisheries **32**, 673 (1966); Chem. Abstr. **69**, 94029 (1968).
602. YOSIOKA, I., M. FUJIO, M. OSAMURA, and I. KITAGAWA: A Novel Cleavage Method of Saponin with Soil Bacteria, Intending to the Genuine Sapogenin: On *Senega* and *Panax* Saponins. Tetrahedron Letters **1966**, 6303.
603. YOSIOKA, I., K. IMAI, and I. KITAGAWA: On the Triterpenic Constituents of the Seeds Saponin of *Aesculus turbinata* Blume. Chem. Pharm. Bull. (Tokyo) **15**, 135 (1967).
604. — — — On Genuine Sapogenins of Horse Chestnut Saponins by means of Soil Bacterial Hydrolysis and a New Minor Sapogenin: 16-Desoxy-Barringtogenol C. Tetrahedron Letters **1967**, 2577.
605. — — — New Steroidal Prosapogenins. Possessing Fully Acetylated Arabinoside Linkage at C-11. Tetrahedron Letters **1971**, 1177.
606. YOSIOKA, I., and I. KITAGAWA: The Structure of Jegosapogenin: 21-Tigloyl-Barringtogenol C. Tetrahedron Letters **1967**, 2353.
607. YOSIOKA, J., A. MATSUDA, K. IMAI, T. NISHIMURA, and I. KITAGAWA: Saponin and Sapogenol, IV. Seeds Sapogenols of *Aesculus turbinata* Blume. On the Configuration of Hydroxyl Functions in Ring E of Aescigenin, Protoaescigenin in Relation to Barringtogenol C and Theasapogenol A. Chem. Pharm. Bull. (Tokyo) **19**, 1200 (1971).
608. YOSIOKA, I., A. MATSUDA, T. NISHIMURA, and I. KITAGAWA: Structure of Theasapogenol E. Chem. Ind. **1966**, 2202.
609. YOSIOKA, I., A. MATSUDA, R. TAKEDA, and I. KITAGAWA: 22α-Hydroxyerythrodiol from *Camellia sasanqua* Thunb. Chem. Ind. **1968**, 745.
610. YOSIOKA, I., T. NISHIMURA, A. MATSUDA, and I. KITAGAWA: Theasopagenol A, the Major Sapogenol of the Seeds Saponin of *Thea sinensis* L. Tetrahedron Letters **1966**, 5979.
611. — — — — Saponin and Sapogenol, III. Seed Sapogenols of *Thea sinensis* L. Theasapogenol E and Minor Sapogenols. Chem. Pharm. Bull. (Tokyo) **19**, 1186 (1971).
612. YOSIOKA, I., T. NISHIMURA, N. WATANI, and I. KITAGAWA: On the Unexpected Identity of Theasapogenol D (Camelliagenin A) a Minor Sapogenol of Tea Seeds, with Dihydropriverogenin A. Tetrahedron Letters **1967**, 5343.

612a. Yosioka, I., S. Saijoh, and I. Kitagawa: Soil Bacterial Hydrolysis Leading to Genuine Aglycons IV. Four Acylated Derivatives of Barringtogenol C from Jego-saponin. Chem. Pharm. Bull. (Tokyo) **20**, 564 (1972).

613. Yosioka, I., T. Sugawara, A. Ohsuka, and I. Kitagawa: Soil Bacterial Hydrolysis Leading to Genuine Aglycone, III. The Structures of Glycosides and Genuine Aglycone of Sanguisorbae Radix. Chem. Pharm. Bull. (Tokyo) **19**, 1700 (1971).

614. Yukhananow, D. Kh., I. N. Sokolskii, E. P. Zinkevich, and V. B. Kuvaev: Plants of the Family *Caryophyllaceae*. Studied to Determine the Presence of the Triterpenoid Saponin Gyposide. Rast. Resur. **7**, 386 (1971); Chem. Abstr. **75**, 121321 (1971).

615. Zinkevich, E. P., and L. P. Vecherko: Triterpenoid Glycosides (Saponins). Tr. Vses. Nauch.-Issled. Inst. Lek. Aromat. Rast. **15**, 640 (1969); Chem. Abstr. **75**, 20804 (1971).

616. Zoltan, O. T., und S. Gorini: Das Verhalten des Liquordruckes beim experimentell durch Triäthylzinnsulfat verursachten Gehirnödem. Untersuchungen über den Zusammenhang zwischen dem durch Triäthylzinnsulfat hervorgerufenem Gehirnödem und dem Liquordruck bei Hunden. Arzneimittelforsch. **20**, 1812 und 1939 (1970).

616a. Zorina, A. D., L. G. Matyukhina, A. G. Chavva, and L. A. Soltikova: Structure of Macedonic Acid, a Pentacyclic Triterpene. Tetrahedron Letters **1972**, 1841.

617. Zubrod, C. G., S. A. Schepartz, J. Leiter, K. M. Endicott, L. M. Carrese, and C. G. Baker: The Chemotherapy Program of the National Cancer Institute. Cancer Chemotherapy Rept. **50**, 349 (1966).

(Eingelaufen am 29. April 1972)

Rizvi, R. H. *310*
Roberts, J. C. *446*
Roberts, J. G. *586*
Roberts, G. *596*
Robertson, A. 109, *147, 148, 203*
Robertson, D. E. 400, *459*
Robertson, J. M. *205*
Robinson, G. A. *99*
Robinson, K. W. *305*
Robinson, R. S. *445*
Rodgers, G. C. *452*
Roffey, P. *306*
Röhm, E. *591*
Roller, P. *99, 596*
Romo, J. *585*
Ronayne, J. *305*
Roncari, G. *452, 457*
Ronchetti, F. *580*
Rondest, J. *596*
Rondest, R. *149*
Rönnquist, O. *57*
Rosen, W. E. *596*
Rosenbrook, W. *454*
Ross, W. C. J. *598*
Rossi, C. *581*
Roswell, D. F. *99*
Roth, W. R. 97, *99*
Rothman, E. S. *596*
Rothschild, Lord *99*
Rothweiler, W. *457*
Routien, J. B. *458*
Row, L. R. *596, 597*
Row, R. L. *596*
Rowley, E. K. *452*
Roy, A. *596*
Roy, S. *596*
Rubalcava, H. E. *583*
Ruffini, G. *582*
Rukmini, C. *596*
Ruof, C. H. *585, 593*
Russo, G. *580, 596, 598*
Ruzicka, L. *596*
Ryhage, R. *447, 449*

Sach, E. *595*
Sackmann, W. *446*
Sado, M. *598*
Saeki, J. *604*
Sagar, B. F. *586*
Saiga, Y. 55, *56, 59*
Saijoh, S. *606*
Saito, M. *205*
Saka, M. *458*

Sakamoto, H. *308*
Sakamoto, J. M. J. *453, 457*
Salazar, J. A. *584, 585*
Saltza, M. H. von *459*
Samulson, O. *593*
Sanchez, C. *585*
Sander, H. *596*
Sandermann, W. 571, *597*
Sandley, F. *597*
Sannié, C. *591, 597*
Sano, T. *587*
Sano, Y. 451, *455, 457*
Sanyal, A. K. *581*
Sanyal, P. K. *581*
Sarangan, S. *586, 595*
Sartori, A. *582*
Sasaki, F. *458*
Sasaki, S. *306, 311*
Sasaki, T. *99*
Sastry, C. S. P. *596, 597*
Satake, K. *459*
Satake, T. *594*
Sato, M. *59*
Savige, W. E. *147, 148*
Savoir, R. *597, 602*
Sawada, J. *455, 456*
Sawada, N. *587*
Sawada, T. 226, *307, 311*
Sawardeker, J. S. *597*
Sax, M. *53*
Scardavi, A. *597*
Schaap, A. P. *60*
Schade, G. *98*
Schafer, H. M. *148*
Schaffner, C. P. *453, 454, 458*
Schaffner, K. *147*
Schaffner, P. C. *446*
Schantz, M. v. *597*
Schartau, O. *97*
Scheidegger, J. J. *597*
Scheidegger, W. *593*
Schemjakin, M. M. *96*
Schepartz, S. A. *606*
Scheuer, P. J. *597*
Schieder, O. 90, 91, *99*
Schiesser, A. *577*
Schimbor, R. F. *457*
Schlegel, R. 363, *458*
Schlemmer, W. *585, 597*
Schlösser, E. *597, 605*
Schmidt, H. 571, *602*
Schmidt, W. *601*
Schneider, D. *99*

Sachverzeichnis. Subject Index

Von · By

A. SIEGEL, Wien

Methymycin 314, 316, 327, 333, 334, 421
Mevalonate, labelled 105
[$^{2-3}$H]Mevalonate 135
2-^{14}C-Mevalonic acid 132
5-^{14}C-Mevalonic acid 132, 137
Mevalonsäure 559, 562, 564
Mexogenin 483
—, -acetat 483
Michael-Addition 367
Michaelis constant 20, 21, 28
Micromonospora sp. 336
Mikrokinematographische Methoden 66
D-Milchsäure 427
L-Milchsäure 440, 441
Minimalagar 396
MM 8 363
Mnemiopsis 4, 15
Molekularsiebe für Chromatographie 468
Molgewichtsbestimmung durch Dampf-
    druckosmometrie 469
Mollugo-Saponin 548
Mollugo spergula 548
Mollusc 41
Molluscizide 572
Molluscogenol A 523
— B 523
— —, -diacetat 523
— C 523
— E 523
— —, -3,12-diacetat 523
Molrotation 477, 478, 532, 534
—, Ergochrome 190
Momordicasäure 510
—, -methylester 510
Monicamycin 363
Monoacetyl-leucomycin A$_3$ 356, 357, 367
—, -3,5-dinitrobenzoat 356, 357
N-Monoacetyl-methylglycosid 425, 426
Monodesmoside 463
Monodesmoside, durch Aglykon saure
    534f
—, durch Aglykon und Zuckeranteil saure
    550f
—, durch Uronsäuren saure 532f
—, saure 539
Monodesmosidische Saponine 477
Monodesmosidische Spirostanolglykoside
    478
Mono-diosphenol 115
Monooxygenase 52, 53
Monoperphthalic acid 140
Monorden → Radicicol 410
Monosporium bonorden 410

Monoterpene 69
Moose 64, 90
Morelloflavone 210, 211, 275, 276, 281, 282, 283, 301
—, -heptamethyl ether 275, 276, 277
—, —, alkali degradation 277
—, —, mass spectra 278, 279
—, —, nmr- spectra 276
—, hexamethyl ether 283
—, -O-methyl ether 282
—, nmr spectrum 276, 277, 282
Morolsäure 517
—, methylester 517
—, monoacetat 517
Mubenin A 535
— B 535
— C 537
Mubenine 505, 534
Mucor 62
Musennin A 537
— B 537
Musennine 505, 534, 539
Mutarotation 433
Mutterkorn 152, 156, 157, 159, 162, 187, 190, 192, 193, 194, 196, 197, 198
Mutterkorn-Alkaloide 152, 194
Mycaminose 338, 342, 343, 344, 345, 346, 413, 421, 424, 432
—, Struktur 414f
—, Synthese 417f, 422
Mycarose 333, 338, 340, 342, 346, 347, 360, 434, 435, 436, 437, 440, 441
—, cyclische Boratkomplexe 440
—, Synthese 442f
Mycarosemonomethyläther 437
3-epi-Mycarose 439, 440, 441
Mycinose 358, 360, 443
—, β-Di-O-acetylderivat 444
Mycosamin 368, 382, 425f
—, -dithioacetal 428, 429
—, -hydrochlorid 428, 429, 430
—, N-monoacetat 425, 429
—, Synthese 430, 431
—, -tetraacetat 368, 425
—, -triacetat 368, 425
Mycoticin → Flavofungin 362, 363, 364, 369, 379, 381
Mycotrienin 363
Myctophid fish 4, 32
Myrsene africana 574
Myrtillogensäure 512
—, -methylester 512
—, -triacetat 512

Samaderin D 128f, 146
Samaderol 128
—, -enolacetate 128
Samensaponine 489, 499
Samogenin 480
—, acetat 480
Sanguisorba officinalis 528, 535, 538, 545
Sanguisorba-Saponin 528
Sanguisorbigenin → Tomentosolsäure 519
—, -acetat 519
—, -methylester 519
Sanicula officinalis 531
Sansevierigenin 483
Sapindosid A 536
— B 536
— C 537
— D 537
— E 544
Sapindus mukorossi 536, 537, 544
Sapogenine
   —, Acetate 470, 480f
   —, Identifizierung 470f
   —, Methylesteracetate 470
   —, Optische Drehung 480f
   —, Persilylderivate 471
   —, Röntgenspektroskopische Untersuchungen 478
   —, Schmelzpunkte 480f
   —, Spektroskopische Methoden 470, 471
   —, Umwandlungen bei Hydrolysen 472
Sapogenin A, genuines 521
Sapogenin aus *Helebordus odorus* 483
—, acetat 483
Saponaria officinalis 538, 555, 557, 558, 574
Saponarosid 527, 538
Saponasid A 555
— D 557
Saponine 461f
—, Acetonfällung 465, 466
—, antibiotische Wirkung 462, 534, 565, 569f
—, antimykotische Wirkung 567
—, antivirale Wirkung 569
—, Ätherfällung 465, 466
—, Auftrenrung von Gemischen 467f
—, Biosynthese 560f
—, cancerostatische Aktivität 565
—, Charakterisierung 469
—, Cholesterinkomplexbildung 462, 466, 494, 565
—, Chromatographie 467, 468, 469
—, Craig-Verteilung 468
—, cytostatische Aktivität 573, 574

Saponine, Dialyse 465
—, Eigenschaften 564f
—, Extraktion 464, 465, 466
—, fungicide Wirkung 558
—, fungistatische Wirkung 565, 569
—, hämolytische Aktivität 462, 463, 478, 489, 502, 524, 534, 546, 549, 558, 564, 565f, 572
—, Hydrolyse 470f
—, in Leguminosen 572
—, Identifizierung 469
—, Isolierung 464f
—, Konfigurationsbestimmung der Glykosidbindung 477f
—, Konformation 478
—, Kristallisation 465, 466, 467
—, Methylierung 473f
—, Molgewichtsbestimmungen 469
—, optische Drehwerte 469
—, Partialhydrolyse 475f
—, Peracetate 469, 476
—, Pharmakologie 463, 573
—, quantitative Bestimmung 575f
—, Reindarstellung 464, 466
—, Schaumbildungsvermögen 478, 502, 564
—, Schmelzpunkte 469
—, Strukturermittlung 469f
—, tierische 558f
—, Toxizität 571f
—, Toxizität für Fische 462, 572
—, Vorkommen 463
—, Zuckerkomponenten 473
Saponin-alkylester 465
Saponin F, genuines 520
Sarsaparillosid → (25S)-5β-Furostan-3β,22α-26-triol 479, 486, 489, 490, 499, 570
—, Chromsäureabbau 488
—, hämolytische Aktivität 566
—, Peracetat 488
—, Umwandlungen 486
Sarsasapogenin 471, 472, 480, 497
—, -acetat 480
—, -β-D-glucosid 500
Scabiosid A 536
— B 535
— C 536
— D 541
— E 541
— F 541
— G 541
Sceptrumgenin 480
—, -acetat 480

Satz: Austro-Filmsatz Richard Gerin, A-1020 Wien
Druck: Paul Gerin, A-1021 Wien

Fortschritte der Chemie organischer Naturstoffe
Progress in the Chemistry of Organic Natural Products

Further volumes see next page

Springer-Verlag Wien · New York

Volume 20: 33 figures. XIII, 509 pages. 1962.
Cloth DM 102,—, S 704,—

Cumulative Index / Generalregister 1–20. 1938–1962. XVI, 369 pages. 1964.
Cloth DM 63,50, S 438,—

Volume 21: 14 figures. VII, 362 pages. 1963.
Cloth DM 80,50, S 556,—

Volume 22: 8 figures. VII, 370 pages. 1964.
Cloth DM 93,50, S 645,—

Volume 23: 58 figures. VIII, 397 pages. 1965.
Cloth DM 99,50, S 687,—

Volume 24: 25 figures. VIII, 475 pages. 1966.
Cloth DM 118,—, S 814,—

Volume 25: 25 figures. VII, 348 pages. 1967.
Cloth DM 84,—, S 580,—

Volume 26: 97 figures. IX, 456 pages. 1968.
Cloth DM 132,—, S 911,—

Volume 27: 47 figures. VIII, 412 pages. 1969.
Cloth DM 120,—, S 830,—

Volume 28: 14 figures. XII, 503 pages. 1970.
Cloth DM 155,—, S 1070,—

Further information will be sent on request.

Volume 29: 18 figures. VIII, 554 pages. 1971.
Cloth DM 181,—, S 1250,—

Contents: **D. Gross,** Vorkommen, Struktur und Biosynthese natürlicher Piperidin-verbindungen. — **W. Rüdiger,** Gallenfarbstoffe und Biliproteide. — **F. Johnson,** The Chemistry of Glutarimide Antibiotics. — **S. Huneck,** Chemie und Biosynthese der Flechtenstoffe. — **D. Lavie** and **E. Glotter,** The Cucurbitanes, a Group of Tetracyclic Triterpenes. — **D. Goldsmith,** Biogenetic-type Synthesis of Terpenoid Systems. — **J. R. Hanson,** The Biosynthesis of the Diterpenes. — **E. Premuzic,** Chemistry of Natural Products Derived from Marine Sources. — Author Index / Namenverzeichnis. — Subject Index / Sachverzeichnis.

Price reduction for subscribers / Preisermäßigung für Subskribenten: 10%.

Special price reduction (20% of the list price) for the set Vols. 1—20 plus Cumulative Index / Vorzugspreis (20% Nachlaß) bei Bezug der Bände 1—20 inklusive Generalregister.